Experientia Supplementum

Volume 112

Experientia Supplementum (EXS) is a multidisciplinary book series originally created as a supplement to the journal *Experientia* which appears now under the cover of *Cellular and Molecular Life Sciences*.

The edited volumes focus on selected topics of biological or biomedical research, discussing current methodologies, technological innovations, novel tools and applications, new developments and recent findings.

* * *

The series is a valuable source of information not only for scientists and graduate students in medical, pharmacological and biological research, but also for physicians as well as practitioners in industry.

EXS is indexed in Medline and Chemical Abstract Service (CAS).

* * *

More information about this series at http://www.springer.com/series/4822

Marija Pezer
Editor

Antibody Glycosylation

Springer

Editor
Marija Pezer
Glycoscience Research Laboratory
Genos Ltd.
Zagreb, Croatia

ISSN 1664-431X   ISSN 2504-3692 (electronic)
Experientia Supplementum
ISBN 978-3-030-76911-6   ISBN 978-3-030-76912-3 (eBook)
https://doi.org/10.1007/978-3-030-76912-3

© The Editor(s) (if applicable) and The Author(s), under exclusive license to Springer Nature Switzerland AG 2021
Chapter "Capillary (Gel) Electrophoresis-Based Methods for Immunoglobulin (G) Glycosylation Analysis" is licensed under the terms of the Creative Commons Attribution 4.0 International License (http://creativecommons.org/licenses/by/4.0/). For further details see license information in the chapter.
This work is subject to copyright. All rights are solely and exclusively licensed by the Publisher, whether the whole or part of the material is concerned, specifically the rights of translation, reprinting, reuse of illustrations, recitation, broadcasting, reproduction on microfilms or in any other physical way, and transmission or information storage and retrieval, electronic adaptation, computer software, or by similar or dissimilar methodology now known or hereafter developed.
The use of general descriptive names, registered names, trademarks, service marks, etc. in this publication does not imply, even in the absence of a specific statement, that such names are exempt from the relevant protective laws and regulations and therefore free for general use.
The publisher, the authors, and the editors are safe to assume that the advice and information in this book are believed to be true and accurate at the date of publication. Neither the publisher nor the authors or the editors give a warranty, expressed or implied, with respect to the material contained herein or for any errors or omissions that may have been made. The publisher remains neutral with regard to jurisdictional claims in published maps and institutional affiliations.

This Springer imprint is published by the registered company Springer Nature Switzerland AG.
The registered company address is: Gewerbestrasse 11, 6330 Cham, Switzerland

Foreword

Antibodies are one of the main weapons in our arsenal for the eternal war against pathogens. They are an elaborate tool that can specifically recognize foreign structures in our body. This is achieved by site-specific recombination of multiple variants of the V, (D), and J sequences of the variable region followed by somatic hypermutation that induces up to a million times higher rate of mutation in this region during antibody maturation. The fact that a completely bizarre process that, against all evolutionary logic, actually induces mutations has been invented during evolution indicates how important antibodies are for our survival. But binding to antigen is only one part of what antibodies do. After binding to a foreign object, antibodies have to activate proper molecular mechanisms to "deal with" this foreign, non-self object. If this non-self antigen is a pathogenic virus, or a bacterium, it has to be eliminated in the most efficient way. If it is on a cell that may be transformed to a tumour cell, or infected by a virus, the entire cell should be eliminated. But if this foreign antigen is a food we eat, dust, or some antigen in the air, then this antigen should be ignored, and the activation of the immune system should be avoided.

The decision of how to react to a foreign antigen is one of the most complex decisions that have to be made, and these decisions have to be made continuously throughout our lifetime. Alternative glycosylation modulates the execution of these decisions by directing IgG to different receptors and in this way activating different branches of our immune system. Fc glycans are an integral part of the CH2 domain of antibodies and as such represent an integral structural component that participates in the interaction with Fc receptors and other proteins. Attaching a different glycan to the polypeptide backbone changes the structure of the antibody and modifies its affinity for different receptors. The best currently known example is the role of core fucose that acts as a "safety switch" against antibody-dependant cellular cytotoxicity (ADCC) by attenuating binding of IgG to Fc-gamma-receptor IIIA.

Contrary to the polypeptide parts of the antibody that can be changed only by inducing changes in the corresponding genes, glycans are encoded in a complex network of at least several dozen genes that are affected by both epigenetics and the environment. This enables flexible and dynamic regulation of antibody function and is extensively used to fine-tune our immune system. More than thirty years ago, the initial discovery of changes in the IgG glycome composition in diseases was made and until now over 100,000 different IgG glycomes have been analysed in different diseases and physiological states. Changes in IgG glycosylation are associated with numerous diseases, often even before any other symptoms of the disease are detectable, indicating that they might be a part of molecular pathophysiology leading to the disease. With ageing IgG glycome converts from a composition that suppresses inflammation to an inflammation-promoting glycome that seems to be an underlying risk factor in many cardiometabolic diseases.

Glycosylation is an essential element in the development of different therapeutic monoclonal antibodies, and glycoengineered drugs are already on the market. Inter-individual differences in glycosylation are large and may be an important underlying element for the response or non-response to a given drug, but this is still understudied. Hopefully, the recent progress in analytical methods will enable more studies in this direction, which would help us to better understand functional aspects of inter-individual differences in antibody glycosylation.

The book *Antibody Glycosylation* edited by Marija Pezer and written by an international team of accomplished scientists from academia and industry provides a comprehensive overview of biosynthesis, regulation, functionality, analytics, and applications of immunoglobulin glycosylation. By covering automatization and bioinformatics in high-throughput analytical settings, it provides new perspectives for research and development in the field of therapeutic antibodies, biomarkers, vaccinations, and immunotherapy.

Glycoscience Research Laboratory,   Gordan Lauc
Genos Ltd., Zagreb, Croatia

Contents

1 **Micro-Heterogeneity of Antibody Molecules** 1
Yusuke Mimura, Radka Saldova, Yuka Mimura-Kimura,
Pauline M. Rudd, and Roy Jefferis

Part I Analytical Methods

2 **Lectin and Liquid Chromatography-Based Methods for
Immunoglobulin (G) Glycosylation Analysis** 29
Tea Petrović and Irena Trbojević-Akmačić

3 **Mass Spectrometry-Based Methods for Immunoglobulin
G *N*-Glycosylation Analysis** 73
Siniša Habazin, Jerko Štambuk, Jelena Šimunović, Toma Keser,
Genadij Razdorov, and Mislav Novokmet

4 **Capillary (Gel) Electrophoresis-Based Methods for
Immunoglobulin (G) Glycosylation Analysis** 137
Samanta Cajic, René Hennig, Robert Burock, and Erdmann Rapp

5 **Automation of Immunoglobulin Glycosylation Analysis** 173
Jenifer L. Hendel, Richard A. Gardner, and Daniel I. R. Spencer

6 **Bioinformatics in Immunoglobulin Glycosylation Analysis** 205
Frédérique Lisacek, Kathirvel Alagesan, Catherine Hayes,
Steffen Lippold, and Noortje de Haan

Part II Biosynthesis and Regulation

7 ***N*-Glycan Biosynthesis: Basic Principles and Factors Affecting
Its Outcome** 237
Teemu Viinikangas, Elham Khosrowabadi, and Sakari Kellokumpu

8   Genetic Regulation of Immunoglobulin G Glycosylation   259
     Azra Frkatovic, Olga O. Zaytseva, and Lucija Klaric

9   Epigenetics of Immunoglobulin G Glycosylation   289
     Marija Klasić and Vlatka Zoldoš

10   Immunoglobulin G Glycosylation Changes in Aging and Other
     Inflammatory Conditions   303
     Fabio Dall'Olio and Nadia Malagolini

11   Estrogen-Driven Changes in Immunoglobulin G
     Fc Glycosylation   341
     Kaitlyn A. Lagattuta and Peter A. Nigrovic

Part III   Effector Functions and Diseases

12   Sweet Rules: Linking Glycosylation to Antibody Function   365
     Falk Nimmerjahn and Anja Werner

13   Immunoglobulin G Glycosylation in Diseases   395
     Marija Pezer

14   Immunoglobulin A Glycosylation and Its Role in Disease   433
     Alyssa L. Hansen, Colin Reily, Jan Novak, and Matthew B. Renfrow

Part IV   Applications

15   Importance and Monitoring of Therapeutic Immunoglobulin G
     Glycosylation   481
     Yusuke Mimura, Radka Saldova, Yuka Mimura-Kimura,
     Pauline M. Rudd, and Roy Jefferis

16   Glycosylation of Plant-Produced Immunoglobulins   519
     Kathrin Göritzer and Richard Strasser

17   The Rapidly Expanding Nexus of Immunoglobulin G N-Glycomics,
     Suboptimal Health Status, and Precision Medicine   545
     Alyce Russell and Wei Wang

18   Glycosylation of Antigen-Specific Antibodies: Perspectives
     on Immunoglobulin G Glycosylation in Vaccination and
     Immunotherapy   565
     Pranay Bharadwaj and Margaret E. Ackerman

Editor and Contributors

About the Editor

Marija Pezer received her PhD in molecular biology from the University of Zagreb, Croatia, in 2013. She currently works at Genos Glycoscience Research Laboratory, Croatia, where she is investigating the biomarker potential of immunoglobulin G glycosylation patterns in different pathological and physiological states. The Genos lab has performed over 150,000 total plasma and IgG N-glycome analyses (world's No. 1). She is also the Head of product development for GlycanAge—the first glycan-based test for biological age.

Contributors

Margaret E. Ackerman Department of Microbiology and Immunology, Geisel School of Medicine, Dartmouth College, Hanover, NH, USA
Thayer School of Engineering, Dartmouth College, Hanover, NH, USA

Kathirvel Alagesan Max Planck Unit for the Science of Pathogens, Berlin, Germany

Pranay Bharadwaj Department of Microbiology and Immunology, Geisel School of Medicine, Dartmouth College, Hanover, NH, USA

Robert Burock glyXera GmbH, Magdeburg, Germany

Samanta Cajic Max Planck Institute for Dynamics of Complex Technical Systems, Magdeburg, Germany

Fabio Dall'Olio Department of Experimental, Diagnostic and Specialty Medicine (DIMES), University of Bologna, Bologna, Italy

Azra Frkatovic Glycoscience Research Laboratory, Genos Ltd., Zagreb, Croatia

Richard A. Gardner Ludger Limited, Culham Science Centre, Abingdon, Oxfordshire, UK

Kathrin Göritzer St. George's University of London, London, UK

Noortje de Haan Copenhagen Center for Glycomics, University of Copenhagen, Copenhagen, Denmark

Siniša Habazin Glycoscience Research Laboratory, Genos Ltd, Zagreb, Croatia

Alyssa L. Hansen Department of Biochemistry and Molecular Genetics, University of Alabama at Birmingham, Birmingham, AL, USA

Catherine Hayes Proteome Informatics Group, SIB Swiss Institute of Bioinformatics, Geneva, Switzerland
Computer Science Department, University of Geneva, Geneva, Switzerland

Jenifer L. Hendel Ludger Limited, Culham Science Centre, Abingdon, Oxfordshire, UK

René Hennig Max Planck Institute for Dynamics of Complex Technical Systems, Magdeburg, Germany
glyXera GmbH, Magdeburg, Germany

Roy Jefferis Institute of Immunology and Immunotherapy, College of Medical and Dental Sciences, University of Birmingham, Birmingham, UK

Sakari Kellokumpu Faculty of Biochemistry and Molecular Medicine, University of Oulu, Oulu, Finland

Toma Keser Faculty of Pharmacy and Biochemistry, University of Zagreb, Zagreb, Croatia

Elham Khosrowabadi Faculty of Biochemistry and Molecular Medicine, University of Oulu, Oulu, Finland

Lucija Klaric MRC Human Genetics Unit, Institute of Genetics and Cancer, University of Edinburgh, Edinburgh, UK

Marija Klasić Faculty of Science, Division of Molecular Biology, Department of Biology, University of Zagreb, Zagreb, Croatia

Kaitlyn A. Lagattuta Harvard-MIT MD-PhD Program, Harvard Medical School, Boston, MA, USA

Steffen Lippold Center for Proteomics and Metabolomics, Leiden University Medical Center, Leiden, The Netherlands

Frédérique Lisacek Proteome Informatics Group, SIB Swiss Institute of Bioinformatics, Geneva, Switzerland
Computer Science Department, University of Geneva, Geneva, Switzerland
Section of Biology, University of Geneva, Geneva, Switzerland

Nadia Malagolini Department of Experimental, Diagnostic and Specialty Medicine (DIMES), University of Bologna, Bologna, Italy

Yuka Mimura-Kimura Department of Clinical Research, National Hospital Organization Yamaguchi Ube Medical Center, Ube, Japan

Yusuke Mimura Department of Clinical Research, National Hospital Organization Yamaguchi Ube Medical Center, Ube, Japan

Peter A. Nigrovic Division of Immunology, Boston Children's Hospital, Boston, MA, USA
Division of Rheumatology, Inflammation, and Immunity, Brigham and Women's Hospital, Harvard Medical School, Boston, MA, USA

Falk Nimmerjahn Chair of Genetics, Department of Biology, Institute of Genetics, University of Erlangen-Nürnberg, Erlangen, Germany
Medical Immunology Campus Erlangen, Erlangen, Germany

Jan Novak Department of Microbiology, University of Alabama at Birmingham, Birmingham, AL, USA

Mislav Novokmet Glycoscience Research Laboratory, Genos Ltd, Zagreb, Croatia

Tea Petrović Glycoscience Research Laboratory, Genos Ltd., Zagreb, Croatia

Marija Pezer Glycoscience Research Laboratory, Genos Ltd., Zagreb, Croatia

Erdmann Rapp Max Planck Institute for Dynamics of Complex Technical Systems, Magdeburg, Germany
glyXera GmbH, Magdeburg, Germany

Genadij Razdorov Glycoscience Research Laboratory, Genos Ltd, Zagreb, Croatia

Colin Reily Departments of Medicine and Microbiology, University of Alabama at Birmingham, Birmingham, AL, USA

Matthew B. Renfrow Department of Biochemistry and Molecular Genetics, University of Alabama at Birmingham, Birmingham, AL, USA

Pauline M. Rudd NIBRT GlycoScience Group, National Institute for Bioprocessing Research and Training, Mount Merrion, Dublin, Ireland
Bioprocessing Technology Institute, Singapore, Singapore

Alyce Russell Centre for Precision Health, Edith Cowan University, Joondalup, Australia
School of Medical and Health Sciences, Edith Cowan University, Joondalup, Australia

Radka Saldova NIBRT GlycoScience Group, National Institute for Bioprocessing Research and Training, Mount Merrion, Dublin, Ireland
UCD School of Medicine, College of Health and Agricultural Science, University College Dublin, Dublin, Ireland

Jelena Šimunović Glycoscience Research Laboratory, Genos Ltd., Zagreb, Croatia

Daniel I. R. Spencer Ludger Limited, Culham Science Centre, Abingdon, Oxfordshire, UK

Jerko Štambuk Glycoscience Research Laboratory, Genos Ltd, Zagreb, Croatia

Richard Strasser University of Natural Resources and Life Sciences Vienna, Vienna, Austria

Irena Trbojević-Akmačić Glycoscience Research Laboratory, Genos Ltd., Zagreb, Croatia

Teemu Viinikangas Faculty of Biochemistry and Molecular Medicine, University of Oulu, Oulu, Finland

Wei Wang Centre for Precision Health, Edith Cowan University, Joondalup, Australia
School of Medical and Health Sciences, Edith Cowan University, Joondalup, Australia

Anja Werner Chair of Genetics, Department of Biology, Institute of Genetics, University of Erlangen-Nürnberg, Erlangen, Germany

Olga O. Zaytseva Glycoscience Research Laboratory, Genos Ltd., Zagreb, Croatia

Vlatka Zoldoš Faculty of Science, Division of Molecular Biology, Department of Biology, University of Zagreb, Zagreb, Croatia

Chapter 1
Micro-Heterogeneity of Antibody Molecules

Yusuke Mimura, Radka Saldova, Yuka Mimura-Kimura, Pauline M. Rudd, and Roy Jefferis

Contents

1.1	Introduction	3
1.2	Disulfide Bond-Related Modifications	5
1.3	N- and C-Terminal Modifications	7
	1.3.1 N-Terminal Modifications	7
	1.3.2 C-Terminal Modifications	8
1.4	Chemical Modifications of Main-Chain Amino Acid Residues	9
	1.4.1 Deamidation	9
	1.4.2 Glycation	10
	1.4.3 Oxidation	11
1.5	Aggregation	12
1.6	Glycosylation	13
	1.6.1 IgG-Fc Oligosaccharide Chain	13
	1.6.2 Fab Oligosaccharide Chain	16
1.7	Conclusion	17
References		18

Y. Mimura (✉) · Y. Mimura-Kimura
Department of Clinical Research, National Hospital Organization Yamaguchi Ube Medical Center, Ube, Japan
e-mail: mimura.yusuke.qy@mail.hosp.go.jp

R. Saldova
NIBRT GlycoScience Group, National Institute for Bioprocessing Research and Training, Mount Merrion, Blackrock, Co Dublin, Ireland

UCD School of Medicine, College of Health and Agricultural Science, University College Dublin, Belfield, Dublin 4, Ireland

P. M. Rudd
NIBRT GlycoScience Group, National Institute for Bioprocessing Research and Training, Mount Merrion, Blackrock, Co Dublin, Ireland

Bioprocessing Technology Institute, Singapore, Singapore

R. Jefferis
Institute of Immunology and Immunotherapy, College of Medical and Dental Sciences, University of Birmingham, Edgbaston, Birmingham, UK

© The Author(s), under exclusive license to Springer Nature Switzerland AG 2021
M. Pezer (ed.), *Antibody Glycosylation*, Experientia Supplementum 112, https://doi.org/10.1007/978-3-030-76912-3_1

Abstract Therapeutic monoclonal antibodies (mAbs) are mostly of the IgG class and constitute highly efficacious biopharmaceuticals for a wide range of clinical indications. Full-length IgG mAbs are large proteins that are subject to multiple posttranslational modifications (PTMs) during biosynthesis, purification, or storage, resulting in micro-heterogeneity. The production of recombinant mAbs in nonhuman cell lines may result in loss of structural fidelity and the generation of variants having altered stability, biological activities, and/or immunogenic potential. Additionally, even fully human therapeutic mAbs are of unique specificity, by design, and, consequently, of unique structure; therefore, structural elements may be recognized as non-self by individuals within an outbred human population to provoke an anti-therapeutic/anti-drug antibody (ATA/ADA) response. Consequently, regulatory authorities require that the structure of a potential mAb drug product is comprehensively characterized employing state-of-the-art orthogonal analytical technologies; the PTM profile may define a set of critical quality attributes (CQAs) for the drug product that must be maintained, employing quality by design parameters, throughout the lifetime of the drug. Glycosylation of IgG-Fc, at Asn297 on each heavy chain, is an established CQA since its presence and fine structure can have a profound impact on efficacy and safety. The glycoform profile of serum-derived IgG is highly heterogeneous while mAbs produced in mammalian cells in vitro is less heterogeneous and can be "orchestrated" depending on the cell line employed and the culture conditions adopted. Thus, the gross structure and PTM profile of a given mAb, established for the drug substance gaining regulatory approval, have to be maintained for the lifespan of the drug. This review outlines our current understanding of common PTMs detected in mAbs and endogenous IgG and the relationship between a variant's structural attribute and its impact on clinical performance.

Keywords Critical quality attributes · Glycoforms · Glycoproteins · Oligosaccharides · Recombinant antibody therapeutics · Posttranslational modifications

Abbreviations

ADA	Anti-drug antibody
ADCC	Antibody-dependent cellular cytotoxicity
AGE	Advanced glycation end product
APR	Aggregation-prone region
ATA	Anti-therapeutic antibody
CDR	Complementarity-determining region
CHO	Chinese hamster ovary
CQA	Critical quality attribute
FcγR	Receptor for Fc portion of IgG
NeuAc	N-acetylneuraminic acid

NeuGc	*N*-glycolylneuraminic acid
PD	Pharmacodynamics
PK	Pharmacokinetics
PTM	Posttranslational modification
QbD	Quality by design

1.1 Introduction

Recombinant monoclonal antibody therapeutics are exemplars of translational medicine and a growing class of biopharmaceuticals (Strohl 2018; Xu et al. 2019; Jefferis 2017b; Lu et al. 2020; Wang et al. 2018). Although the outstanding advantage of mAbs is homogeneity with respect to specificity and affinity, antibody molecules undergo various PTMs during biosynthesis, purification, or storage, and in vivo following administration, resulting in micro-heterogeneity (Figs. 1.1 and 1.2) (Jefferis 2017a, c; Beck and Liu 2019). Micro-heterogeneity of a mAb is a concern for the biopharmaceutical industry because structural modifications can impact efficacy and safety. The sources of heterogeneity identified in early studies included the oligosaccharide attached at each Asn297 residue and N-terminal pyroglutamic acid residues on either the heavy (H) or light (L) chains (Mimura et al. 1995). The development of liquid chromatography/electrospray mass spectrometry and tandem mass spectrometry in the 1990s allowed the identification and quantitation of additional proteoforms due to Met oxidation, Asn deamidation, the presence or absence of H chain C-terminal Lys, etc. (Lewis et al. 1994; Roberts et al. 1995). Recent advances in analytical methods have allowed the identification and localization of additional PTMs in mAbs, both within the complementarity-determining

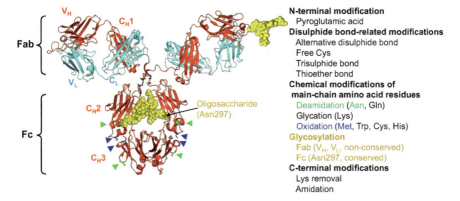

Fig. 1.1 Structure model of a human IgG antibody with the glycans attached to its Fab and Fc regions and common sources of micro-heterogeneity of IgG antibodies. The susceptible sites for oxidation at Met252/Met428 and for deamidation at Asn315/Asn384 are shown by blue and green arrowheads, respectively

Fig. 1.2 Charge heterogeneity of an IgG1 mAb revealed by 2-D electrophoresis. Before (**a**) and after incubation in 10 mM phosphate-buffered saline at pH 7.0, 37 °C for 4 weeks (**b**). Both the H and L chains exhibit heterogeneity in terms of pI, resulting from multiple sources as shown in Fig. 1.1 (**a**). Alterations in the charge properties of the H and L chains took place under physiological conditions (**b**) (Mimura et al. 1998)

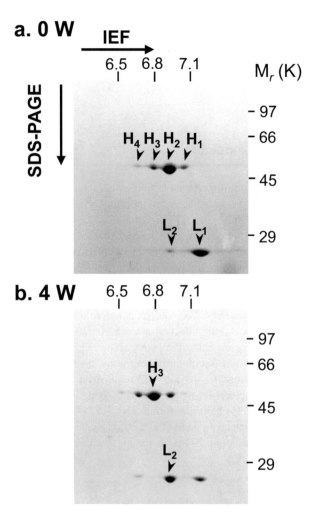

regions (CDRs) of the variable (V) regions and within the constant regions, including Asn/Gln deamidation, Asp isomerization, oxidation of Met, Trp, and His, Cys modification, glycation, and Fc and Fab glycosylation (Harris 2005; Beck and Liu 2019; Jefferis 2016; Chicooree et al. 2015; Ambrogelly et al. 2018; Hmiel et al. 2015; Xu et al. 2019). Micro-heterogeneities generated in vivo over the natural lifespan of individual IgG antibody molecules occur at defined amino acid residues and are "recognized" as self by the individual's immune system, resulting in immune tolerance; however, production of mAbs in nonhuman cell lines followed by downstream processing may introduce additional non-self PTMs resulting in loss of structural fidelity, leading to altered stability and biological activities, including the introduction of immunogenic potential.

The addition and processing of complex oligosaccharide moieties within the IgG-Fc is a principal PTM and is defined as a CQA (Jefferis 2017b; Alt et al. 2016; Kepert et al. 2016). Since the glycoform profile of an approved IgG mAb is an essential CQA, they are necessarily produced in mammalian cell lines such as Chinese hamster ovary (CHO) cells and murine plasmacytoma NS0 and SP2/0 cells. However, given that the oligosaccharide biosynthesis process is species- and cell line-specific, and that the observed addition of nonhuman oligosaccharides terminating in $\alpha(1-3)$-linked galactose ($\alpha(1-3)$-Gal) or N-glycolylneuraminic acid (NeuGc) can result in immunogenicity, the choice and development of a production platform for each mAb constitutes a challenge for the biopharmaceutical industry. One outstanding benefit arising from the above studies has been the demonstration that IgG mAbs bearing oligosaccharides lacking a core fucose residue exhibit increased Fcγ receptor IIIa (FcγRIIIa) binding affinity and consequent increased antibody-dependent cellular cytotoxicity (ADCC) activity; this CQA receives particular attention both within the biopharmaceutical industry and regulators.

Process changes are often introduced within the life cycle of each individual mAb, e.g., optimization for higher productivity and/or increased quality control, etc. All changes must be referred to the regulatory authorities for permission, subject to demonstrating that no adverse responses are experienced by patients. This review outlines the current understanding of common PTMs detected in endogenous IgG and recombinant IgG mAbs and the relationship between a structural variant and its clinical performance.

1.2 Disulfide Bond-Related Modifications

Disulfide bonds result from the coupling of unpaired thiols on Cys residues. The integrity of the disulfide bond is essential to maintain the structure, stability, and biological activities of IgG molecules. The standard structural cartoon for the human IgG1 protein (Eu) exhibits 12 intra-domain disulfide bonds and 4 inter-chain disulfide bonds (Edelman et al. 1969); however, there are variations in the number and pairing of the disulfide bonds between IgG subclasses. IgG1 and IgG4 have two inter-H chain disulfide bonds in the hinge region, while IgG2 and IgG3 have 4 and 11, respectively. For IgG1λ, the L and H chains are connected through the fifth (i.e., the C-terminal) Cys residue of the L chain and the fifth Cys residue of the H chain while for the other IgG subclasses, the C-terminal Cys residue of the L chain is linked to the third Cys residue of the H chain (Liu and May 2012; Lakbub et al. 2016). In addition, there are other Cys-related variants for natural and monoclonal IgG antibodies, including the presence of free thiols, cysteinylation (disulfide bond formation with a free Cys), thioether formation, trisulfide formation, cysteine racemization, and disulfide bond scrambling. Such nonclassical disulfide bond formation can influence the physical stability of IgG and may compromise the safety and efficacy of antibody therapeutics.

Free thiol groups can be detected in natural and recombinant IgG antibodies, although all Cys residues should be involved in disulfide bond formation (–R–S–S–R′–). Free thiols can result from incomplete disulfide bond formation, disulfide bond degradation, and an extra noncanonical Cys residue mostly in the CDRs. Incomplete disulfide bond formation can occur for any intra-domain disulfide bonds and less likely for the inter-chain disulfide bonds (Chumsae et al. 2009). Unpaired Cys residues (Cys 22 and Cys 96) were observed in the V_H domain of a recombinant IgG molecule (Omalizumab, anti-IgE), which substantially reduced its potency (Ouellette et al. 2010; Harris 2005). The unpaired Cys residues were shown to reform a disulfide bond when incubated in serum (Ouellette et al. 2010) or could be prevented by the addition of the oxidizing agent copper sulfate to the cell culture medium (Chaderjian et al. 2005).

Degradation of disulfide bonds can be caused by β-elimination at basic pH, resulting in the generation of dehydroalanine and persulfide. Crosslinking of the resulting Cys and dehydroalanine leads to the formation of a non-reducible thioether linkage (–R–S–R′–), which was found at ~0.4% for recombinant monoclonal IgG1 stored at 4 °C. The L chain—H chain linkage (at L214–H220) was reported to be sensitive for the degradation to the thioether bond formation (Tous et al. 2005), and the rate is higher for IgG1λ than for IgG1κ (Zhang et al. 2013). In addition, the reversibility of the dehydrogenation step can cause racemization of Cys, forming either the L- or D-form (Zhang and Flynn 2013). On the other hand, hydrolysis of the dehydroalanine can also occur, which is an important mechanism that leads to antibody fragmentation in the hinge region.

Insertion of a sulfur atom into a disulfide bond to form trisulfides (–R–S–S–S–R′–) was reported between the H chains of a recombinant human IgG2 (Pristatsky et al. 2009) and later between the L and H chains in all four IgG subclasses and human endogenous IgG (Gu et al. 2010). Trisulfide bonds are formed due to the presence of hydrogen sulfide in the cell culture medium for recombinant mAbs (Kita et al. 2016) and in tissues for endogenous IgG (Gu et al. 2010; Zhao et al. 2001). The presence of trisulfides does not have an adverse effect on antibody function and can be controlled by cysteine feeding in the cell culture medium (Kshirsagar et al. 2012).

Furthermore, an extra noncanonical Cys residue can occur mostly in the CDRs. Cysteinylation of such an unpaired Cys residue is rarely observed in the V regions of IgG, but it can have a significant influence on antibody function (Banks et al. 2008). Cysteinylation and the presence of incomplete disulfide bonds have not been reported in natural human IgGs. Given all the negative impacts of cysteinylation, this modification may have been eliminated from natural IgG during evolution. The same could be true for the presence of a single pair of incomplete disulfide bonds.

A core hinge region sequence of –Cys–Pro–Pro–Cys–, present in IgG1, IgG2, and IgG3, forms a partial helical structure that does not allow for intra-H chain disulfide formation; however, the homologous sequence in the IgG4 subclass is –Cys–Pro–Ser–Cys–, and this does allow for intra-H chain disulfide bond formation. Consequently, natural and recombinant IgG4 antibody populations are a mixture of two hinge isomers, i.e., an inter-chain form with two inter-H chain disulfide bonds and a second isomer with intra-H chain disulfide bonds that are generated in vivo and in vitro (Rispens et al. 2013).

Additional heterogeneity results from "Fab-arm exchange" with the generation of monovalent bispecific antibody molecules in vivo and in vitro. The exchange is facilitated by the presence of an Arg residue at position 409, in place of the Lys409 present in IgG1, IgG2, IgG3, and a polymorphic variant of IgG4 (Rispens et al. 2013, 2014; Davies et al. 2013). For the Arg409 isoform, lateral noncovalent interactions between the two C_H3 domains are reduced such that, under physiological conditions and in the absence of hinge region inter-H chain disulfide bonds, they dissociate to form HL half-molecules. Re-association is nonselective with the formation of monovalent bispecific antibodies.

Later, alternative disulfide bond linkage in the hinge region of IgG2 has been discovered both in recombinant and in natural human IgG2 (Dillon et al. 2008; Wypych et al. 2008; Martinez et al. 2008). Different IgG2 proteoforms showed a subtle difference in structure and thermal stability (Dillon et al. 2008). Inter-chain disulfide bonds within the IgG2 subclass can form more than one pair of combinations in vivo, namely, the A-, A/B-, and B-forms. The classical inter-chain disulfide bond in IgG2 is termed the A-form, with the Cys near or at the C-terminus of the L chain is linked to the C_H1 domain (the third Cys of the H chain). In the B-form, the C-terminus of the L chain is linked to the hinge region, and both the C_H1 domains have linkages to the hinge region, whereas only one C_H1 domain is linked to the hinge in the A/B-form (Liu and May 2012). This heterogeneity in the disulfide bond formation in IgG2 was first reported for recombinant IgG2 but, later, shown to be present in normal serum-derived IgG2 as well (Wypych et al. 2008; Dillon et al. 2008). The interconversion of these proteoforms was shown to be dynamic and promoted by a reducing environment; such an environment can be provided by the presence of thioredoxin/thioredoxin reductase, released into culture media by effector cells; it can be ameliorated by control of dissolved oxygen levels (Koterba et al. 2012; Hutterer et al. 2013; Rispens et al. 2013, 2014; Davies et al. 2013).

An in vitro model revealed that susceptibility to reduction/oxidation differed between IgG subclasses and L chain types with sensitivity being in the order IgG1λ > IgG1κ > IgG2λ > IgG2κ (Koterba et al. 2012).

1.3 N- and C-Terminal Modifications

1.3.1 N-Terminal Modifications

Gln and its cyclized form, pyroglutamate (pyroGlu) or pyrrolidone carboxylic acid (PCA) are commonly observed at the N-termini of the H and λ L chains of human monoclonal antibodies (Kabat et al. 1991). PyroGlu is also observed in recombinant mAbs (Pang et al. 2015; Dick et al. 2007; Yin et al. 2013). PyroGlu formation involves the cyclization of the N-terminal amine and the subsequent loss of ammonia (−17 Da). Not only Gln but also Glu at the N-terminus of a recombinant mAb are shown to cyclize to pyroGlu nonenzymatically in vitro (Liu et al. 2011; Chelius et al. 2006). Nearly complete conversion of the N-terminal Gln to pyroGlu was shown to occur spontaneously inside the bioreactor within 15 days, which was affected by buffer composition and temperature and less dependent on pH (Dick et al. 2007). Sodium phosphate, ammonium carbonate buffers, and cell culture media can increase the rate of the reaction, in contrast to water and *Tris*-HCl buffer. On the other hand, the rate of Glu cyclization is much slower and dependent on pH and temperature, with the half-lives of the N-terminal Glu of the L chain of a mAb at 37 °C being 4.8, 19, and 11 months at pH 4.0, 7.0, and 8.0, respectively (Yu et al. 2006). Although the biological significance of pyroGlu formation has not been investigated, the presence of pyroGlu appears to play a role in the stabilization of proteins and peptides, protecting them from chemical and enzymatic degradation. However, as there is no evidence of a specific benefit attached to the presence of N-terminal pyroGlu, to either the H or L chain, it may be considered a CQA and best avoided during clonal selection for a potential mAb therapeutic.

1.3.2 C-Terminal Modifications

C-terminal Lys residue of the H chain of human serum IgG is generally cleaved in vivo to expose the preceding Gly residue (Edelman et al. 1969). The cleavage of the Lys is thought to be catalyzed by an endogenous carboxypeptidase during secretion or in circulation (Harris 1995). When a recombinant human IgG2 antibody was administered in humans, C-terminal Lys was shown to be lost with a half-life of 62 min (Cai et al. 2011). Recombinant IgG molecules produced in mammalian cells exhibit mixed populations of molecules with Lys present or absent on each H chain, and the presence of the Lys results in the formation of basic charge variants. The level of C-terminal Lys can vary between clones of cells, culture conditions with or without serum, and other parameters of the production platform, contributing to charge heterogeneity (Dick et al. 2008; Antes et al. 2007; Jiang et al. 2016).

Although there is no evidence that the presence or absence of C-terminal Lys is a CQA, it presents an issue for quality by design (QbD) (Tang et al. 2013). The absence of C-terminal Lys is presumed to have no impact on the biological activity, pharmacokinetics (PK) and pharmacodynamics (PD), immunogenicity, and safety (Antes et al. 2007). However, taking a conservative approach, potential C-terminal Lys effects on all antibodies cannot be ruled out. It has been reported that the CHO clones expressing a recombinant IgG gene lacking the codon for C-terminal Lys show lower productivities than those with the codon for the Lys and those lacking the codons for the Lys and the preceding Gly (Hu et al. 2017).

Following cleavage of the H chain C-terminal Gly-Lys sequence, amidation of C-terminal Pro residue has been observed for recombinant IgG antibodies (Johnson et al. 2007; Tsubaki et al. 2013). Although the predominant IgG proteoform separated by weak cation exchange high-performance liquid chromatography terminates in Gly residue as the C-termini of both H chains, high-resolution mass spectrometry of the C-terminal peptide of a minor basic proteoform reveals the presence of amide-terminated Pro ($-CONH_2$) in a single H chain together with carboxylic acid-terminated Gly (–COOH) in the second H chain. C-terminal amidation is catalyzed by peptidyl glycine α-amidating monooxygenase (PAM), and its activity is increased with increasing copper concentrations up to 1 μM in culture media of CHO cells (Kaschak et al. 2011). Pro amidation is considered a common C-terminal modification of the H chain of mAbs, although this modification is undetectable in serum IgG antibodies (Tsubaki et al. 2013). The presence of Pro amidation in mAbs has been shown not to decrease antibody activity, structural stability, in vivo half-life, and subcutaneous bioavailability in rats (Jiang et al. 2016).

1.4 Chemical Modifications of Main-Chain Amino Acid Residues

1.4.1 Deamidation

Deamidation of Asn residues is a major consequence of chemical degradation, contributing to charge heterogeneity of IgG mAbs (Khawli et al. 2010; Wang et al. 2007). Deamidation of Asn residues is observed in vivo, and the position of the Asn residues in the sequence is well defined; however, deamidation during manufacture and storage, etc. may be more randomly distributed (Liu et al. 2008). Deamidation is influenced by the extent of exposure to the external environment (e.g., pH and temperature) and primary structure, with Asn-Gly motif being the most susceptible, followed by Asn-Ser and Asn-Thr motifs. 3-D structures usually impose structural constraints, attenuating deamidation rates (Robinson and Robinson 2001). At neutral to high pH, deamidation of Asn occurs by nucleophilic attack of the backbone nitrogen of the C-terminally flanking amino acid on the side chain carbonyl group of the Asn residue, leading to the formation of a metastable succinimide (cyclic imide) intermediate, which is then hydrolyzed to generate Asp and isoaspartate (isoAsp) in ratios of 1:2–1:3 (Geiger and Clarke 1987).

Deamidation has been detected in normal polyclonal and recombinant IgG molecules at Asn315 and Asn384 of the Fc, with the latter more susceptible (Fig. 1.1, green arrowheads) (Chelius et al. 2005; Liu et al. 2009; Sinha et al. 2009). The relative susceptibility to deamidation at these sites varied between studies; however, the biological significance such as immunogenicity may be ameliorated by the finding that ~23% of normal polyclonal IgG has an Asp384 residue (Liu et al. 2009); thus it may be concluded that healthy humans are constantly exposed to IgG bearing this PTM.

In contrast, deamidation within V regions, particularly within the CDRs, of recombinant antibodies has been shown to compromise antibody specificity and/or binding affinity (Harris et al. 2001; Huang et al. 2005). The kinetics of deamidation at Asn55 in the CDR2 of the H chain of a humanized monoclonal IgG1 antibody was shown in cynomolgus monkeys, with the half-life of 140 h (Huang et al. 2005). Interestingly, although the recombinant anti-HER2 IgG1 Herceptin has hot spots for spontaneous deamidation at Asn55 in the CDR2 of the H chain and Asn30 in the CDR1 of the L chain under in vitro conditions, the approved drug substance did not exhibit these PTMs; therefore, their presence or absence could be used as a lot release criterion (Harris et al. 2001). In vitro incubations of a mAb in phosphate-buffered saline at pH 7.4 and 37 °C were shown to accurately mimic deamidation taking place in vivo when analyzing the Asn384-containing peptide of the mAb by LC/MS/MS. Thus, deamidation of Asn residues in the V regions can be predicted before the selection of chemically stable antibody candidates (Sydow et al. 2014), and this attribute needs routinely be controlled during manufacturing and storage.

In comparison, Gln residues are relatively resistant to deamidation, and the half-lives of deamidation of synthetic peptides at neutral pH are in the range of 100–5000 days for Gln while in the range of 1–500 days for Asn (Robinson and Robinson 2001). Incubation of a mAb at pH 5–9 at 40 °C for 10 weeks revealed pH dependence for deamidation of Gln, showing that seven Gln residues were susceptible to deamidation at alkaline pH, at Gln13 and Gln82 in the V_H region and Gln366 and Gln422 in the Fc and Gln27, Gln100 and Gln199 in the L chain (Liu et al. 2008). As four of the seven Gln residues that are susceptible to deamidation are located in the V regions, deamidation of these Gln residues can have a significant impact on antibody structure and antigen binding.

1.4.2 Glycation

Glycation is a nonenzymatic condensation reaction between reducing sugars (e.g., glucose) and the primary or the secondary amine (e.g., the ε-amines of Lys residues and N-terminal amines on proteins). Glycation occurs under physiological conditions in natural and recombinant antibodies and could alter their charge property to generate acidic species and their impact on the stability and efficacy, including the aggregation propensity (Wei et al. 2017; Banks et al. 2009). The glycation reaction is initiated by the formation of a Schiff's base between susceptible amines and carbonyls such as aldehydes on reducing sugars, which can then undergo an Amadori rearrangement to form a stable covalent bond between the protein and sugar molecules. Glycation of recombinant therapeutic antibodies can occur due to the exposure to glucose during production in glucose-containing culture media, formulation, and administration in patients (Quan et al. 2008; Fischer et al. 2008; Goetze et al. 2012). Although sucrose used as a stabilizing agent in the formulations may cause glycation due to its hydrolysis into glucose when stored at 37 °C, no evidence of glycation was demonstrated after storage at 2–8 °C for 18 months (Gadgil et al. 2007; Banks et al. 2009). With regard to the extent of glycation, endogenous human Fc contains on average 0.045 glucose/Fc, i.e., ~5% of the glycated species (Goetze et al. 2012), but the glycation levels differ substantially between mAbs and individual batches of the same mAb (Fischer et al. 2008; Miller et al. 2011). Glycation in the constant regions of IgG1 and IgG2 has been shown not to affect Fc functionality, including the binding to FcγRIIa, FcRn, and staphylococcal protein A (SpA) (Goetze et al. 2012; Mo et al. 2018); however, glycation of Lys located in or adjacent to a CDR should be monitored more carefully. A model recombinant humanized mAb was found to have a highly glycated site at Lys49 (40–60%) of the V_L region when produced in the culture medium containing 0.8% (49.4 mM) glucose (Zhang et al. 2008). The potential decrease in the antigen-binding capacity by glycation in the V regions may be avoided by the selection of an antibody that does not have a Lys residue within the CDRs or by mutation of the Lys to Arg (Miller et al. 2011). The extent of glycation of antibodies can be controlled by modifying glucose feeding strategies and increasing the total concentration of free amino acids and primary amine-containing compounds during cell culture processes (Zhang et al. 2008).

Glycation modifications can further proceed to form advanced glycation end products (AGEs) under oxidative stress, contributing to product coloration. AGE has been identified on recombinant antibodies such as carboxymethyl lysine (Butko et al. 2014). Whether the glycation and AGE modification affects antibody efficacy and safety has yet to be determined, but the presence of AGE has been shown to be immunogenic in rabbits (Ahmad and Moinuddin 2012).

1.4.3 Oxidation

Oxidation of Met to methionine sulfoxide or even methionine sulfone is commonly observed in the Fc of both endogenous and recombinant IgG. It has been demonstrated, for antibodies of the IgG1 and IgG2 subclasses, that Met252 and Met428 residues are prone to oxidation (Chumsae et al. 2007; Bertolotti-Ciarlet et al. 2009; Wang et al. 2011; Pan et al. 2009; Gaza-Bulseco et al. 2008). Although these residues are distant in linear sequence, they are proximal at the C_H2/C_H3 interface (Fig. 1.1, blue arrowheads). Minimal Met252 oxidation levels of 2–5% are reported for purified IgG antibodies in formulation buffers, while lower levels of oxidation are reported for Met428. Oxidation of Met252 and Met428 increases under conditions of accelerated stability testing and on prolonged storage. The interaction site for FcRn that regulates IgG catabolism and placental passage is formed at the C_H2/C_H3 interface, and Met252 and Met428 oxidation has been shown to reduce both the affinity to FcRn and half-life in vivo in mice transgenic for human FcRn (Bertolotti-Ciarlet et al. 2009; Wang et al. 2011). However, it was shown that oxidation of both Met252 and Met428 on both H chains was required to impact IgG1 half-life, and this was only observed when Met252 oxidation was in excess of ~80% (Wang et al. 2011). Both SpA and streptococcal protein G (SpG) bind IgG-Fc at the C_H2/C_H3 interface, and it has been shown that Met oxidation impacts binding affinity for both (Pan et al. 2009; Gaza-Bulseco et al. 2008); therefore, Met252 and Met428 oxidation levels may be a useful CQA and QbD parameter. The impact of Met252 and Met428 oxidation on FcγR binding has been shown to be minimal, but a subtle decrease in binding to the FcγRIIa-His131 variant was reported (Bertolotti-Ciarlet et al. 2009). Analysis of Herceptin and a potential biosimilar demonstrated that care has to be exercised when resuspending antibody therapeutics since a discrepancy was observed for the level of Met252 oxidation between the innovator product (4.39%) and the proposed biosimilar (10.33%); even so, a note added in proof commented that the value of 4.39% was greater than that determined by the innovator company; no oxidation of Met428 was recorded (Xie et al. 2010). Met oxidation has also been found within CDRs of mouse mAb OKT3 and a human IgG1 mAb (Kroon et al. 1992; Roberts et al. 1995) and may have adverse effects on the potency; therefore, early sequencing and selection for potential antibody clones are employed to eliminate this possibility.

1.5 Aggregation

Control of aggregation of protein pharmaceuticals during manufacture, formulation, storage, and shipping is a major challenge for the biopharmaceutical industry. The level of aggregates in formulated therapeutic antibodies is a CQA as aggregation can often influence the PK/PD of mAbs, adversely impacting safety, efficacy, and the potential for immunogenicity, with the production of ADA/ATA. Protein aggregation is driven by intrinsic factors, including the primary structure, the conformation, and the net charge as well as extrinsic factors, including pH, temperature, protein concentration, ionic strength, and shaking (Li et al. 2016). Various computational algorithms based on the identification of aggregation-prone regions (APRs) in the 3-D protein structure coordinates have been developed to predict the aggregation propensity of a biotherapeutic (Santos et al. 2020). APRs are mainly hydrophobic patches rich in β-branched aliphatic and aromatic residues (Ile, Phe, Val, and Leu) that are exposed to solvent. The propensity for a mAb to aggregate may be determined by the CDRs of the V regions superimposed on an intrinsic susceptibility of the selected H and L chain isotypes. APRs were identified within the C_H1, hinge, C_H2, and C_H3 domains of the IgG1 H chain and the constant regions of both κ and λ L chains (Chennamsetty et al. 2009a). Aggregation-prone motifs in the constant regions of IgG1 are identified in the lower hinge and C_H2/C_H3 interface, including Leu234-Leu235, Met252-Ile253, and Val308-Leu309, which are also important for FcγR and FcRn binding (Wang et al. 2009; Chennamsetty et al. 2009a). The substitution of targeted hydrophobic amino acids with selected hydrophilic residues (e.g., Leu309Lys) has been shown to generate more stable proteins with a diminished tendency to aggregate (Chennamsetty et al. 2009b).

The presence of free thiol is another parameter shown to result in the generation of aggregates through intermolecular disulfide bond formation (Lacy et al. 2008) and is particularly relevant for IgG2 antibodies that contain 50% more free Cys than IgG1 (Huh et al. 2013; Wypych et al. 2008). Additionally, the individual variable sequences can have a profound impact on susceptibilities to aggregation. The germline sequences for V_H, Vκ, and Vλ have been comprehensively reviewed and analyzed for the aggregation potential of their protein products (Ewert et al. 2003; Rouet et al. 2014). These criteria may now be applied to the selection of clones producing potential IgG therapeutics in the early development process and to the optimization of more soluble and stable IgG therapeutics (van der Kant et al. 2017).

It should be noted that incompatibility of diluents with biopharmaceuticals is observed when mixed with human plasma/serum. Instructions to pharmacists for resuspension of Herceptin and Avastin, for intravenous administration, specified resuspension in 0.9% saline and specifically excluded the use of 5% glucose solutions; no reason was given (Arvinte et al. 2013). The consequences for contravening the instruction demonstrated that titration of either Herceptin or Avastin in 5% glucose into human plasma/serum resulted in the formation of insoluble aggregates. The aggregation was shown to result from isoelectric precipitation of complement proteins (C3, C4, and factor H), fibronectin and apolipoproteins bound to the mAb at pH 6.0–6.2 (Luo and Zhang 2015). It was noted that instructions for resuspension of Remicade specified 0.9% saline, pH 7.2, conditions that did not result in the formation of complexes; similarly, neither did Herceptin or Avastin. Thus, the stability/structure of a mAb therapeutic must be considered beyond that of the formulated drug product as aggregates might form in the patient's bloodstream after its administration in the presence of incompatible diluents, possibly resulting in adverse reactions to patients and/or sensitizing them to later production of ADA.

1.6 Glycosylation

1.6.1 IgG-Fc Oligosaccharide Chain

Glycosylation represents the most frequent posttranslational modification, and it is estimated that ~50% of open reading frame genes encode for proteins that may accept the addition of *N*-linked oligosaccharides at Asn-X-Ser/Thr (where X is any amino acid except Pro) (Jefferis 2017c). IgG has a conserved glycosylation site at Asn297 residue [Eu numbering (Edelman et al. 1969)] of each C_H2 domain of the Fc, and differences in the *N*-linked oligosaccharide are a major source of IgG microheterogeneity (Figs. 1.1, 1.3 and 1.4). Fc glycosylation impacts the stability, function, and safety of therapeutic IgG mAbs and is essential for optimal expression of biological activities mediated through Fcγ receptors and the C1q component of complement (Jefferis 2009; Mimura and Jefferis 2021). IgG-Fc effector functions mediated by Fcγ receptors and C1q, including ADCC, phagocytosis, oxidative burst, and complement-dependent cytotoxicity, are abrogated or severely compromised for aglycosylated or deglycosylated forms of IgG (Nose and Wigzell 1983; Pound et al. 1993; Tao and Morrison 1989). Therefore, the presence of the oligosaccharide attached to Asn297 is a CQA of recombinant antibody therapeutics (Jefferis 2017a, b). The oligosaccharide released from human endogenous IgG-Fc is highly heterogeneous and comprised of the core complex diantennary heptasaccharide (GlcNAc$_2$Man$_3$GlcNAc$_2$, designated G0) with the variable addition of fucose,

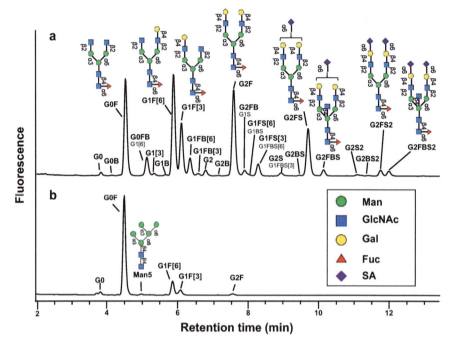

Fig. 1.3 The hydrophilic interaction liquid chromatography profile of oligosaccharides released from (**a**) normal human serum IgG-Fc and (**b**) recombinant IgG1 (bevacizumab). For chromatographic peaks containing multiple oligosaccharide structures, only the predominant components are shown in bold

galactose, bisecting GlcNAc, and *N*-acetylneuraminic (sialic) acid residues (Arnold et al. 2006; Jefferis 2017c) (Figs. 1.3 and 1.4). The heterogeneous oligosaccharides can be classified into three sets (G0, G1, and G2), depending on the number of galactose residues in the outer arms of diantennary oligosaccharides. Within each of these sets are four species that result from the presence or absence of core fucose and bisecting GlcNAc, namely, 16 neutral complex-type structures (Fig. 1.4). Sialylation of the oligosaccharide enhances the heterogeneity, generating 36 possible glycoforms. The oligosaccharides from the Fc fragment of human serum IgG were separated into ~30 structures by hydrophilic interaction liquid chromatography in which fucosylated, monogalactosylated (G1F) glycoforms predominate, with a preference for galactosylation on the α(1–6)-arm (G1F[6]) over the α(1–3)-arm (Fig. 1.3) (Pucic et al. 2011; Mimura et al. 2018). The oligosaccharides of serum IgG-Fc are mostly fucosylated (90–96%), variably galactosylated in aging and diseases and pauci-sialylated (10–20%) (Gudelj et al. 2018; Parekh et al. 1988; Cheng et al. 2020). The outer-arm sugars play important roles in modulating the Fc effector functions, including enhanced ADCC activity of non-fucosylated glycoform. Thus, the rituximab (Rituxan) drug product that gained licensing approval was comprised of ~25% of the G1F oligosaccharide; therefore, regulatory

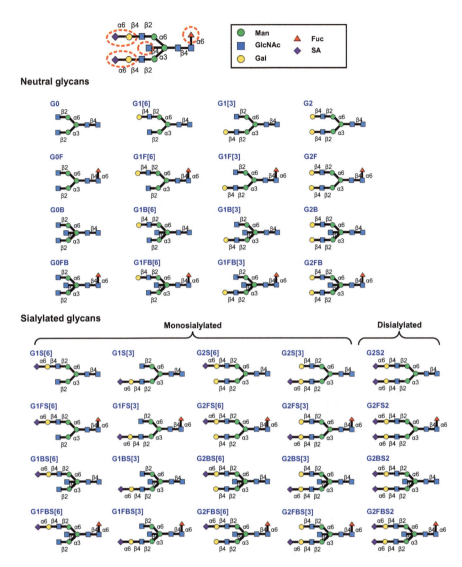

Fig. 1.4 The potential "library" of complex diantennary oligosaccharides associated with human serum IgG-Fc. Heterogeneity arises from variable addition of outer-arm sugars and core fucose (enclosed by red dotted circles) to a "core" heptasaccharide (designated G0)

authorities required that galactosylation of the manufactured product be controlled to within a few percentage points of this value. Analysis of the Fc oligosaccharide of human IgG subclass antibodies revealed unique glycosylation profiles, with preferential galactosylation of the α(1–3)-arm for IgG2 and IgG3 in contrast to that of the α(1–6)-arm for IgG1, IgG4, and polyclonal IgG (Jefferis et al. 1990). The sialic acid of the N-glycan of human IgG-Fc is added preferentially on the α(1–3)-arm of the digalactosylated (G2) glycoforms (van den Eijnden et al. 1980; Grey et al. 1982; Barb et al. 2009). Sialylation occurs in α(2–6)-linkage with *N*-acetylneuraminic acid (NeuAc) in humans, whereas it is in α(2–3)-linkage in CHO-derived recombinant IgG molecules (Takeuchi et al. 1988). Bisecting GlcNAc is present in 10–15% of the *N*-glycans of human IgG-Fc. Currently, licensed therapeutic IgG antibodies are produced in CHO, NS0, and SP2/0 cell lines, and the glycosylation profile of a recombinant glycoprotein is species- and cell type-specific (Raju et al. 2000; de Haan et al. 2020). The oligosaccharide of a CHO-derived therapeutic monoclonal IgG antibody is less heterogeneous than that of human normal polyclonal IgG, with a predominance of non-galactosylated, fucosylated glycoform (G0F) (Fig. 1.3). Bisecting GlcNAc is undetected in recombinant IgG produced in CHO, NS0, and SP2/0 cells (Raju et al. 2000). Terminal α(1–3)-linked galactose (α(1–3)-Gal) and *N*-glycolylneuraminic acid (NeuGc) residues can be attached to the galactosylated oligosaccharides and frequently found in NS0 and SP2/0-derived recombinant IgG antibodies (Sheeley et al. 1997; Stadlmann et al. 2008; Montesino et al. 2012). Such oligosaccharide structures are unnatural and potentially immunogenic in humans. The α-galactosylation and sialylation with NeuGc are reported in cetuximab and infliximab produced from SP2/0 (Qian et al. 2007; Mimura et al. 2009). It has been reported that all humans have IgG antibodies specific for the α(1–3)-Gal epitope (Galili et al. 1993) and that the anti-NeuGc activity is detectable in up to 85% of healthy individuals (Zhu and Hurst 2002; Tangvoranuntakul et al. 2003).

1.6.2 Fab Oligosaccharide Chain

It is established that 15–20% of polyclonal human IgG molecules bear *N*-linked oligosaccharides within the V regions of the κ, λ, or H chains, and sometimes both (Jefferis 2017a). In the immunoglobulin sequence database, ~18% of the V_H sequences contain a potential *N*-linked glycosylation site (Kabat et al. 1991). Very few IgV_H genes have a naturally occurring *N*-linked glycosylation site (Zhu et al. 2002), and the localization of the glycosylation sites in the V_H regions of sequenced proteins is random; therefore, they are likely to be introduced by somatic hypermutation (Dunn-Walters et al. 2000). The functional significance for Fab glycosylation has not been fully evaluated, but studies employing mAbs suggest V region glycosylation can have a neutral, positive, or negative influence on antigen binding (Zhu et al. 2003). On the other hand, in various chronic inflammatory diseases and cancers, including rheumatoid arthritis and follicular lymphoma, high incidence of Fab glycosylation has been observed for anti-citrullinated protein antibodies (ACPAs) (Kempers et al. 2018; Hafkenscheid et al. 2019) and follicular lymphoma-secreted antibodies (Radcliffe et al. 2007; McCann et al. 2008), respectively. The presence of an oligosaccharide in the V region might be associated with the survival of antibody-producing B cells as a consequence of inappropriate activation, selection, and maturation. Analyses of the oligosaccharides of human serum-derived IgG-Fab fragments revealed the presence of diantennary oligosaccharides showing relatively high levels of galactosylation and sialylation or oligomannose, depending on the location of the glycosylation site (Holland et al. 2006; Hafkenscheid et al. 2017; Radcliffe et al. 2007; Gala and Morrison 2004). The licensed antibody cetuximab expressed in SP2/0 cells is glycosylated at Asn88 of the framework region 3 of the V_H region, bearing complex diantennary, triantennary, and hybrid oligosaccharides (Qian et al. 2007). The Fab oligosaccharide pool contains nonhuman oligosaccharides containing both α(1–3)-Gal (30%) and NeuGc (12%) (Qian et al. 2007) and may induce anaphylactic reaction at least in a proportion of patients due to the presence of IgE antibody targeting α(1–3)-Gal epitope (Chung et al. 2008). The immunogenicity of Fab oligosaccharides is required to be controlled by removing the glycosylation sequon in the V_H and/or V_L or changing or engineering host cell lines to add human-like oligosaccharides. On the other hand, Fab glycosylation may contribute positively to solubility and stability and can be beneficial for improved mAb formulation at higher concentrations (>100 mg/ml). Considering the benefit and disadvantages of Fab glycosylation and the essential demand for product consistency, it offers an additional challenge to the biopharmaceutical industry.

1.7 Conclusion

Antibodies are amongst the most structurally and functionally defined glycoproteins, particularly as potential and/or approved biopharmaceuticals. Advanced orthogonal analytical methods have revealed multiple PTMs and consequent structural micro-heterogeneity. The challenge remains to define the functional diversity of each component within the observed structural micro-heterogeneity and the subsequent impact on therapeutic efficacy and consequent patient benefit. Unintended deviations from the primary amino acid sequence of a biopharmaceutical, as defined by the DNA sequence, must be detected since such products may exhibit modulations of essential functions and/or represent non-self-structures and potential immunogenicity. However, purposeful genetic engineering has been employed for the generation of new/novel constructs expressing a selected profile of effector functions that may be of therapeutic advantage. Micro-heterogeneities at defined residues of a given mAb should be compared with that observed for natural IgG, isolated from human serum, or mAb isolated from the serum of patients following treatment. These data inform the selection amongst potential mAbs being developed as therapeutics and particularly to avoid mAbs having amino acid residues within the CDRs that are susceptible to degradation. During early development, the potential to satisfy established CQAs are assessed and employed to establish comparability between batches and before and after process changes. Thus, a thorough understanding of PTMs of mAbs is critical to the establishment of comparability, successful clinical development, and commercial supply.

The oligosaccharides at the Asn297 sites of the Fc are a CQA, and the biopharmaceutical companies have developed several protocols for the production of non-fucosylated IgG mAbs, e.g., to enhance ADCC activity for applications in cancer therapy. Further glycoengineering of mAbs is being pursued to develop next-generation mAbs bearing selected homogeneous oligosaccharides with enhanced effector functions; since they are natural glycoforms, this approach does not render them immunogenic. Importantly, while, in general, the association of Fc glycoforms with clinical efficacy has been established, the further impact of sialylation, galactosylation, and fucosylation of IgG mAbs have not been fully elucidated. Long-standing differences have been reported in the literature on whether agalactosylated IgG is proinflammatory whereas sialylated and/or galactosylated IgG are anti-inflammatory in autoimmune diseases; alternatively, why and how anti-RhD antibodies prepared from hyperimmunized healthy anti-D donors have low fucose contents. These and other issues will be discussed in Chap. 15.

Acknowledgments Pauline M Rudd acknowledges Waters Corporation for research funding and donation of equipment.

Compliance with Ethical Standards

Conflict of Interest Yusuke Mimura, Radka Saldova, Yuka Mimura-Kimura, Pauline M Rudd and Roy Jefferis declare that they have no conflict of interest.

Ethical Approval This article does not contain any studies with human participants or animals performed by any of the authors.

References

Ahmad S, Moinuddin AA (2012) Immunological studies on glycated human IgG. Life Sci 90 (25–26):980–987. https://doi.org/10.1016/j.lfs.2012.05.002

Alt N, Zhang TY, Motchnik P, Taticek R, Quarmby V, Schlothauer T, Beck H, Emrich T, Harris RJ (2016) Determination of critical quality attributes for monoclonal antibodies using quality by design principles. Biologicals 44(5):291–305. https://doi.org/10.1016/j.biologicals.2016.06.005

Ambrogelly A, Gozo S, Katiyar A, Dellatore S, Kune Y, Bhat R, Sun J, Li N, Wang D, Nowak C, Neill A, Ponniah G, King C, Mason B, Beck A, Liu H (2018) Analytical comparability study of recombinant monoclonal antibody therapeutics. MAbs 10(4):513–538. https://doi.org/10.1080/19420862.2018.1438797

Antes B, Amon S, Rizzi A, Wiederkum S, Kainer M, Szolar O, Fido M, Kircheis R, Nechansky A (2007) Analysis of lysine clipping of a humanized Lewis-Y specific IgG antibody and its relation to Fc-mediated effector function. J Chromatogr B Analyt Technol Biomed Life Sci 852(1–2):250–256. https://doi.org/10.1016/j.jchromb.2007.01.024

Arnold JN, Wormald MR, Sim RB, Rudd PM, Dwek RA (2006) The impact of glycosylation on the biological function and structure of human immunoglobulins. Annu Rev Immunol 25:21–50

Arvinte T, Palais C, Green-Trexler E, Gregory S, Mach H, Narasimhan C, Shameem M (2013) Aggregation of biopharmaceuticals in human plasma and human serum: implications for drug research and development. MAbs 5(3):491–500. https://doi.org/10.4161/mabs.24245

Banks DD, Gadgil HS, Pipes GD, Bondarenko PV, Hobbs V, Scavezze JL, Kim J, Jiang XR, Mukku V, Dillon TM (2008) Removal of cysteinylation from an unpaired sulfhydryl in the variable region of a recombinant monoclonal IgG1 antibody improves homogeneity, stability, and biological activity. J Pharm Sci 97(2):775–790. https://doi.org/10.1002/jps.21014

Banks DD, Hambly DM, Scavezze JL, Siska CC, Stackhouse NL, Gadgil HS (2009) The effect of sucrose hydrolysis on the stability of protein therapeutics during accelerated formulation studies. J Pharm Sci 98(12):4501–4510. https://doi.org/10.1002/jps.21749

Barb AW, Brady EK, Prestegard JH (2009) Branch-specific sialylation of IgG-Fc glycans by ST6Gal-I. Biochemistry 48(41):9705–9707. https://doi.org/10.1021/bi901430h

Beck A, Liu H (2019) Macro- and micro-heterogeneity of natural and recombinant IgG antibodies. Antibodies (Basel) 8(1):18. https://doi.org/10.3390/antib8010018

Bertolotti-Ciarlet A, Wang W, Lownes R, Pristatsky P, Fang Y, McKelvey T, Li Y, Li Y, Drummond J, Prueksaritanont T, Vlasak J (2009) Impact of methionine oxidation on the binding of human IgG1 to Fc Rn and Fc gamma receptors. Mol Immunol 46(8–9):1878–1882. https://doi.org/10.1016/j.molimm.2009.02.002

Butko M, Pallat H, Cordoba A, Yu XC (2014) Recombinant antibody color resulting from advanced glycation end product modifications. Anal Chem 86(19):9816–9823. https://doi.org/10.1021/ac5024099

Cai B, Pan H, Flynn GC (2011) C-terminal lysine processing of human immunoglobulin G2 heavy chain in vivo. Biotechnol Bioeng 108(2):404–412. https://doi.org/10.1002/bit.22933

Chaderjian WB, Chin ET, Harris RJ, Etcheverry TM (2005) Effect of copper sulfate on performance of a serum-free CHO cell culture process and the level of free thiol in the recombinant antibody expressed. Biotechnol Prog 21(2):550–553. https://doi.org/10.1021/bp0497029

Chelius D, Rehder DS, Bondarenko PV (2005) Identification and characterization of deamidation sites in the conserved regions of human immunoglobulin gamma antibodies. Anal Chem 77 (18):6004–6011. https://doi.org/10.1021/ac050672d

Chelius D, Jing K, Lueras A, Rehder DS, Dillon TM, Vizel A, Rajan RS, Li T, Treuheit MJ, Bondarenko PV (2006) Formation of pyroglutamic acid from N-terminal glutamic acid in immunoglobulin gamma antibodies. Anal Chem 78(7):2370–2376. https://doi.org/10.1021/ac051827k

Cheng HD, Tirosh I, de Haan N, Stockmann H, Adamczyk B, McManus CA, O'Flaherty R, Greville G, Saldova R, Bonilla FA, Notarangelo LD, Driessen GJ, Holm IA, Rudd PM, Wuhrer M, Ackerman ME, Nigrovic PA (2020) IgG Fc glycosylation as an axis of humoral immunity in childhood. J Allergy Clin Immunol 145(2):710–713.e719. https://doi.org/10.1016/j.jaci.2019.10.012

Chennamsetty N, Helk B, Voynov V, Kayser V, Trout BL (2009a) Aggregation-prone motifs in human immunoglobulin G. J Mol Biol 391(2):404–413. https://doi.org/10.1016/j.jmb.2009.06.028

Chennamsetty N, Voynov V, Kayser V, Helk B, Trout BL (2009b) Design of therapeutic proteins with enhanced stability. Proc Natl Acad Sci U S A 106(29):11937–11942. https://doi.org/10.1073/pnas.0904191106

Chicooree N, Unwin RD, Griffiths JR (2015) The application of targeted mass spectrometry-based strategies to the detection and localization of post-translational modifications. Mass Spectrom Rev 34(6):595–626. https://doi.org/10.1002/mas.21421

Chumsae C, Gaza-Bulseco G, Sun J, Liu H (2007) Comparison of methionine oxidation in thermal stability and chemically stressed samples of a fully human monoclonal antibody. J Chromatogr B Analyt Technol Biomed Life Sci 850(1–2):285–294. https://doi.org/10.1016/j.jchromb.2006.11.050

Chumsae C, Gaza-Bulseco G, Liu H (2009) Identification and localization of unpaired cysteine residues in monoclonal antibodies by fluorescence labeling and mass spectrometry. Anal Chem 81(15):6449–6457. https://doi.org/10.1021/ac900815z

Chung CH, Mirakhur B, Chan E, Le QT, Berlin J, Morse M, Murphy BA, Satinover SM, Hosen J, Mauro D, Slebos RJ, Zhou Q, Gold D, Hatley T, Hicklin DJ, Platts-Mills TA (2008) Cetuximab-induced anaphylaxis and IgE specific for galactose-alpha-1,3-galactose. N Engl J Med 358 (11):1109–1117. https://doi.org/10.1056/NEJMoa074943

Davies AM, Rispens T, den Bleker TH, McDonnell JM, Gould HJ, Aalberse RC, Sutton BJ (2013) Crystal structure of the human IgG4 C(H)3 dimer reveals the role of Arg409 in the mechanism of Fab-arm exchange. Mol Immunol 54(1):1–7. https://doi.org/10.1016/j.molimm.2012.10.029

de Haan N, Falck D, Wuhrer M (2020) Monitoring of immunoglobulin N- and O-glycosylation in health and disease. Glycobiology 30(4):226–240. https://doi.org/10.1093/glycob/cwz048

Dick LW Jr, Kim C, Qiu D, Cheng KC (2007) Determination of the origin of the N-terminal pyroglutamate variation in monoclonal antibodies using model peptides. Biotechnol Bioeng 97(3):544–553. https://doi.org/10.1002/bit.21260

Dick LW Jr, Qiu D, Mahon D, Adamo M, Cheng KC (2008) C-terminal lysine variants in fully human monoclonal antibodies: investigation of test methods and possible causes. Biotechnol Bioeng 100(6):1132–1143. https://doi.org/10.1002/bit.21855

Dillon TM, Ricci MS, Vezina C, Flynn GC, Liu YD, Rehder DS, Plant M, Henkle B, Li Y, Deechongkit S, Varnum B, Wypych J, Balland A, Bondarenko PV (2008) Structural and functional characterization of disulfide isoforms of the human IgG2 subclass. J Biol Chem 283(23):16206–16215. https://doi.org/10.1074/jbc.M709988200

Dunn-Walters D, Boursier L, Spencer J (2000) Effect of somatic hypermutation on potential N-glycosylation sites in human immunoglobulin heavy chain variable regions. Mol Immunol 37(3–4):107–113

Edelman GM, Cunningham BA, Gall WE, Gottlieb PD, Rutishauser U, Waxdal MJ (1969) The covalent structure of an entire gammaG immunoglobulin molecule. Proc Natl Acad Sci U S A 63(1):78–85. https://doi.org/10.1073/pnas.63.1.78

Ewert S, Huber T, Honegger A, Pluckthun A (2003) Biophysical properties of human antibody variable domains. J Mol Biol 325(3):531–553. https://doi.org/10.1016/s0022-2836(02)01237-8

Fischer S, Hoernschemeyer J, Mahler HC (2008) Glycation during storage and administration of monoclonal antibody formulations. Eur J Pharm Biopharm 70(1):42–50. https://doi.org/10.1016/j.ejpb.2008.04.021

Gadgil HS, Bondarenko PV, Pipes G, Rehder D, McAuley A, Perico N, Dillon T, Ricci M, Treuheit M (2007) The LC/MS analysis of glycation of IgG molecules in sucrose containing formulations. J Pharm Sci 96(10):2607–2621. https://doi.org/10.1002/jps.20966

Gala FA, Morrison SL (2004) V region carbohydrate and antibody expression. J Immunol 172(9):5489–5494

Galili U, Anaraki F, Thall A, Hill-Black C, Radic M (1993) One percent of human circulating B lymphocytes are capable of producing the natural anti-Gal antibody. Blood 82(8):2485–2493

Gaza-Bulseco G, Faldu S, Hurkmans K, Chumsae C, Liu H (2008) Effect of methionine oxidation of a recombinant monoclonal antibody on the binding affinity to protein A and protein G. J Chromatogr B Analyt Technol Biomed Life Sci 870(1):55–62. https://doi.org/10.1016/j.jchromb.2008.05.045

Geiger T, Clarke S (1987) Deamidation, isomerization, and racemization at asparaginyl and aspartyl residues in peptides. Succinimide-linked reactions that contribute to protein degradation. J Biol Chem 262(2):785–794

Goetze AM, Liu YD, Arroll T, Chu L, Flynn GC (2012) Rates and impact of human antibody glycation in vivo. Glycobiology 22(2):221–234. https://doi.org/10.1093/glycob/cwr141

Grey AA, Narasimhan S, Brisson JR, Schachter H, Carver JP (1982) Structure of the glycopeptides of a human gamma 1-immunoglobulin G (Tem) myeloma protein as determined by 360-megahertz nuclear magnetic resonance spectroscopy. Can J Biochem 60(12):1123–1131

Gu S, Wen D, Weinreb PH, Sun Y, Zhang L, Foley SF, Kshirsagar R, Evans D, Mi S, Meier W, Pepinsky RB (2010) Characterization of trisulfide modification in antibodies. Anal Biochem 400(1):89–98. https://doi.org/10.1016/j.ab.2010.01.019

Gudelj I, Lauc G, Pezer M (2018) Immunoglobulin G glycosylation in aging and diseases. Cell Immunol 333:65–79. https://doi.org/10.1016/j.cellimm.2018.07.009

Hafkenscheid L, Bondt A, Scherer HU, Huizinga TW, Wuhrer M, Toes RE, Rombouts Y (2017) Structural analysis of variable domain glycosylation of anti-citrullinated protein antibodies in rheumatoid arthritis reveals the presence of highly sialylated glycans. Mol Cell Proteomics 16 (2):278–287. https://doi.org/10.1074/mcp.M116.062919

Hafkenscheid L, de Moel E, Smolik I, Tanner S, Meng X, Jansen BC, Bondt A, Wuhrer M, Huizinga TWJ, Toes REM, El-Gabalawy H, Scherer HU (2019) N-linked glycans in the variable domain of IgG anti-citrullinated protein antibodies predict the development of rheumatoid arthritis. Arthritis Rheumatol 71(10):1626–1633. https://doi.org/10.1002/art.40920

Harris RJ (1995) Processing of C-terminal lysine and arginine residues of proteins isolated from mammalian cell culture. J Chromatogr A 705(1):129–134. https://doi.org/10.1016/0021-9673 (94)01255-d

Harris RJ (2005) Heterogeneity of recombinant antibodies: linking structure to function. Dev Biol (Basel) 122:117–127

Harris RJ, Kabakoff B, Macchi FD, Shen FJ, Kwong M, Andya JD, Shire SJ, Bjork N, Totpal K, Chen AB (2001) Identification of multiple sources of charge heterogeneity in a recombinant antibody. J Chromatogr B Biomed Sci Appl 752(2):233–245. https://doi.org/10.1016/s0378-4347(00)00548-x

Hmiel LK, Brorson KA, Boyne MT 2nd (2015) Post-translational structural modifications of immunoglobulin G and their effect on biological activity. Anal Bioanal Chem 407(1):79–94. https://doi.org/10.1007/s00216-014-8108-x

Holland M, Yagi H, Takahashi N, Kato K, Savage CO, Goodall DM, Jefferis R (2006) Differential glycosylation of polyclonal IgG, IgG-Fc and IgG-Fab isolated from the sera of patients with ANCA-associated systemic vasculitis. Biochim Biophys Acta 1760(4):669–677

Hu Z, Tang D, Misaghi S, Jiang G, Yu C, Yim M, Shaw D, Snedecor B, Laird MW, Shen A (2017) Evaluation of heavy chain C-terminal deletions on productivity and product quality of monoclonal antibodies in Chinese hamster ovary (CHO) cells. Biotechnol Prog 33(3):786–794. https://doi.org/10.1002/btpr.2444

Huang L, Lu J, Wroblewski VJ, Beals JM, Riggin RM (2005) In vivo deamidation characterization of monoclonal antibody by LC/MS/MS. Anal Chem 77(5):1432–1439. https://doi.org/10.1021/ac0494174

Huh JH, White AJ, Brych SR, Franey H, Matsumura M (2013) The identification of free cysteine residues within antibodies and a potential role for free cysteine residues in covalent aggregation because of agitation stress. J Pharm Sci 102(6):1701–1711. https://doi.org/10.1002/jps.23505

Hutterer KM, Hong RW, Lull J, Zhao X, Wang T, Pei R, Le ME, Borisov O, Piper R, Liu YD, Petty K, Apostol I, Flynn GC (2013) Monoclonal antibody disulfide reduction during manufacturing: untangling process effects from product effects. MAbs 5(4):608–613. https://doi.org/10.4161/mabs.24725

Jefferis R (2009) Glycosylation as a strategy to improve antibody-based therapeutics. Nat Rev Drug Discov 8(3):226–234. https://doi.org/10.1038/nrd2804

Jefferis R (2016) Posttranslational modifications and the immunogenicity of biotherapeutics. J Immunol Res 2016:5358272. https://doi.org/10.1155/2016/5358272

Jefferis R (2017a) Antibody posttranslational modifications. In: Liu C, Morrow KJJ (eds) Biosimilars of monoclonal antibodies: a practical guide to manufacturing, preclinical, and clinical development. Wiley, Chichester, pp 155–199. https://doi.org/10.1002/9781118940648

Jefferis R (2017b) Characterization of biosimilar biologics: the link between structure and functions. In: Endrenyi L, Declerck P, Chow S-C (eds) Drugs and the pharmaceutical sciences, vol 216. CRC Press, Boca Raton, pp 109–149

Jefferis R (2017c) Recombinant proteins and monoclonal antibodies. Adv Biochem Eng Biotechnol. https://doi.org/10.1007/10_2017_32

Jefferis R, Lund J, Mizutani H, Nakagawa H, Kawazoe Y, Arata Y, Takahashi N (1990) A comparative study of the N-linked oligosaccharide structures of human IgG subclass proteins. Biochem J 268(3):529–537

Jiang G, Yu C, Yadav DB, Hu Z, Amurao A, Duenas E, Wong M, Iverson M, Zheng K, Lam X, Chen J, Vega R, Ulufatu S, Leddy C, Davis H, Shen A, Wong PY, Harris R, Wang YJ, Li D (2016) Evaluation of heavy-chain C-terminal deletion on product quality and pharmacokinetics of monoclonal antibodies. J Pharm Sci 105(7):2066–2072. https://doi.org/10.1016/j.xphs.2016.04.027

Johnson KA, Paisley-Flango K, Tangarone BS, Porter TJ, Rouse JC (2007) Cation exchange-HPLC and mass spectrometry reveal C-terminal amidation of an IgG1 heavy chain. Anal Biochem 360 (1):75–83. https://doi.org/10.1016/j.ab.2006.10.012

Kabat EA, Wu TT, Perry HM, Gottesman KS, Foeller C (1991) Sequences of proteins of immunological interest, 5th edn. NIH Publication, National Institutes of Health, Bethesda, MD

Kaschak T, Boyd D, Lu F, Derfus G, Kluck B, Nogal B, Emery C, Summers C, Zheng K, Bayer R, Amanullah A, Yan B (2011) Characterization of the basic charge variants of a human IgG1: effect of copper concentration in cell culture media. MAbs 3(6):577–583. https://doi.org/10.4161/mabs.3.6.17959

Kempers AC, Hafkenscheid L, Scherer HU, Toes REM (2018) Variable domain glycosylation of ACPA-IgG: a missing link in the maturation of the ACPA response? Clin Immunol 186:34–37. https://doi.org/10.1016/j.clim.2017.09.001

Kepert JF, Cromwell M, Engler N, Finkler C, Gellermann G, Gennaro L, Harris R, Iverson R, Kelley B, Krummen L, McKnight N, Motchnik P, Schnaible V, Taticek R (2016) Establishing a control system using QbD principles. Biologicals 44(5):319–331. https://doi.org/10.1016/j.biologicals.2016.06.003

Khawli LA, Goswami S, Hutchinson R, Kwong ZW, Yang J, Wang X, Yao Z, Sreedhara A, Cano T, Tesar D, Nijem I, Allison DE, Wong PY, Kao YH, Quan C, Joshi A, Harris RJ, Motchnik P (2010) Charge variants in IgG1: isolation, characterization, in vitro binding properties and pharmacokinetics in rats. MAbs 2(6):613–624. https://doi.org/10.4161/mabs.2.6.13333

Kita A, Ponniah G, Nowak C, Liu H (2016) Characterization of cysteinylation and trisulfide bonds in a recombinant monoclonal antibody. Anal Chem 88(10):5430–5437. https://doi.org/10.1021/acs.analchem.6b00822

Koterba KL, Borgschulte T, Laird MW (2012) Thioredoxin 1 is responsible for antibody disulfide reduction in CHO cell culture. J Biotechnol 157(1):261–267. https://doi.org/10.1016/j.jbiotec.2011.11.009

Kroon DJ, Baldwin-Ferro A, Lalan P (1992) Identification of sites of degradation in a therapeutic monoclonal antibody by peptide mapping. Pharm Res 9(11):1386–1393. https://doi.org/10.1023/a:1015894409623

Kshirsagar R, McElearney K, Gilbert A, Sinacore M, Ryll T (2012) Controlling trisulfide modification in recombinant monoclonal antibody produced in fed-batch cell culture. Biotechnol Bioeng 109(10):2523–2532. https://doi.org/10.1002/bit.24511

Lacy ER, Baker M, Brigham-Burke M (2008) Free sulfhydryl measurement as an indicator of antibody stability. Anal Biochem 382(1):66–68. https://doi.org/10.1016/j.ab.2008.07.016

Lakbub JC, Clark DF, Shah IS, Zhu Z, Go EP, Tolbert TJ, Desaire H (2016) Disulfide bond characterization of endogenous IgG3 monoclonal antibodies using LC-MS: an investigation of IgG3 disulfide-mediated isoforms. Anal Methods 8(31):6046–6055. https://doi.org/10.1039/C6AY01248E

Lewis DA, Guzzetta AW, Hancock WS, Costello M (1994) Characterization of humanized anti-TAC, an antibody directed against the interleukin 2 receptor, using electrospray ionization mass spectrometry by direct infusion, LC/MS, and MS/MS. Anal Chem 66(5):585–595. https://doi.org/10.1021/ac00077a003

Li W, Prabakaran P, Chen W, Zhu Z, Feng Y, Dimitrov DS (2016) Antibody aggregation: insights from sequence and structure. Antibodies (Basel) 5(3):19. https://doi.org/10.3390/antib5030019

Liu H, May K (2012) Disulfide bond structures of IgG molecules: structural variations, chemical modifications and possible impacts to stability and biological function. MAbs 4(1):17–23. https://doi.org/10.4161/mabs.4.1.18347

Liu H, Gaza-Bulseco G, Chumsae C (2008) Glutamine deamidation of a recombinant monoclonal antibody. Rapid Commun Mass Spectrom 22(24):4081–4088. https://doi.org/10.1002/rcm.3831

Liu YD, van Enk JZ, Flynn GC (2009) Human antibody Fc deamidation in vivo. Biologicals 37 (5):313–322. https://doi.org/10.1016/j.biologicals.2009.06.001

Liu YD, Goetze AM, Bass RB, Flynn GC (2011) N-terminal glutamate to pyroglutamate conversion in vivo for human IgG2 antibodies. J Biol Chem 286(13):11211–11217. https://doi.org/10.1074/jbc.M110.185041

Lu RM, Hwang YC, Liu IJ, Lee CC, Tsai HZ, Li HJ, Wu HC (2020) Development of therapeutic antibodies for the treatment of diseases. J Biomed Sci 27(1):1. https://doi.org/10.1186/s12929-019-0592-z

Luo S, Zhang B (2015) Dextrose-mediated aggregation of therapeutic monoclonal antibodies in human plasma: implication of isoelectric precipitation of complement proteins. MAbs 7 (6):1094–1103. https://doi.org/10.1080/19420862.2015.1087636

Martinez T, Guo A, Allen MJ, Han M, Pace D, Jones J, Gillespie R, Ketchem RR, Zhang Y, Balland A (2008) Disulfide connectivity of human immunoglobulin G2 structural isoforms. Biochemistry 47(28):7496–7508. https://doi.org/10.1021/bi800576c

McCann KJ, Ottensmeier CH, Callard A, Radcliffe CM, Harvey DJ, Dwek RA, Rudd PM, Sutton BJ, Hobby P, Stevenson FK (2008) Remarkable selective glycosylation of the immunoglobulin variable region in follicular lymphoma. Mol Immunol 45(6):1567–1572

Miller AK, Hambly DM, Kerwin BA, Treuheit MJ, Gadgil HS (2011) Characterization of site-specific glycation during process development of a human therapeutic monoclonal antibody. J Pharm Sci 100(7):2543–2550. https://doi.org/10.1002/jps.22504

Mimura Y, Jefferis R (2021) Human IgG glycosylation in inflammation and inflammatory disease. In: Barchi JJ (ed) Comprehensive glycoscience, vol 5, 2nd edn. Elsevier, pp 215–232

Mimura Y, Kabat EA, Tanaka T, Fujimoto M, Takeo K, Nakamura K (1995) Microheterogeneity of mouse antidextran monoclonal antibodies. Electrophoresis 16(1):116–123. https://doi.org/10.1002/elps.1150160121

Mimura Y, Nakamura K, Tanaka T, Fujimoto M (1998) Evidence of intra- and extracellular modifications of monoclonal IgG polypeptide chains generating charge heterogeneity. Electrophoresis 19(5):767–775. https://doi.org/10.1002/elps.1150190528

Mimura Y, Jefferis R, Mimura-Kimura Y, Abrahams J, Rudd PM (2009) Glycosylation of therapeutic IgGs. In: An Z (ed) Therapeutic monoclonal antibodies: from the bench to the clinic. Wiley, Hoboken, pp 67–89

Mimura Y, Katoh T, Saldova R, O'Flaherty R, Izumi T, Mimura-Kimura Y, Utsunomiya T, Mizukami Y, Yamamoto K, Matsumoto T, Rudd PM (2018) Glycosylation engineering of therapeutic IgG antibodies: challenges for the safety, functionality and efficacy. Protein Cell 9 (1):47–62. https://doi.org/10.1007/s13238-017-0433-3

Mo J, Jin R, Yan Q, Sokolowska I, Lewis MJ, Hu P (2018) Quantitative analysis of glycation and its impact on antigen binding. MAbs 10(3):406–415. https://doi.org/10.1080/19420862.2018.1438796

Montesino R, Calvo L, Vallin A, Rudd PM, Harvey DJ, Cremata JA (2012) Structural characterization of N-linked oligosaccharides on monoclonal antibody Nimotuzumab through process development. Biologicals 40(4):288–298. https://doi.org/10.1016/j.biologicals.2012.04.005

Nose M, Wigzell H (1983) Biological significance of carbohydrate chains on monoclonal antibodies. Proc Natl Acad Sci U S A 80(21):6632–6636

Ouellette D, Alessandri L, Chin A, Grinnell C, Tarcsa E, Radziejewski C, Correia I (2010) Studies in serum support rapid formation of disulfide bond between unpaired cysteine residues in the VH domain of an immunoglobulin G1 molecule. Anal Biochem 397(1):37–47. https://doi.org/10.1016/j.ab.2009.09.027

Pan H, Chen K, Chu L, Kinderman F, Apostol I, Huang G (2009) Methionine oxidation in human IgG2 Fc decreases binding affinities to protein A and FcRn. Protein Sci 18(2):424–433. https://doi.org/10.1002/pro.45

Pang Y, Wang WH, Reid GE, Hunt DF, Bruening ML (2015) Pepsin-containing membranes for controlled monoclonal antibody digestion prior to mass spectrometry analysis. Anal Chem 87 (21):10942–10949. https://doi.org/10.1021/acs.analchem.5b02739

Parekh R, Roitt I, Isenberg D, Dwek R, Rademacher T (1988) Age-related galactosylation of the N-linked oligosaccharides of human serum IgG. J Exp Med 167(5):1731–1736

Pound JD, Lund J, Jefferis R (1993) Aglycosylated chimaeric human IgG3 can trigger the human phagocyte respiratory burst. Mol Immunol 30(3):233–241

Pristatsky P, Cohen SL, Krantz D, Acevedo J, Ionescu R, Vlasak J (2009) Evidence for trisulfide bonds in a recombinant variant of a human IgG2 monoclonal antibody. Anal Chem 81 (15):6148–6155. https://doi.org/10.1021/ac9006254

Pucic M, Knezevic A, Vidic J, Adamczyk B, Novokmet M, Polasek O, Gornik O, Supraha-Goreta S, Wormald MR, Redzic I, Campbell H, Wright A, Hastie ND, Wilson JF, Rudan I, Wuhrer M, Rudd PM, Josic D, Lauc G (2011) High throughput isolation and glycosylation analysis of IgG-variability and heritability of the IgG glycome in three isolated human populations. Mol Cell Proteomics 10(10):M111 010090. https://doi.org/10.1074/mcp.M111.010090

Qian J, Liu T, Yang L, Daus A, Crowley R, Zhou Q (2007) Structural characterization of N-linked oligosaccharides on monoclonal antibody cetuximab by the combination of orthogonal matrix-assisted laser desorption/ionization hybrid quadrupole-quadrupole time-of-flight tandem mass spectrometry and sequential enzymatic digestion. Anal Biochem 364(1):8–18

Quan C, Alcala E, Petkovska I, Matthews D, Canova-Davis E, Taticek R, Ma S (2008) A study in glycation of a therapeutic recombinant humanized monoclonal antibody: where it is, how it got there, and how it affects charge-based behavior. Anal Biochem 373(2):179–191. https://doi.org/10.1016/j.ab.2007.09.027

Radcliffe CM, Arnold JN, Suter DM, Wormald MR, Harvey DJ, Royle L, Mimura Y, Kimura Y, Sim RB, Inoges S, Rodriguez-Calvillo M, Zabalegui N, de Cerio AL, Potter KN, Mockridge CI, Dwek RA, Bendandi M, Rudd PM, Stevenson FK (2007) Human follicular lymphoma cells contain oligomannose glycans in the antigen-binding site of the B-cell receptor. J Biol Chem 282(10):7405–7415

Raju TS, Briggs JB, Borge SM, Jones AJ (2000) Species-specific variation in glycosylation of IgG: evidence for the species-specific sialylation and branch-specific galactosylation and importance for engineering recombinant glycoprotein therapeutics. Glycobiology 10(5):477–486

Rispens T, Meesters J, den Bleker TH, Ooijevaar-De Heer P, Schuurman J, Parren PW, Labrijn A, Aalberse RC (2013) Fc-Fc interactions of human IgG4 require dissociation of heavy chains and are formed predominantly by the intra-chain hinge isomer. Mol Immunol 53(1–2):35–42. https://doi.org/10.1016/j.molimm.2012.06.012

Rispens T, Davies AM, Ooijevaar-de Heer P, Absalah S, Bende O, Sutton BJ, Vidarsson G, Aalberse RC (2014) Dynamics of inter-heavy chain interactions in human immunoglobulin G (IgG) subclasses studied by kinetic Fab arm exchange. J Biol Chem 289(9):6098–6109. https://doi.org/10.1074/jbc.M113.541813

Roberts GD, Johnson WP, Burman S, Anumula KR, Carr SA (1995) An integrated strategy for structural characterization of the protein and carbohydrate components of monoclonal antibodies: application to anti-respiratory syncytial virus MAb. Anal Chem 67(20):3613–3625. https://doi.org/10.1021/ac00116a001

Robinson NE, Robinson AB (2001) Deamidation of human proteins. Proc Natl Acad Sci U S A 98 (22):12409–12413. https://doi.org/10.1073/pnas.221463198

Rouet R, Lowe D, Christ D (2014) Stability engineering of the human antibody repertoire. FEBS Lett 588(2):269–277. https://doi.org/10.1016/j.febslet.2013.11.029

Santos J, Pujols J, Pallares I, Iglesias V, Ventura S (2020) Computational prediction of protein aggregation: advances in proteomics, conformation-specific algorithms and biotechnological applications. Comput Struct Biotechnol J 18:1403–1413. https://doi.org/10.1016/j.csbj.2020.05.026

Sheeley DM, Merrill BM, Taylor LC (1997) Characterization of monoclonal antibody glycosylation: comparison of expression systems and identification of terminal alpha-linked galactose. Anal Biochem 247(1):102–110

Sinha S, Zhang L, Duan S, Williams TD, Vlasak J, Ionescu R, Topp EM (2009) Effect of protein structure on deamidation rate in the Fc fragment of an IgG1 monoclonal antibody. Protein Sci 18 (8):1573–1584. https://doi.org/10.1002/pro.173

Stadlmann J, Pabst M, Kolarich D, Kunert R, Altmann F (2008) Analysis of immunoglobulin glycosylation by LC-ESI-MS of glycopeptides and oligosaccharides. Proteomics 8 (14):2858–2871. https://doi.org/10.1002/pmic.200700968

Strohl WR (2018) Current progress in innovative engineered antibodies. Protein Cell 9(1):86–120. https://doi.org/10.1007/s13238-017-0457-8

Sydow JF, Lipsmeier F, Larraillet V, Hilger M, Mautz B, Molhoj M, Kuentzer J, Klostermann S, Schoch J, Voelger HR, Regula JT, Cramer P, Papadimitriou A, Kettenberger H (2014) Structure-based prediction of asparagine and aspartate degradation sites in antibody variable regions. PLoS One 9(6):e100736. https://doi.org/10.1371/journal.pone.0100736

Takeuchi M, Takasaki S, Miyazaki H, Kato T, Hoshi S, Kochibe N, Kobata A (1988) Comparative study of the asparagine-linked sugar chains of human erythropoietins purified from urine and the culture medium of recombinant Chinese hamster ovary cells. J Biol Chem 263(8):3657–3663

Tang L, Sundaram S, Zhang J, Carlson P, Matathia A, Parekh B, Zhou Q, Hsieh MC (2013) Conformational characterization of the charge variants of a human IgG1 monoclonal antibody using H/D exchange mass spectrometry. MAbs 5(1):114–125. https://doi.org/10.4161/mabs.22695

Tangvoranuntakul P, Gagneux P, Diaz S, Bardor M, Varki N, Varki A, Muchmore E (2003) Human uptake and incorporation of an immunogenic nonhuman dietary sialic acid. Proc Natl Acad Sci U S A 100(21):12045–12050

Tao MH, Morrison SL (1989) Studies of aglycosylated chimeric mouse-human IgG. Role of carbohydrate in the structure and effector functions mediated by the human IgG constant region. J Immunol 143(8):2595–2601

Tous GI, Wei Z, Feng J, Bilbulian S, Bowen S, Smith J, Strouse R, McGeehan P, Casas-Finet J, Schenerman MA (2005) Characterization of a novel modification to monoclonal antibodies: thioether cross-link of heavy and light chains. Anal Chem 77(9):2675–2682. https://doi.org/10.1021/ac0500582

Tsubaki M, Terashima I, Kamata K, Koga A (2013) C-terminal modification of monoclonal antibody drugs: amidated species as a general product-related substance. Int J Biol Macromol 52:139–147. https://doi.org/10.1016/j.ijbiomac.2012.09.016

van den Eijnden DH, Joziasse DH, Dorland L, van Halbeek H, Vliegenthart JF, Schmid K (1980) Specificity in the enzymic transfer of sialic acid to the oligosaccharide branches of b1- and triantennary glycopeptides of alpha 1-acid glycoprotein. Biochem Biophys Res Commun 92 (3):839–845

van der Kant R, Karow-Zwick AR, Van Durme J, Blech M, Gallardo R, Seeliger D, Assfalg K, Baatsen P, Compernolle G, Gils A, Studts JM, Schulz P, Garidel P, Schymkowitz J, Rousseau F (2017) Prediction and reduction of the aggregation of monoclonal antibodies. J Mol Biol 429 (8):1244–1261. https://doi.org/10.1016/j.jmb.2017.03.014

Wang W, Singh S, Zeng DL, King K, Nema S (2007) Antibody structure, instability, and formulation. J Pharm Sci 96(1):1–26. https://doi.org/10.1002/jps.20727

Wang X, Das TK, Singh SK, Kumar S (2009) Potential aggregation prone regions in biotherapeutics: a survey of commercial monoclonal antibodies. MAbs 1(3):254–267. https://doi.org/10.4161/mabs.1.3.8035

Wang W, Vlasak J, Li Y, Pristatsky P, Fang Y, Pittman T, Roman J, Wang Y, Prueksaritanont T, Ionescu R (2011) Impact of methionine oxidation in human IgG1 Fc on serum half-life of monoclonal antibodies. Mol Immunol 48(6–7):860–866. https://doi.org/10.1016/j.molimm.2010.12.009

Wang X, An Z, Luo W, Xia N, Zhao Q (2018) Molecular and functional analysis of monoclonal antibodies in support of biologics development. Protein Cell 9(1):74–85. https://doi.org/10.1007/s13238-017-0447-x

Wei B, Berning K, Quan C, Zhang YT (2017) Glycation of antibodies: modification, methods and potential effects on biological functions. MAbs 9(4):586–594. https://doi.org/10.1080/19420862.2017.1300214

Wypych J, Li M, Guo A, Zhang Z, Martinez T, Allen MJ, Fodor S, Kelner DN, Flynn GC, Liu YD, Bondarenko PV, Ricci MS, Dillon TM, Balland A (2008) Human IgG2 antibodies display disulfide-mediated structural isoforms. J Biol Chem 283(23):16194–16205. https://doi.org/10.1074/jbc.M709987200

Xie H, Chakraborty A, Ahn J, Yu YQ, Dakshinamoorthy DP, Gilar M, Chen W, Skilton SJ, Mazzeo JR (2010) Rapid comparison of a candidate biosimilar to an innovator monoclonal antibody with advanced liquid chromatography and mass spectrometry technologies. MAbs 2(4):379–394. https://doi.org/10.4161/mabs.11986

Xu Y, Wang D, Mason B, Rossomando T, Li N, Liu D, Cheung JK, Xu W, Raghava S, Katiyar A, Nowak C, Xiang T, Dong DD, Sun J, Beck A, Liu H (2019) Structure, heterogeneity and developability assessment of therapeutic antibodies. MAbs 11(2):239–264. https://doi.org/10.1080/19420862.2018.1553476

Yin S, Pastuskovas CV, Khawli LA, Stults JT (2013) Characterization of therapeutic monoclonal antibodies reveals differences between in vitro and in vivo time-course studies. Pharm Res 30(1):167–178. https://doi.org/10.1007/s11095-012-0860-z

Yu L, Vizel A, Huff MB, Young M, Remmele RL Jr, He B (2006) Investigation of N-terminal glutamate cyclization of recombinant monoclonal antibody in formulation development. J Pharm Biomed Anal 42(4):455–463. https://doi.org/10.1016/j.jpba.2006.05.008

Zhang Q, Flynn GC (2013) Cysteine racemization on IgG heavy and light chains. J Biol Chem 288(48):34325–34335. https://doi.org/10.1074/jbc.M113.506915

Zhang B, Yang Y, Yuk I, Pai R, McKay P, Eigenbrot C, Dennis M, Katta V, Francissen KC (2008) Unveiling a glycation hot spot in a recombinant humanized monoclonal antibody. Anal Chem 80(7):2379–2390. https://doi.org/10.1021/ac701810q

Zhang Q, Schenauer MR, McCarter JD, Flynn GC (2013) IgG1 thioether bond formation in vivo. J Biol Chem 288(23):16371–16382. https://doi.org/10.1074/jbc.M113.468397

Zhao W, Zhang J, Lu Y, Wang R (2001) The vasorelaxant effect of H(2)S as a novel endogenous gaseous K(ATP) channel opener. EMBO J 20(21):6008–6016. https://doi.org/10.1093/emboj/20.21.6008

Zhu A, Hurst R (2002) Anti-N-glycolylneuraminic acid antibodies identified in healthy human serum. Xenotransplantation 9(6):376–381

Zhu D, McCarthy H, Ottensmeier CH, Johnson P, Hamblin TJ, Stevenson FK (2002) Acquisition of potential N-glycosylation sites in the immunoglobulin variable region by somatic mutation is a distinctive feature of follicular lymphoma. Blood 99(7):2562–2568

Zhu D, Ottensmeier CH, Du MQ, McCarthy H, Stevenson FK (2003) Incidence of potential glycosylation sites in immunoglobulin variable regions distinguishes between subsets of Burkitt's lymphoma and mucosa-associated lymphoid tissue lymphoma. Br J Haematol 120(2):217–222

Part I
Analytical Methods

Chapter 2
Lectin and Liquid Chromatography-Based Methods for Immunoglobulin (G) Glycosylation Analysis

Tea Petrović and Irena Trbojević-Akmačić

Contents

2.1	Introduction	32
	2.1.1 Glycosylation	32
	2.1.2 Immunoglobulins	33
2.2	Liquid Chromatography	35
	2.2.1 Immunoglobulin Purification by Liquid Chromatography	37
	2.2.2 N-Glycan Analysis by Liquid Chromatography	40
	2.2.3 O-Glycan Analysis by Liquid Chromatography	47
	2.2.4 Liquid Chromatography Coupled to Mass Spectrometry	49
2.3	Lectin Techniques	53
	2.3.1 Lectin Chromatography	55
	2.3.2 Lectin Microarrays	56
2.4	Perspectives	58
References		59

Abstract Immunoglobulin (Ig) glycosylation has been shown to dramatically affect its structure and effector functions. Ig glycosylation changes have been associated with different diseases and show a promising biomarker potential for diagnosis and prognosis of disease advancement. On the other hand, therapeutic biomolecules based on structural and functional features of Igs demand stringent quality control during the production process to ensure their safety and efficacy. Liquid chromatography (LC) and lectin-based methods are routinely used in Ig glycosylation analysis complementary to other analytical methods, e.g., mass spectrometry and capillary electrophoresis. This chapter covers analytical approaches based on LC and lectins used in low- and high-throughput N- and O-glycosylation analysis of Igs, with the focus on immunoglobulin G (IgG) applications. General principles and practical examples of the most often used LC methods for Ig purification are described, together with typical workflows for N- and O-glycan analysis on the level of free glycans, glycopeptides, subunits, or intact Igs. Lectin chromatography is

---

T. Petrović · I. Trbojević-Akmačić (✉)
Glycoscience Research Laboratory, Genos Ltd., Zagreb, Croatia
e-mail: iakmacic@genos.hr

a historical approach for the analysis of lectin-carbohydrate interactions and glycoprotein purification but is still being used as a valuable tool in Igs purification and glycan analysis. On the other hand, lectin microarrays have found their application in the rapid screening of glycan profiles on intact proteins.

Keywords Critical quality attribute · Glycosylation analysis · Immunoglobulin · Lectins · Lectin chromatography · Lectin microarrays · Liquid chromatography · Mass spectrometry · N-glycans · O-glycans

Abbreviations

2-AA	2-aminobenzoic acid
2-AB	2-aminobenzamide
2-PA	2-aminopyridine
2-PB	2-picoline borane
Ab	Antibody
AC	Affinity chromatography
ADCC	Antibody-dependent cell-mediated cytotoxicity
AEX-RP	Anion-exchange-reverse phase
BEH	Bridged ethylene hybrid, stationary phase
CBMs	Malectin carbohydrate-binding modules
CE	Capillary electrophoresis
CF	Chromatofocusing
CGE-LIF	Capillary gel electrophoresis with laser-induced fluorescence detection
Con A	Concanavalin A
CTLD	C type lectin domain
CTLs	C type lectins
DEAE	Diethylaminoethyl
DMB	1,2-diamino-4,5-methylenedioxybenzene
DTT	Dithiothreitol
ECM	Extracellular matrix
EGF	Epidermal growth factor
ELLA	Enzyme-linked lectin assay
Endo F	Endoglycosidase F
Endo H	Endoglycosidase H
Endo S	Endoglycosidase S
ER	Endoplasmic reticulum
Fab	Fragment antigen binding
Fc	Fragment crystallizable
FcR	Fc receptor
GalNAc	N-acetylgalactosamine
GBP	Glycan binding proteins

GFC	Gel filtration chromatography
GlcNAc	N-acetylglucosamine
GPC	Gel permeation chromatography
GU	Glucose units
HILIC SPE	Hydrophilic interaction liquid chromatography solid-phase extraction
HLPC	High-performance liquid chromatography
HPAEC	High-performance anion-exchange chromatography
IEX	Ion exchange chromatography
Ig	Immunoglobulin (i.e., IgG = immunoglobulin G; IgA, IgM, IgD, IgE)
IVIg	Intravenous immunoglobulin
LC	Liquid chromatography
LC–ESI-MS	Liquid chromatography electrospray ionization mass spectrometry
mAbs	Monoclonal antibodies
MALDI-TOF–MS	Matrix-assisted laser desorption/ionization time-of-flight mass spectrometry
MBL	Mannan binding lectin
MS	Mass spectrometry
NHS	N-hydroxysuccinimide
pAbs	Polyclonal antibodies
PAD	Pulsed amperometric detection
PGC	Porous graphitized carbon
PNGase A	N-glycosidase A
PNGase F	N-glycosidase F
ProA	Procainamide (4-amino-N-[2-(diethylamino)ethyl] benzamide)
QqQ	Triple quadrupole
RP	Reverse phase
SDS	Sodium dodecyl sulfate
SEC	Size-exclusion chromatography
SNA	*Sambucus nigra agglutinin*
TFA	Trifluoroacetic acid
TiO_2	Titanium dioxide
UHPLC-FLR	Ultra-high-performance liquid chromatography with fluorescence detection
ZrO_2	Zirconium dioxide
κ	Kappa
λ	Lambda

2.1 Introduction

Currently, the glycobiology field is working hard to develop new and advance existing rapid, sensitive, and accurate methods for analyzing complex glycan structures in a high-throughput manner. The role of glycans has been studied for decades, and it is known that they play major metabolic, structural, and physical roles in biological systems (Varki et al. 2009). Changes in glycosylation have been observed in aging (Krištić et al. 2014) and different diseases such as alloimmune (Sonneveld et al. 2016; Wuhrer et al. 2009) and autoimmune diseases (Sharapov et al. 2019; Bermingham et al. 2018), cancers (Theodoratou et al. 2016), inflammatory diseases (Trbojevic Akmacic et al. 2015), as well as in infectious diseases (Vadrevu et al. 2018; Larsen et al. 2020), putting glycans into the focus of recent protein and biomarker research. Moreover, the analysis of protein glycosylation has been gaining importance in the pharmaceutical industry since glycans are recognized as critical quality attributes of therapeutic glycoproteins (O'Flaherty et al. 2018).

Glycosylation analysis is still a challenging task, although technological advancements have enabled robust and sensitive glycan analysis on different glycoprotein levels. Several complementary analytical technologies have been routinely used in glycan analysis, and detailed in-depth studies usually require multiple orthogonal approaches (which separate and detect oligosaccharides according to different principles, providing corroboratory qualitative and quantitative information). Liquid chromatography (LC) is a widely applied analytical technique for the detection and purification of (glyco)proteins. One of the fields of LC application is lectin affinity chromatography. It is a common method for the isolation, fractionation, and purification of carbohydrate-containing biomolecules (Mechref et al. 2008). Although LC, capillary electrophoresis (CE), and mass spectrometry (MS) are the most common methods for glycan analysis, lectin-based techniques (e.g., lectin microarrays) can be adopted as a complementary tool for glycan analysis.

2.1.1 Glycosylation

Glycosylation is one of the most common and complex co- and post-translational protein modifications. Glycosylation process, attachment of glycans (oligosaccharides) to proteins or lipids, involves several hundred enzymes, transcriptional factors, and other proteins (Varki et al. 2009). Given the fraction of the mass of the molecule that carries glycan structures, protein glycoconjugates are divided into proteoglycans and glycoproteins. Proteoglycans contain large, branched sugar structures attached to shorter peptide or protein chains, while glycoproteins contain simpler glycans bound to larger protein structures.

The roles of glycoproteins are different. As components of the cell membrane, they participate in cell adhesion and cell recognition, which is especially important in the fertilization process of an egg, embryonic development, and cell differentiation. Glycan structures serve as signaling molecules for proper protein folding and transport to the appropriate cell compartment. Glycans also play a role in modifying the inflammatory response, which is in numerous studies related to changes in glycosylation of immunoglobulin G (IgG) (Maverakis et al. 2015). Likewise, some pathological conditions are caused by and/or manifested as altered protein glycosylation. Glycans are not excluded from the process of carcinogenesis, metastasis, and autoimmune diseases, due to which they have been the subject of many research studies (Varki et al. 2009).

In addition to DNA, proteins, and lipids, glycans are one of the four basic building blocks of life. They are composed of monosaccharides, which are mostly bound to proteins or lipids in the cellular secretory pathway, i.e., in the endoplasmic reticulum (ER) and the Golgi apparatus (Marek et al. 1999). Glycans on a protein are typically bound by an *N*- or *O*-glycosidic bond. *N*-linked glycans are synthesized by the formation of an *N*-glycosidic bond between the oxygen atom of *N*-acetylglucosamine (GlcNAc) and Asn nitrogen in the protein. The glycan-binding Asn must be a part of the amino acid sequence Asn-X-Ser or Asn-X-Thr, where X may be any of the amino acids except proline. *O*-linked glycans are synthesized by *O*-glycosidic bond formation most commonly from the oxygen atom of the hydroxy amino acid Ser or Thr via *N*-acetylgalactosamine (GalNAc), or much less frequently via GlcNAc, mannose, or fucose. Glycan biosynthesis is covered in detail within the corresponding chapter, so we will not focus on it further.

In addition to the complexity resulting from different types of glycosylation, glycans differ depending on the bond type, branching, composition, and length of the oligosaccharide attached to a protein backbone. Identification and characterization of this enormous complexity and repertoire of possible sugar structures represented and still represents a significant analytical challenge.

2.1.2 *Immunoglobulins*

Immunoglobulins (Igs), or antibodies (Abs), are our first line of defense against foreign pathogens. In addition to IgG, which is the most abundant Ab in human plasma (de Haan et al. 2017), there are four more classes of Igs: IgM, IgA, IgE, and IgD, distinguishable by the type of their heavy chains (Schroeder and Cavacini 2010). All Igs are composed of two 50–77 kDa class-specific heavy chains that are joined by one or more disulfide bonds. Also, each heavy chain is joined by a disulfide bond to a 25 kDa light chain. There are two types of light chains, called lambda (λ) and kappa (κ). Certain Ig has either κ or λ light chains, never one of each. So far, no differences in function have been found between these chains, and their ratios are different in different animal species. The reason for this variation is unknown (Janeway and Travers 2001). Fab (F stands for fragment; ab—antigen binding) is the antigen recognition and binding domain, whereas the Fc fragment (named Fc because it showed a crystallization tendency) is comprised of the heavy chain region that interacts with Fc receptors (FcRs) on immune cells triggering different effector functions. While IgM and IgE lack the hinge region and are thus more rigid in the structure, IgG, IgA, and IgD isotypes have a flexible linker containing N- and O-glycans separating the Fab and Fc regions. Igs are heavily glycosylated, and these glycan modifications are critical for the appropriate function of all Igs.

As stated above, while Fab fragments recognize antigens, the Fc fragment has an effector function, which means that it decides what sequence of actions will follow antigen recognition. The specificity of interactions through the regions of the Fab IgG fragment is employed by exogenous therapeutic monoclonal antibodies (mAbs) used in the treatment of cancer, viral diseases, autoimmune diseases, and many others (Adams and Weiner 2005; Chan and Carter 2010; Scott et al. 2012; Singh et al. 2009; Sliwkowski and Mellman 2013). The Fc fragment interacts with FcR on the immune system effector cell to initiate an appropriate sequence of reactions (Bruhns et al. 2009; Jiang et al. 2011; Nimmerjahn and Ravetch 2008; Siberil et al. 2007). The effector functions of Igs can be very different, even opposite, such as inflammation and suppression of the immune response. Since the amino acid sequence of the Fc fragment, unlike the Fab fragment, is immutable, this cannot be the source of different effector activities. Therefore, N-glycans are the ones that enable different responses to antigens (Ahmed et al. 2014).

Glycosylation is of great importance for the proper functioning of the immune system, and one of the most studied glycoproteins is IgG (Rudd et al. 2001). Each CH2 domain of IgG heavy chain carries a covalently bound biantennary N-glycan at the evolutionary conserved Asn residue 297 (Butters 2002; Mimura et al. 2018). In addition to Fc N-glycans, 15–20% of IgG molecules have glycans in the Fab domain (Stadlmann et al. 2010). These Fab glycosylation sites are not evolutionarily conserved but have resulted from somatic hypermutation during the antigen-specific immune response (Dunn-Walters et al. 2000). The fact that there are no conserved glycosylation sites in the Fab domain and that it is only glycosylated in a fraction of IgG molecules makes Fab glycans more challenging to study.

Bound oligosaccharides have structural and functional importance for IgG and its effector functions (Lauc et al. 2014). For example, on Fc glycans, the lack of galactose residues is associated with rheumatoid arthritis (Parekh et al. 1985). The addition of sialic acid dramatically alters the physiological role of IgG, changing it from a pro-inflammatory to an anti-inflammatory mediator (Böhm et al. 2012). Fucose supplementation to the Fc glycan core impairs the binding of IgG to FcγRIIIa and significantly reduces the antibody-dependent cell-mediated cytotoxicity (ADCC) (Scanlan et al. 2008). Age, as well as gender, correlate with specific patterns of IgG glycosylation (Krištić et al. 2014).

Glycosylation of IgA, as the second most abundant Ab in human serum, has been the most studied after IgG, mostly in IgA nephropathy (Coppo and Amore 2004). IgA glycosylation changes have also been reported in rheumatoid arthritis and pregnancy (Bondt et al. 2017). Both isotypes of IgA—IgA1 and IgA2—contain two conserved N-glycan sites per heavy chain, one at Asn263 in the Cα2 domain and one at Asn459, which is the terminal amino acid of its 18-amino acid tail piece. Allotypes of IgA2, IgA2m(1), and IgA2m(2) contain additional evolutionary conserved *N*-glycan sites. IgA2m(1) contains one *N*-glycosylation site in Cα2 and Cα1 domain, while IgA2m(2) contains one site in Cα2 and two *N*-glycosylation sites in Cα1 domain. Additionally, IgA1 has nine potential *O*-glycosylation sites in the 23-amino acid long, proline-rich hinge region. In serum IgA1, three to five of these *O*-glycosylation sites are occupied (Field et al. 1994). IgA2 is not *O*-glycosylated.

Despite the key role of Ig glycosylation in many physiological and pathological processes (Gudelj et al. 2018), the monitoring of Ig glycosylation is often neglected in clinical and immunological research. Historically, glycosylation analysis in large cohorts has been lagging behind other types of analyses due to the underdeveloped methodology that would enable fast, robust, and affordable profiling of glycans in 100 or 1000 of patient and control samples. However, this has been changing with technological advances and the development of high-throughput methods for glycan analysis using several analytical technologies that provide information on glycosylation sites, site occupancy, and content of glycan variants attached to glycoproteins, mainly ultra-high-performance liquid chromatography with fluorescence detection (UHPLC–FLR), liquid chromatography-electrospray mass spectrometry (LC–ESI–MS), capillary gel electrophoresis with laser-induced fluorescence detection (CGE–LIF), and matrix-assisted laser desorption/ionization time-of-flight mass spectrometry (MALDI-TOF–MS) (Huffman et al. 2014). Moreover, lectin-based microarrays have been used to analyze glycan profiles of purified glycoproteins or cell surface proteins (Zhang et al. 2016).

While Ig glycan analysis by MS and CE technologies will be covered in the following chapters, here, we'll describe the robust and sensitive standard approaches in low- and high-throughput *N*-glycosylation analysis of IgG and other Igs based on LC and lectins. Also, new techniques for Ig glycosylation analysis will be mentioned, although some of them are still not routinely used.

2.2 Liquid Chromatography

The most widely used method for the detection and purification of proteins in general is LC. The principle of chromatography is the distribution or partition of individual molecules between two different phases, mobile and stationary phase, based on their relative affinities. Chromatographic techniques can be divided regarding the physical properties of the mobile phase, the nature of the stationary or mobile phase, and the mechanism of separation in the chromatographic system. That is why each method is popularly named after one of its specific features (Wilson and Walker 2010), e.g., ion-exchange chromatography (IEX), chromatofocusing (CF), size-exclusion chromatography (SEC), reverse phase (RP) chromatography, affinity chromatography (AC), etc. (Coskun 2016).

Stationary phases are usually chemically modified small particles packed into a column or a monolith (single piece stationary phases with interconnected large channels). Packed column efficiency depends on the stationary phase particle size, with smaller particles generally resulting in more efficient analyte separation. While high-performance liquid chromatography (HPLC) column particles are usually 3–5 μm in diameter, ultra-high-performance liquid chromatography (UHPLC) columns are <2 μm in diameter. Consequently, UHPLC columns allow more efficient separation, which is a significant advancement for complex mixture separation (e.g., mixture of glycans), but on the other hand, require higher system pressures to move a mobile phase through the system.

In the following sections, we will cover LC applications for Ig purification and *N*- and *O*-glycan analysis on different levels: free glycans, glycopeptides, glycosylation analysis of Ig subunits, and intact proteins. Currently used LC-based approaches for *N*-glycosylation analysis of Igs are summarized on Fig. 2.1.

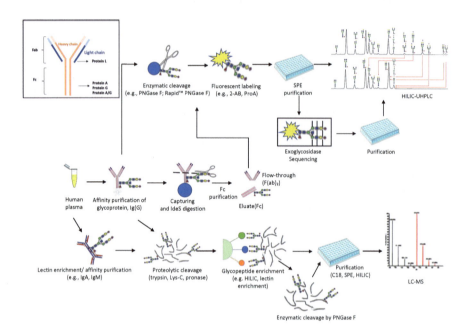

Fig. 2.1 The summary of currently used liquid chromatography-based approaches for N-glycosylation analysis of immunoglobulins (Igs). Free glycans are labeled with a fluorescent dye, e.g., 2-aminobenzamide (2-AB) or procainamide (ProA) and purified or trimmed by exoglycosidases before hydrophilic interaction ultra-high-performance liquid chromatography (HILIC-UHPLC). For analysis of Ig glycosylation on the subunit and whole protein level, IdeS digestion is widely used. Glycopeptide analysis is usually performed using liquid chromatography coupled to mass spectrometry (LC-MS). To obtain glycopeptides, IgG is most often digested with trypsin, enriched with HILIC or lectins, and purified before analysis. Insert: General structure of IgG, with marked heavy and light chains, as well as Fab and Fc fragments. Binding sites for Proteins L, A, and G are shown with arrows

2.2.1 Immunoglobulin Purification by Liquid Chromatography

2.2.1.1 Affinity Chromatography

Diverse approaches exist for the purification of Igs from plasma and other biological fluids. Affinity chromatography is the most widely used separation technique for Ab purification due to its ease of use, speed, yield, and specificity, relying on reversible binding between a protein and its cognate ligand (Walters 1985; Ayyar et al. 2012; Hage et al. 2012). Cuatrecasas et al. in 1968, introduced affinity purification, but over the past few decades, considerable efforts were made in terms of selectivity, specificity, reproducibility, product recovery, storage, maintenance, and economy (Cuatrecasas et al. 1968; Arora et al. 2017). To obtain a streamlined purification process, it is necessary to consider the purification procedure, type of ligand, and the matrix to which it is attached, which may require optimization depending on the type/class of Ab and its ability to recognize the immobilized ligand (Arora et al. 2017).

A variety of bacterial proteins are known to bind mammalian Igs, including protein A (Konrad et al. 2011; Nilsson et al. 1987), G (Akerström and Björck 1986), L (Rodrigo et al. 2015), and their recombinant derivatives (fusion proteins); protein LG (Kihlberg et al. 1996), protein LA (Svensson et al. 1998), and protein AG (Ghitescu et al. 1991). Protein A has a high affinity for the Fc region of different Ig isotypes as well as the Fab region of the human VH3 family (Graille et al. 2000; Naomi et al. 2012). Studies showed that protein A does not bind all subclasses of human IgG equally. While it strongly binds subclasses IgG1, IgG2, and IgG4, it only weakly binds subclass IgG3 (Graille et al. 2000; Björck and Kronvall 1984; Walls et al. 2017). Furthermore, protein A can be used to enrich IgD from preparations of other Ig classes since it binds IgG (strongly), IgA, IgM, and IgE (weakly) (Arora et al. 2017; Rodrigo et al. 2015). Protein A can be covalently bonded to a natural (agarose or cellulose) or a synthetic (polyvinyl ether, pore glass, or polymethacrylate) base matrix (Hilbold et al. 2017). Disadvantages of protein A Ig purification are high costs associated with the support, ligand leaching, and caustic instability (Ramos-de-la-Peña et al. 2019). Despite a lot of advantages of using NaOH for resin cleaning, it was shown that NaOH changes the three-dimensional structure of protein A and G and their affinity towards Ig binding (González-Valdez et al. 2014; Naik et al. 2011). To overcome this, several studies have tried to develop enhanced protein A resins as stationary phases tolerating high NaOH concentrations (up to 0.5 M) (Ramos-de-la-Peña et al. 2019; Linhult et al. 2004). Furthermore, different formats (monoliths, membranes, and microspheres) have been tested to carry out the capture of mAbs via protein A, but none of these formats have replaced chromatographic columns as the standard for biotechnology industry. The main drawbacks are elevated resins costs, their limited lifetime compared to other resins and potential regulatory issues. To develop improved cost-effective solutions for Ig purification, alternative ligands to protein A (e.g., aptamers, artificial binding proteins, engineered Ig-binding proteins, etc.) have been designed and studied (Kruljec and Bratkovič 2017).

Although protein A, protein G, and protein L chromatography are techniques most widely used for the purification of mAbs, protein A chromatography is still considered the golden standard (Hilbold et al. 2017). Furthermore, protein A also interacts with polyclonal antibodies (pAbs) produced by different cell lines with diverse antigen-binding properties (Hilbold et al. 2017). Because of its high binding activity and high purity of isolated Igs, protein A chromatography is commonly used in the pharmaceutical industry (Bolton and Mehta 2016). Although great efforts have been made to decrease the costs and increase the performance of protein A technology, further research is needed to develop and enhance processes, stationary phases, and ligands capable to surpass the simplicity and cost-effectiveness currently offered by protein A-based chromatography.

Protein G and Protein L are ligands with functions similar to that of protein A. However, protein A is not recommended for the isolation of mouse mAbs because it lacks affinity for mouse IgG1, making protein G a more suitable alternative since it binds IgG from most species (Saha et al. 2003). Also, the pH required to dissociate bound IgG is lower, making protein G a commonly used tool for IgG purification. In contrast to protein A, which binds the IgG3 subclass weakly or not at all (depending on the source), protein G binds all subclasses of human IgG equally (GE Healthcare, n.d.). However, protein G has an albumin-binding site which may cause problems with contamination. A recombinant form of protein G that lacks the albumin-binding site is preferred.

Protein L is another Ig-binding protein obtained from the bacteria *Peptostreptococcus magnus* (Myhre and Erntell 1985). It does not bind to the Ig Fc domain but instead interacts with all Ig classes (i.e., IgG, IgM, IgA, IgE, and IgD) with κ light chains (De Chateau et al. 1993), enabling their purification (Rodrigo et al. 2015). Since it binds with high affinity to a large number of Igs with κ1, κ3, and κ4 light chains, protein L is also suitable for the purification of Fab, scFv (short-chain variable region), F(ab)2, and Ab derivatives (Grodzki and Berenstein 2010a). Since protein L binds to a wider range of Igs than either protein A or G, it is a useful tool in affinity chromatography and for Ab immobilization.

To improve the binding characteristics of the bacterial proteins, several fusion proteins were developed. In general, these fusion proteins combine the Ig-binding domains of the three main bacterial Ig-binding proteins: A, G, and L. For instance, protein AG is a recombinant fusion protein that includes four Ig-binding domains from protein A and two from protein G. Protein AG binds to the broadest range of IgG subclasses from human, mouse, and rabbit. It binds to all human IgG subclasses and to IgA, IgE, IgM, and slightly to IgD (Grodzki and Berenstein 2010a).

The major disadvantage of the use of bacterial Fc-binding proteins in chromatographic affinity purification of Abs, in addition to high production costs, is Abs elution in acidic conditions. These low-pH elution conditions may cause some Abs to aggregate or denature and alter the subclass distribution or glycan composition (Ramos-de-la-Peña et al. 2019; Gagnon 1996). Also, low-pH exposure of IgG has been found to cause Cγ2 unfolding associated with protonation of specific acidic residues (McMahon and O'Kennedy 2000). Furthermore, low-pH IgG exposure induces other conformational changes resulting in increased aggregation and hydrophobicity, which significantly alter FcγR-binding behavior and biological activity (Lopez et al. 2019).

Mannan binding lectin (MBL) and jacalin (extracted from jackfruit) have also been used in the affinity purification of Abs since they specifically bind human IgM and IgA, respectively (Roque-Barreira et al. 1986; Nevens et al. 1992).

2.2.1.2 Melon Gel Chromatography

Besides affinity purification on immobilized protein A, G, or L, Melon gel is a relatively new approach for IgG isolation (Thermo Fisher Scientific, USA). Melon gel, in contrast to protein A and protein G, binds all non-γ-globulin and plasma proteins while allowing purified IgG to be collected in the flow-through fraction. As mentioned before, low pH which is used during IgG immunoaffinity purification using, e.g., protein G can alter its antigen-binding behavior, lead to aggregation and cause denaturation (Nevens et al. 1992; Gagnon et al. 2015; Gagnon and Nian 2016). Furthermore, it is demonstrated that low-pH IgG purification approaches can dramatically alter F(ab')2 antigen recognition (McMahon and O'Kennedy 2000; Djoumerska-Alexieva et al. 2010). Melon gel purification, in contrast, does not require low-pH elution conditions.

Moreover, it was shown that low pH causes IgG aggregation and enhances binding to Fc receptors, impacting the Ab-binding kinetics and affinity, compared to IgG samples purified via Melon gel (Lopez et al. 2019; Dorion-Thibaudeau et al. 2014). It is known that the binding affinity of IgG for the Fc receptors can be modulated by the IgG subclass (Jefferis et al. 1994), and each of the IgG subclasses has a unique binding profile to FcγRs (Vidarsson et al. 2014). Lopez et al. compared protein G and Melon gel purification method in a manner of FcγR binding. For the FcγRIIa and FcγRIIIa receptors, binding affinity follows the hierarchy IgG3 > IgG1 > > IgG2 = IgG4, and it is the strongest for IgG3 in both purification methods. Additionally, IgG1, which follows IgG3 in binding affinity, was significantly more abundant in the Melon gel-purified IgG samples (Lopez et al. 2019).

2.2.1.3 Size-Exclusion Chromatography (SEC)

Size-exclusion chromatography (SEC) is a liquid column chromatographic technique used for the separation of molecules according to their size when a solution flows through a column filled with porous packing. The method appeared in the late 1950s and was named gel permeation chromatography (GPC) (Moore 1996) or gel filtration chromatography (GFC) (Porath and Flodin 1959). GPC usually refers to the chromatographic separation of synthetic macromolecules with the use of porous gels or rigid inorganic packing particles, while GFC or simply gel filtration refers to a process of biological macromolecules (biopolymers) separation (Kostanski et al.

2004). Size separation is achieved by differential pore permeation. The accessible volume of a pore is greater for a small molecule than for a large one. Because of that, larger molecules have shorter retention times in the pores than smaller ones and are eluted from the column earlier. Over the years, much effort has been spent on the design and manufacture of a wide range of gels compatible with a variety of polymers and mobile phases (Aust et al. 2001). SEC is often used for Ig desalting and buffer exchange, as well as an additional step of purification.

2.2.1.4 Ion-Exchange Chromatography (IEX)

Ion-exchange chromatography (IEX) relies on the binding of charged sample molecules to oppositely charged groups attached to an insoluble matrix (ion exchanger). Such binding is electrostatic and reversible (Grodzki and Berenstein 2010b). IEX can be used for the purification of Igs, but it should be preceded by ammonium sulfate fractionation or followed by affinity chromatography or gel filtration due to its lower selectivity (Grodzki and Berenstein 2010b). Igs can be purified both by cation- (binds positively charged groups) or anion-exchange chromatography (binds negatively charged groups) (Coskun 2016). However, they are most frequently purified by anion-exchange chromatography with diethyl aminoethyl (DEAE) resins (GE Healthcare, n.d.; Grodzki and Berenstein 2010b). The Igs in the sample bind to the ion-exchange matrix and after binding, an equilibration buffer is used to wash the column and remove all molecules that do not bind under the conditions. Elution is done by increasing the ionic strength of the buffer and by changing the pH. At the selected pH, proteins with the lowest net charge will be the first to elute, while the proteins with the highest net charge will be the last to elute. Also, by changing the ionic strength, the proteins bound to the matrix are eluted differentially in a purified and concentrated form (Grodzki and Berenstein 2010b; Yang and Harrison 1996; Fishman and Berg 2019).

2.2.2 *N-Glycan Analysis by Liquid Chromatography*

Liquid chromatography, UHPLC specifically, is being widely used in high-throughput *N*-glycan analysis of isolated glycoproteins (e.g., IgG) and total glycoproteins from plasma/serum samples or tissues because of the relatively low cost of equipment, very good, reliable, and robust quantification, and the ability to separate glycan isomers (Trbojević-Akmačić et al. 2015). The standard procedure for UHPLC analysis includes deglycosylation of proteins, labeling of released glycans with a fluorescent dye, a clean-up procedure to wash out excess reagents, and finally fluorescent detection of labeled and purified glycans.

Additionally, nanoLC systems coupled to MS have been widely used for IgG (Selman et al. 2012; Chandler et al. 2019; Liu et al. 2020), and more recently, IgA analysis (Momčilović et al. 2020) of subclass-specific glycopeptides, where the chromatographic dimension is utilized to separate glycopeptides into subclass-specific clusters predominantly based on the peptide part. Glycopeptides are further separated and analyzed in MS based on their m/z ratio. This technique is described in more detail in the chapter on MS-based methods for Ig glycome analysis.

2.2.2.1 Glycan Release

N-glycans can be released from glycoproteins using either enzymatic or chemical methods. In contrast to chemical glycan release, enzymatic cleavage of N-glycans is more straightforward and specific; this is the reason why it is the most popular way to release N-glycans. Several enzymes have been successfully used and optimized for cleaving N-linked glycans, such as peptide-N-glycosidase F (PNGase F) (Plummer et al. 1984; Tarentino et al. 1985), Rapid™ PNGase F (van de Bovenkamp et al. 2019), peptide-N-glycosidase A (PNGase A), endoglycosidases F, H, D (Freeze and Kranz 2010), and S (Collin and Olsén 2001). PNGase F is a widely used glycoamidase due to its broad substrate specificity and high activity. This enzyme will remove oligomannose, hybrid, and complex N-glycans attached to Asn by cleaving the bond between a protein and the innermost GlcNAc and converting Asn to Asp. However, it will not remove N-glycans with certain modifications of the N-glycan core (e.g., α-1,3 fucose) found so far only in slime molds, plants, insects, and parasites (Plummer et al. 1984). It is important to note that although most commercially available PNGase F enzymes today are cloned from *Elizabethkingia miricola* and overexpressed in *Escherichia coli*, different specificities dependent on reaction conditions have been observed (Vilaj et al. 2020). In comparison to PNGase F, PNGase A allows the release of N-glycans containing core α-1,3 fucose (Fan and Lee 1997). However, PNGase A shows lower efficiency in N-glycan release from glycoproteins (Tarentino and Plummer 1982) and cannot cleave highly complex glycans (Taga et al. 1984). Furthermore, PNGase A is a glycoprotein itself and therefore can be self-deglycosylated, which can bias the outcome of the N-glycan analysis by causing contamination from endogenous PNGase A glycan structures (Altmann et al. 1998). Considering the rapid development of N-glycoproteome studies, there is a need for PNGases that combine the advantages of both PNGase A and PNGase F (Fan and Lee 1997).

Other used enzymes cleave between the two core GlcNAc residues, leaving one GlcNAc attached to Asn. These endoglycosidases are more specific in terms of the N-glycan structures they will cleave. Endoglycosidase H will release oligomannose and hybrid N-glycans, but not complex N-glycans. Endoglycosidase F will release simple biantennary N-glycans, but not oligomannose or hybrid N-glycans (Weng et al. 2015). On the other hand, endoglycosidase S (EndoS) has a specific endoglycosidase activity on native IgG by hydrolyzing the glycans attached to the conserved Asn on the heavy chains of IgG (Collin and Olsén 2001; Collin et al. 2002). EndoS hydrolysis has been shown to modulate human IgG/FcγR interactions by influencing the binding/dissociation of IgG to soluble and cell-bound FcγR (Allhorn et al. 2008).

To release glycans, first, it is necessary to denature proteins and break disulfide bonds to make glycans more available to the enzyme. One of the typical approaches is denaturation by high temperature (Akazawa-Ogawa et al. 2018) or sodium dodecyl sulfate (SDS) in combination with β-mercaptoethanol or dithiothreitol (DTT) for reduction of protein disulfide bonds. Moreover, deglycosylation is usually performed at 37 °C for a few hours to overnight to increase the accessibility of glycans to PNGase F and facilitate complete glycan release.

Although denaturation of glycoproteins prior to PNGase F digestion increases deglycosylation efficiency, the digestion process is still time-consuming. Methods such as microwave-assisted deglycosylation have been shown to lead to complete deglycosylation in 10 min (Zhou et al. 2012). In addition, there have been reports on the use of pressure cycling to accelerate enzyme-catalyzed digestion (Szabo et al. 2010). However, these approaches are generally not suited for high-throughput Ig N-glycan analyses and may lead to significant errors in quantification due to differences in the release kinetics of N-glycans with different structures (Huang and Orlando 2017).

Another approach is a chemical release technique called hydrazinolysis (Patel et al. 1993; Fischler and Orlando 2019). Hydrazinolysis uses anhydrous hydrazine to cleave the complete glycan from the peptide backbone. Unfortunately, the use of hydrazine causes degradation of a peptide backbone and can lead to unwanted modifications of released glycans, such as loss of N-acetylation and loss of the free reducing end (Fischler and Orlando 2019). For these reasons, chemical deglycosylation by hydrazinolysis is barely used for glycan analysis.

The main disadvantage of glycan release from whole Ig is the inability to distinguish between Fab and Fc glycans, site-specific glycosylation, and glycosylation originating from different Ig subclasses.

2.2.2.2 Fluorescent Labeling Methods

Labeling of released N-glycans plays a crucial role in both detection and characterization by various analytical techniques. Free glycans lack chromophore or fluorophore properties to enable straightforward UV or fluorescent detection; and do not ionize well to result in high-quality mass spectra. Although techniques like high-performance anion-exchange chromatography (HPAEC) coupled to pulsed amperometric detection (PAD), which do not require glycan derivatization, have been used in the past for glycan analysis (Lee 1990), today glycans are often derivatized during sample preparation, not only to enable their detection but also to enhance the sensitivity of the analysis.

A large number of fluorescent labels have been routinely applied. The easiest way to label released N-glycans is via reductive amination, where the amine group of the label reacts with the aldehyde group of a glycan resulting in Schiff base, which is reduced by a reducing agent to yield a secondary amine. In addition to reductive amination, other derivatization methods such as Michael addition, hydrazide labeling, and permethylation have been employed (Ruhaak et al. 2010a).

Many tags have been historically used for the reductive amination of IgG *N*-glycans prior to chromatographic analysis, e.g., 2-aminobenzoic acid (2-AA) (Anumula and Dhume 1998) and 2-aminopyridine (PA) (Takegawa et al. 2006), and some of them, like 2-aminobenzamide (2-AB) (Bigge et al. 1995) and procainamide (4-amino-*N*-[2-(diethylamino)ethyl] benzamide, ProA) (Klapoetke et al. 2010), are still mostly used today.

The label 2-AB has been widely applied in chromatographic analysis of IgG *N*-glycans. Consequently, extensive databases of 2-AB-labeled *N*-glycans have been developed over the years, the most extensive one being GlycoStore (www.glycostore.org) (Klapoetke et al. 2010; Abrahams et al. 2018; Campbell et al. 2008), which contains standardized elution positions of 2-AB- and ProA-labeled *N*-glycans in hydrophilic interaction liquid chromatography (HILIC) with fluorescence detection.

Labeling with 2-AB gives 1:1 stoichiometry, thus allowing relative quantification of different glycans based on fluorescence intensity. It is also very convenient for high-throughput glycan analysis since the sample preparation requires low volumes of solvents, it is relatively fast and simple. Despite these advantages, the major drawback of 2-AB *N*-glycan labeling is poor ionization efficiency, which hinders MS analysis from the same sample making other labels like ProA more attractive.

Having all advantages of 2-AB, ProA additionally shows increased fluorescence and ionization performance (Keser et al. 2018). ProA-labeled glycans are suitable for both (U)HPLC–FLR and ESI–MS analysis, providing more efficient ionization compared to 2-AB, facilitating glycan identification and structure confirmation (Kozak et al. 2015).

Recently, more multifunctional labels, like InstantPC and RapiFluor-MS, which contain both a fluorophore for sensitive chromatographic analysis and an easily ionizable group to facilitate sensitive MS detection, have been developed. InstantPC (instant procaine) rapidly (within 5 min) modifies glycosylamine-bearing *N*-glycans after their enzymatic release, yielding a stable urea linkage. As with ProA, InstantPC contains a tertiary amine to enhance positive mode ESI (Segu et al. 2020). RapiFluor-MS is another commercially available label similar to ProA which is comprised of an *N*-hydroxysuccinimide carbamate reactive group, a quinoline fluorophore, and a basic tertiary amine, an amine that generates a high MS signal in the positive mode and, thus, enhances sensitivity (Lauber et al. 2015). Other "instant" labels used in IgG *N*-glycan analysis are InstantAB (Reusch et al. 2015) and 6-aminoquinoyl-*N*-hydroxysuccinimidyl carbamate (AQC) (Stöckmann et al. 2013).

Advantages of these new commercially available "instant" labels are rapid analysis and increased sensitivity. Labeling can be done in just a few minutes, compared to labeling with the abovementioned traditional labels that have a reaction time of 2–4 h (Stöckmann et al. 2013). Despite these advantages, the major drawback in terms of high-throughput glycan analysis is their price.

2.2.2.3 Reducing Agent

The most used reducing agent in glycan labeling reaction used to be sodium cyanoborohydride, resulting in high yields of labeled oligosaccharides (Bigge et al. 1995). A major drawback of this reagent is that upon hydrolysis, it readily forms the toxic, volatile compound hydrogen cyanide. Today the most used reducing agent is 2-picoline borane (2-PB) (Ruhaak et al. 2010a; Trbojević-Akmačić et al. 2017). 2-PB is an efficient and non-toxic alternative that serves to reductively aminate oligosaccharides with 2-AA, 2-AB, or ProA and is thus less harmful to researchers and the environment, especially considering the analysis of a large number of samples in high-throughput studies. Furthermore, 2-PB can be used in both aqueous and non-aqueous conditions and shows equal efficacies for reductive amination using various fluorescent labels (Ruhaak et al. 2010b). Robust conditions for the reductive amination of IgG N-glycans with 2-AB label and 2-PB as the reducing agent that were used in many high-throughput studies were reported in (Trbojević-Akmačić et al. 2017).

Glycan derivatization is enhanced by the addition of glacial acetic acid up to a content of 30% (v/v). Usually, a reaction temperature of 60–65 °C for 2–3 h was found to be optimal. In these conditions, most glycans are derivatized and glycan degradation reactions such as acid-catalyzed loss of sialic acid are minimized.

2.2.2.4 Clean-Up Strategies

After derivatization, labeled glycans have to be purified prior to analysis. Not only does the excess of salts have to be removed (e.g., for MALDI analysis), but also the concentration of remaining free labeling reagent, normally present in large excess during the labeling step, should be reduced. Although different clean-up techniques, like solid-phase extraction (SPE) (Ruhaak et al. 2010b), liquid-liquid extraction (Ciucanu and Kerek 1984), gel filtration (Nakagawa et al. 2007), paper chromatography (Royle et al. 2002), and precipitation (Pabst et al. 2009), have been used for purification, SPE is still the most widely used. Also, SPE is the method of choice for analyzing larger cohorts as it can be adapted to high-throughput setups (Ruhaak et al. 2010b). The diversity of SPE stationary phases and their use in glycan purification is relatively extensive, ranging from RP (Wilhelm et al. 2019), porous graphitized carbon (PGC) (Kolarich et al. 2015; Ashwood et al. 2019), and HILIC (Szabo et al. 2018; Kim et al. 2019), to anion-exchange chromatography (Szabo et al. 2018). A widely applied technique to capture and separate hydrophilic glycans that are released from glycoproteins but not retained by RP chromatography is PGC (Lam et al. 2011). It has successfully been used for the purification of fluorescently labeled glycans and even permethylated glycans (Ruhaak et al. 2010b; Costell et al. 2007). Unfortunately, this glycan-targeting approach has disadvantages due to the high costs of the PGC and the fact that the excess of label may not be removed (Ruhaak et al. 2010b).

The most widely used purification method for the removal of proteins, salts, and the excess of reagents after fluorescent labeling, especially for the analysis of larger cohorts, is HILIC SPE (Trbojević-Akmačić et al. 2016). With this method, excess of label, which is less hydrophilic than glycans, may be removed, and glycans are retained based on their hydrophilic properties (Ruhaak et al. 2008). There is an increasing number of SPE sorbent materials in the market: pipette tips (e.g., ZIPTIP®) (Schmelter et al. 2018; Poole 2003), cartridges (e.g., LudgerClean™T1 cartridge) (Kozak et al. 2015), discs (Poole 2003), filter plates (Berna et al. 2002), magnetics beads (Madhwani and McBain 2016), and sorbent materials (Augusto et al. 2013).

In 2013, an automated platform for high-throughput UHPLC IgG *N*-glycan analysis was developed utilizing an automated liquid handling workstation. That platform allowed the preparation of 96 samples in only 22 h, and its major advantage was rapid and low-cost analysis (Stöckmann et al. 2013). In the meantime, other automatic systems have been successfully used for sample processing in IgG *N*-glycans analysis (Reusch et al. 2014; Reed et al. 2018; Haxo et al. 2016). Nowadays, automated IgG *N*-glycan analysis platforms are increasingly being used to allow 96 samples to be prepared in just 2 h. Moreover, commercially available fast IgG *N*-glycan analysis kits enable rapid enzymatic release and fluorescent labeling of IgG *N*-glycans. Those products facilitate the release of IgG *N*-glycans in minutes. Such methods are faster than conventional protocols and have very good reproducibility.

2.2.2.5 Detection of Labeled Glycans with (U)HPLC

A method that provides robust separation and quantification of fluorescently labeled glycans is HILIC-(U)HPLC. The main advantage of this method over others is the ability to separate structural isomers, as well as neutral and acidic glycans in the same run. The retention time of a glycan depends on its hydrophilic potential, which is influenced by glycan size, charge and structure, and the bonds and branches within the glycan. Analysis of glycans by HILIC is steadily progressing, with the advancements of instrumentation and column chemistry (Mariño et al. 2010). UHPLC over HPLC provides increased resolution and sensitivity due to a stationary phase with a particle diameter of <2 μm, which can function at higher pressures and increased flow rates without decreasing resolution and column efficiency (Ahn et al. 2010a). This reduces the analysis time and mobile phase solvents consumption. The typical mobile phase for HILIC analysis of IgG *N*-glycans is a mixture of acetonitrile and 50 or 100 mM ammonium formate pH 4.4, with glycan separation in a linear gradient of an increasing percentage of the water phase. Ethylene bridged hybrid (BEH) stationary phase has been developed in which methyl silica gel bundles are bridged, ensuring mechanical stability (Alden et al. 2012). The fluorescence detector is equipped with a xenon lamp, and excitation and emission wavelengths are chosen depending on the used fluorescent label to obtain high sensitivity. Dextran hydrolysate, a mixture of labeled glucose oligomers, is used as the external standard for system calibration. The retention time of each oligomer is converted to glucose units

(GU), which are used as reference standard values and for data comparability (Guile et al. 1996). Today, there are bioinformatics tools that contain normalized retention data expressed as GU values. One of those tools is GlycoStore (www.glycostore.org), which contains data on around 850 unique glycan structures (Klapoetke et al. 2010; Abrahams et al. 2018; Campbell et al. 2008). GlycoBase and autoGU (Klapoetke et al. 2010), as well as GlycoProfileAssigner (Duffy and Rudd 2015), are databases and analytical tools developed to assist the interpretation and assignment of HPLC-glycan profiles.

Several orthogonal LC and LC/MS(MS) approaches for the characterization of therapeutic glycoproteins have been recently compared by Largy et al. (2017). It was shown that mixed modes of LC, anion-exchange-reverse phase (AEX-RP), and anion-exchange-hydrophilic interaction liquid chromatography (AEX-HILIC), can be applied for separation of 2-AB and RapiFluor-MS-labeled free N-glycans to assess the sialylation profile and as an orthogonal approach to separate N-glycans coeluting in HILIC mode (Largy et al. 2017). The mixed-mode AEX-RP column in a gradient of ammonium formate pH 4.5 and acetonitrile enables glycan separation based on their charge (AEX mode) and hydrophobicity (RP mode). Glycans are consequently separated according to their sialylation level, e.g., non-sialylated, monosialylated, disialylated, etc. In contrast to HPAEC-based methods, AEX-RP separation using the abovementioned mobile phases allows straightforward coupling of the LC system to ESI-QTOF-MS, facilitating subsequent MS analysis and glycan structure elucidation. On the other hand, the AEX-HILIC mode enables fast sialylation profiling (13 min run) without MS identification of glycan peaks (Largy et al. 2017). Moreover, RP chromatography has been applied for the quantification of sialic acids after their release in acetic acid and labeling with 1,2-diamino-4,5-methylenedioxybenzene (DMB) (Largy et al. 2017).

2.2.2.6 N-Glycan Sequencing by Exoglycosidases

As mentioned earlier, all human IgGs have N-glycans attached to conserved N-glycosylation sites in the Fc region, the Asn-297 residues, capped with Man, GlcNAc, Gal, NeuAc (in the α-2,3/6 linkage) or Fuc (α-1,6-linked to the innermost GlcNAc residue—the core fucose). Although N-glycan analysis is often performed by using endoglycosidases such as PNGase F and Endo H to cleave off the glycans and identify N-glycosylation sites, exoglycosidases are usually employed to cleave monosaccharide units from the nonreducing end of glycans on glycoproteins and glycolipids (Bourne and Henrissat 2001). Sequential N-glycan trimming by exoglycosidases has been routinely used in the glycobiology field to obtain structural information about N-glycan species (Royle et al. 2006). Application of a set of exoglycosidases results in specific HILIC-(U)HPLC chromatographic glycan peak retention time shifts providing structural information about underlying N-glycans. Although exoglycosidase sequencing of fluorescently labeled N-glycans is a fast and powerful tool for structural elucidation, not all monosaccharides in a glycan structure are equally accessible to the corresponding glycosidase. For example, several fucosidases are capable of hydrolyzing core α-1,6-linked fucose, but all of them rely on the previous removal of the majority of the glycan chain (e.g., by endoglycosidases) (Tsai et al. 2017) to enable enzyme binding and core fucose hydrolysis.

Chen et al. recently presented an approach named NGlyoReduction, which trims complex glycans on N-glycopeptides using a set of exoglycosidases. Unlike deglycosylation with endoglycosidases, the reduction of glycans generated an oligomannosylated N-glycopeptidome, which shares similar retention time and ionization efficiency with intact glycopeptidome (Chen et al. 2016). Oligomannosylated glycopeptidome produced by NGlycoReduction approach can be analyzed by HPLC–ESI–MS to enable the identification of intact glycopeptide structure, peptide sequence, and glycosylation site. The glycan structure of intact glycopeptides can be further identified from their own MS/MS spectra. Advantages of this approach are low cost, simple processing of data, and potential application for characterizing site-specific N-glycosylation involving complex N-glycans.

2.2.3 O-Glycan Analysis by Liquid Chromatography

Analysis of O-glycosylation in many cases represents a difficult analytical problem. Contrary to N-glycosylation, it is difficult to predict which Ser or Thr presents a potential O-glycosylation site just by knowing the amino acid sequence. Moreover, O-glycosylation is, in general, far less explored than N-glycosylation, mainly due to the underdeveloped methodology for sample preparation. In terms of Ig O-glycosylation analysis, another aspect is the fact that only low-abundant human Ig (sub)-classes—IgA1, IgD, and IgG3—contain O-glycans (Arnold et al. 2007), making technologies like LC less applicable in Ig O-glycan profiling due to its lower sensitivity.

Another aggravating aspect of routine O-glycan profiling is the non-existence of specific enzymes that would cleave all types of O-glycans. Although an O-glycosidase is commercially available, it releases only Core-1 or Core-3 disaccharide O-glycans. Recently, the mucin-selective protease StcE has been used for the cleavage of peptide/glycan domains (Malaker et al. 2019). Another enzyme, O-endopeptidase, has been used for the digestion of mucin-type glycoproteins and glycopeptides. This enzyme can cleave the N-terminus of O-glycosylated Ser or Thr but does not digest at Ser or Thr unless occupied by an O-glycan (Yang et al. 2018). However, O-endopeptidase is not fully active toward sialylated glycoproteins and therefore, it is suggested that this acidic residue is removed prior to digestion (Yang et al. 2020).

Therefore, a general approach to free O-glycan analysis is their release by chemical methods, e.g., β-elimination (Fukuda 1989; Zheng et al. 2009) or hydrazinolysis (Patel et al. 1993). The reaction of β-elimination is usually performed using an ammonia solution. This reaction results in the incorporation of one NH_3 into the amino acid residues to which the glycans are attached. The resulting protein contains a modified amino acid residue with a distinct mass (Zheng et al. 2009;

Rademaker et al. 1998). However, a β-elimination reaction can be difficult to control, is not very robust, and causes side reactions on proteins (Hanisch et al. 2009). It was shown that hydrazinolysis with anhydrous hydrazine at 60 °C for 5 h selectively releases *O*-glycans, while reaction at 95 °C for 4 h releases both *N*- and *O*-glycans (Patel et al. 1993). Released glycans can then be labeled in a reductive amination reaction, e.g., with 2-AB, and analyzed by LC. This approach has been successfully used for *O*-glycan analysis of human secretory IgA (Royle et al. 2003).

Due to analytical challenges in *O*-glycan analysis, most often hyphenated techniques, where a separation technique is coupled with an online spectroscopic detection technology (e.g., LC-MS), have been used for their characterization. Hoffmann et al. (2016) have performed site-specific *O*-glycosylation analysis of Ig glycopeptides in human plasma using RP LC coupled online to ion trap MS after proteinase K digestion, precipitation, and glycopeptide enrichment and fractionation via HILIC. In this study, two *O*-glycopeptides carrying monosialylated T-antigens that might correspond to the J chain were detected and suggested *O*-glycosylation at Thr97. However, IgA1 *O*-glycopeptides have not been detected in this study, most likely due to hydrophilicity of the *O*-glycopeptide region causing these fragments to end up in late eluting HILIC wash fraction or inability to unambiguously identify proteinase K generated glycopeptides. On the other hand, the IgA1 peptide containing the Ser105 *O*-glycosylation site has been detected in a non-glycosylated form. The fact that some other human IgA studies have not been able to detect IgA1 *O*-glycopeptides (Bai et al. 2015) implicates that currently employed *O*-glycan analysis approaches are still not sufficiently sensitive and robust.

The approach by Bai et al. (2015) that enabled novel human IgD Ser121 *O*-glycosylation site identification consisted of albumin depletion using an albumin-affinity column, trypsin digestion, deglycosylation with PNGase F and exoglycosidases, followed by jacalin-affinity-chromatography *O*-glycopeptides enrichment. Collected fractions were then analyzed by LC-MS/MS.

O-glycosylation analysis of therapeutic glycoprotein etanercept (fusion protein of two naturally occurring soluble human TNF receptors linked to an Fc portion of an IgG1), on the level of free *O*-glycans has been successfully performed by their overnight release at 45 °C in sodium hydroxide and sodium borohydride. After neutralization with glacial acetic acid at 0 °C, purification on Dowex 50WX8 hydrogen form column and removal of borates, reduced unlabeled *O*-glycans have been analyzed using PGC column and detected with MS. This alkaline β-elimination reaction in reducing conditions was shown to minimize *O*-glycan peeling, a common problem in *O*-glycan analysis that results in their degradation following release (Largy et al. 2017).

2.2.4 Liquid Chromatography Coupled to Mass Spectrometry

Sometimes information obtained in glycan analysis by only using LC is not sufficient to address specific research questions, e.g., glycan structure elucidation/confirmation or site-specific glycan analysis. In these cases, coupling LC to MS system can provide an additional dimension in glycan analysis. Glycan analysis by MS enables highly sensitive high-throughput analysis of the enriched glycoproteome and offers site-specific qualitative and quantitative profiling of glycoproteins, e.g., subclass-specific analysis of IgG Fc *N*-glycans by LC–MS (Selman et al. 2012). On the other hand, it provides mass information of free glycans after their separation in, e.g., PGC or HILIC mode (Ashwood et al. 2019; Vreeker and Wuhrer 2017).

Liquid chromatography allows the separation of isomeric molecules, which then reach the ionization site in the analyzer at different times. Ionization can be performed in several ways: inductively coupled plasma, laser-assisted matrix ionization, chemical ionization, the rapid atomic bombardment of analytes, and ESI. During ionization, the analyte, which was previously in the liquid state, is converted to the ionized gaseous state. After ionization, the analyte enters a mass spectrometer formed by the system quadrupoles (Ho et al. 2003).

Over the past two decades, softer MALDI and ESI techniques (imparting little residual energy onto the analyzed molecule resulting in little fragmentation) have been developed that provide greater sensitivity and generation of intact ions of high-molecular-mass compounds. While ESI often results in doubly charged $[M + 2H]^{2+}$ ions (Wuhrer et al. 2007; Harazono et al. 2008; Rehder et al. 2006) and/or triply charged $[M + 3H]^{3+}$ ions (Wuhrer et al. 2007; Stadlmann et al. 2008; Olivova et al. 2008), MALDI of tryptic IgG glycopeptides typically results in singly charged $[M + H]^{+}$ ions (Takegawa et al. 2006; Kroon et al. 1995). In MS analysis, the ionization efficiency of glycans (especially sialylated glycans) is low, so a suitable derivatization method is required (Mariño et al. 2010). Permethylation (Kang et al. 2008), methyl esterification of sialic acids (Powell and Harvey 1996), or reducing end labeling before MS analysis are typical derivatization methods. To avoid suppression of glycopeptide ionization, it is useful to enrich them from coexisting ionizing peptides.

2.2.4.1 Proteolytic Cleavage

Although recent advances in MS analysis offer many advantages in glycoproteomics, direct analysis of glycoproteins in complex samples by MS is a significant challenge. The low abundance of glycopeptides, the heterogeneity of glycan composition at each glycosylation site, the complexity of glycan structures, and a low ionization are still barriers to overcome (An et al. 2009). The most important steps in Ig glycopeptide LC-MS analysis is the conversion of glycoprotein to glycopeptides and RP clean-up of glycopeptides from the excess reagents. Enzymes such as trypsin, lysyl endopeptidase (Lys-C) (Fernández et al. 2001;

Ongay et al. 2012), endoproteinase Glu-C (V8 protease), and pronase are available for digestion of glycoproteins to glycopeptides (An et al. 2009). Proteolytic cleavage is improved by adding denaturing agents, e.g., guanidine hydrochloride, urea, and ACN (Wuhrer et al. 2007). Also, by using reducing agents such as DTT and β-mercaptoethanol, the accessibility of the proteolytic cleavage site is further improved.

Trypsin is the most widely used proteolytic enzyme in proteomics and glycoproteomics. Trypsin is a serine protease that specifically cleaves proteins at the carboxyl side of Lys and Arg in the preferred mass range for MS, producing interpretable peptide fragmentation mass spectra. Trypsin digestion is generally done at 37 °C overnight in a reaction buffer of 10–100 mM (Tris-HCl or ammonium bicarbonate, pH 8) (Vaezzadeh et al. 2010). As mentioned earlier, denaturing agents such as guanidine-HCl or urea are used to improve trypsin digestion (Wuhrer et al. 2007).

IgG trypsin digestion results in specific peptide moieties for each IgG subclass allowing subclass-specific glycopeptide analysis: EEQYNSTYR for IgG1, EEQFNSTFR for IgG2, and EEQYNSTYR for IgG4. Due to allotype variation in the amino acid at the position N-terminal of the Asn297, a tryptic digest of IgG3 results in a mass identical to either IgG2 peptide (EEQFNSTFR; predominant in Caucasian populations) or IgG4 peptide (EEQYNSTYR; predominant in Asian and African populations) (Dard et al. 2001). Although trypsin digestion is the standard approach for IgG Fc glycopeptide analysis, it was shown that incomplete IgG denaturation and digestion could cause biases in obtained glycopeptide profiles (Falck et al. 2015).

Lyc C is an alkaline protease that cleaves peptides on the carboxyl side of Lys. As for IgG, during a short digestion time and with lower protease/protein ratios, partial cleavage occurs, while almost complete digestion is achieved overnight by digestion at 37 °C and 1:200–1:20 protease/protein ratio (Hirayama et al. 1998).

Endoproteinase Glu-C is a serine protease, which cleaves peptide bonds at the carboxyl-terminal side of Glu. Depending on the pH of the digestion buffer, peptide bonds are further cleaved at the terminal carboxylic side of Asp. Also, endoproteinase Glu-C digestion of human IgG enables the characterization and quantification of plasma-purified IgG1 and IgG4 by LC–ESI–MS and LC–ESI–MS/MS (Huhn et al. 2009).

Pronase is a commercially available mixture of endo- and exopeptidases, nonspecifically cleaving proteins into their individual amino acids. Pronase was applied to elucidate the structural and heterogeneity data of IgG glycans (Rothman et al. 1989).

2.2.4.2 Glycopeptide and Glycan Enrichment

One of the major challenges in the field of MS glycan analysis is the relatively low abundance of specific glycopeptides (Peterman and Mulholland 2006). This can be alleviated by selective enrichment of glycopeptides from peptide mixtures before MS detection.

Accordingly, various glycoprotein enrichment methods have been developed and discussed in several reviews (Chen et al. 2014). These developed methods are based on HILIC (Selman et al. 2011), SPE (Pezer et al. 2016), and titanium dioxide affinity chromatography (Larsen et al. 2007). Moreover, lectin affinity chromatography is widely used to enrich glycoproteins/glycopeptides (Chen et al. 2017). Lectins will be described in the corresponding section.

Enrichment by HILIC has been commonly used because it is relatively cheap and easy to operate. Different HILIC materials have been used for glycopeptide enrichment. Chemistry of the chosen HILIC stationary phase, composition of the mobile phase, and the properties of the sample itself can affect the efficiency of enrichment (Jensen et al. 2013). Also, SPE enrichment techniques are generally used because they are cheap, simpler than gel-based techniques, and feasible for designing high-throughput MS-based analytical systems. One method of negatively charged sialylated glycopeptides enrichment is chelation interaction using titanium dioxide (TiO_2), zirconium dioxide (ZrO_2), and hybrid metal oxide-based materials, which also allow the purification of neutral glycopeptides (Palmisano et al. 2012; Wan et al. 2011). Hydrophilic interactions between the metal surface and glycopeptides play an important role in the enrichment of neutral glycopeptides (Kayili et al. 2019).

2.2.4.3 Glycopeptide and Glycan Analysis by LC-MS

One of the widely applied analytical techniques for glycopeptide analysis involves a direct analysis of proteolytic digests by RP-HPLC coupled to ESI–MS (Wuhrer et al. 2009). Separation of proteolytic IgG digests by RP-HPLC is mostly performed on a C18 analytical column, but as an alternative to RP-HPLC, graphitized carbon chromatography can be applied (Wagner-Rousset et al. 2008). Trifluoroacetic acid (TFA) or formic acid (FA) are commonly used mobile phase additives (Wuhrer et al. 2007; Harazono et al. 2008; Huang et al. 2005). Although TFA has a positive influence on analyte retention, it can form a gas-phase ion that potentially increases the ESI suppression (Chakraborty and Berger 2005). Therefore, FA is used for the analysis of IgG glycosylation. nanoLC-RP-ESI-ion trap (IT)-MS has been successfully used for O-glycopeptide analysis of IgG3 hinge region after proteinase K and trypsin in-gel digestion of protein bands (Plomp et al. 2015).

Another powerful technique for the separation and analysis of glycopeptides is HILIC coupled to ESI–MS, HILIC-ESI–MS. Retention in HILIC is mainly achieved via hydrophilic interactions between the stationary phase and the glycan moieties of the glycopeptides, which result in the separation of glycopeptides from peptides. Separation of glycopeptides can be achieved according to the degree of sialylation (Boersema et al. 2008).

Free mAb N-glycans have been successfully analyzed using HILIC-UPLC-QTOF-MS setup after labeling with $^{12}C_6$ and $^{13}C_6$ stable isotope analogs (Δmass = 6 Da) of 2-AA (Millán Martín and Iglesias 2015). This kind of a twoplex method is ideally suited for comparability studies of mAbs (e.g., lot-to-lot analysis, or innovator and biosimilar similarity assessment), enabling quantitative structural characterization of glycans, including isomeric species.

2.2.4.4 Analysis of Ig Glycosylation on the Subunit and Whole Protein Level

Contrary to bottom-up methods, where proteins are subjected to proteolytic digestion and characterized by MS based on their amino acid sequences and post-translational modifications, middle-up and middle-down approaches rely on the analysis of larger fragments or subunits, e.g., heavy and light Ig chains. These techniques are highly sensitive and informative and reduce sample preparation time and data complexity. Stationary phase based on wide-pore (300 Å) hybrid silica bonded with amide ligand was introduced in 2016 and successfully utilized to resolve glycoforms at the middle-up level of trastuzumab analysis (fragments of 25–100 kDa) after cleavage with IdeS protease (immunoglobulin-degrading enzyme of *Streptococcus pyogenes*) (Periat et al. 2016). Enzyme IdeS is a cysteine protease that cleaves all IgG subclasses at the hinge region resulting in F(ab')2 and Fc fragments and after reduction of disulfide bonds in six 25 kDa domains (a duplicate of each—LC, Fd, and Fc/2). Biochemical properties of this highly specific protease have been studied by Vincents et al. (2004). Additionally, after domain separation, IdeS allows Fab- and Fc-specific released glycan analysis (Anumula 2012) as already described in the previous sections. HILIC separation of Ig fragments using the abovementioned wide-pore stationary phase has been most effective with TFA ion-pairing instead of FA or ammonium formate (usually used in HILIC separation of free glycans). The addition of TFA reduces protein retention and improves resolution and peak shape. Chromatographic resolution can also be improved by the use of several in-line columns (Periat et al. 2016). Since mobile phases containing TFA are volatile, this mode of separation can easily be coupled to MS. Major drawbacks of this approach are required enzymatic digestion and chemical reactions before analysis, allowing the possibility of artifact formation, and limited analysis of protein modifications due to partial sequence coverage (percentage of the protein detected as peptides) (Fornelli et al. 2014). Larger precursor mass makes MS fragmentation and data processing more challenging (Lermyte et al. 2019).

Another approach, not suffering from disadvantages related to sample preparation reactions, is a top-down approach, enabling fast analysis of intact glycoproteins. However, this approach also has a limitation of lower sequence coverage, as well as poor detection of low-abundant glycoforms. The top-down approach is orthogonal to other MS methods based on protein proteolysis. Low-abundant mAb glycoforms were successfully analyzed on intact Ab level using high-resolution nanoLC-chip/MS(/MS) technology (Jacobs et al. 2016). The advantage of this approach is the ability to detect hypoglycosylation (the lack of glycans), the possibility of high-throughput applications due to minimal sample processing, and the quantitative analysis of glycans on each IgG glycosylation site. This technology can easily be employed for application in lot-to-lot glycoform heterogeneity evaluation of therapeutic Abs, while on the other hand, it is less suitable (compared to subunit analysis) for more detailed characterization due to poorer separation of same molecular weight glycans. To the best of our knowledge, these techniques have been applied so far only for Fc-glycosylation profiling.

2.3 Lectin Techniques

Lectins are glycan-binding proteins (GBPs) that selectively recognize glycan epitopes of carbohydrates or glycoproteins through the reversible binding between the anomeric hydroxyl groups on carbohydrates and the hydrophilic groups from amino acid residues in lectin protein (Hang et al. 2015; Pröpster et al. 2016). This heterogeneous group of proteins with at least one non-catalytic domain reversibly binds to specific accessible glycans present on glycoproteins and glycolipids without altering the structure of the carbohydrates (Lannoo and Van Damme 2014; Wu and Liu 2019). Lectins were initially discovered in plants, and they have been the most extensively studied. Moreover, they were found in other organisms, ranging from viruses (Van Breedam et al. 2014) to humans (Wesener et al. 2017).

Most lectins are multivalent and capable of agglutinating cells and thus are frequently designated as agglutinins. The first alternative name originated from the ability of proteins or glycoproteins to agglutinate the blood group ABO (H) glycotopes (epitopes) and sialic acid on red blood cells (Lannoo and Van Damme 2014). Lectins can be divided into classes based on their amino acid sequences and biochemical properties: C-type lectins (CTLs, require calcium ions for carbohydrate binding); G-type lectins (*Galanthus nivalis* agglutinin-related lectins); L-type lectins (Legume seeds and LysMs in bacterial autolysins); malectin carbohydrate-binding modules (CBMs) (bind glucose oligomers; maltose) (Berg and Tymoczko 2002; Bellande et al. 2017). While G-type and L-type lectin families are mostly found in plants (Bellande et al. 2017), LysM, CBMs are spread in bacteria (Bellande et al. 2017), and C-type lectins are spread in animals (Berg and Tymoczko 2002).

CTLs are the largest family of known GBPs, which comprises 16 different groups, defined by their phylogenetic relationships and domain structures (Drickamer et al. 2002; Zelensky and Gready 2005; Mayer et al. 2017). During evolution, CTLs have interacted with a large range of glycan ligands, although some also bind proteins, lipids, and inorganic molecules. CTLs include selectins, collectins, proteoglycans with C-type lectin domain (CTLD), and endocytic receptors (Cummings and McEver 2009). Selectins are divided into L-selectins (CD62L), E-selectins (CD62E) and P-selectins (CD162) (André et al. 2015; McEver 2015). While L-selectins are expressed by leukocytes, P-selectins are predominantly expressed by platelets and endothelial cells, and E-selectins are found on endothelial cells. These proteins function as adhesion and signaling receptors in many pathways, including homeostasis and innate immunity, and are crucial in inflammatory responses and leukocyte and platelet trafficking (Mayer et al. 2017).

Collagen-containing C-type lectins, known as collectins, are oligomeric proteins that are characterized by a collagen-like domain with a short Cys rich N-terminus (Drickamer et al. 1986). So far, nine CTLs have been discovered, but MBL was the first one, and it is the most characterized collectin to date (Van De Wetering et al. 2004; Howard et al. 2018). Collectins stimulate in vitro phagocytosis by recognizing surface glycans on pathogens, stimulate the production of cytokines and reactive oxygen species by immune cells, and act before the induction of an Ab-mediated response (Cummings and McEver 2009).

Among the lectin receptors in innate immunity, siglecs and galectins are well described (Manning et al. 2017; Bhide and Colley 2017). While galectins bind β-galactose-containing glycoconjugates, siglecs represent cell surface proteins that bind sialic acid (Pillai et al. 2012).

Proteoglycans with CTLD (lecticans or hyalectins) include aggregan, brevican, versican, and neurocan and they exist in the extracellular matrix (ECM) (Yanagishita 1993; Schmitt 2016). Like the selectins, each of these core proteins contains a CTLD, an epidermal growth factor (EGF)-like domain, and a complement-regulatory protein domain. They mediate leukocyte–endothelium adhesion through various carbohydrate ligands (Nelson et al. 1995; Cummings and McEver 2009; Yamaguchi 2000). Recent studies have shown that the CTLD of aggrecan activates classical and, to a lesser extent, the alternative pathway of complement (Melin Fürst et al. 2013).

Lectins play many key roles in the control of various physiological and pathological processes in living organisms, including cell migration, inflammation, immune defense, fertilization, embryogenesis, infection, and cancer formation (Sharon and Lis 1989, 2003). Also, they mediate cell-cell interactions by binding with complementary carbohydrates on opposing cells (Cummings and McEver 2009; André et al. 2015; Sharon and Lis 1989). Given the vast complexity and diversity of glycan structures, as well as different types of interactions with proteins, it is not surprising that the range and respective biological activities of lectins are substantial. Conveniently, this glycan-recognition and -binding properties have been used as a tool in a wide range of glycoscience applications: detection, isolation of glycoproteins, mapping of neuronal glyco-functions, investigation of carbohydrates on cells and subcellular organelles, selection of lectin-resistant mutants, etc. (Gabius et al. 2011).

Lectins found in nature are mostly purified directly from different organisms. On the other hand, lectins can also be produced by recombinant techniques. To this date, the most reported applications of recombinant lectins are in cancer diagnosis and as anti-microbial, anti-viral, and anti-insect molecules. Furthermore, lectins have been used in the field of functional as well as structural glycomics (Wu and Liu 2019; Oliveira et al. 2013; Hu et al. 2015). The sensitivity of lectins makes them a valuable tool in disease diagnosis compared to instrumental techniques (Pihíková et al. 2015). Additionally, in glycoconjugate analyses where the amount of purified glycoprotein is insufficient for instrumental techniques, including mass spectrometry, techniques relying on lectin-glycan recognition can be very helpful. The choice of proper lectin for such studies and its quality as a tool highly depends on well-defined lectin specificity. Here we will cover some applications of lectins as stationary phases in chromatography techniques used for Ig N- and O-glycan analysis, as well as their utilization in increasingly used microarrays.

Lectin analytical methods are based on lectin-carbohydrate interactions and can be divided into traditional and modern ones. Precipitation analysis, originally developed for the detection of antigen-Ab complexes (Wu et al. 1997) and hemagglutination inhibition, based on a lectin-mediated cell agglutination process (Sano and Ogawa 2014), belong to the traditional methods. Today, lectins are used in several immunoassay-like techniques, including enzyme-linked lectin assay (ELLA), lectin-western blot, lectin affinity chromatography, lectin microarrays, biosensor technologies, and as a tool to analyze glycoconjugates (Lauc et al. 2002). Since lectins do not share a common structural feature, there is no universal secondary Abs to lectins. Therefore, to be used in immunoassay-like techniques lectins must be labeled with a suitable tag, e.g., biotin, digoxigenin, or digoxin, which will enable their identification and detection (Lauc et al. 2002).

2.3.1 Lectin Chromatography

One of the main fields of lectin application is lectin affinity chromatography, first developed by Donnelly and Goldstein (Donnelly and Goldstein 1970). It is a common method for the isolation, fractionation, and purification of carbohydrate-containing structures (Mechref et al. 2008; Fanayan et al. 2012). For instance, concanavalin A (Con A) is the most widely used lectin and has been employed for glycoprotein/glycopeptide enrichment due to its affinity in identifying N-glycan high mannose core (Feng et al. 2009). Because of the low affinity between lectins and carbohydrates, glycans can be competitively displaced from the complex by a competitor compound. Also, due to the low specificity of some lectins, two or more glycans may bind to a carrier during affinity chromatography (Durham and Regnier 2006). On the other hand, other lectins have narrow specificity for unique glycan structures, such as L-phytohemagglutinin (L-PHA) for the targeted β-1,6-branched N-linked glycan enrichment (Ahn et al. 2010b) and jacalin for galactosyl-β-1,3-N-acetylgalactosamine or O-linked GalNAc core structures in certain glycosylated proteins (Roque-Barreira and Campos-Neto 1985).

Durham and Regnier used the combination of Con A and lectin from *Artocarpus altilis* to isolate O-glycosylated peptides in the study of O-glycosylation sites of human serum proteins (Durham and Regnier 2006). Totten et al. (2018) used lectin affinity chromatography to separate core fucosylated and highly branched protein glycoforms. Another group applied lectin affinity chromatography on a column with sepharose-immobilized *Artocarpus incisa* lectin to study the correlation between changes in protein glycosylation and the progression of breast cancer (Lobo et al. 2017). Lectin chromatography has also been used for the purification of IgA glycoforms from sera of IgA nephropathy patients and healthy controls (Amore et al. 2001).

Due to its affinity for core 1 type *O*-glycans, a jackfruit (*Artocarpus integrifolia*) lectin jacalin has been used for binding and purification of *O*-glycosylated proteins such as IgA1 (Hortin and Trimpe 1990), and several animal IgGs that are suspected of carrying *O*-glycosylation, such as rabbit IgG (Kabir and Gerwig 1997) and bovine IgG (Porto et al. 2007). Moreover, it enables the separation of IgA1 from the IgA2 subclass. Binding to the jacalin agarose column does not require divalent cations and is not significantly affected by ionic strength and pH over a wide range. However, it was observed it has a lower affinity for disialylated and higher affinity to monosialylated and non-sialylated IgG3 *O*-glycans (Plomp et al. 2015). Moreover, it was shown that jacalin could also bind to mannose and galactose residues in *N*-glycans (Bourne et al. 2002), so it is advisable to remove *N*-glycans with PNGase F and/or exoglycosidases before *O*-glycopeptide enrichment (Bai et al. 2015).

Sambucus nigra agglutinin (SNA) lectin fractionation was used by Kaneko et al. to enrich intravenous immunoglobulin G (IVIg) for sialylated IgG and to determine the role of sialylation in its anti-inflammatory activity (Kaneko et al. 2006). IVIg is a therapeutic IgG preparation initially developed as a replacement agent for treating primary and secondary Ab deficiencies (Guhr et al. 2011) and was shown to be beneficial to some patients with acute and chronic autoimmune diseases (Kumar et al. 2006; Imbach 1991). Equivalent approach of IVIg enrichment for sialylated IgG by SNA lectin fractionation was used by Gurh et al. (2011), whose results indicated that this approach is not a suitable method to enrich IVIg specific for Fc-sialylated IgG, because the IVIg is predominantly enriched for Fab-sialylated IgG.

2.3.2 Lectin Microarrays

Lectin-based microarrays have been used to analyze glycan profiles of purified glycoproteins, as well as for the characterization of therapeutic glycoproteins, including mAb (Hirabayashi et al. 2013, 2015; Syed et al. 2016) (Fig. 2.2). In this technology, several lectins with known specificity are immobilized as microdots on a

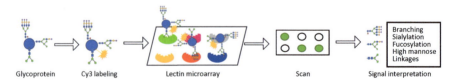

Fig. 2.2 Lectin-based microarrays have been used to analyze glycan profiles of purified glycoproteins. Glycoproteins are labeled with a fluorescent dye (Cy3) and then applied to the lectin chips. The binding signal at each lectin spot is measured, and the presence or absence of glycan variants in the testing sample detected based on the known selectivity of lectins toward glycan structures. Adapted from Zhang et al. (2016)

solid glass surface which is activated by chemical (e.g., epoxy, N-hydroxysuccinimide (NHS), amino, gold) or biochemical (e.g., streptavidin) derivatization procedures. After immobilization, residual activated groups are blocked (e.g., with amine or glycan-free serum albumin). Interaction of carbohydrate residues with the corresponding lectins can be detected either directly through their prior labeling with fluorescent reagents (e.g., Cy3 monoreactive dye) (Zhang et al. 2016) or indirectly—e.g., by overlaying a fluorescently labeled Ab (if available) against the target glycoprotein. After the extensive washing of the unbound probe, specific lectin-glycan interactions are detected with very high accuracy (Hirabayashi et al. 2013). To detect such interactions, confocal fluorescence and evanescent-field activated fluorescence can be used (Hirabayashi et al. 2013; Kuno et al. 2005). Methods such as bimolecular fluorescence quenching and recovery detection do not require the preliminary labeling of target carbohydrates (Koshi et al. 2006). Lectin microarrays are robust in generating glycan profiles that are generally consistent with the known glycan characteristics of an individual glycoprotein. They enable direct glycoprotein analysis, both as isolated glycoproteins or from body fluids (Rosenfeld et al. 2007), and are often used for rapid screening of Ig drugs.

Zhang et al. (2016), using commercial lectin chips, were able to determine glycan profiles of therapeutic mAbs and perform glycan profiling of proteins produced by different host cell systems. Furthermore, they evaluated the utility of lectin microarray in monitoring terminal galactosylation and sialylation of glycoproteins. Lectins that selectively bind core fucose, N-acetyl-D-lactosamine (Galβ1-4GlcNAc), mannose, sialic acid, or GlcNAc, which are structures commonly found in recombinant glycoproteins, were used (Hirabayashi et al. 2013).

Lectin microarray is sensitive to alterations in the terminal glycan structures, i.e., galactosylation vs. sialylation, and can effectively distinguish glycan isomers containing different sialic acid linkages (Zhang et al. 2016). Furthermore, such a platform demonstrates the usefulness of the lectin microarray in screening glycan patterns of protein samples and increased full coverage of all glycan variants of a glycoprotein. Despite all advantages, assay performance could be improved using lectins with advanced selectivity and binding affinity to distinct glycan species. One of the advantages is the possibility of lectin microarray customization by including lectins relevant to the glycan species that are possibly present in the testing sample. Lectin microarrays are commercially available by different producers.

Nowadays, carbohydrate-binding proteins like lectins play an important role in the structural and functional elucidation of glycoproteins, as well as in the study of their binding affinities and interactions with other proteins (Roucka et al. 2017). In contrast to Abs, lectins are generally more stable, more affordable, better characterized and address a broader spectrum of glycoproteins. As mentioned before, MS, UHPLC, and CE are the most common methods for Ig glycan analysis. While MS analysis mostly involves enzymatic digestion of a glycoprotein, UHPLC and CE usually require the release of glycans from a glycoprotein. In contrast, lectin microarrays enable rapid and direct measurement of glycan profiles on an intact protein (including O-glycans) without the need for protein digestion and glycan release. Therefore, the lectin microarray platform could be adopted as a complementary tool for the high-throughput screening of glycan profiles of therapeutic glycoproteins.

2.4 Perspectives

In the last decade, the glycobiology field has significantly advanced in terms of the analytical capabilities of available technologies that facilitated our understanding of the various biological roles of Igs (and other glycoproteins). A correlation between changes in Ig N- and O-glycan moieties and the pathogenesis of many diseases, including inflammatory diseases, immune deficiencies, cardiovascular diseases, and cancer, has fostered a continuing interest in the elucidation of their functions.

Thus, the development of new and optimization of existing high-throughput approaches is crucial. Most lectin- and LC-based glycan sample preparation workflows developed to date are impacted considerably by the increase in robustness, throughput, sensitivity, integration of different analytical strategies, and hyphenation of orthogonal technologies (e.g., LC and MS).

Regarding the analytical techniques, the requirement for rapid and more detailed characterization of Ig glycosylation on all levels (free glycans, glycopeptides, subunit level, and intact Ig level) will lead to an increase in the application of lectin- and LC-based methods combined with mass spectrometric techniques. Development of fast and robust purification and enrichment techniques that would enable low-abundant Igs isolation and their in-depth glycan analysis remains a challenge. Lectin microarrays could prove to be useful here since they do not require additional steps of protein digestion or glycan cleaving that usually lead to sample losses and make glycan analysis of low-abundant Igs even more demanding.

Employment of rapid, sensitive protocols enabling Ig glycosylation analysis from low amounts of starting material already allows glycan detection and quantification using multiple analytical technologies without the necessity for separate sample preparation, significantly decreasing hands-on time. Due to the higher degree of automation and throughput increase, separation techniques coupled with mass spectrometric detection will dominate the field. With the development of automated, coupled techniques, the development of bioinformatics tools is essential.

To summarize, the analytical techniques applied for Igs glycosylation analysis will further develop toward a more detailed and sensitive characterization of N- and O-glycosylation but also toward fast, automated high-throughput methods for glycosylation monitoring.

Compliance with Ethical Standards

Conflict of Interest T. Petrović and I. Trbojević-Akmačić are employees of Genos Ltd., a private research organization that specializes in high-throughput glycomic analysis.

Ethical Approval This article does not contain any studies with human participants or animals performed by any of the authors.

References

Abrahams JL, Campbell MP, Packer NH (2018) Building a PGC-LC-MS N-glycan retention library and elution mapping resource. Glycoconj J 35(1):15–29

Adams GP, Weiner LM (2005) Monoclonal antibody therapy of cancer. Nat Biotechnol 23 (9):1147–1157

Ahmed AA, Giddens J, Pincetic A, Lomino JV, Ravetch JV, Wang LX, Bjorkman PJ (2014) Structural characterization of anti-inflammatory immunoglobulin G Fc proteins. J Mol Biol 426 (18):3166–3179

Ahn J, Bones J, Yu YQ, Rudd PM, Gilar M (2010a) Separation of 2-aminobenzamide labeled glycans using hydrophilic interaction chromatography columns packed with 1.7 μm sorbent. J Chromatogr B 878(3-4):403–408

Ahn YH, Kim YS, Ji ES, Lee JY, Jung JA, Ko JH, Yoo JS (2010b) Comparative quantitation of aberrant glycoforms by lectin-based glycoprotein enrichment coupled with multiple-reaction monitoring mass spectrometry. Anal Chem 82(11):4441–4447

Akazawa-Ogawa Y, Nagai H, Hagihara Y (2018) Heat denaturation of the antibody, a multi-domain protein. Biophys Rev 10(2):255–258

Akerström B, Björck L (1986) A physicochemical study of protein G, a molecule with unique immunoglobulin G-binding properties. J Biol Chem 261(22):10240–10247

Alden B, Iraneta P, Hudalla C, Wyndham K, Walter T, Lawrence N (2012) Synthesis and applications of BEH particles in liquid chromatography. LGCG 30:20–29

Allhorn M, Olin AI, Nimmerjahn F, Collin M (2008) Human IgG/FcγR interactions are modulated by streptococcal IgG glycan hydrolysis. PLoS One 3(1):e1413

Altmann F, Paschinger K, Dalik T, Vorauer K (1998) Characterisation of peptide-N4-(N-acetyl-β-glucosaminyl) asparagine amidase A and its N-glycans. Eur J Biochem 252 (1):118–123

Amore A, Cirina P, Conti G, Brusa P, Peruzzi L, Coppo R (2001) Glycosylation of circulating IgA in patients with IgA nephropathy modulates proliferation and apoptosis of mesangial cells. J Am Soc Nephrol 12(9):1862–1871

An HJ, Froehlich JW, Lebrilla CB (2009) Determination of glycosylation sites and site-specific heterogeneity in glycoproteins. Curr Opin Chem Biol 13(4):421–426

André S, Kaltner H, Manning JC, Murphy PV, Gabius HJ (2015) Lectins: getting familiar with translators of the sugar code. Molecules 20(2):1788–1823

Anumula KR (2012) Quantitative glycan profiling of normal human plasma derived immunoglobulin and its fragments Fab and Fc. J Immunol Methods 382(1-2):167–176

Anumula KR, Dhume ST (1998) High resolution and high sensitivity methods for oligosaccharide mapping and characterization by normal phase high performance liquid chromatography following derivatization with highly fluorescent anthranilic acid. Glycobiology 8(7):685–694

Arnold JN, Wormald MR, Sim RB, Rudd PM, Dwek RA (2007) The impact of glycosylation on the biological function and structure of human immunoglobulins. Annu Rev Immunol 25:21–50

Arora S, Saxena V, Ayyar BV (2017) Affinity chromatography: a versatile technique for antibody purification. Methods 116:84–94

Ashwood C, Pratt B, MacLean BX, Gundry RL, Packer NH (2019) Standardization of PGC-LC-MS-based glycomics for sample specific glycotyping. Analyst 144(11):3601–3612

Augusto F, Hantao LW, Mogollón NG, Braga SC (2013) New materials and trends in sorbents for solid-phase extraction. TrAC Trends Anal Chem 43:14–23

Aust N, Parth M, Lederer K (2001) SEC of ultra-high molar mass polymers: optimization of experimental conditions to avoid molecular degradation in the case of narrow polystyrene standards. Int J Polym Anal Charact 6(3–4):245–260

Ayyar BV, Arora S, Murphy C, O'Kennedy R (2012) Affinity chromatography as a tool for antibody purification. Methods 56(2):116–129

Bai X, Li D, Zhu J, Guan Y, Zhang Q, Chi L (2015) From individual proteins to proteomic samples: characterization of O-glycosylation sites in human chorionic gonadotropin and human-plasma proteins. Anal Bioanal Chem 407(7):1857–1869

Bellande K, Bono JJ, Savelli B, Jamet E, Canut H (2017) Plant lectins and lectin receptor-like kinases: how do they sense the outside? Int J Mol Sci 18(6):1164

Berg JM, Tymoczko JLSL (2002) Biochemistry. In: Section 11.4. Lectins are specific carbohydrate-binding proteins, 5th edn. W. H. Freeman, New York

Bermingham ML, Colombo M, McGurnaghan SJ, Blackbourn LA, Vučković F, Baković MP, Trbojević-Akmačić I, Lauc G, Agakov F, Agakova AS, Hayward C (2018) N-glycan profile and kidney disease in type 1 diabetes. Diabetes Care 41(1):79–87

Berna M, Murphy AT, Wilken B, Ackermann B (2002) Collection, storage, and filtration of in vivo study samples using 96-well filter plates to facilitate automated sample preparation and LC/MS/MS analysis. Anal Chem 74(5):1197–1201

Bhide GP, Colley KJ (2017) Sialylation of N-glycans: mechanism, cellular compartmentalization and function. Histochem Cell Biol 147(2):149–174

Bigge JC, Patel TP, Bruce JA, Goulding PN, Charles SM, Parekh RB (1995) Nonselective and efficient fluorescent labeling of glycans using 2-amino benzamide and anthranilic acid. Anal Biochem 230(2):229–238

Björck L, Kronvall G (1984) Purification and some properties of streptococcal protein G, a novel IgG-binding reagent. J Immunol 133(2):969–974

Boersema PJ, Mohammed S, Heck AJ (2008) Hydrophilic interaction liquid chromatography (HILIC) in proteomics. Anal Bioanal Chem 391(1):151–159

Böhm S, Schwab I, Lux A, Nimmerjahn F (2012) The role of sialic acid as a modulator of the anti-inflammatory activity of IgG. Semin Immunopathol 34(3):443–453

Bolton GR, Mehta KK (2016) The role of more than 40 years of improvement in protein A chromatography in the growth of the therapeutic antibody industry. Biotechnol Prog 32 (5):1193–1202

Bondt A, Nicolardi S, Jansen BC, Kuijper TM, Hazes JM, Van Der Burgt YE, Wuhrer M, Dolhain RJ (2017) IgA N-and O-glycosylation profiling reveals no association with the pregnancy-related improvement in rheumatoid arthritis. Arthritis Res Ther 19(1):1–8

Bourne Y, Henrissat B (2001) Glycoside hydrolases and glycosyltransferases: families and functional modules. Curr Opin Struct Biol 11(5):593–600

Bourne Y, Astoul CH, Zamboni V, Peumans WJ, Menu-Bouaouiche L, Van Damme EJ, Barre A, Rougé P (2002) Structural basis for the unusual carbohydrate-binding specificity of jacalin towards galactose and mannose. Biochem J 364(1):173–180

Bruhns P, Iannascoli B, England P, Mancardi DA, Fernandez N, Jorieux S, Daëron M (2009) Specificity and affinity of human Fcγ receptors and their polymorphic variants for human IgG subclasses. Blood 113(16):3716–3725

Butters TD (2002) Control in the N-linked glycoprotein biosynthesis pathway. Chem Biol 9 (12):1266–1268

Campbell MP, Royle L, Radcliffe CM, Dwek RA, Rudd PM (2008) GlycoBase and autoGU: tools for HPLC-based glycan analysis. Bioinformatics 24(9):1214–1216

Chakraborty AB, Berger SJ (2005) Optimization of reversed-phase peptide liquid chromatography ultraviolet mass spectrometry analyses using an automated blending methodology. J Biomol Tech 16(4):327

Chan AC, Carter PJ (2010) Therapeutic antibodies for autoimmunity and inflammation. Nat Rev Immunol 10(5):301–316

Chandler KB, Mehta N, Leon DR, Suscovich TJ, Alter G, Costello CE (2019) Multi-isotype glycoproteomic characterization of serum antibody heavy chains reveals isotype-and subclass-specific N-glycosylation profiles. Mol Cell Proteomics 18(4):686–703

Chen CC, Su WC, Huang BY, Chen YJ, Tai HC, Obena RP (2014) Interaction modes and approaches to glycopeptide and glycoprotein enrichment. Analyst 139(4):688–704

Chen R, Cheng K, Ning Z, Figeys D (2016) N-Glycopeptide reduction with Exoglycosidases enables accurate characterization of site-specific N-Glycosylation. Anal Chem 88 (23):11837–11843

Chen M, Shi X, Duke RM, Ruse CI, Dai N, Taron CH, Samuelson JC (2017) An engineered high affinity Fbs1 carbohydrate binding protein for selective capture of N-glycans and N-glycopeptides. Nat Commun 8(1):1–5

Ciucanu I, Kerek F (1984) A simple and rapid method for the permethylation of carbohydrates. Carbohydr Res 131(2):209–217

Collin M, Olsén A (2001) EndoS, a novel secreted protein from *Streptococcus pyogenes* with endoglycosidase activity on human IgG. EMBO J 20(12):3046–3055

Collin M, Svensson MD, Sjöholm AG, Jensenius JC, Sjöbring U, Olsén A (2002) EndoS and SpeB from *Streptococcus pyogenes* inhibit immunoglobulin-mediated opsonophagocytosis. Infect Immun 70(12):6646–6651

Coppo R, Amore A (2004) Aberrant glycosylation in IgA nephropathy (IgAN). Kidney Int 65 (5):1544–1547

Coskun O (2016) Separation techniques: chromatography. North Clin Istanb 3(2):156

Costell CE, Contado-Miller JM, Cipollo JF (2007) A glycomics platform for the analysis of permethylated oligosaccharide alditols. J Am Soc Mass Spectrom 18(10):1799–1812

Cuatrecasas P, Wilchek M, Anfinsen CB (1968) Selective enzyme purification by affinity chromatography. Proc Natl Acad Sci U S A 61(2):636

Cummings RD, McEver RP (2009) C-type lectins. In: Essentials of glycobiology, 2nd edn. Cold Spring Harbor Laboratory Press, Cold Spring Harbor

Dard P, Lefranc MP, Osipova L, Sanchez-Mazas A (2001) DNA sequence variability of IGHG3 alleles associated to the main G3m haplotypes in human populations. Eur J Hum Genet 9 (10):765–772

De Chateau M, Nilson BH, Erntell M, Myhre E, Magnusson CG, Åkerström B, Björck L (1993) On the interaction between protein L and immunoglobulins of various mammalian species. Scand J Immunol 37(4):399–405

de Haan N, Reiding KR, Krištić J, Ederveen ALH, Lauc G, Wuhrer M (2017) The N-glycosylation of mouse immunoglobulin G (IgG)-fragment crystallizable differs between IgG subclasses and Strains. Front Immunol 8:608

Djoumerska-Alexieva IK, Dimitrov JD, Voynova EN, Lacroix-Desmazes S, Kaveri SV, Vassilev TL (2010) Exposure of IgG to an acidic environment results in molecular modifications and in enhanced protective activity in sepsis. FEBS J 277(14):3039–3050

Donnelly EH, Goldstein IJ (1970) Glutaraldehyde-insolubilized concanavalin A: an adsorbent for the specific isolation of polysaccharides and glycoproteins. Biochem J 118(4):679–680

Dorion-Thibaudeau J, Raymond C, Lattová E, Perreault H, Durocher Y, De Crescenzo G (2014) Towards the development of a surface plasmon resonance assay to evaluate the glycosylation pattern of monoclonal antibodies using the extracellular domains of CD16a and CD64. J Immunol Methods 408:24–34

Drickamer K, Dordal MS, Reynolds L (1986) Mannose-binding proteins isolated from rat liver contain carbohydrate-recognition domains linked to collagenous tails. Complete primary structures and homology with pulmonary surfactant apoprotein. J Biol Chem 261(15):6878–6887

Drickamer K, Dell A, Fadden AJ (2002) Genomic analysis of C-type lectins. Biochem Soc Symp 69:59–72

Duffy FJ, Rudd PM (2015) GlycoProfileAssigner: automated structural assignment with error estimation for glycan LC data. Bioinformatics 31(13):2220–2221

Dunn-Walters D, Boursier L, Spencer J (2000) Effect of somatic hypermutation on potential N-glycosylation sites in human immunoglobulin heavy chain variable regions. Mol Immunol 37(3-4):107–113

Durham M, Regnier FE (2006) Targeted glycoproteomics: serial lectin affinity chromatography in the selection of O-glycosylation sites on proteins from the human blood proteome. J Chromatogr A 1132(1-2):165–173

Falck D, Jansen BC, Plomp R, Reusch D, Haberger M, Wuhrer M (2015) Glycoforms of immunoglobulin G based biopharmaceuticals are differentially cleaved by trypsin due to the glycoform influence on higher-order structure. J Proteome Res 14(9):4019–4028

Fan JQ, Lee YC (1997) Detailed studies on substrate structure requirements of glycoamidases A and F. J Biol Chem 272(43):27058–27064

Fanayan S, Hincapie M, Hancock WS (2012) Using lectins to harvest the plasma/serum glycoproteome. Electrophoresis 33(12):1746–1754

Feng S, Yang N, Pennathur S, Goodison S, Lubman DM (2009) Enrichment of glycoproteins using nanoscale chelating concanavalin A monolithic capillary chromatography. Anal Chem 81 (10):3776–3783

Fernández LE, Kalume DE, Calvo L, Mallo MF, Vallin A, Roepstorff P (2001) Characterization of a recombinant monoclonal antibody by mass spectrometry combined with liquid chromatography. J Chromatogr B Biomed Sci Appl 752(2):247–261

Field MC, Amatayakul-Chantler SU, Rademacher TW, Rudd PM, Dwek RA (1994) Structural analysis of the N-glycans from human immunoglobulin A1: comparison of normal human serum immunoglobulin A1 with that isolated from patients with rheumatoid arthritis. Biochem J 299(1):261–275

Fischler DA, Orlando R (2019) N-linked Glycan Release Efficiency: a Quantitative Comparison between NaOCl and PNGase F Release Protocols. J Biomol Tech 30(4):58

Fishman JB, Berg EA (2019) Purification of antibodies: diethylaminoethyl (DEAE) chromatography. Cold Spring Harb Protoc 2019(1):pdb-rot099135

Fornelli L, Ayoub D, Aizikov K, Beck A, Tsybin YO (2014) Middle-down analysis of monoclonal antibodies with electron transfer dissociation orbitrap Fourier transform mass spectrometry. Anal Chem 86(6):3005–3012

Freeze HH, Kranz C (2010) Endoglycosidase and glycoamidase release of N-linked glycans. Curr Protoc Mol Biol 89(1):17–13

Fukuda M (1989) Characterization of O-Linked saccharides from cell surface glycoproteins. Methods Enzymol 179:17–29

Gabius HJ, André S, Jiménez-Barbero J, Romero A, Solís D (2011) From lectin structure to functional glycomics: principles of the sugar code. Trends Biochem Sci 36(6):298–313

Gagnon P (1996) Purification tools for monoclonal antibodies. Validated Biosystems, Tucson

Gagnon P, Nian R (2016) Conformational plasticity of IgG during protein A affinity chromatography. J Chromatogr A 1433:98–105

Gagnon P, Nian R, Leong D, Hoi A (2015) Transient conformational modification of immunoglobulin G during purification by protein A affinity chromatography. J Chromatogr A 1395:136–142

GE Healthcare. Antibody purification handbook. GE Health Available from: https://www.cytivalifesciences.com/en/us/support/Handbooks

Ghitescu LU, Galis Z, Bendayan M (1991) Protein AG-gold complex: an alternative probe in immunocytochemistry. J Histochem Cytochem 39(8):1057–1065

González-Valdez J, Yoshikawa A, Weinberg J, Benavides J, Rito-Palomares M, Przybycien TM (2014) Toward improving selectivity in affinity chromatography with PEG ylated affinity ligands: the performance of PEG ylated protein A. Biotechnol Prog 30(6):1364–1379

Graille M, Stura EA, Corper AL, Sutton BJ, Taussig MJ, Charbonnier JB, Silverman GJ (2000) Crystal structure of a *Staphylococcus aureus* protein A domain complexed with the Fab fragment of a human IgM antibody: structural basis for recognition of B-cell receptors and superantigen activity. Proc Natl Acad Sci 97(10):5399–5404

Grodzki AC, Berenstein E (2010a) Antibody purification: affinity chromatography–protein A and protein G Sepharose. In: Immunocytochemical methods and protocols. Humana Press, New York, pp 33–41

Grodzki AC, Berenstein E (2010b) Antibody purification: ion-exchange chromatography. In: Immunocytochemical methods and protocols. Humana Press, New York, pp 27–32

Gudelj I, Lauc G, Pezer M (2018) Immunoglobulin G glycosylation in aging and diseases. Cell Immunol 333:65–79

Guhr T, Bloem J, Derksen NI, Wuhrer M, Koenderman AH, Aalberse RC, Rispens T (2011) Enrichment of sialylated IgG by lectin fractionation does not enhance the efficacy of immunoglobulin G in a murine model of immune thrombocytopenia. PLoS One 6(6):e21246

Guile GR, Rudd PM, Wing DR, Prime SB, Dwek RA (1996) A rapid high-resolution high-performance liquid chromatographic method for separating glycan mixtures and analyzing oligosaccharide profiles. Anal Biochem 240(2):210–226

Hage DS, Anguizola JA, Bi C, Li R, Matsuda R, Papastavros E, Pfaunmiller E, Vargas J, Zheng X (2012) Pharmaceutical and biomedical applications of affinity chromatography: recent trends and developments. J Pharm Biomed Anal 69:93–105

Hang Y, He XP, Yang L, Hua J (2015) Probing sugar–lectin recognitions in the near-infrared region using glyco-diketopyrrolopyrrole with aggregation-induced-emission. Biosens Bioelectron 65:420–426

Hanisch FG, Teitz S, Schwientek T, Müller S (2009) Chemical de-O-glycosylation of glycoproteins for application in LC-based proteomics. Proteomics 9(3):710–719

Harazono A, Kawasaki N, Itoh S, Hashii N, Matsuishi-Nakajima Y, Kawanishi T, Yamaguchi T (2008) Simultaneous glycosylation analysis of human serum glycoproteins by high-performance liquid chromatography/tandem mass spectrometry. J Chromatogr B 869(1-2):20–30

Haxo T, Jones A, Kimzey M, Dale E, Vlasenko S, Mast S. Automated N-glycan sample preparation with an instant glycan labeling dye for mass spectrometry. Prozyme Application Note. 2016.

Hilbold NJ, Le Saoût X, Valery E, Muhr L, Souquet J, Lamproye A, Broly H (2017) Evaluation of several protein a resins for application to multicolumn chromatography for the rapid purification of fed-batch bioreactors. Biotechnol Prog 33(4):941–953

Hirabayashi J, Yamada M, Kuno A, Tateno H (2013) Lectin microarrays: concept, principle and applications. Chem Soc Rev 42(10):4443–4458

Hirabayashi J, Kuno A, Tateno H (2015) Development and applications of the lectin microarray. Top Curr Chem 367:105–124

Hirayama K, Yuji R, Yamada N, Kato K, Arata Y, Shimada I (1998) Complete and rapid peptide and glycopeptide mapping of mouse monoclonal antibody by LC/MS/MS using ion trap mass spectrometry. Anal Chem 70(13):2718–2725

Ho CS, Lam CW, Chan MH, Cheung RC, Law LK, Lit LC, Ng KF, Suen MW, Tai HL (2003) Electrospray ionisation mass spectrometry: principles and clinical applications. Clin Biochem Rev 24(1):3

Hoffmann M, Marx K, Reichl U, Wuhrer M, Rapp E (2016) Site-specific O-glycosylation analysis of human blood plasma proteins. Mol Cell Proteomics 15(2):624–641

Hortin GL, Trimpe BL (1990) Lectin affinity chromatography of proteins bearing O-linked oligosaccharides: application of jacalin-agarose. Anal Biochem 188(2):271–277

Howard M, Farrar CA, Sacks SH (2018) Structural and functional diversity of collectins and ficolins and their relationship to disease. Semin Immunopathol 40(1):75–85

Hu D, Tateno H, Hirabayashi J (2015) Lectin engineering, a molecular evolutionary approach to expanding the lectin utilities. Molecules 20(5):7637–7656

Huang Y, Orlando R (2017) Kinetics of N-glycan release from human immunoglobulin G (IgG) by PNGase F: all glycans are not created equal. J Biomol Tech 28(4):150

Huang L, Lu J, Wroblewski VJ, Beals JM, Riggin RM (2005) In vivo deamidation characterization of monoclonal antibody by LC/MS/MS. Anal Chem 77(5):1432–1439

Huffman JE, Pučić-Baković M, Klarić L, Hennig R, Selman MH, Vučković F, Novokmet M, Krištić J, Borowiak M, Muth T, Polašek O (2014) Comparative performance of four methods for high-throughput glycosylation analysis of immunoglobulin G in genetic and epidemiological research. Mol Cell Proteomics 13(6):1598–1610

Huhn C, Selman MH, Ruhaak LR, Deelder AM, Wuhrer M (2009) IgG glycosylation analysis. Proteomics 9(4):882–913

Imbach P (1991) Immune thrombocytopenic purpura and intravenous immunoglobulin. Cancer 68 (S6):1422–1425
Jacobs JF, Wevers RA, Lefeber DJ, van Scherpenzeel M (2016) Fast, robust and high-resolution glycosylation profiling of intact monoclonal IgG antibodies using nanoLC-chip-QTOF. Clin Chim Acta 461:90–97
Janeway CA Jr, Travers PWM (2001) Immunobiology: the immune system in health and disease, 5th edn. Garland Science, New York
Jefferis R, Pound J, Lund J, Goodall M (1994) Effector mechanisms activated by human IgG subclass antibodies: clinical and molecular aspects. Ann Biol Clin 52(1):57–65
Jensen PH, Mysling S, Højrup P, Jensen ON (2013) Glycopeptide enrichment for MALDI-TOF mass spectrometry analysis by hydrophilic interaction liquid chromatography solid phase extraction (HILIC SPE). In: Mass spectrometry of glycoproteins. Humana Press, Totowa, NY, pp 131–144
Jiang XR, Song A, Bergelson S, Arroll T, Parekh B, May K, Chung S, Strouse R, Mire-Sluis A, Schenerman M (2011) Advances in the assessment and control of the effector functions of therapeutic antibodies. Nat Rev Drug Discov 10(2):101–111
Kabir S, Gerwig GJ (1997) The structural analysis of the O-glycans of the jacalin-bound rabbit immunoglobulin G. IUBMB Life 42(4):769–778
Kaneko Y, Nimmerjahn F, Ravetch JV (2006) Anti-inflammatory activity of immunoglobulin G resulting from Fc sialylation. Science 313(5787):670–673
Kang P, Mechref Y, Novotny MV (2008) High-throughput solid-phase permethylation of glycans prior to mass spectrometry. Rapid Commun Mass Spectrom 22(5):721–734
Kayili HM, Ertürk AS, Elmacı G, Salih B (2019) Poly (amidoamine) dendrimer-coated magnetic nanoparticles for the fast purification and selective enrichment of glycopeptides and glycans. J Sep Sci 42(20):3209–3216
Keser T, Pavić T, Lauc G, Gornik O (2018) Comparison of 2-aminobenzamide, procainamide and RapiFluor-MS as derivatizing agents for high-throughput HILIC-UPLC-FLR-MS N-glycan analysis. Front Chem 6:324
Kihlberg BM, Sjöholm AG, Björck L, Sjöbring U (1996) Characterization of the binding properties of protein LG, an immunoglobulin-binding hybrid protein. Eur J Biochem 240(3):556–563
Kim W, Kim J, You S, Do J, Jang Y, Kim D, Lee J, Ha J, Kim HH (2019) Qualitative and quantitative characterization of sialylated N-glycans using three fluorophores, two columns, and two instrumentations. Anal Biochem 571:40–48
Klapoetke S, Zhang J, Becht S, Gu X, Ding X (2010) The evaluation of a novel approach for the profiling and identification of N-linked glycan with a procainamide tag by HPLC with fluorescent and mass spectrometric detection. J Pharm Biomed Anal 53(3):315–324
Kolarich D, Windwarder M, Alagesan K, Altmann F (2015) Isomer-specific analysis of released N-glycans by LC-ESI MS/MS with porous graphitized carbon. In: Glyco-engineering. Humana Press, New York, NY, pp 427–435
Konrad A, Eriksson Karlström A, Hober S (2011) Covalent immunoglobulin labeling through a photoactivable synthetic Z domain. Bioconjug Chem 22(12):2395–2403
Koshi Y, Nakata E, Yamane H, Hamachi I (2006) A fluorescent lectin array using supramolecular hydrogel for simple detection and pattern profiling for various glycoconjugates. J Am Chem Soc 128(32):10413–10422
Kostanski LK, Keller DM, Hamielec AE (2004) Size-exclusion chromatography—a review of calibration methodologies. J Biochem Biophys Methods 58(2):159–186
Kozak RP, Tortosa CB, Fernandes DL, Spencer DI (2015) Comparison of procainamide and 2-aminobenzamide labeling for profiling and identification of glycans by liquid chromatography with fluorescence detection coupled to electrospray ionization–mass spectrometry. Anal Biochem 486:38–40
Krištić J, Vučković F, Menni C, Klarić L, Keser T, Beceheli I, Pučić-Baković M, Novokmet M, Mangino M, Thaqi K, Rudan P (2014) Glycans are a novel biomarker of chronological and biological ages. J Gerontol A Biol Sci Med Sci 69(7):779–789

Kroon DJ, Freedy J, Burinsky DJ, Sharma B (1995) Rapid profiling of carbohydrate glycoforms in monoclonal antibodies using MALDI/TOF mass spectrometry. J Pharm Biomed Anal 13 (8):1049–1054

Kruljec N, Bratkovič T (2017) Alternative affinity ligands for immunoglobulins. Bioconjug Chem 28(8):2009–2030

Kumar A, Teuber SS, Gershwin ME (2006) Intravenous immunoglobulin: striving for appropriate use. Int Arch Allergy Immunol 140(3):185–198

Kuno A, Uchiyama N, Koseki-Kuno S, Ebe Y, Takashima S, Yamada M, Hirabayashi J (2005) Evanescent-field fluorescence-assisted lectin microarray: a new strategy for glycan profiling. Nat Methods 2(11):851–856

Lam MP, Lau E, Siu SO, Ng DC, Kong RP, Chiu PC, Yeung WS, Lo C, Chu IK (2011) Online combination of reversed-phase/reversed-phase and porous graphitic carbon liquid chromatography for multicomponent separation of proteomics and glycoproteomics samples. Electrophoresis 32(21):2930–2940

Lannoo N, Van Damme EJ (2014) Lectin domains at the frontiers of plant defense. Front Plant Sci 5:397

Largy E, Cantais F, Van Vyncht G, Beck A, Delobel A (2017) Orthogonal liquid chromatography–mass spectrometry methods for the comprehensive characterization of therapeutic glycoproteins, from released glycans to intact protein level. J Chromatogr A 1498:128–146

Larsen MR, Jensen SS, Jakobsen LA, Heegaard NH (2007) Exploring the sialiome using titanium dioxide chromatography and mass spectrometry. Mol Cell Proteomics 6(10):1778–1787

Larsen MD, de Graaf EL, Sonneveld ME, Plomp HR, Nouta J, Hoepel W, Chen HJ, Linty F, Visser R, Brinkhaus M, Šuštić T (2020) Afucosylated IgG characterizes enveloped viral responses and correlates with COVID-19 severity. Science 371(6532):eabc8378

Lauber MA, Yu YQ, Brousmiche DW, Hua Z, Koza SM, Magnelli P, Guthrie E, Taron CH, Fountain KJ (2015) Rapid preparation of released N-glycans for HILIC analysis using a labeling reagent that facilitates sensitive fluorescence and ESI-MS detection. Anal Chem 87 (10):5401–5409

Lauc G, Dumić J, Supraha S, Flögel M (2002) Lectins labelled with digoxin as a novel tool to study glycoconjugates. Food Technol Biotechnol 40(4):289–292

Lauc G, Krištić J, Zoldoš V (2014) Glycans–the third revolution in evolution. Front Genet 5:145

Lee YC (1990) High-performance anion-exchange chromatography for carbohydrate analysis. Anal Biochem 189(2):151–162

Lermyte F, Tsybin YO, O'Connor PB, Loo JA (2019) Top or middle? Up or down? Toward a standard lexicon for protein top-down and allied mass spectrometry approaches. J Am Soc Mass Spectrom 30(7):1149–1157

Linhult M, Gülich S, Gräslund T, Simon A, Karlsson M, Sjöberg A, Nord K, Hober S (2004) Improving the tolerance of a protein a analogue to repeated alkaline exposures using a bypass mutagenesis approach. Proteins 55(2):407–416

Liu S, Fu Y, Huang Z, Liu Y, Liu BF, Cheng L, Liu X (2020) A comprehensive analysis of subclass-specific IgG glycosylation in colorectal cancer progression by nanoLC-MS/MS. Analyst 145(8):3136–3147

Lobo MD, Moreno FB, Souza GH, Verde SM, Moreira RD, Monteiro-Moreira AC (2017) Label-free proteome analysis of plasma from patients with breast cancer: stage-specific protein expression. Front Oncol 7:14

Lopez E, Scott NE, Wines BD, Hogarth PM, Wheatley AK, Kent SJ, Chung AW (2019) Low pH exposure during Immunoglobulin G purification methods results in aggregates that avidly bind Fcγ Receptors: implications for measuring Fc dependent antibody functions. Front Immunol 10:2415

Madhwani T, McBain AJ (2016) The application of magnetic bead selection to investigate interactions between the oral microbiota and salivary immunoglobulins. PLoS One 11(8): e0158288

Malaker SA, Pedram K, Ferracane MJ, Bensing BA, Krishnan V, Pett C, Yu J, Woods EC, Kramer JR, Westerlind U, Dorigo O (2019) The mucin-selective protease StcE enables molecular and functional analysis of human cancer-associated mucins. Proc Natl Acad Sci 116(15):7278–7287

Manning JC, Romero A, Habermann FA, Caballero GG, Kaltner H, Gabius HJ (2017) Lectins: a primer for histochemists and cell biologists. Histochem Cell Biol 147(2):199–222

Marek KW, Vijay IK, Marth JD (1999) A recessive deletion in the GlcNAc-1-phosphotransferase gene results in peri-implantation embryonic lethality. Glycobiology 9(11):1263–1271

Mariño K, Bones J, Kattla JJ, Rudd PM (2010) A systematic approach to protein glycosylation analysis: a path through the maze. Nat Chem Biol 6(10):713–723

Maverakis E, Kim K, Shimoda M, Gershwin ME, Patel F, Wilken R et al (2015) Glycans in the immune system and the altered glycan theory of autoimmunity: a critical review. J Autoimmun 57:1–3

Mayer S, Raulf MK, Lepenies B (2017) C-type lectins: their network and roles in pathogen recognition and immunity. Histochem Cell Biol 147(2):223–237

McEver RP (2015) Selectins: initiators of leucocyte adhesion and signalling at the vascular wall. Cardiovasc Res 107(3):331–339

McMahon MJ, O'Kennedy R (2000) Polyreactivity as an acquired artefact, rather than a physiologic property, of antibodies: evidence that monoreactive antibodies may gain the ability to bind to multiple antigens after exposure to low pH. J Immunol Methods 241(1–2):1–10

Mechref Y, Madera M, Novotny MV (2008) Glycoprotein enrichment through lectin affinity techniques. In: 2D page: sample preparation and fractionation, vol 424. Humana Press, Totowa, pp 373–396

Melin Fürst C, Mörgelin M, Vadstrup K, Heinegård D, Aspberg A, Blom AM (2013) The C-type lectin of the aggrecan G3 domain activates complement. PLoS One 8(4):e61407

Millán Martín S, Iglesias N (2015) Comparative analysis of monoclonal antibody N-glycosylation using stable isotope labelling and UPLC-fluorescence-MS. Analyst 140(5):1442–1447

Mimura Y, Katoh T, Saldova R, O'Flaherty R, Izumi T, Mimura-Kimura Y, Utsunomiya T, Mizukami Y, Yamamoto K, Matsumoto T, Rudd PM (2018) Glycosylation engineering of therapeutic IgG antibodies: challenges for the safety, functionality and efficacy. Protein Cell 9(1):47–62

Momčilović A, de Haan N, Hipgrave Ederveen AL, Bondt A, Koeleman CA, Falck D, de Neef LA, Mesker WE, Tollenaar R, de Ru A, van Veelen P (2020) Simultaneous immunoglobulin A and G glycopeptide profiling for high-throughput applications. Anal Chem 92(6):4518–4526

Moore JC (1996) Gel permeation chromatography. I. A new method for molecular weight distribution of high polymers. J Polym Sci A Polym Chem 34(10):1833–1841

Myhre EB, Erntell M (1985) A non-immune interaction between the light chain of human immunoglobulin and a surface component of a Peptococcus magnus strain. Mol Immunol 22(8):879–885

Naik AD, Menegatti S, Gurgel PV, Carbonell RG (2011) Performance of hexamer peptide ligands for affinity purification of immunoglobulin G from commercial cell culture media. J Chromatogr A 1218(13):1691–1700

Nakagawa H, Hato M, Takegawa Y, Deguchi K, Ito H, Takahata M, Iwasaki N, Minami A, Nishimura SI (2007) Detection of altered N-glycan profiles in whole serum from rheumatoid arthritis patients. J Chromatogr B 853(1-2):133–137

Naomi L, Gustavsson P, Michael R, Lindgren J, Nørskov-lauritsen L, Lund M et al (2012) Novel peptide ligand with high binding capacity for antibody purification. J Chromatogr A 1225:158–167

Nelson RM, Venot A, Bevilacqua MP, Linhardt RJ, Stamenkovic I (1995) Carbohydrate-protein interactions in vascular biology. Annu Rev Cell Dev Biol 11(1):601–631

Nevens JR, Mallia AK, Wendt MW, Smith PK (1992) Affinity chromatographic purification of immunoglobulin M antibodies utilizing immobilized mannan binding protein. J Chromatogr A 597(1–2):247–256

Nilsson B, Moks T, Jansson B, Abrahmsen L, Elmblad A, Holmgren E, Henrichson C, Jones TA, Uhlen M (1987) A synthetic IgG-binding domain based on staphylococcal protein A. Protein Eng Des Sel 1(2):107–113

Nimmerjahn F, Ravetch JV (2008) Fcγ receptors as regulators of immune responses. Nat Rev Immunol 8(1):34–47.26

O'Flaherty R, Trbojević-Akmačić I, Greville G, Rudd PM, Lauc G (2018) The sweet spot for biologics: recent advances in characterization of biotherapeutic glycoproteins. Expert Rev Proteomics 15(1):13–29

Oliveira C, Teixeira JA, Domingues L (2013) Recombinant lectins: an array of tailor-made glycan-interaction biosynthetic tools. Crit Rev Biotechnol 33(1):66–80

Olivova P, Chen W, Chakraborty AB, Gebler JC (2008) Determination of N-glycosylation sites and site heterogeneity in a monoclonal antibody by electrospray quadrupole ion-mobility time-of-flight mass spectrometry. Rapid Commun Mass Spectrom 22(1):29–40

Ongay S, Boichenko A, Govorukhina N, Bischoff R (2012) Glycopeptide enrichment and separation for protein glycosylation analysis. J Sep Sci 35(18):2341–2372

Pabst M, Kolarich D, Pöltl G, Dalik T, Lubec G, Hofinger A, Altmann F (2009) Comparison of fluorescent labels for oligosaccharides and introduction of a new postlabeling purification method. Anal Biochem 384(2):263–273

Palmisano G, Parker BL, Engholm-Keller K, Lendal SE, Kulej K, Schulz M, Schwämmle V, Graham ME, Saxtorph H, Cordwell SJ, Larsen MR (2012) A novel method for the simultaneous enrichment, identification, and quantification of phosphopeptides and sialylated glycopeptides applied to a temporal profile of mouse brain development. Mol Cell Proteomics 11 (11):1191–1202

Parekh RB, Dwek RA, Sutton BJ, Fernandes DL, Leung A, Stanworth D, Rademacher TW, Mizuochi T, Taniguchi T, Matsuta K, Takeuchi F (1985) Association of rheumatoid arthritis and primary osteoarthritis with changes in the glycosylation pattern of total serum IgG. Nature 316(6027):452–457

Patel T, Bruce J, Merry A, Bigge C, Wormald M, Parekh R, Jaques A (1993) Use of hydrazine to release in intact and unreduced form both N-and O-linked oligosaccharides from glycoproteins. Biochemistry 32(2):679–693

Periat A, Fekete S, Cusumano A, Veuthey JL, Beck A, Lauber M, Guillarme D (2016) Potential of hydrophilic interaction chromatography for the analytical characterization of protein biopharmaceuticals. J Chromatogr A 1448:81–92

Peterman SM, Mulholland JJ (2006) A novel approach for identification and characterization of glycoproteins using a hybrid linear ion trap/FT-ICR mass spectrometer. J Am Soc Mass Spectrom 17(2):168–179

Pezer M, Stambuk J, Perica M, Razdorov G, Banic I, Vuckovic F, Gospic AM, Ugrina I, Vecenaj A, Bakovic MP, Lokas SB (2016) Effects of allergic diseases and age on the composition of serum IgG glycome in children. Sci Rep 6(1):1–10

Pihíková D, Kasák P, Tkac J (2015) Glycoprofiling of cancer biomarkers: label-free electrochemical lectin-based biosensors. Open Chem 13(1):636–655

Pillai S, Netravali IA, Cariappa A, Mattoo H (2012) Siglecs and immune regulation. Annu Rev Immunol 30:357–392

Plomp R, Dekkers G, Rombouts Y, Visser R, Koeleman CA, Kammeijer GS, Jansen BC, Rispens T, Hensbergen PJ, Vidarsson G, Wuhrer M (2015) Hinge-region O-glycosylation of human immunoglobulin G3 (IgG3). Mol Cell Proteomics 14(5):1373–1384

Plummer TH, Elder JH, Alexander S, Phelan AW, Tarentino AL (1984) Demonstration of peptide: N-glycosidase F activity in endo-beta-N-acetylglucosaminidase F preparations. J Biol Chem 259(17):10700–10704

Poole CF (2003) New trends in solid-phase extraction. TrAC Trends Anal Chem 22(6):362–373

Porath J, Flodin P (1959) Gel filtration: a method for desalting and group separation. Nature 183 (4676):1657–1659

Porto AC, Oliveira LL, Ferraz LC, Ferraz LE, Thomaz SM, Rosa JC, Roque-Barreira MC (2007) Isolation of bovine immunoglobulins resistant to peptic digestion: new perspectives in the prevention of failure in passive immunization of neonatal calves. J Dairy Sci 90(2):955–962

Powell AK, Harvey DJ (1996) Stabilization of sialic acids in N-linked oligosaccharides and gangliosides for analysis by positive ion matrix-assisted laser desorption/ionization mass spectrometry. Rapid Commun Mass Spectrom 10(9):1027–1032

Pröpster JM, Yang F, Rabbani S, Ernst B, Allain FH, Schubert M (2016) Structural basis for sulfation-dependent self-glycan recognition by the human immune-inhibitory receptor Siglec-8. Proc Natl Acad Sci 113(29):E4170–E4179

Rademaker GJ, Pergantis SA, Blok-Tip L, Langridge JI, Kleen A, Thomas-Oates JE (1998) Mass spectrometric determination of the sites of O-glycan attachment with low picomolar sensitivity. Anal Biochem 257(2):149–160

Ramos-de-la-Peña AM, González-Valdez J, Aguilar O (2019) Protein A chromatography: challenges and progress in the purification of monoclonal antibodies. J Sep Sci 42(9):1816–1827

Reed CE, Fournier J, Vamvoukas N, Koza SM (2018) Automated preparation of MS-sensitive fluorescently labeled N-glycans with a commercial pipetting robot. SLAS Technol 23 (6):550–559

Rehder DS, Dillon TM, Pipes GD, Bondarenko PV (2006) Reversed-phase liquid chromatography/mass spectrometry analysis of reduced monoclonal antibodies in pharmaceutics. J Chromatogr A 1102(1-2):164–175

Reusch D, Haberger M, Kailich T, Heidenreich AK, Kampe M, Bulau P, Wuhrer M (2014) High-throughput glycosylation analysis of therapeutic immunoglobulin G by capillary gel electrophoresis using a DNA analyzer. MAbs 6(1):185–196

Reusch D, Haberger M, Maier B, Maier M, Kloseck R, Zimmermann B, Hook M, Szabo Z, Tep S, Wegstein J, Alt N (2015) Comparison of methods for the analysis of therapeutic immunoglobulin G Fc-glycosylation profiles—part 1: separation-based methods. MAbs 7(1):167–179

Rodrigo G, Gruvegård M, Van Alstine JM (2015) Antibody fragments and their purification by protein L affinity chromatography. Antibodies 4(3):259–277

Roque-Barreira MC, Campos-Neto AN (1985) Jacalin: an IgA-binding lectin. J Immunol 134 (3):1740–1743

Roque-Barreira MC, Praz F, Halbwachs-Mecarelli L, Greene LJ, Campos-Neto A (1986) IgA-affinity purification and characterization of the lectin jacalin. Braz J Med Biol Res = Rev Bras Pesqui Med Biol 19(2):149

Rosenfeld R, Bangio H, Gerwig GJ, Rosenberg R, Aloni R, Cohen Y, Amor Y, Plaschkes I, Kamerling JP, Maya RB (2007) A lectin array-based methodology for the analysis of protein glycosylation. J Biochem Biophys Methods 70(3):415–426

Rothman RJ, Warren L, Vliegenthart JF, Hard KJ (1989) Clonal analysis of the glycosylation of immunoglobulin G secreted by murine hybridomas. Biochemistry 28(3):1377–1384

Roucka M, Zimmermann K, Fido M, Nechansky A (2017) Application of lectin array technology for biobetter characterization: Its correlation with FcγRIII binding and ADCC. Microarrays 6 (1):1

Royle L, Mattu TS, Hart E, Langridge JI, Merry AH, Murphy N, Harvey DJ, Dwek RA, Rudd PM (2002) An analytical and structural database provides a strategy for sequencing O-glycans from microgram quantities of glycoproteins. Anal Biochem 304(1):70–90

Royle L, Roos A, Harvey DJ, Wormald MR, Van Gijlswijk-Janssen D, Redwan ER, Wilson IA, Daha MR, Dwek RA, Rudd PM (2003) Secretory IgA N-and O-glycans provide a link between the innate and adaptive immune systems. J Biol Chem 278(22):20140–20153

Royle L, Radcliffe CM, Dwek RA, Rudd PM (2006) Detailed structural analysis of N-glycans released from glycoproteins in SDS-PAGE gel bands using HPLC combined with exoglycosidase array digestions. In: Glycobiology protocols. Humana Press, New York, pp 125–143

Rudd PM, Elliott T, Cresswell P, Wilson IA, Dwek RA (2001) Glycosylation and the immune system. Science 291(5512):2370–2376

Ruhaak LR, Huhn C, Waterreus WJ, De Boer AR, Neusüss C, Hokke CH, Deelder AM, Wuhrer M (2008) Hydrophilic interaction chromatography-based high-throughput sample preparation method for N-glycan analysis from total human plasma glycoproteins. Anal Chem 80 (15):6119–6126

Ruhaak LR, Steenvoorden E, Koeleman CA, Deelder AM, Wuhrer M (2010a) 2-Picoline-borane: a non-toxic reducing agent for oligosaccharide labeling by reductive amination. Proteomics 10 (12):2330–2336

Ruhaak LR, Zauner G, Huhn C, Bruggink C, Deelder AM, Wuhrer M (2010b) Glycan labeling strategies and their use in identification and quantification. Anal Bioanal Chem 397 (8):3457–3481

Saha K, Bender F, Gizeli E (2003) Comparative study of IgG binding to proteins G and A: nonequilibrium kinetic and binding constant determination with the acoustic waveguide device. Anal Chem 75(4):835–842

Sano K, Ogawa H (2014) Chapter 4: Hemagglutination (inhibition) Assay. Methods Mol Biol 1200:47–52

Scanlan CN, Burton DR, Dwek RA (2008) Making autoantibodies safe. Proc Natl Acad Sci 105 (11):4081–4082

Schmelter C, Funke S, Treml J, Beschnitt A, Perumal N, Manicam C, Pfeiffer N, Grus FH (2018) Comparison of two solid-phase extraction (spe) methods for the identification and quantification of porcine retinal protein markers by lc-ms/ms. Int J Mol Sci 19(12):3847

Schmitt M (2016) Versican vs versikine: tolerance vs attack. Blood 128(5):612–613

Schroeder HW, Cavacini L (2010) Structure and function of immunoglobulins. J Allergy Clin Immunol 125(2):S41–S52

Scott AM, Wolchok JD, Old LJ (2012) Antibody therapy of cancer. Nat Rev Cancer 12(4):278–287

Segu Z, Stone T, Berdugo C, Roberts A, Doud E, Li Y (2020) A rapid method for relative quantification of N-glycans from a therapeutic monoclonal antibody during trastuzumab biosimilar development. Mabs 12(1):1750794

Selman MH, Hemayatkar M, Deelder AM, Wuhrer M (2011) Cotton HILIC SPE microtips for microscale purification and enrichment of glycans and glycopeptides. Anal Chem 83 (7):2492–2499

Selman MH, Derks RJ, Bondt A, Palmblad M, Schoenmaker B, Koeleman CA, van de Geijn FE, Dolhain RJ, Deelder AM, Wuhrer M (2012) Fc specific IgG glycosylation profiling by robust nano-reverse phase HPLC-MS using a sheath-flow ESI sprayer interface. J Proteome 75 (4):1318–1329

Sharapov SZ, Tsepilov YA, Klaric L, Mangino M, Thareja G, Shadrina AS, Simurina M, Dagostino C, Dmitrieva J, Vilaj M, Vuckovic F (2019) Defining the genetic control of human blood plasma N-glycome using genome-wide association study. Hum Mol Genet 28 (12):2062–2077

Sharon N, Lis H (1989) Lectins as cell recognition molecules. Science 246(4927):227–234

Sharon N, Lis H (2003) Lectins, 2nd edn. Kluwer Academic, Dordrecht

Siberil S, Dutertre CA, Fridman WH, Teillaud JL (2007) FcγR: the key to optimize therapeutic antibodies? Crit Rev Oncol Hematol 62(1):26–33

Singh N, Pirsch J, Samaniego M (2009) Antibody-mediated rejection: treatment alternatives and outcomes. Transplant Rev 23(1):34–46

Sliwkowski MX, Mellman I (2013) Antibody therapeutics in cancer. Science 341 (6151):1192–1198

Sonneveld ME, Natunen S, Sainio S, Koeleman CA, Holst S, Dekkers G, Koelewijn J, Partanen J, van der Schoot CE, Wuhrer M, Vidarsson G (2016) Glycosylation pattern of anti-platelet IgG is stable during pregnancy and predicts clinical outcome in alloimmune thrombocytopenia. Br J Haematol 174(2):310–320

Stadlmann J, Pabst M, Kolarich D, Kunert R, Altmann F (2008) Analysis of immunoglobulin glycosylation by LC-ESI-MS of glycopeptides and oligosaccharides. Proteomics 8 (14):2858–2871

Stadlmann J, Pabst M, Altmann F (2010) Analytical and functional aspects of antibody sialylation. J Clin Immunol 30(1):15–19

Stöckmann H, Adamczyk B, Hayes J, Rudd PM (2013) Automated, high-throughput IgG-antibody glycoprofiling platform. Anal Chem 85(18):8841–8849

Svensson HG, Hoogenboom HR, Sjöbring U (1998) Protein LA, a novel hybrid protein with unique single-chain Fv antibody-and Fab-binding properties. Eur J Biochem 258(2):890–896

Syed P, Gidwani K, Kekki H, Leivo J, Pettersson K, Lamminmäki U (2016) Role of lectin microarrays in cancer diagnosis. Proteomics 16(8):1257–1265

Szabo Z, Guttman A, Karger BL (2010) Rapid release of N-linked glycans from glycoproteins by pressure-cycling technology. Anal Chem 82(6):2588–2593

Szabo Z, Thayer JR, Reusch D, Agroskin Y, Viner R, Rohrer J, Patil SP, Krawitzky M, Huhmer A, Avdalovic N, Khan SH (2018) High performance anion exchange and hydrophilic interaction liquid chromatography approaches for comprehensive mass spectrometry-based characterization of the N-glycome of a recombinant human erythropoietin. J Proteome Res 17(4):1559–1574

Taga EM, Waheed A, Van Etten RL (1984) Structural and chemical characterization of a homogeneous peptide N-glycosidase from almond. Biochemistry 23(5):815–822

Takegawa Y, Deguchi K, Ito H, Keira T, Nakagawa H, Nishimura SI (2006) Simple separation of isomeric sialylated N-glycopeptides by a zwitterionic type of hydrophilic interaction chromatography. J Sep Sci 29(16):2533–2540

Tarentino AL, Plummer TH Jr (1982) Oligosaccharide accessibility to peptide: N-glycosidase as promoted by protein-unfolding reagents. J Biol Chem 257(18):10776–10780

Tarentino AL, Gomez CM, Plummer TH Jr (1985 Aug 1) Deglycosylation of asparagine-linked glycans by peptide: N-glycosidase F. Biochemistry 24(17):4665–4671

Theodoratou E, Thaçi K, Agakov F, Timofeeva MN, Štambuk J, Pučić-Baković M, Vučković F, Orchard P, Agakova A, Din FV, Brown E (2016) Glycosylation of plasma IgG in colorectal cancer prognosis. Sci Rep 6(1):1–2

Totten SM, Adusumilli R, Kullolli M, Tanimoto C, Brooks JD, Mallick P, Pitteri SJ (2018) Multi-lectin affinity chromatography and quantitative proteomic analysis reveal differential glycoform levels between prostate cancer and benign prostatic hyperplasia sera. Sci Rep 8(1):1–3

Trbojevic Akmacic I, Ventham NT, Theodoratou E, Vučković F, Kennedy NA, Krištić J et al (2015) Inflammatory bowel disease associates with proinflammatory potential of the immunoglobulin G glycome. Inflamm Bowel Dis 21(6):1237–1247

Trbojević-Akmačić I, Ugrina I, Štambuk J, Gudelj I, Vučković F, Lauc G, Pučić-Baković M (2015) High-throughput glycomics: optimization of sample preparation. Biochem Mosc 80(7):934–942

Trbojević-Akmačić I, Vilaj M, Lauc G (2016) High-throughput analysis of immunoglobulin G glycosylation. Expert Rev Proteomics 13(5):523–534

Trbojević-Akmačić I, Ugrina I, Lauc G (2017) Comparative analysis and validation of different steps in glycomics studies. Methods Enzymol 586:37–55

Tsai TI, Li ST, Liu CP, Chen KY, Shivatare SS, Lin CW, Liao SF, Lin CW, Hsu TL, Wu YT, Tsai MH (2017) An effective bacterial fucosidase for glycoprotein remodeling. ACS Chem Biol 12(1):63–72

Vadrevu SK, Trbojevic-Akmacic I, Kossenkov AV, Colomb F, Giron LB, Anzurez A, Lynn K, Mounzer K, Landay AL, Kaplan RC, Papasavvas E (2018) Frontline science: plasma and immunoglobulin G galactosylation associate with HIV persistence during antiretroviral therapy. J Leukoc Biol 104(3):461–471

Vaezzadeh AR, Deshusses JM, Waridel P, François P, Zimmermann-Ivol CG, Lescuyer P, Schrenzel J, Hochstrasser DF (2010) Accelerated digestion for high-throughput proteomics analysis of whole bacterial proteomes. J Microbiol Methods 80(1):56–62

Van Breedam W, Pöhlmann S, Favoreel HW, de Groot RJ, Nauwynck HJ (2014) Bitter-sweet symphony: glycan–lectin interactions in virus biology. FEMS Microbiol Rev 38(4):598–632 van de Bovenkamp FS, Derksen NI, Ooijevaar-de Heer P, Rispens T (2019) The enzymatic removal of immunoglobulin variable domain glycans by different glycosidases. J Immunol Methods 467:58–62

Van De Wetering JK, Van Golde LMG, Batenburg JJ (2004) Collectins Players of the innate immune system. Euro J Biochem 271(7):1229–1249

Varki A, Cummings RD, Esko JD, Freeze HH, Stanley P, Bertozzi CR, Hart GW, Etzler ME (2009) Essentials of glycobiology, 2nd edn. Cold Spring Harbor Laboratory Press, Cold Spring Harbor

Vidarsson G, Dekkers G, Rispens T (2014) IgG subclasses and allotypes: from structure to effector functions. Front Immunol 5:520

Vilaj M, Lauc G, Trbojević-Akmačić I (2020) Evaluation of different PNGase F enzymes in immunoglobulin G and total plasma N-glycans analysis. Glycobiology 31(1):2–7

Vincents B, von Pawel-Rammingen U, Björck L, Abrahamson M (2004) Enzymatic characterization of the streptococcal endopeptidase, IdeS, reveals that it is a cysteine protease with strict specificity for IgG cleavage due to exosite binding. Biochemistry 43(49):15540–15549

Vreeker GC, Wuhrer M (2017) Reversed-phase separation methods for glycan analysis. Anal Bioanal Chem 409(2):359–378

Wagner-Rousset E, Bednarczyk A, Bussat MC, Colas O, Corvaïa N, Schaeffer C, Van Dorsselaer A, Beck A (2008) The way forward, enhanced characterization of therapeutic antibody glycosylation: comparison of three level mass spectrometry-based strategies. J Chromatogr B 872(1–2):23–37

Walls D, Loughran ST, Madadlou A, O'Sullivan S (2017) Sheehan D. Protein chromatography, Springer New York

Walters RR (1985) Affinity chromatography. Anal Chem 57(11):1099A–1114A

Wan H, Yan J, Yu L, Sheng Q, Zhang X, Xue X, Li X, Liang X (2011) Zirconia layer coated mesoporous silica microspheres as HILIC SPE materials for selective glycopeptide enrichment. Analyst 136(21):4422–4430

Weng Y, Sui Z, Jiang H, Shan Y, Chen L, Zhang S, Zhang L, Zhang Y (2015) Releasing N-glycan from peptide N-terminus by N-terminal succinylation assisted enzymatic deglycosylation. Sci Rep 5(1):1–5

Wesener DA, Dugan A, Kiessling LL (2017) Recognition of microbial glycans by soluble human lectins. Curr Opin Struct Biol 44:168–178

Wilhelm JG, Dehling M, Higel F (2019) High-selectivity profiling of released and labeled N-glycans via polar-embedded reversed-phase chromatography. Anal Bioanal Chem 411(3):735–743

Wilson K, Walker J (eds) (2010) Principles and techniques of biochemistry and molecular biology. Cambridge University Press, Cambridge

Wu AM, Liu JH (2019) Lectins and ELLSA as powerful tools for glycoconjugate recognition analyses. Glycoconj J 36(2):175–183

Wu AM, Song SC, Sugii S, Herp A (1997) Differential binding properties of Gal/GalNAc specific lectins available for characterization of glycoreceptors. Indian J Biochem Biophys 34(1–2):61–71

Wuhrer M, Stam JC, van de Geijn FE, Koeleman CA, Verrips CT, Dolhain RJ, Hokke CH, Deelder AM (2007) Glycosylation profiling of immunoglobulin G (IgG) subclasses from human serum. Proteomics 7(22):4070–4081

Wuhrer M, Porcelijn L, Kapur R, Koeleman CA, Deelder AM, de Haas M, Vidarsson G (2009) Regulated glycosylation patterns of IgG during alloimmune responses against human platelet antigens. J Proteome Res 8(2):450–456

Yamaguchi Y (2000) Lecticans: organizers of the brain extracellular matrix. Cell Mol Life Sci 57(2):276–289

Yanagishita M (1993) Function of proteoglycans in the extracellular matrix. Pathol Int 43(6):283–293

Yang YB, Harrison K (1996) Influence of column type and chromatographic conditions on the ion-exchange chromatography of immunoglobulins. J Chromatogr A 743(1):171–180

Yang S, Onigman P, Wu WW, Sjogren J, Nyhlen H, Shen RF, Cipollo J (2018) Deciphering protein O-glycosylation: solid-phase chemoenzymatic cleavage and enrichment. Anal Chem 90 (13):8261–8269

Yang S, Wu WW, Shen R, Sjogren J, Parsons L, Cipollo JF (2020) Optimization of O-GIG for O-glycopeptide characterization with sialic acid linkage determination. Anal Chem 92 (16):10946–10951

Zelensky AN, Gready JE (2005) The C-type lectin-like domain superfamily. FEBS J 272 (24):6179–6217

Zhang L, Luo S, Zhang B (2016) The use of lectin microarray for assessing glycosylation of therapeutic proteins. MAbs 8(3):524–535

Zheng Y, Guo Z, Cai Z (2009) Combination of β-elimination and liquid chromatography/quadrupole time-of-flight mass spectrometry for the determination of O-glycosylation sites. Talanta 78 (2):358–363

Zhou H, Briscoe AC, Froehlich JW, Lee RS (2012) PNGase F catalyzes de-N-glycosylation in a domestic microwave. Anal Biochem 427(1):33–35

Chapter 3
Mass Spectrometry-Based Methods for Immunoglobulin G *N*-Glycosylation Analysis

Siniša Habazin, Jerko Štambuk, Jelena Šimunović, Toma Keser, Genadij Razdorov, and Mislav Novokmet

Contents

3.1	Basic Principles of Mass Spectrometry	76
	3.1.1 Ionization	77
	3.1.2 Gas-Phase Separation and Detection	77
	3.1.3 Tandem MS	80
3.2	Levels of IgG *N*-Glycosylation Analysis	81
3.3	Sample Preparation for IgG Glycosylation Analysis	86
	3.3.1 Protein A, G, and L Affinity Chromatography for IgG Enrichment from Biological Samples	86
	3.3.2 Sample Preparation for Released IgG Glycan Analysis	91
	3.3.3 Fluorescent and Isotopic Glycan Labeling	95
	3.3.4 Sample Preparation for Glycopeptide Mapping and Subclass-Specific IgG Fc Glycosylation Analysis	97
3.4	Deciphering the IgG Glycan Structure	100
	3.4.1 Fragmentation	101
	3.4.2 Fragmentation Nomenclature	102
	3.4.3 Fragmentation Candidates	103
	3.4.4 Ionization Polarity of MS Analysis	104
	3.4.5 Exoglycosidase Digestion Monitored by MS	108
3.5	Selected Approaches for IgG Glycosylation Analysis	109
	3.5.1 MALDI–MS	109
	3.5.2 LC–MS for IgG Glycosylation Analysis	112
	3.5.3 Capillary Electrophoresis-Mass Spectrometry	119
3.6	Perspectives	120
References		121

S. Habazin · J. Štambuk · J. Šimunović · G. Razdorov · M. Novokmet (✉)
Glycoscience Research Laboratory, Genos Ltd., Zagreb, Croatia
e-mail: mnovokmet@genos.hr

T. Keser
Faculty of Pharmacy and Biochemistry, University of Zagreb, Zagreb, Croatia

© The Author(s), under exclusive license to Springer Nature Switzerland AG 2021
M. Pezer (ed.), *Antibody Glycosylation*, Experientia Supplementum 112,
https://doi.org/10.1007/978-3-030-76912-3_3

Abstract Mass spectrometry and its hyphenated techniques enabled by the improvements in liquid chromatography, capillary electrophoresis, novel ionization, and fragmentation modes are truly a cornerstone of robust and reliable protein glycosylation analysis. Boost in immunoglobulin G (IgG) glycan and glycopeptide profiling demands for both applied biomedical and research applications has brought many new advances in the field in terms of technical innovations, sample preparation, improved throughput, and confidence in glycan structural characterization. This chapter summarizes mass spectrometry basics, focusing on IgG and monoclonal antibody *N*-glycosylation analysis on several complexity levels. Different approaches, including antibody enrichment, glycan release, labeling, and glycopeptide preparation and purification, are covered and illustrated with recent breakthroughs and examples from the literature omitting excessive theoretical frameworks. Finally, selected highly popular methodologies in IgG glycoanalytics such as liquid chromatography–mass spectrometry and matrix-assisted laser desorption ionization are discussed more thoroughly yet in simple terms making this text a practical starting point either for the beginner in the field or an experienced clinician trying to make sense out of the IgG glycomic or glycoproteomic dataset.

Keywords Mass spectrometry · Liquid chromatography · Solid-phase extraction · Ion fragmentation · Glycomics · Glycoproteomics · Biomarker discovery · Antibodies · Immunoglobulins

Abbreviations

2-AA	2-aminobenzoic acid (anthranilic acid)
2-AB	2-aminobenzamide (anthranilamide)
2-AP	2-aminopyridine
4-HCCA	α-cyano-4-hydroxycinnamic acid
Abs	Antibodies
ACN	Acetonitrile
ANTS	8-aminonaphthalene-1,3,6-trisulfonate
APS-PEG	Aminopropylsilane-polyethylene glycol
APTS	8-aminopyrene-1,3,6-trisulfonate
CE	Capillary electrophoresis
CEC	Capillary electrochromatography
CE–LIF	Capillary electrophoresis with laser-induced fluorescence detection
CE–MS	Capillary electrophoresis coupled to mass spectrometry
CFG	Consortium for Functional Glycomics
CGE	Capillary gel electrophoresis
CID	Collision-induced dissociation
CIEF	Capillary isoelectric focusing
Cl-CCA	4-chloro-α-cyanocinnamic acid
CNBr	Cyanogen bromide

CZE	Capillary zone electrophoresis
DHB	2,5-dihydroxybenzoic acid (gentisic acid)
DTT	Dithiothreitol
EACA	ε-aminocaproic acid
ECD	Electron capture dissociation
EDC	1-ethyl-3(−3-dimethylaminopropyl) carbodiimide hydrochloride
EDTA	Ethylenediaminetetraacetic acid
ESI	Electrospray ionization
ESI–MS	Electrospray ionization mass spectrometry
ETD	Electron transfer dissociation
EThcD	Electron transfer/higher energy collision dissociation
Fab	Fragment antigen-binding
FASP	Filter-aided sample preparation
Fc	Fragment crystallizable
Fmoc-Cl	Fluorenylmethoxycarbonyl chloride
FT-ICR	Fourier-transform ion cyclotron resonance
GlcNAc	N-acetylglucosamine
HCD	Higher-energy collision dissociation
HILIC	Hydrophilic interaction liquid chromatography
HILIC–SPE	Solid-phase extraction using hydrophilic interaction liquid chromatography
HILIC–UPLC–FLR	Hydrophilic interaction ultraperformance liquid chromatography with fluorescence detection
HOBt	1-hydroxybenzotriazole
HPAEC	High-pH anion exchange chromatography
IAA	Iodoacetamide
IgE	Immunoglobulin E
IgG	Immunoglobulin G
IgM	Immunoglobulin M
Igs	Immunoglobulins
IM-MS	Ion mobility-mass spectrometry
ISD	In-source decay
ITP	Isotachophoresis
K_a	Equilibrium association constant
LC–MS	Liquid chromatography coupled with mass spectrometry
LDI	Laser desorption ionization
m/z	Mass-to-charge ratio
mAbs	Monoclonal antibodies
MALDI	Matrix-assisted laser desorption/ionization
MALDI–MS	MALDI coupled with mass spectrometry
MALDI-TOF–MS	MALDI coupled with time-of-flight mass spectrometry
MECC/MEKC	Micellar electrokinetic capillary chromatography
MEEKC	Microemulsion electrokinetic chromatography
MRM	Multiple reaction monitoring

MS	Mass spectrometry
MS/MS	Tandem mass spectrometry
NACE	Nonaqueous capillary electrophoresis
nano-LC	Nano-liquid chromatography
Neu5Gc	N-glycolylneuraminic acid
NP-HPLC	Normal-phase high-performance liquid chromatography
pAbs	Polyclonal antibodies
PGC	Porous graphitized carbon
PGC-LC	Porous graphitized carbon liquid chromatography
PMP	1-phenyl-3-methyl-5-pyrazolone
PNGase F	Peptide:N-glycosidase F
PpL	Protein L
ProA	Procainamide
PTM	Posttranslational modification
Q-TOF	Quadrupole time-of-flight
RP	Reversed-phase
RP-LC	Reversed-phase liquid chromatography
SpA	Protein A
SpG	Protein G
TETA	Triethylenetetramine
TFA	Trifluoroacetic acid
TMT	Tandem mass tag
TOF	Time-of-flight
UPLC®	Ultra performance liquid chromatography
(U)HPLC–FLR–MS	(ultra)high-performance liquid chromatography with fluorescence detection coupled to mass spectrometry
ZIC-HILIC	Zwitterionic hydrophilic liquid interaction chromatography

3.1 Basic Principles of Mass Spectrometry

Since the beginning of the twentieth century, it is possible to manipulate individual gas-phase ions in a vacuum using electric and magnetic fields (Griffiths 2008). The study of matter that relies on individual ion gas-phase manipulation is called *mass spectrometry* (MS) (Gross 2004). Central to MS is the ability to separate gas-phase ions based on their mass and charge. The mass spectrometer is an instrument by which matter is: (1) *ionized* into gas-phase, (2) *manipulated and separated* on the level of individual ions, and eventually (3) *detected*. There are several techniques for all three phases of MS, with specific strengths and weaknesses, resulting in conceptually different instruments applicable to various analytical problems.

Results of the MS measurement are usually presented as a mass spectrum. The mass spectrum is a graph showing the intensity (proportional to the strength of detected signal) versus *m/z* (ratio of mass to charge) relationship. Intensity is always scaled relatively, often normalized to the base peak (the most intense peak) in the mass spectrum. The m/z ratio is a dimensionless quantity defined as the ratio between an ion's mass and the unified atomic mass unit, divided by its charge number (regardless of the sign). To determine mass from m/z, one has to be aware of the measured ion charge state. Since most chemical elements on Earth are, in fact, mixtures of isotopes, differing only by their number of neutrons and consequently by their mass, almost any molecule measured by MS is separated into several isotopologues called the isotopic envelope. The nominal mass difference between neighboring isotopologues is $1u$ (1 neutron), and consequently, the m/z difference between them is equal to $1/z$. Due to its high sensitivity, specificity, speed, and almost universal applicability, MS is a central analytical technique in many *omics* fields, including glycomics and proteomics (Aebersold and Mann 2003; Ruhaak et al. 2018).

3.1.1 Ionization

To study matter by MS, first, molecules have to be ionized into a gas-phase. Several techniques can do this. At the beginning of the 1990s, the first soft ionization techniques, matrix-assisted laser desorption/ionization (MALDI) and electrospray ionization (ESI) were invented (Tanaka et al. 1988; Fenn et al. 1989). These ionization techniques enabled MS of the large and fragile biomolecules, such as proteins, glycans, and glycoconjugates (Zaia 2010).

The ESI process involves a conductive spray needle under electrical potential containing molecules of interest in a solution (Smith et al. 1990). At the tip of the ESI needle, a Taylor cone is formed out of the liquid meniscus under the influence of an electric field gradient. A jet of highly charged droplets flows towards the counter electrode at atmospheric pressure. Charged droplets are further spliced into a plume of smaller ones, from which desolvated molecules of interest enter the gas-phase as adducts with small ions (proton, sodium cation, etc.). Adducts can contain single or multiple small ions, which define the charge state of the molecule of interest.

MALDI involves the co-crystallization of molecules of interest with matrix molecules from a solvent, usually in an array of spots on a metal plate (Karas and Krüger 2003). Crystals are irradiated with a laser beam. The matrix absorbs the energy from the laser beam, which promotes desorption into the gas-phase and ionization of the molecules of interest by the formation of adducts with small ions. As opposed to ESI, MALDI produces almost exclusively singly charged ions.

3.1.2 Gas-Phase Separation and Detection

Ions of interest enter the low-vacuum ion transfer region individually and are ready for the next phases of MS, which take place in the quadrupoles, ion traps, or collision cells supported with an array of ion optics for electrostatic manipulation of charged particles. Individual gas-phase ions formed are subjected to Lorentz force if immersed in an electromagnetic field (Maher et al. 2015):

$$\vec{F} = q\left(\vec{E} + \vec{v} \times \vec{B}\right) \qquad (3.1)$$

where F represents force, q charge, E electric field, v speed, and B magnetic field.

Since force is proportional to mass (m) and acceleration (a), which is equal to time (t) derivative of speed (v) or second time derivative of the path (r),

$$\vec{F} = m\vec{a} = m\frac{d\vec{v}}{dt} = m\frac{d^2\vec{r}}{dt^2} \qquad (3.2)$$

it is possible to describe ion motion in space and time by these classical equations:

$$\left(\frac{m}{q}\right)\frac{d^2\vec{r}}{dt^2} = \vec{E} + \frac{d\vec{r}}{dt} \times \vec{B} \qquad (3.3)$$

The previous equation shows that the ion mass to charge ratio is a principal concept for the MS.

The separation of individual ions is done by a mass analyzer (Haag 2016). There are a number of mass analyzer types for ion separation. The main difference between them is the mass-resolving power, where mass resolution is defined as a mass to mass difference ratio (m represents m/z):

$$\frac{m}{\Delta m'} \qquad (3.4)$$

Briefly, by convention, the mass resolution is defined as a measure of how well the two ions with close m/z can be distinguished. Only one isotopic composition then corresponds to each spectral peak. Mass accuracy, on the other hand, describes the difference between measured and theoretical m/z.

Mass resolution spans several orders of magnitude and ranges from 10^3 up to over 10^6. For an ion with the m/z of 1000, this means complete signal separation at m/z 1001 (resolution 10^3) and 1,000,001 (resolution 10^6). Inversely proportional to mass resolution, mass accuracy is essential for analyte identification based on accurate mass measurements.

Other performance parameters of different mass analyzers are m/z range (mostly limited to around 2000), speed (ranging from microseconds to seconds), duty cycle (or efficiency), and dynamic range (max-to-min signal ratio).

Low-resolving mass analyzers with a mass resolution of over 10^3 are transmission quadrupole (Q) and quadrupole ion trap (QIT), both belonging to the scanning type (Miller and Denton 1986; Nolting et al. 2019). To measure a broad range of m/z signals, Q needs to scan over the stability region. For a given m/z, a quadrupole operates within the stability region only when the applied radio frequency (RF) and direct current (DC) voltages combination enable the ions to pass through. The efficiency of scanning mode is usually very low and can be expressed as single m/z over the total number of m/z in the range scanned (for example, 1/2000 for m/z range of 2000). Both of these are rarely used for analyte identification based on accurate mass measurements. The quadrupole is mostly used as a mass filter for ion selection based on m/z, while QIT is used for structural studies due to its ability to select and fragment ions at multiple levels (see Sect. 3.1.3 on tandem MS). These analyzers are very affordable, almost maintenance-free, and require only a modest vacuum relieving the need for complex mass spectrometer vacuum systems.

Next in line are time-of-flight (TOF) analyzers with a middle mass-resolving power ($>10^4$) (Boesl 2017). This resolution is usually enough for the charge state determination of multiply charged ions produced by ESI. Time-of-flight mass analyzers are the fastest (a microsecond scan time), with the broadest mass range ($m/z > 10^4$).

Kingdon trap (KT), commercially called Orbitrap™, performs measurements with a high-resolving power of over 10^5 (Nolting et al. 2019; Eliuk and Makarov 2015). It is predominately used in proteomics and other -omics studies, where analytes from complex mixtures are identified by accurate mass measurement.

However, to achieve the high mass resolution needed for this type of analysis, KT needs long scan times. Since the mass resolution is inversely proportional to scan time, a compromise has to be made between the duration of the scan and the required mass resolution.

The very same trade-off is imminent when using Fourier transform ion cyclotron resonance (FT-ICR), a mass analyzer with an ultra-high mass-resolving power (Nolting et al. 2019). With the resolution of over 10^6, it is possible to separate individual isotopologues with the same mass number (for example, ^{15}N, ^{33}S, ^{13}C, and ^2H for $A + 1$ mass number, where A is the atomic mass number) and unambiguously determine the molecular formula (Nikolaev et al. 2012). This analyzer tends to be expensive and complex for usage, making it less popular in the omics field.

Detection of ions in MS occurs by charge counting or by charge-induced image current measurement. Charge counting is performed by the electron multiplier. When a beam of measured ions emerges from a mass analyzer and strikes a detector, it causes a release of an electron from the surface layer. During the process called secondary emission, a single electron induces the emission of additional ones, resulting in the amplification of the input current. All MS analyzers, except for FT-ICR and KT, are coupled to this type of detector. FT-ICR and KT use the measurement of the image current, where ions flying between the detector plates induce an image current, which can easily be transformed from the time domain to the frequency domain by the Fourier transform algorithm. The calculated frequency is proportional to m/z (Gross 2004).

3.1.3 Tandem MS

Since molecular isomers have the same molecular formula and consequently the same mass, they cannot be separated by MS directly. However, by combining two mass analyzers with an ion activation technique in between, it is possible to select an ion of interest by the first mass analyzer, activate and fragment it, and measure the resulting fragments by the second mass analyzer. This is, in essence, tandem MS (MS/MS), which enables different isomer separation by their fragmentation pattern (de Hoffmann 1996).

Ions can be activated by different techniques, but only collision-induced dissociation (CID) is used in all tandem instruments. Ions are activated by collisions with inert gas, usually in a collision cell. A few collisions are required to fragment a single molecular bond, which translates into ion fragmentation at the least stable bonds—a peptide bond in peptides and a glycosidic bond in the case of the glycans (Wuhrer et al. 2007a). CID fragmentation mass spectrum contains a ladder of peptide fragments, usually single amino acid apart. Due to the linearity of the peptide's primary structure, this is usually enough to decipher the amino acid sequence. Annotation of glycan fragmentation mass spectrum is demanding because of branched glycan structures, and it is discussed in more detail later in the chapter.

Another popular activation method, at least for glycopeptides, is electron transfer dissociation (ETD) (Wuhrer et al. 2007a). It is based on the transfer of electrons from electron donor anions to analyte ions, which induces fragmentation of the glycosidic or peptide bonds (Syka et al. 2004; Kim and Pandey 2012).

Even though it is possible to combine any mass analyzers with the collision cell (q) in the tandem instrument, some combinations are more popular: triple quadrupole (QqQ), quadrupole time-of-flight (Q-TOF and Qq-TOF), quadrupole-Orbitrap (Q-KT), tandem time-of-flight (TOF-TOF) (Eliuk and Makarov 2015; Yost and Enke 1979; Chernushevich et al. 2001; Vestal and Campbell 2005). The transmission quadrupole (Q) is the ideal first mass analyzer due to its mass filtering capabilities.

There are two approaches to tandem experiments: *tandem-in-space* and *tandem-in-time* (de Hoffmann 1996). In tandem-in-space, ion isolation, activation, and fragment separation are being performed separately in space. On the other hand, the quadrupole ion trap instrument can only perform *tandem-in-time* experiments. By this approach, ions are isolated, activated, and fragments separated in the same space but sequentially in time. Another capability of the *tandem-in-time* instrument is the application of multiple levels of fragmentation or MS^n. The first stage of MS analysis is typically denoted as MS^1, and it involves only an ion selection process without any fragmentation. After the first level of fragmentation (MS^2), a single fragment can be isolated, activated, and fragmented in the second level of fragmentation (MS^3). The process can be performed in several cycles, achieving higher and higher fragmentation levels (e.g., MS^4, MS^5). This multilevel fragmentation can differentiate isomers with substructural differences. Quadrupole ion trap is the only analyzer capable of *tandem-in-time*, with up to 10 levels of fragmentation.

Each of the mass spectrometry solutions described in this introduction is applicable for antibody glycosylation analysis with certain advantages and disadvantages in throughput, user-friendliness, mass resolution and accuracy, ion fragmentation capabilities, and level of structural information obtained. In this way, ESI–MS(/MS) and MALDI–MS(/MS) systems are most commonly used for IgG glycoanalysis due to their moderate cost and broader availability. Both approaches are discussed more thoroughly in Sects. 3.5.1 and 3.5.2.

3.2 Levels of IgG *N*-Glycosylation Analysis

Immunoglobulin G has one conserved *N*-glycosylation site on each heavy chain in the fragment crystallizable (Fc) region at the Asn297. Additionally, only about 15–25% of the protein molecules have an *N*-glycosylation site at the IgG variable fragment antigen-binding (Fab) region (Holland et al. 2006; Stadlmann et al. 2010). The analysis of released IgG glycans (glycomics approach) gives an overall picture of *N*-glycome present on the protein, covering glycosylation of all IgG subclasses and both Fc and Fab fragments combined. Glycans still attached to the peptide backbone (glycopeptides) carry information about the site of glycosylation. Site-specific analysis reveals glycosylation site occupancy by different glycan structures. In the analysis of IgG glycopeptides (glycoproteomics approach), information about subclass-specific Fc glycosylation is obtained. However, insight on Fab glycosylation is missing, although it can still be analyzed on released glycan level. Site-specific glycosylation analysis of the Fab region of polyclonal IgG or IgG isolated from human serum is currently not viable. However, site-specific analysis of the Fc portion can give valuable information since its tryptic peptides enclose sequence differences depending on the IgG subclass (Selman et al. 2012a). Substantial heterogeneity, coming from both the constant and the variable region and the wide repertoire of possible glycan structures attached to it (Fig. 3.1) (Pučić et al. 2011), is masking the signals of possible Fab glycopeptides. Whatever approach is chosen, mass spectrometry-based techniques can help resolve glycan composition and its structural features.

Enzymatically released *N*-glycans can be analyzed in their unmodified form but are more often derivatized in different ways before the analysis. When analyzed in combination with separation-based methods such as liquid chromatography or capillary electrophoresis, glycans are primarily modified on their reducing end with one of the many different fluorescent dyes available (see Sect. 3.3.3). Other approaches include modification of hydroxyl and *N*-acetyl groups (permethylation), modification of sialic acid residues (esterification or methylation of carboxyl groups). Permethylation of released glycans also stabilizes sialic acids and enables efficient detection of a mixture of acidic and neutral glycans (Atwood et al. 2008; Kang et al. 2007). Esterification of sialic acids reveals the difference in the linkage types to underlying galactose residues (Fig. 3.1). Masses of esterified glycans will be

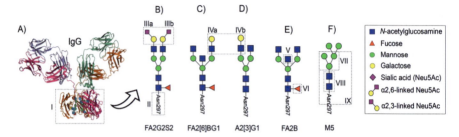

Fig. 3.1 The multi-layered IgG *N*-glycan diversity and structural motifs determine the choice of the MS analysis methods. A conserved IgG Fc region (**A**) *N*-glycosylation site (**I**) at Asn297 (**II**) is mainly occupied by biantennary glycans with structures spanning from disialylated (**B**), bisected (**C, E**), and monogalactosylated (**C, D**) to high-mannose (**F**). The sialic acid can be either α(2,3)- or α(2,6)-linked (**IIIa** and **IIIb**), while the antennary galactose residues can be present on 6-arm (**IVa**), 3-arm (**IVb**) or both in digalactosylated glycans (**B**). The bisecting GlcNAc (**V**) is β(1,4)-linked to the core mannose, while the core fucose (**VI**) is α(1,6)-linked to the innermost GlcNAc residue. About 90% of IgG glycans are core fucosylated. Minor glycan structures include, e.g., high-mannose structures (**F**). Further information on IgG glycosylation can be found in (Gudelj et al. 2018). All these glycans share some common features: the trimannosyl core (**VII**) and *N,N*'-diacetylchitobiose core (**VIII**), together with the asparagine residue from the peptide backbone, form the Man3GlcNAc2Asn *core* (**IX**). *For glycan depictions, rules proposed by the Consortium for Functional Glycomics (CFG) are followed. A textual IgG glycan nomenclature (below depictions* **B–F**) *can also be used:* F—α1,6-linked core fucose, A_n—number (*n*) of antennas (GlcNAcs) on a trimannosyl *N*-glycan core, G_n—number of β1,4-linked galactoses on 3- (Aebersold and Mann 2003) or 6-antenna (Fenn et al. 1989), B—bisecting GlcNAc, S_n—number of sialic acids (Neu5Ac) linked to galactoses, M_n—number (n) of mannoses

different if sialic acid residues are linked to galactose by α(2,6) versus α(2,3) linkage (Reiding et al. 2014).

Glycoproteomics is more challenging than traditional proteomics due to the high complexity and variability of glycan structures. Different MS-based protein and glycoprotein structural characterization approaches remain similar as the analysis can be performed on multiple structural levels. When an intact glycoprotein of interest is analyzed directly without fragmentation, the approach is called a native MS. Intact glycoprotein analysis gives information about its molecular weight but provides little structural insight. More recently, ion mobility-mass spectrometry (IM-MS) has seen considerable progress in new techniques for intact glycoprotein analysis, especially with the discovery of collision-induced protein unfolding in the gas-phase (Hernandez-Alba et al. 2018; Tian et al. 2015). Still, for structurally resolved information, the glycoprotein is usually cleaved into smaller fragments before or during the MS analysis. This fragmentation process can be performed either in the gas-phase inside a mass spectrometer or in-solution before the MS analysis by proteolytic cleavage. When the fragmentation happens primarily in the gas-phase, the mass spectrometer is used to analyze the entire molecule directly—this is called a "down"—type approach.

On the other hand, when the fragmentation happens primarily in solution, and the MS is used to analyze each fragment, the molecular structure is deduced on the information obtained from the corresponding fragmentation spectra in an "up"-type approach.

The "up" and "down" methods are further divided into (1) top-down, (2) middle-up, (3) middle-down, and (4) bottom-up subtypes (Zhang et al. 2009). In a top-down method, structural information is obtained from the fragmentation pattern of the intact glycoprotein inside the MS. When the glycoprotein is cleaved into a few large fragments before being introduced into the MS, the approach involves a middle-up method if the masses of these fragments are determined directly or a middle-down method if further gas-phase ion fragmentation is performed. In a bottom-up method, a glycoprotein is first digested into smaller peptides and glycopeptides before being introduced into the MS, regardless of whether these peptides are fragmented during the analysis or not (Zhang et al. 2009; Lermyte et al. 2019). Glycoproteins can be analyzed at one additional level—MS of the released glycans. The glycan part can be enzymatically or chemically released from the protein part and then analyzed separately to explore the distribution and quantify and/or confirm the glycan structures by MS/MS. Both "up" and "down" approaches have their advantages and disadvantages in terms of structural resolution, glycosylation information, sequence coverage, sample consumption, cost, speed, and effort required for the analysis. Therefore, different techniques are often combined to provide an in-depth glycoprotein characterization (Ayoub et al. 2013; Tran et al. 2016). Figure 3.2 summarizes the

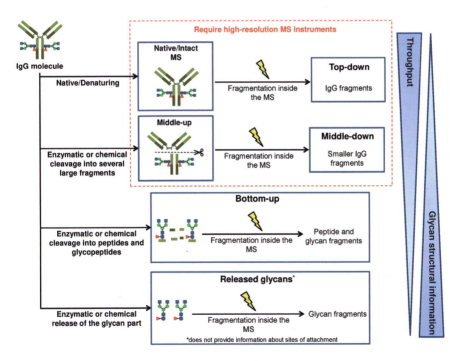

Fig. 3.2 Different levels of IgG structural characterization by MS-based methods

application of each of these approaches for the glycobiological structural characterization of immunoglobulins.

Analysis of IgG by a top-down approach consumes little sample, results in improved throughput, and can provide useful structural information in a very short time by reducing the number of sample preparation steps (Todoroki et al. 2020). The major drawback of a top-down MS for large proteins such as IgG is its limited structural resolution. The isotopic peaks of an intact IgG are usually not well resolved due to many different charge states. Therefore, the intact protein mass is difficult to determine, and high-resolution MS instruments are required. However, significant progress has been made recently in both top-down and native protein MS due to technological innovations, including improvements in ionization techniques and high-resolution detectors (Lermyte et al. 2019). Therefore, accurate MS analysis of intact antibodies is increasingly conducted in monoclonal antibodies (mAbs) production to rapidly detect major glycoforms (Tissot et al. 2009). The high-resolution instruments enable monitoring the degree of glycosylation of the intact molecule and the site-specific distribution of the major glycans, although the detection of minor glycoforms remains challenging (Zhang et al. 2009).

Middle-up IgG analysis involves cleaving the glycoprotein into several large fragments before the MS analysis. It provides a simple way to detect structural changes within a specific region of the molecule but cannot pinpoint the change to a specific amino acid residue (Zhang et al. 2009). Obviously, the site-specific information cannot be obtained this way. Yet, as each IgG subclass exists in several amino acid sequence variants (allotypes), it is possible to explore allotype-specific glycosylation IgG Fc glycosylation using the middle-up approach (Sénard et al. 2020). The IgG digestion can be achieved either chemically or by using structure-specific enzymes that are selective for only one or a handful of cleavage sites in the glycoprotein. For example, reduction of disulfide bridges can be employed to separate light and heavy IgG chains—this approach can be used either alone or in combination with proteolysis (Lermyte et al. 2019). On the other hand, bacterial IgG-degrading enzymes (see Sect. 3.3.4 and Fig. 3.3 for more details) (Ayoub et al. 2013) allow cleavage in the lower and upper hinge region, respectively (Lermyte et al. 2019). The high-resolution MS can provide accurate and often isotopically resolved mass spectra of these large fragments for further conclusions about both protein integrity and the presence of post-translational modifications, including glycosylation.

In the middle-down approach, as in the middle-up, chemical or enzymatic digestion is used to cleave the IgG into large fragments. However, instead of just measuring the mass of these fragments, they are also subsequently fragmented in the gas-phase during the MS analysis. This enables increased structural resolution with the benefit of simple sample preparation as separating the heavy and light chain prior to MS/MS analysis facilitates spectral assignment, while reduction of disulfide bridges allows for complete unfolding and improves dissociation efficiency (Lermyte et al. 2019). A middle-down proteomics approach using IdeS digestion has been used to assess the complexity, long-term storage stability, and biosimilarity of therapeutic antibody formulations (Todoroki et al. 2020).

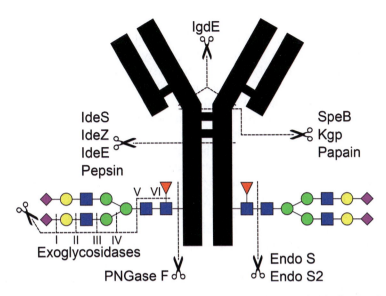

Fig. 3.3 Enzymology of IgG digestion strategies for glycoanalytics. Aside from endo- and exoglycosidases, different proteases are sometimes employed to fragment Ab molecules, drug-Ab conjugates, or Fc-fusion proteins for middle-up or middle-down glycoproteomics or to analyze Fc and Fab glycosylation separately (Sjögren et al. 2016). Bacterial proteases IdeS (von Pawel-Rammingen et al. 2002), IdeZ (Faid et al. 2018; Eriksson and Norgren 2003) and IdeE (Lannergård and Guss 2006) cleave IgG below hinge region, yielding F(ab')$_2$ and Fc fragments; similar to pepsin partially degrading Fc to pFc'. The IgdE (Spoerry et al. 2016) digests human IgG1 above the hinge and releases two Fab fragments (Faid et al. 2018). Streptococcal pyrogenic exotoxin B (SpeB) (Eriksson and Norgren 2003) and lysine-gingipain (Kgp) from *Porphyromonas gingivalis* (Vincents et al. 2011) are Cys-proteases cleaving in the hinge region above disulfide bridges. Papain similarly releases two Fab fragments from Fc, although when activated with 1–2 mM cysteine, F(ab')$_2$ can be obtained. Enzymes like neuraminidases (**I**), galactosidases (**II**), hexosaminidases (**III**), mannosidases (**IV, V**), and fucosidases (**VI**) of different linkage specificities are used for exoglycosidase glycan sequencing. Endoglycosidases cleave entire glycans either within the chitobiose core (Endo S, Endo S2) or between the innermost GlcNAc and Asn residue (PNGase F). Glycan depictions are for illustration purposes only and are unrelated to the enzymes' specificities

A bottom-up approach is the most frequently used for detailed structural characterization of IgG and its glycosylation as it can reveal the exact protein sequence and characterize glycosylation at the residue level simultaneously (Hinneburg et al. 2016). Contrary to the mass analysis of larger antibody fragments, the glycopeptide mass can be determined with much higher accuracy for greater confidence in the glycan composition assignment. In bottom-up methods, the IgG is cleaved into small peptides by a protease, followed by the MS or, more frequently MS/MS analysis. Bottom-up methods provide the most detailed structural information compared to the other approaches. However, on the downside, they are labor-intensive, lengthy, and often suffer from problems such as large sample consumption and artifacts introduced during digestion. Proteolytic IgG digestion is usually performed with trypsin or Lys-C after reduction and alkylation under denaturing conditions. Chemical cleavage methods, such as cyanogen bromide (CNBr) digestion, are also used occasionally (Zhang et al. 2009). Bottom-up IgG glycosylation analysis provides information on the glycan structure, heterogeneity, and attachment sites (Ayoub et al. 2013). In cases where glycans are present at sites beyond the Fc region, such as *N*-glycosylation in the antibody Fab region, Fc glycoforms are typically analyzed at the glycopeptide level (Zhang et al. 2009).

IgG glycosylation profile can also be explored by releasing glycans from the protein part of the molecule. This approach enables complete coverage of glycan components, but it completely ignores the protein part and information about sites of attachment and site-specific glycan distribution. Glycans can be released from the antibody either chemically (e.g., hydrazinolysis) or enzymatically (e.g., peptide:N-glycosidase F or endoglycosidase F) (Zhang et al. 2009). Released glycans are usually chemically derivatized before the MS analysis in order to enhance the sensitivity of the analysis since free glycans lack chromophores or fluorophores and do not ionize well. Released glycans are most often analyzed with online fluorescence detection and MS/MS because it enables high sensitivity, relative quantitation, and structural information within a single chromatographic analysis (Zhang et al. 2009; Keser et al. 2018). Furthermore, by using antibody-based affinity capture and IgG-degrading enzymes (IdeS), it is also possible to obtain IgG Fab and Fc released glycan profiles separately (Bondt et al. 2014). Analysis of IgG glycosylation on a released glycan level is most quantitative and highly sensitive and can provide detailed structural information, but it needs to be combined with other approaches to include information about sites of attachment and site-specific glycan distribution.

3.3 Sample Preparation for IgG Glycosylation Analysis

3.3.1 *Protein A, G, and L Affinity Chromatography for IgG Enrichment from Biological Samples*

Fast, specific and simple enrichment of natural immunoglobulins (Ig) or engineered antibodies (Ab) is a prerequisite for streamlined glycomic and glycoproteomic high-throughput workflows. Affinity chromatography is typically used for enrichment. It utilizes reversible non-covalent—mainly electrostatic, van der Waals and hydrophobic—interactions between biomolecules and immobilized ligands. It facilitates the enrichment of, among others, monoclonal and polyclonal (pAbs) antibodies or their fragments from complex biological matrices (blood plasma and serum, ascites fluid, cell culture supernatants) for subsequent downstream characterization, additional purification, or their removal as potential contaminants. Samples are applied to chromatographic media, washed, and eluted under controlled conditions (pH, ionic strength, solvent polarity) that ensure optimal binding, minimal loss of the desired analyte(s) during the washing step and eventually high recovery without unwanted

Ab degradation (Moser and Hage 2010). Traditionally, purification of IgG and other immunoglobulins involved sequential precipitation with high concentrations of ammonium, sodium, or dextran sulfate (Lermyte et al. 2019; Ayoub et al. 2013). This approach is still used for large-scale production and, if followed by size-exclusion chromatography, offers a cost-effective alternative to affinity purification. IgG recovered in this way is contaminated with other co-precipitated proteins, which now makes affinity chromatography using recombinant bacterial protein A, G, or L as a ligand generally accepted and widely exploited in both preparative and analytical protocols. For convenience, protein ligands are typically immobilized to beaded matrices such as crosslinked agarose-derived Sepharose™, gold, or membrane surfaces using different bioconjugation chemistries (Faccio 2018). Increased availability and simplified synthesis of nanomaterials, polymeric substrates, and functionalized microparticles have also brought innovation to antibody affinity purification and immunoassays (Table 3.1).

3.3.1.1 Protein A

Protein A (SpA) is a 42 kDa surface protein produced by *Staphylococcus aureus*, possessing a high affinity (K_a 1.4 × 10^8 M^{-1}) (Choe et al. 2016) for the human IgG Fc region with a binding ratio of 1:2 (Yang et al. 2003). Its five binding domains (A–E) target the Ab between the constant heavy chain domains C_H2 and C_H3 (Choe et al. 2016).

Regarding human subclass specificity, protein A does not bind IgG$_3$ (except allotypes with His435 instead of Arg435) (Rispens and Vidarsson 2014), while it binds strongly to the rabbit, pig, and guinea pig IgG. Rat, chicken, and goat IgG do not bind to protein A at all (Page and Thorpe 2002). As with other immobilized bacterial proteins used for Ab purification, SpA requires low-pH for elution. This often leads to Ab denaturation or aggregation and can be detrimental to the protein's functional and structural properties, so alternative ways are sought through protein engineering to allow for milder elution conditions. One interesting approach was demonstrated by Koguma et al. by developing a thermoresponsive mutant of SpA that binds IgGs at low temperatures and releases them upon heating (Koguma et al. 2013). Another strategy involved the introduction of a calcium-dependent protein A-derived domain with high dynamic binding capacity, where complete elution was carried out at neutral pH (Kanje et al. 2018). A simple double-point mutation of His to Ser and Asn to Ala resulted in milder elution conditions compared to commercially available SpA, making it attractive to produce a novel affinity chromatography media with enhanced properties (Pabst et al. 2014). Other cost-effective options like mimetic SpA-like peptide ligands and peptidomimetics have also emerged (Faccio 2018; Choe et al. 2016).

Table 3.1 Selected recent examples of innovative protein-ligand immobilization strategies and applications for IgG enrichment and analysis

No.	Protein	Support	Sample(s)	Summary	References
1	A, AG	SiO$_2$ magnetic microspheres	Rabbit serum	Ligand immobilized to aldehyde-functionalized magnetic SiO$_2$ meso- and macroporous microbeads with the fast magnetic response, synthesized by hydrolysis-condensation approach. Reusable. Enriched IgG to >95% purity. Incubation time: 1 h at room temperature	Salimi et al. (2018)
2	AG	β-D-glucan microspheres	Mouse and rabbit serum	Carbodiimide-linked protein AG conjugate microparticles based on β-D-glucan. Enriched IgG to 92% purity. Suitable for the enrichment of fusion proteins by co-immunoprecipitation for protein-protein interaction studies	Yang et al. (2020)
3	A	PP fibers	CHO cell culture supernatant spiked with human IgG1	Polypropylene capillary-channeled polymer fibers with absorbed SpA, packed into a microbore column. Samples loaded in phosphate buffer pH 7.4 (0.1 mL min^{-1}) and eluted with 4.5 mM H$_3$PO$_4$ (1.0 mL min^{-1})	Trang and Marcus (2017)
4	A	Agarose beads	Human serum, IgG-spiked human saliva	Innovative micro-bead injection technique (μ-BIS) adapted for a lab-on-valve platform for reproducible packing of microcolumns between two optic fibers. Suitable for microscale quantification of IgG from biological matrices using HRP-labeled detection antibody	Ramos et al. (2019)
5	A	Au nanodots	IgG, human plasma	Photoluminescent gold nanodots prepared by 11-mercaptoundecanoic acid-mediated self-assembly of gold nanoparticles were conjugated with protein SpA for capture and luminescence IgG assay in human plasma. Results are comparable with ELISA	Shiang et al. (2011)

(continued)

Table 3.1 (continued)

No.	Protein	Support	Sample(s)	Summary	References
6	G	96-well plate wells	Anti-EGFR antibody	Nunc-Immuno™ MicroWell™ MaxiSorp™ 96-well plate with on-plate immobilized antibody-protein G complex crosslinked with bis(sulfosuccinimidyl)suberate and dimethyl pimelimidate. Potentially reusable chemically active surface for antigen binding-based immunoassays or extraction of antibody-specific proteins	Korodi et al. (2020)
7	A, G	Au nanoparticles	Human plasma	Citrate-capped gold nanoparticles conjugated with Cys-tagged protein A or G for the depletion of IgG from human plasma or cell culture supernatants	Liu et al. (2019a)
8	G	SiO_2 and carboxylated PS microparticles	–	In-depth optimization study on bead material, size, surface area, incubation time, ligand density, and conjugation chemistry influencing antibody binding and isolation efficiency	Ramirez et al. (2019)
9	G	PMMA monoliths	Human plasma	Poly(glycidyl methacrylate-co-ethylene dimethacrylate) epoxy-activated monolithic supports with immobilized protein G in the form of a 96-well filter plate. Reusable. Suitable for IgG enrichment in high-throughput glycomic and glycoproteomic workflows	Pučić et al. (2011)
10	L	PMMA monoliths	Rat serum	Poly(glycidyl methacrylate-co-ethylene dimethacrylate) epoxy-activated monolithic supports with immobilized recombinant protein L in the form of a 96-well filter plate. Reusable. Used for high-throughput IgG enrichment from Wistar rat serum	Habazin et al. (2019)

Abbreviations: *CHO* Chinese hamster ovary, *HRP* horseradish peroxidase, *EGFR* epidermal growth factor receptor, *PMMA* polymethacrylate, *PP* polypropylene, *PS* polystyrene

3.3.1.2 Protein G

Protein G (SpG) is another 30 kDa surface protein derived from group G *streptococci*, involved in bacterial pathogenesis (Fahnestock et al. 1986). It has three binding domains (C1–C3), and, unlike SpA, whose interaction with Ab Fc-domain is governed primarily by hydrophobic interactions, SpG exhibits pronounced hydrogen bonding and salt bridges as a dominant mode of binding of its C2 domain to the Fc (Yang et al. 2003; Rispens and Vidarsson 2014). Distinctive residues of Fc are also responsible for competitive binding to both SpA and SpG, which is not surprising since both proteins share a similar function on the surface of the bacterial cell wall (Sauer-Eriksson et al. 1995). SpG exhibits excellent binding capacity (K_a 6.7 × 10^9 M^{-1} towards human IgG) (Tran et al. 2016; Page and Thorpe 2002) for IgGs from a broad range of different species. All human and mouse IgG subclasses bind well to SpG (Sheng and Kong 2012). It is of note that wild-type SpG has an albumin-binding site, and for Ab purification and enrichment, a recombinant variant with suppressed albumin-binding capacity should be used to preclude co-enrichment of serum albumin (Choe et al. 2016).

3.3.1.3 Protein L

Multi-domain peptostreptococcal protein L (PpL) binds immunoglobulins via the interactions with the variable domain (V_L) of the κ-light chain ($V_κ$) rather than with the Fc-domain (Nilson et al. 1992). The affinity of PpL is not restricted to a particular immunoglobulin class but only to κ-light chain subgroups. The interaction is achieved through the two binding sites on each of five Ig-binding domains of PpL (B1-B5) and can be used to isolate various Ab fragments lacking Fc (Rodrigo et al. 2015). Hot spots for PpL interaction with the Ser-rich β-strand of $V_κ$ are Tyr34 from site 1 and Tyr36 from site 2, as confirmed by X-ray crystallography (Svensson et al. 2004).

Protein L domains are capable of binding $V_{κI}$, $V_{κIII}$, and $V_{κIV}$ light chains of IgG (K_a 5.7 × 10^7 M^{-1}) (Choe et al. 2016) with no evident affinity for $V_{κII}$ or λ subgroups, immunoglobulins heavy chains or constant domains (C_L) of a κ-light chain. The broad specificity and usefulness of this ligand have led to the development of engineered mutants capable of binding all κ-chains, with the potential to capture and purify any Ab fragment containing this light chain type (Lakhrif et al. 2016). Furthermore, by using salt additives in the elution buffers, the avidity of PpL can be fine-tuned to separate, e.g., high molecular weight Ab aggregates from monomers (Chen et al. 2020; Kihlberg et al. 1996; Nilsson et al. 1987).

3.3.1.4 Recombinant Fusion Proteins and Alternative Scaffolds

The Ig-binding domains of bacterial proteins can be combined into fusion proteins to expand binding properties (Choe et al. 2016). Thus, e.g., protein LG (Kihlberg et al. 1996), LA (Nilsson et al. 1987), and AG (Ghitescu et al. 1991) have been engineered and successfully employed as versatile novel ligands for Ab enrichment with broader specificity. Exhaustive reviews of less common immunoglobulin ligands were given by Mouratou et al. (2015) as well as Kruljec and Bratkovič (2017), covering natural and artificial proteins, aptamers, peptide ligands, and other small molecules used for Ab enrichment, depletion, and chromatography polishing steps.

3.3.2 Sample Preparation for Released IgG Glycan Analysis

Antibody *N*-glycosylation macro- and microheterogeneity manifest in the number of glycosylation sites and several glycan structural diversity levels (monosaccharide composition, size, branching, linkage, and charge). Two highly conserved Asn-297 glycosylation sites of the IgG heavy chains C_H2 domain are mostly occupied with covalently linked biantennary glycans (Fig. 3.1), affecting its stability, half-life, and role in the immune response, primarily through Fc-receptor (FcγR) interaction modulation (Gudelj et al. 2018). Together with the understudied and often neglected *N*-glycosylation within variable Fab domains (van de Bovenkamp et al. 2016), overall changes in IgG glycosylation profile such as increase or decrease in galactosylation, sialylation, bisection, and fucosylation reflect inflammatory status, biological age, and disease progression or regression of the individual. This makes contemporary multi-approach IgG glycoanalytics a valuable source of prospective biomarkers and renders it an indispensable tool in precision medicine and large-scale clinical cohort studies. IgG glycosylation analysis can be performed at different levels, targeting (1) intact glycoprotein, (2) cleaved Ab subunits, (3) released glycans, and (4) glycopeptides. The choice of the approach depends on the desired level of information, including Ab conformation, its subunit glycoforms, total released glycan diversity (glycome), glycan charge profile, and site-specific glycosylation microheterogeneity on glycopeptide level (see Sect. 3.2 and Fig. 3.2 for more details). Of the four aforementioned approaches, released IgG glycan and glycopeptide analyses are by far the most common as they mostly do not require complex state-of-the-art analytical instrumentation yet provide sufficient throughput, robustness, and versatility in glycobiomarker discovery, mAbs quality control, and diagnostic procedures.

Typically, all glycomic workflows include glycan release, labeling, enrichment, subsequent chromatographic or capillary electrophoresis separation coupled with mass spectrometry and/or fluorescence detection. All IgG *N*-glycans are of GlcNAcβ1-Asn type, assuming that innermost *N*-acetylglucosamine (GlcNAc) monosaccharide residue is attached to the polypeptide backbone asparagine (Asn) within an Asn-Xaa-Ser/Thr sequon, where Xaa is any amino acid except proline (Stanley et al. 2017). The N-glycosidic bond between the Asn and the innermost GlcNAc can be hydrolyzed enzymatically using different endoglycosidases, commonly peptide-N-glycosidase F (PNGase F) (Huhn et al. 2009), by alkaline hydrolysis in the presence of hydroxylamine (Kameyama et al. 2018), or by oxidative release with sodium hypochlorite (Song et al. 2016). Released glycans are then enriched and purified either prior to or after labeling. Fluorescent and isotopic labels enhance glycan ionization and chromatographic behavior, facilitating their detection and separation. In this subsection, standard glycan release, labeling, and enrichment methods employed for IgG glycosylation profiling will be discussed and illustrated by recent advances.

3.3.2.1 Chemical and Enzymatic Glycan Release

A classical approach for chemical deglycosylation of both N- and O-glycans is hydrazinolysis (Kamerling and Gerwig 2007). It is a high-speed and reproducible option for IgG glycosylation analysis, outperforming enzymatic release with PNGase F (Kotsias et al. 2019). This reaction yields de-N-acetylated glycan hydrazone derivatives with simultaneous degradation of peptide backbone and formation of amino acid hydrazides. Upon hydrazine removal, free hydrazide amino groups are then re-N-acetylated using acetic anhydride, and the resulting β-acetohydrazide is acid-hydrolyzed to obtain free glycan with restored reducing terminus (Kamerling and Gerwig 2007). Since anhydrous hydrazine is a regulated and highly toxic substance, hydrazine monohydrate was tested as a safer replacement and shown to be almost equally efficient, even for large-scale preparations (Nakakita et al. 2007). Besides special hydrazine handling precautions, hydrazinolysis must be performed under strictly anhydrous conditions. The glycan "peeling" side-reaction reported for this approach can be precluded by the addition of ethylenediaminetetraacetic acid (EDTA) (Kozak et al. 2014) or malonic acid (Goso 2016). Of note, undesirable immunogenic N-glycolylneuraminic acid (Neu5Gc) residues on therapeutic mAbs glycans for human use are entirely degraded in this reaction, masking their presence (Kameyama et al. 2018). With the advent of new, fast-acting recombinant N- and O-glycosidases and enzyme-compatible detergents, hydrazinolysis tends to be outdated or replaced with safer modifications, especially for routine Ab glycosylation analyses.

Alternative chemical deglycosylation methods based on aqueous ammonia treatment with (Wang et al. 2018a) or without (Wang et al. 2018b) the presence of reducing agents, followed by derivatization with 9-fluorenylmethoxycarbonyl chloride (Fmoc-Cl) or Girard's reagent P, respectively, also exist. These workflows elegantly employ the formation of unstable glycosylamines and their subsequent conversion to the open-ring form of 1-amino alditols (Wang et al. 2018a) or simply back to glycans with reducing end termini upon ammonia removal (Wang et al. 2018b). Feasibility of these two new approaches was not directly demonstrated for IgG glycomics, although modified alkaline hydrolysis of monoclonal human IgG1 glycans with lithium hydroxide in the presence of hydroxylamine at 80 °C was published by Kameyama et al. Here, reducing glycans are further converted to rather more stable oximes and fluorescently labeled with 2-aminobenzamide (2-AB) tag, followed by 2-picoline borane reduction (Kameyama et al. 2018). The reaction causes some sialic acids and GlcNAc to deacylate even at lower temperatures, and the overall sialoglycan recovery was found to be half of that achieved with PNGase F. Care should be taken here if glycans with substantial quantities with sialic acids are to be analyzed.

Seeking a cost-effective protocol for producing large quantities of glycans from biological samples, the oxidative release of natural glycans (ORNG) using common household bleach was recently introduced (Song et al. 2016). Sodium hypochlorite (NaClO) used in ORNG degrades N- or O-glycans very slowly compared to peptides if reaction conditions are carefully controlled. The proposed mechanism involves chlorination of glycosylation site amide bond to give N-chloroamide, which is further converted to glycosylamines via glycan-isocyanate intermediates and finally hydrolyzed resulting in free reducing N-glycans. Bovine IgG and human plasma glycoproteins were successfully N-deglycosylated in this way, with results comparable to PNGase F treatment and a slightly better yield in fucosylated glycans, obviously more resistant to enzymatic release.

PNGase F from *Elizabethkingia meningoseptica* or *E. miricola* (Flavobacteriaceae) and peptide-N-glycosidase A (PNGase A) from almonds or rice are both asparagine amidases cleaving the linkage between the innermost GlcNAc and the Asn residue, converting it to aspartate (Wang and Voglmeir 2014) (Fig. 3.3). PNGase A can cleave N-glycans with or without α(1,3)-linked core fucose, while PNGase F can cleave only glycans without α(1,3) core fucosylation. PNGase F activity requires a minimal N,N'-diacetylchitobiose motif with the GlcNAc 2-acetamido group linked to Asn for recognition and is a widespread enzyme of choice for routine N-glycosylation analysis, of human IgG in particular (Table 3.2), as well as plasma or serum proteins (Huhn et al. 2009). A detailed investigation of PNGase F-mediated glycan release kinetics from human IgG tryptic glycopeptides (Huang and Orlando 2017) revealed that peptide backbone composition has a barely noticeable influence on the deglycosylation reaction rate. On the contrary, different glycan structures (fucosylated, bisected, mono-, and di-galactosylated, or sialylated) have been proven to influence and govern the PNGase F-catalyzed deglycosylation reaction rate. Thus, for example, core fucosylated glycoforms exhibit the fastest release rates while those sialylated and bisected are released slower. Following this, caution should be exercised and enzyme performance thoroughly tested when performing rapid IgG deglycosylation protocols of growing popularity (Cook et al. 2012). Novel bacterial N-glycosidases such as PNGase H$^+$ from *Terriglobus roseus* (Wang et al. 2014) with the combined activity of PNGase A and PNGase F or PNGase F-II, also from *E. meningoseptica* (Sun et al. 2015), with broader combined specificities towards glycans with α(1,3)- and/or α(1,6)-linked core fucose, were also described. The PNGase F is highly effective for complete and cost-effective glycan release in reducing and non-reducing conditions from different antibodies, regardless of their structure, and thus remains an enzyme of choice for IgG glycosylation analysis.

Table 3.2 Recent literature review on antibody PNGase F deglycosylation protocols and technical improvements

No.	Sample	Highlight	Summary	References
1	Commercial mAbs	Enzyme immobilized on SiO$_2$ beads and integrated into an LC chip	On-chip 6-s deglycosylation, glycan trapping, separation on C$_8$ stationary phase, and detection of released unlabeled glycosylamines. Enzyme microreactor, columns, and nano-ESI emitter tip were combined into a single microfluidic chip	Bynum et al. (2009)
2	Human IgG	Reusable magnetic beads with immobilized PNGase F	PNGase F-GST conjugated to glutathione-functionalized paramagnetic beads efficiently deglycosylated proteins in 10 min at 50 °C. Results are comparable with an in-solution reaction	Bodnar et al. (2016)
3		Comparison of different PNGases F for high-throughput glycomics	A detailed reproducibility and efficiency testing of enzymes from different producers for IgG and human plasma UPLC-FLR glycan analysis was performed.	Vilaj et al. (2020)
4		Rapid deglycosylation in a domestic microwave oven	A 20 min irradiation at 20% oven maximum power effectively catalyzed IgG enzymatic deglycosylation without notable loss of labile sialic acids	Zhou et al. (2012)
5		Integrated platform for online deglycosylation, HILIC enrichment, and labeling	An elaborate fluidic platform constructed from two 10-port valves for automated HILIC glycopeptide enrichment, N$_2$-assisted buffer exchange, hIMER deglycosylation, and C$_{18}$ solid phase-assisted heavy/light dimethyl labeling is described	Weng et al. (2014)
6		PCT for rapid PNGase F deglycosylation	PCT device alternates cyclically between atmospheric and elevated pressure and enhances enzyme conformational changes and digestion site accessibility. IgG glycans were successfully released in 5 min with 1:2500 enzyme:Substrate molar ratio at 30 kPsi. Up to 12 samples can be processed simultaneously	Szabo et al. (2010)
7		PNGase F-GST immobilized on glutathione affinity resin	Model glycoproteins can be deglycosylated in 10–20 min at 50 °C by continuous aspiration/dispensing cycles using an automated protein purification platform	Szigeti et al. (2016)
8	–	Overview of different N-glycosidase immobilization approaches	A critical review of enzyme immobilization strategies (entrapment, aggregation, adsorption, covalent binding, microencapsulation), matrices, and substrates	Karav et al. (2017)

Abbreviations: *PCT* pressure-cycling technology, *F-GST* glutathione S-transferase, *hIMER* hydrophilic PNGase F immobilized enzymatic reactor, *nano-ESI* nano-electrospray ionization, *UPLC–FLR* ultraperformance liquid chromatography with fluorescence detection

Another attractive endoglycosidase for IgG glycosylation analysis is endoglycosidase S (Endo S) (Collin and Olsén 2001; Trastoy et al. 2018). This endo-β-N-acetylglucosaminidase from *Streptococcus pyogenes* hydrolyzes the β(1,4) linkage within the chitobiose core of complex glycans (Trastoy et al. 2018) and leaves the innermost GlcNAc residue—regardless of core fucosylation—attached to the peptide backbone (Fig. 3.3).

The structural background of Endo S interaction with different oligosaccharide moieties was recently studied in detail, applying X-ray crystallography, small-angle X-ray scattering, and molecular docking simulations (Trastoy et al. 2018). This enzyme was also proven to be useful for therapeutic mAb glycosylation engineering. In a study by Tong et al., transglycosylation of rituximab with glycan oxazolines upon Endo S digestion was performed, paving the way for more efficient antibody chemoenzymatic remodeling (Tong et al. 2018). A closely related *S. pyogenes* serotype M49 enzyme is Endo S2 (Sjögren et al. 2013), exhibiting pronounced specificity towards high-mannose and hybrid IgG glycoforms (Sjögren et al. 2015).

Overviews of sequential exoglycosidase digestion for antibody glycan analysis using capillary electrophoresis or liquid chromatography are given elsewhere (Sjögren et al. 2016; von Pawel-Rammingen et al. 2002), and this topic will not be discussed here in detail. An array of these glycoside hydrolases exist, e.g., α-neuraminidase, α-fucosidase, β-galactosidase, α-mannosidase, β-mannosidase, β-N-acetylglucosaminidase; with different linkage specificities (Fig. 3.3). The course of the sequencing reaction is usually monitored through the chromatographic retention time shift upon monosaccharide residue cleavage or using MS (see Sects. 3.4.5 and 3.5.1.2).

3.3.3 Fluorescent and Isotopic Glycan Labeling

Glycans are bulky hydrophilic biomolecules of low ionization efficiency, lacking chromophores. Numerous glycan derivatization strategies alleviate a bottleneck in detecting and quantifying these protein post-translational modifications (PTM) products. Typically, released glycan analysis uses a fluorogenic small-molecule reducing end tagging via (1) reductive amination of imines (Schiff bases) (Bigge et al. 1995), (2) Michael addition (Ruhaak et al. 2010a), (3) hydrazone, and (4) oxime formation with hydrazide or aminooxy reagents. Immunoglobulin G glycomic workflows still favor simple reductive amination labeling due to the low cost and high stability of secondary amines obtained. The reaction is typically performed in dimethyl sulfoxide in the presence of acetic acid and sodium cyanoborohydride ($NaBH_3CN$) as a reducing agent, overnight at 37 °C or 2–4 h at 60–65 °C. At higher temperatures, acid hydrolysis and loss of labile sialic acids can easily occur. Less toxic reducing agents, e.g., 2-methylpyridine (picoline) borane complex, are now of widespread use in IgG glycan labeling (Ruhaak et al. 2010b). Most of the labels used are aniline-, pyridine-, naphthalene-, or pyrene-derived (Table 3.3) and can be either neutral or negatively charged for charge-based

Table 3.3 Common fluorescent labels used in IgG released glycan analysis.

Label	Structure	Charge	Separation	Comments	References
2-aminobenzoic acid (2-AA)		(−)	LC, CE	Negatively charged label. It can also be used for negative mode MALDI-TOF-MS.	Jeong et al. (2018)
2-aminobenzamide (2-AB)		0	LC	Most widely used fluorescent label for IgG glycosylation analysis with FLR and/or MS detection.	Keser et al. (2018)
2-aminopyridine (2-AP)		0	LC	Used for separation and MS detection of isomeric IgG glycans on zwitterionic HILIC column.	Gong et al. (2013) and Albrecht et al. (2017)
Procainamide (ProA, DEAEAB)		0	LC	Protonation of the tertiary amine group greatly enhances signal intensities in positive mode ESI-MS. highly suitable for MS/MS structural glycan analysis.	Harvey (2000)
8-aminopyrene-1,3,6-trisulfonate (APTS)		(−)	LC, CE	Negative mode ESI–MS detection is also possible after HILIC-UPLC separation.	Reusch et al. (2014)
8-aminonaphthalene-1,3,6-trisulfonate (ANTS)		(−)	LC, CE	RP separation of ANTS-labeled glycans can also be achieved with the addition of ion-pairing reagents.	Takegawa et al. (2006) and Harvey (2000)
GlycoWorks™ *Rapi*Fluor-MS™		0	LC	NHS-carbamate dye. Labeling achieved in 5 min after digestion with rapid™ PNGase F. enhanced fluorescence and ionization properties. Available commercially only in the form of a kit. Higher MS sensitivity than 2-AB and ProA.	Tian et al. (2015) and Karav et al. (2017)

Abbreviations: *CE* capillary electrophoresis, *DEAEAB* N-(2-diethylamino)ethyl-4-aminobenzamide, *ESI-MS* electrospray ionization mass spectrometry, *FLR* fluorescence detection, *HILIC–UPLC* hydrophilic interaction ultraperformance liquid chromatography, *LC* liquid chromatography, *MALDI-TOF–MS* matrix-assisted laser desorption/ionization coupled with time-of-flight mass spectrometry, *NHS* N-hydroxysuccinimide, *RP* reversed-phase separations. A new generation of fluorescent glycan labels is predominantly based on the *N*-hydroxysuccinimidyl (NHS) carbamate reactive group for fast labeling of *N*-glycosylamines (Lauber et al. 2015). Therefore, 'instant' 2-AB and ProA NHS-ester analogs have been synthesized (Vainauskas et al. 2018) as well as a very popular quinoline-based *GlycoWorks™ Rapi*Fluor-MS™ reagent (Waters Corporation, Milford, MA).

Modern proteomics heavily relies on stable heavy isotope (^{2}H, ^{18}O, ^{13}C, ^{15}N) labeling tags as such approach allows for multiplexing, i.e., simultaneous MS analysis of combined complex samples with enhanced ionization properties and accurate quantification by high-resolution MS (Gevaert et al. 2008). Among many variations, the most popular are isobaric tandem mass tags (TMT), based on the measurement of the ratio of diagnostic heavy and light reporter ion pairs upon TMT tag fragmentation (Thompson et al. 2003). Accordingly, glyco-TMTs were developed and used for mAb glycosylation analysis (Gong et al. 2013). Reductive amination duplex labeling with [^{12}C$_6$] and [^{13}C$_6$] aniline or 2-AA is also possible (Vainauskas et al. 2018; Gevaert et al. 2008). Internal standardization for MS-based glycomics has also been greatly improved by Zhou et al. with the development of the iGlycoMab (Zhou et al. 2016)—a murine mAb with integral ^{15}*N*-labeled glycans, which can be released, mixed with the sample, and analyzed to compensate for variations in sample preparation. For MALDI- and LC–MS, glycan permethylation with iodomethane (CH$_3$I) in the presence of sodium or potassium hydroxide is also used and is further described in Sects. 3.4 and 3.5.

3.3.4 Sample Preparation for Glycopeptide Mapping and Subclass-Specific IgG Fc Glycosylation Analysis

3.3.4.1 Enzymatic Digestion

Representative workflows that analyze site- and subclass-specific IgG glycosylation comprise in-gel or in-solution protease digestion of Ab, glycopeptide enrichment, chromatographic separation, and MS detection. Trypsin is typically used as a digestion enzyme, although protein conformation and glycoforms themselves can bias its activity. Falck et al. have pointed out that there is a digestion preference of trypsin for high mannose, bisected, and α(2,3)-sialylated IgG glycoforms over the abundant core fucosylated neutral ones and that it decreases when a chemical denaturation step is introduced before digestion (Falck et al. 2015). For high-throughput glycoproteomics, the denaturation step can be omitted for convenience. If needed, simple formic acid denaturation is sufficient (Falck et al. 2015), while MS-compatible RapiGest™ SF surfactant or traditional dithiothreitol (DTT) and iodoacetamide (IAA) reduction and alkylation can also be used (Yan et al. 2012). Microwave-assisted rapid trypsin digestion is an attractive alternative for mAb tryptic digestion, yet it can cause Asn deamidation artifacts (Formolo et al. 2014). Limited sequence coverage, multidimensional separations, and peptide fractionation requirements in bottom-up mAb reductionist (glyco)proteomics and PTMs identification can be bypassed with the middle-up approaches made possible by an array of Ab proteases (Fig. 3.3). Allotype-specific HILIC– and CE–MS analysis of IgG glycoforms after on-beads IdeS digestion and isolation of Fc/2 subunits was achieved in this manner (Sénard et al. 2020). An excellent overview of alternative endoproteases such as Lys-C, Lys-N, Glu-C, or Asp-N is given by Tsiatsiani and Giansanti and in references therein (Giansanti et al. 2016; Kurogochi and Amano 2014). The different cleavage specificity of these enzymes can occasionally be utilized in case of too large tryptic peptides.

3.3.4.2 Stable Isotope Glycopeptide Labeling

Analytical benefits arising out of the stable isotope tagging in glycomics (see Sect. 3.3.3) can be translated to therapeutic and natural IgG glycoproteomics. Human IgG and IgG1 glycoforms from plasma cell myeloma were quantified with high sensitivity by MALDI-TOF-MS using benzoic acid light (D_0) and heavy (D_5) N-succinimidyl esters (Kurogochi and Amano 2014), while a similar approach with amino group-reactive succinic anhydride and its $D_4{}^{13}C_4$ isotopologue on IgG glycopeptides was demonstrated (Pabst et al. 2016).

More elaborate approaches include synthetic, isotopically labeled Fc glycopeptides for standard addition quantitative analysis (Roy et al. 2018) or SILu™MAb *s*table *i*sotope-*l*abeled *u*niversal *m*onoclonal *a*nti*b*ody for internal standardization combined with on-bead digestion and multiple reaction monitoring (MRM) mode MS analysis (Shiao et al. 2020).

Due to the high cost and limited commercial availability of these reagents, glycoproteomics still depends on intact unlabeled glycopeptide MS analysis and relative quantitation.

For glycosylation site occupancy analysis, PNGase F-mediated incorporation of ^{18}O into aspartic acid (+3 Da mass shift) upon digestion in $H_2{}^{18}O$ is a tool of the trade (Angel et al. 2007). Care should be taken here to avoid (or promote) residual trypsin variable incorporation of one or two ^{18}O atoms into peptides C-termini carboxyl groups. An interesting example of the MRM method developed to monitor serum IgG2, and IgA1 glycosylation site occupancy changes in the patients with congenital disorders of glycosylation and non-alcoholic fatty liver disease were developed by Hülsmeier et al. (2016).

3.3.4.3 Glycan and Glycopeptide Enrichment and Purification Strategies

As with other instrumental approaches, glycan or glycopeptide sample preparation is inseparable from technical advances in the LC-MS field. It destines the analytical workflow success, allows preset method validation criteria to be met, and further pushes the limits of detection, quantification, and structural characterization. The use of different fluorescent or isotopic tags, low analyte concentrations, and presence of residual digestion enzymes or labeling reagents within a complex matrix of non-glycosylated peptides and proteins further complicate the choice of appropriate enrichment protocol. In their essence, all interaction modes target either the glycan or the peptide part via covalent (e.g., boronate and hydrazide chemistries) or non-covalent (e.g., HILIC, RP, ion exchange, lectins, PGC, TiO_2) interactions (Chen et al. 2014). The methods for both released IgG glycan and glycopeptide enrichment can frequently be used interchangeably as these analytes are of similar physicochemical properties.

A single exception is a reversed-phase chromatography, typically performed using C_8 or C_{18} functionalized silica, where hydrophobic interactions with the peptide moiety are utilized to desalt and enrich glycopeptides only, whereas released glycans cannot be purified in this manner. By tuning elution conditions, non-glycosylated peptides and contaminants are retained on or washed away from octyl- or octadecyl-modified silica. This enrichment mode is widely employed in high-throughput IgG glycopeptide enrichment for cohort studies due to low cost, speed, similar peptide hydrophobicities across subclasses, and low bias towards certain glycan types (Zaytseva et al. 2018; Alagesan et al. 2017).

Hydrophilic interaction liquid chromatography (HILIC) stationary phases and their zwitterionic modifications (ZIC®-HILIC sulfobetaine) (Alagesan et al. 2017) have seen the fastest expansion with dozens of innovative materials of broad or narrow affinity ranges and binding capacities developed.

An overview of the field, including discussion on retention mechanisms, recent breakthroughs, and future perspectives in HILIC-based materials for glycoanalytics, has been published recently (Chen et al. 2014; Selman et al. 2011). This form of normal-phase liquid chromatography has now been improved almost to a level of molecular recognition for specific carbohydrate structures. It can be technically adapted for comprehensive non-targeted protein glycosylation analysis or high-throughput screening in the forms of high-performance liquid chromatography, filter-aided sample preparation (FASP), 96-well filter plates, and pipette tip-based devices.

A lineup of new adsorbents and (nano)composites proved to be a successful alternative to the traditional IgG and mAbs glycopeptide purification approaches. Here, of particular interest are environmentally friendly (ligno)cellulosic materials, including cellulose microspheres (Sha et al. 2018), unmodified (Selman et al. 2011), and boric acid functionalized TiO_2-modified (Liu et al. 2019b) cotton wool, delignified balsa wood (Zhou et al. 2020), and boronic acid@fibrous cellulose (Sajid et al. 2020). Carrageenan on graphene oxide/poly(ethylenimine) support (Chen et al. 2019), freeze-casted honeycomb chitosan membrane (Zhang et al. 2019), Ti^{4+}-immobilized dendritic polyglycerol-coated chitosan (Zou et al. 2017), as well as carboxymethyl chitosan (Bodnar and Perreault 2015) are more elaborate examples of chemically remodeled linear polysaccharides with demonstrated excellent binding capacity, stability, and specificity for IgG glycopeptide enrichment.

Porous graphitic carbon (PGC)—a versatile stationary phase embracing several different binding interactions—is also effective for microscale purification and chromatography of native and tagged glycans for biosimilars *N*-glycosylation profiling (Liu et al. 2017). Most enrichment modes can be conveniently combined for enhanced results. Ohta and coworkers have, accordingly, constructed a simple stop-and-go HILIC/RP-SPE tip (StageTip) for IgG1 glycopeptide capture (Ohta et al. 2017).

Surface engineering can greatly expand the chemical space of two-dimensional (2D) materials, and it has expectedly transformed sample preparation for protein glycosylation analysis. Several outstanding illustrations of new 2D material platforms for IgG glycoproteomics involve hydrophilic magnetic graphene coated with phytic acid for combined phospho- and glycopeptide enrichment (Hong et al. 2018), polydopamine-coated graphene decorated with magnetic Fe_3O_4 and chitosan (Bi et al. 2020), MoS_2/Au-nanoparticle-L-cysteine nanocomposite and, its variant with Au-nanowires grown on SiO_2 coating offering larger surface area for generous L-cysteine grafting and enhanced hydrophilicity. Unfortunately, novel techniques are typically limited to the labs where they were developed, and seldom are these materials and reagents widely available commercially.

For large-scale glycoanalysis, automation of robust and cost-effective glycan and glycopeptide enrichment is essential as it allows for the full diagnostic potential of protein glycosylation to be utilized. In this setting, glycoblotting for on-bead free reducing end glycan enrichment and fluorescent labeling seems to be promising (Amano and Nishimura 2010). Glycans are first ligated to hydrazide beads for either subsequent release and aminooxy-fluorophore tagging via *trans*-imination (BlotGlyco H®) or reduction of disulfide spacer after hydrazone bond formation with a predesignated fluorophore already coupled to a bead (BlotGlyco ABC®) (Miura et al. 2008). Sialic acids may also be on-bead methyl-esterified using 3-methyl-1-(*p*-tolyl)triazene for linkage-specific analysis if desired. A dedicated liquid handling platforms are fully compatible with BlotGlyco workflows (Nishimura 2011). Recently, glycoblotting was shown to fit the needs of large-scale human serum and IgG glycomics for cancer glycobiomarker discovery (Gebrehiwot et al. 2019).

Hydrophilic polypropylene 96-well filter plates are a straightforward option for HILIC purification of 2-AB- or ProA-tagged released glycans before (Ultra) High-performance Liquid Chromatography analysis coupled with mass spectrometry as they are compatible with vacuum manifolds and pipetting robots (Tian et al. 2015; Menni et al. 2013; Goh and Ng 2018).

3.4 Deciphering the IgG Glycan Structure

N-glycans found on IgG are oligomeric carbohydrate structures of different monosaccharides linked to each other by a glycosidic bond in a complex, branched manner. A glycosidic bond can be formed between an anomeric carbon of one monosaccharide and potentially any other hydroxyl group of the other sugar. Moreover, more than one hydroxyl group on any monosaccharide can be occupied simultaneously, adding even more to the glycan complexity (Varki et al. 2015). When compared to linear biomolecules targeted by omics studies, such as DNA or proteins, glycans exhibit several orders of magnitude higher complexity and therefore require a more comprehensive approach to identify their compositional and structural features. There are several different methodological approaches to identify the structure of a glycan successfully. Here we will focus on the opportunities MS approaches are offering in this demanding endeavor.

Mass spectrometry is a powerful tool for glycan analysis, and in combination with several separation techniques (HPLC, nano-LC, UHPLC, CE) coupled to a mass analyzer, it can provide an almost complete picture of the analyzed sugar molecule. High-resolution instruments offer information about the glycan candidates already on the MS^1 level. Given the known information about the analyzed sample, such as biological type/origin, or biosynthetic pathways of analyzed glycans, even the m/z value alone can be very informative. In the case of IgG, a repertoire of possible constituent monosaccharides is very well known and defined by human biology, i.e., its sugar donors, biosynthetic pathway, glycosyltransferases, and glycosidases. Even more, IgG from human plasma has been characterized in detail (Pučić et al. 2011; Kobata 2008), which makes the elucidation of its glycan structures in further epidemiological and clinical studies rather simpler. Although this is true for the IgG from human plasma, the analysis of recombinant antibodies or biofluids with low IgG concentrations is not as straightforward. Glycosylation is a post-translational modification governed by a very complex system of enzymes, and it is dependent on a number of environmental factors (Menni et al. 2013). All these factors play an important role in the production of recombinant proteins in different cell lines under different conditions. Hence, IgG produced in different setups can yield a repertoire of possible glycan species attached to the consensus N-glycosylation sequence (Goh and Ng 2018). It is therefore important to use analytical techniques that can undoubtedly confirm the composition and structure of analyzed N-glycans.

3.4.1 Fragmentation

Even though a number of different analytical techniques are developed for the analysis of N-glycans, mass spectrometry-based techniques remain the most used analytical approach (Harvey 2020). As mentioned above, MS spectra obtained with high-resolution mass analyzers can provide good enough clues about the glycan identity. However, due to oligosaccharide complexity, a number of isobaric structures can have the same m/z value. A standard technique to overcome the issue of isomers in the characterization of glycan structures has become MS/MS (tandem MS or MS^n), where a single ion of interest can be selected (precursor ion) and later fragmented. Tandem MS becomes exceptionally informative in the ion trapping instrumental setup as it enables sequential fragmentation events. In the multiple stages of MS^n experiments, any fragment ion can be selected and subjected to another round of fragmentation (Ashline et al. 2007). With the appropriate fragmentation method in the MS/MS experiment, each sugar molecule undergoes a series of linkage cleavages and shows a specific pattern of fragments. Numerous fragments obtained can be very informative about the original sugar composition and the type of linkages between constituent monosaccharides. Several tandem MS methods are developed and used for the analysis of oligosaccharides: collision-induced dissociation (CID), higher energy collisional dissociation (HCD), and electron-based dissociations: electron capture dissociation (ECD) and electron-transfer dissociation (ETD). CID is one of the first approaches that have been developed, and it is the most widely used method for the fragmentation of glycan structures (Sobott et al. 2009). During a CID event, the kinetic energy of the selected parent ion is increased as the ion enters a collision cell of the instrument. The ion collides with inert gas molecules within the cell, and the kinetic energy of the excited ion is transferred to vibrational energy, which results in bond cleavages (An and Lebrilla 2011). Collision gases commonly used in fragmentation experiments are nitrogen (N_2), helium (He), and argon (Ar). HCD is a form of CID, designed explicitly for Orbitrap mass analyzers (Jedrychowski et al. 2011). Levels of fragmentation energies are similar to CID. ECD is a fragmentation method introduced relatively recently (Zubarev et al. 1998). It is based on the irradiation of multiply charged ions with low-energy electrons. The electrons are captured by the irradiated ions resulting in dominantly c- and z-fragments in peptide fragmentation events. The c- and z-ion series are generated by the fragmentation of the C_α-N bond. The ETD shares the same mechanism of dissociation as ECD, with the main difference being the source of electrons. In ETD, the electron transfer occurs from singly charged reagent anions to multiply charged protonated peptides (Wuhrer et al. 2007a). A fluoranthene is the most common ETD reagent used.

3.4.2 Fragmentation Nomenclature

For the interpretation of generated fragmentation spectra, Domon and Costello proposed nomenclature of glycan fragments to simplify the communication of the often complicated MS/MS spectra (Domon and Costello 1988). The glycan fragmentation nomenclature is distinguished from the peptide by capital letters used in carbohydrate fragment spectra annotation. After a fragmentation event, ions can retain charge on either the reducing or non-reducing end of the oligosaccharide. Fragments with the charge at the reducing end are labeled X, Y, and Z, while ions with the charge at the non-reducing end are labeled A, B, and C. A- and X-ions are a result of a cross-ring fragmentation, while Y-, Z-, B-, and C-ions represent the cleavage of the glycosidic bond (Domon and Costello 1988). Subscript and superscript numbers are used to designate the exact location of glycosidic bond cleavage and the bond that has been cleaved following the cross-ring fragmentation, respectively, counting from the reducing end of the glycan chain. Figure 3.4 shows a

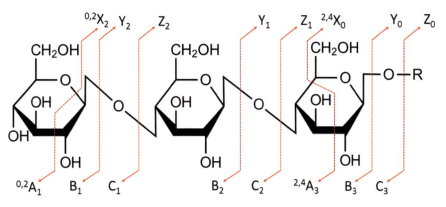

Fig. 3.4 Carbohydrate fragmentation nomenclature as proposed by Domon and Costello in 1988

schematic representation of nomenclature used in the interpretation of oligosaccharide fragmentation spectra introduced by Domon and Costello.

3.4.3 Fragmentation Candidates

Glycans can be fragmented either released or still attached to the peptide backbone asparagine. The choice of the approach is dependent on the information needed.

3.4.3.1 Fragmentation of Glycopeptides

Fragmentation of the tryptic glycopeptides can provide confirmation of both the peptide sequence and the glycan conjugate composition and its structural features. The CID experiments result dominantly in the cleavage of glycosidic bonds while the amide bonds of the peptide remain mostly intact (Conboy and Henion 1992; Huddleston et al. 1993). Fragmentation spectra are dominated by carbohydrate B- and Y-ions. Although CID favors glycosidic bond cleavage, information about peptide and glycan moiety can be obtained in a single MS/MS experiment by optimizing the collision energy stepping (Hinneburg et al. 2016). With the instrumental setup that allows multiple sequential fragmentation steps, tandem MS experiments can resolve both glycan and peptide sequences. After the MS^2 experiment, one of the most abundant fragments is a complete peptide with the first chitobiose core GlcNAc still attached to the N-glycosylation site. By choosing this fragment for further MS^3 experiments, we can obtain the amino acid sequence of the glycopeptide (Demelbauer et al. 2004). Another approach is to use complementary fragmentation events of ETD and HCD (EThcD). A diagnostic ion generated by neutral loss in the HCD event can be set to trigger sequential fragmentation of the parent ion by ETD. Both spectra are then recorded, with the HCD spectra dominated by B- and Y-ions generated by cleavage of glycosidic bonds and the ETD spectra dominated by b- and y-ions generated by cleavage of amide bonds (Singh et al. 2012). However, the downside of ETD is the requirement of multiply charged precursor ions (Zhurov et al. 2013). Experiences from different research groups report that useful spectra for IgG glycopeptides are obtained for precursors below m/z 1400 (Alley et al. 2009) or even lower for precursors below 850 Da (Hinneburg et al. 2016). Since the majority of IgG glycopeptide ion signals are in the range around m/z 850 or higher (doubly and triply charged), complementing CID with ETD is limited due to the lack of multiply charged precursor ions.

3.4.3.2 Fragmentation of Released Glycans

In tandem MS experiments, glycan modifications reveal additional structural features that would otherwise go unnoticed. Reducing end labeling introduces a chemical tag by which we can easily differentiate between GlcNAc originating from the branches and the GlcNAc from the chitobiose core. This is very informative when assigning fucosylation as core versus antennary. Permethylation of hydroxyl and *N*-acetyl groups improves the ionization efficiency of glycans and enables a more straightforward determination of branching as well as the position of the glycosidic linkage. Major fragments observed in tandem MS experiments of permethylated glycans are ions generated by cleavages of glycosidic bonds at GlcNAc and sialic acid residues. Nevertheless, a significant number of fragments arising from cross-ring fragmentation can also be detected, which allows for a more detailed assignation of linkages between constituent monosaccharides (Morelle et al. 2004).

3.4.4 Ionization Polarity of MS Analysis

Glycans can be analyzed in positive (positively charged ions are measured) or negative ionization mode (negatively charged ions are measured). So far, positive ion mode is used more extensively for glycan analysis as it tends to give better signal intensities, and it is more compatible with the buffered mobile phases and fluorescent dyes (2-AB, ProA) of the widespread HILIC–UPLC–FLR method (Keser et al. 2018; Pabst et al. 2009). Negative ion mode, still not dominantly used for glycan analysis, has a major advantage in the structural analysis of glycans. Fragmentation spectra of negatively charged ions provide far more informative patterns and diagnostic ions. There is no need for any derivatization to gain additional data about glycan structure as cross-ring fragmentations dominate in the negative mode tandem MS experiments of unmodified glycans (Harvey 2020). Cross-ring fragmentations, in contrast to cleavage of glycosidic bonds typical for positive ion mode, produce a number of diagnostic ions highly specific for the branch occupancy (e.g., galactose positioned on 6-arm versus 3-arm) and for the type of linkages between constituent monosaccharides [$\alpha(2,6)$ versus $\alpha(2,3)$ sialic acid linkage] (Harvey 2020).

Nevertheless, negative ion mode is also feasible, if needed, for derivatized glycans where some dyes even increase the ionization efficacy (Pabst et al. 2009). However, some fluorescent tags might change the fragmentation patterns and reduce the level of information in the corresponding fragmentation spectra (Harvey 2020). Negative ion spectra of glycans are more complex and require substantial experience for proper interpretation, making this MS approach somewhat disadvantageous for routine IgG glycosylation analyses.

3.4.4.1 Example of Positive Ion Mode Fragmentation

The main event of positively charged ion fragmentation is the dissociation of glycosidic bond (Conboy and Henion 1992; Huddleston et al. 1993). Figure 3.5 shows a typical tandem MS spectrum of the 2-AB-labeled FA2BG1 structure analyzed by HILIC-UPLC-FLR-MS. Several structural features were successfully deciphered from the combined separation- and MS-based approach. The most abundant ion in Fig. 3.5 at m/z 366.13, $[M + H]^+$, is a typical diagnostic ion for galactosylation of the antennae. Additionally, HILIC can successfully separate monogalactosylated isomers with galactose linked to 3-arm versus 6-arm. With these two sets of information, it is possible to assign a galactose linked to 6-arm of the glycan. The ion with the m/z value of 488.21, $[M + H]^+$, locates the fucose to the chitobiose core. Reducing end derivatization of the glycan structure with the fluorescent dye marks the innermost GlcNAc and produces a diagnostic ion that annotates the structure as core fucosylated. During the CID of antennary fucosylated structures, it was observed that in certain conditions, an intramolecular fucose transfer between antennae could occur. It is an event characterized by the migration of fucose from its original position during fragmentation, whereby fucose creates linkages with adjacent monosaccharides, producing artifacts rather than diagnostic ions (Hsu and Turk 2004; Kováčik et al. 1995; Wuhrer et al. 2006; Harvey et al. 2002). It is, therefore, important to be cautious when assigning a structure to glycans carrying fucose and to back up the analysis with the appropriate standards. An additional informative fragment is detected at the m/z 1056.451, $[M + H]^+$, pointing to the presence of a bisected GlcNAc rather than a third antenna. The rest of the fragments supply information about the sequence and, thereby, about the composition of the analyzed glycan. With all the information from the retention time of the glycan, its fragmentation spectra, and biological origin (human IgG glycan), it is possible to assign the structure as the abovementioned FA2BG1.

3.4.4.2 Example of Negative Ion Mode Fragmentation

As mentioned above, the main advantage of negative ion mode fragmentation is the generation of cross-ring cleavages. It is a type of reaction where two bonds across a single monosaccharide unit are cleaved, yielding A- and X-ions (Fig. 3.4). Figure 3.6 shows the CID MS spectrum of a 2-AB labeled FA2B glycan recorded in negative

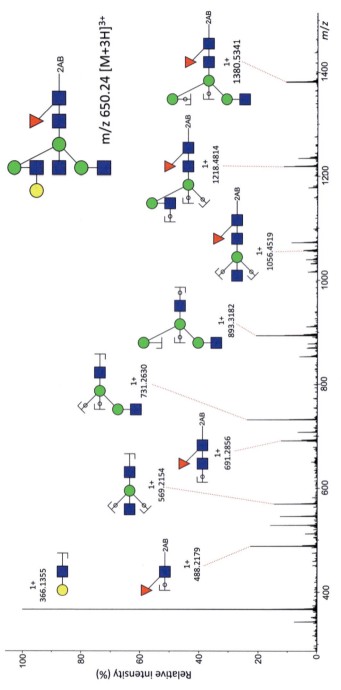

Fig. 3.5 Collision-induced fragmentation (CID) of the triply charged 2-AB labeled core fucosylated, monogalactosylated, biantennary glycan with bisecting GlcNAc recorded in positive mode. Diagnostic [M + H]+ ions: m/z 366.13 galactosylation; 488.21 core fucosylation; 1056.45 bisecting N-acetylglucosamine (GlcNAc). The rest of the fragments represent proposed fragmentation events. Symbols for monosaccharide units of fragment ion compositions are represented as defined by CFG: blue square (N-acetylglucosamine), green circle (mannose), yellow circle (galactose), and red triangle (fucose)

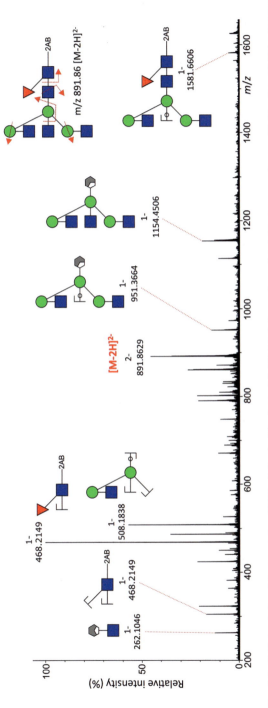

Fig. 3.6 Collision-induced fragmentation (CID) of the doubly charged 2-AB labeled core fucosylated, biantennary glycan with bisecting GlcNAc (FA2B) recorded in negative mode. Diagnostic [M-H]⁻ ions: m/z 468.21 core fucosylation; m/z 508.18 bisected N-acetylglucosamine (GlcNAc)—[D-221]⁻. The rest of the fragments represent proposed fragmentation events. Symbols for monosaccharide units of fragment ion compositions are represented as defined by CFG: blue square (N-acetylglucosamine), green circle (mannose), and red triangle (fucose)

ion mode. The most abundant ion in the spectrum is the [M-H]⁻ Z-ion with the m/z 468.221. It originates from the cleavage of the glycosidic bond between chitobiose core GlcNAcs. Since the reducing end is derivatized with the 2-AB fluorescent dye, we can assign the fucose to the core of the structure. In the fragment spectrum of glycans carrying bisecting GlcNAc, the otherwise present so-called D-ion is missing a result of cleavages between Manβ1-4GlcNAc and Manα1-3Man. The resulting loss of chitobiose core and the 3-arm yields the D-ion neighboured by [D-18]⁻ formed by neutral loss of water (Harvey et al. 2008; Everest-Dass et al. 2013). Instead, one of the most prominent signals arises from the [D-221]⁻ ion. It represents the 6-arm occupancy, with the loss of bisecting GlcNAc. In the case of the structure in Fig. 3.5, it is the [M-H]⁻ fragment with the m/z value of 508.18. The main ambiguity in the analyzed FA2B structure arises from its differentiation from the potential isomer FA3. It has the same mass and the same composition as FA2B, and the retention times of two isomers in the HILIC–UPLC analysis are too close to resolve the issue. The only difference between the two isomeric structures is the position of one GlcNAc: is it the third antenna or bisecting GlcNAc? This is exactly the question that can be answered unequivocally with the information from fragmentation in negative ion mode, where the detection of abundant [D-221]⁻ ion clearly resolves the issue (Fig. 3.6).

3.4.5 Exoglycosidase Digestion Monitored by MS

Glycan sequencing by exoglycosidases uses the specificity of a broad range of enzymes for different monosaccharide types, linkage anomericity, linkage position, aglycon component (proteins, lipids, etc.), and monosaccharide stereoisomerism. Enzymes that remove particular monosaccharides from the non-reducing end of a glycan can be added sequentially to the enzymatic reaction, or they can be premixed and added simultaneously (Dwek et al. 1993). The most accurate and simplest approach is to analyze the structure of a single fractionated and purified N-glycan. The approach is also used for the analysis of complex mixtures, for example, a complete IgG N-glycome library. The method requires a supporting analytical technique that allows recognition of any possible changes introduced in the sequence of analyzed glycan. Techniques most often used in this type of analysis are separation-based ones which detect signals from fluorescent labels (capillary electrophoresis, ultra-high-performance liquid chromatography) and monitor shifts in retention time reflecting potential digestions of glycosidic bonds. Although not applied as often as the previously mentioned combination, MS techniques can also successfully be applied to monitor the enzymatic reactions. MALDI is an interesting option as it allows, with its speed and simplicity of sample preparation and efficient evaluation of the assay (Matsumoto et al. 2000; Kannicht et al. 2019). There is no need for chromatographic separation, and the combination of the information about applied exoglycosidase with the accompanied monosaccharide mass loss can provide valuable information about sequence and the type of glycosidic linkage.

3.5 Selected Approaches for IgG Glycosylation Analysis

3.5.1 MALDI–MS

Matrix-assisted laser desorption/ionization is a type of laser desorption ionization and one of the most widely used soft ionization techniques. Here, little energy is passed over to the molecule preventing uncontrolled fragmentation (Gross 2004). Compared to the second most widely used soft ionization technique—electrospray ionization (ESI), MALDI mass spectra are less complex to analyze since multiple charges are not common for MALDI ionization (Saba et al. 2002). Studies have demonstrated MALDI application in intact proteins, subclass site-specific, and IgG released *N*-glycans analysis (Huhn et al. 2009). The MS instruments with MALDI source are ideally combined with TOF and FT-ICR mass analyzers, although the latter is better suited for laser desorption ionization (LDI) MS (Gross 2004).

3.5.1.1 Intact Proteins and Glycopeptides

In the top-down and the middle-down approaches, the MALDI in-source decay (ISD) setup was reported as beneficial in the sequence analysis of undigested proteins. The ISD occurs upon transfer of hydrogen radical from the matrix to the radical-sites of the neutral analytes (Takayama 2016). This is achieved by fragmentation of the entire polypeptide chain as a result of hydrogen radical transfer from the matrix (Ayoub et al. 2013). Using MALDI–ISD, IgG can be analyzed in its denatured and reduced form after detergent, DTT, and IAA treatment. Following detergent removal, samples can be spotted on a MALDI target plate and analyzed with an FT-ICR mass spectrometer. Another approach combined IdeS digestion and separate glycosylation analysis of Fab and Fc portion (Tran et al. 2016). In this case, MALDI–ISD also enables the analysis of the N- and C-terminus protein sequence of the heavy chain as an advantage to the classical method, the Edman sequencing, which provides information only about the N-terminus. Additionally, Edman sequencing is limited to up to 50 amino acid residues, but with this method, coverage can be greater than 50 even without protein hydrolysis (Ayoub et al. 2013; Tran et al. 2016).

In the bottom-up approach, prior to MALDI-TOF IgG subclass-specific glycosylation analysis, IgG is usually digested with trypsin. Samples are then desalted, followed by HILIC–SPE glycopeptide enrichment (Selman et al. 2012b). One of the challenges generally arising in the MS analysis, including the MALDI approach, is the difference in the ionization efficiency of peptides and glycopeptides, so relative abundances of IgG subclasses do not always reflect their real abundances in biological samples (Selman et al. 2012b).

3.5.1.2 Released *N*-Glycans

Among other abovementioned approaches aimed at the analysis of released IgG *N*-glycans, MALDI–MS is widely used for this purpose due to its speed, simple sample preparation, and absence of sample separation. This ionization technique offers a possibility to analyze unmodified glycan moieties as well as glycans modified and labeled at the reducing end. Unmodified structures are usually observed in spectra as sodium adducts (Harvey 2001). The reducing end of glycans can be fluorescently labeled to avoid the appearance of intense metal adducts signals in spectra, which can usually be observed after chemical or enzymatic release (Gil et al. 2010). Since the detailed structure and linkage information about glycan species are available only by analysis of native or labeled glycans by MALDI–TOF–MS, separation by normal-phase high-performance liquid chromatography (NP–HPLC) can be introduced before the MALDI–MS analysis. Collected peaks can be treated with high purity exoglycosidases and subsequently analyzed by MS to elucidate the sequence and linkage of glycan moieties (Qian et al. 2007).

Acidic glycans such as sialylated ones are more prone to the production of sodium and potassium adducts, so it is crucial to remove additional salts from the sample before MS analysis (Harvey 2001). One option is resin beads that can be mixed with the matrix (Harvey 2001). Alternatively, cation-exchange resins can also be used for purification before sample application onto the MALDI target (Saba et al. 2002). Dialysis on microfiltration membrane and drop-dialysis are also methods of choice for desalting the samples before applying them onto the target plate and MALDI-Q-TOF analysis (Qian et al. 2007). Besides salts, excess derivatization reagents can be removed in a post-reaction work-up using C_{18} or HILIC cartridges (Gil et al. 2010).

3.5.1.3 Sialic Acid Stability

The sialic acids glycosidic bonds are more prone to hydrolysis than those of other monosaccharides, and chemical derivatization is readily used for stabilization. There are methods reported for both the derivatization of sialic acids for released *N*-glycans and glycopeptides. The different ionization efficiencies of neutral and acidic glycans are the major problem in the analysis of complex glycan mixtures. Various methods can be employed to neutralize charged glycans, such as amidation in mildly acidic conditions or permethylation (Krenkova et al. 2013). With permethylation, sensitivity can also be improved (Harvey 1996). Methyl esterification and amidation have the disadvantage of incomplete modification of α(2,3)-linked sialic acid. As an alternative, acetohydrazide and 1-ethyl-3(-3-dimethylaminopropyl) carbodiimide hydrochloride (EDC) can be used to treat glycoproteins since aldehydes of the reducing ends of glycans are prone to react with hydrazide. Another problem could be the precipitation and aggregation due to low protein isoelectric point, but so far, that wasn't reported with recombinant mAbs since they have a basic pI (Gil et al. 2010). A carboxylic activator EDC and 1-hydroxybenzotriazole (HOBt) in methanol or ethanol as a catalyst results in ethyl-esterification and lactonization of sialic acids. This derivatization method was applied on released N-glycans resulting in better sialic acid stability and additional information about their linkage since ethyl ester was formed with structures carrying α(2,6)-linked sialic acid (+14.016 Da) and lactone was the result of the reaction of α(2,3)-linked sialic acid with the neighboring galactose (−18.011 Da). For glycopeptides, the situation is different. Dimethylamine was introduced to additionally stabilize sialylated species and also to produce a larger mass difference. The reaction products are dimethylamide with α(2,6)-linked sialic acids (+27.047 Da) and a lactone with α(2,3)-linked sialic acids (−18.011 Da) (De Haan et al. 2015). Besides sialic acid stabilization, these derivatization methods enable a distinction between α(2,3) and α(2,6) linkages, which adds value to the characterization of IgG sialoglycans.

3.5.1.4 Matrix Substances for MALDI–MS

A proper matrix choice plays an essential role in ISD prevention and sialic acid loss (Selman et al. 2012b). Sample and the UV-absorbing matrix are both applied to the target and allowed to co-crystallize. In order to reduce crystal size and improve glycan solubility, recrystallization from ethanol is often applied (Harvey 2001). Among a wide variety of matrices, some of them are commonly used in IgG N-glycans analysis, such as 2,5-dihidroxybenzoic acid (DHB) or super-DHB, which contains 10% 2-hydroxy-5-methoxybenzoic acid (Harvey 2001). For glycopeptide spotting, a good performance regarding signal intensity is obtained with α-cyano-4-hydroxycinnamic acid (4-HCCA), either alone or mixed with nitrocellulose. For large proteins and glycoproteins, sinapinic acid is a matrix of choice (Harvey 2001). 4-HCCA has characteristics of a "hot" matrix, such as high proton affinity and extensive analyte fragmentation, and can cause a complete degradation of sialylated glycopeptides (Zauner et al. 2013). Therefore, two "cold" matrices—DHB and 4-chloro-α-cyanocinnamic acid (Cl-CCA)—were tested and enabled detection of intact sialylated glycopeptides, especially in a negative mode. In addition, Cl-CCA showed a better performance in the detection of highly sialylated species compared to DHB. These findings revealed significant benefits of Cl-CCA use in glycopeptides analysis (Selman et al. 2012b).

3.5.1.5 MALDI–MS for High-Throughput and Quantitative Analysis

Protocols for IgG glycosylation analysis by MALDI–MS were optimized for high-throughput applications, and some were also automated (Shubhakar et al. 2016; Bailey et al. 2005). In 2005, a high-throughput platform for MALDI glycopeptide analysis of recombinant mAbs was developed on microplates (Bailey et al. 2005). Hence, purification with C_{18} cartridges can be easily translated to a high-throughput mode, and the MALDI-TOF instrument can be coupled with an MS^n instrument in order to perform fragmentation on multiple levels and obtain as many details about the analyzed structure as possible (Huang et al. 2017). Furthermore, the high-throughput protocol for *N*-glycan analysis can be automated using HILIC–SPE enrichment, permethylation, and liquid–liquid extraction, followed by MALDI-TOF–MS (Shubhakar et al. 2016).

Processing a large amount of data is always a challenge, especially in high-throughput settings. Data analysis, extraction, and quantification can be done using a number of commercially available software packages or with specifically developed tools distributed under free software licenses, like MALDIquant, Mass-Up, or MassyTools (Jansen et al. 2015; Gibb and Strimmer 2012; López-Fernández et al. 2015).

MALDI–MS techniques are considered semi-quantitative since glycan species ionization is structure-dependent (Saba et al. 2002).

3.5.2 LC–MS for IgG Glycosylation Analysis

3.5.2.1 Coupling LC to MS for Enhanced Separation and Structural Characterization

Glycans show a high degree of structural heterogeneity with various linkage arrangements and monosaccharide branching within the molecule. Such complexity causes difficulties in isomer separation and makes structural analysis challenging. Often a single analytical technique cannot untangle complex mixtures of glycans and define their structural properties (Wuhrer et al. 2009). In order to unravel the complete glycan structure, multiple approaches should be combined to obtain complete information on linkage specificity, composition, isobaric, and isomeric structures (Mendez-Huergo et al. 2014; Miura and Endo 2016; Abrahams et al. 2018).

Although mass spectrometry proved to be a powerful tool for protein glycosylation analyses due to its high sensitivity, speed, and robustness, methods that rely only on accurate mass are unable to distinguish structural isomers (Zaia 2010). Besides structural isomers of released glycans, protease digested glycopeptides can also form various isomeric peptide-glycan combinations. Since antibodies can have one (IgG) to six (IgE) Fc glycosylation sites, there is a large number of possible peptide-glycan combinations that produce the same mass (Arnold et al. 2007).

Mass spectrometers equipped with MALDI sources often require the derivatization of sialic acids to avoid their in-source decay (Luo et al. 2009). On the contrary, in-source fragmentation of glycans can be easily controlled on ESI instruments (Wuhrer et al. 2009). Possible loss of sialic acids or a complete arm from the glycan core can be efficiently diagnosed and prevented by adjusting ESI source voltages (Huhn et al. 2009; Stadlmann et al. 2008). Another way to decrease in-source fragmentation is the use of buffered solutions. However, this also decreases signal intensities when compared to acidic aqueous solvents (Huhn et al. 2009). Negatively charged glycans containing sialic acids have a low ionization efficiency and are underrepresented in positive mode MALDI analysis. Similar holds true for native or glycans tagged with 2-AB in a negative ion MALDI mode. In this case, again, neutral and charged glycans in both ESI–MS polarities show a higher ionization efficiency and give more representative MS spectra when compared to MALDI–MS (Wuhrer et al. 2005). Ionization processes within ESI sources used for LC–MS predominantly form multiply protonated species, which, in positive mode collision-induced dissociation analysis, do not form cross-ring fragments required for detailed structure determination. Higher-order tandem MS methods should be used to untangle the structural features of glycans or glycopeptides.

Released glycans have 1–2 orders of magnitude lower ionization efficiency when compared to peptides (Karas et al. 2000). For this reason, glycans are usually derivatized to increase ionization efficiency. Furthermore, in complex mixtures, ionization of glycopeptides or glycans tends to be suppressed by peptides through competitive ionization and detector saturation (Luo et al. 2009; Hu and Mechref 2012). Therefore, using LC coupled to MS is beneficial for separating samples with potentially limited availability (Luo et al. 2009).

An efficient way to distinguish isomeric structures is by coupling various liquid chromatography methods with mass spectrometry (LC–MS). LC–MS plays an important role in the separation of isomeric structures indistinguishable for MS. Accordingly, a common liquid chromatography method for separating enzymatically released glycan mixtures includes HILIC and PGC (Mauko et al. 2012; Michael and Rizzi 2015; Ruhaak et al. 2009; Saldova et al. 2014). Reversed-phase chromatography is routinely used for the separation of IgG glycopeptides (Selman et al. 2012a), antibody fragments produced by complete or incomplete proteolysis with various proteases, heavy chains obtained by reduction of antibody, and complete intact antibodies (Zhang et al. 2014). Derivatized glycans released from protein can also be analyzed using RP–LC–MS (Abrahams et al. 2018; Sinha et al. 2008). Various mass analyzers such as time-of-flight detectors, triple quadrupoles, ion traps, Orbitraps, and IT-FCRs can be coupled with LC.

3.5.2.2 HILIC–UHPLC–MS

By this approach, a mixture of charged, highly hydrophilic, and uncharged glycans can be easily retained and separated on the column. Glycans can be analyzed either underivatized or derivatized with fluorescent dyes or other reagents. This technique was introduced by Alpert in 1990 and is widely used in various glycomic studies (Reusch et al. 2015a, b; Alpert 1990). HILIC is considered a variant of normal-phase liquid chromatography (NP-LC) because it contains a stationary phase that is more polar than the mobile phase (Buszewski and Noga 2012). Nevertheless, mobile phases used for HILIC differ from solvents used in NP-LC (Hemström and Irgum 2006). Glycans are separated in a gradient changing from a high percentage organic solvent used for binding glycans to the column to a high percentage of aqueous phase (buffer) to separate and elute glycans (Buszewski and Noga 2012). Retention times of analytes on HILIC stationary phases are affected by various interactions, which include hydrogen bonding, ionic interactions, and dipole-dipole interactions (Wuhrer et al. 2009). Separation is achieved through the interactions between the thin layer of water coating the polar adsorbent and the mobile phase that contains an organic solvent. The salt concentration and pH are important factors governing charged stationary phase surface properties (Veillon et al. 2017).

There are various stationary phases for HILIC columns, of which non-ionic amide-modified columns are mostly used for glycan separation (Wuhrer et al. 2009). Since there is no mass difference between structural isomers, coupling HILIC to MS can further resolve a number of isobaric structures (Butler et al. 2003). Mobile phases for HILIC separations are compatible with mass spectrometers because elution of glycans from the column occurs at a low percentage of buffer containing volatile salts and a high ratio of organic solvent, which enables its relatively easy evaporation and ionization (Wuhrer et al. 2009).

HILIC–ESI–LC–MS is suitable for the analysis of native glycans as well as glycans reduced into alditols or labeled with fluorescent or UV-absorbing dye (Wuhrer et al. 2005). Besides the benefits of high sensitivity for fluorescence detection, fluorophores (see Table 3.3) also increase ionization efficiency and thus the sensitivity of MS detection. Detailed HILIC-MS analysis of commercially available mAb (IgG4) and IgG from human serum was described by Zhao et al. (2016). Fab and Fc fragments were enzymatically separated using specific proteases such as IdeS and further individually analyzed to obtain information on Fab and Fc-specific glycan profiles (Janin-Bussat et al. 2013).

Besides amide/amine HILIC columns, zwitterionic HILIC (ZIC–HILIC) columns can also be used for the separation of released glycans (Mauko et al. 2012). ZIC–HILIC columns showed capability for separation of 2-AP derivatized glycans from human serum IgG, while ZIC–HILIC coupled with ESI–MS can be used for separation and structural elucidation of reduced glycans from mAbs (Takegawa et al. 2006; Mauko et al. 2011). Mauko et al. showed that the ZIC–HILIC column displayed similar results when compared to amide HILIC separation but with a shorter separation time. Also, the use of acetic acid instead of ammonium formate buffer enabled easier coupling to MS (Mauko et al. 2012).

Besides released antibody glycans analysis, HILIC columns can be used for the separation of glycopeptide digests. Although tandem MS analysis of released glycans can give complete information on glycan structure, it does not give any information about protein and glycosylation site. HILIC-MS of glycopeptides provides information on peptide backbone carrying glycan and site-specificity of glycosylation (Wuhrer et al. 2009). Historically, HILIC separation has been traditionally part of 2D chromatography, where the first dimension was RP separation of peptides followed by HILIC separation, which enabled enrichment and separation of glycopeptides (Wuhrer et al. 2009; Zhang and Wang DI c. 1998). HILIC-MS can be performed at the nanoscale and capillary scale, providing femtomole range sensitivity (Wuhrer et al. 2005). IgG tryptic digests separated on ZIC–HILIC column results in the separation of glycopeptides based both on peptide backbone and charge of attached glycans (Takegawa et al. 2006).

3.5.2.3 RP–LC–MS

The most common RP–LC–MS setup includes an RP pre-column (trapping column) that enables online desalting and sample preconcentration, an analytical column, and a Q-TOF instrument equipped with an ESI ion source (Stadlmann et al. 2008). Formic acid is a typical additive in aqueous solvents in gradient separations with increasing organic solvent content (acetonitrile, ACN) (Wuhrer et al. 2007b; Hirayama et al. 1998; Takakura et al. 2014). Trifluoroacetic acid (TFA) can be used instead of formic acid to prevent separation of the same peptides carrying charged and neutral glycans; however, it has a strong suppressive effect on ionization (Shou and Naidong 2005). In order to overcome this problem, propionic acid can be introduced after the analytical column, either using a tee connector or sheath flow ESI sprayer (Selman et al. 2012a).

Endopeptidases, such as pronase, Lys-C, Glu-C, and the most widely used trypsin, are applied for the digestion of immunoglobulins (Huhn et al. 2009). The resulting glycopeptides are combinations of various peptide sequences and attached glycan structures yielding sometimes indistinguishable mass isomers. In the case of trypsin-digested human IgG, three Fc glycopeptides can be separated using RP-LC based on differences in their peptide sequences: IgG1 (EEQY*N*STYR), IgG2 (EEQF*N*STFR), and IgG4 (EEQF*N*STYR). Tryptic peptides from IgG3 allotypes present in Caucasian populations (EEQF*N*STFR) have amino acid composition and sequence that are identical to the IgG2 peptide. On the other hand, tryptic peptides from IgG3 allotypes predominant in Asian and African populations (EEQY*N*STFR) have a different sequence but the same amino acid composition as the IgG4 peptide. In both cases, RP–LC–MS is unable to separate peptides with the same amino acid content, which is the major drawback of this approach (Dard et al. 2001). The subclasses IgG2 and IgG4 or IgG3 and IgG4 thus have to be quantified together.

Source settings considerably affect the analyte charge state. Tryptic IgG glycopeptides typically ionize as doubly, $[M + 2H]^{2+}$, and triply, $[M + 3H]^{3+}$, charged protonated species (Stadlmann et al. 2008; Wuhrer et al. 2007b; Olivova et al. 2008). High concentrations of coeluting peptides can suppress the ionization of targeted glycopeptides, while the measurement of glycopeptides in complex samples is complicated by different ionization efficiencies. Also, protein digests can include peptides or incompletely digested (glyco)peptides with masses overlapping with targeted glycopeptides which additionally complicates the analysis. For these reasons, it is common to use purified antibodies, which are enriched and desalted before digestion (Huhn et al. 2009).

Glycopeptide analysis of other antibody classes is more challenging than that of IgG. The presence of multiple glycosylation sites often requires the use of more than one protease to obtain glycopeptides of optimal length. An LC–ESI–MS/MS method for the analysis of recombinant secretory IgA complexes using digestion with trypsin followed by Glu-C has been described by Paul et al. (Paul et al. 2014; Huang et al. 2015). Associations of effector function IgA subclasses with their glycosylation profiles were recently demonstrated by Steffen et al. (2020). Similar glycosylation analysis of IgM showed the presence of high-mannose and complex glycans on the heavy chain of the molecule (Loos et al. 2014). Immunoglobulin E, an antibody with six glycosylation sites on each heavy chain mainly occupied by high-mannose and complex-type glycans, can be treated with trypsin, proteinase K, or chymotrypsin to obtain optimal glycopeptides from all glycosylation sites (Plomp et al. 2014).

Reversed-phase liquid chromatography coupled to mass spectrometry is an alternative method for analysis of released glycans, in which separation is achieved through non-covalent interactions with non-polar stationary phases such as C_8 and C_{18} (Vreeker and Wuhrer 2017). There are numerous C_{18} stationary phases with various influences on retention times. One of the advantages of RP-LC is the high compatibility of solvents with MS because they include water, acid (typically 0.1% formic acid), and organic solvent (ACN). Glycans are bound to the stationary phase in the high aqueous phase, with separation occurring at a slightly increased percentage of organic solvent. Native glycans, due to their hydrophilic nature, are not retained on RP columns even under high-salt and low-pH conditions (Fan et al. 1994). Therefore, binding glycans to alkyl chains of the C_{18} column requires derivatization to increase their hydrophobicity. Retention times and separation are highly dependent on the stationary phase, mobile phase, and hydrophobicity of derivatives used. Permethylation is a commonly used derivatization technique preceding glycan analysis on RP–LC due to a significant increase in hydrophobicity. Also, permethylated glycans tend to have higher ionization efficiencies in ESI sources and increased stability of attached sialic acids (Ruhaak et al. 2010a; Vreeker and Wuhrer 2017). Various studies demonstrated the use of RP–LC separation of permethylated IgG glycans (Zhou et al. 2016; Ritamo et al. 2013). However, the presence of side reactions, which are especially favored with low sample amounts, causes problems with the reproducibility of this method (de Haan et al. 2020; Kang et al. 2005). Various labels can be used in RP–MS analysis of IgG *N*-glycans. The most commonly used label for glycans, 2-AB, slightly increases hydrophobicity and requires long runs to achieve separation (Chen and Flynn 2007; Adamczyk et al. 2014). This means that charged glycans labeled with 2-AB are poorly separated on RP-LC columns, which can be fixed by the addition of ion-pairing agents (Higel et al. 2013; Melmer et al. 2011). On the other hand, 2-AA provides a good ionization efficiency, especially for negative mode MS, but also enables separation of glycans based on their type, presence of core fucose, and sialic acids (Higel et al. 2013, 2014). Besides 2-AA, 8-aminonaphthalene-1,3,6-trisulfonic acid (ANTS) is another negatively charged dye convenient for a negative mode MS analysis (Higel et al. 2013). Finally, 2-aminopyridine (2-AP) has the lowest hydrophobicity resulting in rapid eluting glycan peaks. However, it is less compatible with MS since higher amounts of hydrophilic solvent are required (Pabst et al. 2009).

The separation of structural isomers using RP represents a challenging task. Isomeric complex glycan structures generate similar interactions with the stationary phase. The differences between such isomers are not large enough to cause shifts in hydrophobicity which could significantly impact retention properties and achieve baseline separation. Therefore, the most common way to separate glycans on RP is by using shallow linear gradients or even isocratic runs, which take a significant amount of time and ultimately provide lower peak capacities, i.e., the largest number of equally resolved peaks that can be fit into the separation interval (Melmer et al. 2011). Because of the abovementioned issues, the separation of IgG *N*-glycan isomers is mainly done using HILIC and PGC stationary phases (Veillon et al. 2017).

3.5.2.4 PGC for Enhanced Isomeric Glycan Separation

Porous graphitized carbon liquid chromatography (PGC–LC) is a high-resolution method for the separation of complex glycan mixtures because of its ability to efficiently separate isomeric glycan structures (Abrahams et al. 2018). Glycans bind to PGC through various hydrophobic, polar, and ionic interactions (Fan et al. 1994; Pabst and Altmann 2008a). Additionally, the planarity of graphite type carbon used for the stationary phase in PGC columns enables the separation based on the three-dimensional structure of the molecule (West et al. 2010). On the contrary to RP–LC, polar analytes are strongly retained on the column, which enables the analysis of native, reduced, labeled, or permethylated glycans. Mobile phases used for PGC include organic solvents and aqueous solutions containing acids, bases, or volatile buffers, which makes this separation method compatible with MS (Ruhaak et al. 2009).

PGC–LC–ESI–MS is commonly used for detailed structural analysis of underivatized glycans due to its ability to separate anomers (Abrahams et al. 2018). Although glycan permethylation can be used to increase ionization of sialylated glycans, this derivatization results in inadequate separation, which can be solved by increasing column temperature (Costello et al. 2007; Zhou et al. 2017). The mobile phase choice is crucial for proper separation since its pH and buffer capacity affect the retentivity on the column by determining the ionization state of both the stationary phase and the analyte in solution (Bapiro et al. 2016). During the PGC–LC run, larger high-mannose glycans elute first, while multi-sialylated glycans are strongly retained. Better retention of sialylated glycans can be achieved by using ion-pairing agents such as TFA (Melmer et al. 2011). Native and derivatized glycans will dominantly produce proton adducts, followed by less abundant sodium, potassium, and ammonium adducts (Pabst and Altmann 2008a; Zhou et al. 2017). However, PGC columns are susceptible to contamination which causes column deterioration and affects the retention of analytes (Bapiro et al. 2016). There are also reported interferences between the electric field of the ESI source with the separation on PGC (Carrasco et al. 2001). The solution for this problem was proposed by Pabst et al., who reduced it by the electrical grounding of the column. This research group also presented the importance of ionic strength on the peak shape of charged glycans and studied the effects of solvent, temperature, and ion polarity on the detection and elution of charged glycans. Changes in pH did not affect ionization efficiency in both polarities, although increased retention time of charged glycans was observed. Finally, a high concentration of organic solvent was found to be required to obtain better intensities for charged glycans in negative mode (Pabst and Altmann 2008b). Use of PGC–LC–MS for glycosylation analysis of polyclonal human IgG and commercial mAb was described by Stadlmann et al., demonstrating separation of neutral and sialylated glycans released from antibodies within a single run (Stadlmann et al. 2008). Abrahams et al. obtained glycosylation profiles of various glycoproteins, including IgG and IgA, and demonstrated the need for a PGC retention time library (Abrahams et al. 2018; Stadlmann et al. 2008). Glycopeptides can also be separated on PGC. Nevertheless, there are reports of difficulties eluting sialylated glycopeptides from the column (de Haan et al. 2020). Antibody glycopeptide analysis using PGC published by Huang et al. included comprehensive site-specific glycosylation analysis of Pronase E-derived glycopeptides originating from human colostrum IgA complexes (Huang et al. 2015).

3.5.2.5 Anion Exchange LC–MS

Anion exchange separation techniques can be utilized for the separation of glycans since they are weak acids (Cataldi et al. 2000). High-pH anion exchange chromatography (HPAEC) coupled with MS can be used to obtain glycan structural information. Stationary phases used in HPAEC columns are polymer-based and are stable in a wide pH range. Advantages of using this method were presented by Maier et al., who described structural assignment and quantification of IgG glycans using HPAEC coupled to MS (Maier et al. 2016). Usually, glycans are separated under alkaline conditions, and there is no need for derivatization. Retention on the column is primarily affected by analyte formal charge, size, linkages, and composition (Veillon et al. 2017). The coupling of HPAEC with MS can be problematic due to the high-salt concentrations, which significantly suppress the ionization of glycans. A desalter such as a microfluidic membrane suppressor can be coupled to an alkaline mobile phase to enable sodium-free ionization conditions. Although it is not as widely used as the other methods, its separation mechanism and selectivity make it an interesting alternative to PGC-, RP-, and HILIC-based chromatography (Maier et al. 2016).

3.5.2.6 Challenges of Miniaturization

Separation, throughput, and sensitivity can be drastically altered by changing column variables such as diameter, length, and particle size of stationary phase (Vreeker and Wuhrer 2017). C_{18} nano-LC–MS is a commonly used method for peptide separation. Even in the case of glycan analysis, this method gives higher sensitivity, separation efficiency, and resolution when compared to conventional LC-columns-based methodology (Veillon et al. 2017). Compared to traditional LC–MS, analysis of 2-AB labeled glycans analyzed on nano-LC–MS systems were reported to have ten times higher sensitivity. When compared to MALDI–TOF–MS of NP-LC fractionated and collected peaks, the increase in sensitivity is 100-fold. Nevertheless, nano-LC systems display higher instability of retention times between injections. In the case of fluorescently labeled glycans, this can be easily corrected using a dextran ladder retention time recalibration (Wuhrer et al. 2005; Royle et al. 2002; Rudd et al. 2001). In addition, nano-LC chromatography runs are significantly longer and, due to smaller inner diameters of capillaries and columns fluidic parts, are more susceptible to clogging. Moreover, tight fluidic connections must be considered because even the smallest dead volume can severely impact retention times and peak shape. A limited amount of sample can be injected, but this can be solved by using trapping columns on which higher flow rates and, therefore, larger sample volumes can be applied (Vreeker and Wuhrer 2017). Due to the aforementioned issues limiting nanoscale chromatography, antibody glycosylation analyses are often performed on micro and capillary-LC scale (Huhn et al. 2009; Stadlmann et al. 2008; Rehder et al. 2006; Huang et al. 2005; Harazono et al. 2008).

3.5.3 Capillary Electrophoresis-Mass Spectrometry

Capillary electrophoresis (CE) as a separation technique for IgG glycosylation analysis can be used independently or coupled to MS detection (Volpi and Maccari 2013). The separation of analytes is based on charge, size, and shape properties (Volpi and Maccari 2013), and the main advantage is the separation of positional isomers (Zaia 2013). Several CE techniques are available: capillary zone electrophoresis (CZE), capillary gel electrophoresis (CGE), capillary isoelectric focusing (CIEF), isotachophoresis (ITP), micellar electrokinetic capillary chromatography (MECC/MEKC), microemulsion electrokinetic chromatography (MEEKC), nonaqueous capillary electrophoresis (NACE), and capillary electrochromatography (CEC) (Volpi and Maccari 2013). However, the number of CE techniques that can be coupled to MS is somewhat limited since solvents that can be used for efficient CE separation are, in that case, restricted to volatile buffer systems. Therefore, CZE is the technique that is most often a part of CE–MS (Huang et al. 2017). CE–MS coupling can be done in an online or an offline setup. The two most common ionization techniques used are ESI (positive and negative ion mode) preferentially for the online coupling and MALDI for the off-line setup.

The online coupling has been done with different types of interfaces: coaxial sheath flow interface, the liquid junction, or sheathless interface (Zhong et al. 2014; Gennaro and Salas-Solano 2008). Coaxial sheath flow is used at the distal end of the capillary with a flow of a few μL/min which can cause a dilution of analytes migrating from the capillary. To overcome this issue and provide sufficient flow (below 0.5 nL/min) for the MS source at the same time, a liquid junction interface can be used (Zaia 2013). Sheathless interface is used in normal polarity separation, where the use of buffer at high-pH values promotes ionization. On the other hand, the introduction of acidic buffers is beneficial for reverse polarity analysis in combination with the coating of capillary and pressure assistance (Bindila et al. 2005). Bare fused silica capillaries are usually used to reduce the possibility of capillary coating bleeding into the MS source. Interaction between glycopeptides and the walls of uncoated capillaries can occur, so a washing procedure can be introduced after each CE–MS sample injection (Zaia 2013).

Analyses of both underivatized and derivatized glycans were reported in the literature. Jayo et al. optimized conditions for underivatized glycan analysis in reversed polarity mode and negative mode ESI–MS (Jayo et al. 2014). Besides APTS as the most common label used for derivatization via reductive amination, Fmoc-Cl is also an attractive option for derivatization prior to MS analysis. Excess of labeling reagent is usually removed by solid-phase extraction (SPE) prior to the analysis (Nakano et al. 2009). SPE can also be part of the inline SPE–CE–MS system mainly used for cations analysis. It can also be beneficial in the enrichment and analysis of the anions of strong acids like APTS-labeled glycans. These analytes showed 1000 times higher sensitivity in CE–LIF analysis compared to CE–MS. When applying an in-line SPE-CE-MS system for separation of APTS-labeled glycans in acidic background electrolyte (BGE), 800 times lower concentration of analytes can be used and still provide comparable intensities to standard CE–MS (Jooß et al. 2014). For ESI, derivatized glycans have the advantage of better ionization compared to native neutral oligosaccharides lacking easily ionizable functional groups. Specifically, APTS-labeled glycans can maintain a negative charge at low-pH values (Gennaro et al. 2006). Even though TOF mass analyzers were preferentially used in CE–MS coupling, a method using drift tube ion mobility (DTIM) was developed to analyze derivatized and native N-glycans. Accordingly, the CZE–DTIM–MS setup showed excellent separation and reproducibility (Jooß et al. 2019).

For the analysis of intact mAbs, a microfluidic CE–ESI–MS strategy was applied and was successful in the separation of intact mAbs charge variants. The main challenges arising here were to prevent analyte adsorption and to enhance electrophoretic mobility differences. High ionic strength buffers, surfactants, and additives in BGE have that role, and often triethylenetetramine and ε-aminocaproic acid are added for that purpose. However, the problem of their incompatibility with ESI remains. In a microfluidic strategy, uniform and stable aminopropyl silane-polyethylene glycol surface coating was used, which eliminated additives, enabled MS-compatible BGE, and resulted in the highly efficient separation of intact mAbs (Redman et al. 2015). Capillary electrophoresis offers many opportunities for glycan and glycopeptide analysis, even more as new solutions emerge regarding the compatibility and improvement in coupling CE techniques to MS detection.

3.6 Perspectives

An accelerating transition from traditional to precision medicine has transformed how fundamental research translates from lab bench to patient bedside care and the other way round. The increasingly popular comprehensive multi-omics approaches for studying biological fingerprints require the fast development of novel analytical methods followed by big data mining, integration, and network analysis. These breakthroughs can reveal hidden insights into patterns and relationships entangling all biological organization levels from the molecular and single-cell to the population level on the pathway to individualized therapy.

Mass spectrometry is no exception here, although its implementation into the clinic was slow due to complex instrumentation, lengthy protocols, and challenging interpretation of the results with shadowy meaning to the medical practitioners. Miniaturization, more robust analytical workflows, sample multiplexing, and technical advancements in the field of chromatography made mass spectrometry more tempting for widespread use in protein post-translational modifications analysis, including glycosylation. In this chapter, a concise methodological overview of IgG—a central immune system glycoprotein—N-glycosylation analysis was given. In the past decade, extensive large-cohort studies revealed changes in IgG glycosylation to be correlating with numerous pathological states. This promisingly positions immunoglobulin glycans as potential biomarker candidates and paves the way for the advent of new disciplines such as pharmacoglycomics or glycoimmunology, all made possible by the invention of a mass spectrometer.

Compliance with Ethical Standards

Funding Conflict of Interest: Siniša Habazin, Jerko Štambuk, Jelena Šimunović, Genadij Razdorov and Mislav Novokmet are employees of Genos Ltd., a privately held company specialized in commercial high-throughput glycan analysis. Toma Keser declares that he has no conflict of interest.

Ethical Approval: This book chapter does not contain any studies with human participants or animals performed by any of the authors.

References

Abrahams JL, Campbell MP, Packer NH (2018) Building a PGC-LC-MS N-glycan retention library and elution mapping resource. Glycoconj J 35:15–29

Adamczyk B, Tharmalingam-Jaikaran T, Schomberg M, Szekrényes Á, Kelly RM, Karlsson NG, Guttman A, Rudd PM (2014) Comparison of separation techniques for the elucidation of IgG N-glycans pooled from healthy mammalian species. Carbohydr Res 389:174–185

Aebersold R, Mann M (2003) Mass spectrometry-based proteomics. Nature 422:198–207

Alagesan K, Khilji SK, Kolarich D (2017) It is all about the solvent: on the importance of the mobile phase for ZIC-HILIC glycopeptide enrichment. Anal Bioanal Chem 409:529–538

Albrecht S, Mittermayr S, Smith J, Martín SM, Doherty M, Bones J (2017) Twoplex 12/13C6 aniline stable isotope and linkage-specific sialic acid labeling 2D-LC-MS workflow for quantitative N-glycomics. Proteomics 17:1600304

Alley WR, Mechref Y, Novotny MV (2009) Characterization of glycopeptides by combining collision-induced dissociation and electron-transfer dissociation mass spectrometry data. Rapid Commun Mass Spectrom 23:161–170

Alpert AJ (1990) Hydrophilic-interaction chromatography for the separation of peptides, nucleic acids and other polar compounds. J Chromatogr A 499:177–196

Amano M, Nishimura S-I (2010) Large-scale glycomics for discovering cancer-associated N-glycans by integrating glycoblotting and mass spectrometry. Methods Enzymol 478:109–125

An HJ, Lebrilla CB (2011) Structure elucidation of native N- and O-linked glycans by tandem mass spectrometry (tutorial). Mass Spectrom Rev 30:560–578

Angel PM, Lim J-M, Wells L, Bergmann C, Orlando R (2007) A potential pitfall in 18O-based N-linked glycosylation site mapping. Rapid Commun Mass Spectrom 21:674–682

Arnold JN, Wormald MR, Sim RB, Rudd PM, Dwek RA (2007) The impact of glycosylation on the biological function and structure of human immunoglobulins. Annu Rev Immunol 25:21–50

Ashline DJ, Lapadula AJ, Liu Y-H, Lin M, Grace M, Pramanik B, Reinhold VN (2007) Carbohydrate structural isomers analyzed by sequential mass spectrometry. Anal Chem 79:3830–3842

Atwood JA, Cheng L, Alvarez-Manilla G, Warren NL, York WS, Orlando R (2008) Quantitation by isobaric labeling: applications to glycomics. J Proteome Res 7:367–374

Ayoub D, Jabs W, Resemann A, Evers W, Evans C, Main L, Baessmann C, Wagner-Rousset E, Suckau D, Beck A (2013) Correct primary structure assessment and extensive glyco-profiling of cetuximab by a combination of intact, middle-up, middle-down and bottom-up ESI and MALDI mass spectrometry techniques. mAbs 5:699–710

Bailey MJ, Hooker AD, Adams CS, Zhang S, James DC (2005) A platform for high-throughput molecular characterization of recombinant monoclonal antibodies. J Chromatogr B Anal Technol Biomed Life Sci 826:177–187

Bapiro TE, Richards FM, Jodrell DI (2016) Understanding the complexity of porous graphitic carbon (PGC) chromatography: modulation of mobile-stationary phase interactions overcomes loss of retention and reduces variability. Anal Chem 88:6190–6194

Bi C, Yuan Y, Tu Y, Wu J, Liang Y, Li Y, He X, Chen L, Zhang Y (2020) Facile synthesis of hydrophilic magnetic graphene nanocomposites via dopamine self-polymerization and Michael addition for selective enrichment of N-linked glycopeptides. Sci Rep 10:71. https://doi.org/10.1038/s41598-019-56944-4

Bigge JC, Patel TP, Bruce JA, Goulding PN, Charles SM, Parekh RB (1995) Nonselective and efficient fluorescent labeling of glycans using 2-amino benzamide and anthranilic acid. Anal Biochem 230:229–238

Bindila L, Peter-Katalinić J, Zamfir A (2005) Sheathless reverse-polarity capillary electrophoresis-electrospray-mass spectrometry for analysis of underivatized glycoconjugates. Electrophoresis 26:1488–1499

Bodnar ED, Perreault H (2015) Synthesis and evaluation of carboxymethyl chitosan for glycopeptide enrichment. Anal Chim Acta 891:179–189

Bodnar J, Szekrenyes A, Szigeti M, Jarvas G, Krenkova J, Foret F, Guttman A (2016) Enzymatic removal of N-glycans by PNGase F coated magnetic microparticles. Electrophoresis 37:1264–1269

Boesl U (2017) Time-of-flight mass spectrometry: introduction to the basics. Mass Spectrom Rev 36:86–109

Bondt A, Rombouts Y, Selman MHJ, Hensbergen PJ, Reiding KR, Hazes JMW, Dolhain RJEM, Wuhrer M (2014) Immunoglobulin G (IgG) fab glycosylation analysis using a new mass spectrometric high-throughput profiling method reveals pregnancy-associated changes. Mol Cell Proteomics 13:3029–3039

Buszewski B, Noga S (2012) Hydrophilic interaction liquid chromatography (HILIC)-a powerful separation technique. Anal Bioanal Chem 402:231–247

Butler M, Quelhas D, Critchley AJ et al (2003) Detailed glycan analysis of serum glycoproteins of patients with congenital disorders of glycosylation indicates the specific defective glycan processing step and provides an insight into pathogenesis. Glycobiology 13:601–622

Bynum MA, Yin H, Felts K, Lee YM, Monell CR, Killeen K (2009) Characterization of IgG N-glycans employing a microfluidic chip that integrates glycan cleavage, sample purification, LC separation, and MS detection. Anal Chem 81:8818–8825

Carrasco B, Garcia De La Torre J, Davis KG, Jones S, Athwal D, Walters C, Burton DR, Harding SE (2001) Crystallohydrodynamics for solving the hydration problem for multi-domain proteins: open physiological conformations for human IgG. Biophys Chem 93:181–196

Cataldi TRI, Campa C, De Benedetto GE (2000) Carbohydrate analysis by high-performance anion-exchange chromatography with pulsed amperometric detection: the potential is still growing. Fresenius J Anal Chem 368:739–758

Chen X, Flynn GC (2007) Analysis of N-glycans from recombinant immunoglobulin G by on-line reversed-phase high-performance liquid chromatography/mass spectrometry. Anal Biochem 370:147–161

Chen C-C, Su W-C, Huang B-Y, Chen Y-J, Tai H-C, Obena RP (2014) Interaction modes and approaches to glycopeptide and glycoprotein enrichment. Analyst 139:688–704

Chen Y, Sheng Q, Hong Y, Lan M (2019) Hydrophilic nanocomposite functionalized by carrageenan for the specific enrichment of glycopeptides. Anal Chem 91:4047–4054

Chen SW, Tan D, Yang YS, Zhang W (2020) Investigation of the effect of salt additives in Protein L affinity chromatography for the purification of tandem single-chain variable fragment bispecific antibodies. MAbs 12(1):1718440. https://doi.org/10.1080/19420862.2020.1718440

Chernushevich IV, Loboda AV, Thomson BA (2001) An introduction to quadrupole-time-of-flight mass spectrometry. J Mass Spectrom 36:849–865

Choe W, Durgannavar TA, Chung SJ (2016) Fc-binding ligands of immunoglobulin G: an overview of high affinity proteins and peptides. Materials 9:994

Collin M, Olsén A (2001) EndoS, a novel secreted protein from *Streptococcus pyogenes* with endoglycosidase activity on human IgG. EMBO J 20:3046–3055

Conboy JJ, Henion JD (1992) The determination of glycopeptides by liquid chromatography/mass spectrometry with collision-induced dissociation. J Am Soc Mass Spectrom 3:804–814

Cook KS, Bullock K, Sullivan T (2012) Development and qualification of an antibody rapid deglycosylation method. Biologicals 40:109–117

Costello CE, Contado-Miller JM, Cipollo JF (2007) A glycomics platform for the analysis of permethylated oligosaccharide alditols. J Am Soc Mass Spectrom 18:1799–1812

Dard P, Lefranc MP, Osipova L, Sanchez-Mazas A (2001) DNA sequence variability of IGHG3 alleles associated to the main G3m haplotypes in human populations. Eur J Hum Genet 9:765–772

De Haan N, Reiding KR, Haberger M, Reusch D, Falck D, Wuhrer M (2015) Linkage-specific sialic acid derivatization for MALDI-TOF-MS profiling of IgG glycopeptides. Anal Chem 87:8284–8291

de Haan N, Falck D, Wuhrer M (2020) Monitoring of immunoglobulin N- and O-glycosylation in health and disease. Glycobiology 30:226–240

de Hoffmann E (1996) Tandem mass spectrometry: a primer. J Mass Spectrom 31:129–137

Demelbauer UM, Zehl M, Plematl A, Allmaier G, Rizzi A (2004) Determination of glycopeptide structures by multistage mass spectrometry with low-energy collision-induced dissociation: comparison of electrospray ionization quadrupole ion trap and matrix-assisted laser desorption/ionization quadrupole ion trap reflectron time-of-flight approaches. Rapid Commun Mass Spectrom 18:1575–1582

Domon B, Costello CE (1988) A systematic nomenclature for carbohydrate fragmentations in FAB-MS/MS spectra of glycoconjugates. Glycoconj J 5:397–409

Dwek RA, Edge CJ, Harvey DJ, Wormald MR, Parekh RB (1993) Analysis of glycoprotein-associated oligosaccharides. Annu Rev Biochem 62:65–100

Eliuk S, Makarov A (2015) Evolution of orbitrap mass spectrometry instrumentation. Annu Rev Anal Chem (Palo Alto Calif) 8:61–80

Eriksson A, Norgren M (2003) Cleavage of antigen-bound immunoglobulin G by SpeB contributes to streptococcal persistence in opsonizing blood. Infect Immun 71:211–217

Everest-Dass AV, Abrahams JL, Kolarich D, Packer NH, Campbell MP (2013) Structural feature ions for distinguishing N- and O-linked glycan isomers by LC-ESI-IT MS/MS. J Am Soc Mass Spectrom 24:895–906

Faccio G (2018) From protein features to sensing surfaces. Sensors (Basel) 18(4):1204. https://doi.org/10.3390/s18041204

Fahnestock SR, Alexander P, Nagle J, Filpula D (1986) Gene for an immunoglobulin-binding protein from a group G streptococcus. J Bacteriol 167:870–880

Faid V, Leblanc Y, Bihoreau N, Chevreux G (2018) Middle-up analysis of monoclonal antibodies after combined IgdE and IdeS hinge proteolysis: investigation of free sulfhydryls. J Pharm Biomed Anal 149:541–546

Falck D, Jansen BC, Plomp R, Reusch D, Haberger M, Wuhrer M (2015) Glycoforms of immunoglobulin G based biopharmaceuticals are differentially cleaved by trypsin due to the glycoform influence on higher-order structure. J Proteome Res 14:4019–4028

Fan JQ, Kondo A, Kato I, Lee YC (1994) High-performance liquid chromatography of glycopeptides and oligosaccharides on graphitized carbon columns. Anal Biochem 219:224–229

Fenn JB, Mann M, Meng CK, Wong SF, Whitehouse CM (1989) Electrospray ionization for mass spectrometry of large biomolecules. Science (New York, NY) 246:64–71

Formolo T, Heckert A, Phinney KW (2014) Analysis of deamidation artifacts induced by microwave-assisted tryptic digestion of a monoclonal antibody. Anal Bioanal Chem 406:6587–6598

Gebrehiwot AG, Melka DS, Kassaye YM, Gemechu T, Lako W, Hinou H, Nishimura S-I (2019) Exploring serum and immunoglobulin G N-glycome as diagnostic biomarkers for early detection of breast cancer in Ethiopian women. BMC Cancer 19:588

Gennaro LA, Salas-Solano O (2008) On-line CE-LIF-MS technology for the direct characterization of N-linked glycans from therapeutic antibodies. Anal Chem 80:3838–3845

Gennaro LA, Salas-Solano O, Ma S (2006) Capillary electrophoresis-mass spectrometry as a characterization tool for therapeutic proteins. Anal Biochem 355:249–258

Gevaert K, Impens F, Ghesquière B, Van Damme P, Lambrechts A, Vandekerckhove J (2008) Stable isotopic labeling in proteomics. Proteomics 8:4873–4885

Ghitescu L, Galis Z, Bendayan M (1991) Protein AG-gold complex: an alternative probe in immunocytochemistry. J Histochem Cytochem 39:1057–1065

Giansanti P, Tsiatsiani L, Low TY, Heck AJR (2016) Six alternative proteases for mass spectrometry–based proteomics beyond trypsin. Nat Protoc 11:993–1006

Gibb S, Strimmer K (2012) MALDIquant: a versatile R package for the analysis of mass spectrometry data. Bioinformatics 28:2270–2271

Gil G-C, Iliff B, Cerny R, Velander WH, Van Cott KE (2010) High throughput quantification of N-glycans using one-pot sialic acid modification and matrix assisted laser desorption ionization time of flight mass spectrometry. Anal Chem 82:6613–6620

Goh JB, Ng SK (2018) Impact of host cell line choice on glycan profile. Crit Rev Biotechnol 38:851–867

Gong B, Hoyt E, Lynaugh H, Burnina I, Moore R, Thompson A, Li H (2013) N-glycosylamine-mediated isotope labeling for mass spectrometry-based quantitative analysis of N-linked glycans. Anal Bioanal Chem 405:5825–5831

Goso Y (2016) Malonic acid suppresses mucin-type O-glycan degradation during hydrazine treatment of glycoproteins. Anal Biochem 496:35–42

Griffiths J (2008) A brief history of mass spectrometry. Anal Chem 80:5678–5683

Gross JH (2004) Mass spectrometry. A textbook. Springer, Berlin

Gudelj I, Lauc G, Pezer M (2018) Immunoglobulin G glycosylation in aging and diseases. Cell Immunol 333:65–79

Haag AM (2016) Mass analyzers and mass spectrometers. Adv Exp Med Biol 919:157–169

Habazin S, Novokmet M, Štambuk J, Razdorov G, Keser T, Lauc G (2019) A sweet glimpse of rat immunoglobulin G glycosylation: towards comprehensive rodent animal model glyco(proteo)mics. Glycoconj J 36:267–397

Harazono A, Kawasaki N, Itoh S, Hashii N, Matsuishi-Nakajima Y, Kawanishi T, Yamaguchi T (2008) Simultaneous glycosylation analysis of human serum glycoproteins by high-performance liquid chromatography/tandem mass spectrometry. J Chromatogr B Anal Technol Biomed Life Sci 869:20–30

Harvey DJ (1996) Identification of cleaved oligosaccharides by matrix-assisted laser desorption/ionization. In: Methods in molecular biology. Humana Press, Totowa, pp 243–253

Harvey DJ (2000) N-(2-diethylamino)ethyl-4-aminobenzamide derivative for high sensitivity mass spectrometric detection and structure determination of N-linked carbohydrates. Rapid Commun Mass Spectrom 14:862–871

Harvey DJ (2001) Identification of protein-bound carbohydrates by mass spectrometry. Proteomics 1:311–328

Harvey DJ (2020) Negative ion mass spectrometry for the analysis of N-linked glycans. Mass Spectrom Rev 39:586–679. https://doi.org/10.1002/mas.21622

Harvey DJ, Mattu TS, Wormald MR, Royle L, Dwek RA, Rudd PM (2002) "Internal residue loss": rearrangements occurring during the fragmentation of carbohydrates derivatized at the reducing terminus. Anal Chem 74:734–740

Harvey DJ, Royle L, Radcliffe CM, Rudd PM, Dwek RA (2008) Structural and quantitative analysis of N-linked glycans by matrix-assisted laser desorption ionization and negative ion nanospray mass spectrometry. Anal Biochem 376:44–60

Hemström P, Irgum K (2006) Hydrophilic interaction chromatography. J Sep Sci 29:1784–1821

Hernandez-Alba O, Wagner-Rousset E, Beck A, Cianférani S (2018) Native mass spectrometry, ion mobility, and collision-induced unfolding for conformational characterization of IgG4 monoclonal antibodies. Anal Chem 90:8865–8872

Higel F, Demelbauer U, Seidl A, Friess W, Sörgel F (2013) Reversed-phase liquid-chromatographic mass spectrometric N-glycan analysis of biopharmaceuticals. Anal Bioanal Chem 405:2481–2493

Higel F, Seidl A, Demelbauer U, Sörgel F, Frieß W (2014) Small scale affinity purification and high sensitivity reversed phase nanoLC-MS N-glycan characterization of mAbs and fusion proteins. mAbs 6:894–903

Hinneburg H, Stavenhagen K, Schweiger-Hufnagel U, Pengelley S, Jabs W, Seeberger PH, Silva DV, Wuhrer M, Kolarich D (2016) The art of destruction: optimizing collision energies in quadrupole-time of flight (Q-TOF) instruments for glycopeptide-based glycoproteomics. J Am Soc Mass Spectrom 27:507–519

Hirayama K, Yuji R, Yamada N, Kato K, Arata Y, Shimada I (1998) Complete and rapid peptide and glycopeptide mapping of mouse monoclonal antibody by LC/MS/MS using ion trap mass spectrometry. Anal Chem 70:2718–2725

Holland M, Yagi H, Takahashi N, Kato K, Savage COS, Goodall DM, Jefferis R (2006) Differential glycosylation of polyclonal IgG, IgG-Fc and IgG-Fab isolated from the sera of patients with ANCA-associated systemic vasculitis. Biochim Biophys Acta Gen Subj 1760:669–677

Hong Y, Zhao H, Pu C, Zhan Q, Sheng Q, Lan M (2018) Hydrophilic phytic acid-coated magnetic graphene for titanium(IV) immobilization as a novel hydrophilic interaction liquid chromatography–immobilized metal affinity chromatography platform for glyco- and phosphopeptide enrichment with controllable selectivity. Anal Chem 90:11008–11,015

Hsu F-F, Turk J (2004) Studies on sulfatides by quadrupole ion-trap mass spectrometry with electrospray ionization: structural characterization and the fragmentation processes that include an unusual internal galactose residue loss and the classical charge-remote fragmentation. J Am Soc Mass Spectrom 15:536–546

Hu Y, Mechref Y (2012) Comparing MALDI-MS, RP-LC-MALDI-MS and RP-LC-ESI-MS glycomic profiles of permethylated N-glycans derived from model glycoproteins and human blood serum. Electrophoresis 33:1768–1777

Huang Y, Orlando R (2017) Kinetics of N-Glycan release from human immunoglobulin G (IgG) by PNGase F: all glycans are not created equal. J Biomol Tech 28:150–157

Huang L, Lu J, Wroblewski VJ, Beals JM, Riggin RM (2005) In vivo deamidation characterization of monoclonal antibody by LC/MS/MS. Anal Chem 77:1432–1439

Huang J, Guerrero A, Parker E, Strum JS, Smilowitz JT, German JB, Lebrilla CB (2015) Site-specific glycosylation of secretory immunoglobulin a from human colostrum. J Proteome Res 14:1335–1349

Huang C, Liu Y, Wu H, Sun D, Li Y (2017) Characterization of IgG glycosylation in rheumatoid arthritis patients by MALDI-TOF-MSn and capillary electrophoresis. Anal Bioanal Chem 409:3731–3739

Huddleston MJ, Bean MF, Carr SA (1993) Collisional fragmentation of glycopeptides by electrospray ionization LC/MS and LC/MS/MS: methods for selective detection of glycopeptides in protein digests. Anal Chem 65:877–884

Huhn C, Selman MHJ, Ruhaak LR, Deelder AM, Wuhrer M (2009) IgG glycosylation analysis. Proteomics 9:882–913

Hülsmeier AJ, Tobler M, Burda P, Hennet T (2016) Glycosylation site occupancy in health, congenital disorder of glycosylation and fatty liver disease. Sci Rep 6:33927. https://doi.org/10.1038/srep33927

Janin-Bussat MC, Tonini L, Huillet C, Colas O, Klinguer-Hamour C, Corvaïa N, Beck A (2013) Cetuximab fab and Fc N-glycan fast characterization using IdeS digestion and liquid chromatography coupled to electrospray ionization mass spectrometry. Methods Mol Biol 988:93–113

Jansen BC, Reiding KR, Bondt A, Hipgrave Ederveen AL, Palmblad M, Falck D, Wuhrer M (2015) MassyTools: a high-throughput targeted data processing tool for relative quantitation and quality control developed for glycomic and glycoproteomic MALDI-MS. J Proteome Res 14:5088–5098

Jayo RG, Thaysen-Andersen M, Lindenburg PW, Haselberg R, Hankemeier T, Ramautar R, Chen DDY (2014) Simple capillary electrophoresis-mass spectrometry method for complex glycan analysis using a flow-through microvial interface. Anal Chem 86:6479–6486

Jedrychowski MP, Huttlin EL, Haas W, Sowa ME, Rad R, Gygi SP (2011) Evaluation of HCD- and CID-type fragmentation within their respective detection platforms for murine phosphoproteomics. Mol Cell Proteomics 10(12):M111.009910. https://doi.org/10.1074/mcp.M111.009910

Jeong YR, Kim SY, Park YS, Lee GM (2018) Simple and robust N-glycan analysis based on improved 2-aminobenzoic acid labeling for recombinant therapeutic glycoproteins. J Pharm Sci 107:1831–1841

Jooß K, Sommer J, Bunz SC, Neusüß C (2014) In-line SPE-CE using a fritless bead string design-Application for the analysis of organic sulfonates including inline SPE-CE-MS for APTS-labeled glycans. Electrophoresis 35:1236–1243

Jooß K, Meckelmann SW, Klein J, Schmitz OJ, Neusüß C (2019) Capillary zone electrophoresis coupled to drift tube ion mobility-mass spectrometry for the analysis of native and APTS-labeled N-glycans. Anal Bioanal Chem 411:6255–6264

Kamerling JP, Gerwig GJ (2007) 2.01—Strategies for the structural analysis of carbohydrates. In: Kamerling H (ed) Comprehensive glycoscience. Elsevier, Oxford, pp 1–68

Kameyama A, Dissanayake SK, Thet Tin WW (2018) Rapid chemical de-N-glycosylation and derivatization for liquid chromatography of immunoglobulin N-linked glycans. PLoS One 13(5):e0196800. https://doi.org/10.1371/journal.pone.0196800

Kang P, Mechref Y, Klouckova I, Novotny MV (2005) Solid-phase permethylation of glycans for mass spectrometric analysis. Rapid Commun Mass Spectrom 19:3421–3428

Kang P, Mechref Y, Kyselova Z, Goetz JA, Novotny MV (2007) Comparative glycomic mapping through quantitative permethylation and stable-isotope labeling. Anal Chem 79:6064–6073

Kanje S, Venskutonytė R, Scheffel J, Nilvebrant J, Lindkvist-Petersson K, Hober S (2018) Protein engineering allows for mild affinity-based elution of therapeutic antibodies. J Mol Biol 430(18 Pt B):3427–3438. https://doi.org/10.1016/j.jmb.2018.06.004

Kannicht C, Grunow D, Lucka L (2019) Enzymatic sequence analysis of N-glycans by exoglycosidase cleavage and mass spectrometry: detection of lewis X structures. Methods Mol Biol 1934:51–64

Karas M, Krüger R (2003) Ion formation in MALDI: the cluster ionization mechanism. Chem Rev 103:427–440

Karas M, Bahr U, Dülcks T (2000) Nano-electrospray ionization mass spectrometry: addressing analytical problems beyond routine. Fresenius J Anal Chem 366:669–676

Karav S, Cohen JL, Barile D, de Moura BJMLN (2017) Recent advances in immobilization strategies for glycosidases. Biotechnol Prog 33:104–112

Keser T, Pavić T, Lauc G, Gornik O (2018) Comparison of 2-aminobenzamide, procainamide and RapiFluor-MS as derivatizing agents for high-throughput HILIC-UPLC-FLR-MS N-glycan analysis. Front Chem 6:324

Kihlberg B-M, Sjöholm AG, Björck L, Sjöbring U (1996) Characterization of the binding properties of protein LG, an immunoglobulin-binding hybrid protein. Eur J Biochem 240:556–563

Kim M-S, Pandey A (2012) Electron transfer dissociation mass spectrometry in proteomics. Proteomics 12:530–542

Kobata A (2008) The N-linked sugar chains of human immunoglobulin G: their unique pattern, and their functional roles. Biochim Biophys Acta 1780:472–478

Koguma I, Yamashita S, Sato S, Okuyama K, Katakura Y (2013) Novel purification method of human immunoglobulin by using a thermo-responsive protein A. J Chromatogr A 1305:149–153

Korodi M, Rákosi K, Baibarac M, Fejer SN (2020) Reusable on-plate immunoprecipitation method with covalently immobilized antibodies on a protein G covered microtiter plate. J Immunol Methods 483:112812

Kotsias M, Blanas A, van Vliet SJ, Pirro M, Spencer DIR, Kozak RP (2019) Method comparison for N-glycan profiling: towards the standardization of glycoanalytical technologies for cell line analysis. PLoS One 14(10):e0223270. https://doi.org/10.1371/journal.pone.0223270

Kováčik V, Hirsch J, Kováč P, Heerma W, Thomas-Oates J, Haverkamp J (1995) Oligosaccharide characterization using collision-induced dissociation fast atom bombardment mass spectrometry: evidence for internal monosaccharide residue loss. J Mass Spectrom 30:949–958

Kozak RP, Royle L, Gardner RA, Bondt A, Fernandes DL, Wuhrer M (2014) Improved nonreductive O-glycan release by hydrazinolysis with ethylenediaminetetraacetic acid addition. Anal Biochem 453:29–37

Krenkova J, Szekrenyes A, Keresztessy Z, Foret F, Guttman A (2013) Oriented immobilization of peptide-N-glycosidase F on a monolithic support for glycosylation analysis. J Chromatogr A 1322:54–61

Kruljec N, Bratkovič T (2017) Alternative affinity ligands for immunoglobulins. Bioconjug Chem 28:2009–2030

Kurogochi M, Amano J (2014) Relative quantitation of glycopeptides based on stable isotope labeling using MALDI-TOF MS. Molecules 19:9944–9961

Lakhrif Z, Pugnière M, Henriquet C, di Tommaso A, Dimier-Poisson I, Billiald P, Juste MO, Aubrey N (2016) A method to confer Protein L binding ability to any antibody fragment. MAbs 8:379–388

Lannergård J, Guss B (2006) IdeE, an IgG-endopeptidase of *Streptococcus equi* ssp. equi. FEMS Microbiol Lett 262:230–235

Lauber MA, Yu Y-Q, Brousmiche DW, Hua Z, Koza SM, Magnelli P, Guthrie E, Taron CH, Fountain KJ (2015) Rapid preparation of released N-glycans for HILIC analysis using a labeling reagent that facilitates sensitive fluorescence and ESI-MS detection. Anal Chem 87:5401–5409

Lermyte F, Tsybin YO, O'Connor PB, Loo JA (2019) Top or middle? Up or down? Toward a standard lexicon for protein top-down and allied mass spectrometry approaches. J Am Soc Mass Spectrom 30:1149–1157

Liu S, Gao W, Wang Y, He Z, Feng X, Liu B-F, Liu X (2017) Comprehensive N-glycan profiling of cetuximab biosimilar candidate by NP-HPLC and MALDI-MS. PLos One 12:e0170013

Liu S, Haller E, Horak J, Brandstetter M, Heuser T, Lämmerhofer M (2019a) Protein A- and Protein G-gold nanoparticle bioconjugates as nano-immunoaffinity platform for human IgG depletion in plasma and antibody extraction from cell culture supernatant. Talanta 194:664–672

Liu L, Jin S, Mei P, Zhou P (2019b) Preparation of cotton wool modified with boric acid functionalized titania for selective enrichment of glycopeptides. Talanta 203:58–64

Loos A, Gruber C, Altmann F, Mehofer U, Hensel F, Grandits M, Oostenbrink C, Stadlmayr G, Furtmüller PG, Steinkellner H (2014) Expression and glycoengineering of functionally active heteromultimeric IgM in plants. Proc Natl Acad Sci U S A 111:6263–6268

López-Fernández H, Santos HM, Capelo JL, Fdez-Riverola F, Glez-Peña D, Reboiro-Jato M (2015) Mass-Up: an all-in-one open software application for MALDI-TOF mass spectrometry knowledge discovery. BMC Bioinformatics 16:318

Luo Q, Rejtar T, Wu SL, Karger BL (2009) Hydrophilic interaction 10 μm I.D. porous layer open tubular columns for ultratrace glycan analysis by liquid chromatography-mass spectrometry. J Chromatogr A 1216:1223–1231

Maher S, Jjunju FP, Taylor S (2015) Colloquium: 100 years of mass spectrometry: perspectives and future trends. Rev Mod Phys 87:113

Maier M, Reusch D, Bruggink C, Bulau P, Wuhrer M, Mølhøj M (2016) Applying mini-bore HPAEC-MS/MS for the characterization and quantification of Fc N -glycans from heterogeneously glycosylated IgGs. J Chromatogr B 1033–1034:342–352

Matsumoto A, Shikata K, Takeuchi F, Kojima N, Mizuochi T (2000) Autoantibody activity of IgG rheumatoid factor increases with decreasing levels of galactosylation and sialylation. J Biochem 128:621–628

Mauko L, Nordborg A, Hutchinson JP, Lacher NA, Hilder EF, Haddad PR (2011) Glycan profiling of monoclonal antibodies using zwitterionic-type hydrophilic interaction chromatography coupled with electrospray ionization mass spectrometry detection. Anal Biochem 408:235–241

Mauko L, Lacher NA, Pelzing M, Nordborg A, Haddad PR, Hilder EF (2012) Comparison of ZIC-HILIC and graphitized carbon-based analytical approaches combined with exoglycosidase digestions for analysis of glycans from monoclonal antibodies. J Chromatogr B Anal Technol Biomed Life Sci 911:93–104

Melmer M, Stangler T, Premstaller A, Lindner W (2011) Comparison of hydrophilic-interaction, reversed-phase and porous graphitic carbon chromatography for glycan analysis. J Chromatogr A 1218:118–123

Mendez-Huergo SP, Maller SM, Farez MF, Mariño K, Correale J, Rabinovich GA (2014) Integration of lectin-glycan recognition systems and immune cell networks in CNS inflammation. Cytokine Growth Factor Rev 25:247–255

Menni C, Keser T, Mangino M et al (2013) Glycosylation of immunoglobulin g: role of genetic and epigenetic influences. PLoS One 8:e82558

Michael C, Rizzi AM (2015) Quantitative isomer-specific N-glycan fingerprinting using isotope coded labeling and high performance liquid chromatography-electrospray ionization-mass spectrometry with graphitic carbon stationary phase. J Chromatogr A 1383:88–95

Miller PE, Denton MB (1986) The quadrupole mass filter: basic operating concepts. J Chem Educ 63:617

Miura Y, Endo T (2016) Glycomics and glycoproteomics focused on aging and age-related diseases—glycans as a potential biomarker for physiological alterations. Biochim Biophys Acta Gen Subj 1860:1608–1614

Miura Y, Hato M, Shinohara Y et al (2008) BlotGlycoABC™, an integrated glycoblotting technique for rapid and large scale clinical glycomics. Mol Cell Proteomics 7(2):370–377. https://doi.org/10.1074/MCP.M700377-MCP200

Morelle W, Slomianny M-C, Diemer H, Schaeffer C, van Dorsselaer A, Michalski J-C (2004) Fragmentation characteristics of permethylated oligosaccharides using a matrix-assisted laser desorption/ionization two-stage time-of-flight (TOF/TOF) tandem mass spectrometer. Rapid Commun Mass Spectrom 18:2637–2649

Moser AC, Hage DS (2010) Immunoaffinity chromatography: an introduction to applications and recent developments. Bioanalysis 2:769–790

Mouratou B, Béhar G, Pecorari F (2015) Artificial affinity proteins as ligands of immunoglobulins. Biomolecules 5:60–75

Nakakita S, Sumiyoshi W, Miyanishi N, Hirabayashi J (2007) A practical approach to N-glycan production by hydrazinolysis using hydrazine monohydrate. Biochem Biophys Res Commun 362:639–645

Nakano M, Higo D, Arai E, Nakagawa T, Kakehi K, Taniguchi N, Kondo A (2009) Capillary electrophoresis-electrospray ionization mass spectrometry for rapid and sensitive N-glycan analysis of glycoproteins as 9-fluorenylmethyl derivatives. Glycobiology 19:135–143

Nikolaev EN, Jertz R, Grigoryev A, Baykut G (2012) Fine structure in isotopic peak distributions measured using a dynamically harmonized Fourier transform ion cyclotron resonance cell at 7 T. Anal Chem 84:2275–2283

Nilson BH, Solomon A, Björck L, Akerström B (1992) Protein L from Peptostreptococcus magnus binds to the kappa light chain variable domain. J Biol Chem 267:2234–2239

Nilsson B, Moks T, Jansson B, Abrahmsén L, Elmblad A, Holmgren E, Henrichson C, Jones TA, Uhlén M (1987) A synthetic IgG-binding domain based on staphylococcal protein A. Protein Eng 1:107–113

Nishimura S-I (2011) Toward automated glycan analysis. Adv Carbohydr Chem Biochem 65:219–271

Nolting D, Malek R, Makarov A (2019) Ion traps in modern mass spectrometry. Mass Spectrom Rev 38:150–168

Ohta Y, Kameda K, Matsumoto M, Kawasaki N (2017) Rapid glycopeptide enrichment using cellulose hydrophilic interaction/reversed-phase stagetips. Mass Spectrom 6:A0061–A0061

Olivova P, Chen W, Chakraborty AB, Gebler JC (2008) Determination of N-glycosylation sites and site heterogeneity in a monoclonal antibody by electrospray quadrupole ion-mobility time-of-flight mass spectrometry. Rapid Commun Mass Spectrom 22:29–40

Pabst M, Altmann F (2008a) LC-MS/MS analysis of permethylated N-glycans facilitating isomeric characterization. Anal Chem 80:7534–7542

Pabst M, Altmann F (2008b) Influence of electrosorption, solvent, temperature, and ion polarity on the performance of LC-ESI-MS using graphitic carbon for acidic oligosaccharides. Anal Chem 80:7534–7542

Pabst M, Kolarich D, Pöltl G, Dalik T, Lubec G, Hofinger A, Altmann F (2009) Comparison of fluorescent labels for oligosaccharides and introduction of a new postlabeling purification method. Anal Biochem 384:263–273

Pabst TM, Palmgren R, Forss A, Vasic J, Fonseca M, Thompson C, Wang WK, Wang X, Hunter AK (2014) Engineering of novel Staphylococcal Protein A ligands to enable milder elution pH and high dynamic binding capacity. J Chromatogr A 1362:180–185

Pabst M, Benešová I, Fagerer SR et al (2016) Differential isotope labeling of glycopeptides for accurate determination of differences in site-specific glycosylation. J Proteome Res 15:326–331

Page M, Thorpe R (2002) Purification of IgG by precipitation with sodium sulfate or ammonium sulfate. In: Walker JM (ed) The protein protocols handbook. Humana Press, Totowa, NJ, pp 983–984

Paul M, Reljic R, Klein K et al (2014) Characterization of a plant-produced recombinant human secretory IgA with broad neutralizing activity against HIV. mAbs 6:1585–1597

Plomp R, Hensbergen PJ, Rombouts Y, Zauner G, Dragan I, Koeleman CAM, Deelder AM, Wuhrer M (2014) Site-specific N-glycosylation analysis of human immunoglobulin e. J Proteome Res 13:536–546

Pučić M, Knežević A, Vidič J et al (2011) High throughput isolation and glycosylation analysis of IgG–variability and heritability of the IgG glycome in three isolated human populations. Mol Cell Proteomics 10(10):M111.010090. https://doi.org/10.1074/mcp.M111.010090

Qian J, Liu T, Yang L, Daus A, Crowley R, Zhou Q (2007) Structural characterization of N-linked oligosaccharides on monoclonal antibody cetuximab by the combination of orthogonal matrix-assisted laser desorption/ionization hybrid quadrupole-quadrupole time-of-flight tandem mass spectrometry and sequential enzy. Anal Biochem 364:8–18

Ramirez K, Campbell E, Han S-Y, Buehler J, Phan T, Young Yoon H, Lee YL, Suresh T, Sulchek T (2019) Optimization of microparticle reagents to collect and detect antibody. Langmuir 35:11717–11,724

Ramos II, Marques SS, Magalhães LM, Barreiros L, Reis S, Lima JLFC, Segundo MA (2019) Assessment of immunoglobulin capture in immobilized protein A through automatic bead injection. Talanta 204:542–547

Redman EA, Batz NG, Mellors JS, Ramsey JM (2015) Integrated microfluidic capillary electrophoresis-electrospray ionization devices with online ms detection for the separation and characterization of intact monoclonal antibody variants. Anal Chem 87:2264–2272

Rehder DS, Dillon TM, Pipes GD, Bondarenko PV (2006) Reversed-phase liquid chromatography/mass spectrometry analysis of reduced monoclonal antibodies in pharmaceutics. J Chromatogr A 1102:164–175

Reiding KR, Blank D, Kuijper DM, Deelder AM, Wuhrer M (2014) High-throughput profiling of protein N-glycosylation by MALDI-TOF-MS employing linkage-specific sialic acid esterification. Anal Chem 86:5784–5793

Reusch D, Haberger M, Kailich T, Heidenreich A-K, Kampe M, Bulau P, Wuhrer M (2014) High-throughput glycosylation analysis of therapeutic immunoglobulin G by capillary gel electrophoresis using a DNA analyzer. MAbs 6:185–196

Reusch D, Haberger M, Falck D et al (2015a) Comparison of methods for the analysis of therapeutic immunoglobulin G Fc-glycosylation profiles-part 2: mass spectrometric methods. mAbs 7:732–742

Reusch D, Haberger M, Maier B et al (2015b) Comparison of methods for the analysis of therapeutic immunoglobulin G Fc-glycosylation profiles—part 1: separation-based methods. mAbs 7:167–179

Rispens T, Vidarsson G (2014) Human IgG subclasses. In: Antibody Fc. Elsevier, Amsterdam, pp 159–177

Ritamo I, Räbinä J, Natunen S, Valmu L (2013) Nanoscale reversed-phase liquid chromatography–mass spectrometry of permethylated N-glycans. Anal Bioanal Chem 405:2469–2480

Rodrigo G, Gruvegård M, Van Alstine JM (2015) Antibody fragments and their purification by protein L affinity chromatography. Antibodies 4:259–277

Roy R, Ang E, Komatsu E, Domalaon R, Bosseboeuf A, Harb J, Hermouet S, Krokhin O, Schweizer F, Perreault H (2018) Absolute quantitation of glycoforms of two human IgG subclasses using synthetic Fc peptides and glycopeptides. J Am Soc Mass Spectrom 29:1086–1098

Royle L, Mattu TS, Hart E, Langridge JI, Merry AH, Murphy N, Harvey DJ, Dwek RA, Rudd PM (2002) An analytical and structural database provides a strategy for sequencing O-glycans from microgram quantities of glycoproteins. Anal Biochem 304:70–90

Rudd PM, Colominas C, Royle L, Murphy N, Hart E, Merry AH, Hebestreit HF, Dwek RA (2001) A high-performance liquid chromatography based strategy for rapid, sensitive sequencing ofN-linked oligosaccharide modifications to proteins in sodium dodecyl sulphate polyacrylamide electrophoresis gel bands. Proteomics 1:285–294

Ruhaak LR, Deelder AM, Wuhrer M (2009) Oligosaccharide analysis by graphitized carbon liquid chromatography-mass spectrometry. Anal Bioanal Chem 394:163–174

Ruhaak LR, Zauner G, Huhn C, Bruggink C, Deelder AM, Wuhrer M (2010a) Glycan labeling strategies and their use in identificartion and quantification. Anal Bioanal Chem 397:3457–3481

Ruhaak LR, Steenvoorden E, Koeleman CAM, Deelder AM, Wuhrer M (2010b) 2-picoline-borane: a non-toxic reducing agent for oligosaccharide labeling by reductive amination. Proteomics 10:2330–2336

Ruhaak LR, Xu G, Li Q, Goonatilleke E, Lebrilla CB (2018) Mass spectrometry approaches to glycomic and glycoproteomic analyses. Chem Rev 118:7886–7930

Saba JA, Kunkel JP, Jan DCH, Ens WE, Standing KG, Butler M, Jamieson JC, Perreault H (2002) A study of immunoglobulin g glycosylation in monoclonal and polyclonal species by electrospray and matrix-assisted laser desorption/ionization mass spectrometry. Anal Biochem 305:16–31

Sajid MS, Jabeen F, Hussain D, Gardner QA, Ashiq MN, Najam-ul-Haq M (2020) Boronic acid functionalized fibrous cellulose for the selective enrichment of glycopeptides. J Sep Sci 43:1348–1355

Saldova R, Asadi Shehni A, Haakensen VD, Steinfeld I, Hilliard M, Kifer I, Helland Å, Yakhini Z, Børresen-Dale AL, Rudd PM (2014) Association of N-glycosylation with breast carcinoma and systemic features using high-resolution quantitative UPLC. J Proteome Res 13:2314–2327

Salimi K, Usta DD, Koçer İ, Çelik E, Tuncel A (2018) Protein A and protein A/G coupled magnetic SiO2 microspheres for affinity purification of immunoglobulin G. Int J Biol Macromol 111:178–185

Sauer-Eriksson AE, Kleywegt GJ, Uhlén M, Jones TA (1995) Crystal structure of the C2 fragment of streptococcal protein G in complex with the Fc domain of human IgG. Structure 3:265–278

Selman MHJ, Hemayatkar M, Deelder AM, Wuhrer M (2011) Cotton HILIC SPE microtips for microscale purification and enrichment of glycans and glycopeptides. Anal Chem 83:2492–2499

Selman MHJ, Derks RJE, Bondt A, Palmblad M, Schoenmaker B, Koeleman CAM, van de Geijn FE, Dolhain RJEM, Deelder AM, Wuhrer M (2012a) Fc specific IgG glycosylation profiling by robust nano-reverse phase HPLC-MS using a sheath-flow ESI sprayer interface. J Proteome 75:1318–1329

Selman MHJ, Hoffmann M, Zauner G, McDonnell LA, Balog CIA, Rapp E, Deelder AM, Wuhrer M (2012b) MALDI-TOF-MS analysis of sialylated glycans and glycopeptides using 4-chloro-α-cyanocinnamic acid matrix. Proteomics 12:1337–1348

Sénard T, Gargano AFG, Falck D, de Taeye SW, Rispens T, Vidarsson G, Wuhrer M, Somsen GW, Domínguez-Vega E (2020) MS-based allotype-specific analysis of polyclonal IgG-Fc N-glycosylation. Front Immunol 11:2049. https://doi.org/10.3389/fimmu.2020.02049

Sha Q, Wu Y, Wang C, Sun B, Zhang Z, Zhang L, Lin Y, Liu X (2018) Cellulose microspheres-filled pipet tips for purification and enrichment of glycans and glycopeptides. J Chromatogr A 1569:8–16

Sheng S, Kong F (2012) Separation of antigens and antibodies by immunoaffinity chromatography. Pharm Biol 50:1038–1044

Shiang Y-C, Lin C-A, Huang C-C, Chang H-T (2011) Protein A-conjugated luminescent gold nanodots as a label-free assay for immunoglobulin G in plasma. Analyst 136:1177–1182

Shiao J-Y, Chang Y-T, Chang M-C, Chen MX, Liu L-W, Wang X-Y, Tsai Y-J, Kuo T-C, Tsai I-L (2020) Development of efficient on-bead protein elution process coupled to ultra-high performance liquid chromatography–tandem mass spectrometry to determine immunoglobulin G subclass and glycosylation for discovery of bio-signatures in pancreatic disease. J Chromatogr A 1621:461039

Shou WZ, Naidong W (2005) Simple means to alleviate sensitivity loss by trifluoroacetic acid (TFA) mobile phases in the hydrophilic interaction chromatography-electrospray tandem mass spectrometric (HILIC-ESI/MS/MS) bioanalysis of basic compounds. J Chromatogr B Anal Technol Biomed Life Sci 825:186–192

Shubhakar A, Kozak RP, Reiding KR, Royle L, Spencer DIR, Fernandes DL, Wuhrer M (2016) Automated high-throughput permethylation for glycosylation analysis of biologics using MALDI-TOF-MS. Anal Chem 88:8562–8569

Singh C, Zampronio CG, Creese AJ, Cooper HJ (2012) Higher energy collision dissociation (HCD) product ion-triggered electron transfer dissociation (ETD) mass spectrometry for the analysis of N-linked glycoproteins. J Proteome Res 11:4517–4525

Sinha S, Pipes G, Topp EM, Bondarenko PV, Treuheit MJ, Gadgil HS (2008) Comparison of LC and LC/MS methods for quantifying N-glycosylation in recombinant IgGs. J Am Soc Mass Spectrom 19:1643–1654

Sjögren J, Struwe WB, Cosgrave EFJ, Rudd PM, Stervander M, Allhorn M, Hollands A, Nizet V, Collin M (2013) EndoS2 is a unique and conserved enzyme of serotype M49 group A

Streptococcus that hydrolyses N-linked glycans on IgG and α1-acid glycoprotein. Biochem J 455:107

Sjögren J, Cosgrave EFJ, Allhorn M, Nordgren M, Björk S, Olsson F, Fredriksson S, Collin M (2015) EndoS and EndoS2 hydrolyze Fc-glycans on therapeutic antibodies with different glycoform selectivity and can be used for rapid quantification of high-mannose glycans. Glycobiology 25:1053–1063

Sjögren J, Olsson F, Beck A (2016) Rapid and improved characterization of therapeutic antibodies and antibody related products using IdeS digestion and subunit analysis. Analyst 141:3114–3125

Smith RD, Loo JA, Edmonds CG, Barinaga CJ, Udseth HR (1990) New developments in biochemical mass spectrometry: electrospray ionization. Anal Chem 62:882–899

Sobott F, Watt SJ, Smith J, Edelmann MJ, Kramer HB, Kessler BM (2009) Comparison of CID versus ETD based MS/MS fragmentation for the analysis of protein ubiquitination. J Am Soc Mass Spectrom 20:1652–1659

Song X, Ju H, Lasanajak Y, Kudelka MR, Smith DF, Cummings RD (2016) Oxidative release of natural glycans for functional glycomics. Nat Methods 13:528–534

Spoerry C, Seele J, Valentin-Weigand P, Baums CG, von Pawel-Rammingen U (2016) Identification and characterization of IgdE, a novel IgG-degrading protease of *Streptococcus suis* with unique specificity for porcine IgG. J Biol Chem 291:7915–7925

Stadlmann J, Pabst M, Kolarich D, Kunert R, Altmann F (2008) Analysis of immunoglobulin glycosylation by LC-ESI-MS of glycopeptides and oligosaccharides. Proteomics 8:2858–2871

Stadlmann J, Pabst M, Altmann F (2010) Analytical and functional aspects of antibody sialylation. J Clin Immunol 30(Suppl 1):S15–S19

Stanley P, Taniguchi N, Aebi M (2017) N-Glycans. In: Varki A, Cummings RD, Esko JD Essentials of glycobiology [Internet]. 3rd. Cold Spring Harbor (NY): Cold Spring Harbor Laboratory Press

Steffen U, Koeleman CA, Sokolova MV et al (2020) IgA subclasses have different effector functions associated with distinct glycosylation profiles. Nat Commun 11(1):120. https://doi.org/10.1038/s41467-019-13992-8

Sun G, Yu X, Bao C, Wang L, Li M, Gan J, Qu D, Ma J, Chen L (2015) Identification and characterization of a novel prokaryotic peptide N-glycosidase from Elizabethkingia meningoseptica. J Biol Chem 290:7452–7462

Svensson HG, Wedemeyer WJ, Ekstrom JL, Callender DR, Kortemme T, Kim DE, Sjöbring U, Baker D (2004) Contributions of amino acid side chains to the kinetics and thermodynamics of the bivalent binding of protein L to Ig kappa light chain. Biochemistry 43:2445–2457

Syka JEP, Coon JJ, Schroeder MJ, Shabanowitz J, Hunt DF (2004) Peptide and protein sequence analysis by electron transfer dissociation mass spectrometry. Proc Natl Acad Sci U S A 101:9528–9533

Szabo Z, Guttman A, Karger BL (2010) Rapid release of N-linked glycans from glycoproteins by pressure-cycling technology. Anal Chem 82:2588–2593

Szigeti M, Bondar J, Gjerde D, Keresztessy Z, Szekrenyes A, Guttman A (2016) Rapid N-glycan release from glycoproteins using immobilized PNGase F microcolumns. J Chromatogr B Analyt Technol Biomed Life Sci 1032:139–143

Takakura D, Harazono A, Hashii N, Kawasaki N (2014) Selective glycopeptide profiling by acetone enrichment and LC/MS. J Proteome 101:17–30

Takayama M (2016) MALDI in-source decay of protein: the mechanism of c-Ion formation. Mass Spectrom (Tokyo) 5(1):A0044. https://doi.org/10.5702/massspectrometry.A0044

Takegawa Y, Deguchi K, Keira T, Ito H, Nakagawa H, Nishimura SI (2006) Separation of isomeric 2-aminopyridine derivatized N-glycans and N-glycopeptides of human serum immunoglobulin G by using a zwitterionic type of hydrophilic-interaction chromatography. J Chromatogr A 1113:177–181

Tanaka K, Waki H, Ido Y, Akita S, Yoshida Y, Yoshida T, Matsuo T (1988) Protein and polymer analyses up to m/z 100 000 by laser ionization time-of-flight mass spectrometry. Rapid Commun Mass Spectrom 2:151–153

Thompson A, Schäfer J, Kuhn K, Kienle S, Schwarz J, Schmidt G, Neumann T, Hamon C (2003) Tandem mass tags: a novel quantification strategy for comparative analysis of complex protein mixtures by MS/MS. Anal Chem 75:1895–1904

Tian Y, Han L, Buckner AC, Ruotolo BT (2015) Collision induced unfolding of intact antibodies: rapid characterization of disulfide bonding patterns, glycosylation, and structures. Anal Chem 87:11509–11,515

Tissot B, North SJ, Ceroni A, Pang PC, Panico M, Rosati F, Capone A, Haslam SM, Dell A, Morris HR (2009) Glycoproteomics: past, present and future. FEBS Lett 583:1728–1735

Todoroki K, Mizuno H, Sugiyama E, Toyo'oka T (2020) Bioanalytical methods for therapeutic monoclonal antibodies and antibody–drug conjugates: a review of recent advances and future perspectives. J Pharm Biomed Anal 179:112991

Tong X, Li T, Orwenyo J, Toonstra C, Wang L-X (2018) One-pot enzymatic glycan remodeling of a therapeutic monoclonal antibody by endoglycosidase S (Endo-S) from *Streptococcus pyogenes*. Bioorg Med Chem 26:1347–1355

Tran BQ, Barton C, Feng J et al (2016) Comprehensive glycosylation profiling of IgG and IgG-fusion proteins by top-down MS with multiple fragmentation techniques. J Proteome 134:93–101

Trang HK, Marcus RK (2017) Application of protein A-modified capillary-channeled polymer polypropylene fibers to the quantitation of IgG in complex matrices. J Pharm Biomed Anal 142:49–58

Trastoy B, Klontz E, Orwenyo J, Marina A, Wang L-X, Sundberg EJ, Guerin ME (2018) Structural basis for the recognition of complex-type N-glycans by Endoglycosidase S. Nat Commun 9:1874

Vainauskas S, Kirk CH, Petralia L et al (2018) A novel broad specificity fucosidase capable of core α1-6 fucose release from N-glycans labeled with urea-linked fluorescent dyes. Sci Rep 8 (1):9504. https://doi.org/10.1038/s41598-018-27797-0

van de Bovenkamp FS, Hafkenscheid L, Rispens T, Rombouts Y (2016) The emerging importance of IgG Fab glycosylation in immunity. J Immunol 196:1435–1441

Varki A, Cummings RD, Esko JD et al (eds) (2015) Essentials of glycobiology, 3rd edn. Cold Spring Harbor Laboratory Press, Cold Spring Harbor, NY

Veillon L, Huang Y, Peng W, Dong X, Cho BG, Mechref Y (2017) Characterization of isomeric glycan structures by LC-MS/MS. Electrophoresis 38:2100–2114

Vestal ML, Campbell JM (2005) Tandem time-of-flight mass spectrometry. Methods Enzymol 402:79–108

Vilaj M, Lauc G, Trbojević-Akmačić I (2020) Evaluation of different PNGase F enzymes in immunoglobulin G and total plasma N-glycans analysis. Glycobiology. https://doi.org/10.1093/glycob/cwaa047

Vincents B, Guentsch A, Kostolowska D, von Pawel-Rammingen U, Eick S, Potempa J, Abrahamson M (2011) Cleavage of IgG1 and IgG3 by gingipain K from *Porphyromonas gingivalis* may compromise host defense in progressive periodontitis. FASEB J 25:3741–3750

Volpi N, Maccari F (2013) Capillary Electrophoresis of Biomolecules. Human Press, Springer Science+Business Media

von Pawel-Rammingen U, Johansson BP, Björck L (2002) IdeS, a novel streptococcal cysteine proteinase with unique specificity for immunoglobulin G. EMBO J 21:1607–1615

Vreeker GCM, Wuhrer M (2017) Reversed-phase separation methods for glycan analysis. Anal Bioanal Chem 409:359–378

Wang T, Voglmeir J (2014) PNGases as valuable tools in glycoprotein analysis. Protein Pept Lett 21:976–985

Wang T, Cai ZP, Gu XQ, Ma HY, Du YM, Huang K, Voglmeir J, Liu L (2014) Discovery and characterization of a novel extremely acidic bacterial N-glycanase with combined advantages of PNGase F and A. Biosci Rep 34(6):e00149. https://doi.org/10.1042/BSR20140148

Wang C, Qiang S, Jin W, Song X, Zhang Y, Huang L, Wang Z (2018a) Reductive chemical release of N-glycans as 1-amino-alditols and subsequent 9-fluorenylmethyloxycarbonyl labeling for MS and LC/MS analysis. J Proteome 187:47–58

Wang C, Yang M, Gao X, Li C, Zou Z, Han J, Huang L, Wang Z (2018b) The ammonia-catalyzed release of glycoprotein N-glycans. Glycoconj J 35:411–420

Weng Y, Qu Y, Jiang H, Wu Q, Zhang L, Yuan H, Zhou Y, Zhang X, Zhang Y (2014) An integrated sample pretreatment platform for quantitative N-glycoproteome analysis with combination of on-line glycopeptide enrichment, deglycosylation and dimethyl labeling. Anal Chim Acta 833:1–8

West C, Elfakir C, Lafosse M (2010) Porous graphitic carbon: a versatile stationary phase for liquid chromatography. J Chromatogr A 1217:3201–3216

Wuhrer M, Deelder AM, Hokke CH (2005) Protein glycosylation analysis by liquid chromatography-mass spectrometry. J Chromatogr B Anal Technol Biomed Life Sci 825:124–133

Wuhrer M, Koeleman CAM, Hokke CH, Deelder AM (2006) Mass spectrometry of proton adducts of fucosylated N-glycans: fucose transfer between antennae gives rise to misleading fragments. Rapid Commun Mass Spectrom 20:1747–1754

Wuhrer M, Catalina MI, Deelder AM, Hokke CH (2007a) Glycoproteomics based on tandem mass spectrometry of glycopeptides. J Chromatogr B Analyt Technol Biomed Life Sci 849:115–128

Wuhrer M, Stam JC, van de Geijn FE, Koeleman CAMM, Verrips CT, Dolhain RJEMEM, Hokke CH, Deelder AM (2007b) Glycosylation profiling of immunoglobulin G (IgG) subclasses from human serum. Proteomics 7:4070–4081

Wuhrer M, De Boer AR, Deelder AM (2009) Structural glycomics using Hydrophilic interaction chromatography (HILIC) with mass spectrometry. Mass Spectrom Rev 28:192–206

Yan X, Tchekhovskoi D, Mirokhin Y, Stein S, Kilpatrick L (2012) Optimization of tryptic digestion methods for LC-MS/MS analysis of chimeric immunoglobulin G. J Biomol Tech 23:S44

Yang L, Biswas ME, Chen P (2003) Study of binding between protein A and immunoglobulin G using a surface tension probe. Biophys J 84:509–522

Yang Z, Sun A, Zhao X et al (2020) Preparation and application of a beta-d-glucan microsphere conjugated protein A/G. Int J Biol Macromol 151:878–884

Yost RA, Enke CG (1979) Triple quadrupole mass spectrometry for direct mixture analysis and structure elucidation. Anal Chem 51:1251–1264

Zaia J (2010) Mass spectrometry and glycomics. Omics 14:401–418

Zaia J (2013) Capillary electrophoresis-mass spectrometry of carbohydrates. In: Methods in molecular biology. Humana Press, Totowa, NJ, pp 139–151

Zauner G, Selman MHJ, Bondt A, Rombouts Y, Blank D, Deelder AM, Wuhrer M (2013) Glycoproteomic analysis of antibodies. Mol Cell Proteomics 12:856–865

Zaytseva OO, Jansen BC, Hanić M et al (2018) MIgGGly (mouse IgG glycosylation analysis)—a high-throughput method for studying Fc-linked IgG N-glycosylation in mice with nanoUPLC-ESI-MS. Sci Rep 8:13688

Zhang J, Wang DI c. (1998) Quantitative analysis and process monitoring of site-specific glycosylation microheterogeneity in recombinant human interferon-γ from Chinese hamster ovary cell culture by hydrophilic interaction chromatography. J Chromatogr B Biomed Appl 712:73–82

Zhang Z, Pan H, Chen X (2009) Mass spectrometry for structural characterization of therapeutic antibodies. Mass Spectrom Rev 28:147–176

Zhang H, Cui W, Gross ML (2014) Mass spectrometry for the biophysical characterization of therapeutic monoclonal antibodies. FEBS Lett 588:308–317

Zhang L, Ma S, Chen Y, Wang Y, Ou J, Uyama H, Ye M (2019) Facile fabrication of biomimetic chitosan membrane with honeycomb-like structure for enrichment of glycosylated peptides. Anal Chem 91:2985–2993

Zhao J, Li S, Li C, Wu SL, Xu W, Chen Y, Shameem M, Richardson D, Li H (2016) Identification of low abundant isomeric N-glycan structures in biological therapeutics by LC/MS. Anal Chem 88:7049–7059

Zhong X, Zhang Z, Jiang S, Li L (2014) Recent advances in coupling capillary electrophoresis based separation techniques to ESI and MALDI MS. Electrophoresis 35:1214–1225

Zhou H, Briscoe AC, Froehlich JW, Lee RS (2012) PNGase F catalyzes de-N-glycosylation in a domestic microwave. Anal Biochem 427:33–35

Zhou S, Tello N, Harvey A, Boyes B, Orlando R, Mechref Y (2016) Reliable LC-MS quantitative glycomics using iGlycoMab stable isotope labeled glycans as internal standards. Electrophoresis 37:1489–1497

Zhou S, Dong X, Veillon L, Huang Y, Mechref Y (2017) LC-MS/MS analysis of permethylated N-glycans facilitating isomeric characterization. Anal Bioanal Chem 409:453–466

Zhou Y, Sheng X, Garemark J, Josefsson L, Sun L, Li Y, Emmer Å (2020) Enrichment of glycopeptides using environmentally friendly wood materials. Green Chem 22:5666–5676

Zhurov KO, Fornelli L, Wodrich MD, Laskay ÜA, Tsybin YO (2013) Principles of electron capture and transfer dissociation mass spectrometry applied to peptide and protein structure analysis. Chem Soc Rev 42:5014–5030

Zou X, Jie J, Yang B (2017) Single-step enrichment of N-glycopeptides and phosphopeptides with novel multifunctional Ti4 + -immobilized dendritic polyglycerol coated chitosan nanomaterials. Anal Chem 89:7520–7526

Zubarev RA, Kelleher NL, McLafferty FW (1998) Electron capture dissociation of multiply charged protein cations. A nonergodic process. J Am Chem Soc 120(13):3265–3266. https://pubs.acs.org/doi/10.1021/ja973478k. Accessed 6 Aug 2020

Chapter 4
Capillary (Gel) Electrophoresis-Based Methods for Immunoglobulin (G) Glycosylation Analysis

Samanta Cajic , René Hennig , Robert Burock , and Erdmann Rapp

Contents

4.1	Historical Background	139
4.2	Background: Principles of Capillary (Gel) Electrophoresis (C(G)E)	140
4.3	Performance, Benefits, and Potentials of Capillary (Gel) Electrophoresis C(G)E	142
4.4	Data Analysis and Interpretation	146
4.5	Exoglycosidase Sequencing of Glycans	148
4.6	Coupling Capillary Electrophoresis with Mass Spectrometry	151
4.7	Latest Developments: Miniaturization of CE Systems—Microchip CE	153
4.8	Application of C(G)E for Immunoglobulin Analysis	154
4.9	Conclusion	162
References		162

Abstract The in-depth characterization of protein glycosylation has become indispensable in many research fields and in the biopharmaceutical industry. Especially knowledge about modulations in immunoglobulin G (IgG) *N*-glycosylation and their effect on immunity enabled a better understanding of human diseases and the development of new, more effective drugs for their treatment. This chapter provides a deeper insight into capillary (gel) electrophoresis-based (C(G)E) glycan analysis, addressing its impressive performance and possibilities, its great potential regarding real high-throughput for large cohort studies, as well as its challenges and limitations. We focus on the latest developments with respect to miniaturization and mass

---

Samanta Cajic and René Hennig contributed equally with all other contributors.

S. Cajic
Max Planck Institute for Dynamics of Complex Technical Systems, Magdeburg, Germany

R. Hennig (✉) · E. Rapp
Max Planck Institute for Dynamics of Complex Technical Systems, Magdeburg, Germany

glyXera GmbH, Magdeburg, Germany
e-mail: r.hennig@glyxera.com

R. Burock
glyXera GmbH, Magdeburg, Germany

© The Authors(s) 2021
M. Pezer (ed.), *Antibody Glycosylation*, Experientia Supplementum 112,
https://doi.org/10.1007/978-3-030-76912-3_4

spectrometry coupling, as well as data analysis and interpretation. The use of exoglycosidase sequencing in combination with current C(G)E technology is discussed, highlighting possible difficulties and pitfalls. The application section describes the detailed characterization of *N*-glycosylation, utilizing multiplexed CGE with laser-induced fluorescence detection (xCGE-LIF). Besides a comprehensive overview on antibody glycosylation by comparing species-specific IgGs and human immunoglobulins A, D, E, G, and M, the chapter comprises a comparison of therapeutic monoclonal antibodies from different production cell lines, as well as a detailed characterization of Fab and Fc glycosylation. These examples illustrate the full potential of C(G)E, resolving the smallest differences in sugar composition and structure.

Keywords Capillary gel electrophoresis · *N*-glycosylation · Biopharmaceuticals · Immunoglobulin · IgA · IgD · IgE · IgG · IgM · Monoclonal antibody · mAb · APTS · xCGE–LIF · CE–MS · Chip–CE · Exoglycosidase

Abbreviations

APTS	8-aminopyrene-1,3,6-trisulfonic-acid
CE	Capillary electrophoresis
CGE	Capillary gel electrophoresis
CHO	Chinese hamster ovary
CQA	Critical quality attribute
CZE	Capillary zone electrophoresis
DB	Database
EOF	Electroosmotic flow
ESI	Electrospray ionization
Fab	Fragment antigen binding
Fc	Fragment crystallizable
Fuc	Fucose
Gal	Galactose
GlcNAc	*N*-acetylglucosamine
GU	Glucose units
HAGR	Hepatic asialo-glycoprotein receptor
HILIC	Hydrophilic interaction liquid chromatography
HMOS	Human milk oligosaccharides
HPLC	High-performance liquid chromatography
Ig	Immunoglobulin
LIF	Laser-induced fluorescence detection
mAb	Monoclonal antibody
MALDI-TOF-MS	matrix-assisted laser desorption ionization with time-of-flight MS
Man	Mannose

MS	Mass spectrometry
MS/MS	Tandem mass spectrometry
MTU	Migration time unit
MTU″	Double aligned migration time units
Neu5Ac	*N*-acetyl-neuraminic acid
Neu5Gc	*N*-glycolyl-neuraminic acid
NMR	Nuclear magnetic resonance
Sia	Sialic acid
SNFG	Symbol nomenclature for glycans
UPLC	Ultra-performance liquid chromatography
xCGE-LIF	Multiplexed capillary gel electrophoresis with laser-induced fluorescence detection

4.1 Historical Background

Electrophoresis was born more than 200 years ago (Ruess 1809), even long before the concept of chromatography was described. However, it took about 150 more years before the use of capillaries was introduced (Hjertén 1967). From the late 1960s, it took an additional decade to establish capillary electrophoresis (*CE*) as it is most widely known today, enabling separations that seemed unattainable at that time (Jorgenson and DeArman Lukacs 1981; Jorgenson and Lukacs 1981a, b). The increased efficiency and the amazing separation capabilities together with short analysis times, induced a growing interest among the scientific community. Since then, CE has constantly been improved and has become an important tool in the analysis of a wide class of compounds, from small ions over amino acids and peptides/proteins to large DNA fragments (Saraswathy and Ramalingam 2011). However, the employment of CE for glycan analysis lagged behind other commonly used analytical techniques. Attempts to use CE for glycan analysis started in the 1990s, but they were rather humble and not broadly embraced—neither by the glycoscientific community nor by industrial laboratories. Today, 40 years since the potential of CE was recognized, the advantages of the method for analysis of glycans are becoming increasingly obvious and appreciated. Ready-to-use CE methods and kits on the market (glyXera 2020; Thermo Fisher Scientific 2018; Ludger 2018; ProZyme 2018; SCIEX 2018; PerkinElmer 2018) enable fast and robust comparison of glycan profiles, which makes the technology more routine in academic research and industry. In the following sections, we will emphasize some of the reasons why CE is gaining popularity and underline why this technology is attractive for the biopharmaceutical industry, especially regarding the analysis of immunoglobulin (*Ig*) glycosylation. Additionally, some current and future challenges, as well as possible future methodological developments, will be addressed.

4.2 Background: Principles of Capillary (Gel) Electrophoresis (C(G)E)

Capillary electrophoresis is the umbrella term for various capillary electrophoretic separation techniques and methods, such as capillary zone electrophoresis (*CZE*) and capillary gel electrophoresis (*CGE*). The differential migration of charged molecular species (ions) in a narrow capillary (25–75μm) containing an electrolyte solution under the influence of high voltage (usually 10–30 kV, generating an electric field) is the basis for separation by CE (Watson 2012). An analyte is attracted to either the anode (positive electrode) or the cathode (negative electrode), and its movement (electrophoretic mobility) in an electric field results from both electrical and frictional force contributions. The movement of a charged analyte through a conductive solution is dependent on the charge of the analyte and the magnitude of the applied electric field. Additionally, a frictional force will impede the movement induced by the electrical field to a certain extent, dependent on the viscosity of the electrolyte solution and the analyte radius/molecular size or shape. Thus, the mobility and, therefore, the separation is based on the size-to-charge ratio of the analytes. Besides electrophoretic velocity, electroosmotic flow (*EOF*) is influencing the movement of the analyte through the capillary. EOF is the bulk liquid motion in a capillary when a high voltage (electric field) is applied. In a capillary composed of fused silica, the surface possesses negative charges (SiO^-) over a wide pH range (pH 4–12). This negatively charged surface attracts positively charged electrolytes from the running buffer. Along an electric field, these cations (enriched at the capillary wall) move towards the cathode. Due to solvation and liquid viscosity, these cations drag the surrounded bulk solvent with them along the capillary axis, creating a net flow toward the cathode, the EOF. Thus, all analytes, irrespective of their charge, are pushed in one direction, enabling simultaneous separation and detection of cationic (EOF + ion migration = fastest), neutral (EOF = middle) and anionic (EOF − ion migration = slowest) analytes (Holland et al. 1997). Although this is rather a rough summary of the separation mechanism, the reader should recognize that the parameters involved in optimizing the technique to produce separation are very complex. However, the effort pays off as very small changes in molecular structure can lead to quite marked differences in migration if variables are carefully balanced. There is a large volume of literature regarding CE separations of glycans exploiting a variety of possible setups, methods, modes, and parameters (e.g., capillary material, dimensions, and surface coating, voltage, current, polarity, temperature, buffer type and concentration, pH, additives/modifiers, injection technique, detection mode, etc.) that is beyond the scope of this review (Campa and Rossi 2008; Ruhaak et al. 2010a; Lu et al. 2018). However, due to a boom of applications in the field of glycan analysis and particular interest in the biopharmaceutical industry, one technique deserves to be further dissected—capillary gel electrophoresis (CGE).

After providing the complete genetic blueprint of human life (Collins et al. 2004), CGE-based Genetic Analyzers (DNA analyzers) found their application in the field of glycan analysis, with a more complex challenge but a new hope of deciphering the glycome (Schwarzer et al. 2008; Laroy et al. 2006; Callewaert et al. 2001; Ruhaak et al. 2010b). These instruments combine four great advantageous features, which make them very powerful analytical tools in glycoscience. First, in CGE, the abovementioned EOF is typically completely suppressed by employing a neutral capillary (permanently or dynamically coated) filled with a gel mesh polymer. The use of such a gel buffer increases the viscosity of the electrophoresis medium. Consequently, due to the role of viscosity in frictional drag, this increases the time analytes spend in the capillary (decrease of electrophoretic mobility), while decreasing diffusion and thereby improving the separation (Watson 2012; Guttman et al. 1994; Luo et al. 2010). Thanks to this first advantageous feature, glycans are separated based on mass-to-charge ratio and molecular size/shape (hydrodynamic volumes) with high resolution (Guttman et al. 1996a; Guttman and Herrick 1996). Second, the large surface-to-volume ratio of narrow capillaries enables a very good heat transfer. This in turn raises the possibility to apply very high electric fields, which increases the electrophoretic mobility and consequently decreases the analysis time. Third, besides its high separation efficiency in a short time, this method is also attractive due to the possibility to employ laser-induced fluorescence (*LIF*) detection, known for its impressive sensitivity (low attomole range) (Guttman 1996; Hennig et al. 2011a). The fourth advantage is the possibility to employ a multi-capillary format (multiplexing), incorporating up to 96 capillaries in parallel, so that hundreds of samples can be measured by CGE per day (Mittermayr et al. 2013). As a result, modern multiplexed capillary gel electrophoresis with laser-induced fluorescence detection (*xCGE-LIF*) instruments offer an amazing high-throughput, separation performance and sensitivity. Another attractive option provided, e.g., by Applied Biosystems Genetic Analyzers, is a recording of up to six different fluorescent dyes (in six independent detection traces/channels) simultaneously within one run (all excited by only one laser). As described later, this special feature has been exploited for internal alignment of migration times (glycans and internal base pair standard labeled with different dyes), giving long-time stable migration times (Callewaert et al. 2001; Laroy et al. 2006; Schwarzer et al. 2008; Reusch et al. 2014). The basic setup of a Genetic Analyzer ((x)CGE-LIF system) and the principle of glycan separation is depicted in Fig. 4.1.

The majority of glycans are uncharged and therefore would neither be electrokinetically injected nor migrate in an electric field when EOF is suppressed. Additionally, glycans lack a chromophore/fluorophore necessary for optical detection (Guttman et al. 1996b). Thus, an ideal label for C(G)E-LIF needs to have high fluorescent yield, carry (sufficient) negative charges stable over a broad pH range (for electric field-mediated mobilization), and it needs to have an excitation wavelength corresponding to the output wavelength of commercially available lasers (commonly argon-ion laser) (Briggs et al. 2009). All these requirements are met in 8-aminopyrene-1,3,6-trisulfonic-acid (*APTS*) fluorescent dye (Evangelista et al. 1995; Ruhaak et al. 2010b), which is almost universally employed in CE-based glycan analysis today. High absorptivity and quantum yield of APTS enable a highly sensitive LIF detection, thus making CE an appropriate method for high-sensitivity glycan analysis, even with low sample amounts (Guttman et al. 1996a).

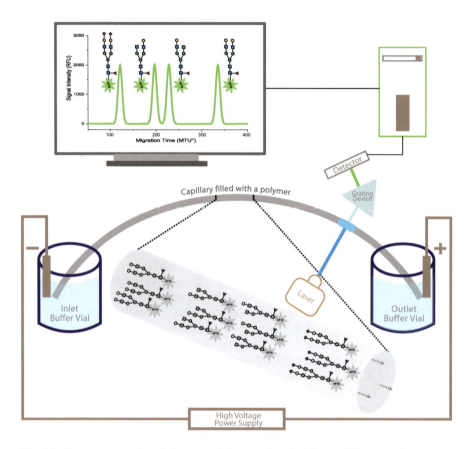

Fig. 4.1 Common separation of glycans utilizing an xCGE-LIF-based DNA sequencing system. High injection voltage is applied for a brief period, causing sample (glycans labeled with a negatively charged fluorescent dye, e.g., APTS) to enter the inlet of the gel-filled capillary by electromigration (electrokinetic injection). The electric field is applied under reverse polarity (cathode at injection and anode on the detection side) so that anions (i.e., negatively charged labeled glycans) migrate towards the detection window. As demonstrated, fluorescently labeled glycans are separated, based not only on the differences in their mass-to-charge ratios but also on their hydrodynamic volumes. Although sequencing systems benefit from multi-capillary format, only one capillary is presented here for simplicity. The x-axis of the schematic electropherogram is given in double aligned migration time units (MTU″). The signal intensity of the y-axis is given in relative fluorescent units (RFU). Symbolic representation of *N*-glycan structures follows the guidelines of Symbol Nomenclature for Glycans (SNFG) (Varki et al. 2015)

4.3 Performance, Benefits, and Potentials of Capillary (Gel) Electrophoresis C(G)E

CE provides remarkable separation efficiencies and resolution in a very short time, which makes this technology appealing for the analysis of structurally complex and diverse molecules like glycans. Recent work has shown that CE can separate even challenging positional and linkage isomers in one single analysis run. For example, the method is capable of distinguishing even carbohydrate position in a glycan structure [e.g., nonreducing terminal residue on α1–3 arm from terminal residue on α1–6 arm of the core structure (see Fig. 4.2b, c) (Hennig et al. 2016; Schwedler et al. 2014a, b; Chen et al. 2017; Huang et al. 2017; Guttman et al. 2015)], along with position and linkage of fucose (*Fuc*) [e.g., α1–6 core Fuc from α1–3/1–4 Fuc on antenna *N*-acetylglucosamine (*GlcNAc*) or α1–2 Fuc on galactose (Hennig et al. 2016; Konze et al. 2017; Thiesler et al. 2016; Weiz et al. 2016; Schwedler et al. 2014a)], type and linkage of sialic acids (*Sia*) [e.g., *N*-acetylneuraminic acid (*Neu5Ac*) from *N*-glycolylneuraminic acid (*Neu5Gc*), see Fig. 4.3d (Abeln et al. 2017), and α2–3 from α2–6 (Hennig et al. 2016; Konze et al. 2017; Thiesler et al. 2016; Donczo et al. 2017; Meininger et al. 2016)], and linkage of galactose (*Gal*) [e.g., β1–3 Gal from β1–4 Gal (Konze et al. 2017; Thiesler et al. 2016; Schwedler et al. 2014a; Muñoz et al. 2019)]. This feature becomes especially advantageous when thinking about the importance of determining immunogenic α-Gal and Neu5Gc on glycoprotein therapeutics (Chung et al. 2008; Teranishi et al. 2002; Van der Linden et al. 2000). Additionally, the fact that only α2–6, and not α2–3 sialic acid, affects the anti-inflammatory activity of IgG antibody (Anthony et al. 2008) makes it crucial to have a method capable of their differentiation.

Besides speed, resolution, sensitivity, and simplicity, compatibility with a wide range of samples is one of the biggest advantages of C(G)E. Due to the robustness of the method, only a modest sample purification is necessary. Large amounts of cell debris (Konze et al. 2017; Thiesler et al. 2016; Abeln et al. 2017) or other impurities like large excess of fluorescent labeling dye (Croset et al. 2012) will not prevent a successful analysis. For comparison, in LC-based analytics, such cell debris can block the column or at least accumulate inside of it, resulting in a significant decrease of separation power. Consequently, a costly column exchange or a time-consuming column cleaning procedure is needed. In contrast, in CGE, the separation matrix (polymeric gel) is exchanged after each run; thus impure samples are not problematic in terms of clogging or sample carryover. Even sample clean-up after exoglycosidase digest—for LC an obligatory step—can be omitted, further reducing analysis time and workload (Szigeti and Guttman 2017).

Moreover, CE analysis is beneficial when enormous differences in analyte concentrations are present in one sample. For example, it could be shown by Kottler et al. that lactose, which represents up to 85% of the free oligosaccharides in human milk, did not interfere with the analysis of other more complex but low abundant human milk oligosaccharides (*HMOS*) (Coppa et al. 1993). While in xCGE-LIF the lactose peak appears without any peak shape deformation, even when injected in amounts far above the linear detection range, for LC-based methods, lactose needs to be depleted from the milk samples before HMOS analysis to avoid massive peak broadening and tailing as a result of column overload (Kottler et al. 2013). Similar problems might be faced when using mass spectrometry (*MS*).

After reviewing all the given advantages, an important aspect of the *N*-glycan analysis is the costs. Resources in academia and industry are often quite limited, especially if hundreds or thousands of samples need to be analyzed in a short time—

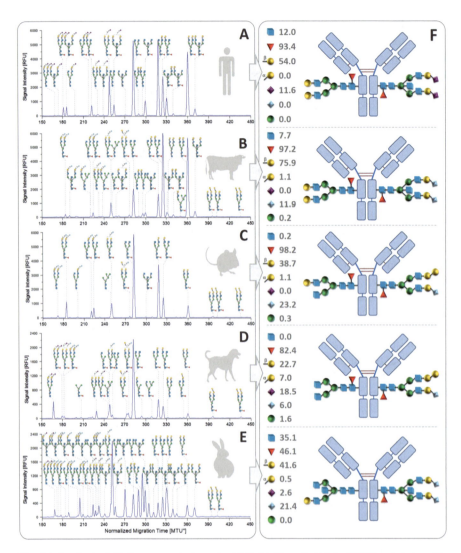

Fig. 4.2 Bovine IgG *N*-glycome analysis on the (**a**) PerkinElmer LabChip GXII Touch (microchip CE), (**b** and **c**) glyXera glyXboxCE™ built on Applied Biosystems Genetic Analyzer 3130 (xCGE-LIF), (**d**) Dionex Ultimate 3000RS UPLC system equipped with Waters ACQUITY UPLC Glycan BEH Amide column. For (**a**), labeled bovine IgG *N*-glycans were prepared using the PerkinElmer Glycan Profiling Assay Reagent Kit (PerkinElmer 2018). For (**b**) and (**c**), APTS-labeled *N*-glycans were prepared from bovine IgG following a published procedure (Hennig et al. 2016; Huffman et al. 2014). For (**d**), *N*-glycans were enzymatically released from bovine IgG, AB-labeled (Ruhaak et al. 2010c) and subsequently purified (Ruhaak et al. 2008, 2012) before measurement by HILIC-UPLC-FLD. The signal intensity of the y-axis is given in relative fluorescence units (RFU). Symbolic representation of *N*-glycan structures follows the guidelines of Symbol Nomenclature for Glycans (SNFG) (Varki et al. 2015)

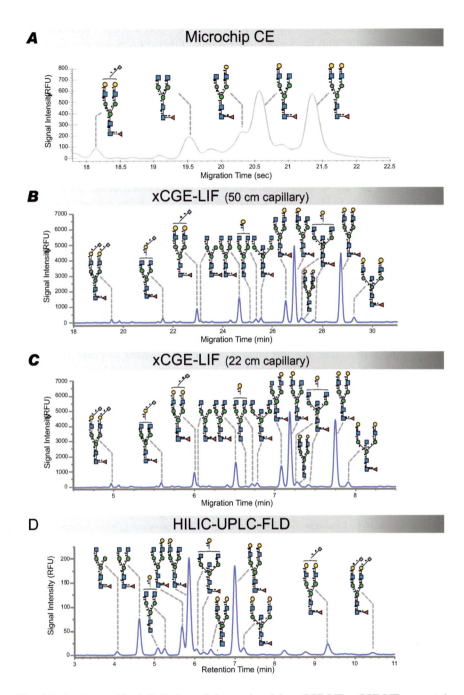

Fig. 4.3 Species-specific IgG *N*-glycosylation analyzed by xCGE-LIF. xCGE-LIF generated fingerprints of APTS-labeled *N*-glycans derived from human IgG (**a**), bovine IgG (**b**), mouse IgG (**c**), dog IgG (**d**) and rabbit IgG (**e**) (Vendors: Dog IgG—Abcam, other IgGs—Sigma-Aldrich). APTS-labeled *N*-glycans were prepared using the glyXprep[48] kit (glyXera 2020) by carefully

i.e., in real high-throughput. Here glycan analysis by xCGE-LIF, utilizing DNA analyzers originally built for Sanger sequencing, outperforms conventional LC and MS approaches. With low material costs (due to minimal sample preparation) and low operating costs (low maintenance, no organic solvents and solvent disposal costs), the overall cost are significantly lower for xCGE-LIF (Huffman et al. 2014; Mahan et al. 2015) than for LC or LC–MS. Moreover, data analysis and interpretation became user-friendly with available software solutions (see section "Data analysis and interpretation"), further reducing expenses for hands-on time and expertise.

Finally, C(G)E-based methods were also successfully applied for the absolute quantification of carbohydrates (Sarkozy et al. 2021; Eussen et al. 2021), which completes the wide field of applications. Taken together, unprecedented separation together with low costs, robustness, speed of analysis, multiplexing capability, high sensitivity, and nano- to femtoliter injection volumes make CE a veritable competitor to other more traditionally used techniques for glycan analysis.

4.4 Data Analysis and Interpretation

One way to obtain structural information and assign glycans to peaks is through co-injection of glycan standards, which are commercially available and fully characterized (Guttman et al. 1996a, b; Guttman and Herrick 1996; Reusch et al. 2014). In a so-called spiking experiment, the glycan is indirectly identified by the height increase of an often perfectly Gaussian-shaped CE peak. Unfortunately, only a very limited number of glycan standards are available on the market. A more general approach for the identification of glycans in a separation-based method with spectroscopic detection relies on comparing migration times (for CE separations)

Fig. 4.3 (continued) following the kit instructions. Data processing was performed using glyXtoolCE™ (glyXera 2021). Data processing comprised alignment of migration times to two orthogonal internal standards, resulting in a double aligned x-axis in migration time units (MTU″). The signal intensity of the y-axis is given in relative fluorescence units (RFU). *N*-glycan structures were assigned via database matching using glyXtoolCE™ (in combination with glyXbase™) and confirmed by exoglycosidase sequencing as published by Thiesler et al. (2016). (**f**) Overview of IgG *N*-glycome characteristics with: blue square for bisected *N*-glycans, red triangle for core fucosylated *N*-glycans, yellow circle for glycans with terminal galactose in β1–4 or α1–3 linkage, purple diamond for glycans with terminal *N*-acetylneuraminic acid (Neu5Ac), light blue diamond for glycans with terminal *N*-glycolylneuraminic acid (Neu5Gc) and green circle for oligo-mannosidic *N*-glycans. Numbers indicate their approximated relative abundances in percent. IgG illustrations indicate the broad characteristics of the *N*-glycan structures attached to the species-specific IgGs. IgG illustrations are for visualization purposes only and do not reflect a real combination of *N*-glycans or their linkage position to the protein backbone. Symbolic representation of *N*-glycan structures follows the guidelines of Symbol Nomenclature for Glycans (SNFG) (Varki et al. 2015)

between a sample and a glycan database (*DB*). In this case, a fully characterized glycan standard (pure or inside of a complex mixture) is analyzed once, and its migration time is stored inside a glycan DB. The DB enables identification for all further analyses without the need for spiking the glycan standard again. Accordingly, only glycan structures with known migration times can be identified. However, the buildup of a glycan DB requires a very reliable CE setup and method with very stable and reproducible migration times.

A long-term stable migration time, independent from instrument, operator, and lab, in combination with a DB, comprising a broad variety of glycan structures, eases the structural annotation. Several research groups tackled this problem in a similar way. By running an accompanying oligosaccharide standard (e.g., a glucose ladder or a set of single oligosaccharides) for each sample or set of samples, the migration time can be aligned to this standard, resulting in a standardized time axis. Hence, the migration time of individual glycan peaks can be given in standardized migration time units (*MTU*). The MTU of a peak can now be used to search inside a dedicated DB to assign the corresponding glycan structure. Often a glucose ladder is used as an external alignment standard (in a separate run), resulting in a standardized migration time axis in Glucose Units (*GU*) (Guttman et al. 1996a; Guttman and Herrick 1996; Mittermayr et al. 2013; Laroy et al. 2006; Mittermayr and Guttman 2012; Liu et al. 2007). Recently, it was shown that a co-injection of a bracketing triple internal standard (maltose, maltotriose, and maltopentadecaose; by Agilent) could negate the need for additional glucose ladder run (Jarvas et al. 2016)—at least for simple samples and regarding short term repeatability. As bigger variations of migration times in-between the bracketing standard can occur (region of interest), Hennig et al. use a patented orthogonal double alignment, combining the standard bracketing approach and the ladder approach (Hennig et al. 2015, 2016; Huffman et al. 2014), resulting in correspondingly double aligned migration time units (MTU''). An alignment to the bracketing standard, labeled with the same fluorescent dye as the sample, is complemented with an orthogonal alignment to a DNA base pair standard, labeled with a different dye. Both standards are co-injected with the sample and detected in separate spectral traces. The additional DNA ladder-supporting points for the alignment significantly improved the (long-term) stability of the migration times, which allowed to build up a large *N*-glycan DB with more than 400 structures (glyXbase™ by glyXera). Recently, the glycan analysis software glyXtoolCE™ (by glyXera) was developed (Hennig et al. 2011b, 2016; glyXera 2021), which automates migration time alignment together with an instant structural assignment and furthermore provides the background adjustment raw data smoothing, peak picking, integration, relative quantification and sample comparison (Hennig et al. 2011b, 2016; Behne et al. 2013). Similar logic is behind the recent approach from Feng et al. (2017); however, alignment was performed in a non-automated fashion and with only a single normalization point to an internal glycan standard. Related bioinformatics tools such as GUcal were developed that can carry out GU value calculation in an automated fashion and concomitantly search through the database (at the moment comprised of 92 structures, available at www.glycostore.org) (GU database) for structural assignment (Jarvas et al. 2015, 2018). All these developments facilitate data processing and interpretation and make CE an easy-to-use high-throughput tool.

Nevertheless, while spiking experiments and database comparison give a strong indication of the glycan structures, they do not fully confirm them. Also, alternative approaches, when applied alone, do not allow full structural elucidation of glycans. For example, with single-stage MS, glycan composition can be estimated based upon the addition of monosaccharide constituent masses. However, glycans with different structures but identical monosaccharide composition result in identical mass values and cannot be distinguished with MS. Even with tandem mass spectrometry (*MS/MS*), isobaric stereoisomers [like hexoses, galactose, and mannose (*Man*)], positional isoforms, and the different types of glycosidic linkages are difficult or impossible to determine. Here chromatographic and electrophoretic separations are advantageous with their ability to resolve closely related positional and linkage glycan isomers. Nevertheless, the structural variety of glycans is enormous, and resolution capabilities are not indefinite. Thus, what appears as a single peak often comprises a mixture of glycans (multi-structure peak). For that reason, also glycan analytical methods based on chromatographic and electrophoretic separation need a second dimension to provide correct and complete sample structural information. Thus, they are often complemented with an additional technique, such as MS or exoglycosidase sequencing. Principles, benefits, applications, and some limitations of exoglycosidase sequencing are outlined in the section below. A description of different CE and MS coupling possibilities, together with the overview of recent CE–MS-based glycomics studies, is provided in section "Coupling capillary electrophoresis with mass spectrometry."

4.5 Exoglycosidase Sequencing of Glycans

Exoglycosidases are enzymes that cleave the terminal carbohydrate monomers on the non-reducing end of a glycan. They can be highly selective for specific monosaccharide types (e.g., Gal, Fuc, or Man), linkage orientation (α or β) or the position of the glycosidic linkage (e.g., 1–3, 1–4, or 2–3). Apparently, the specific enzymatic cleavage of monosaccharide residues serves as the most efficient approach for glycan structural elucidation in CE- and LC-based analysis because no additional equipment is needed. The conversion of a glycan by an exoglycosidase is harnessed with CE by analyzing the sample before and after the treatment. Digest-induced charge, size, and shape changes can be observed in CE as a change in migration time. Depending on enzyme specificity, information on monosaccharide type and the number of cleaved residues, sequence, or even linkage and anomericity can be obtained, as detailed shown by Cajic et al. (Thiesler et al. 2016) (in the supplemental material). Analysis of peak positions and relative peak areas before and after an exoglycosidase digest is therefore a means of exhaustive structural annotation. These digests can be conducted following four different strategies. First option is sequencing in a parallel fashion (Hennig et al. 2016; Thiesler et al. 2016; Feng et al. 2017), with the sample being evenly split into one aliquot per enzyme and simultaneous analysis of enzyme-treated aliquots. Second option is sequencing with parallel exoglycosidase treatments, mediated by (different) carefully designed mixtures of exoglycosidases (Szigeti and Guttman 2017; Guttman and Ulfelder 1997; Guttman 1997). Through multiple combinations of exoglycosidases (parallel enzyme array), glycans are sequenced down to their Man3-core structure (Prime and Merry 1998; Rudd and Dwek 1997). Instead of splitting the sample, it is possible to apply individual enzymes (third option) or mixtures of enzymes (fourth option) sequentially to the same sample, followed by an analysis of each step (Szigeti and Guttman 2017; Guttman et al. 2015; Mechref et al. 2005; Ma and Nashabeh 1999). This sequential treatment of a single sample is often performed when the sample amount is limited or when the sample complexity is too high. It is common practice, e.g., to reduce sample complexity by sialidase treatment before peaks are annotated (Muñoz et al. 2019; Vanderschaeghe et al. 2010; Zhuang et al. 2007, 2011). Each strategy has its advantages and drawbacks, parts of which have been already addressed elsewhere (Mittermayr et al. 2013; Prime and Merry 1998; Holland et al. 2017) and are consequently not discussed here.

Approaches to enzymatically elucidate a glycan structure in a glycan mixture are quite diverse. For instance, it was shown that the reaction of a sequential enzyme treatment could be performed inside the CE autosampler, with the sample being injected from the enzyme reaction vial directly into CE (Szigeti and Guttman 2017). Another interesting alternative to offline methods is the incorporation of enzymes into the capillary (online digest). In-capillary enzymatic digests are a rapid option, with incubation times down to only a few minutes. By passing an enzyme plug inside the capillary, the digest of the samples is accomplished during the separation process itself (mixing via polarity switching, stopped flow or low flow incubations) (Luo et al. 2010; Holland et al. 2017; Archer-Hartmann et al. 2011a, b; Yagi et al. 2011; Gattu et al. 2017; Yamagami et al. 2017). The thermally tunable phospholipid nanogels are especially attractive due to the reported enhancement of the stability and performance of exoglycosidase enzymes (Holland et al. 2017; Yamagami et al. 2017). However, in-capillary endeavors are sometimes incompatible with certain enzymes (Yagi et al. 2011; Yamagami et al. 2017) and often accompanied by a loss in separation efficiency (Archer-Hartmann et al. 2011a, b).

Although exoglycosidases can be combined with many other methods, due to already praised advantageous features of CE, combining the specificity of exoglycosidase enzymes with the strengths of CE seems most appropriate to determine glycan structures unambiguously and with minimum effort. Because of the high resolving power of CE, multi-structure sequencing of a complex glycan pool can be performed requiring no prior isolation of the individual glycans, resulting in significant labor and time savings. Even if enzymes are cleaving terminal residues without (much) specificity regarding linkage or position, CE can still provide this information due to its high separation efficiency. Thanks to the very good reproducibility of the peak areas, even small differences in glycan abundance can be detected reliably so that a digest-induced shift of a minor structure under a peak (less than 5% of total peak area) can be successfully tracked. Because of the electrokinetic loading system and highly sensitive LIF detection, only a very low amount of sample is required per injection, which means that most of the sample remains for other analyses. High-throughput capabilities obtained from multi-capillary CE formats and fast separations qualify CE as an ideal screening method of, for example, optimal reaction conditions or batch-to-batch variations. Excellent migration time stability of some CE methods allows migration time-database matching for original and exoglycosidase digested products, making analysis much faster and more efficient. All these characteristics make CE in conjunction with exoglycosidase sequencing a valuable tool for detailed glycan analysis.

Despite all acknowledged benefits, the use of enzymes for determining the glycan structure requires great care and expertise and should involve a large dose of healthy skepticism about any deduced structure (Jacob and Scudder 1994). Even if certain sugar residues are exposed at the reducing end of the glycan, these residues are not always removed by the exoglycosidases with the corresponding specificities. This resistance can be due to the strict linkage specificities of the exoglycosidases, due to steric hindrance of neighboring sugar residues or due to the attached (fluorescent) label. For example, hydrolysis of bisecting GlcNAc in human IgG N-glycans upon hexosaminidase treatment or the core Fuc upon bovine kidney α-fucosidase treatment might be incomplete or completely impeded (Mittermayr et al. 2013; Laroy et al. 2006; Guttman and Ulfelder 1997; Jacob and Scudder 1994; Kamerling and Gerwig 2007; O'Flaherty et al. 2017). In addition, used enzymes may have additional selectivity for other structural features, such as local and nonlocal branching. For example, β-galactosidase can hydrolyze β1–3,4,6-linked Gal exposed at the non-reducing end of an antenna without Fuc attached to the subterminal GlcNAc. If Fuc is attached to the antenna GlcNAc, the β-galactosidase will not hydrolyze the Gal (Yu et al. 2011). Hence, a particular exoglycosidase will cleave a terminal monosaccharide only if all its specificity requirements are met. However, the purity and composition of both glycan material and used reagents can greatly affect enzyme activity and pose an additional difficulty for accurate and reliable structural assignment (Jacob and Scudder 1994). For example, variation in enzyme activity has been observed between different batches from one vendor and different sources (vendors) of the same enzyme. Enzyme activities and side activities are often tested only on artificial p-nitrophenyl glycosides or other simple substrates, even though activity on natural complex glycans can significantly differ (Kobata 2013). Additionally, vendor-added salts or other additives have the potential to interfere with different aspects of the analysis. For instance, electrokinetic injection in CE is sensitive to high salt content originating from the non-volatile digestion media (Mittermayr et al. 2013; Laroy et al. 2006), which is usually unavoidable with commercial enzymes. Finally, a major issue can be the purity of the exoglycosidase. Many of the enzyme preparations contain a certain amount of contaminant enzymes, ranging from very low to unacceptably high (Jacob and Scudder 1994; Kamerling and Gerwig 2007). Therefore, even when a positive or negative result is obtained by digestion with exoglycosidases, it does not necessarily confirm the presence or absence of the corresponding sugar residues at the non-reducing end of glycan, respectively

(Kobata and Takasaki 1992). Thus, using dedicated positive and negative controls for each enzyme reaction is unavoidable.

Despite the abovementioned considerations, it was shown that, when used cautiously, exoglycosidase can effectively elucidate even subtle changes in glycan structures, including linkage type (e.g., α2–3 versus α2–6 Sia or β1–3 versus β1–4 Gal) (Callewaert et al. 2001; Hennig et al. 2016; Konze et al. 2017; Thiesler et al. 2016), anomericity (e.g., immunogenic α-Gal versus β-Gal) (Abeln et al. 2017; Szabo et al. 2012; Yagi et al. 2012) and position on glycan (e.g., core Fuc versus Fuc on antenna GlcNAc or Gal) (Konze et al. 2017; Thiesler et al. 2016; Meininger et al. 2016; Liu et al. 2007; Zhao et al. 2014)—a complexity which is often inaccessible by methods other than nuclear magnetic resonance (NMR) spectroscopy.

4.6 Coupling Capillary Electrophoresis with Mass Spectrometry

Combining CE, one of the most effective isomer separation tools, with the information-rich MS technique is a mutually advantageous and powerful alliance for in-depth glycan analysis. Two setups are widely applied: the direct linking of CE to MS, the so-called online CE–MS, or a time-separated analysis, the offline CE–MS, in serial or parallel mode. For a serial offline approach, a CE instrument can be modified to spot the eluent from the capillary directly onto a target-plate for matrix-assisted laser desorption ionization with time-of-flight MS (*MALDI-TOF–MS*) detection (Suzuki et al. 1997). Consequently, separation is done by CE and identification by MS. Furthermore, parallel offline approaches are gaining popularity, especially for MS-based glycopeptide or intact glycoprotein analysis (glycoproteomics). Here, the vast diversity of protein/peptide–glycan combinations (driven by the complexity of glycans) is dramatically increasing the computing time for the interpretation of MS spectra (Thaysen-Andersen and Packer 2014; Yang et al. 2017). For reduction of computing time, an initial global characterization of the *N*-glycome can be performed by CE. The resulting glycan list is used to search against a targeted set of defined *N*-glycan structures as variable protein/peptide modifications (Thaysen-Andersen and Packer 2014; Parker et al. 2013; Lebede et al. 2021; Pralow et al. 2021; Pioch et al. 2018). This pre-knowledge of attached glycan structures greatly reduces the search space, which in turn significantly decreases the computing time. This allows relatively fast and in-depth analysis of quite complex samples, even with demanding glycoproteomic approaches.

The major advantage of online CE–MS is that glycans are identified by both their differential migration times and their molecular masses and/or fragmentation patterns in one analysis. The most common interface used for the transfer of glycans from the liquid phase of CE to the gas phase of MS is electrospray ionization (*ESI*) since this soft ionization method allows the direct transfer of the glycans from separation capillary to MS and is rather easy to implement. A detailed description of the advantages and drawbacks of the different interfacing options is beyond the scope of this chapter and has been already provided elsewhere (Simó and Cifuentes 2005; Maxwell and Chen 2008; Zhong et al. 2014; Zhang et al. 2017), together with the comprehensive overview of CE–MS-based glycomics studies (Pioch et al. 2012; Nakano et al. 2011; Mechref and Novotny 2009; Mechref 2013; Lindenburg et al. 2015; Dotz et al. 2015).

Although the CE–MS technology was introduced already 30 years ago (Olivares et al. 1987; Smith et al. 1988a, b) and continues to advance ever since, there has been relatively little work performed on merging CE with online electrospray MS for analysis of released glycans, in particular, when compared to the hyphenated LC–MS techniques. Coupling the high resolving power of CE and structural information of MS in one system often comes at the expense of resolution, sensitivity, analysis time, reproducibility and/or robustness. Very often, the separation conditions giving the unmatched separation efficiency to CE are detrimental to MS performance and vice versa. For example, gels and buffers often used in CE because of resolution improvement are not volatile and are not suitable for CE–ESI–MS since they often suppress the ionization of the analyte, yielding poor MS sensitivity or even clog the system. On the other hand, the choice of volatile, "MS-friendly" buffers can not only affect reproducibility and analysis time but even negatively impact separations. This necessary compromise between optimal MS performance and elevated CE separation efficiency, plus its lacking robustness, are the major reasons that CE–MS has still not been widely adopted as a routine method for glycan analysis. Another concern to be addressed is that with all the limitations (small sample loads and consequent low sample concentration, sample dilution by the sheath-liquid introduction, labeling incompatibility issues, and ionization efficiency considerations) inherent in CE–MS analysis, obtaining detailed linkage and/or positional information by MS/MS is often challenging.

Even though we cannot yet talk in terms of widespread acceptance, still some efforts to make CE–MS a more viable approach in the field are ongoing. Applications are mainly limited to the use of capillary zone electrophoresis (CZE) coupled to MS for protein characterization and glycan identification. Besides the analysis of intact glycoproteins, CZE–MS is used to characterize side-specific microheterogeneity on glycopeptide level (Pioch et al. 2012; Lindenburg et al. 2015; Dotz et al. 2015). The analysis of native or derivatized glycans is often performed on complex samples (e.g., plasma *N*-glycomes), taking advantage of the good separation performance of CE (Snyder et al. 2017; Lageveen-Kammeijer et al. 2019; Huhn et al. 2012; Jayo et al. 2012). CZE–MS-based characterization of single protein glycosylation is performed less frequently (Jayo et al. 2014) and focuses mainly on monoclonal antibodies (*mAbs*) (Pioch et al. 2012; Mechref 2013; Dotz et al. 2015; Gennaro and Salas-Solano 2008; Bunz et al. 2013a, b).

4.7 Latest Developments: Miniaturization of CE Systems—Microchip CE

Easy-to-use technologies that allow rapid and efficient screening in a compact format at the best cost are in great demand, especially in the biopharmaceutical industry. Miniaturization of the electrophoretic process onto microchips (microchip CE) holds a great promise to meet all these needs. No other technology seems better suited for miniaturization owing to two main hallmarks of CE. Firstly, sophisticated pumping systems are not necessary since the separation is driven by an applied electric field. Secondly, separation efficiency depends primarily on the strength of the electric field applied along the separation capillary and not on its length (Holland et al. 1997). Most of the commercial microchip-based systems employ relatively short separation channels (<10 cm, down to few millimeters) and modest electric field strengths (<500 V/cm). However, in practice, microchip CE systems show lower separation efficiencies compared with standard CE capillary separations, primarily due to their shorter channel lengths. Even with this apparent disadvantage, these devices provide sufficient resolution to separate and compare all major N-glycans found on IgG and mAbs in 60 s or less (Vanderschaeghe et al. 2010, 2013; Smejkal et al. 2010; Primack et al. 2011; LabChip Microfluidics 2018), illustrated in Fig. 4.2a.

The misconception that complex N-glycan samples are not amenable to separation by microchip CE has been proven wrong by recent research work. To increase the component resolution for these microfluidic separations, longer separation channels need to be fabricated on microchips. To keep the overall dimension of a chip turns had to be integrated into the channels without introducing significant sample dispersion. With these longer separation channels (>20 cm) and higher electric field strengths (>1000 V/cm), microfluidic devices are now able to rapidly and efficiently separate N-glycans derived even from complex samples in less than 3 min (Zhuang et al. 2007, 2011; Snyder et al. 2016, 2017; Mitra et al. 2012, 2013, 2016). Although they have great resolving power, there is often higher variability in migration times when higher separation field strengths are applied (Mitra et al. 2012).

Figure 4.2 shows a direct comparison of separations of bovine IgG-derived N-glycans performed by microchip CE, xCGE-LIF equipped with 50 and 22 cm capillary-arrays, and hydrophilic interaction high-performance liquid chromatography (*HILIC–HPLC*). Miniaturization of the entire electrophoretic process resulted in a reasonably satisfactory separation of the major N-glycan peaks in seconds, as opposed to a much better separation for xCGE-LIF and HILIC-HPLC, but in minutes. However, for glycoproteins with a low glycan complexity, the capillary length in conventional CE and CGE systems can be further shortened without considerable loss of separation power, as shown for bovine IgG analyzed by xCGE-LIF with 50 and 22 cm capillaries (Fig. 4.2b, c). An xCGE-LIF instrument equipped with a 16-channel array with capillaries of 22 cm in length has an effective separation time of 32 s per sample (8.5 min/16 samples), which is close to 22.5 s obtained by the chip-CE instrument, but with a significantly better separation of the

N-glycan peaks. Keeping the multiplexing aspect of xCGE-LIF in mind, with its possibility to run up to 96 samples in parallel, it becomes obvious that there is potential to further increase the throughput of microchip CE.

Thus, despite the mentioned advancements, "Lab-on-a-chip" technology—incorporating both sample preparation and analysis onto the same microfluidic devices with a minimum hands-on time and being sufficiently simple for non-experts—still remains an ultimate goal. But first, the current generation of miniaturized systems has yet to demonstrate benefits in cost and performance, compelling enough to make them seriously competitive with conventional benchtop-scale CE technology. This is reflected in the current market volatility. The reliable availability of equipment, consumables, or support can often not be guaranteed. Therefore, low-cost, reliable and highly efficient microchip CE systems are expected to encourage greater use of this technology in the glycomics field.

4.8 Application of C(G)E for Immunoglobulin Analysis

Compared to common LC- and MS-based approaches, CE-based glycan analysis approaches are lagging far behind with respect to the number of applications and publications. This is largely due to the existing obstacles that needed to be tackled, including missing kits for sample preparation and non-flexible, difficult to handle or often unstable instruments. However, as discussed in detail in the previous sections, all these initial difficulties are now solved. Since stable and easy-to-handle DNA analyzers were adopted for *N*-glycan analyses by Callewaert and coworkers (Callewaert et al. 2001; Laroy et al. 2006), this approach gained popularity in the field of glycomics. Moreover, the availability of commercial analysis kits and instrumental solutions [e.g., by glyXera, PerkinElmer, Prozyme/Agilent, Sciex and Thermo Fisher Scientific (glyXera 2020; Thermo Fisher Scientific 2018; ProZyme 2018; SCIEX 2018; PerkinElmer 2018)], including software and glycan databases, made C(G)E-based methods appealing to the scientific community, as well as to the biopharmaceutical industry. This broader acceptance can be additionally attributed to the separation power of C(G)E and the ability to resolve complex glycan mixtures and to separate positional and linkage isomers. The following paragraphs are intended to highlight significant applications of C(G)E to the analyses of *N*-glycans in the science and biopharmaceutical industry and some new, exciting possibilities with a special emphasis on immunoglobulins.

IgG The C(G)E-based analysis of IgG *N*-glycosylation is well established (Reusch et al. 2014, 2015a; Routier et al. 1998) and was already applied to big cohorts (Huffman et al. 2014). Small sample quantities required for the xCGE-LIF-based analysis turned out to be quite beneficial for studies with limited sample availability, e.g., for the analysis of mouse IgG *N*-glycosylation (Patenaude et al. 2020; Schaffert et al. 2020). Consequently, this method can be applied to all kinds of mammalian samples, from large to small. As an example, Fig. 4.3 shows a comparison of the IgG

N-glycans derived from various mammalian species, namely human, cow, mouse, dog, and rabbit. Human and bovine (cow) IgG-derived *N*-glycans show quite some similarities, despite the absence of the Neu5Gc sialic acid type in humans, respectively of Neu5Ac in cow (Raju et al. 2000). However, more dominant are the differences between the species. Their IgG *N*-glycomes vary in composition and abundance of different glycans; from quite simple mixtures of glycans, with only a few structures (like the murine/mouse IgG in Fig. 4.3c), to extremely complex mixtures, with a wide range of structures, as shown in Fig. 4.3e for leporid (rabbit) IgG. Due to the big proportion of bisected *N*-glycans, in combination with a low degree of galactosylation and fucosylation, leporid IgG-derived *N*-glycans show one of the most complex glycosylation patterns (Taniguchi et al. 1985), which is still well resolved by xCGE-LIF, compared to results achieved by HILIC-UPLC analysis (Vainauskas et al. 2016). Additionally, the ability to separate Neu5Ac from Neu5Gc, as shown for canine (dog) IgG in Fig. 4.3d at about 180 MTU", exposes the big potential of C(G)E-based glycan analysis techniques. As recombinant IgGs produced in different host cells is one of the major therapeutic agents to treat life-threatening diseases (Raju et al. 2000), fast and effective analysis of species-specific glycosylation differences by methods like C(G)E is becoming invaluable in this field.

Other Igs Humans have five distinct classes of Igs, namely IgA, IgD, IgE, IgG, and IgM. All Igs are comprised of two heavy and two light chains that are joined together by disulfide bonds. The glycosylation of Igs greatly varies between the different classes, ranging from only one conserved *N*-glycosylation site in IgG up to various *N*- and *O*-glycosylation sites for IgD (Maverakis et al. 2015; Clerc et al. 2016). It has long been known that aberrant IgA glycosylation causes IgA nephropathy and Henoch-Schoenlein purpura nephritis (Allen et al. 1995; Novak et al. 2007), and it was recently discovered that IgE sialylation is one regulator of allergic reactions (Shade et al. 2020). But, although Igs are the major component of the adaptive immune system, only little is known about the influence of their glycosylation, apart from the extensively studied IgG. This is partially due to current technological limitations and the complexity of their glycosylation (Maverakis et al. 2015).

Here, we show how to overcome these limitations by giving a comprehensive overview on *N*-glycosylation of all plasma immunoglobulins, analyzed by xCGE-LIF as shown in Fig. 4.4. In contrast to human IgG with incomplete sialylation (Fig. 4.4d), human IgA, IgD, IgE, and IgM show a high degree of sialylation—often in combination with a big variety of oligo-mannosidic *N*-glycan structures (Fig. 4.4a–c, e). In comparison to HILIC-HPLC results published in the early 2000s (Arnold et al. 2004, 2005; Mattu et al. 1998), the xCGE-LIF results display markedly better resolution, resolving structures appearing as multi-structure peaks in HILIC-HPLC. Guttman and coworkers achieved similar results using a CE-based approach for the analysis of IgA from blood and saliva (Meszaros et al. 2020). They applied the method to biomedically relevant samples and found a link between altered IgA *N*-glycosylation and oral mucositis (Gebri et al. 2020). As the role of Ig glycosylation (besides IgG) is still not fully understood, xCGE-LIF might

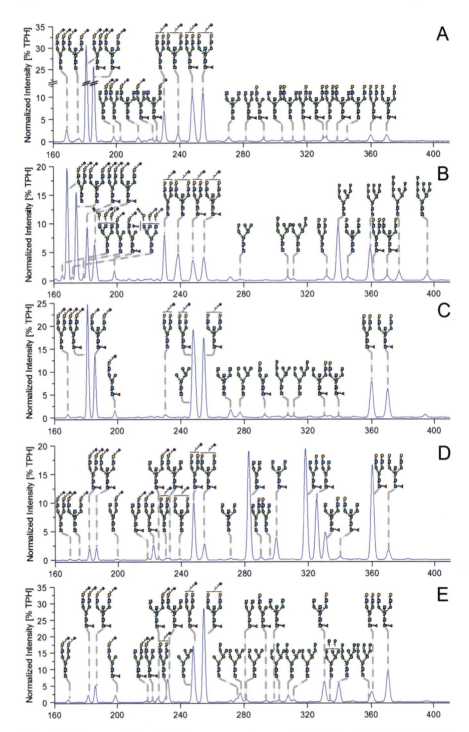

Fig. 4.4 *N*-glycan analysis of all human serum immunoglobulins by xCGE-LIF. xCGE-LIF generated fingerprints of APTS-labeled *N*-glycans derived from human IgA (**a**), IgD (**b**), IgE (**c**), IgG (**d**) and IgM (**e**) (Vendor: Abcam for IgA, IgD, IgE, and IgM; Sigma-Aldrich for IgG). APTS-

therefore simplify and speed up data generation and interpretation—thus, in the future promote interesting new findings when applied to big cohort studies.

Fab and Fc Antibody Domains A general approach in the field of glycobiology is to analyze released *N*-glycans of the entire IgG (the IgG *N*-glycome) (Huffman et al. 2014). However, *N*-glycans can appear at two different positions on IgG. The majority of *N*-glycans are attached to the conserved *N*-glycosylation site inside the constant (Fragment crystallizable, *Fc*) domain, while only a minor portion of *N*-glycans (up to 20%) might originate from the variable (Fragment antigen binding, *Fab*) domain of IgG (van de Bovenkamp 2019). The so-called Fab *N*-glycans differ from those attached to the Fc part, as the Fab portion possesses primarily highly sialylated and bisected complex type *N*-glycans (Clerc et al. 2016; van de Bovenkamp 2019), as shown in Fig. 4.5b (black curve), compared with the Fc-associated glycans with lower sialylation, as shown in Fig. 4.5b (red curve). Having a global look at glycosylation of other human immunoglobulins (Fig. 4.4), the Fab glycosylation shows high similarity. The higher degree of sialylation in Fab-associated glycans of IgG might originate in part from selective removal of non-sialylated structures by the hepatic asialo-glycoprotein receptor (*HAGR*) (as for IgA, IgD, IgE, IgM, and all other blood proteins), resulting in a plasma glycoprotein typical glycosylation (Dalziel et al. 1999). Fc-associated glycans are less accessible ("hidden"), and accordingly not cleared from the bloodstream by HAGR. Furthermore, the presumably better accessibility of the IgG Fab glycosylation site to glycosyltransferases results in better processing compared to Fc glycans that are spatially localized inside the constant domain and not accessible to the same enzymes (van de Bovenkamp 2019).

Unfortunately, many big cohort studies of IgG glycosylation are focusing on released *N*-glycans and are not differentiating between IgG Fab- and Fc *N*-glycosylation (Huffman et al. 2014; Wang et al. 2017; Russell et al. 2017; Barrios et al. 2016; Trbojevic Akmacic et al. 2015; Lauc et al. 2013; Pučić et al. 2011), although recent studies suggest that the prevalence and complexity of Fab glycans might change for certain physiological and pathological conditions. On the one hand, sialylated Fab glycans are presumed to be protective during pregnancy (Bondt et al. 2014), while on the other hand, an increase in Fab glycans is associated with several autoimmune diseases, like rheumatoid arthritis and Sjögren's syndrome, or cancer like multiple myeloma (Kinoshita et al. 1991). These aspects show that a separate analysis of Fab and Fc glycosylation might be beneficial for a better

---

Fig. 4.4 (continued) labeled *N*-glycans were prepared using the glyXprep^{48} kit (glyXera 2020), by carefully following the kit instructions. Data processing was performed using glyXtoolCE^{TM} (glyXera 2021). Data processing comprised alignment of migration times to two orthogonal internal standards, resulting in a double aligned x-axis in migration time units (MTU″). Peak heights were normalized to the sum of all peaks, resulting in a normalized intensity in % of total peak height (TPH). *N*-glycan structures were assigned via database matching using glyXtoolCE^{TM} (in combination with glyXbase^{TM}) and confirmed by exoglycosidase sequencing as published by Cajic et al. (Thiesler et al. 2016). Symbolic representation of *N*-glycan structures follows the guidelines of Symbol Nomenclature for Glycans (SNFG) (Varki et al. 2015)

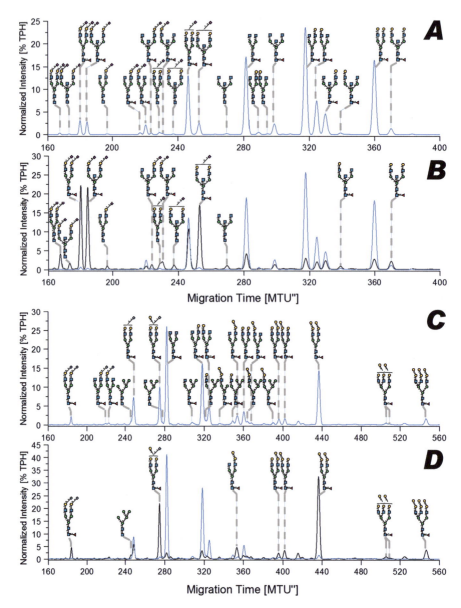

Fig. 4.5 Fab and Fc glycosylation analysis by xCGE-LIF, exemplarily shown for human IgG and the SP2/0 cell-derived mAb cetuximab (Erbitux). xCGE-LIF generated fingerprints of APTS-labeled *N*-glycans derived from: whole human IgG (**a**), human IgG Fab (**b**, black curve), human IgG Fc (**b**, blue curve), whole mAb cetuximab (**c**), cetuximab Fab (**d**, black curve) and cetuximab Fc (**d**, blue curve) (human IgG was purified from normal control plasma, purchased at Affinity Biologicals; cetuximab was purchased at Evidentic). Fab and Fc part of human IgG and mAb cetuximab were purified and generated as published by Bondt et al. (2014). Release and APTS-labeling of *N*-glycans were performed, using the glyXprep[48] kit (glyXera 2020), by carefully following the kit instructions. Data processing was performed using glyXtoolCE™ (glyXera 2021). Data processing comprised alignment of migration times to two orthogonal internal standards, resulting in a double aligned x-axis in migration time units (MTU″). Peak heights were

understanding of immunological processes and that Fab and Fc glycans might be a promising biomarker for the early detection of various diseases. Here, especially C(G)E shows big potential because of its high sensitivity—enabling detection of even the lowest amounts of Fab glycans (Fig. 4.5b), and its speed (by multiplexing)—enabling higher sample throughput if needed.

Therapeutic Glycoproteins Since glycosylation of monoclonal antibodies (mAb) is a critical quality attribute (*CQA*), the detailed characterization and control of antibody glycosylation during the development process and later product life cycle is essential. Despite the ever-growing use of C(G)E-based methods for biopharmaceutical characterization, the number of publications in the field is still quite limited (Reusch et al. 2014; Croset et al. 2012; Bunz et al. 2013a; Bielser et al. 2020; Borza et al. 2018). Nevertheless, several studies showed that the CGE-based analysis results are quite comparable to the results obtained with conventional MS- and LC-based methods (Reusch et al. 2015a, b; De Leoz et al. 2020). As shown for the SP2/0 cell-derived mAb cetuximab in Fig. 4.5c, d, the overall results are quite comparable to the earlier analysis results achieved with HILIC-HPLC and MALDI-TOF MS (Qian et al. 2007), or ESI-TOF MS (Janin-Bussat et al. 2013). Here xCGE-LIF could also resolve the isomeric structures containing the immunogenic α-Gal from non-immunogenic β-Gal for Fc, Fab (Fig. 4.5d) and complete mAb *N*-glycosylation (Fig. 4.5c). After the discovery that the α1–3 Gal epitope can cause an anaphylactic shock via an anti-oligosaccharide IgE-mediated reaction (Chung et al. 2008; Chinuki and Morita 2019), the absence of this epitope is intended for all newly developed therapeutic glycoproteins and biosimilars. Nevertheless, several previously commercialized mAbs are produced in murine hybridoma cell lines like NS0 and SP2/0, which do express the α1–3 Gal epitope (Qian et al. 2007; Uçaktürk 2012; Stadlmann et al. 2008), as shown in Fig. 4.6a for the comparison of NS0-derived mAb ramucirumab (Cyramza) (blue line) and SP2/0-derived mAb cetuximab (Erbitux) (black line). In contrast, the current dominant production cell lines are derived from Chinese Hamster Ovary (*CHO*) cells, which do not express the immunogenic α1–3 Gal epitope, as shown in Fig. 4.6b. Accordingly, the α1–3 Gal epitope can be used to identify the production cell line of a mAb in a straightforward manner, as shown for comparison of the CHO-derived mAb rituximab (Rituxan®) (black line) and the NS0 derived mAb ramucirumab (blue line) in Fig. 4.6c (zoom-in around 360 MTU″). However, CHO cells are differing from NS0 and SP2/0 cells in two additional important *N*-glycosylation properties: CHO cells incorporate mainly the non-immunogenic Neu5Ac in α2–3 linkage, as representatively shown for the mAbs bevacizumab (Avastin®) (blue line) and rituximab

---

Fig. 4.5 (continued) normalized to the sum of all peaks, resulting in a normalized intensity in % of total peak height (TPH). *N*-glycan structures were assigned via database matching using glyXtoolCE™ (in combination with glyXbase™) and confirmed by exoglycosidase sequencing as published by Cajic et al. (Thiesler et al. 2016). Symbolic representation of *N*-glycan structures follows the guidelines of Symbol Nomenclature for Glycans (SNFG) (Varki et al. 2015)

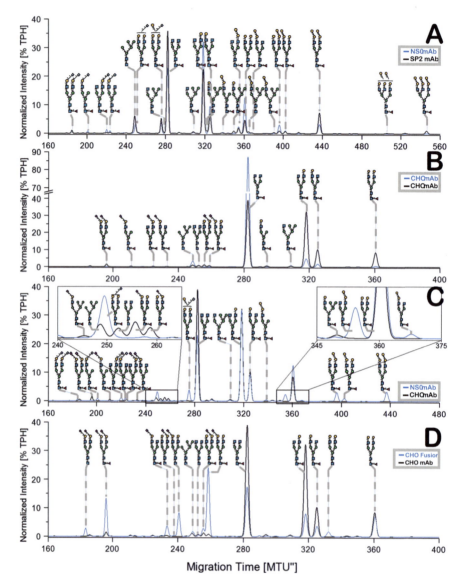

Fig. 4.6 xCGE-LIF-based *N*-glycan analysis comparing the therapeutic proteins ramucirumab (NS0), cetuximab (SP2/0), bevacizumab (CHO), rituximab (CHO) and etanercept (CHO). xCGE-LIF generated fingerprints of APTS-labeled *N*-glycans derived from mAb ramucirumab (Cyramza) produced in NS0 cells (**a** and **c**; blue curve), mAb cetuximab (Erbitux) produced in SP2/0 cells (**a**; black curve), mAb bevacizumab (Avastin®) produced in CHO cells (**b**; blue curve), mAb rituximab produced in CHO cells (Rituxan®) (**b**, **c** and **d**; black curve) and fusion protein etanercept (Enbrel®) produced in CHO cells (**d**; blue curve) (all purchased at Evidentic). Release and APTS-labeling of *N*-glycans were performed using the glyXprep[48] kit (glyXera 2020), by carefully following the kit instructions. Data processing was performed using glyXtoolCE™ (glyXera 2021). Data processing comprised alignment of migration times to two orthogonal internal standards, resulting in a double aligned x-axis in migration time units (MTU″). Peak heights were normalized to the sum of all peaks, resulting in a normalized intensity in % of total peak height (TPH). *N*-glycan structures were assigned via database matching using glyXtoolCE™ (in combination with glyXbase™) and

(black line) in Fig. 4.6b and the fusion protein etanercept (Enbrel®) (blue line) in Fig. 4.6d (Borza et al. 2018; Mcleod 2013). In contrast, NS0 and SP2/0 cells integrate Neu5Gc in α2–6 linkage (Qian et al. 2007; Stadlmann et al. 2008; Beck et al. 2008), as shown for SP2/0-derived cetuximab in Fig. 4.5c, d and ramucirumab in Fig. 4.6a. This enables an easy distinction between originator and biosimilar mAb, as both structural N-glycan properties (Neu5Gc versus Neu5Ac and α2–3 versus α2–6 linkage) can be directly resolved by xCGE-LIF, without the need for an additional exoglycosidase treatment, as shown in detail for the zoom-in around 250 MTU″ in Fig. 4.6c. Furthermore, the high resolution of xCGE-LIF enables the monitoring of structural N-glycan properties of glycoengineered CHO cells, like the missing core fucose of FUT8 knockout CHO cells (Yamane-Ohnuki et al. 2004), the additional bisected GlcNAc for GnTIII over-expression CHO cells (Umaña et al. 1999), or the emerging α2–6-linked Neu5Ac in CHO cells with stable ST6GAL1 expression (Houeix and Cairns 2019).

Besides the resolution of structural properties of N-glycans, the relative quantities (like the degree of galactosylation or sialylation) are easily accessible, too. As demonstrated for the mAbs bevacizumab and rituximab in Fig. 4.6b, remarkable differences in N-glycosylation can be observed. While bevacizumab has a considerably low degree of galactosylation (with about 90% of all N-glycans gathered in only one peak comprising of FA2G0), rituximab shows a more complex N-glycosylation pattern (with di-antennary N-glycans bearing up to two terminal sialic acids). In the case of sialylation, both mAbs are clearly lagging behind the fusion protein etanercept (tumor necrosis factor receptor linked to the Fc portion of human IgG1) with a sialylation degree of more than 50%. These differences in the N-glycosylation are often intentionally induced, as they affect the proper function of the protein (Kanda et al. 2007), its stability and thus the product potency and quality.

Moreover, xCGE-LIF is a powerful tool for high-throughput screening of hundreds of cell clones during early-stage discovery, as well as post-discovery characterization, with minimal material consumption and costs. Here, xCGE-LIF-based methods help to reduce the screening time, greatly accelerating the development of biosimilars, biobetters, and new biotherapeutics. In combination with commercially available enzymes like IdeZ (by New England Biolabs) or FabRICATOR® (Z) (by Genovis), "site"-specific characterization of IgG-based therapeutic proteins can be achieved, ensuring a more detailed level of protein characterization, and preventing surprises during pharmacokinetic testing. Consequently, xCGE-LIF holds potential toward detailed analysis of mAb glycosylation in real high throughput.

Fig. 4.6 (continued) confirmed by exoglycosidase sequencing as published by Cajic et al. (Thiesler et al. 2016). Symbolic representation of N-glycan structures follows the guidelines of Symbol Nomenclature for Glycans (SNFG) (Varki et al. 2015)

4.9 Conclusion

CE has matured to the point that it can stay side by side with other more commonly used techniques for glycan analysis. Because of its unbeatable speed, resolution, sensitivity, and simplicity, the method is gaining more and more attention. Ongoing efforts to bring CE-based analysis kits and software solutions on the market (together with the increasing availability, size, and quality of glycan databases) will make the method attractive to the biopharmaceutical industry, helping to reveal the full potential of therapeutic glycoproteins like mAbs. Application of CE-based workflows for Ig analysis will enable more detailed characterization of their glycosylation and fast, automated high-throughput monitoring of their glycosylation patterns, especially when employed for large cohort studies. The recent and upcoming developments in miniaturization and analysis toolboxes show that there is an exciting future in glycan analysis for this recently revived technology.

Compliance with Ethical Standards

Funding Erdmann Rapp acknowledges support by Deutsche Forschungsgemeinschaft (DFG, German Research Foundation)—Project-ID RA2992/1-1—Forschungsgruppe FOR 2509. Robert Burock acknowledges support by funds of the ESF.

Conflict of Interest Author Erdmann Rapp is the founder, CEO and CSO of glyXera GmbH. Authors René Hennig and Robert Burock are employees of glyXera GmbH. glyXera provides high-performance glycoanalytical products and services and holds several patents for xCGE-LIF based glycoanalysis. Author Samanta Cajic declares no competing interests.

Ethical Approval This article does not contain any studies with human participants or animals performed by any of the authors.

References

Abeln M, Borst KM, Cajic S, Thiesler H, Kats E, Albers I et al (2017) Sialylation is dispensable for early murine embryonic development in vitro. ChemBioChem 18(13):1305–1316

Allen AC, Harper SJ, Feehally J (1995) Galactosylation of N- and O-linked carbohydrate moieties of IgA1 and IgG in IgA nephropathy. Clin Exp Immunol 100(3):470–474

Anthony RM, Nimmerjahn F, Ashline DJ, Reinhold VN, Paulson JC, Ravetch JV (2008) Recapitulation of IVIG anti-inflammatory activity with a recombinant IgG Fc. Science 320 (5874):373–376

Archer-Hartmann SA, Sargent LM, Lowry DT, Holland LA (2011a) Microscale exoglycosidase processing and lectin capture of glycans with phospholipid assisted capillary electrophoresis separations. Anal Chem 83(7):2740–2747

Archer-Hartmann SA, Crihfield CL, Holland LA (2011b) Online enzymatic sequencing of glycans from Trastuzumab by phospholipid-assisted capillary electrophoresis. Electrophoresis 32 (24):3491–3498

Arnold JN, Radcliffe CM, Wormald MR, Royle L, Harvey DJ, Crispin M et al (2004) The glycosylation of human serum IgD and IgE and the accessibility of identified oligomannose structures for interaction with mannan-binding lectin. J Immunol 173(11):6831–6840

Arnold JN, Wormald MR, Suter DM, Radcliffe CM, Harvey DJ, Dwek RA et al (2005) Human serum IgM glycosylation: identification of glycoforms that can bind to Mannan-binding lectin. J Biol Chem 280(32):29080–29087

Barrios C, Zierer J, Gudelj I, Stambuk J, Ugrina I, Rodríguez E et al (2016) Glycosylation profile of IgG in moderate kidney dysfunction. J Am Soc Nephrol 27(3):933–941

Beck A, Wagner-Rousset E, Bussat M-C, Lokteff M, Klinguer-Hamour C, Haeuw J-F et al (2008) Trends in glycosylation, glycoanalysis and glycoengineering of therapeutic antibodies and Fc-fusion proteins. Curr Pharm Biotechnol 9(6):482–501

Behne A, Muth T, Borowiak M, Reichl U, Rapp E (2013) glyXalign: high-throughput migration time alignment preprocessing of electrophoretic data retrieved via multiplexed capillary gel electrophoresis with laser-induced fluorescence detection-based glycoprofiling. Electrophoresis 34(16):2311–2315

Bielser JM, Kraus L, Burgos-Morales O, Broly H, Souquet J (2020) Reduction of medium consumption in perfusion mammalian cell cultures using a perfusion rate equivalent concentrated nutrient feed. Biotechnol Prog 36(5):1–8

Bondt A, Rombouts Y, Selman MHJ, Hensbergen PJ, Reiding KR, Hazes JMW et al (2014) Immunoglobulin G (IgG) fab glycosylation analysis using a new mass spectrometric high-throughput profiling method reveals pregnancy-associated changes. Mol Cell Proteomics 13 (11):3029–3039

Borza B, Szigeti M, Szekrenyes A, Hajba L, Guttman A (2018) Glycosimilarity assessment of biotherapeutics 1: quantitative comparison of the N-glycosylation of the innovator and a biosimilar version of etanercept. J Pharm Biomed Anal 153:182–185. https://doi.org/10.1016/j.jpba.2018.02.021

Briggs JB, Keck RG, Ma S, Lau W, Jones AJS (2009) An analytical system for the characterization of highly heterogeneous mixtures of N-linked oligosaccharides. Anal Biochem 389(1):40–51

Bunz S-C, Cutillo F, Neusüß C (2013a) Analysis of native and APTS-labeled N-glycans by capillary electrophoresis/time-of-flight mass spectrometry. Anal Bioanal Chem 405 (25):8277–8284

Bunz SC, Rapp E, Neusüss C (2013b) Capillary electrophoresis/mass spectrometry of APTS-labeled glycans for the identification of unknown glycan species in capillary electrophoresis/laser- induced fluorescence systems. Anal Chem 85(21):10218–10224

Callewaert N, Geysens S, Molemans F, Contreras R (2001) Ultrasensitive profiling and sequencing of N-linked oligosaccharides using standard DNA-sequencing equipment. Glycobiology 11 (4):275–281

Campa C, Rossi M (2008) Capillary electrophoresis of neutral carbohydrates: mono-, oligosaccharides, glycosides. Methods Mol Biol 384:247–305

Chen J, Fang M, Chen X, Yi C, Ji J, Cheng C et al (2017) N-glycosylation of serum proteins for the assessment of patients with IgD multiple myeloma. BMC Cancer 17(1):881. Available from: https://bmccancer.biomedcentral.com/articles/10.1186/s12885-017-3891-3

Chinuki Y, Morita E (2019) Alpha-Gal-containing biologics and anaphylaxis. Allergol Int 68 (3):296–300. https://doi.org/10.1016/j.alit.2019.04.001

Chung CH, Mirakhur B, Chan E, Le Q-T, Berlin J, Morse M et al (2008) Cetuximab-induced anaphylaxis and IgE specific for galactose-α-1,3-galactose. N Engl J Med 358(11):1109–1117

Clerc F, Reiding KR, Jansen BC, Kammeijer GSM, Bondt A, Wuhrer M (2016) Human plasma protein N-glycosylation. Glycoconj J 33(3):309–343

Collins FS, Lander ES, Rogers J et al (2004) International human genome sequencing consortium, finishing the euchromatic sequence of the human genome. Nature 431(7011):931–945

Coppa GV, Gabrielli O, Pierani P, Catassi C, Carlucci A, Giorgi PL (1993) Changes in carbohydrate composition in human milk over 4 months of lactation. Pediatrics 91(3):637–641. Available from: http://pediatrics.aappublications.org/content/91/3/637.abstract

Croset A, Delafosse L, Gaudry JP, Arod C, Glez L, Losberger C et al (2012) Differences in the glycosylation of recombinant proteins expressed in HEK and CHO cells. J Biotechnol 161 (3):336–348. https://doi.org/10.1016/j.jbiotec.2012.06.038

Dalziel M, McFarlane I, Axford JS (1999) Lectin analysis of human immunoglobulin G N-glycan sialylation. Glycoconj J 16(12):801–807

De Leoz MLA, Duewer DL, Fung A, Liu L, Yau HK, Potter O et al (2020) NIST interlaboratory study on glycosylation analysis of monoclonal antibodies: comparison of results from diverse analytical methods. Mol Cell Proteomics 19(1):11–30

Donczo B, Szarka M, Tovari J, Ostoros G, Csanky E, Guttman A (2017) Molecular glycopathology by capillary electrophoresis: analysis of the N-glycome of formalin-fixed paraffin-embedded mouse tissue samples. Electrophoresis 38(12):1602–1608

Dotz V, Haselberg R, Shubhakar A, Kozak RP, Falck D, Rombouts Y et al (2015) Mass spectrometry for glycosylation analysis of biopharmaceuticals. TrAC Trends Anal Chem 73:1–9

Eussen SRBM, Mank M, Kottler R et al (2021) Presence and levels of galactosyllactoses and other oligosaccharides in human milk and their variation during lactation and according to maternal phenotype. Nutrients 13:2324. https://doi.org/10.3390/nu13072324

Evangelista RA, Liu M-S, Chen F-TA (1995) Characterization of 9-aminopyrene-1,4,6-trisulfonate derivatized sugars by capillary electrophoresis with laser-induced fluorescence detection. Anal Chem 67(13):2239–2245

Feng H, Li P, Rui G, Stray J, Khan S, Chen S-M et al (2017) Multiplexing N-glycan analysis by DNA analyzer. Electrophoresis 38(13–14):1788–1799

Gattu S, Crihfield CL, Holland LA (2017) Microscale measurements of michaelis-menten constants of neuraminidase with nanogel capillary electrophoresis for the determination of the sialic acid linkage. Anal Chem 89(1):929–936

Gebri E, Kovács Z, Mészáros B, Tóth F, Simon Á, Jankovics H et al (2020) N-glycosylation alteration of serum and salivary immunoglobulin A is a possible biomarker in oral mucositis. J Clin Med 9(6):1747

Gennaro LA, Salas-Solano O (2008) On-line CE−LIF−MS technology for the direct characterization of N-linked glycans from therapeutic antibodies. Anal Chem 80(10):3838–3845

glyXera (2020) glyXprep™ kit for N-glycan analysis [cited 2020 Mar 30]. Available from: https://www.glyxera.com/product/glyxprep/

glyXera (2021) glyXtoolCE. [cited 2021 Mar 30]. Available from: https://www.glyxera.com/product/glyxtool-ce/

GU database

Guttman A (1996) High-resolution carbohydrate profiling by capillary gel electrophoresis. Nature 380(6573):461–462

Guttman A (1997) Multistructure sequencing of N-linked fetuin glycans by capillary gel electrophosesis and enzyme matrix digestion. Electrophoresis 18(7):1136–1141

Guttman A, Herrick S (1996) Effect of the quantity and linkage position of mannose(α1,2) residues in capillary gel electrophoresis of high-mannose-type oligosaccharides. Anal Biochem 235(2):236–239

Guttman A, Ulfelder KW (1997) Exoglycosidase matrix-mediated sequencing of a complex glycan pool by capillary electrophoresis. J Chromatogr A 781(1–2):547–554

Guttman A, Cooke N, Starr CM (1994) Capillary electrophoresis separation of oligosaccharides: I. Effect of operational variables. Electrophoresis 15(12):1518–1522

Guttman A, Chen F-TA, Evangelista RA (1996a) Separation of 1-aminopyrene-3,6,8-trisulfonate-labeled asparagine-linked fetuin glycans by capillary gel electrophoresis. Electrophoresis 17(2):412–417

Guttman A, Chen F-TA, Evangelista RA, Cooke N (1996b) High-resolution capillary gel electrophoresis of reducing oligosaccharides labeled with 1-aminopyrene-3,6,8-trisulfonate. Anal Biochem 233(2):234–242

Guttman M, Váradi C, Lee KK, Guttman A (2015) Comparative glycoprofiling of HIV gp120 immunogens by capillary electrophoresis and MALDI mass spectrometry. Electrophoresis 36(11–12):1305–1313

Hennig R, Borowiak M, Ruhaak LR, Wuhrer M, Rapp E (2011a) High-throughput CGE-LIF based analysis of APTS-labeled N-glycans, utilizing a multiplex capillary DNA sequencer. Glycoconj J 28(5):331

Hennig R, Reichl U, Rapp E (2011b) A software tool for automated high-throughput processing of CGE-LIF based glycoanalysis data, generated by a multiplexing capillary DNA sequencer. Glycoconj J 28(5):331

Hennig R, Rapp E, Kottler R, Cajic S, Borowiak M, Reichl U (2015) N-Glycosylation fingerprinting of viral glycoproteins by xCGE-LIF. Methods Mol Biol 1331:123–143

Hennig R, Cajic S, Borowiak M, Hoffmann M, Kottler R, Reichl U et al (2016) Towards personalized diagnostics via longitudinal study of the human plasma N-glycome. Biochim Biophys Acta Gen Subj 1860(8):1728–1738

Hjertén S (1967) Free zone electrophoresis. Chromatogr Rev 9(2):122–219

Holland LA, Chetwyn NP, Perkins MD, Lunte SM (1997) Capillary electrophoresis in pharmaceutical analysis. Pharm Res 14(4):372–387

Holland LA, Gattu S, Crihfield CL, Bwanali L (2017) Capillary electrophoresis with stationary nanogel zones of galactosidase and Erythrina cristagalli lectin for the determination of β(1–3)-linked galactose in glycans. J Chromatogr A 1523:90–96

Houeix B, Cairns MT (2019) Engineering of CHO cells for the production of vertebrate recombinant sialyltransferases. Peer J 7:e5788. https://doi.org/10.7717/peerj.5788

Huang C, Liu Y, Wu H, Sun D, Li Y (2017) Characterization of IgG glycosylation in rheumatoid arthritis patients by MALDI-TOF-MSn and capillary electrophoresis. Anal Bioanal Chem 409 (15):3731–3739

Huffman JE, Pučić-Baković M, Klarić L, Hennig R, Selman MHJ, Vučković F et al (2014) Comparative performance of four methods for high-throughput glycosylation analysis of immunoglobulin G in genetic and epidemiological research. Mol Cell Proteomics 13(6):1598–1610

Huhn C, Ruhaak LR, Mannhardt J, Wuhrer M, Neusüß C, Deelder AM et al (2012) Alignment of laser-induced fluorescence and mass spectrometric detection traces using electrophoretic mobility scaling in CE-LIF-MS of labeled N-glycans. Electrophoresis 33(4):563–566

Jacob GS, Scudder P (1994) Glycosidases in structural analysis. Methods Enzymol 230:280–299

Janin-Bussat M-C, Tonini L, Huillet C, Colas O, Klinguer-Hamour C, Corvaïa N et al (2013) Cetuximab Fab and Fc N-glycan fast characterization using IdeS digestion and liquid chromatography coupled to electrospray ionization mass spectrometry. In: Beck A (ed) Glycosylation engineering of biopharmaceuticals: methods and protocols. Humana Press, Totowa, NJ, pp 93–113. https://doi.org/10.1007/978-1-62703-327-5_7

Jarvas G, Szigeti M, Guttman A (2015) GUcal: an integrated application for capillary electrophoresis based glycan analysis. Electrophoresis 36(24):3094–3096

Jarvas G, Szigeti M, Chapman J, Guttman A (2016) Triple-internal standard based glycan structural assignment method for capillary electrophoresis analysis of carbohydrates. Anal Chem 88 (23):11364–11367

Jarvas G, Szigeti M, Guttman A (2018) Structural identification of N-linked carbohydrates using the GUcal application: a tutorial. J Proteome 171:107–115

Jayo RG, Li J, Chen DDY (2012) Capillary electrophoresis mass spectrometry for the characterization of O-Acetylated N-glycans from fish serum. Anal Chem 84(20):8756–8762

Jayo RG, Thaysen-Andersen M, Lindenburg PW, Haselberg R, Hankemeier T, Ramautar R et al (2014) Simple capillary electrophoresis-mass spectrometry method for complex glycan analysis using a flow-through microvial interface. Anal Chem 86(13):6479–6486

Jorgenson JW, DeArman Lukacs K (1981) Zone electrophoresis in open-tubular glass capillaries: preliminary data on performance. J High Resolut Chromatogr 4(5):230–231

Jorgenson JW, Lukacs KD (1981a) Free-zone electrophoresis in glass capillaries. Clin Chem 27 (9):1551–1553

Jorgenson JW, Lukacs KD (1981b) Zone electrophoresis in open-tubular glass capillaries. Anal Chem 53(8):1298–1302

Kamerling JP, Gerwig GJ (2007) Strategies for the structural analysis of carbohydrates. Compr Glycosci From Chem to Syst Biol 2–4:1–68

Kanda Y, Yamada T, Mori K, Okazaki A, Inoue M, Kitajima-Miyama K et al (2007) Comparison of biological activity among nonfucosylated therapeutic IgG1 antibodies with three different N-linked Fc oligosaccharides: the high-mannose, hybrid, and complex types. Glycobiology 17(1):104–118

Kinoshita N, Ohno M, Nishiura T, Fujii S, Nishikawa A, Kawakami Y et al (1991) Glycosylation at the Fab portion of myeloma immunoglobulin G and increased fucosylated biantennary sugar chains: structural analysis by high-performance liquid chromatography and antibody-lectin enzyme immunoassay using *Lens culinaris* agglutinin. Cancer Res 51(21):5888–5892

Kobata A (2013) Exo- and endoglycosidases revisited. Proc Jpn Acad Ser B Phys Biol Sci 89 (3):97–117

Kobata A, Takasaki S (1992) Structural characterization of oligosaccharides from glycoproteins. 4A Glycosidase treatment and other methods, including methylation analysis. In: Glycobiology: a practical approach. IRL Press, Oxford, pp 165–185

Konze SA, Cajic S, Oberbeck A, Hennig R, Pich A, Rapp E et al (2017) Quantitative assessment of sialo-glycoproteins and N-glycans during cardiomyogenic differentiation of human induced pluripotent stem cells. ChemBioChem 18(13):1317–1331

Kottler R, Mank M, Hennig R, Müller-Werner B, Stahl B, Reichl U et al (2013) Development of a high-throughput glycoanalysis method for the characterization of oligosaccharides in human milk utilizing multiplexed capillary gel electrophoresis with laser-induced fluorescence detection. Electrophoresis 34(16):2323–2336

LabChip Microfluidics (2018) Rapid analysis of N-glycans on theLabChip GXII touch microchip-CE platform [cited 2018 May 22]. Available from: https://www.perkinelmer.com/lab-solutions/resources/docs/APP_Rapid_Analysis_N-Glycans_on_LabChip_GXII_Touch.pdf

Lageveen-Kammeijer GSM, de Haan N, Mohaupt P, Wagt S, Filius M, Nouta J et al (2019) Highly sensitive CE-ESI-MS analysis of N-glycans from complex biological samples. Nat Commun [cited 2019 May 20] 10(1):2137. Available from: http://www.nature.com/articles/s41467-019-09910-7

Laroy W, Contreras R, Callewaert N (2006) Glycome mapping on DNA sequencing equipment. Nat Protoc 1(1):397–405

Lauc G, Huffman JE, Pučić M, Zgaga L, Adamczyk B, Mužinić A et al (2013) Loci associated with N-glycosylation of human immunoglobulin G show pleiotropy with autoimmune diseases and haematological cancers. PLoS Genet 9(1):e1003225

Lebede M, Di Marco F, Esser-Skala W, Hennig R, Wohlschlager T, Huber CG (2021) Exploring the chemical space of protein glycosylation in non-covalent protein complexes: an expedition along different struc-tural levels of human chorionic gonadotropin employing mass spectrometry. J Am Chem Soc. https://doi.org/10.1021/acs.analchem.1c02199

Lindenburg PW, Haselberg R, Rozing G, Ramautar R (2015) Developments in interfacing designs for CE–MS: towards enabling tools for proteomics and metabolomics. Chromatographia 78 (5–6):367–377

Liu XE, Desmyter L, Gao CF, Laroy W, Dewaele S, Vanhooren V et al (2007) N-glycomic changes in hepatocellular carcinoma patients with liver cirrhosis induced by hepatitis B virus. Hepatology 46(5):1426–1435

Lu G, Crih CL, Gattu S, Veltri LM, Holland LA, Bennett CE et al (2018) Capillary electrophoresis separations of glycans. Chem Rev 118(17):7867–7885. https://doi.org/10.1021/acs.chemrev.7b00669

Ludger (2018) Glycan labeling kits [cited 2018 May 16]. Available from: https://www.ludger.com/products/glycan_labeling_kits.php

Luo R, Archer-Hartmann SA, Holland LA (2010) Transformable capillary electrophoresis for oligosaccharide separations using phospholipid additives. Anal Chem 82(4):1228–1233

Ma S, Nashabeh W (1999) Carbohydrate analysis of a chimeric recombinant monoclonal antibody by capillary electrophoresis with laser-induced fluorescence detection. Anal Chem 71 (22):5185–5192

Mahan AE, Tedesco J, Dionne K, Baruah K, Cheng HD, De Jager PL et al (2015) A method for high-throughput, sensitive analysis of IgG Fc and Fab glycosylation by capillary electrophoresis. J Immunol Methods 417:34–44. https://doi.org/10.1016/j.jim.2014.12.004

Mattu TS, Pleass RJ, Willis AC, Kilian M, Wormald MR, Lellouch AC et al (1998) The glycosylation and structure of human serum IgA1, Fab, and Fc regions and the role of N-glycosylation on Fcα receptor interactions. J Biol Chem 273(4):2260–2272

Maverakis E, Kim K, Shimoda M, Gershwin ME, Patel F, Wilken R et al (2015) Glycans in the immune system and the altered glycan theory of autoimmunity: a critical review. J Autoimmun 57(January):1–13. https://doi.org/10.1016/j.jaut.2014.12.002

Maxwell EJ, Chen DDY (2008) Twenty years of interface development for capillary electrophoresis–electrospray ionization–mass spectrometry. Anal Chim Acta 627(1):25–33

Mcleod B (2013) Characterization of glycans from erbitux®, Rituxan® and Enbrel® using PNGase F (glycerol-free), recombinant. Glycobiol Protein Tools—New Engl Biolabs:2–5. Available from: https://www.neb.com/-/media/catalog/application-notes/characterization-of-glycans-from-erbitux-rituxan-and-enbrel-using-pngase-f-glycerol-free-recombinant.pdf

Mechref Y (2013) Glycomic profiling through capillary electrophoresis and microchip capillary electrophoresis. In: Capillary electrophoresis and microchip capillary electrophoresis. Wiley, Hoboken, NJ, pp 367–383

Mechref Y, Novotny MV (2009) Glycomic analysis by capillary electrophoresis-mass spectrometry. Mass Spectrom Rev 28(2):207–222

Mechref Y, Muzikar J, Novotny MV (2005) Comprehensive assessment of N-glycans derived from a murine monoclonal antibody: a case for multimethodological approach. Electrophoresis 26 (10):2034–2046

Meininger M, Stepath M, Hennig R, Cajic S, Rapp E, Rotering H et al (2016) Sialic acid-specific affinity chromatography for the separation of erythropoietin glycoforms using serotonin as a ligand. J Chromatogr B 1012–1013:193–203

Meszaros B, Kovacs Z, Gebri E, Jankovics H, Vonderviszt F, Kiss A et al (2020) N-glycomic analysis of Z(IgA1) partitioned serum and salivary immunoglobulin A by capillary electrophoresis. Curr Mol Med 20:781–788. Available from: http://www.eurekaselect.com/node/180847/article

Mitra I, Zhuang Z, Zhang Y, Yu C-Y, Hammoud ZT, Tang H et al (2012) N-glycan profiling by microchip electrophoresis to differentiate disease states related to esophageal adenocarcinoma. Anal Chem 84(8):3621–3627

Mitra I, Alley WR, Goetz JA, Vasseur JA, Novotny MV, Jacobson SC et al (2013) Comparative profiling of N-glycans isolated from serum samples of ovarian cancer patients and analyzed by microchip electrophoresis. J Proteome Res 12(10):4490–4496

Mitra I, Snyder CM, Zhou X, Campos MI, Alley WR, Novotny MV et al (2016) Structural characterization of serum N-glycans by methylamidation, fluorescent labeling, and analysis by microchip electrophoresis. Anal Chem 88(18):8965–8971

Mittermayr S, Guttman A (2012) Influence of molecular configuration and conformation on the electromigration of oligosaccharides in narrow bore capillaries. Electrophoresis 33 (6):1000–1007

Mittermayr S, Bones J, Guttman A (2013) Unraveling the glyco-puzzle: glycan structure identification by capillary electrophoresis. Anal Chem 85(9):4228–4238

Muñoz RI, Kähne T, Herrera H, Rodríguez S, Guerra MM, Vío K et al (2019) The subcommissural organ and the Reissner fiber: old friends revisited. Cell Tissue Res 375(2):507–529

Nakano M, Kakehi K, Taniguchi N, Kondo A (2011) Capillary electrophoresis and capillary electrophoresis–mass spectrometry for structural analysis of N-glycans derived from glycoproteins. In: Capillary electrophoresis of carbohydrates. Humana Press, Totowa, NJ, pp 205–235

Novak J, Moldoveanu Z, Renfrow MB, Yanagihara T, Suzuki H, Raska M et al (2007) IgA nephropathy and henoch-schoenlein purpura nephritis: aberrant glycosylation of IgA1, formation of IgA1-containing immune complexes, and activation of mesangial cells. Contrib Nephrol 157:134–138

O'Flaherty R, Harbison AM, Hanley PJ, Taron CH, Fadda E, Rudd PM (2017) Aminoquinoline fluorescent labels obstruct efficient removal of N-glycan core α(1-6) fucose by bovine kidney α-L-fucosidase (BKF). J Proteome Res 16(11):4237–4243

Olivares JA, Nguyen NT, Yonker CR, Smith RD (1987) On-line mass spectrometric detection for capillary zone electrophoresis. Anal Chem 59(8):1230–1232

Parker BL, Thaysen-Andersen M, Solis N, Scott NE, Larsen MR, Graham ME et al (2013) Site-specific glycan-peptide analysis for determination of N-glycoproteome heterogeneity. J Proteome Res 12(12):5791–5800

Patenaude AM, Erhardt J, Hennig R, Rapp E, Lauc G, Pezer M (2020) N-glycosylation analysis of mouse immunoglobulin G isolated from dried blood spots. Electrophoresis:1–11. https://doi.org/10.1002/elps.202000249

PerkinElmer (2018) Glycan profiling assay release and labeling kit [cited 2018 May 22]. Available from: http://www.perkinelmer.de/product/glycan-release-and-labeling-kit-760523

Pioch M, Bunz SC, Neusüß C (2012) Capillary electrophoresis/mass spectrometry relevant to pharmaceutical and biotechnological applications. Electrophoresis 33(11):1517–1530

Pioch M, Hoffmann M, Pralow A, Reichl U, Rapp E (2018) GlyXtoolMS: an open-source pipeline for semiautomated analysis of glycopeptide mass spectrometry data. Anal Chem 90 (20):11908–11916

Pralow A, Hoffmann M, Nguyen-Khuong T, Pioch M, Hennig R, Genzel Y, Rapp E, Reichl U (2021) Comprehensive N-glycosylation analysis of the influenza A virus proteins HA and NA from adherent and suspension MDCK cells. FEBS J. https://doi.org/10.1111/febs.15787

Primack J, Flynn GC, Pan H (2011) A high-throughput microchip-based glycan screening assay for antibody cell culture samples. Electrophoresis 32(10):1129–1132

Prime S, Merry T (1998) Exoglycosidase sequencing of N-linked glycans by the reagent array analysis method (RAAM). Methods Mol Biol 76:53–69

ProZyme (2018) GlykoPrep® rapid N-glycan preparation with APTS (24-ct) [GP24NG-APTS]. ProZyme-product [cited 2018 May 16]. Available from: https://prozyme.com/products/gp24ng-apts

Pučić M, Knežević A, Vidič J, Adamczyk B, Novokmet M, Polašek O et al (2011) High throughput isolation and glycosylation analysis of IgG-variability and heritability of the IgG glycome in three isolated human populations. Mol Cell Proteomics 10(10):1–15

Qian J, Liu T, Yang L, Daus A, Crowley R, Zhou Q (2007) Structural characterization of N-linked oligosaccharides on monoclonal antibody cetuximab by the combination of orthogonal matrix-assisted laser desorption/ionization hybrid quadrupole-quadrupole time-of-flight tandem mass spectrometry and sequential enzy. Anal Biochem 364(1):8–18

Raju TS, Briggs JB, Borge SM, Jones AJS (2000) Species-specific variation in glycosylation of Igc: evidence for the species-specific sialylation and branch-specific galactosylation and importance for engineering recombinant glycoprotein therapeutics. Glycobiology 10(5):477–486

Reusch D, Haberger M, Kailich T, Heidenreich A-K, Kampe M, Bulau P et al (2014) High-throughput glycosylation analysis of therapeutic immunoglobulin G by capillary gel electrophoresis using a DNA analyzer. MAbs 6(1):185–196

Reusch D, Haberger M, Maier B, Maier M, Kloseck R, Zimmermann B et al (2015a) Comparison of methods for the analysis of therapeutic immunoglobulin G Fc-glycosylation profiles—part 1: separation-based methods. MAbs 7(1):167–179. https://doi.org/10.4161/19420862.2014.986000

Reusch D, Haberger M, Falck D, Peter B, Maier B, Gassner J et al (2015b) Comparison of methods for the analysis of therapeutic immunoglobulin G Fc-glycosylation profiles-part 2: mass spectrometric methods. MAbs 7(4):732–742

Routier FH, Hounsell EF, Rudd PM, Takahashi N, Bond A, Hay FC et al (1998) Quantitation of the oligosaccharides of human serum IgG from patients with rheumatoid arthritis: a critical evaluation of different methods. J Immunol Methods 213(2):113–130

Rudd PM, Dwek RA (1997) Rapid, sensitive sequencing of oligosaccharides from glycoproteins. Curr Opin Biotechnol 8(4):488–497

Ruess FF (1809) Notice sur un nouvel effet de l'électricité galvanique. In: Mémoires de la Société Impériale des Naturalistes de l'Université Impérial e de Moscou. Impr. de l'Université impériale, Moscou, pp 327–336

Ruhaak LR, Huhn C, Waterreus W-J, de Boer AR, Neusüss C, Hokke CH et al (2008) Hydrophilic interaction chromatography-based high-throughput sample preparation method for N-glycan analysis from total human plasma glycoproteins. Anal Chem 80(15):6119–6126

Ruhaak LR, Zauner G, Huhn C, Bruggink C, Deelder AM, Wuhrer M (2010a) Glycan labeling strategies and their use in identification and quantification. Anal Bioanal Chem 397(8):3457–3481

Ruhaak LR, Hennig R, Huhn C, Borowiak M, Dolhain RJEM, Deelder AM et al (2010b) Optimized workflow for preparation of APTS-labeled N-glycans allowing high-throughput analysis of human plasma glycomes using 48-channel multiplexed CGE-LIF. J Proteome Res 9(12):6655–6664

Ruhaak LR, Steenvoorden E, Koeleman CAM, Deelder AM, Wuhrer M (2010c) 2-Picoline-borane: a non-toxic reducing agent for oligosaccharide labeling by reductive amination. Proteomics 10(12):2330–2336

Ruhaak LR, Huhn C, Koeleman CAM, Deelder AM, Wuhrer M (2012) Robust and high-throughput sample preparation for (semi-)quantitative analysis of N-glycosylation profiles from plasma samples. Methods Mol Biol 893:371–385

Russell AC, Šimurina M, Garcia MT, Novokmet M, Wang Y, Rudan I et al (2017) The N-glycosylation of immunoglobulin G as a novel biomarker of Parkinson's disease. Glycobiology 27(5):501–510

Saraswathy N, Ramalingam P (2011) DNA sequencing methods, chapter5. In: Concepts and techniques in genomics and proteomics. Woodhead, Oxford, p 57

Sarkozy D, Borza B, Domokos A, Varadi E, Szigeti M, Meszaros-Matwiejuk A et al (2021) Ultrafast high-resolution analysis of human milk oligosaccharides by multicapillary gel electrophoresis. Food Chem 341:128200. https://doi.org/10.1016/j.foodchem.2020.128200

Schaffert A, Hanić M, Novokmet M, Zaytseva O, Krištić J, Lux A et al (2020) Minimal B cell extrinsic IgG glycan modifications of pro- and anti-inflammatory IgG preparations in vivo—supplementary material. Front Immunol 10:3024

Schwarzer J, Rapp E, Reichl U (2008) N-glycan analysis by CGE-LIF: profiling influenza A virus hemagglutinin N-glycosylation during vaccine production. Electrophoresis 29(20):4203–4214

Schwedler C, Kaup M, Weiz S, Hoppe M, Braicu EI, Sehouli J et al (2014a) Identification of 34 N-glycan isomers in human serum by capillary electrophoresis coupled with laser-induced fluorescence allows improving glycan biomarker discovery. Anal Bioanal Chem 406(28):7185–7193

Schwedler C, Kaup M, Petzold D, Hoppe B, Braicu EI, Sehouli J et al (2014b) Sialic acid methylation refines capillary electrophoresis laser-induced fluorescence analyses of immunoglobulin G N-glycans of ovarian cancer patients. Electrophoresis 35(7):1025–1031

SCIEX (2018) Fast glycan analysis and labeling for the PA 800 plus [cited 2018 May 16]. Available from: https://sciex.com/products/consumables-and-standards/fast-glycan-analysis-and-labeling-for-the-pa-800-plus

Shade KTC, Conroy ME, Washburn N, Kitaoka M, Huynh DJ, Laprise E et al (2020) Sialylation of immunoglobulin E is a determinant of allergic pathogenicity. Nature 582(7811):265–270. https://doi.org/10.1038/s41586-020-2311-z

Simó C, Cifuentes A (2005) Mass spectrometry detection in capillary electrophoresis. In: Analysis and detection by capillary electrophoresis. Elsevier, Amsterdam, pp 441–517

Smejkal P, Szekrényes Á, Ryvolová M, Foret F, Guttman A, Bek F et al (2010) Chip-based CE for rapid separation of 8-aminopyrene-1,3,6-trisulfonic acid (APTS) derivatized glycans. Electrophoresis 31(22):3783–3786

Smith RD, Olivares JA, Nguyen NT, Udseth HR (1988a) Capillary zone electrophoresis-mass spectrometry using an electrospray ionization interface. Anal Chem 60(5):436–441

Smith RD, Baringa CJ, Udseth HR (1988b) Improved electrospray ionization interface for capillary zone electrophoresis-mass spectrometry. Anal Chem 60(18):1948–1952

Snyder CM, Alley WR, Campos MI, Svoboda M, Goetz JA, Vasseur JA et al (2016) Complementary glycomic analyses of sera derived from colorectal cancer patients by MALDI-TOF-MS and microchip electrophoresis. Anal Chem 88(19):9597–9605

Snyder CM, Zhou X, Karty JA, Fonslow BR, Novotny MV, Jacobson SC (2017) Capillary electrophoresis–mass spectrometry for direct structural identification of serum N-glycans. J Chromatogr A 1523:127–139

Stadlmann J, Pabst M, Kolarich D, Kunert R, Altmann F (2008) Analysis of immunoglobulin glycosylation by LC-ESI-MS of glycopeptides and oligosaccharides. Proteomics 8 (14):2858–2871

Suzuki H, Muller O, Guttman A, Karger BL (1997) Analysis of 1-aminopyrene 3,6,8-trisulfonate-derivatized oligosaccharides by capillary electrophoresis with matrix-assisted laser desorption/ionization time-of-flight mass spectrometry. Anal Chem 69(22):4554–4559

Szabo Z, Guttman A, Bones J, Shand RL, Meh D, Karger BL (2012) Ultrasensitive capillary electrophoretic analysis of potentially immunogenic carbohydrate residues in biologics: galactose-α-1,3-galactose containing oligosaccharides. Mol Pharm 9(6):1612–1619

Szigeti M, Guttman A (2017) Automated N-glycosylation sequencing of biopharmaceuticals by capillary electrophoresis. Sci Rep 7(1):11663

Taniguchi T, Mizuochi T, Kobata A, Beale M, Dwek RA, Rademacher TW (1985) Structures of the sugar chains of rabbit immunoglobulin G: occurrence of asparagine-linked sugar chains in Fab fragment. Biochemistry 24(20):5551–5557

Teranishi K, Manez R, Awwad M, Cooper DKC (2002) Anti-Galα1-3Gal IgM and IgG antibody levels in sera of humans and old world non-human primates. Xenotransplantation 9(2):148–154

Thaysen-Andersen M, Packer NH (2014) Advances in LC-MS/MS-based glycoproteomics: getting closer to system-wide site-specific mapping of the N- and O-glycoproteome. Biochim Biophys Acta Proteins Proteomics 1844(9):1437–1452. https://doi.org/10.1016/j.bbapap.2014.05.002

Thermo Fisher Scientific (2018) GlycanAssure APTS Kit [cited 2018 May 16]. Available from: http://www.thermofisher.com/order/catalog/product/A28676

Thiesler CT, Cajic S, Hoffmann D, Thiel C, Van Diepen L, Hennig R et al (2016) Glycomic characterization of induced pluripotent stem cells derived from a patient suffering from phosphomannomutase 2 congenital disorder of glycosylation (PMM2-CDG). Mol Cell Proteomics 15(4):1435–1452

Trbojevic Akmacic I, Ventham NT, Theodoratou E, Vučković F, Kennedy NA, Krištić J et al (2015) Inflammatory bowel disease associates with proinflammatory potential of the immunoglobulin G glycome. Inflamm Bowel Dis 21(6):1237–1247

Uçaktürk E (2012) Analysis of glycoforms on the glycosylation site and the glycans in monoclonal antibody biopharmaceuticals. J Sep Sci 35(3):341–350

Umaña P, Jean-Mairet J, Moudry R, Amstutz H, Bailey JE (1999) Engineered glycoforms of an antineuroblastoma IgG1 with optimized antibody-dependent cellular cytotoxic activity. Nat Biotechnol 17(2):176–180

Vainauskas S, Duke RM, McFarland J, McClung C, Ruse C, Taron CH (2016) Profiling of core fucosylated N-glycans using a novel bacterial lectin that specifically recognizes α1,6 fucosylated chitobiose. Sci Rep 6(September):1–12. https://doi.org/10.1038/srep34195

van de Bovenkamp F (2019) The role of variable domain glycosylation of antibodies in immunity. PhD Thesis of the University of Amsterdam. Available at: https://dare.uva.nl/search?identifier=af243516-5114-4f95-af99-ba0648757e13

Van der Linden ECB, Sjoberg ER, Juneja LR, Crocker PR, Varki N, Varki A (2000) Loss of N-glycolylneuraminic acid in human evolution. Implications for sialic acid recognition by siglecs. J Biol Chem 275(12):8633–8640

Vanderschaeghe D, Szekrényes A, Wenz C, Gassmann M, Naik N, Bynum M et al (2010) High-throughput profiling of the serum N-glycome on capillary electrophoresis microfluidics systems: toward clinical implementation of GlycoHepatoTest. Anal Chem 82(17):7408–7415

Vanderschaeghe D, Guttman A, Callewaert N (2013) High-throughput profiling of the serum N-glycome on capillary electrophoresis microfluidics systems. Methods Mol Biol 919:87–96

Varki A, Cummings RD, Aebi M, Packer NH, Seeberger PH, Esko JD et al (2015) Symbol nomenclature for graphical representations of glycans. Glycobiology 25(12):1323–1324

Wang JR, Gao WN, Grimm R, Jiang S, Liang Y, Ye H et al (2017) A method to identify trace sulfated IgG N-glycans as biomarkers for rheumatoid arthritis. Nat Commun 8(1):1–13. https://doi.org/10.1038/s41467-017-00662-w

Watson DG (2012) High-performance capillary electrophoresis. In: Pharmaceutical analysis: a textbook for pharmacy students and pharmaceutical chemists. Elsevier Health Sciences, Edinburgh, pp 376–397

Weiz S, Wieczorek M, Schwedler C, Kaup M, Braicu EI, Sehouli J et al (2016) Acute-phase glycoprotein N-glycome of ovarian cancer patients analyzed by CE-LIF. Electrophoresis 37 (11):1461–1467

Yagi Y, Yamamoto S, Kakehi K, Hayakawa T, Ohyama Y, Suzuki S (2011) Application of partial-filling capillary electrophoresis using lectins and glycosidases for the characterization of oligosaccharides in a therapeutic antibody. Electrophoresis 32(21):2979–2985

Yagi Y, Kakehi K, Hayakawa T, Ohyama Y, Suzuki S (2012) Specific detection of N-glycolylneuraminic acid and Galα1–3Gal epitopes of therapeutic antibodies by partial-filling capillary electrophoresis. Anal Biochem 431(2):120–126

Yamagami M, Matsui Y, Hayakawa T, Yamamoto S, Kinoshita M, Suzuki S (2017) Plug-plug kinetic capillary electrophoresis for in-capillary exoglycosidase digestion as a profiling tool for the analysis of glycoprotein glycans. J Chromatogr A 1496:157–162

Yamane-Ohnuki N, Kinoshita S, Inoue-Urakubo M, Kusunoki M, Iida S, Nakano R et al (2004) Establishment of FUT8 knockout Chinese hamster ovary cells: an ideal host cell line for producing completely defucosylated antibodies with enhanced antibody-dependent cellular cytotoxicity. Biotechnol Bioeng 87(5):614–622

Yang Y, Franc V, Heck AJR (2017) Glycoproteomics: a balance between high-throughput and in-depth analysis. Trends Biotechnol 35(7):598–609. https://doi.org/10.1016/j.tibtech.2017.04.010

Yu T, Guo C, Wang J, Hao P, Sui S, Chen X et al (2011) Comprehensive characterization of the site-specific N-glycosylation of wild-type and recombinant human lactoferrin expressed in the milk of transgenic cloned cattle. Glycobiology 21(2):206–224

Zhang W, Hankemeier T, Ramautar R (2017) Next-generation capillary electrophoresis–mass spectrometry approaches in metabolomics. Curr Opin Biotechnol 43:1–7

Zhao Y-P, Xu X-Y, Fang M, Wang H, You Q, Yi C-H et al (2014) Decreased core-fucosylation contributes to malignancy in gastric cancer. Pizzo S V., editor. PLoS One 9(4):e94536

Zhong X, Zhang Z, Jiang S, Li L (2014) Recent advances in coupling capillary electrophoresis-based separation techniques to ESI and MALDI-MS. Electrophoresis 35(9):1214–1225

Zhuang Z, Starkey JA, Mechref Y, Novotny MV, Jacobson SC (2007) Electrophoretic analysis of N-glycans on microfluidic devices. Anal Chem 79(18):7170–7175

Zhuang Z, Mitra I, Hussein A, Novotny MV, Mechref Y, Jacobson SC (2011) Microchip electrophoresis of N-glycans on serpentine separation channels with asymmetrically tapered turns. Electrophoresis 32(2):246–253

Open Access This chapter is licensed under the terms of the Creative Commons Attribution 4.0 International License (http://creativecommons.org/licenses/by/4.0/), which permits use, sharing, adaptation, distribution and reproduction in any medium or format, as long as you give appropriate credit to the original author(s) and the source, provide a link to the Creative Commons license and indicate if changes were made.

The images or other third party material in this chapter are included in the chapter's Creative Commons license, unless indicated otherwise in a credit line to the material. If material is not included in the chapter's Creative Commons license and your intended use is not permitted by statutory regulation or exceeds the permitted use, you will need to obtain permission directly from the copyright holder.

Chapter 5
Automation of Immunoglobulin Glycosylation Analysis

Jenifer L. Hendel, Richard A. Gardner, and Daniel I. R. Spencer

Contents

5.1	Introduction	175
	5.1.1 Biopharmaceutical Glycomics	176
	5.1.2 Clinical Glycomics	177
	5.1.3 Towards High-Throughput Glycomics	177
	5.1.4 Robotics: The Ultimate High-Throughput Solution?	180
5.2	Automation of Glycomics Sample Preparation	182
	5.2.1 Sample Origins and Protein Purification	182
	5.2.2 Preparing Glycans for Analysis: Glycan Release, Derivatization and Clean-Up	185
	5.2.3 Automated Methods for Glycopeptide Preparation	196
5.3	Commentary	197
5.4	Future Perspectives	200
5.5	Conclusions	200
References		201

Abstract The development of reliable, affordable, high-resolution glycomics technologies that can be used for many samples in a high-throughput manner are essential for both the optimization of glycosylation in the biopharmaceutical industry as well as for the advancement of clinical diagnostics based on glycosylation biomarkers. We will use this chapter to review the sample preparation processes that have been used on liquid-handling robots to obtain high-quality glycomics data for both biopharmaceutical and clinical antibody samples. This will focus on glycoprotein purification, followed by glycan or glycopeptide generation, derivatization and enrichment. The use of liquid-handling robots for glycomics studies on other sample types beyond antibodies will not be discussed here. We will summarize our thoughts on the current status of the field and explore the benefits and challenges associated with developing and using automated platforms for sample preparation. Finally, the

---

J. L. Hendel · R. A. Gardner · D. I. R. Spencer (✉)
Ludger Limited, Culham Science Centre, Abingdon, Oxfordshire, UK
e-mail: Jenifer.hendel@ludger.com; Richard.gardner@ludger.com;
Daniel.spencer@ludger.com

© The Author(s), under exclusive license to Springer Nature Switzerland AG 2021
M. Pezer (ed.), *Antibody Glycosylation*, Experientia Supplementum 112,
https://doi.org/10.1007/978-3-030-76912-3_5

future outlook for the automation of glycomics will be discussed along with a projected impact on the field in general.

Keywords Automation · Robotization · Sample preparation · Glycan analysis · High-throughput strategies · Glycomics · Antibody

Abbreviations

2AA	2-aminobenzoic acid
2AB	2-aminobenzamide
2-PB	2-picoline borane
A1AT	Alpha-1-anti-trypsin
APTS	8-aminopyrene-1,3,6-trisulfonic Acid
AQC	Aminoquinoline carbamate
BOA	O-benzyloxyamine hydrochloride
CE	Capillary electrophoresis
CE-LIF	Capillary electrophoresis-laser-induced fluorescence
CHO	Chinese hamster ovary
DMSO	Dimethyl sulfoxide
EDTA	Ethylenediaminetetraacetic acid
ESI	Electrospray ionization
FAb	Fragment antigen-binding
GHP	GH polypro
HILIC	Hydrophilic interaction liquid chromatography
HPLC	High performance liquid chromatography
Hpt	Haptoglobin
IBD	Inflammatory bowel disease
IgG/A/M	Immunoglobulin G/A/M
LC	Liquid chromatography
mAb	Monoclonal antibody
MALDI MS	Matrix assisted laser desorption ionization mass spectrometry
MALDI-TOF–MS	Matrix assisted laser desorption ionization-time of flight mass spectrometry
MS	Mass spectrometry
MSn	Tandem mass spectrometry (MSn with $n = 2$ or 3)
NMWL	Nominal molecular weight limit
PBM	Protein binding membrane
PCR	Polymerase chain reaction
PNGaseF	Peptide:N-glycosidase F
RFMS	Rapifluor-MS
rhEPO	Recombinant human erythropoietin
SDS	Sodium dodecyl sulphate
SPE	Solid phase extraction

TFA	Trifluoroacetic acid
Trf	Transferrin
UHPLC or UPLC	Ultra-High Performance Liquid Chromatography
QbD	Quality by design

5.1 Introduction

Glycans are implicated in virtually all physiological processes (Varki 2017). Whilst there are a wide variety of important glycoproteins covering a wide range of functions, one of the main and widely studied groups are the antibodies. Antibodies are a group of bioactive glycoproteins with significance in both biology and the biopharmaceutical industry (Dalziel et al. 2014). As has been introduced in earlier chapters, antibodies are important in therapeutic development, understanding disease progression and provide opportunities for medical diagnostics. In this chapter, the emphasis will be on high-throughput processing of glycomics activities to support the study of antibodies in two specific areas, namely glycoprotein biopharmaceuticals and glycan biomarkers of disease for clinical diagnostics. In both of these areas, considerable advancement has been made owing to the vast improvement in analytical platforms and glycomics technologies (O'Flaherty et al. 2018; Yamamoto et al. 2016). These advancements have contributed to both a greater understanding and interest in the monitoring and optimization of glycosylation in biopharmaceutical realization and the incredible potential for clinical glycomics. However, this has also brought with it a new challenge; large sample cohorts. Therefore, this chapter is aimed at glycoscientists who are dealing with large sample sets and are interested in automation.

We will review the current status of automated sample handling and preparation of antibodies for glycomics analysis. For detailed information on biosynthesis, function, and application of antibody glycosylation, the other chapters of this book and the references therein are a good resource of knowledge. Firstly, we will focus on sample origin considerations and automated methods for purification of antibodies prior to glycan/glycopeptide generation. This section will be divided into the preparation of biological samples (specifically serum and plasma samples) and also the preparation of biopharmaceutical therapeutic antibodies. Secondly, we will look at automated methods for both glycan generation and derivatization and glycopeptide generation. Thirdly, we will discuss the advantages and drawbacks of implementing and using automation within glycomics laboratories for sample handling. Finally, we will finish off by looking at what the future holds for this subsection of the glycomics field. It should be noted that the automation of sample analysis, data acquisition and characterization using any of the common analytical platforms (LC, MS, and CE) as well as automated glycomics for other types of samples is out of scope for this review.

5.1.1 Biopharmaceutical Glycomics

Glycoprotein drugs span a range of structure-activity relationship classes. These encompass human-engineered versions of naturally occurring glycoproteins; hormones (e.g., Follicle-Stimulating Hormone; infertility), cytokines (e.g., erythropoietin; kidney dysfunction) and blood-clotting factors (e.g., Factor VIII; hemophilia). The largest class of glycoprotein drugs are monoclonal antibodies (mAbs), and the majority of approved mAbs are of the IgG1 isotype—these IgG mAbs target serious inflammatory conditions, cancers, autoimmune, cardiovascular, and infectious diseases. mAb sales are expected to reach US$130–200 billion in 2022; this is motivated by several components of the market, which include a healthy pipeline, the increasing roles for biosimilars, and also emerging economies (Grilo and Mantalaris 2019).

For mAb drugs, glycans are the greatest source of within-batch and batch-to-batch variability. Drug glycosylation impacts clinical performance (safety and efficacy), manufacturability and cost per dose. Glycosylation patterns can influence both pharmacodynamics and pharmacokinetics. Furthermore, the presence of various glycosylation features, including alpha 1–3 linked galactose and N-glycolylneuraminic acid can lead to adverse immunogenic reactions if they are present in biologic therapeutics. It is for these many reasons that regulatory authorities mandate that glycosylation is properly designed, measured, and controlled throughout the entire drug lifecycle. In many cases, certain glycosylation features are indicated as critical quality attributes required for therapeutic release (Costa et al. 2014).

To remain competitive, companies are seeking to optimize biomanufacturing and develop new technologies for faster drug discovery, including high-throughput screening, selection of the best clones and culturing in miniaturized bioreactor systems. Regardless of which biomanufacturing pathway is pursued, the distribution of glycan structures present on mAbs can vary. Bioprocessing parameters including; cell line, dissolved oxygen, nutrients in the feed and bioreactor type have all been shown (Costa et al. 2014; Shubhakar et al. 2016; Hossler et al. 2009) to impact glycosylation profiles, and as a result, these need to be monitored characterized and understood to satisfy regulatory scrutiny.

Striving to identify the most beneficial glycosylation patterns for drugs will lead to better and safer therapeutics but will also lead to an increase in the need for additional sample analysis. The biopharmaceutical industries' focus on obtaining optimal glycosylation in every stage during the drug lifecycle is creating a significant demand for high-throughput analysis of large sample sets.

5.1.2 Clinical Glycomics

Most proteins are glycosylated, largely as a result of post-translational modification, and these glycans play a vital role in the regulation of key biological processes, including brain development, immunity, and growth. Antibodies, which are glycoproteins belonging to the immunoglobulin superfamily, are mainly produced by plasma cells, and changes in their glycosylation profiles have been studied in response to both normal physiological processes like aging (Bonté et al. 2018) and the onset of various diseases. Indeed, abnormal glycosylation has been found as a hallmark of many human diseases like cancer, neurodegenerative, and inflammatory diseases (Dube and Bertozzi 2005; Adamczyk et al. 2012; Dennis et al. 1999). Considering this, the potential for using glycosylation signatures in biological fluids, such as plasma/serum from blood, as disease biomarkers or as a diagnostic tool for patient stratification and precision medicine is steadily becoming more attractive (Liu et al. 2019; Peng et al. 2018). The knowledge of these altered glycosylation features in plasma immunoglobulins is already leading to a greater understanding of disease pathways. However, it is yet to be determined how these features will be exploited for clinical purposes. It is possible that a new generation of clinical diagnostics could be established or alternatively, that serum glycoanalysis could be an orthogonal method to support a diagnosis by more traditional approaches (e.g. genomics).

However, to successfully implement glycan markers in clinical diagnosis, validation studies on large biological cohorts need to be performed (Shipman et al. 2020). Serum glycan profiles from different cohorts of individuals (e.g., healthy versus disease samples) need to be compared on statistically significant sample sets to identify changes in the abundance of individual glycan species accurately.

Thus, in order for glycan biomarkers to be used in clinical prognosis and diagnosis of many diseases in the future, there needs to be an analytical strategy available to characterize glycosylation on a larger-scale. However, even with the expansion of commercially available glycan analysis kits, most of the glycan biomarkers discovered so far have been studied using glycoanalytical technologies that would not be suitable for use in routine clinical diagnostic labs. The methods are limited in sample throughput, resolution, and affordability. These problems highlight the need for robust, reliable, and high-throughput sample processing methods for glycomics studies.

5.1.3 Towards High-Throughput Glycomics

Given the information discussed above, the development of reliable, affordable, high-resolution glycomics technologies which can be used for processing many samples in a high-throughput fashion is essential for both the optimization of glycosylation in the biopharmaceutical industry as well as for the advancement of clinical diagnostics based on glycosylation biomarkers.

The status quo for glycomics studies is largely manual techniques geared towards low-throughput sample handling, often still requiring specialized expertise. In spite of the significant improvements in various chromatographic and mass spectrometric methods, the most challenging aspect remains the tedious and lengthy sample preparation steps. The most common strategies for antibody glycosylation analysis are usually either (1) enzymatic or chemical glycan release followed by chemical derivatization and measurement of the glycans by mass spectrometry, capillary electrophoresis or liquid chromatography with fluorescence detection; (2) proteolytic cleavage of the glycoprotein to produce glycopeptides which are measured by various mass spectrometry methods (Yamamoto et al. 2016).

Regardless of which of the above methods are used for glycoanalysis, additional multi-step purification processes are needed to obtain samples that will provide clear and consistent data. This means that purification is often needed to obtain a suitably pure antibody glycoprotein before using enzymatic treatment to provide glycans or glycopeptides, and further clean-up and enrichment are often needed before analysis. These technical issues make it more challenging to achieve high-throughput glycomics.

In addition to the technical issues, another challenge in glycomics is that multiple orthogonal analysis methods are often needed for the complete detailed characterization of all glycan species in a particular sample; a standard protocol may include the combination of various techniques (commonly LC, MS, and exoglycosidase sequencing) to have confidence in the structural assignments. Since characterization and analysis are so demanding, it would be ideal if the sample preparation was also not as arduous.

In an attempt to improve sample preparation and processing time, a number of high-throughput manual methods have been reported for the glycoanalysis of antibody glycoproteins from both clinical and biopharmaceutical sources (Bondt et al. 2014, 2016; Wang et al. 2016; Kumpel et al. 2020; Shajahan et al. 2019; Royle et al. 2008; Trbojević Akmačić et al. 2015). A common feature among each of these methods is that they all leverage the 96-well microtiter plate format, which has been a major enabler in the lab when handling liquid samples. In these studies, the standard low-throughput protocol for glycan analysis is often adapted to the plate format with little modification to the overall method itself, and multichannel pipettes are used to ease the amount of repetitive manual labor for the analyst.

If required, manual high-throughput purification of the antibodies is achieved by using 96-well filter plates functionalized with or containing beads functionalized with selective capture agents such as protein A or protein G.

For *N*-glycan analysis, many of the traditional methods show compatibility with processing in 96-well plates. For example, Royle et al. developed an in-gel method adapted to a 96-well plate for PNGaseF release, which was followed by 2-AB labeling and clean-up before HPLC analysis (Royle et al. 2008). This technique was first used to analyze serum IgG from a rheumatoid arthritis patient cohort and continues to be widely used for manual high-throughput antibody glycosylation analysis applications [selected citations (Wang et al. 2016; Kumpel et al. 2020; Adamczyk et al. 2014)]. Alternatively, Bondt et al. developed a manual high-throughput method for studying FAb glycosylation at the level of released glycans obtained from serum-derived polyclonal IgG using affinity capturing beads and enzymes in 96-well plates (Bondt et al. 2014). The released glycans were subjected to ethyl esterification derivatization, resulting in linkage-specific modification of sialic acids, followed by HILIC sample purification with cotton-packed pipette tips and analysis on MALDI-TOF–MS. In addition, the focus has been given to specific parts of the process while developing manual methods for high-throughput sample preparation. For example, Trbojević Akmačić et al. found that the clean-up of labeled glycans was the biggest source of variation. As a result, they tested multiple high-throughput clean-up approaches (cellulose, silica gel, BioGel, and a hydrophilic GHP filter) for solid-phase extraction. All stationary phases were suitable for efficient purification of labeled glycans, but the GHP filter plate proved the easiest to handle and provided the most reproducible data (Trbojević Akmačić et al. 2015).

Some manual high-throughput methods have been developed with the intention of being more automation ready by slight modifications to the original methods. Shajahan et al. reported small changes to the traditional permethylation procedure for *N*-glycan derivatization in plasma, cell lines and purified glycoproteins to make the process more automation friendly (Shajahan et al. 2019). In this case, the micropermethylation reaction was conducted in a 96-deep-well polypropylene plate, and the permethylated glycans were purified by C18 tips mounted on a multichannel pipette before being analyzed by MALDI–MS and ESI–MSn. The major change to this protocol was the introduction of the C18 tip clean-up instead of the more traditional liquid-liquid extraction technique that is often used.

Likewise, Aich et al. reported on what they have called an integrated solution-based procedure for the analysis of *N*-glycans from therapeutic monoclonal antibodies (Aich et al. 2016). Their focus was to limit the number of purification, concentration, and manipulation steps to minimize the time required for sample processing and to also employ the non-toxic reducing reagent 2-PB. Therefore, all reactions starting from denaturing of proteins to the 2-AA labeling reactions were performed in the same 96-well deep plate. After completion of all reactions, the samples were purified once using a HILIC–SPE 96-well microplate using a positive pressure manifold.

Lastly, an ultrafast method for antibody glycopeptide analysis was developed by Yang et al. (2016). Here the generation and purification of tryptic glycopeptides are performed in an aqueous buffer followed by label-free quantification using matrix-assisted laser desorption/ionization-time of flight mass spectrometry. The assay time is less than 15 min, and the authors indicate that it is automation ready because the process has been designed for 96-well PCR plates using a multichannel pipette for minimal sample handling. This method was evaluated for glycoprofiling of mAbs expressed under various cell culture conditions, as well as for the evaluation of antibody culture clones and various production batches. The innovation in this method was the optimization of the trypsin digest; by employing a short digestion time with a high temperature (70 °C) denaturing step in the presence of urea prior to trypsin addition, the generation of glycopeptides was complete in less than 10 min.

From the examples given, it is clear that the development of manual high-throughput methods for glycomics facilitates the study of large sample cohorts; however, performing manual sample preparation and manipulation is still a time-consuming process that requires significant hands-on labor. Additionally, whilst the use of manual methods is suitable for small sample sets, scaling up to larger sizes increases the likelihood of operator error leading to inconsistency and poor data repeatability. These challenges have led to increasing demand for the automation of glycomics workflows; laboratories are looking to automation as the ultimate solution for simple and scalable sample preparation methods in order to generate data that is both repeatable and reliable while also reducing the laboratory burden on their analysts.

5.1.4 Robotics: The Ultimate High-Throughput Solution?

Multipurpose liquid-handling robotic workstations have been designed to do much of the sampling, mixing, and combining of liquid samples automatically. Biology research labs and drug development labs are among the types of laboratories that have implemented these tools as solutions to limiting sample contamination and freeing up personnel to do other tasks (Alexovič et al. 2020; Kong et al. 2012).

Robotic workstations are used in glycomics studies to perform unattended actions including transport of objects (glassware, plates, racks etc.), aspiration and injection of liquids, mixing and extraction. The deck can also be expanded to include other equipment, including a centrifuge, vacuum manifolds, storage devices and incubators. As mentioned previously, the benefit of using these workstations is that the actions are all completed in a uniform fashion. When introducing an automated system into a laboratory setting, one should consider the volume that the robot is capable of handling and the corresponding precision. Other features to factor in are the footprint of the workstation (i.e., space required to accommodate the tool) and the ease of use of its software interface.

The following is a non-exhaustive list of companies who offer liquid handling robots; Agilent Technologies, Analytik Jena, Andrew Alliance, Apogent Discoveries, Beckman Coulter, Dynamic Devices, Gilson, Hamilton Company, LEAP Technologies, Opentrons, Perkin Elmer, PhyNexus, Qiagen, Tecan, and Thermo Fisher Scientific. Each of these companies offers a range of automated solutions that meet various different specifications required for each size and type of laboratory. We will not provide commentary on their comparative effectiveness or use as we are not experts on the entire range of offerings and hence would encourage the reader to seek out each individual robotics platform. We will, however, mention the robots which are used in many of the literature reports and provide the corresponding references for the readers to explore further if desired. Each lab will have a variable set-up of the robot it is using based on the equipment available and the method requirements.

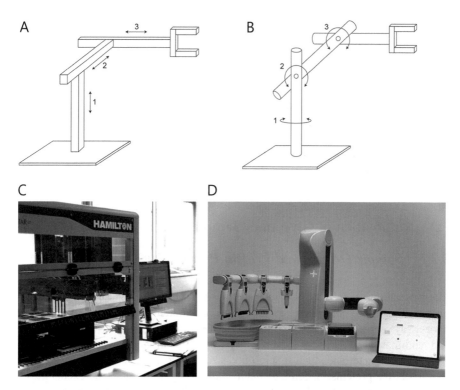

Fig. 5.1 (**a**) Representation of an anthropomorphic/articulated robotic workstation; (**b**) Representation of a Cartesian robotic workstation; (**c**) Commercial example of an anthropomorphic/articulated robotic workstation, the Andrew from Andrew Alliance; (**d**) Commercial example of a Cartesian robotic workstation, the Hamilton Microlab STARlet from Hamilton Company

The two main types of automated liquid-handling workstations which are used in glycomics laboratories fit into one of two robotic configurations: (1) the Cartesian configuration and (2) the anthropomorphic or articulated configuration (Fig. 5.1a, b respectively).

The main feature of the Cartesian robot is the programmed positioning of a portable single-probe or multi-probe injector in x-, y-, z-axis (Alexovič et al. 2018). The x-y-z drives control the translational motion of the end-effector, which holds the liquid-dispensing devices or plate gripper in the Cartesian coordinates. A liquid-handling robot with this structure is beneficial for its rigidity, stability and therefore, the positioning precision (repeatability), which is usually up to 10–100 microns (Kong et al. 2012). This configuration is seen, for example, in the Hamilton Star and StarLet robots, as well as the Beckmann Coulter Biomek and Tecan liquid-handling workstations (Fig. 5.1c).

The anthropomorphic/articulated robotic workstation is based on the artificial mimicking of a human arm. The core of the system represents a multi-jointed robotic arm (5- or 6-joints) with gripping fingers of various sizes at the end. Similar to Cartesian configurations, the articulated robotic arm attends to the various units located within the working bench-area at predefined positions and levels. Here, the fingers at the end of the robotic arm are used to grip either pipettes or plastic ware (Alexovič et al. 2018). The liquid-handling robot supplied by Andrew Alliance is an example of an articulated workstation (Fig. 5.1d).

We will use this chapter to review the sample preparation processes that have been used on liquid-handling robots to obtain high-quality glycomics data of both biopharmaceutical and clinical antibody samples. This will include a focus on glycoprotein purification, glycan, or glycopeptide generation, derivatization and enrichment. The use of liquid-handling robots for glycomics studies on other sample types beyond antibodies has also been reported but will not be discussed here.

5.2 Automation of Glycomics Sample Preparation

5.2.1 Sample Origins and Protein Purification

The generation of accurate and reproducible glycomics data can only be accomplished if glycoprotein samples are free from contaminants that lead to artifacts and poor assay reproducibility (Colhoun et al. 2018); this is important for both manual and automated methods. Recent methods for glycoprotein purification exploit molecular characteristics such as solubility, size, charge, and specific binding affinity to immobilize proteins. Specific techniques include gel filtration, ion-exchange, or affinity chromatography, all of which allow the efficient removal of contaminants (buffers, reagents, and other carbohydrates and proteins) by washing the immobilized glycoproteins prior to enzymatic glycan release. However, not all options are amenable to automation for high-throughput processing. For example, although the aforementioned reproducible and robust method for glycoprotein immobilization in polyacrylamide "gel-block" (Royle et al. 2008) was used extensively in manual preparations, the gel-block hindered adaptation onto an automated platform.

Automation friendly methods for glycoprotein purification prior to enzymatic processing are divided into two categories; plasma/serum samples for biomarker and diagnostic studies and glycoprotein therapeutic antibodies.

5.2.1.1 Serum and Plasma

Glycoprofiling of the serum or plasma N-glycome is challenging due to the high complexity and heterogeneity of the glycoproteins present in these samples. Serum and plasma fractions are taken from blood samples that have undergone various biochemical protocols after collection (Tuck et al. 2010). In the case of serum, coagulation factors (i.e., fibrinogen) and blood cells are removed by centrifugation, while plasma is typically obtained from blood samples by adding an anticoagulant agent (i.e., EDTA). The glycoproteins found in serum samples are immunoglobulins

(IgG, IgM, IgA) and acute phase proteins (transferrin, alpha-1-anti-trypsin, haptoglobin). Plasma, on the other hand, contains all of the glycoproteins which are present in serum plus fibrinogen glycoprotein.

Although glycoprofiling of complete plasma or serum samples provides a comprehensive assessment of released glycans from all glycoproteins present in the liquid sample, it lacks information about the origin of the glycan and hence the identity of the carrier protein. Hence, to focus in on the glycomic information from each of the immunoglobulin antibodies, a strategy is required to selectively capture these desired glycoproteins from the mixture. This is often accomplished by affinity purification, wherein a solid surface is coated with an anti-glycoprotein-specific antibody, which is a method that has been incorporated in automated workflows. The two strategies that have been adapted for automation of antibody purification are (1) the use of 96-well affinity-protein-containing plates and (2) pipette tips packed with affinity resin.

96-Well affinity purification plates are cited most often in automated glycoprotein purification workflows. This is presumably because any plate-based tool is easily implemented onto a robot workstation deck. For the isolation of IgG or IgA from serum or plasma, plates containing either protein G or protein A are employed. These are often purchased from commercial suppliers pre-loaded with the required resin and are accompanied by documentation that clearly indicates their binding capacity. Stöckmann et al. reported the affinity purification of IgG from 20 to 50 µL of human or animal serum using a robotics-compatible 96-well filter plate containing solid-supported Protein G (Stöckmann et al. 2013). In this case, a Hamilton Robotics StarLet liquid-handling platform was used which was equipped with eight software-controlled pipettes, a vacuum manifold, and an automated heater shaker. In this example, a Thermo Scientific™ Pierce™ "Protein G Spin Plate for IgG Screening" was used, which contains 50 µL Protein G agarose resin per well capable of binding ≥ 0.5 mg of human IgG/well (ThermoFisher Scientific, n.d.). In brief, the glycoprotein is purified by binding onto the plate, washing several times with a washing buffer and then treating the plate with an elution buffer to release the purified glycoproteins. This protocol has been used to purify serum IgG from galactosemia patients (Maratha et al. 2016), Juvenile idiopathic arthritis affected children (Cheng et al. 2017) and pooled serum samples from 100 healthy male and female adults (Stöckmann et al. 2015a).

As an extension of the plate-based method discussed above, it is also feasible to combine resins to remove multiple glycoprotein antibodies of interest from serum or plasma at one time. This is a strategy that was adopted by Momčilović et al. for combined enrichment of IgG and IgA (Momčilović et al. 2020). Here, samples were prepared in 96-well plate format from minimal (~5 µL) amounts of serum using affinity purification on a Microlab STAR liquid-handling robot. Due to the difference in serum concentrations of IgG and IgA (IgG has a higher serum concentration than IgA), their simultaneous enrichment was optimized by using a mixture of IgG and IgA affinity beads in a ratio that allowed for complete capture of IgA and only partial capture of IgG. This method successfully allowed the detection and analysis of the highly occupied glycosylation sites from both IgG and IgA in a single analytical run.

Utilizing modified pipette tips is the other automation friendly approach for affinity purification of glycoprotein antibodies. In this case, robot compatible tips are modified by packing them with the desired affinity resin for capture. Although preparation of these tips can be done manually, there are also several commercial sources of pre-packed tips for robotic platforms. O'Flaherty et al. reported the use of affinity resin-modified tips for capturing and glycoprofiling six abundant individual glycoproteins from human serum [which included the immunoglobulins, IgG, IgM, IgA, and the acute phase proteins, transferrin (Trf), alpha-1-anti-trypsin (A1AT), haptoglobin (Hpt)] by serial extraction using a Hamilton robotics StarLet liquid-handling platform (O'Flaherty et al. 2019). They used a series of both commercially available and manually packed tips containing different anti-glycoprotein capture resins (PhyNexus phytips; 20 μL of each resin). Glycoproteins from 50 μL of whole serum sample were then passed through the tips, and glycoproteins were captured in the following sequence: Trf, IgG, IgM, IgA, Hpt, and A1AT. This was optimized using a sample of pooled human serum on an automated liquid-handling station (Hamilton Starlet) in a 96-well format. This strategy proved quite successful with the captured proteins measuring >98% pure as determined by 1D-SDS page. After optimization, this method was applied to fractionate human serum glycoproteins from patients with ovarian cancer. One of the advantages of this approach is that glycomics data can be obtained for various glycoproteins from a single clinical source. The one caveat being that affinity resins with high selectivity and specificity for the targeted glycoproteins must be available.

5.2.1.2 Therapeutic Antibody Glycoproteins

The majority of therapeutic antibodies for biopharmaceuticals are produced in bioreactor-based mammalian cell cultures (e.g., Chinese hamster ovary (CHO) or murine myeloma transfectomas), and a select few are produced in other expression systems (e.g., *Escherichia coli*) (Kelley 2009). The monitoring of relevant glycan characteristics of biopharmaceuticals is required throughout the drug-life cycle, which means that these glycoproteins require testing during process development, medium development, clone selection and for final product release. To ensure reliable and reproducible glycomics data is generated, purification techniques are required to provide purified therapeutic antibodies from a variety of mixtures, some more complex than others, ranging from final formulations to fermentation broths.

In general, the same high-throughput purification strategies are applied to therapeutic glycoproteins as are applied to serum and plasma samples; the most common method used for automated therapeutic glycoprotein purification is 96-well plate-based affinity purification. For therapeutic antibodies, the trend is that 96-well affinity plates containing protein A are implemented for purification. Protein A has been chosen to exploit the strong affinity of protein A to bind to the Fc portion of IgG molecules and efficiently purify the immunoglobulin from complex mixtures such as cell culture media. The two methods for obtaining affinity plates are to either purchase plates which already contain a specified amount of affinity resin or to prepare them in-house by adding affinity resin to standard 96-well filter plates.

Both Doherty et al. and Stöckmann et al. chose to use commercial protein A plates for purification. Doherty et al. used an automated purification method to purify IgG directly from bioreactor cell culture supernatants. IgG from the clarified medium was captured in each of the wells of a commercial Protein A filter plate (Pierce) using a Hamilton Microlab STAR liquid-handling platform. Here, glycans were successfully released from the immunoglobulin using PNGaseF while immobilized on protein A; this technique will be elaborated on further in the following section (Doherty et al. 2013). Likewise, Stöckmann et al. used preconditioned 96-well IgG affinity purification plates (Thermo Scientific, 50 µL Protein A agarose resin per well) to isolate IgG from Chinese hamster ovary cell cultures using a Hamilton Robotics StarLet liquid-handling platform. However, in contrast, they chose to elute the IgG from the resin prior to glycan processing (Stöckmann et al. 2013).

Reusch et al. opted for capturing IgG from fermentation broth by adding a protein A-Sepharose slurry (Protein A-Sepharose from GE Healthcare) to samples in a standard 96-well plate. This approach may be favorable when capturing amounts of IgG that are outside the capacity of the commercially available affinity plates. As an example of how methods are adapted to suit the capability of the automation platform, here they chose to incubate the samples and the slurry of protein A in the robot without shaking. Instead, the beads were repeatedly resuspended by pipetting up and down intermittently throughout the incubation time using a 96-channel pipetting head. The captured antibodies were eluted from the beads on a robot-mounted vacuum manifold. This procedure allowed efficient capturing and purification of IgG from a complex fermentation matrix which was confirmed by quantifying the protein content (Reusch et al. 2013).

In many of the examples mentioned, for both biological samples and biopharmaceutical samples, a cited benefit of using IgG-binding affinity resins is that they can be regenerated after use. It is important to note that this should only be recommended after a thorough validation. Tests should be done to ensure that the full binding affinity is restored after cleaning and also to check that the complete removal of all previously bound proteins is sufficient so as to not skew or contaminate the next sample set (Stöckmann et al. 2013).

5.2.2 Preparing Glycans for Analysis: Glycan Release, Derivatization and Clean-Up

The most common method for the investigation of glycosylation is the analysis of liberated intact glycans. As a result, the majority of automation reports describe some variation of N-glycan analysis. Within the realm of N-glycan analysis, there are several complementary multi-step analytical approaches. The workflows tend to include the following; generally, the first step is to cleave the glycan moiety from the protein backbone using an enzymatic or chemical reaction. The gold standard method for the release of *N*-linked glycans is enzymatic treatment with peptide-*N*-glycosidase F (PNGase F); Secondly, due to the lack of inherent chromophores, it is common to derivatize the glycans with a fluorescent label after release and prior to analysis. Alternatively, the glycans can be chemically modified by permethylation or esterification to facilitate detection. Thirdly, most of these approaches require clean-up steps throughout the process, and these can be after enzymatic treatment and/or after labeling.

The following sections will detail the current literature on automated *N*-glycan sample preparation. As before, the sections will be divided by what type of sample is being processed; biological serum or plasma samples or therapeutic glycoproteins. While this helps to categorize the studies, it should be noted that it is possible for these methods to be applied across sample types, and multiple studies indicate that their processes were optimized for both.

5.2.2.1 Plasma and Serum

Automated Methods for *N*-Glycan Preparation Employing Anomeric Fluorescent Labeling Strategies

2-Aminobenzamide (2-AB) is the current gold standard fluorescent label used in glycan analysis. Therefore, it is not surprising that multiple laboratories have developed automated methods to perform this type of *N*-glycan analysis. Translation of the commonly used manual method for *N*-glycan release, 2-AB labeling and clean-up to an automated high-throughput method in a 96-well plate-based format was described by Ventham et al. (2015) In this protocol, various steps were optimized and adapted to suit the automated platform, however, the process itself closely mirrors a typical manual high-throughput method. A Hamilton Microlab STARlet liquid-handling robot was used for the automated sample processing steps, including liquid transfer and vacuum manifold mediated clean-up while all incubation steps, vacuum drying and plate sealing were carried out offline. This report is a good example of automating what is possible given the capabilities of certain laboratories. Glycoprotein denaturation, using both heat and chemical treatment, was carried out in a skirted 96 well PCR plate. A foil plate seal was employed in each of the incubation steps to ensure minimal solvent loss and to protect from contamination. To the same plate was added the PNGaseF and the respective buffers. The deglycosylation step was allowed to proceed overnight, and the solvents were removed under vacuum. *N*-glycan purification was accomplished on an automation compatible 96-well protein binding membrane plate using the integrated Hamilton vacuum manifold. The *N*-glycan solutions were transferred to a non-skirted PCR plate and dried under vacuum to enable the samples to be 2-AB labelled using reductive amination with non-toxic 2-PB reductant. A HILIC (hydrophilic liquid interaction chromatography) SPE type clean-up was implemented to remove excess labeling reagents using LC-T1 cartridges (commercially available from Ludger Ltd.) placed into a 96-well base plate which fits onto the vacuum manifold on the robot deck. After elution from the T1 cartridges, the samples were ready for analysis by UHPLC. The repeatability of this automated N-glycan processing protocol was assessed by processing 48 replicates of a pooled sample of human serum IgG. The method showed excellent repeatability with a Pearson's coefficient of 0.9998 for the normalized peak areas in the IgG glycan data. This specific workflow has been used on various sample sets, which include; the study of whole serum N-glycosylation from IBD patients (Ventham et al. 2015), to test if different collection tubes for serum samples impact N-glycosylation data (Ventham et al. 2015) and also to study the whole serum N-glycosylation changes in pregnant women (Reiding et al. 2019).

In 2013 Stöckmann et al. reported what they called the first example of a low-cost, fully automated high-throughput assay for N-glycomics (Stöckmann et al. 2013).Here they used a liquid-handling robot to prepare fluorescent 2-AB labeled N-glycans from serum samples and indicate that the process is versatile enough to be used to determine the glycosylation pattern of individual glycoproteins or classes of glycoproteins, such as immunoglobulin G (IgG). Indeed, this protocol was used to study IgG in both biopharmaceutical samples and biological samples, including serum IgG (Stöckmann et al. 2013) as well as whole serum from healthy participants (Stöckmann et al. 2015b), plasma IgG from juvenile idiopathic arthritis patients (Cheng et al. 2017) and whole serum from pregnant women (Reiding et al. 2019). The automated process was optimized on a Hamilton Robotics StarLet liquid-handling platform and included all required steps for N-glycan sample preparation, affinity purification of the glycoprotein, denaturation, N-glycan release, fluorescent labeling and clean-up. Their strategy to optimize the automated process to ensure that all steps were completely robot compatible was elegantly described, and a few highlights of the method are; (1) They chose to use an ultrafiltration plate with a nominal molecular weight limit (NMWL) to prepare for and perform the PNGaseF digestion. This permitted the removal of excess small molecule reagents such as detergents and alkylating agents after the protein denaturation. Additionally, upon their enzymatic release from the protein, IgG N-glycans were easily recovered by simple filtration because their size was well below the molecular weight cut-off of the ultrafiltration membrane. (2) The released N-glycans were immobilized on hydrazide beads to facilitate clean-up before labeling. This was introduced because 2-AB glycan labeling after solvent removal by evaporation frequently led to inconsistent data, presumably from the presence of contaminants (buffer salts or residual detergent) interfering with the labeling reaction. To avoid time-consuming solid-phase extraction and aqueous solvent evaporation, solid-supported hydrazide beads were chosen to selectively react with reducing carbohydrates via the formation of a stable covalent hydrazone bond so that any noncarbohydrate species could be removed by a simple washing step. Once washed, the glycans could then be released from the solid support by incubation in water and catalytic amounts of acid. (3) Solid-phase extraction cartridges were used for clean-up after labeling. After the labeling reaction was quenched, to remove excess labeling reagents, the entire reaction mixture was transferred to solid-phase extraction cartridges (normal phase) set into a 96-well plate format. Interestingly, although the glycans had been released from the hydrazide resin and are now 2-AB labeled, they found that reproducible results could only be obtained upon quantitative transfer of the entire reaction mixture, including the hydrazide resin, to the solid-phase extraction plate. Water-mediated glycan elution provided the samples, which were concentrated and ready for UPLC. The optimization of this method was successful which is reflected in its reproducibility; samples prepared on different days have coefficients of variation (CVs) that are generally below 10%. The processing time of up to 96 samples, including glycoprotein affinity purification, was around 22 h and is completely automated, built to run overnight without human intervention.

Building on the success of their previous fully automated N-glycan processing and 2-AB labeling platform, Stockman et al. developed an improved method (Stöckmann et al. 2015a). They chose to increase the throughput and adapted the method to a 384 well plate workflow. Additionally, they replaced the reductive amination 2-AB glycan labeling step with the "instant" aminoquinoline carbamate (AQC) labeling reaction. These changes allowed them to streamline the process significantly. The process followed the same overall workflow; glycoprotein affinity purification, glycoprotein denaturation, N-glycan release, and fluorescent labeling. However, multiple clean-up steps could be removed due to the nature of fluorescent labeling chemistry. In the first step, the glycoproteins were denatured and washed, and then the N-glycans were released from the protein using PNGase F. These steps were all carried out in the same 384-well ultrafiltration plate (10 kDa molecular weight cut-off) with vacuum filtration performed by the Hamilton Robotics StarLet liquid-handling platform. This allowed for easy washing during the denaturing step, and then the N-glycans could be simply separated from the protein after PNGaseF treatment. The previously automated method required N-glycan purification after deglycosylation, where N-glycans were extracted from the solution using a solid-supported hydrazide resin. In this case, residual impurities and buffer salts do not interfere with the AQC labeling reaction, hence the elimination of the need for time-consuming SPE prior to labeling. Also, no buffer exchanges were required because the deglycosylation and the subsequent AQC labeling reaction both occur at the same pH. Finally, after ACQ labeling, no final clean-up steps or solvent concentration are needed because the labeling reaction proceeds in a mixture of (30:70) buffer:acetonitrile, which is an appropriate solvent composition for injection of the sample onto the UPLC HILIC column.

It is worth mentioning that the quality of the data obtained for AQC is comparable to the 2-AB labeling method with similar N-glycan profiles. However, the fluorescence emission of the 2-AB profile is about 30-fold less intense in comparison to the AQC profile. This indicates that AQC labeling could be more favorable for analyzing and profiling IgG samples with a minimal amount of material. One disadvantage of the method is that migration of O-acetyl groups commonly found on sialic acids may occur under the conditions of release and AQC labeling. Overall, the modifications made to their previous method to develop this new fully automated AQC labeling N-glycan processing method had a significant impact on the throughput of their workflow, increasing the number of samples processed in a single automated run from 96 to 768. In addition, the elimination of the SPE steps and the increased throughput significantly lowers the cost per sample, making this method an attractive one for processing large patient sample cohorts. Indeed, this technique was used by O'Flaherty et al. to perform the detailed characterization of N-glycosylation of six serum glycoproteins (O'Flaherty et al. 2019). They were able to study the antibodies (IgG, IgM and IgA) and acute phase proteins (Trf, Hpt, and A1AT) from a single small (50 µL) serum sample. The strength of this study and technique is that they are using one human biological fluid (serum) to obtain data on six glycoproteins in detail—this undertaking without the aid of automation would be very laborious. The utility of this method for biomarker identification in ovarian cancer was also demonstrated in the same study where the glycomics data indicates that N-glycosylation of Trf and Hpt glycoproteins may be suitable targets to be exploited.

As an alternative approach to the automation of a standard manual N-glycomics method, Nishimura et al. developed an "all-in-one" solution for automated and high-throughput N-glycan enrichment (Nishimura 2011). This is an integrated system wherein the selective capturing of total glycans, methyl esterification of sialic acids, and fluorescent tagging are all carried out using a hydrazide-functionalized bead handled in a multi-well filter plate. The automation of N-glycan release, purification, labeling, and MALDI–MS spotting were performed on the SweetBlot 7 automated system from System Instruments Co. In this study, the affinity-purified serum from 115 Ethiopian breast cancer patients and 33 healthy volunteers were studied to identify biomarkers of disease (Gebrehiwot et al. 2019). Affinity-purified IgG fractions were transferred via automated liquid-handling pipette into 96 well polymerase chain reaction (PCR) plates for denaturation, trypsin digestion and deglycosylation. At the same time, BlotGlyco H beads (Sumitomo Bakelite Co., Ltd., 10 mg/mL suspension with water) are loaded into the wells of a multi-Screen Solvinert filter plate (Millipore). The released N-glycan mixture is then transferred into the 96-well filter plate packed with BlotGlyco H beads. These BlotGlyco H beads are the key element in the all-in-one protocol and are stable hydrazide-functionalized polymer support. The hydrazide group on the bead reacts selectively with aldehyde or ketone groups that are present at the reducing terminus of glycans (reactive aldehyde and ketone groups are very rare in biological samples). The formation of the hydrazone bond between the BlotGlyco H bead and glycan is reversible, ensuring that the glycans can be released when needed. The unreacted hydrazide functional groups on BlotGlyco H beads are then capped by incubation with acetic anhydride. The beads are then washed to remove any impurities. This is conveniently performed in the same filter plate. The following step is on-bead methyl esterification of the sialic acid residues. This was performed to stabilize the sialylated glycans. The final steps were the transiminization with O-benzyloxyamine hydrochloride (BOA) fluorescent dye and mild acid hydrolysis of the hydrazone bond. The released and BOA-labeled N-glycans were then eluted from the filter plate under vacuum and were ready for subsequent MALDI-TOF–MS analysis. Currently, this solid-phase protocol is the only example of "all-in-one" glycoblotting technique in a single automatable workflow where all steps are completed on the bead. The automated protocol was validated by processing replicates of the same human serum digests, and its reliability was confirmed by good repeatability (Miura et al. 2008). This all-in-one approach has been used with multiple clinical projects, including the differentiation of whole serum *N*-glycan profiles in subjects with congenital disorders of glycosylation and hepatocellular carcinoma and healthy donors (Miura et al. 2008) and the inter-ethnic differences in whole serum *N*-glycome among US origin, South Indian, Japanese, and Ethiopian ethnic populations (Gebrehiwot et al. 2018).

Automated Methods for *N*-Glycan Preparation Employing Permethylation Derivatization Strategies

The permethylation of released glycans is an alternative strategy to fluorescent labeling, and it is routinely performed prior to MALDI-TOF–MS analysis. There are a number of reasons that an analyst might choose to implement glycan permethylation, including (1) the improvement of ionization efficiency of the glycans when compared to nonderivatized oligosaccharides, (2) the stabilization of the sialic acid moieties, (3) the detection of both neutral and acidic glycans in positive ion mode on MS, (4) the ability to determine branching and glycosidic linkage positions, (5) increased glycan hydrophobicity enabling reverse-phase chromatography analysis, and (6) fast profiling and analysis times (Ciucanu 2006). However, the conventional manual permethylation techniques are laborious for large sample sets. Hence, this was the main motivation for various research groups to automate the permethylation workflow to enable high-throughput processing.

The standard permethylation method involves the following steps: *N*-glycan release, enrichment, permethylation, and liquid–liquid extraction. Shubhakar et al. were the first to adapt the majority of this method to a liquid-handling robot, specifically a Hamilton Microlab Starlet (Shubhakar et al. 2016). PNGaseF deglycosylation was performed in 96-well PCR plates with liquid handling completed on the robotic workstation. Following the enzymatic digestion, the released *N*-glycan samples were purified using a 96-well format robot compatible HILIC SPE filter plate (Ludger Clean Pre-Permethylation Clean-up Plate) followed by vacuum-mediated solvent removal. The HILIC SPE filter plate was found to provide better data after permethylation than a protein binding membrane (PBM) plate. Off-deck incubation, plate sealing and centrifugal evaporation steps were required in this process. However, liquid-handling robots are available with higher specifications which do have integrated plate sealers, incubators, and centrifuges. The use of these more sophisticated robots would allow for the total automation of the process of sample preparation and derivatization without any manual handling. The next step is the permethylation reaction. This step is carried out by dissolving the glycans in DMSO and transferring them into a 96-well plate containing solid sodium hydroxide, followed by the addition of methyl iodide to the mixture. After incubation, the permethylated glycans are isolated using liquid-liquid extraction. The most challenging aspects of this process to automate were the permethylation reaction and the liquid-liquid extraction. The liquid handling of highly volatile organic liquids, such as methyl iodide and dichloromethane which are required for these steps, is notoriously difficult. To solve this problem, the robot was programmed to pre-wet the tips to avoid the loss of the volatile liquids and also to use the liquid level detection feature to ensure good sample recovery after four cycles of liquid-liquid extraction. One of the major benefits of using this automated procedure is that it is quick; the permethylation and liquid-liquid extraction process for 96 samples can be performed within 5 h. A second major benefit of analyzing permethylated glycans is that data acquisition using the MALDI-TOF–MS takes less than 1 min per sample. The automated permethylation method was developed and fully validated on IgG from human serum. The authors also performed a comparison study to show that the permethylated glycan data from MALDI-TOF–MS were similar to those obtained using HILIC UHPLC data and 2-AB labeling. This method has also been used to analyze other glycoprotein standards (fetuin O-glycans), N-glycan standards, and a range of biopharmaceutical samples (IgG1 mAb standard, IgG4 mAbs, and rhEPO) (Shubhakar et al. 2016).

As mentioned previously, one of the challenging steps in the standard permethylation process is the liquid–liquid extraction required for the isolation of the permethylated glycans. In addition to volatile solvents being tricky to handle, liquid-liquid extraction also comes with the risk of forming emulsions during the process as it can sometimes be difficult to obtain a clear phase separation. The process also uses known toxic solvent, dichloromethane. Additionally, liquid-liquid extraction is not suitable for recovering permethylated glycans bearing polar substituents, such as sulfates. As a way to improve the permethylation process and eliminate the need for liquid-liquid extraction, Shajahan et al. developed a high through put method which uses solid-phase extraction (SPE) to isolate the permethylated glycans instead (Shajahan et al. 2019). They chose to use this approach for the following reasons; it can be performed on smaller volumes, it does not require additional solvents, it allows for higher N-glycan recovery. Most significantly, SPE is easily adapted to automated workflows since it can be done in a pipet tip and C18 pipet tips are widely available from commercial suppliers.

The process follows the same steps as the standard workflow beginning with glycan release. The liberated glycans are dissolved in DMSO in a polypropylene microplate and are treated with a gel solution of sodium hydroxide in DMSO followed by iodomethane. The permethylation reaction is terminated by the addition of water which results in a bi-phasic solution due to the excess iodomethane forming a separate lower layer. In contrast to other methods where the excess iodomethane is removed during evaporation under vacuum, the iodomethane is removed by repeatedly pushing air through the mixture using a pipet to encourage evaporation directly from the plate. The final step in the process is the binding of the permethylated glycans to a C18 resin by repeatedly passing the sample solutions through C18 filled SPE tips (Thermo Fisher Scientific). The resin is washed with water, and the permethylated glycans are eluted with a small volume of methanol. The resulting solution is then ready for analysis. The full process on 96 samples from glycan release to the acquisition of MALDI-MS can be accomplished in less than 1 day. The method is very sensitive, with good N-glycan spectra being obtained when only 1 microgram of glycoprotein was used. This method was used to study immunoglobulins (IgG) and transferrin isolated from human serum, along with standard glycoproteins such as bovine fetuin and κ-casein.

The authors indicate that since the entire permethylation process is carried out in conventional polypropylene 96-well plates, it can be performed manually or by using any degree of automation and that sophisticated robotics-based sample handling equipment are not necessarily required as is the case with the other automated permethylation methods. However, one caveat is that while this method has been designed for high-throughput processing in 96-well plate format and would presumably transfer to a robotic platform without issue, it is not clear that the process has been explicitly trialed and tested on an automated platform as of yet.

5.2.2.2 Therapeutic Antibody Glycoproteins

Automated Methods for *N*-Glycan Preparation Employing Anomeric Fluorescent Labeling Strategies

The biopharmaceutical industry could benefit from automated high-throughput methods designed for a variety of activities, including final product release, QC and batch-to-batch comparison. Additionally, many tasks in the development process could also be streamlined with the help of automation. These tasks include both high-throughput screening of clones for cell line selection and quality by design (QbD) studies to understand the impact of process parameters on product quality. To address the need to rapidly generate data for bioprocesses that are under development, Doherty et al. developed an automated workflow that is capable of glycosylation sample analysis directly from a bioreactor (Doherty et al. 2013). The protocol follows the standard *N*-glycan release, labeling, and clean-up strategy but with some clever optimization to reduce processing times and allow efficient automation. All steps were carried out using a Hamilton MicroLab STAR Robot, and liquid-handling steps were performed employing 8-channels, whereas sample transfer was accomplished by operating the 96-channel head. mAbs were purified from the cell culture media by capture on protein A resin in a 96-well plate format. Instead of eluting the captured mAbs from the protein A resin after washing, the glycans were removed enzymatically with PNGaseF while still immobilized. This allowed for easy isolation of the *N*-glycans without any additional clean-up steps. They were able to optimize the deglycosylation time as well, reducing the incubation time down to 60 min without a loss in yield. Complete automation of these steps was made possible by the modification of the robot deck with a custom-made integrated incubator. In this automated protocol, the reduction and alkylation steps that often precede PNGaseF release were eliminated. Not every IgG will require reduction and alkylation for efficient *N*-glycan release, but in order to see if these steps could be eliminated, they compared samples that had undergone reduction and alkylation with those that were directly treated with PNGaseF. In this case, for the samples they analyzed, there were no differences in *N*-glycan profiles. The released glycans were fluorescently labeled with 2-AB. For the final clean-up step, the excess chemical reagents were removed using solid-phase extraction (SPE) with a bulk synthetic polyamide stationary phase (DPA-6S, 25 mg/well, Sigma–Aldrich) packed into 96-well plates and run in hydrophilic interaction mode. The advantage of this particular SPE clean-up method was that it outperformed other commercially available options, and as an added benefit, it could also be applied to the purification of underivatized glycans for complementary analytical techniques. Processing samples using this automated method takes approximately 5 h as opposed to multiple days for classical analysis and provided reproducible data for therapeutics taken directly from a bioreactor.

Although 2-AB remains the gold standard label used in the biopharmaceutical industry, an alternate fluorescent label that has been introduced recently is Rapifluor-MS (RFMS). RFMS is described as an instant label, similar to the AQC label that was mentioned previously in this chapter, that reacts "instantly" with the glycosylamine that is generated directly after PNGaseF treatment without the need of additional chemical reagents. The process of adapting the RFMS workflow to an Andrew Alliance semi-automation platform was reported by Reed et al. (2018). They used both a murine IgG1 mAb as well as the biopharmaceutical Cetuximab to optimize their automated procedure. The optimization of this protocol centered on time and temperature requirements and presented several challenges. The first challenge was that the automated platform is unable to transfer reaction vials to other locations on the robot deck. According to the manual method, sample transfer is required into and out of heating blocks set at 90 and 50 °C during the denaturation and deglycosylation steps. Here, the denaturation step requires 3 min of heating followed by 3 min at room temperature, and deglycosylation requires 5 min of heating with 3 min at room temperature. In an effort to automate this step and limit manual intervention, the robot deck was modified to include a computer-controlled Peltier effect heating block that would be used for both steps. Unfortunately, the time required to reach the desired temperatures on the heating block from room temperature was significantly longer than the required incubation times resulting in the samples being exposed to elevated temperatures for much longer than under normal conditions. This posed several problems; (1) the samples are generally uncapped and over time, evaporation becomes a concern, and (2) while the time for deglycosylation was comparable to the manual timing, the automated heating block was 10 °C off of the optimal temperature for PNGaseF digestion which can cause issues with the integrity of the glycosylamine. This was a problem because it is known that the glycosylamine produced directly following PNGase F release is not stable and converts slowly to a hemi-acetal sugar. Therefore, a prolonged delay between the enzymatic release of the glycosylamine and the labeling step will decrease the glycan labeling yield as the hemi-acetal cannot react with RFMS. In order to ensure that the manual and automated methods produced similar results, reoptimization of the denaturation and deglycosylation temperatures was performed and resulted in changing the target temperatures to 75 and 55 °C, respectively and increasing the time required for these steps. After the liberation of the *N*-glycans, RFMS labeling reagent was added to the reaction vials via a robotic pipette, which then also transferred the samples to the clean-up cartridges (GlycoWorks HILIC μElution Plate). The liquid handling is performed by the robotic arm; however, the analyst must be present at this stage to manually manage the vacuum, turning it on and off when required. Additionally, after clean-up the waste tray must be manually replaced with sample collection tubes, and then the vacuum managed during the elution of the purified and labeled N-glycans.

Adapting this N-glycan processing strategy to an automated platform resulted in an overall increase in experiment time, with the automated protocol taking from 1 to 3 h depending on the number of samples being processed. The authors indicated that this is a significant increase in experiment time over the manually performed protocol. However, this can be justified as the manual intervention is limited to three user interactions over the whole experiment and relieves the analyst of all repetitive pipetting actions. The data obtained using the automated method was comparable to the manually performed protocol. Acknowledging the limitations of the automated method, the authors suggest this as a cost-effective benchtop solution for medium- to low-throughput laboratories. In another publication, this method was tested for its applicability in the manufacturing environment of mAb-based therapeutics by Chen et al., who used it to analyze 48 samples of infliximab derived from six different batches (Zhang et al. 2020). Based on the historical knowledge of the infliximab product, the levels of critical glycan species such as high mannose and sialylated glycans were evaluated among the samples. Indeed, using this method, they were able to identify the out-of-specification results and were able to report them with the accompanying detailed glycomics information for further investigation.

The automated method employing the AQC label (also in the instant label family), which has been described in detail for processing human IgG samples, was also used to process therapeutic glycoproteins. Stöckmann et al. applied the same automated platform for glycan release, AQC labeling and clean-up on a series of cell-culture-derived IgG N-glycans (Stöckmann et al. 2013). The authors were able to process antihuman IL-8 IgG successfully, showing that they could rapidly screen cell-culture-derived IgG to identify clones that produce the desired glycosylation pattern.

In many biopharmaceutical testing labs, the alternative to liquid chromatography analysis of fluorescently labeled glycans is high-performance capillary electrophoresis with laser-induced fluorescent (CE-LIF) detection. In this process, the released glycans are labelled via reductive amination with 8-aminopyrene-1,3,6-trisulfonic acid (APTS). An optimized automated method for glycan release, glycan partitioning, APTS labeling and sample clean-up for biotherapeutic mAb samples is described by Szigeti et al. (2016). In this method, the main area of innovation was the use of a magnetic bead-based protocol to avoid the centrifugation and vacuum drying steps which are often a challenge to automate. Magnetic bead-mediated techniques were first introduced in the proteomics field as a method to clean up and enrich samples using interaction-based capture mechanisms (Molares Vila et al. 2010). This technology can also be leveraged for glycans as carboxyl-coated magnetic beads can be used for the selective enrichment of glycans via solid-phase reversible immobilization. In this case, organic solvents (acetonitrile) encourage the glycans to interact with the surface of the magnetic beads allowing for the residual peptide material left after glycan release, as well as excess ATPS reagent, to be removed from the reaction mixtures. The bound glycans can then be eluted off of the magnetic beads by treatment with aqueous media.

The fully automated N-glycan sample preparation platform from therapeutic glycoprotein antibodies was performed using a Biomek FXP Laboratory Automation Workstation (Beckman Coulter). The robot deck was equipped with 96-well plate holders, a magnetic stand, various pipette tip holders, solvent reservoirs, a vortex heating block and slots for sample and reagent vials. All actions were performed by the robot. The workflow for the method starts with protein denaturation and N-glycan release using PNGase F in a 96-well plate containing commercial magnetic beads (CleanSeq). After 1 h incubation, the N-glycans were captured by increasing the acetonitrile content of each well. Once the glycans were captured, the magnetic beads were held in place using the magnetic stand allowing for the removal of the supernatant. The APTS labeling of the captured glycans was then performed by reductive amination. After a 2 h incubation, the acetonitrile content of each well was increased to ensure the glycans were once again bound to the magnetic beads. The excess APTS dye was removed by washing the beads three times, and in every wash step, the following process was performed; the ATPS labeled glycans were resuspended in water, the acetonitrile content of the solution was increased to capture all glycans, and the supernatant was removed. Finally, the APTS labeled glycans were eluted from the beads by the addition of water and were ready for CE-LIF analysis.

Applying this fully automated magnetic bead-based glycan sample preparation protocol, 96 samples of commercially available murine IgG1 mAb were analyzed in less than 4 h processing time with excellent yield and good repeatability. They also showed that the method could be successfully applied to the automated sample preparation and analysis of multiple marketed therapeutic antibodies, including both the innovator and a biosimilar of adalimumab (commercial-name Humira) and trastuzumab (commercial-name Herceptin). The benefit of this method is that it is fully automated and requires no human intervention from the beginning to the end of the sample preparation process. Provided that the analytical platform of choice is CE-LIF, this method is well suited for rapid, large-scale sample processing for glycosylation analysis of biopharmaceutical antibodies.

Automated Methods for N-Glycan Preparation Employing Permethylation Derivatization Strategies

The automated N-glycan processing method using permethylation derivatization was described in Sect. 2.2.1.2 reported by Shubhakar et al. was also used to study mAbs and therapeutic glycoprotein samples (Shubhakar et al. 2016). They analyzed both a standard IgG1 mAb and also a series of IgG4 mAbs that were produced by glutamine synthetase Chinese hamster ovary (CHO) cell lines grown in stirred tank bioreactors. These cells producing IgG4 mAbs were cultured under five different bioreactor conditions varying in temperature and type of aeration. Using the automated and high-throughput method, they were able to screen the samples from each bioreactor condition quickly and were able to identify an impact in the percentage abundance of galactosylation in the N-glycan profiles. This information is essential in the biopharmaceutical industry to help with clone selection and cell culture process optimization for biopharmaceutical realization.

5.2.3 Automated Methods for Glycopeptide Preparation

As mentioned in the previous section, the study of N-glycans is the current method of choice for glycoanalysis. However, when glycans are released from their parent antibody, the information on site-specific glycosylation is lost. On the other hand, mass spectrometric analysis of glycopeptides allows for the in-depth analysis of IgG glycosylation in a site-specific manner. These studies are required for large sample sets, especially during drug development and biomarker identification, when site-specific information is vital to the understanding and characterization of biologically relevant glycoproteins, particularly when the target antibody contains Fab as well as Fc glycosylation. The literature available on the automation of high-throughput methods for glycopeptide processing of antibodies is limited; indeed, only one publication was found that addresses this topic.

The standard method for preparing glycopeptides often includes three stages; affinity purification, tryptic digestion and clean-up. Reusch et al. chose to translate the manual method for glycopeptide analysis and adapt it to a robotic workstation (Reusch et al. 2013). This process required multiple optimization steps to make the protocol automation friendly; in this case, the biggest change was the elimination of vacuum centrifugation steps used to dry samples by evaporation. The workflow was designed on an extensively equipped Hamilton Microlab Star Robot containing: a 96-channel pipetting head and 4 single pipetting channels; grippers designed for moving labware on the robot deck; 14 cooled 96-well plate format positions for plates and troughs; 24 cooled 1.5–2.0-mL cup-holders; seven 96-well plate positions; five 96-well filter plate positions; four heated shakers; and a vacuum manifold. With this protocol, 8–384 samples could be handled in parallel (up to four 96-well plates). For the method development, they started with un-purified IgG1 and IgG4 containing fermentation supernatants and purified IgG1 in formulated bulk. The method started with the affinity purification of the glycoproteins; the authors chose to skip the vacuum centrifugation drying step that is usually directly after elution of the samples from the affinity resin. With this small change, the robotization of the entire procedure became more feasible. Trypsin was added directly to the purified IgG solutions. To increase the speed of proteolytic cleavage and to avoid incomplete digestion, two aliquots of trypsin were needed, and these were added to the reaction mixture at the beginning of the experiment and after 3 h of incubation. In addition, the 96-well plate was covered with a "robo" lid (Corning Glassware, Corning, NY, USA) to minimize evaporation during the incubation time. The purification and concentration of glycopeptides were accomplished using HILIC. Filter plates

(96-well polyethylene frit Orochem plates) were loaded with CL-4B Sepharose. This HILIC purification proved to be a crucial step for efficient glycopeptide enrichment. In order for the exclusive retention of glycopeptides on the HILIC resin, the washing buffer had to contain trifluoroacetic acid (TFA). If the TFA was not present in the buffer, both peptides and glycopeptides were detected in the mass spectra. Retained glycopeptides were eluted from the HILIC resin with water under vacuum into a 96-well collection plate. The glycopeptides were analyzed on MS directly from the elution mixture. This method for automated glycopeptide preparation showed good repeatability, and all peaks could be assigned to glycopeptide structures for IgGs produced using CHO cells. The authors were able to develop and optimize the method to allow for a high degree of automation. The optimized method was applied to measure the glycosylation state in the course of a fermentation run. Here, samples were taken from the fermentation media each day and were analyzed. They were able to establish the optimal fermentation time to obtain the desired glycosylation and could also investigate the impact of alterations in the fermentation conditions on glycosylation. The overall benefit of this method is that by analyzing glycopeptides, site-specific information on glycosylation is obtained. The authors speculated that this method would be applicable to wider glycomics applications. For example, if used with plasma samples, the method should enable the analyst to distinguish between the various IgG subtypes.

5.3 Commentary

One of the limitations of glycosylation analysis is the hands-on time required for sample preparation. As a solution to this problem, a relatively new approach that has been adopted in some laboratories is the use of automated liquid-handling robots. The status quo for the automated sample preparation of antibodies was reviewed herein. To remain in scope, only automated antibody processing for glycomics was reviewed. There are numerous reports on the automated processing of other biologically important glycoprotein samples (e.g., cells, tissues, and other bodily fluids), but they were not investigated. In general, there are a select group of researchers actively working on automated processes for glycosylation analysis workflows for antibodies. The automated preparation of antibody samples for glycosylation studies often starts with affinity purification regardless of whether the sample is a biological fluid or a therapeutic antibody. The purification of an antibody is needed to remove contaminants, other proteins or reagents, which could negatively impact the subsequent sample processing and resulting glycomic data. The affinity purification processes that have been successfully automated employ either 96-well filter plates or tips that have been packed with antibody-specific binding resins. These are either purchased from commercial suppliers or are prepared in-house. The most widely used affinity resins for glycosylated antibody purification are protein A and protein G.

The automation procedures for sample preparation of *N*-glycans, for both biopharmaceutical and biological antibodies, all use enzymatic digestion (PNGaseF) for glycan release, which is followed by a chemical derivatization step that falls into three main categories; reductive amination with a fluorescent dye, reaction with an "instant" fluorescent dye, or permethylation. Although most strategies for total automation aim to limit the total number of steps required, clean-up steps are invariably needed throughout the workflow and many groups have looked to employ different strategies to make these steps more automatable, including the use of SPE plates, immobilization of the glycans on insoluble beads and filter plates with size exclusion molecular weight cut-offs. If site-specific information about an antibody is needed, then glycopeptide analysis is the technique of choice. The automation of this process, which involves proteolytic digestion, purification, and concentration of the formed glycopeptides, has surprisingly only been reported once. In this case, the process was adapted to total automation by carrying out the proteolytic digestion in the elution mixture after affinity purification. The glycopeptides were purified and enriched using a HILIC plate and were ready for analysis upon elution.

After reviewing the literature on the feasibility and overall success of robotizing glycomics workflows for antibody glycosylation analysis, there were several trends emerged. The first is that plasticware is a pivotal item for translating manual processes to automated processes. Plates are a key point of automation friendly methods; whether they are filter plates packed with various affinity resins, clean-up resins or size exclusion membranes, the commercial availability of these products makes a huge difference in the ease of translation. Likewise, pipette tips are just as integral to the success of an automated process. Secondly, transferring a manual high-throughput procedure onto an automated platform requires a good amount of optimization and ingenuity from the analyst. The analyst must also be able to program the corresponding software and run simulation experiments to ensure that the computer and liquid-handling robot are properly "talking." Furthermore, out-of-the-box solutions are often needed to overcome roadblocks; for instance, translating heating or shaking step might demand an alteration to the method to allow automation—the analyst may ask: Do I really need to do this step? Or can it be done in a different way? Instead of shaking, will be repeated up and down pipetting result in the same outcome? Likewise, the analyst may consider if a certain type of antibody glycoprotein requires denaturation or if the numerous centrifugation and vacuum—centrifugation steps that make full automation a challenge could be eliminated. In some cases, the resolution is to automate what is possible given the equipment and resources available. Finally, it is common practice for test methods to be run in simulation mode before initiating the actual run to maximize productivity and minimize any potential errors.

There are several key advantages of using automated methods. (1) Automation eliminates most human errors resulting in more reliable data. The use of a precision device increases confidence in the accuracy of sample preparation, and this is then reflected in the quality of the resulting data. (2) Automation removes the physical stress that repetitive manual pipetting creates for scientists and also reduces the risk of exposure to hazardous materials used in certain experimental procedures.

(3) Many more samples can be processed at the same time. Automation allows for higher throughput, and many platforms can facilitate the queuing of samples for continuous or parallel processing. In addition, once a process is automated, a further increase in throughput can be achieved by converting the method to a 384-well plate format. (4) Fully automated processes mean that there is no human intervention required at any point in sample processing, allowing analysts to focus on other tasks, freeing up people resources. (5) The computerized control of the robot means that protocols can be easily transferred between labs by sharing the programming code. (6) Robots are able to provide sophisticated computational features such as; monitored air displacement pipetting, independently spreadable pipetting channels and program logs indicating errors in the process. The disposable tips can have integrated conductivity probes that allow for liquid level detection, and they can also be fitted with anti-droplet control for pipetting volatile organic solvents. In addition, when the process is semi-automated, the robot programming can prompt the analyst. For example, if a robot was not equipped with an on-deck incubator at this step, an on-screen instruction can be set to instruct the user to manually perform the incubation and to press "OK" when the required action was completed. (7) Lower costs per sample for processing. Typical consumable costs are less than the manual low-throughput equivalents. Also, the increase in precision may allow for smaller volumes of initial starting material, which translates to smaller amounts of reagents as well. Specifically, for the biopharmaceutical industry, the high-throughput screening capabilities can allow for faster data generation allowing for quicker decision making on the optimal cell processing parameters or clone selection and thereby reducing overall expenditure.

There are various challenges associated with developing and implementing robotized methods. (1) One of the biggest roadblocks is that not all aspects of sample preparation are readily automated, and the ease of translation often depends on the sophistication of the equipment. Full automation is difficult to achieve. (2) Another challenge when using a liquid-handling platform is that dead volumes must be considered in cost analysis. The solvent and reagent reservoirs must have sufficient volume for liquid transfer, and this will always mean that excess is required. For cheap reagents, this is not a concern but for expensive reagents, dead volume can lead to increased cost. (3) Sample stability and evaporation of solvents is a concern on robotic workstations—although some now come with plate sealing options or are capable of maneuvering lids—the time without this is still a potential problem. (4) The start-up costs for implementing robotics are high. Indeed, a huge limiting factor to the adaptation and expansion of automated glycomics research is that acquiring a liquid-handling robot requires a significant upfront capital investment for the machine itself and additionally, there is the cost of expert personnel for method programming and development. Therefore, a laboratory may require a high sample demand to make automation worthwhile. (5) While hands-on time will be limited, the automated process does not always offer any time savings in comparison to the manual method. (6) Robotic platforms require users who are comfortable and proficient in the programming language or operating system provided with the machine.

5.4 Future Perspectives

The future outlook for the advancement of automated methods to support both biopharmaceutical realization and medical glycomics is promising. The ever-growing demand for higher throughput studies will keep the field moving; the increasing size of sample sets coming from both biopharmaceutical and clinical sources has been a relatively recent challenge to the glycomics field, and this will inevitably drive continuous improvement and further development of automation. The biopharmaceutical pipeline is currently dominated by mAbs, and in the future, this will only increase as more innovator mAbs are developed, and new biosimilars of off-patent therapeutics are released. Likewise, there has been tremendous growth in the clinical glycomics field, with many glycosylation-based disease biomarkers being discovered. In the future, the translation of these biomarkers to clinical diagnostic tools will require high-throughput solutions for a large number of patient samples. Progress continues to be made to adapt automation to all aspects of glycomics sample preparation. Indeed, out of the publications reviewed herein regarding the automated preparation of antibodies for glycoanalysis, three were published in the first quarter of 2020. The advancement in automation equipment and capabilities will be a key factor in the continued expansion of the available automated methods for glycosylation analysis of antibodies. This will be impacted by the improvement in both robot compatible plastic ware as well as improvement in solid-phase separation, affinity and size exclusion resins and their commercialization in 96- or 384-well formats. Finally, there is hope that the cost burden of implementing automation will be driven down by both its wider use and also by the publication of literature supporting its effectiveness for high-throughput glycomics studies.

5.5 Conclusions

Antibodies are glycosylated proteins that are key molecules in two major fields; the biopharmaceutical industry and medical glycomics. Scientific progress in both of these areas has steadily increased the demand for higher throughput sample processing. In the biopharmaceutical industry, the importance of glycosylation on the efficacy and safety of mAbs, which have proven to be hugely successful for treating illness, has made glycomics monitoring vital at all stages of drug development. At the same time, significant developments showing that the identification and diagnosis of the disease can be facilitated by the study of human antibodies is driving interest in the use of clinical medical glycomics.

From the examples provided in this chapter it is apparent that automation offers a viable solution to minimize variability due to human error, provide greater consistency, and reduce the effort required for sample preparation in glycomics studies. Multiple research groups have successfully been able to optimize and adapt glycomics protocols for both *N*-glycan and glycopeptide preparation to liquid-handling robot platforms. While there are limitations to the uptake of automated methods in many laboratories, the main advantage is that sample preparation protocols that were once extremely labor-intensive are now adapted to robotic platforms, which allow IgG *N*-glycan analysis with throughput capabilities that are much greater than before. These developments will help stimulate the interest in large-scale glycan analysis, which continues to grow and become more popular in life science as researchers realize its importance in biological processes. This growing interest should lead to high throughput glycan analysis becoming more common in academic, industrial and clinical centers and has a great potential to impact wider society, ultimately leading to an improvement in human health and well-being.

Compliance with Ethical Standards

Funding RAG, DIRS, and JLH were supported by the European Union (Seventh Framework Programme HighGlycan project, grant number 278535).

Conflict of Interest All authors (RAG, DIRS, and JLH) are employed by Ludger Ltd.

Ethical Approval This article does not contain any new studies with human participants or animals performed by any of the authors. All studies referenced herein were subjected to their own ethical approval detailed in the original publications.

References

Adamczyk B, Tharmalingam T, Rudd PM (2012) Glycans as cancer biomarkers. Biochim Biophys Acta 1820:1347–1353

Adamczyk B, Tharmalingam-Jaikaran T, Schomberg M, Szekrényes Á, Kelly RM, Karlsson NG, Guttman A, Rudd PM (2014) Comparison of separation techniques for the elucidation of IgG N-glycans pooled from healthy mammalian species. Carbohydr Res 389:174–185

Aich U, Liu A, Lakbub J, Mozdzanowski J, Byrne M, Shah N, Galosy S, Patel P, Bam N (2016) An integrated solution-based rapid sample preparation procedure for the analysis of N-glycans from therapeutic monoclonal antibodies. J Pharm Sci 105:1221–1232

Alexovič M, Dotsikas Y, Bober P, Sabo J (2018) Achievements in robotic automation of solvent extraction and related approaches for bioanalysis of pharmaceuticals. J Chromatogr B Anal Technol Biomed Life Sci 1092:402–421

Alexovič M, Urban PL, Tabani H, Sabo J (2020) Recent advances in robotic protein sample preparation for clinical analysis and other biomedical applications. Clin Chim Acta 507:104–116

Bondt A, Rombouts Y, Selman MHJ, Hensbergen PJ, Reiding KR, Hazes JMW, Dolhain RJEM, Wuhrer M (2014) Immunoglobulin G (IgG) fab glycosylation analysis using a new mass spectrometric high-throughput profiling method reveals pregnancy-associated changes. Mol Cell Proteomics 13:3029–3039

Bondt A, Nicolardi S, Jansen BC et al (2016) Longitudinal monitoring of immunoglobulin A glycosylation during pregnancy by simultaneous MALDI-FTICR-MS analysis of N-and O-glycopeptides. Sci Rep 6:1–12

Bonté F, Girard D, Archambault J, Desmoulière A (2018) Biochemistry and cell biology of ageing: part I biomedical science, vol 90. Springer, Singapore, pp 249–280

Cheng HD, Stöckmann H, Adamczyk B, McManus CA, Ercan A, Holm IA, Rudd PM, Ackerman ME, Nigrovic PA (2017) High-throughput characterization of the functional impact of IgG Fc glycan aberrancy in juvenile idiopathic arthritis. Glycobiology 27:1099–1108

Ciucanu I (2006) Per- O -methylation reaction for structural analysis of carbohydrates by mass spectrometry. Anal Chim Acta 576:147–155

Colhoun HO, Treacy EP, MacMahon M, Rudd PM, Fitzgibbon M, O'Flaherty R, Stepien KM (2018) Validation of an automated ultraperformance liquid chromatography IgG N-glycan analytical method applicable to classical galactosaemia. Ann Clin Biochem 55:593–603

Costa AR, Rodrigues ME, Henriques M, Oliveira R, Azeredo J (2014) Glycosylation: impact, control and improvement during therapeutic protein production. Crit Rev Biotechnol 34:281–299

Dalziel M, Crispin M, Scanlan CN, Zitzmann N, Dwek RA (2014) Emerging principles for the therapeutic exploitation of glycosylation. Science 343(6166):1235681. https://doi.org/10.1126/science.1235681

Dennis JW, Granovsky M, Warren CE (1999) Protein glycosylation in development and disease. BioEssays 21:412–421

Doherty M, Bones J, McLoughlin N, Telford JE, Harmon B, DeFelippis MR, Rudd PM (2013) An automated robotic platform for rapid profiling oligosaccharide analysis of monoclonal antibodies directly from cell culture. Anal Biochem 442:10–18

Dube DH, Bertozzi CR (2005) Glycans in cancer and inflammation—potential for therapeutics and diagnostics. Nat Rev Drug Discov 4:477

Gebrehiwot AG, Melka DS, Kassaye YM, Rehan IF, Rangappa S, Hinou H, Kamiyama T, Nishimura SI (2018) Healthy human serum N-glycan profiling reveals the influence of ethnic variation on the identified cancer-relevant glycan biomarkers. PLoS One 13:1–23

Gebrehiwot AG, Melka DS, Kassaye YM, Gemechu T, Lako W, Hinou H, Nishimura SI (2019) Exploring serum and immunoglobulin G N-glycome as diagnostic biomarkers for early detection of breast cancer in Ethiopian women. BMC Cancer 19:1–18

Grilo AL, Mantalaris A (2019) The increasingly human and profitable monoclonal antibody market. Trends Biotechnol 37:9–16

Hossler P, Khattak SF, Li ZJ (2009) Optimal and consistent protein glycosylation in mammalian cell culture. Glycobiology 19:936–949

Kelley B (2009) Industrialization of mAb production technology: the bioprocessing industry at a crossroads. MAbs 1:443–452

Kong F, Yuan L, Zheng YF, Chen W (2012) Automatic liquid handling for life science: a critical review of the current state of the art. J Lab Autom 17:169–185

Kumpel BM, Saldova R, Koeleman CAM et al (2020) Anti-D monoclonal antibodies from 23 human and rodent cell lines display diverse IgG Fc-glycosylation profiles that determine their clinical efficacy. Sci Rep 10:1–14

Liu D, Li Q, Zhang X et al (2019) Systematic review: immunoglobulin G N-glycans as next-generation diagnostic biomarkers for common chronic diseases. OMICS 23:607–614

Maratha A, Stockmann H, Coss KP et al (2016) Classical galactosaemia: novel insights in IgG N-glycosylation and N-glycan biosynthesis. Eur J Hum Genet 24:976–984

Miura Y, Hato M, Shinohara Y et al (2008) BlotGlycoABC™, an integrated glycoblotting technique for rapid and large scale clinical glycomics. Mol Cell Proteomics 7:370–377

Molares Vila A, Rupérez Pérez de Arrilucea P, Caso Peláez E, Gago-Martínez A (2010) Development of a new magnetic beads-based immunoprecipitation strategy for proteomics analysis. J Proteome 73:1491–1501

Momčilović A, De Haan N, Hipgrave Ederveen AL et al (2020) Simultaneous immunoglobulin A and G glycopeptide profiling for high-throughput applications. Anal Chem 92:4518–4526

Nishimura SI (2011) Toward automated glycan analysis, 1st ed. Adv Carbohydr Chem Biochem 65:219–271. https://doi.org/10.1016/B978-0-12-385520-6.00005-4

O'Flaherty R, Trbojević-Akmačić I, Greville G, Rudd PM, Lauc G (2018) The sweet spot for biologics: recent advances in characterization of biotherapeutic glycoproteins. Expert Rev Proteomics 15:13–29

O'Flaherty R, Muniyappa M, Walsh I, Stöckmann H, Hilliard M, Hutson R, Saldova R, Rudd PM (2019) A robust and versatile automated glycoanalytical technology for serum antibodies and acute phase proteins: ovarian cancer case study. Mol Cell Proteomics 18:2191–2206

Peng W, Zhao J, Dong X, Banazadeh A, Huang Y, Hussien A, Mechref Y (2018) Clinical application of quantitative glycomics. Expert Rev Proteomics 15(12):1007–1031. https://doi.org/10.1080/14789450.2018.1543594

Reed CE, Fournier J, Vamvoukas N, Koza SM (2018) Automated preparation of MS-sensitive fluorescently labeled N-glycans with a commercial pipetting robot. SLAS Technol 23:550–559

Reiding KR, Bondt A, Hennig R et al (2019) High-throughput serum N-glycomics: method comparison and application to study rheumatoid arthritis and pregnancy-associated changes. Mol Cell Proteomics 18:3–15

Reusch D, Haberger M, Selman MHJ, Bulau P, Deelder AM, Wuhrer M, Engler N (2013) High-throughput work flow for IgG Fc-glycosylation analysis of biotechnological samples. Anal Biochem 432:82–89

Royle L, Campbell MP, Radcliffe CM et al (2008) HPLC-based analysis of serum N-glycans on a 96-well plate platform with dedicated database software. Anal Biochem 376:1–12

Shajahan A, Supekar N, Heiss C, Azadi P (2019) High-throughput automated micro-permethylation for glycan structure analysis. Anal Chem 91:1237–1240

Shipman JT, Nguyen HT, Desaire H (2020) So you discovered a potential glycan-based biomarker; now what? We developed a high-throughput method for quantitative clinical glycan biomarker validation. ACS Omega 5:6270–6276

Shubhakar A, Kozak RP, Reiding KR, Royle L, Spencer DIR, Fernandes DL, Wuhrer M (2016) Automated high-throughput permethylation for glycosylation analysis of biologics using MALDI-TOF-MS. Anal Chem 88:8562–8569

Stöckmann H, Adamczyk B, Hayes J, Rudd PM (2013) Automated, high-throughput IgG-antibody glycoprofiling platform. Anal Chem 85:8841–8849

Stöckmann H, Duke RM, Millán Martín S, Rudd PM (2015a) Ultrahigh throughput, ultrafiltration-based N-glycomics platform for ultraperformance liquid chromatography (ULTRA3). Anal Chem 87:8316–8322

Stöckmann H, O'Flaherty R, Adamczyk B, Saldova R, Rudd PM (2015b) Automated, high-throughput serum glycoprofiling platform. Integr Biol (Camb) 7:1026–1032

Szigeti M, Lew C, Roby K, Guttman A (2016) Fully automated sample preparation for ultrafast N-glycosylation analysis of antibody therapeutics. J Lab Autom 21:281–286

ThermoFisher Scientific (n.d.) Pierce^{TM} protein G spin plate for igg screening. https://www.thermofisher.com/order/catalog/product

Trbojević Akmačić I, Ugrina I, Štambuk J, Gudelj I, Vučković F, Lauc G, Pučić-Baković M (2015) High-throughput glycomics: optimization of sample preparation. Biochemist 80:934–942

Tuck MK, Chan DW, Chia D et al (2010) NIH Public Access 8:113–117

Varki A (2017) Biological roles of glycans. Glycobiology 27:3–49

Ventham NT, Gardner RA, Kennedy NA, Shubhakar A, Kalla R, Nimmo ER, Fernandes DL, Satsangi J, Spencer DIR (2015) Changes to serum sample tube and processing methodology does not cause inter-individual variation in automated whole serum N-glycan profiling in health and disease. PLoS One 10:1–16

Wang Y, Klarić L, Yu X et al (2016) The association between glycosylation of immunoglobulin G and hypertension. Medicine (Baltimore) 95:1–11

Yamamoto S, Kinoshita M, Suzuki S (2016) Current landscape of protein glycosylation analysis and recent progress toward a novel paradigm of glycoscience research. J Pharm Biomed Anal 130:273–300

Yang X, Kim SM, Ruzanski R et al (2016) Ultrafast and high-throughput N-glycan analysis for monoclonal antibodies. MAbs 8:706–717

Zhang X, Reed CE, Birdsall RE, Yu YQ, Chen W (2020) High-throughput analysis of fluorescently labeled N-glycans derived from biotherapeutics using an automated LC-MS-based solution. SLAS Technol 25:380–387

Chapter 6
Bioinformatics in Immunoglobulin Glycosylation Analysis

Frédérique Lisacek, Kathirvel Alagesan, Catherine Hayes, Steffen Lippold, and Noortje de Haan

Contents

6.1	Introduction	206
6.2	Glycomic Data Collection and Processing	208
	6.2.1 Reference Databases	208
	6.2.2 Identification and Quantification Software Tools	210
6.3	Glycoproteomics Data Collection and Processing	210
	6.3.1 Reference Databases	210
	6.3.2 Identification Software Tools	211
	6.3.3 Quantification Software Tools	212
6.4	Data Integration with Other Omics	212
6.5	Practical Examples	214
	6.5.1 Glycomic Data Processing	214
	6.5.2 Glycopeptide Data Processing for Enriched Immunoglobulins	218
	6.5.3 Visualizing Profiles	224
6.6	Conclusion	228
References		231

F. Lisacek (✉)
Proteome Informatics Group, SIB Swiss Institute of Bioinformatics, Geneva, Switzerland

Computer Science Department, University of Geneva, Geneva, Switzerland

Section of Biology, University of Geneva, Geneva, Switzerland
e-mail: frederique.lisacek@sib.swiss

K. Alagesan
Max Planck Unit for the Science of Pathogens, Berlin, Germany

C. Hayes
Proteome Informatics Group, SIB Swiss Institute of Bioinformatics, Geneva, Switzerland

Computer Science Department, University of Geneva, Geneva, Switzerland

S. Lippold
Center for Proteomics and Metabolomics, Leiden University Medical Center, Leiden, The Netherlands

N. de Haan
Copenhagen Center for Glycomics, University of Copenhagen, Copenhagen, Denmark

© The Author(s), under exclusive license to Springer Nature Switzerland AG 2021
M. Pezer (ed.), *Antibody Glycosylation*, Experientia Supplementum 112,
https://doi.org/10.1007/978-3-030-76912-3_6

Abstract Analytical methods developed for studying immunoglobulin glycosylation rely heavily on software tailored for this purpose. Many of these tools are now used in high-throughput settings, especially for the glycomic characterization of IgG. A collection of these tools, and the databases they rely on, are presented in this chapter. Specific applications are detailed in examples of immunoglobulin glycomics and glycoproteomics data processing workflows. The results obtained in the glycoproteomics workflow are emphasized with the use of dedicated visualizing tools. These tools enable the user to highlight glycan properties and their differential expression.

Keywords Glycomics · Glycoproteomics · Workflow · Software · Database · Glycoinformatics

Abbreviations

ETD	Electron-transfer dissociation
Fc	Fragment crystallizable
HTP	High throughput
Ig	Immunoglobulin
IMGT	ImMunoGeneTics information system
LC	Liquid chromatography
MIRAGE	Minimum Information Required for A Glycomics Experiment
MS	Mass spectrometry
MS/MS	Tandem mass spectrometry
NMR	nuclear magnetic resonance
PDB	Protein Data Bank
PNGase F	Peptide-*N*-glycosidase F
RP	Reverse phase
RT	Retention time
SNFG	Symbol Nomenclature for Glycans

6.1 Introduction

Glycosylation plays a major role in creating proteoform diversity reaching the magnitude of 10^6 in the human proteome. The mapping of this diversity is far from being charted, particularly in the case of glycoforms, mainly because of the limitations and the very broad variety of experimental techniques needed to capture information. Mass spectrometry (MS) and nuclear magnetic resonance (NMR) as part of a fuller range of analytical methods (Gray et al. 2019) are used to solve the structure of glycans. In recent years, the improved accuracy of mass spectrometers and sophisticated fragmentation techniques continue to refine the identification of intact glycopeptides (peptide with attached glycan) and new glycoproteomics datasets are regularly published (Ye et al. 2019; Riley et al. 2019).

Glycomics and glycoproteomics have benefitted from past experience in proteomics and the principles of the Proteome Standards Initiative (Orchard et al. 2003) and ProteomeXchange (Vizcaíno et al. 2014) are being gradually applied. In the past decade, reporting guidelines have been issued through the MIRAGE (Minimum Information Required for A Glycomics Experiment) initiative (https://www.beilstein-institut.de/en/projects/mirage/) as part of the many minimum information standards defined to facilitate data checking and processing (Kolarich et al. 2013).

Determining a glycosylation profile may entail either glycomics or glycoproteomics studies or both, depending on experimental design. In most cases, mass spectrometry data analysis software is used. Despite the expansion of glycan (Fujita et al. 2020) and glycoconjugate databases and repositories (Alocci et al. 2019; York et al. 2020), direct use of such resources is limited for glycan identification as there is no template that can be used to infer structures detectable in an organism. Some of the commonly applied strategies rely on established proteomics search tools that are refined to identify multiple modifications or exploiting idiosyncratic fragmentation patterns of glycans, while others use dedicated software for glyco(proteo)mics data, as exemplified in this chapter.

Glycosylation analyses of proteins can be performed either at intact glycoprotein, at glycopeptide, or released glycan level. Irrespective of the level of analytical method, structural analysis of glycoproteins remains elusive due to lack of high-throughput analytical tools and data analysis pipeline capable of analyzing high-volume spectral data to determine the glycan composition, structure and/or glycopeptide sequence information (Pralow et al. 2020; Alagesan et al. 2020). Immunoglobulins (Igs) and Fc receptors are complex glycoproteins and key components of both the innate and adaptive immune systems. The specific glycosylation of both immunoglobulin and its receptors is well known to be crucial for maintaining and modulating effector functions. In humans, there are five distinct antibody isotypes: IgM, IgD, IgE, IgA, and IgG. All of them are glycosylated (Arnold et al. 2007; de Haan et al. 2020). IgG provides an ideal case scenario for the development of high-throughput analytical tools and data analysis pipeline as it (1) is one of the most abundant proteins in human serum, accounting for about 10–20% of plasma protein, and (2) mostly bears a single N-linked glycan attached to Asn-297 of each heavy chain, which can have a wide variety of different structures. These unique IgG features allow for fine-tuning and extensive benchmarking of high-throughput (HTP) glycoanalytical methods (Trbojević-Akmačić et al. 2016; Wuhrer et al. 2007; Gstöttner et al. 2020; Amez-Martín et al. 2020).

This chapter briefly surveys the landscape of glycan and intact glycopeptide identification and quantification bioinformatics resources and focuses on profiling of antibodies using glycomics or glycoproteomics approaches as well as using visualization tools to easily compare these profiles. The coverage of resources is depicted in Fig. 6.1.

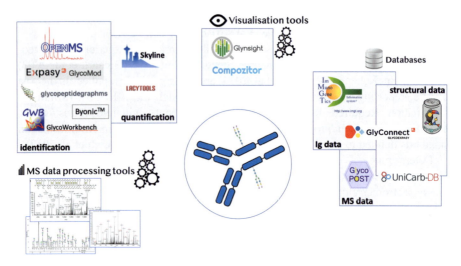

Fig. 6.1 Summary of resources. The glycoinformatics resources that are cited and used in this chapter are categorized as either database or tool. The tools are further distinguished according to their purpose: tools processing mass spectrometry (MS) data (identification or quantification) or used to visualize processed data. Databases store different types of data

6.2 Glycomic Data Collection and Processing

A recently published chapter covers widely the database and tool collection available for glycomics studies (Lisacek et al. 2017) presented as exhaustive lists of the various databases and software tools. The present section is an update on recent additions to the collection that is not reproduced here but complemented by the partially overlapping content of Table 6.1.

6.2.1 Reference Databases

In recent years, several community-driven moves in glycan bioinformatics have been initiated, the first of which is the agreement to resort to a single repository for glycan structures. This resulted in the creation of GlyTouCan in 2016, only briefly mentioned in (Lisacek et al. 2017) as version 1.0 (Aoki-Kinoshita et al. 2016). GlyTouCan is an uncurated registry for glycan structures that assigns globally unique accession numbers. It is increasingly adopted as a reference (Tiemeyer et al. 2017) and reached version 3.0 in 2021 to include a substantially reshaped registration flow improving data submission (Fujita et al. 2020). Manually annotated and experimentally verified spectra collected in UniCarb-DB (Hayes et al. 2011) are cross-referenced to GlyTouCan in order to increase the speed and accuracy of glycan assignment.

Table 6.1 Summary of cited glycoinformatics resources

Type	Resource name	Availability	References
Databases			
Glycoproteins	GlyConnect	glyconnect.expasy.org	Alocci et al. (2019)
Immunoglobulins	IMGT	imgt.org	Lefranc et al. (2015)
Glycans	UniCarb-DB	unicarb-db.expasy.org	Hayes et al. (2011)
3D structures	Protein Data Bank	www.wwpdb.org	wwPDB consortium et al. (2019)
Proteins	UniProt	uniprot.org	The UniProt Consortium (2019)
Repositories (Enabled data submission)			
MS data	GlycoPOST	glycopost.glycosmos.org	Watanabe et al. (2020)
Glycans	GlyTouCan	glytoucan.org	Fujita et al. (2020)
Software			
Generic MS Processing	ProteoWizard	proteowizard.sourceforge.net/tools.shtml	Kessner et al. (2008)
	OpenMS	www.openms.de	Pfeuffer et al. (2017)
Glycan MS Processing	GlycoMod	web.expasy.org/glycomod	Cooper et al. (2001)
	GlycoWorkbench	code.google.com/archive/p/glycoworkbench	Ceroni et al. (2008)
	GRITS toolbox	www.grits-toolbox.org	Weatherly et al. (2019)
Glycopeptide MS processing	Byonic	Licensed	Bern et al. (2012)
	GlycopeptideGraphMS	bitbucket.org/glycoaddict/glycopeptidegraphms/src/master/	Choo et al. (2019)
Quantitative MS processing	Skyline	skyline.ms/project/home/begin.view?	Pino et al. (2020)
	LacyTools	git.lumc.nl/cpm/lacytools	Jansen et al. (2016)
Visualization	Glynsight	glycoproteome.expasy.org/glynsight	Alocci et al. (2018)
	Compozitor	glyconnect.expasy.org/compozitor	Robin et al. (2020)

Glycomics mass spectrometry data have been taken one step further with the introduction of a MIRAGE-compliant experimental data repository named GlycoPOST (Watanabe et al. 2020) and the implementation of a pipeline for collecting such data (Rojas-Macias et al. 2019). Some publishers have started imposing data deposition and the field of glycoscience is slowly (re)connecting to life sciences.

6.2.2 Identification and Quantification Software Tools

GlycoMod (Cooper et al. 2001) and GlycoWorkbench (Ceroni et al. 2008) are still in use in much the same way for the identification of glycan structures, described in (Lisacek et al. 2017). These earlier protocols (1) from analytical MS data to monosaccharide composition using GlycoMod, (2) from structure to predicted MS/MS data using GlycoWorkbench, and (3) from analytical MS and MS/MS data to structure using UniCarb-DB, are still valid. A more recent identification platform is the GRITS toolbox (Weatherly et al. 2019) that can be considered as an upgrade of GlycoWorkBench. In particular, managing adducts is made more flexible and larger datasets composed of thousands of spectra are more easily handled.

Quantification was barely addressed in (Lisacek et al. 2017) mainly due to the limited number of quantitative glycomics datasets a few years back. The situation is rapidly changing and the need for related software is increasing. Skyline is accurately described by its authors as a software ecosystem for quantitative mass spectrometry informatics (Pino et al. 2020). This platform efficiently tackles issues such as large dataset management, integration with other commonly used tools, data visualization, independently of the experimental workflow. This approach made the tool popular in many -omics using mass spectrometry irrespective of the size of identified molecules, that is, ranging from metabolites to large peptides, encompassing glycans and glycopeptides. Application to serum glycoprotein site occupancy using Skyline was reported early in (Hülsmeier et al. 2016). Since then, the use of the tool has spread both in glycomics and glycoproteomics.

6.3 Glycoproteomics Data Collection and Processing

6.3.1 Reference Databases

As hinted in the introduction, glycoproteomics is fast-growing and three appropriate databases have been released in recent years. In fact, the GlycoPOST repository (Watanabe et al. 2020) was designed for mass spectrometry data submission in relation to JPOST, the Japanese member of the ProteomeXchange Consortium. A database of curated glycoproteins and their associated glycans called GlyConnect (Alocci et al. 2019) includes many entries describing glycosylated immunoglobulins. This database is reciprocally cross-linked with UniProt (The UniProt Consortium 2019), the reference protein sequence database, and feeds glycomics data to GlyGen (York et al. 2020), the recently released US portal for glycoscience.

6.3.2 Identification Software Tools

The topic of automatic intact glycopeptide identification is regularly reviewed (Thaysen-Andersen and Packer 2014; Hu et al. 2017; Cao et al. 2020; Abrahams et al. 2020), following the abundant production of new tools that warrant frequent updates of the catalog. This chapter is not destined to add to this set of reviews.

One of the key points in running glycopeptide identification software is the selection of the glycan composition file upon which the identification of the glycan moiety is based. In the majority of cases, the composition file can be qualified as "empirical" since its definition relies on glycan data collected from the literature or from databases/repositories. It is, however, an unevenly flexible definition. In some software, it is a modifiable parameter while in others it is not. Alternatively, the knowledge of glycan biosynthesis can be used to generate expected glycans. For example, MAGIC (Lih et al. 2016) starts with 29 monosaccharides to be combined according to enzymatic rules and in a similar way, 19 groups of seed structures are defined in GlycoPAT (Liu et al. 2017). In this way, the composition file is "theoretical" as the result of potential enzymatic activity. This definition is also flexible since both the enzymes or the monosaccharides can be removed or constrained.

Most classical proteomics search engines accommodate the selection of glyco-based modifications from a collection they provide. For example, in principle, Mascot (Savitski et al. 2011) relies on the UniMod subset of glycans (Creasy and Cottrell 2004). The user may also use in-house definitions as in (Bollineni et al. 2018), where a few hundred compositions were built from a customized and systematic addition of monosaccharides prior to using the Mascot engine. Note that instrument vendors also provide a glycopeptide identification component in the software suites that are usually available for proteomics data analysis such as ProteomeDiscover by Thermo or ProteinScape at Bruker, but the information on glycan composition files is not easily accessible.

Proteomics software developers, in particular when posttranslational modifications are accounted for, have already acknowledged that the performance of search engines correlates with limiting the search space (Schwämmle et al. 2015). Similar observations have been made in glycoproteomics and this viewpoint, also promoted here, is strongly advocated in (Khatri et al. 2017).

6.3.3 Quantification Software Tools

Most glycopeptide identification tools have some sort of built-in quantification module, either based on a spectral count or MS1 feature intensity or integrated area. While these modules are often sufficient to obtain a rough estimation of (relative) glycopeptide abundance, more sophisticated software tools have been reported for accurate glycopeptide quantification in a high-throughput manner. As described above, Skyline is well suitable for the quantification of glycopeptide data and has excellent visualization capability that allows for data curation and representation (Pino et al. 2020). LacyTools is an automated data processing tool for high-throughput analysis of LC-MS glycoproteomics data and is suitable for the simultaneous analysis and quality control of thousands of samples (Jansen et al. 2016). Both tools feature important options such as quality control based on isotopic pattern matching, separate integration of different charge states, and a flexible inclusion of isotope signals.

6.4 Data Integration with Other Omics

In the early years of bioinformatics in the 1980s when the term itself was hardly used, the main focus was on collecting and comparing gene and protein sequences. A handful of resources created then are still available now and this is the case of the international ImMunoGeneTics information system (IMGT), created in 1989 by Marie-Paule Lefranc in Montpellier, France. This resource remains a unique gene-centric reference in immunogenetics and immunoinformatics (Lefranc et al. 2015). IMGT/3Dstructure-DB is a subsection of the IMGT platform where Ig three-dimensional information is included. This data is cross-referencing the Protein Data Bank (PDB) (wwPDB consortium et al. 2019), where glycan information is symbolically represented in protein structures using a 3D version (Sehnal and Grant 2019) of the Symbol Nomenclature for Glycans (SNFG) depiction increasingly adopted in publications and textbooks (Varki et al. 2015).

Since the PDB has long considered glycans decorating glycoproteins as "ligands," IMGT lists "glucid" as potential ligands of immunoglobulins. IMGT/3Dstructure-DB (version 4.12.3) contains 21 entries where 3D structures include information on attached glycans. Figure 6.2 summarizes the steps of the search and illustrates the possible visualization of attached glycans with the example of the PDB entry 5 K65 of VEGF binding IgG1-Fc (Fcab CT6), as recorded from (Lobner et al. 2017).

The recently introduced feature of the PDB clearly showing the presence of a glycan on the 3D representation of the protein is the first step toward bringing together different data sources that cover complementary information on immunoglobulins. Further work in this direction is needed to reach a better level of integration between genomics and glycomics.

6 Bioinformatics in Immunoglobulin Glycosylation Analysis 213

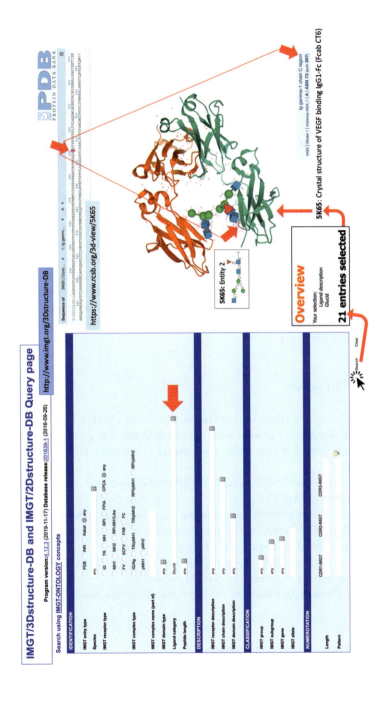

Fig. 6.2 Potential data integration with IMGT. The IMGT platform integrates genetic, structural, and functional information about immunoglobulins. It is split into several sections, one of which is devoted to 3D structures. Querying this section by ligand category with the "glucid" term (central red arrow) outputs a list of 21 entries with cross-links to the Protein Data Bank (PDB). The new interface of the PDB highlights the glycan in the 3D structure with the recommended SNFG notation

6.5 Practical Examples

This part describes the stepwise processing of glycomic and glycoproteomic Ig MS data, as well as the visualization of selected glycoproteomic identification results. Supplementary information is provided in the Notes section (notes 9–12), at the end of the chapter.

6.5.1 Glycomic Data Processing

The glycomic approach to Ig N-glycan characterization is traditionally performed after isolating the Ig of interest from its complex matrix by affinity chromatography. Subsequently, the N-glycans are released from their carrier protein using the enzyme Peptide-N-Glycosidase F (PNGase F). Prior to MS analysis, glycans may be subjected to permethylation, reduction, or reducing end labeling to increase sensitivity and specificity. Analysis of released glycans allows for in-depth structural elucidation and isomer separation, but it lacks information on the site of glycosylation. Therefore, for Ig-specific glycan analysis at the glycomics level, protein purity is of utmost importance. The glycomics data analysis protocol outlined below allows for the structural determination of reduced glycans released from human IgG. The workflow is illustrated in Fig. 6.3.

6.5.1.1 Software Required

1. *Manual inspection of raw data*: Bruker Daltonic Software (Bruker DALTONIK GmbH, Bremen, Germany) was used since the example data was generated with a Bruker instrument (amaZon speed Electron-transfer dissociation (ETD) ion trap mass spectrometer).
2. *Data format conversion*: MSConvertGUI (ProteoWizard 3.0 suite, http://proteowizard.sourceforge.net/)
3. *Determination of glycan composition from the experimentally determined precursor mass*: GlycoMod (https://web.expasy.org/glycomod/) that works well for both free and derivatized glycan as well as for glycopeptides.
4. *Interpretation of tandem MS spectra* (MS/MS spectra): GlycoWorkbench (https://code.google.com/archive/p/glycoworkbench/)
5. *Quantification*: Skyline (64-Bit) 20.2.0.343 (https://skyline.ms/project/home/begin.view?)

6.5.1.2 Glycan Composition Determination

1. Determine the experimental glycan masses. Convert any doubly and triply charged m/z values to singly charged or neutral mass. Bruker Data Analysis

6 Bioinformatics in Immunoglobulin Glycosylation Analysis

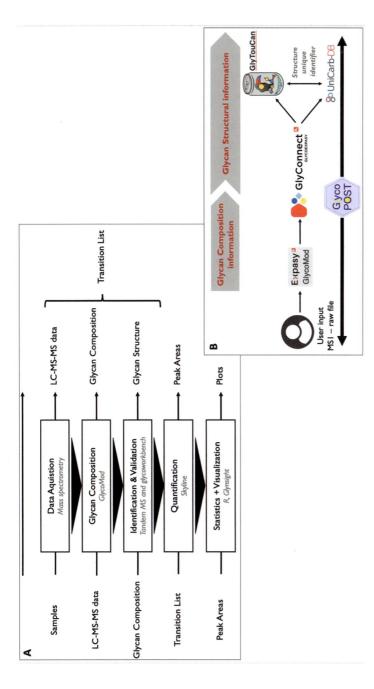

Fig. 6.3 Glycomics workflow. (**a**) The glycomics data analysis was divided into five major procedures. The raw data is acquired on the LC-MS instrument resulting in tandem MS spectra and retention time for each precursor mass selected for fragmentation. The MS1 masses allow for the determination of glycan composition followed by glycan structure identification and validation. The quantification can be performed using Skyline and peak areas can be exported and readily visualized using either R or various other visualization tools such as Glynsight available on the Expasy server. (**b**) MS raw files can be submitted to GlycoPOST including experimental metadata, glycan identification, and quantification results. Glycan composition information can be processed to obtain structural information

software allows for easy deconvolution of the precursor masses. For other instruments, a vendor-specific software package can be used for this purpose. As described below, KNIME (KNIME GmbH) with OpenMS 2.3 can be used for MS1 peak picking and deconvolution.

2. Open the web-based GlycoMod tool with the following parameters:
   (a) All mass values: Monoisotopic
   (b) Mass tolerance: ±0.2 Da (can be adjusted depending upon the mass accuracy of the instrument)
   (c) Ion mode and adduct: Neutral [M]
   (d) *N*-linked oligosaccharide: Reduced oligosaccharides
   (e) Monosaccharide residues present (if known): Underivatized and input the range if known
3. Enter the list of experimental masses.
4. At the bottom, click start GlycoMod.
5. A list of possible glycan compositions is calculated for each mass imputed with a link to GlyConnect.
6. Click on GlyConnect to view possible isomers where each structure is cross-referenced to GlyTouCan and UniCarb-DB.

6.5.1.3 Annotation of MS and MS/MS Glycomics Spectra

1. Launch GlycoWorkbench.
2. Draw the proposed glycan structure using the GlycanBuilder for each precursor mass. Theoretical m/z values can be calculated by configuring the parameters in the "Mass options" available under the "Edit" tab.
3. In silico fragmentation can be computed for selected structures using "compute fragments for selected structure" available in the "Tools" tab.
4. In the "Fragmentation options" pop-up window offers options to select specifications for fragmentation.
5. The results are displayed in a tabular form in the side panel containing all predicted fragments. Different fragments with the same exact mass are represented separately within the details tab. Whereas, the Summary tab contains a condensed form of the information presented within the detail tab.
6. Glycan structures can then be assigned based on glycan fragmentation pathways, in silico fragmentation, observed fragmentation, and precursor mass.
7. Additionally, experimental MS/MS peak lists can also be associated with the computed peak list.
8. Copy the list of fragments m/z and intensities directly from the MS software or export the list first to a spreadsheet using, for example, Microsoft Excel.
9. Next, go to Tools → Annotation → Annotate peaks with fragments from selected structures. A Fragment option window will appear. After setting the parameter, click OK. Then the associated peak list will appear in the right panel.

6.5.1.4 Convert Raw Mass Spectrometry Data to mzXML

1. Open MSconvertGUI.
2. Select the list of files that needs conversion and select the Output directory for the converted files.
3. In the options box below the output directory, adjust the settings to output format = "mzXML," Binary encoding precision = "64-bit," and check the boxes next to "write index," "use zlib compression," and "TPP compatibility."
4. In the filter box, select the subset option and do not change the settings that pop up (default setting).
5. At the bottom right corner, click "start," and wait for your files to finish converting to mzXML.

6.5.1.5 Skyline for Glycomics Quantitation

1. Open Skyline and select either the molecule interface or the mixed interface.
2. Go to the "Setting" tab and select "Transition settings."
3. In the pop-up window, select "Instrument" tab and input Min and Max m/z.
4. Next, in the "Full-Scan" tab, input/select the following parameters depending on the type of instrument used: (a) Isotope peak included (count), (b) Precursor mass analyzer (QIT), (c) Peaks (3), and (d) Resolution (0.3 m/z). The values in parentheses indicate the value used for the analysis.
5. At the bottom, select "Include all matching scans" and click OK.

6.5.1.6 Setting Up Transition List

1. Go to the "Edit" tab, and select Insert → "Transition list."
2. In the pop-up window, select "Molecules" at the bottom and click on the "Columns" button.
3. Select the required columns to set up the transition list. Be sure to include at least (a) Molecule List Name, (b) Precursor Name, (c) Precursor m/z, (d) precursor charge, (e) Explicit retention time, and (f) Explicit retention time window.
4. For large lists of glycans, it is advisable to create the "Transition list" in Excel.
5. The "Transition list" can be easily copied/imported to Skyline using File → Import → "Transition list." Once successful, you will see the Transitions in the Target menu.
6. Now save the File before importing the results (converted mzXML files).
7. Wait for the import to complete. Skyline integrates the area under the curve based on the information provided in the "Transition List." Check if the correct peaks are integrated for each glycan. If required, manual integration can be performed by clicking and dragging the integration window beneath the x-axis.

8. Skyline offers various parameters for the quality control of the integrated signals, including mass accuracy, isotopic pattern matching, and retention time which can be conveniently plotted for each sample.
9. Results can be exported using Files → Export → Report function as CSV file.

6.5.2 Glycopeptide Data Processing for Enriched Immunoglobulins

The analysis of Ig glycosylation at the glycopeptide level has the advantage that protein-, site-, Ig isotype- and often even subclass-specificity are maintained (Momčilović et al. 2020). General workflows include the affinity purification of the Igs of interest, their tryptic cleavage, and the analysis of the resulting (glyco)-peptide mixture by reverse-phase (RP) liquid chromatography (LC) coupled to high-resolution mass spectrometry (MS). The RP-LC separation of (Ig) glycopeptides results in the retention time clustering of analytes with the same peptide portion but different glycan moieties. Often, distinct retention time clusters are also obtained for glycopeptides with varying degrees of sialylation. Based on this very characteristic and predictable elution behavior of glycopeptides, software packages were developed for the identification [GlycopeptideGraphMS (Choo et al. 2019)] and targeted quantification and quality control [LaCyTools (Jansen et al. 2016)] of glycopeptides derived from purified proteins. Combining these tools with MS2-aided glycopeptide validation resulted in a workflow for the semiautomated identification and quantification of glycopeptides that is highly suitable for high-throughput Ig glycosylation analysis (Lippold et al. 2020). The protocol outlined below is adjusted for data [publicly available via (Bern et al. 2012)] obtained by the RP LC-Orbitrap-MS/MS analysis of tryptic IgG and IgA glycopeptides, Fig. 6.4 (Glycoproteomics workflow).

6.5.2.1 Software Required

1. *Manual inspection of the raw data*: instrumentation proprietary software. In this case, Xcalibur (v. 2.2, Thermo Fisher Scientific).
2. *MS2-based glycopeptide identification*: software package for automated MS/MS assignment of glycopeptides [example can be found in (Abrahams et al. 2020)]. In this case, PMI-Byonic (v. 3.7.13, Protein Metrics), (Bern et al. 2012) Fig. 6.4a.
3. *Data format conversion*: MSConvertGUI (ProteoWizard 3.0 suite, http://proteowizard.sourceforge.net/) (Kessner et al. 2008), Fig. 6.4b.
4. *MS1 peak picking and deconvolution*: KNIME (KNIME GmbH) with OpenMS 2.3 (https://www.openms.de) (Pfeuffer et al. 2017). Details on the installation of these, along with a dedicated workflow to prepare GlycopeptideGraphMS inputs are available as part of the GlycopeptideGraphMS download.
(workflow: KNIME_OPENMS_GraphMS_Preprocessing_120318, Fig. 6.4c; https://bitbucket.org/glycoaddict/glycopeptidegraphms/src/master/).

6 Bioinformatics in Immunoglobulin Glycosylation Analysis   219

5. *MS1-based glycopeptide identification*: GlycopeptideGraphMS, Fig. 6.4c (v. 2.06, https://bitbucket.org/glycoaddict/glycopeptidegraphms/src/master/).
6. *Glycopeptide quantification and quality control* (QC): LaCyTools, Fig. 6.4d (v 2.0.1, https://git.lumc.nl/cpm/lacytools).
7. Python 3 with the Anaconda package (https://www.anaconda.com/).

6.5.2.2 Glycosylation Site Identification Based on MS2

Byonic can be used to identify the peptides that are present in samples as well as validating the presence of glycopeptides (Fig. 6.4a). There are three inputs: raw files, a protein database, and a glycan database (in this case, Byonic proprietary *N*-linked database).

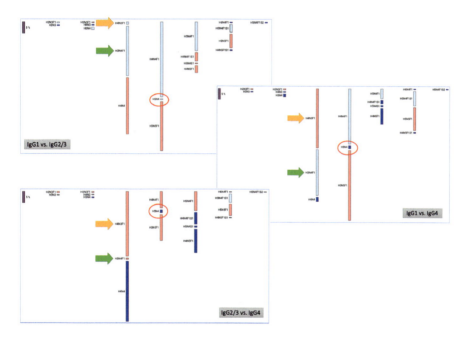

Fig. 6.4 Glycoproteomics workflow. The glycoproteomics dataset was subjected to (**a**) glycosylation site identification by MS2 and (**b**) MS1 feature selection and deconvolution. (**c**) The outcomes of these processes were combined using GlycopeptideGrapMS for extensive glycoform identification and visualization. (**d**) The identified glycopeptides were subjected to a QC step and quantified using LaCyTools

1. Run PMI-Byonic on the raw LC-MS/MS data files with the Homo sapiens UniProt database including 71,591 protein sequences (20,205 from Swiss-Prot and 51,386 from TrEMBL at time of writing). Byonic has an inbuilt glycan database for putative glycan assignments. In this example the "N-glycan 309 mammalian no sodium" glycan list is used. (see Note 1). Individual options are set to:

   (a) C-terminal cleavage of lysine and arginine: true
   (b) Maximum number of missed cleavages: 2
   (c) Precursor mass tolerance: 10 ppm
   (d) Fragment ion mass tolerance: 20 ppm
   (e) Fixed modification: cysteine carbamidomethylation
   (f) Variable modification: methionine oxidation

2. Output from this step is a list of polypeptide cleavage products and glycopeptide identifications. Identify those products of the protein(s) of interest. Include glycopeptide identifications with a Byonic score above 200 (see Note 2) (Fig. 6.4a).

6.5.2.3 Glycoform Identification Based on MS1

GlycopeptideGraphMS can be further used in the identification process to cluster the identified glycopeptides by retention time. These nodes can reveal information on the underlying proteins and glycosylation sites. Advanced (and/or customized) options can be investigated using GlycopeptideGraphMS manual (https://bitbucket.org/glycoaddict/glycopeptidegraphms/).

1. A preprocessing step is required to create input for GlycopeptideGraphMS. Convert the raw data to the mzML format using MSConvertGUI with the following settings:

   (a) Binary encoding precision: 64-bit
   (b) Filters: subset – MS level: 1-1
   (c) Check: write index, use zlib compression, TPP compatibility

2. Load the obtained mzML file in the KNIME OpenMS workflow and define an output directory. Adjust the default settings in the feature finder to account for the m/z range (400–3500) and expected charge states (2–7). Run the file filter, peak picking, feature finder, and de-charging node (Fig. 6.4b). Adjust these settings according to instrumentation.

3. Load the obtained .csv file from the KNIME OpenMS workflow into GlycopeptideGraphMS and format it into the consensus format needed for further processing using the "Format consensusXML" function.

4. Perform the GlycopeptideGraphMS analysis with the following settings (only values deviating from the default are mentioned):

(a) Mass deviation tolerance: 0.02 Da
(b) Maximum subgroup degree: 1 (the lower the value, the higher the noise)
(c) Composition search blocks (see Note 3):
   (i) Hexose (Hex, 162.0528 Da, max. 30 s retention time (RT) difference)
   (ii) N-Acetylhexosamine (HexNAc, 203.0794 Da, max. 30 s RT difference)
   (iii) Hexose + N-acetylhexosamine (HexHexNAc, 365.1322 Da, max. 30 s RT difference)
   (iv) Deoxyhexose (Fuc, 146.0579 Da, max. 20 s RT difference)
   (v) N-Acetylneuraminic acid (NeuAc, 291.0954 Da, max. 120 s RT difference)

5. The obtained output.csv provides a list of all nodes identified in so-called "subgraphs" (glycopeptides that share the same peptide portion and differ in their glycan content). Entries (rows) with the same value under "refnode" belong to the same subgraph. Use the m/z (mz_node) and retention time (rt_node) of each node to match them to glycopeptides identified during the MS2-based identification (Sect. 5.2.2).

6. Open the "reference node XLSX" which lists one node per identified subgraph. Populate each entry by one MS2-confirmed node per subgraph (see Note 4 and Note 5), indicating the node number (node), m/z (mz_node), retention time (rt_node), the number of the subgraph node with the lowest mass with its m/z (refnode; mz_refnode), the number of the respective monosaccharides present in the identified glycopeptide and the peptide sequence of the identified glycopeptide, as illustrated in Table 6.2.

7. Use the graphical representation option in GlycopeptideGraphMS to visualize the identified subgraphs. Closely related clusters help to identify (unexpected) peptide modifications and glycoforms, Fig. 6.4c which can be used to re-run the glycopeptide (Byonic) search and optimize the output. It also allows the identification of unexpected peptide modifications, such as partial cysteine oxidation (identified on the IgA data in this example).

8. Run the composition predictor in GlycopeptideGraphMS (input: reference node XLSX and output.csv) to obtain a full list of glycoforms, based on the MS1 LC-MS features.

9. Remove the glycoforms from the predicted composition output of GlycopeptideGraphMS that contain negative values in the final glycoform composition for one or more of the composition search blocks (illogical nodes). Use the graphical representation to remove the glycoforms with logical compositions that were only connected to the subgraph via an illogical node.

Table 6.2 Description of an MS2-confirmed node per subgraph, following the GlycopeptideGraphMS analysis

Subgraph	node	mz_node	rt_node	Refnode	mz_refnode	N	H	F	S	Pep
6	989	2658.155	1769.709	989	2658.155	3	4	1	1	SLHVPGLNK

6.5.2.4 Targeted Glycopeptide Quantification

1. Convert the raw data (see Note 6) to the mzXML format using MSConvertGUI with the following settings:

    (a) Binary encoding precision: 64-bit
    (b) Filters: subset – MS level: 1-1
    (c) Check: write index, use zlib compression, TPP compatibility

2. Place the LaCyTools master folder in the working directory.
3. Based on the peptide fragments identified in the previous steps, create text files for the peptide "building blocks" representing the peptide potions, following the format below (example for the IgA2 glycopeptide covering site Asn337). Save the text file as [peptidename].block file and place it in the blocks folder in the LaCyTools master folder (Fig. 6.4d).
   TPL.block format:

    (a) mass
        957.5495
    (b) available_for_charge_carrier 0
    (c) carbons 42
    (d) hydrogens 75
    (e) nitrogens 11
    (f) oxygens 14
    (g) sulfurs 0

4. When there is more than one data file to quantify, pick one as the reference file. Create a tab-delimited text file (LaCyTools alignment file) listing the m/z of a minimum of five highly confident (see Note 7) glycopeptide compositions covering the entire retention time range in the first column and the retention time of these features in the reference file in the second column.
5. Create a tab-delimited text file (LaCyTools analyte list) containing the following:

    (a) Column 1: all identified glycopeptides, using the annotation TPL1H5N4F1S1 for the peptide covering IgA2 site Asn337 (TPL), carrying a glycan with five hexoses (H), four N-acetylhexosamines (N), one fucose (F), and one sialic acid (S)
    (b) Column 2: the retention time of the center of the LaCyTools glycopeptide cluster the analyte belongs to
    (c) Column 3: the integration m/z window
    (d) Column 4: the glycopeptide cluster retention time window
    (e) Column 5: the minimum charge state for the analyte
    (f) Column 6: the maximum charge state for the analyte
        A dedicated python script is available from [4] to create the LaCyTools analyte list based on the GlycopeptideGraphMS output.

6. In LaCyTools, open the Settings tab and adjust as follows:

    (a) Alignment time window and Alignment mass window: match the variation in retention time over the different runs and the mass accuracy of the analysis, respectively
    (b) Sum spectrum resolution: the average number of data points per 1 m/z in the MS1 spectra
    (c) Charge carrier: proton
    (d) Extraction window padding: 0
    (e) Minimum isotopic fraction: 0.75
    (f) Background detection window: 10 (see Note 8).

7. Open the Batch process tab, select the created LaCyTools alignment file under Alignment file and the LaCyTools analyte list under Reference file. Choose the directory where the .mzXML files are stored and select the following output format: Analyte Intensity, per charge state, background subtracted; Alignment QC; Analyte QC. Run the batch process.
8. The obtained Summary file contains the absolute intensities and QC parameters (mass accuracy, isotopic pattern QC; IPQ, and signal-to-noise; S/N) of all targeted glycopeptides in the selected charge states. In Excel: remove charge states of analytes with an S/N < 9, an IPQ > 0.25 and an absolute ppm error > 10. Sum the areas of the remaining charge states per analyte and correct the total for the exact fraction of the isotope pattern integrated for the respective analyte.
9. To obtain the final data, perform total area normalization per glycopeptide cluster (subgraph). For each data file, sum the absolute intensities of all glycoforms per peptide portion and divide each individual glycopeptide of this cluster by the sum. The obtained values can be used for data visualization and statistical analysis (Fig. 6.4d).

Using this protocol glycopeptide data was quantified for IgG1, IgG2/3, IgG4, IgA2 Asn337 and the joining chain Asn49 (Lippold et al. 2020).

6.5.3 Visualizing Profiles

We rely here on a so-called condensed notation: H for hexose, N for hexosamine, F for fucose, and S for sialic acid. Additionally, small letters designate modifications such as "a" for acetylation, "p" for phosphorylation, and "s" for sulfation, for the most frequent.

6.5.3.1 Structural Dependencies Brought Out by GlyConnect Compozitor

The output of most search engines identifying glycans or intact glycopeptides is usually provided as a list or in a table where the items or the rows are independent. However, the glycan structures or the compositions associated with peptides share common substructures. The GlyConnect Compozitor tool was developed to visualize these dependencies (Robin et al. 2020). Moreover, this web interface provides the option of comparing experimental results with the content of the GlyConnect database.

The glycoproteomics results of Sect. 5.2 for Ig gamma 4 were selected. Nine compositions were identified on peptide EEQFNSTYR (Asn177 in the corresponding UniProt P01861 entry):

H2N3F1
H3N3F1
H4N3F1
H3N4F1
H4N4F1
H3N5F1
H5N4F1
H5N5F1
H5N4F1S1

To bring out dependencies, Compozitor maps compositions related by shared monosaccharide counts into a graph that eases glycome visualization and comparison. When available, each composition is associated with defined structures. Here is how to generate graphs:

1. Go to glyconnect.expasy.org.
2. On the homepage, click on the orange PROTEIN button to open the Protein View of the database.
3. In the Protein View page, type "P01861" in the search window.
4. In the output of two entries, select/click on entry 770 (Id column), i.e., human immunoglobulin heavy constant gamma 4, for display.
5. The protein page shows all glycan structures reported to be N-linked to Asn-177 in ten different references. On the right side of the page, click on the Compozitor link.
6. The Compozitor search fields are pre-filled with the details of the 770 entry. In this case, the protein name is "immunoglobulin heavy constant gamma 4" and the Species is "Homo sapiens," shown in the "Protein" tab.
7. Click on the "Add to selection" button and observe that 33 compositions are recorded in the GlyConnect database, ready for display as "Selection A." Then, move to the "Custom" tab by clicking on it.
8. Copy-paste the nine compositions listed above in the input window. The glycan type is N-linked by default. Type "experimental" in the "selection label" window in order to name your input set of compositions.
9. Click again on the "Add to selection" button and observe that the nine compositions are now saved as "Selection B."
10. Display the graph of connected compositions by clicking on the "Compute graph" button.

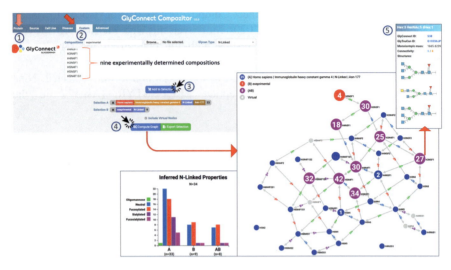

Fig. 6.5 Compozitor output of IgG4 data. The GlyConnect Compozitor tool is accessed directly from the database when querying it to visualize the *N*-glycome of human IgG4. This step is numbered 1 and the data is visualized in the Protein tab. It is labeled "Selection A." Step #2 consists of selecting the Custom tab to input the nine experimentally identified compositions in the glycoproteomics screening of IgG4. During step #3 the input data is recorded as "Selection B" and in step #4, the Compozitor graph is output. Magenta nodes correspond to the match between experimental data and the database content. A window with links to potential structures pops up by mousing over a node in step #5

The graph appears below and shows a well-connected set of 33 + 9 composition nodes that are distinguished by color: blue for the 33 compositions recorded in the database (Selection A), red for the nine input experimental set (Selection B), and magenta for the overlap between the two, as shown in Fig. 6.5. A number contained in a node represents how many glycan structure entries match this composition in GlyConnect. For example, H3N5F1 matched 27 partially or fully defined structures. This information is accessible by mousing over the node to prompt a caption showing the potential structures as shown in Fig. 6.5. The user can then explore the possibilities by clicking on the structure thereby opening the corresponding GlyConnect glycan structure page. The graph shows that the experimental compositions match exactly the mostly documented structures in the database (nodes with highest numbers).

Compozitor allows for "virtual nodes" (in gray) when only one intermediary step is needed to connect isolated nodes. If not for these, the graph would be more scattered.

6.5.3.2 Comparing Profiles with Glynsight

Again, the results of glycopeptide quantification generated by the workflow of Sect. 5.2 were recorded and input in the Glynsight tool (Alocci et al. 2018) for comparing

6   Bioinformatics in Immunoglobulin Glycosylation Analysis   227

the expression profiles of each of the IgGs. An example of file in the correct .csv (comma separated values) input format is shown below for IgG4:

```
composition,quantification,
H2N3F1,0.55,
H3N3F1,1.74,
H3N4F1,0.59,
H3N5F1,37.19,
H4N3F1,28.24,
H4N4F1,6.99,
H5N4F1,12.01,
H5N4F1S1,1.04,
H5N5F1,11.64,
```

1. Go to https://glycoproteome.expasy.org/glynsight;
2. In the left menu, click on "Manage experiments" and click on "Import experiments" in the list of tasks that is displayed. Use the "upload files" button to import the .csv files such as the one shown above. The easiest is to give a meaningful name to the file, such as IgG4.csv for the example above.
3. The name of each of the uploaded files is shown in the left menu under "Experiment list." Click on each name of experiment to see the corresponding profile displayed as a bar chart, each bar representing a composition. These are incrementally ordered and structured in columns. All compositions featuring in a column have a constant overall number of monosaccharides. The first column of the profile here contains H3N3 and H2N3F1 (six monosaccharides in total), the next column contains compositions with an overall count of seven monosaccharides, etc. Changing from one experiment to the next immediately brings out the variations.
4. The default display mode is "individual," showing one experiment at a time. To compare profiles, select "differential display" in the Display mode section of the left menu. In this mode, the list of experiments in the left side menu is duplicated to provide the user with the option of selecting an experiment in each of the two lists.
5. Compare IgG1 with IgG2/3, then with IgG4, etc.

Figure 6.6 shows the profile comparisons. Glynsight shows red and blue bars whose height for each composition represents the result of the subtraction. For any given composition, if its intensity in the first experiment selected is higher than in the second, then the bar is blue. In the opposite situation, the bar is red. Brighter blue (resp. red) highlights differences arising from compositions unique to the first (resp. second) experiment while toned down blue (resp. red) shows differences arising from compositions expressed in both experiments but higher in the first (resp. second) than in the second. Grey corresponds to glycans present in other experiments but not in those currently displayed.

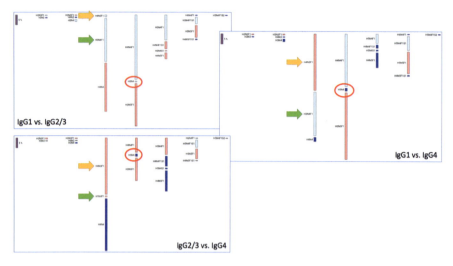

Fig. 6.6 Glynsight comparison of profiles. The comparative profiles of IgG1, IgG2/3, and IgG4 with respect to one another as displayed by Glynsight. Blue (resp. red) bars indicate that expression is higher (resp. lower) in the first profile than in the second. Brighter blue (resp. red) shows differences arising from compositions unique to the first (resp. second) profile while toned down blue (resp. red) shows differences arising from compositions expressed in both profiles but higher in the first (resp. second) than in the second. Green arrows single out the expression of H3N4F1 underexpressed in IgG1, but roughly the same in IgG2/3 and IgG4. Yellow arrows single out the expression of H4N3F1 overexpressed in IgG4 but roughly the same in IgG1 and IgG2/3. Red circles show the constant expression of H5N4 across all sets

The large bars reflect significant differences between the two profiles. The screen capture of the figure is not as visual as when using the software live, since the dynamics of change are not transcribed. Nonetheless, variations are observable. For example, H3N4F1 is overexpressed in IgG1 with respect to both IgG2/3 and IgG4, while it is the same between IgG2/3 and IgG4 (see green arrows in Fig. 6.6). In contrast, H4N3F1 is over overexpressed in IgG4 with respect to both IgG1 and IgG2/3, while it is the same IgG1 and IgG2/3 (see yellow arrows). Another example of constant expression across all sets is that of H5N4 (red circles).

6.6 Conclusion

This chapter has demonstrated the extent of available bioinformatics resources for studying the glycosylation of immunoglobins. On the one hand, both generic and specific databases are interconnected to reflect the growing body of literature on Ig glycosylation and its functional influence. The therapeutic importance of these proteins warrants a special section soon to be developed in the GlyConnect database. On the other hand, the cited software tools have enabled high-throughput and robust analysis, especially for IgG and IgA. In fact, Ig(G)s have quite a head start, compared to other glycoproteins and many publications already reveal IgG N-glycome variations as a reflection of disease states (see recent (Clerc et al. 2018; Martin et al. 2020) for examples). Although the same basic principles apply for the analysis of other glycoproteins, more developments and streamlining of resources are warranted, especially when shifting from isolated proteins to complex samples.

Notes

Glycopeptide data analysis

1. This *N*-glycan library is a good start for Ig samples but should be used with caution as it is far from complete regarding reported *N*-glycan structures. The glycan library used in Byonic can be fine-tuned in an iterative manner after Step 7 in "Glycoform identification based on MS1." In addition, GlyConnect (https://glyconnect.expasy.org/) can be used to construct a protein-specific library based on literature knowledge.
2. According to the Byonic user manual (https://www.proteinmetrics.com/support-information/), the Byonic score ranges from 0 to 1000. A score of 300 is considered good, 400 is very good and scores over 500 are almost sure to be correct. In our experience, lower scores may reflect confident glycopeptide identification, and as further QC will be performed both during the MS1-based identification and the quantification step, using a relatively low threshold score of 200 is justified here.
3. The max. RT differences are highly dependent on the chromatographic setup used and should be optimized based on the specific data analyzed. In general, when using C18 chromatography, the RT difference between two neutral monosaccharides is smaller than between a neutral and a charged monosaccharide.
4. In the likely scenario that more than one node per cluster could be assigned to a specific glycopeptide by MS2, choose the one with the highest confidence. Confidence can be based on Byonic score, manual evaluation of the MS2 data, and prior (literature) knowledge on the expected products.
5. In the scenario that no MS2 confirmed glycopeptide was obtained for any of the nodes in one subgraph, leave the row empty. This can indicate multiple things, including (1) no MS2 spectra were obtained for this glycopeptide cluster, in this case, a targeted MS2 run can be considered; (2) an unknown glycopeptide modification occurred, when this modification is identified Byonic can be run again including it in the search; (3) these nodes were erroneously identified as glycopeptide cluster, exclude the subgraph.
6. To obtain superior quantification precision, data integration should be performed on MS1-only data featuring a high number of data points per chromatographic feature. [4] However, data files including MS2 scans are also suitable, provided that there is a minimum of five MS1 data points per chromatographic feature.
7. While subgraphs or glycopeptide clusters are defined as a group of glycopeptides that share the same peptide portion and differ in their glycan content, in the current context separate LaCyTools glycopeptide clusters were defined for

glycopeptides with the same peptide portion but different sialic acid contents. In general, the more clusters defined, the longer LaCyTools processing time, but the lower the chance of analyte overlaps that interfere with the quantification.
8. Please find an in-depth explanation of all settings in [3] and [5]. The definition of the isotopic pattern QC (IPQ) changed between the initially published version of LaCyTools and the current and is now defined as the sum of the relative intensity deviation per isotope.

Visualization tools

9. Several of the images showing a Compozitor graph in figures have been slightly modified from the raw output to disentangle the network which is generated by the D3.js (https://d3js.org) library. This transformation is made easy since any node can be dragged wherever a user wishes in the space of the browser window by maintaining the mouse over it. Consequently, paths can be shortened or stretched upon user actions.
10. Compositions can be exported in a text file as soon as they have been added to the selection and before computing the graph ("Export selection" button). They can be selectively exported once the graph is computed ("Export" button). The graph can be exported in the .svg format.
11. The "Zoom in" and "Zoom out" buttons have obvious purposes and the former operation is supplemented by the "Zoom on" window located in the top right corner; typing a specific composition in this window will trigger zooming in and centering the graph on the corresponding node.
12. Mousing over a bar of the bar plot of glycan properties highlights all nodes that are counted in that bar to show where they are located in the graph.

Compliance with Ethical Standards

Funding The development of resources of the glycomics@ExPASy initiative has been supported by the European Union FP7 Innovative Training Network (grant 316929) and the Swiss Federal Government through the State Secretariat for Education, Research and Innovation (SERI) and are currently supported by the Swiss National Science Foundation (SNSF) (grant 31003A_179249). ExPASy is maintained by the Swiss Institute of Bioinformatics and hosted at the Vital-IT Competency Center. KA gratefully acknowledges the generous financial support of the Max Planck Society to the Max Planck Unit for the Science of Pathogens. CH is supported by supplement #SUB00002508FE to the NIH grant 1U01GM125267 (GlyGen). SL was funded by the European Commission H2020 (Analytics for Biologics project; 765502). NH was funded by the European Research Council (ERC) under the European Union's Horizon 2020 research and innovation program (GlycoSkin H2020-ERC; 772735).

Ethical Approval This chapter does not contain any studies with human participants performed by any of the authors.

Conflict of Interest Frédérique Lisacek declares that she has no conflict of interest.
Kathirvel Alagesan declares that he has no conflict of interest.
Catherine Hayes declares that she has no conflict of interest.
Steffen Lippold declares that he has no conflict of interest.
Noortje de Haan declares that she has no conflict of interest.

References

Abrahams JL, Taherzadeh G, Jarvas G et al (2020) Recent advances in glycoinformatic platforms for glycomics and glycoproteomics. Curr Opin Struct Biol 62:56–69

Alagesan K, Hoffmann M, Rapp E, Kolarich D (2020) Glycoproteomics technologies in glycobiotechnology. In: Advances in biochemical engineering/biotechnology. Springer, Heidelberg. https://doi.org/10.1007/10_2020_144

Alocci D, Ghraichy M, Barletta E et al (2018) Understanding the glycome: an interactive view of glycosylation from glycocompositions to glycoepitopes. Glycobiology 28:349–362

Alocci D, Mariethoz J, Gastaldello A et al (2019) GlyConnect: glycoproteomics goes visual, interactive, and analytical. J Proteome Res 18:664–677

Amez-Martín M, Wuhrer M, Falck D (2020) Immunoglobulin G glycoprofiles are unaffected by common bottom-up sample processing. J Proteome Res 19:4158–4162

Aoki-Kinoshita K, Agravat S, Aoki NP et al (2016) GlyTouCan 1.0—The international glycan structure repository. Nucleic Acids Res 44:D1237–D1242

Arnold JN, Wormald MR, Sim RB et al (2007) The impact of glycosylation on the biological function and structure of human immunoglobulins. Annu Rev Immunol 25:21–50

Bern M, Kil YJ, Becker C (2012) Byonic: advanced peptide and protein identification software. In: Baxevanis AD, Petsko GA, Stein LD et al (eds) Current protocols in bioinformatics. Wiley, Hoboken, NJ, pp 13.20.1–13.20.14

Bollineni RC, Koehler CJ, Gislefoss RE et al (2018) Large-scale intact glycopeptide identification by Mascot database search. Sci Rep 8:2117

Cao W, Liu M, Kong S et al (2020) Recent advances in software tools for more generic and precise intact glycopeptide analysis. Mol Cell Proteomics. https://doi.org/10.1074/mcp.R120.002090

Ceroni A, Maass K, Geyer H et al (2008) GlycoWorkbench: a tool for the computer-assisted annotation of mass spectra of glycans. J Proteome Res 7:1650–1659

Choo MS, Wan C, Rudd PM et al (2019) GlycopeptideGraphMS: improved glycopeptide detection and identification by exploiting graph theoretical patterns in mass and retention time. Anal Chem 91:7236–7244

Clerc F, Novokmet M, Dotz V et al (2018) Plasma N-glycan signatures are associated with features of inflammatory bowel diseases. Gastroenterology 155:829–843

Cooper CA, Gasteiger E, Packer NH (2001) GlycoMod—a software tool for determining glycosylation compositions from mass spectrometric data. Proteomics 1:340–349

Creasy DM, Cottrell JS (2004) Unimod: protein modifications for mass spectrometry. Proteomics 4:1534–1536

de Haan N, Falck D, Wuhrer M (2020) Monitoring of immunoglobulin N- and O-glycosylation in health and disease. Glycobiology 30:226–240

Fujita A, Aoki NP, Shinmachi D et al (2020) The international glycan repository GlyTouCan version 3.0. Nucleic Acids Res. 49(D1):D1529–D1533

Gray CJ, Migas LG, Barran PE et al (2019) Advancing solutions to the carbohydrate sequencing challenge. J Am Chem Soc 141:14463–14479

Gstöttner C, Nicolardi S, Haberger M et al (2020) Intact and subunit-specific analysis of bispecific antibodies by sheathless CE-MS. Anal Chim Acta 1134:18–27

Hayes CA, Karlsson NG, Struwe WB et al (2011) UniCarb-DB: a database resource for glycomic discovery. Bioinformatics 27:1343–1344

Hu H, Khatri K, Zaia J (2017) Algorithms and design strategies towards automated glycoproteomics analysis: algorithms and design strategies. Mass Spectrom Rev 36:475–498

Hülsmeier AJ, Tobler M, Burda P et al (2016) Glycosylation site occupancy in health, congenital disorder of glycosylation and fatty liver disease. Sci Rep 6:33927

Jansen BC, Falck D, de Haan N et al (2016) LaCyTools: a targeted liquid chromatography–mass spectrometry data processing package for relative quantitation of glycopeptides. J Proteome Res 15:2198–2210

Kessner D, Chambers M, Burke R et al (2008) ProteoWizard: open source software for rapid proteomics tools development. Bioinformatics 24:2534–2536

Khatri K, Klein JA, Zaia J (2017) Use of an informed search space maximizes confidence of site-specific assignment of glycoprotein glycosylation. Anal Bioanal Chem 409:607–618

Kolarich D, Rapp E, Struwe WB et al (2013) The minimum information required for a glycomics experiment (MIRAGE) Project: improving the standards for reporting mass-spectrometry-based glycoanalytic data. Mol Cell Proteomics 12:991–995

Lefranc M-P, Giudicelli V, Duroux P et al (2015) IMGT®, the international ImMunoGeneTics information system® 25 years on. Nucleic Acids Res 43:D413–D422

Lih TM, Choong W-K, Chen C-C et al (2016) MAGIC-web: a platform for untargeted and targeted N-linked glycoprotein identification. Nucleic Acids Res 44:W575–W580

Lippold S, de Ru AH, Nouta J et al (2020) Semiautomated glycoproteomics data analysis workflow for maximized glycopeptide identification and reliable quantification. Beilstein J Org Chem 16:3038–3051

Lisacek F, Mariethoz J, Alocci D et al (2017) Databases and associated tools for glycomics and glycoproteomics. In: Lauc G, Wuhrer M (eds) High-throughput glycomics and glycoproteomics. Springer, New York, NY, pp 235–264

Liu G, Cheng K, Lo CY et al (2017) A comprehensive, open-source platform for mass spectrometry-based glycoproteomics data analysis. Mol Cell Proteomics 16:2032–2047

Lobner E, Humm A-S, Mlynek G et al (2017) Two-faced Fcab prevents polymerization with VEGF and reveals thermodynamics and the 2.15 Å crystal structure of the complex. mAbs 9:1088–1104

Martin TC, Šimurina M, Ząbczyńska M et al (2020) Decreased immunoglobulin G core fucosylation, a player in antibody-dependent cell-mediated cytotoxicity, is associated with autoimmune thyroid diseases. Mol Cell Proteomics 19:774–792

Momčilović A, de Haan N, Hipgrave Ederveen AL et al (2020) Simultaneous immunoglobulin A and G glycopeptide profiling for high-throughput applications. Anal Chem 92:4518–4526

Orchard S, Hermjakob H, Apweiler R (2003) The proteomics standards initiative. Proteomics 3:1374–1376

Pfeuffer J, Sachsenberg T, Alka O et al (2017) OpenMS—a platform for reproducible analysis of mass spectrometry data. J Biotechnol 261:142–148

Pino LK, Searle BC, Bollinger JG et al (2020) The Skyline ecosystem: Informatics for quantitative mass spectrometry proteomics. Mass Spectrom Rev 39:229–244

Pralow A, Cajic S, Alagesan K, Kolarich D, Rapp E (2020) State-of-the-art glycomics technologies in glycobiotechnology. In: Advances in biochemical engineering/biotechnology. Springer, Heidelberg. https://doi.org/10.1007/10_2020_143

Riley NM, Hebert AS, Westphall MS et al (2019) Capturing site-specific heterogeneity with large-scale N-glycoproteome analysis. Nat Commun 10:1311

Robin T, Mariethoz J, Lisacek F (2020) Examining and fine-tuning the selection of glycan compositions with GlyConnect compozitor. Mol Cell Proteomics 19:1602–1618

Rojas-Macias MA, Mariethoz J, Andersson P et al (2019) Towards a standardized bioinformatics infrastructure for N- and O-glycomics. Nat Commun 10:3275

Savitski MM, Lemeer S, Boesche M et al (2011) Confident phosphorylation site localization using the mascot delta score. Mol Cell Proteomics 10:M110.003830

Schwämmle V, Verano-Braga T, Roepstorff P (2015) Computational and statistical methods for high-throughput analysis of post-translational modifications of proteins. J Proteome 129:3–15

Sehnal D, Grant OC (2019) Rapidly display glycan symbols in 3D structures: 3D-SNFG in LiteMol. J Proteome Res 18:770–774

Thaysen-Andersen M, Packer NH (2014) Advances in LC–MS/MS-based glycoproteomics: Getting closer to system-wide site-specific mapping of the N- and O-glycoproteome. Biochim Biophys Acta 1844:1437–1452

The UniProt Consortium (2019) UniProt: a worldwide hub of protein knowledge. Nucleic Acids Res 47:D506–D515

Tiemeyer M, Aoki K, Paulson J et al (2017) GlyTouCan: an accessible glycan structure repository. Nucleic Acids Res 27:915–919

Trbojević-Akmačić I, Vilaj M, Lauc G (2016) High-throughput analysis of immunoglobulin G glycosylation. Expert Rev Proteomics 13:523–534

Varki A, Cummings RD, Aebi M et al (2015) Symbol nomenclature for graphical representations of glycans. Glycobiology 25:1323–1324

Vizcaíno JA, Deutsch EW, Wang R et al (2014) ProteomeXchange provides globally coordinated proteomics data submission and dissemination. Nat Biotechnol 32:223–226

Watanabe Y, Aoki-Kinoshita KF, Ishihama Y et al (2020) GlycoPOST realizes FAIR principles for glycomics mass spectrometry data. Nucleic Acids Res 49(D1):D1523–D1528

Weatherly DB, Arpinar FS, Porterfield M et al (2019) GRITS Toolbox—a freely available software for processing, annotating and archiving glycomics mass spectrometry data. Glycobiology 29:452–460

Wuhrer M, Stam JC, van de Geijn FE et al (2007) Glycosylation profiling of immunoglobulin G (IgG) subclasses from human serum. Proteomics 7:4070–4081

wwPDB consortium, Burley SK, Berman HM et al (2019) Protein Data Bank: the single global archive for 3D macromolecular structure data. Nucleic Acids Res 47:D520–D528

Ye Z, Mao Y, Clausen H et al (2019) Glyco-DIA: a method for quantitative O-glycoproteomics with in silico-boosted glycopeptide libraries. Nat Methods 16:902–910

York WS, Mazumder R, Ranzinger R et al (2020) GlyGen: computational and informatics resources for glycoscience. Glycobiology 30:72–73

Part II
Biosynthesis and Regulation

Chapter 7
N-Glycan Biosynthesis: Basic Principles and Factors Affecting Its Outcome

Teemu Viinikangas, Elham Khosrowabadi, and Sakari Kellokumpu

Contents

7.1	Introduction	240
7.2	Biosynthesis of N-Glycans in the Endoplasmic Reticulum	241
	7.2.1 Building Blocks for N-Glycan Synthesis	241
	7.2.2 Precursor Synthesis and Its Attachment to Nascent Polypeptide Chains	241
	7.2.3 N-Glycan Processing in the ER and Quality Control	244
7.3	N-Glycan Processing in the Golgi Apparatus	246
	7.3.1 N-Glycosylation of Immunoglobulins	249
7.4	Golgi Microenvironment Is Important for Normal Processing and Maturation of N-Glycans	249
	7.4.1 Golgi pH Homeostasis	250
	7.4.2 Golgi Ion Homeostasis	251
	7.4.3 Golgi Redox State	251
7.5	Concluding Remarks	252
References		254

Abstract Carbohydrate chains are the most abundant and diverse of nature's biopolymers and represent one of the four fundamental macromolecular building blocks of life together with proteins, nucleic acids, and lipids. Indicative of their essential roles in cells and in multicellular organisms, genes encoding proteins associated with glycosylation account for approximately 2% of the human genome. It has been estimated that 50–80% of all human proteins carry carbohydrate chains—glycans—as part of their structure. Despite cells utilize only nine different monosaccharides for making their glycans, their order and conformational variation in glycan chains together with chain branching differences and frequent post-synthetic modifications can give rise to an enormous repertoire of different glycan structures of which few thousand is estimated to carry important structural or functional information for a cell. Thus, glycans are immensely versatile encoders of multicellular life. Yet, glycans do not represent a random collection of unpredictable structures but rather,

---

T. Viinikangas · E. Khosrowabadi · S. Kellokumpu (✉)
Faculty of Biochemistry and Molecular Medicine, University of Oulu, Oulu, Finland
e-mail: sakari.kellokumpu@oulu.fi

© The Author(s), under exclusive license to Springer Nature Switzerland AG 2021
M. Pezer (ed.), *Antibody Glycosylation*, Experientia Supplementum 112,
https://doi.org/10.1007/978-3-030-76912-3_7

a collection of predetermined but still dynamic entities that are present at defined quantities in each glycosylation site of a given protein in a cell, tissue, or organism.

In this chapter, we will give an overview of what is currently known about *N*-glycan synthesis in higher eukaryotes, focusing not only on the processes themselves but also on factors that will affect or can affect the final outcome—the dynamicity and heterogeneity of the *N*-glycome. We hope that this review will help understand the molecular details underneath this diversity, and in addition, be helpful for those who plan to produce optimally glycosylated antibody-based therapeutics.

Keywords Endoplasmic reticulum · Golgi apparatus · *N*-glycosylation · Organelle homeostasis

List of Abbreviations

Alg	Yeast Asparagine-linked glycosyltransferase
ALS	Amyotrophic lateral sclerosis
Asn	Asparagine
BiP	Binding protein
CDG	Congenital disorders of glycosylation
CEA	Carcinoembryonic antigen
CMAH	Cytidine monophospho-*N*-acetylneuraminic acid hydroxylase
CMP-Sia	Cytidine-5′-monophospho-*N*-acetylneuraminic acid
COPII	Coat protein complex II
Dol	Dolichol
Dol-P	Dolichol phosphate
Dol-P-Glc	Dolichol monophosphate glucose
Dol-P-Man	Dolichol monophosphate mannose
Dpagt1	Dolichyl-phosphate *N*-acetylglucosamine-phosphotransferase 1
EDEM	ER degradation enhancing mannosidase-like protein
ER	Endoplasmic reticulum
ERAD	ER-associated degradation
ERGIC	ER-Golgi intermediate compartment
ERGIC-53	ER-Golgi intermediate compartment 53 kDa protein
ERManI	ER mannosidase I
ERp57	Endoplasmic reticulum resident disulfide isomerase
Fuc	Fucose
FUT	Fucosyltransferase
Gal	Galactose
GalNAc	*N*-acetyl-D-galactosamine
GCNT2	*N*-acetyl-lactosamine-β-1,6-*N*-acetylglucosaminyltransferase
GDP-Man	Guanosine-5′-diphosphate-α-D-mannose
Glc	Glucose
GlcNAc	*N*-acetyl-D-glucosamine

GLUT-2	Glucose transporter type 2
HIF	Hypoxia-inducible factor
LacNAc	N-Acetyl-D-lactosamine
LLO	Lipid-linked oligosaccharide
LMAN1	Lectin mannose-binding 1
M6P	Mannose-6-phosphate
Man	Mannose
MCFD2	Multiple coagulation factor deficiency protein 2
MGAT	Mannosyl-glycoprotein-N-acetylglucosaminyltransferase
MPR	Mannose-6-phosphate receptor
NAGPA	N-acetylglucosamine-1-phosphodiester α-N-acetylglucosaminidase
Neu5Ac	5-N-Acetylneuraminic acid (sialic acid)
Neu5Gc	5-N-Glycolylneuraminic acid (sialic acid)
OS-9	Protein OS-9; amplified in osteosarcoma 9
OST	Oligosaccharyltransferase complex
ROS	Reactive oxygen species
Rft1	A transmembrane protein encoded by the yeast Rft1 gene
Sec61	A three-subunit (Sec61α, Sec61β, and Sec61γ) protein translocation complex
Ser	Serine
SERCA2	Sarcoplasmic/endoplasmic reticulum calcium ATPase 2
Sia	Sialic acid
SPCA1/2	Secretory pathway Ca(2+)-ATPase type 1 and 2
ST3Gal1	Beta-galactoside α-2,3-sialyltransferase 1
ST3Gal3	Beta-galactoside α-2,3-sialyltransferase 3
ST6Gal1	Beta-galactoside alpha-2,6-sialyltransferase 1
STT3A/B	Catalytic A and -B subunits of the oligosaccharyltransferase complex
TGN	*Trans*-Golgi network
Thr	Threonine
TMEM165	Transmembrane protein 165
UDP-GlcNAc	Uridine-5′-diphosphate-N-acetyl-α-D-glucosamine
UDP-Gal	Uridine-5′-diphosphate-α-D-galactose
UDP-Glc	Uridine-5′-diphosphate-α-D-glucose
UGGT	UDP-glucose-glycoprotein glucosyltransferase
VIP 36	Vesicular integral-membrane protein 36
XTP3-B	XTP3-transactivated gene B lectin

7.1 Introduction

Asparagine-linked (*N*-linked) glycosylation is an essential protein modification, affecting a number of basic cellular processes such as protein folding, its half-life, trafficking and immunogenicity as well as its interactions between cells, cells and extracellular matrix components or pathogens (Varki and Gagneux 2017). In all eukaryotic cells, *N*-glycans are synthesized in two specialized organelles, the endoplasmic reticulum (ER) and the Golgi apparatus. Together, these organelles harbor dozens of functionally distinct glycosyltransferases and glycosidases that sequentially modify the growing oligosaccharide chain (Kornfeld and Kornfeld 1985; Dunphy 1985; Spiro 2002; Rabouille et al. 1995). Yet, it is much less clear how this sequence of enzymatic reactions is orchestrated to guarantee faithful synthesis of *N*-glycans, considering that enzymes do not use any template, can compete with each other for the same substrate and/or the acceptor, and even localize in the same Golgi compartment. Another puzzling issue is an intrinsic microheterogeneity of glycans made by the cell. For example, an *N*-glycan attached to a specific asparagine of a given protein can be different from an *N*-glycan attached to the same site in another protein molecule. Distinguishing this "background noise" from dynamic changes that are functionally important, e.g., during embryonic development, cell differentiation, and aging can sometimes be problematic. Nevertheless, unlike other polymerization events in the cell, glycosylation apparently need not be a high-fidelity system, since cells normally tolerate such microheterogeneity without facing problems in cell survival or proliferation.

The other side of the coin is that this variation can sometimes lead to a devastating disease. Congenital disorders of glycosylation (CDGs) are a rare, yet diverse group of serious, often multiorgan diseases characterized by defects in glycosylation (Freeze et al. 2014; Francisco et al. 2020). More than 140 different CDG syndromes are known as of today, the severity of which varies from prenatal death to survival into adulthood with a relatively normal life span. Disturbed *N*-glycosylation forms the largest group of the CDGs. It is divided into two groups (Type I and Type II) based primarily on the genetic defect and the site it is affecting. Type I CDGs are characterized by defects in the synthesis of *N*-glycans in the endoplasmic reticulum (ER), while the Type II CDGs have problems in their processing in the Golgi apparatus. In addition to CDGs, glycosylation changes play an important role in many other human diseases including autoimmune diseases, inflammation, tumorigenesis, and its progression (Reily et al. 2019). Yet, the underlying mechanistic details that cause these changes are incompletely understood, as are also the reasons why some changes lead to disease and some do not. Partly, this is due to the dynamic and variable nature of the glycan themselves, their cell- and tissue-specific expression (Medzihradszky et al. 2015) as well as the lack of tools that would allow glycan editing at will in a specified glycosylation site or protein itself.

7.2 Biosynthesis of *N*-Glycans in the Endoplasmic Reticulum

7.2.1 Building Blocks for N-Glycan Synthesis

The early steps in *N*-glycan biosynthesis in the endoplasmic reticulum (ER) are conserved in all three domains of life (Dell et al. 2010), whereas their processing and maturation differ markedly. All *N*-glycans share a common core structure (asn-GlcNAc2Man3-) which is further elongated in a species- and tissue-specific manner (Medzihradszky et al. 2015) by adding a few other subterminal or terminal sugar residues to the core structure. Depending on the sugar residue and the linkage type used, these additions can significantly influence the structure of the *N*-glycan (Medzihradszky et al. 2015). The main sugar residues utilized as building blocks are *N*-acetylglucosamine (GlcNAc), mannose (Man), galactose (Gal), fucose (Fuc), and sialic acid (*N*-acetylneuraminic acid (Neu5Ac) being the predominant form), of which the latter two act as chain-capping residues. In some instances, *N*-acetylgalactosamine (GalNAc) residues can be used to construct an *N*-glycan. Glucose (Glc) residues are also temporarily incorporated into the growing *N*-glycan during its synthesis in the ER, yet they are invariably removed as glucose residues have not been detected in a mature *N*-glycan isolated from cultured cells or tissues (Zuber et al. 2000). Occasionally, mature *N*-glycans can also be modified by the addition of sulfate or phosphate, generating determinants that modulate cell adhesion or glycoprotein localization in the cells. Interestingly, despite the deletion of the CMAH gene (needed for *N*-glycolylneuraminic acid (Neu5Gc) synthesis) 3 million years ago, this neuraminic acid variant is still regularly detected in trace amounts in human glycans (Angata and Varki 2002). This is due to dietary consumption of Neu5Gc-containing animal products (e.g., red meat and dairy products) and its incorporation into newly synthesized glycans (Banda et al. 2012). Perhaps unsurprisingly, the highest Neu5Gc levels are detected in epithelial and endothelial cells that line the intestine and blood (and lymph) vessels, respectively.

7.2.2 Precursor Synthesis and Its Attachment to Nascent Polypeptide Chains

The *N*-glycosylation of nascent polypeptides in the ER lumen relies on the prior assembly of a lipid-linked oligosaccharide (LLO) precursor (Glc$_3$Man$_9$-GlcNAc$_2$) onto a membrane-embedded dolichyl phosphate (Dol-P) carrier (Fig. 7.1). This set of events is orchestrated by the Alg-family of ER-localized, membrane-associated glycosyltransferases (Kelleher et al. 2007). They stepwise assemble the LLO using nucleotide sugars (UDP-GlcNAc, GDP-Man, Dol-P-Man, and Dol-P-Glc) as donor substrates. The LLO assembly begins on the cytoplasmic face of the ER membrane by the formation of a GlcNAc$_2$-PP-Dol intermediate from GlcNAc-1-phosphate and

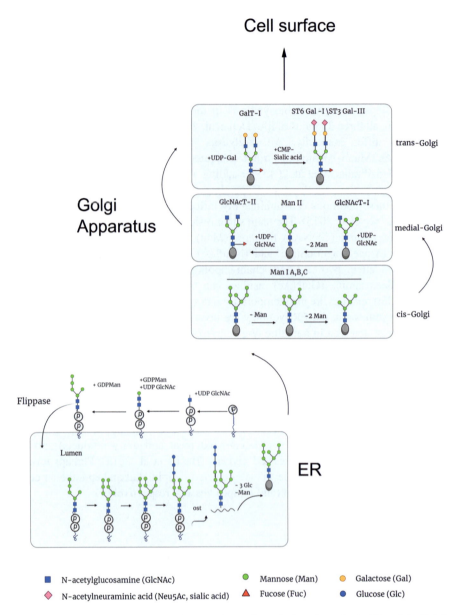

Fig. 7.1 A schematic representation of the *N*-glycan biosynthetic pathway in the ER and the Golgi apparatus. The figure shows the gradual maturation of an *N*-glycan and the various steps involved. For more details, please see the text

GlcNAc. These additions are catalyzed by Dpagt1 (Alg7) and Alg13p/Alg14p UDP-GlcNAc-transferases, respectively. The three enzymes exist as hexamers with a stoichiometry of 2:2:2 (Noffz et al. 2009). Alg14 appears to be the central unit, capable of recruiting other enzymes to the complex (Lu et al. 2012). Next, ER mannosyltransferases (Alg1, Alg2, and Alg11) that also form complexes with each other (Gao 2004) add five mannose residues from GDP-Man donors to form a $Man_5GlcNAc_2$-PP-Dol intermediate. Thus, LLO precursor synthesis on the cytoplasmic face of the ER membrane involves three main enzyme complexes, one formed by Dpagt1/Alg13/Alg14 and the other two either by Alg1/Alg2 and Alg1/Alg11. This arrangement likely ensures that each mannose residue will be linked correctly to the precursor despite the coexistence of several competing enzymes on the same membrane.

The next step involves translocation of the $Man_5GlcNAc_2$-PP-Dol intermediate into the ER lumen, a process that is thought to be mediated by a protein termed as the Rft1, but it is still uncertain whether it acts as a bonafide flippase protein (Helenius et al. 2002). In the ER lumen, mannosyltransferases (Alg3/Alg9/Alg12) and glucosyltransferases (Alg6/Alg8/Alg10) further elongate the LLO precursor by attaching four additional mannose residues and three glucose residues, respectively. This completes the precursor synthesis and yields the $Glc_3Man_9GlcNAc_2$-PP-Dol structure, which will be used later as the donor substrate for *en bloc* transfer of an *N*-glycan to a suitable polypeptide chain. It is noteworthy that unlike the initial catalytic steps on the cytosolic face of ER, the completion of the LLO precursor synthesis in the ER lumen does not use nucleotide sugars as donors. Rather, membrane-embedded Dol-P-Man and Dol-P-Glc are used as sugar donors in this case. Their synthesis takes place also on the cytoplasmic side of the ER membrane (from GDP-Man and UDP-Glc, respectively) before they are translocated (flipped) to the luminal side (Helenius et al. 2002).

The most preferred acceptor asparagine residues for *N*-glycosylation are the ones within the Asn-X-Ser/Thr motif (where $X \neq$ proline) (Zielinska et al. 2010). Of these two, the Asn-X-Thr sequon is preferred over Asn-X-Ser, mainly because the interaction between the side chain methyl group of threonine and the asparagine-lysine (NK) motif in the binding pocket of the oligosaccharyltransferase (OST) increases the stability of the complex (Kasturi et al. 1995, 1997; Medus et al. 2017). The identity of the amino acid X and flanking amino acids also contribute to the glycosylation of a given sequon. In addition, the position of the sequon within the polypeptide, the secondary and tertiary structure of the protein, and its final destination in a cell can impair or enhance the likelihood of whether that site becomes glycosylated or not (Rao and Wollenweber 2010). Thus, the presence of sequons alone cannot be used as an adequate predictor of *N*-glycosylation. Indeed, roughly one-third of the identified sequons in secreted glycoproteins remain non-glycosylated (Schulz 2012).

The transfer of the completed precursor oligosaccharide is catalyzed by ER membrane-localized OST complex. It is an octamer consisting of a single catalytic subunit and seven accessory subunits, each important for optimal glycosylation efficiency. Most multicellular animals (sponges are an exception) possess two such complexes due to an ancient duplication of the gene encoding the catalytic subunit. The STT3A and STT3B (for OST-A and OST-B complexes, respectively) have different kinetic properties, acceptor substrate preferences, and partially non-overlapping roles in glycosylation (Shrimal et al. 2013a). The accessory subunit compositions between the two complexes also differ. OST-A complex associates with Sec61 core components of the ER translocon complex and co-translationally glycosylates the nascent polypeptide in accessible sequons during polypeptide chain translocation into the ER lumen. Sequons within the last ~50–55 residues of the C-terminus are, however, inside the translocon and hence inaccessible for STT3A. Instead, STT3B transferase in OST-B complexes can posttranslationally add N-glycans to such sequons. It also can use internal sequons that are skipped by the STT3A as acceptors (Lu et al. 2018; Shrimal et al. 2013b). Often, these include closely spaced sequons adjacent to signal cleavage site or sequons with cysteine residues nearby or inside the motif (i.e., the N-C-T/S motif (Shrimal et al. 2013a).

7.2.3 N-Glycan Processing in the ER and Quality Control

The newly attached $Glc_3Man_9GlcNAc_2$ N-glycan structure is further modified once the polypeptide is translocated to the ER lumen and begins to fold. The first step involves the removal of the terminal glucose residue by a transmembrane enzyme α-glucosidase I. The second glucose is then rapidly removed by the soluble α-glucosidase II (Grinna and Robbins 1979; Janssen et al. 2010) The resulting mono-glucosylated glycan is a preferred ligand for the carbohydrate-recognizing molecular chaperones calnexin and calreticulin. These chaperones readily associate also with the protein ERp57 (Ruddock and Molinari 2006), a disulfide isomerase that catalyzes the formation of inter- and intramolecular disulfide bonds, thereby helping proper folding of the nascent glycoprotein. Calnexin binding appears to happen irrespective of the folding state of the glycoprotein (Zapun et al. 1997), suggesting that it most likely interacts with the nascent polypeptide as soon as it arrives in the ER lumen. During the folding process (Fig. 7.2), the last glucose residue is removed by the α-glucosidase II. Unfolded or misfolded proteins display exposed hydrophobic patches that are recognized by UDP-glucose-glycoprotein glucosyl-transferase (UGGT) (Caramelo et al. 2003), an enzyme that can glucosylate the same mannose residue again in that N-glycan. By doing so, it recreates the $Glc_1Man_9GlcNAc_2$ structure that is again acted upon by the chaperone-disulfide isomerase complex. This removal and re-addition of glucose residues can continue for several cycles until the protein is properly folded. Once this is achieved, the glycan is finally trimmed by ER mannosidase I (ERMan1) that removes the terminal mannose residue from the middle branch of the N-linked oligosaccharide. The resulting $Man_8GlcNAc_2$ structure then can be recognized by the ERGIC-53 (LMAN1), a mannose-specific lectin of the LMAN1/MCFD2 cargo receptor complex (Zheng et al. 2010), thereby facilitating packaging and transport of the native N-glycosylated glycoprotein into COPII (Coat Protein complex II)-coated vesicular carriers that ferry cargo from the ER to the Golgi via the ER-Golgi intermediate compartment (ERGIC) (Hanna et al. 2018; Peotter et al. 2019).

7 N-Glycan Biosynthesis: Basic Principles and Factors Affecting Its Outcome

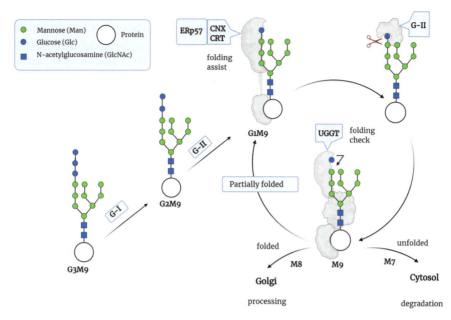

Fig. 7.2 A glycan-based quality control system in the ER that distinguishes correctly folded glycoproteins from unfolded or misfolded ones. *CNX* calnexin, *CRT* calreticulin, *G-I-II* α-glucosidases I and II

An ER stress caused by various factors (e.g., altered calcium homeostasis, redox state and glucose deprivation or mutations) is characterized by accumulation of misfolded or unassembled proteins in the ER and can be detrimental to cell viability. Metazoan cells can, however, cope with this stress by launching an ER stress response that suppresses the rate of translation and increases the expression of molecular chaperones to ease protein folding in the ER. In acase these maneuvers still fail, terminally misfolded glycoproteins will be directed to degradation via an ER-associated degradation (ERAD) pathway (Benyair et al. 2015). It starts when ERMan1 mannosidase and EDEM (ER degradation enhancing mannosidase-like) proteins (EDEM1/2/3 in mammals) are recruited to cleave off two mannose residues (instead of one) from an *N*-glycan. As a result, the glycan becomes unrecognizable by ERGIC-53, thereby preventing glycoprotein transport to later secretory compartments. The exposed α(1,6)-linked mannose in the $Man_7GlcNAc_2$ structure is now recognized by the OS-9/XTP3-B lectin complex that directs the bound glycoprotein to the transient ERAD (ER-associated degradation) protein complex at the ER membrane with ubiquitin ligase activity. ERAD complex then ensures that the glycoprotein is returned back to the cytoplasmic side of the ER membrane by tagging it for proteasomal degradation through ubiquitination (Benyair et al. 2015).

The other route for degrading of misfolded glycoproteins relies on malectin, a membrane-associated, ER stress-induced lectin first identified in 2008 (Schallus et al. 2008). It is highly conserved in metazoans (Yang et al. 2018) and it shows high specificity toward di-glucosylated N-glycans ($Glc_2Man_9GlcNAc_2$) (Schallus et al. 2008, 2010). Malectin forms a stable complex with ribophorin I, a subunit of the OST complex with proposed chaperone activity based on its ability to recognize misfolded protein backbones (Qin et al. 2012; Galli et al. 2011) This complex seems to act as an early intervention mechanism for detecting and capturing nascent non-native N-glycoproteins before it delivers them to proteasomal degradation if initial attempts to fold will fail (Stanley 2016). Whether this malectin-ribophorin I-mediated removal mechanism involves a unique retro-translocation machinery different from the ERAD machinery is currently unclear.

7.3 N-Glycan Processing in the Golgi Apparatus

Correctly folded glycoproteins entering *cis*-Golgi compartment carry typically an N-glycan with eight mannose residues left ($Man_8GlcNAc_2$). While some N-glycans may exit the Golgi without being modified, their proportion is normally low in humans (Lee et al. 2014). Partly, this is due to a presence of a quality control mechanism that is present in the Golgi. Golgi membranes harbor a mannose-binding lectin VIP36, that can recycle high mannose type N-glycans back to the ER (Lee et al. 2014). In support of this, VIP36 also interacts with the ER-localized BiP chaperone (Nawa et al. 2007). By doing so, VIP36 can halt the secretion of improperly glycosylated glycoproteins to post-Golgi compartments. Another mechanism to prevent high mannose type N-glycoproteins from passing through the Golgi takes over when a mono-glycosylated N-glycan ($Glc_1Man_9GlcNAc_2$) carrying glycoprotein arrives in the Golgi. The glycan part is cleaved internally by the Golgi endo-α-mannosidase between the two Man residues of the Glcα1–3Manα1–2-Manα1–2 moiety, thereby yielding a $Man_8GlcNAc_2$ isomer that is different from that produced by ERMan1 in the ER (Thompson et al. 2012). Interestingly, experimental evidence also suggests that the calreticulin-based glycoprotein quality control may be functional also in the Golgi compartment, as calreticulin was found to co-localize with endo-α-mannosidase in the ERGIC and *cis*/medial-Golgi compartment at least in cultured rat liver cells (Zuber et al. 2000).

Normally, the vast majority of N-glycans are processed in the Golgi to complex and/or hybrid type N-glycans by a distinct set of glycosidases and N-acetylglucosaminyltransferases, also termed as MGAT1–5 (Kellokumpu et al. 2016; Khoder-Agha et al. 2019a). The processing involves complex mutual interplay between the MGAT homomers and heteromers, mannosidase II (ManII) acting as a central hub (Khoder-Agha et al. 2019a). Thus, upon arriving in the Golgi, ER-derived MGAT homomers form heteromeric complexes not only with other MGATs but also with relevant UDP-N-acetylglucosamine transporters. Thereby, they organize into multienzyme/multi-transporter assemblies in the Golgi membranes. Their interplay likely involves either distinct or dynamic complexes (Khoder-Agha et al. 2019a) to facilitate efficient processing and branching of N-glycans in the *cis*- and medial-Golgi.

7 N-Glycan Biosynthesis: Basic Principles and Factors Affecting Its Outcome

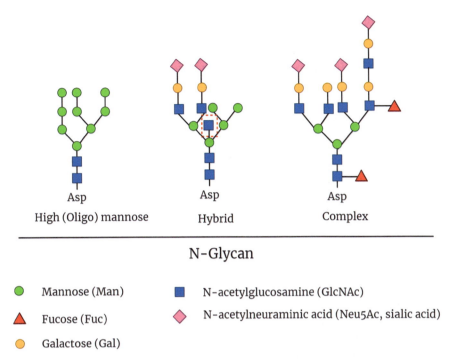

Fig. 7.3 Three examples depicting the main *N*-glycan types present on the cell surface in higher eukaryotes. High-mannose type *N*-glycan is characterized by having not undergone any processing in the Golgi compartment. Hybrid-type *N*-glycan typically has only one branch that has been processed in the Golgi. Bulky complex-type *N*-glycans in turn have two to five branches that are made and terminated in the medial and *trans*-Golgi cisternae

The processing begins in the *cis*-Golgi by the removal of three mannose residues to yield the Man$_5$GlcNAc$_2$ structure. Golgi mannosidases IA-C are responsible for the cleavages. Then, the first GlcNAc is added by MGAT1, using nucleotide-activated *N*-acetylglucosamine as a donor substrate. Once this GlcNAc is added, two additional mannose residues are removed by the Golgi α-mannosidase II. This creates a scaffold for MGAT2 to add a second GlcNAc to the exposed mannose residue, yielding a precursor for all complex-type *N*-glycans. MGAT4 and MGAT5 can then initiate the synthesis of third and fourth GlcNAc branches, respectively. Alternatively, MGAT3 can add a bisecting GlcNAc at the tri-mannosyl core structure (Fig. 7.3, middle). If this bisecting GlcNAc is added before MGAT4 and MGAT5 have added theirs, the synthesis of the third and fourth GlcNAc branches by MGAT4 and MGAT5 is halted (Kizuka and Taniguchi 2018). Bisecting GlcNAc also cannot be further elongated with any other sugar residue. Its addition also significantly alters the conformation of an *N*-glycan and suppresses the addition of terminal sugar residues such as sialic acid and fucose. The human natural killer-1 epitope (HSO3-3GlcAβ1-3Galβ1-4GlcNAc), a sulfated trisaccharide structure that is extensively expressed in the nervous system, is another terminal epitope suppressed by the bisecting GlcNAc (Nakano et al. 2019). Bisecting GlcNAc is also known to inhibit α-mannosidase II, suggesting that this addition may be one reason for the synthesis of hybrid type N-glycans. Yet, not all hybrid-type N-glycans contain a bisecting GlcNAc. It is, therefore, possible that rapid elongation of the first GlcNAc branch by galactose can also inhibit the necessary removal of two terminal mannoses by α-mannosidase II and thus, the build-up of the other GlcNAc branches.

Normally, the GlcNAc branches are further elongated by adding galactose, N-acetylglucosamine, and sialic acid. Galactose is added to the GlcNAC nearly always with the β(1,4)-linkage. This structure, termed N-acetyllactosamine (LacNAc), can be repeated several times in one branch, forming poly-LacNAc structures. Poly-LacNAc motifs in turn can act as substrates for making additional branches to the antennae. This is done by a set of special enzymes called GCNT2s A-C (Dimitroff 2019) by adding extra GlcNAc residues with the β(1,6)- linkage to internal galactose residues. These GlcNAc residues can also be subsequently elongated by β(1,4)-galactosyltransferases to form additional LacNAc structures. This kind of branched N-glycan is termed an I-branched glycan. They are most frequently found in adult erythrocytes, mucosal epithelia, and cells of the eye and olfactory bulb (Dimitroff 2019). GalNAc residues are also occasionally found in N-glycans of mammals forming LacdiNAc (GalNAcβ(1,4)GlcNAc) type structures.

The antennae are often capped with sialic acid by various sialyltransferases. This blocks further elongation of the branches except in the case of polysialylation. Polysialylated N-glycans are commonly detected in neural cell adhesion molecules (NCAMs) of the nervous system (Kiss and Rougon 1997). Fucose is another residue that cannot be elongated further. It can be added by specific Golgi fucosyltransferases either to the asparagine-linked GlcNAc to produce the "core fucosylated" N-glycan, or to GlcNAc residues of the antennae.

In specific cases, sugar residues of the antennae can undergo further modifications such as sulphation, phosphorylation, and O-acetylation (Klein and Roussel 1998; Wang et al. 2017). For example, lysosomal acid hydrolases carry N-glycans with a phosphate that directs the enzymes to lysosomes. Lysosomal enzymes share common conformational lysine-containing motifs that are recognized by the *cis*-Golgi-localized GlcNAc-1-phosphotransferase enzyme. In the first catalytic step, GlcNAc-1-phosphotransferase transfers GlcNAc-1-P from UDP-GlcNAc to the C6 hydroxyl group of selected mannose residue present in the high mannose-type N-glycan (Oh 2015; Qian et al. 2010). In the second step, N-acetylglucosamine-1-phosphodiester α-N-acetyl-glucosaminidase (NAGPA) cleaves the GlcNAc, leaving only the phosphate group linked to the mannose. Man-6-phosphate (M6P) tag is the ligand for transmembrane Man-6-P receptors (MPRs) residing in the *trans*-Golgi network (TGN). Once recognized by the MPR, the receptor escorts the lysosomal hydrolase with its ligand to endosomes and eventually to lysosomes in clathrin-coated vesicles. In lysosomes, the enzyme is released at low pH and the receptor is recycled back to the *trans*-Golgi. Mutations that impair tagging of mannose with phosphate lead to lysosomal storage diseases, a group of over 70 rare diseases characterized by accumulation of macromolecules in lysosomes (Xu et al. 2016).

7.3.1 N-*Glycosylation of Immunoglobulins*

N-glycosylation is also an important modification of all immunoglobulin isotypes and contributes affecting their binding characteristics and effector functions. Although their synthesis might not be different in any way from other *N*-glycans, there are some special issues that are worth discussing. *N*-glycans attached to immunoglobulin G (IgG) are best characterized owing to IgG abundance in the serum and successful production of many IgG-based therapeutic antibodies by the biopharma industry. IgG *N*-glycans are typically found in the Fc region but a minor proportion (15–25%) of serum IgG can contain *N*-glycans within their variable domains. These so-called "Fab glycans" differ from the Fc region *N*-glycans by having a higher proportion of terminally galactosylated and sialylated *N*-glycans with a bisecting GlcNAc, while having a lower abundance of core-fucosylated *N*-glycans (van de Bovenkamp et al. 2016). Yet, it is not clear why the number of antennae in IgG *N*-glycans seems to be limited to only two antennae (or three if the bisecting GlcNAc is considered also as an own branch). One possibility that may explain this is that antibody-producing plasma cells do not express the MGAT4 or MGAT5 enzymes needed for further branching. Another explanation could be that the addition of bisecting GlcNAc (or some other regulatory system) will prevent further branching of IgG *N*-glycans. The existence of such a system would be logical, given that an increase in *N*-glycan "bulkiness" brought about by additional branching might interfere with the folding and pairing of the Fc regions in the ER, and thereby alter its conformation known to be important for its binding to Fc receptors and antibody effector functions. Similarly, it is unclear why the Fab *N*-glycans display a higher proportion of more mature (more completely processed) *N*-glycans than those of Fc *N*-glycans. Whether this difference stems from better accessibility of the Fab glycans over Fc glycans, or something else such as increased extracellular glycosylation or decreased degradation of glycosidases, remains to be explored.

7.4 Golgi Microenvironment Is Important for Normal Processing and Maturation of *N*-Glycans

Despite the rather homogenous nature of high mannose type *N*-glycans arriving in the Golgi, *N*-glycans leaving the Golgi are much less so. For example, a single glycoprotein can carry complex, hybrid, and high-mannose *N*-glycans on the same polypeptide. Hybrid and complex type *N*-glycans can also display a variable number of antennae in their structure that may, or may not, carry sialic acid and/or fucose. While we do not have a clear picture at the molecular level of what determines the outcome in each case, this heterogeneity reflects both protein- and cell-specific processing of *N*-glycans brought about for example by epigenetic changes that determine what enzymes are expressed by the cell. Other factors that also modulate *N*-glycan biosynthesis are discussed below.

7.4.1 Golgi pH Homeostasis

The environmental cues outside or inside the cells can also contribute to *N*-glycan diversity. Unlike the ER, the other secretory pathway compartments, including the ER-Golgi intermediate compartment (ERGIC), the Golgi apparatus itself, and secretory vesicles, have uniquely acidic lumens, their pH decreasing along the pathway toward the plasma membrane (Paroutis et al. 2004). This pH gradient is crucial for their efficient functioning in membrane trafficking, glycosylation, proteolysis, protein sorting, or cargo transport (Kellokumpu 2019). Altered glycosylation due to abnormal Golgi pH is also responsible for several human disorders identified recently (Khosrowabadi and Kellokumpu 2020). Proper pH in the Golgi lumen appears to be important especially for the activity and assembly of the glycosyltransferase complexes in the Golgi. Previously, we have shown that the *trans*-Golgi β(1,4)GalT1 galactosyltransferase not only forms homomers in the ER but also heteromers with either ST3Gal3 or ST6Gal1 sialyltransferases upon its arrival in the Golgi (Hassinen et al. 2011; Hassinen and Kellokumpu 2014). Interestingly, these two heteromeric complexes assemble only in the acidic pH of the Golgi lumen (pH < 6.5). In the former case, complex formation could be prevented by increasing Golgi pH only by 0.2 pH units (Hassinen et al. 2011; Hassinen and Kellokumpu 2014). This increase was also sufficient to redirect the ST3Gal3 enzyme from the Golgi to post-Golgi compartments, consistent with their oligomerization-driven retention in the organelle (Rivinoja et al. 2009). The loss of the enzyme heteromers and enzyme mislocalization also coincided with reduced α(2,3)-sialylation and increased α(2,6)-sialylation of carcinoembryonic antigen (CEA) (Rivinoja et al. 2009). A similar decrease and increase in α(2,3)- and α(2,6)-sialylation, respectively, in CEA *N*-glycans has also been observed in cancer tissues in vivo (Kobata et al. 1995). Since Golgi resting pH is often elevated in cancer cells (Rivinoja et al. 2006), these findings suggest that Golgi resting pH may be used to regulate what linkage type will be used to link sialic acids to an *N*-glycan. This kind of switch from one linkage type to another one can have dramatic effects on cell behavior. For example, increased expression of α(2,6)-linked sialic acid in *N*-glycans can inhibit tumor cell apoptosis and activate growth factor pathways (Francisco et al. 2020; references therein).

Interestingly, the formation of the β(1,4)GalT1/ST6Gal heteromer was shown to increase markedly the catalytic activity of the β(1,4)GalT1 perhaps via substrate channeling (Hassinen et al. 2011). Alternatively, heteromer formation may also increase the accessibility of the donor or acceptor substrates to the active site of the β(1,4)GalT1, even though it is not directly involved in homodimer formation (Harrus et al. 2018). Yet, the active site is more exposed in the β(1,4)GalT1/ST6Gal heterodimers than it is in homodimers (Khoder-Agha et al. 2019b). In addition to 3D structures, this view is supported by the observation that a single mutation in the active site (H243) was able to abolish homodimer formation but not heterodimer formation.

Acidic Golgi resting pH is also needed the keep certain glycosyltransferases active. Accordingly, Golgi acidity (pH < 6.5) is essential for the full catalytic activity of ST6Gal1 sialyltransferase but not for β(1,4)GalT1, nor the MGATs (Hassinen et al. 2011). Partly, this can be explained by the pH-sensitive interactions between the β(1,4)GalT1 and the two sialyltransferases acting on *N*-glycans (Hassinen and Kellokumpu 2014). Yet, it is likely that pH-dependent conformational changes in the tertiary structure of ST6Gal1 also contribute to the activity loss if the Golgi resting pH is close to neutral. Collectively, these data suggest that the main role of the decreasing pH gradient from the *cis*-to-*trans* side of the Golgi compartments (pH 6.7–pH 6.3) is to orchestrate mutual interactions between glycosyltransferases, to promote their active conformation, and to get them correctly localized, in accord with their suggested oligomerization-mediated retention in the Golgi (Nilsson et al. 2009).

7.4.2 Golgi Ion Homeostasis

Golgi lumen contains high amounts of calcium, magnesium, and manganese ions (Van Baelen et al. 2004; Pizzo et al. 2010; Vangheluwe et al. 2009). The presence of these divalent cations is important for cargo concentration and sorting (Chanat and Huttner 1991) as well as for glycosylation (Vanoevelen et al. 2007). The cations are transported into the Golgi lumen by the SERCA2 and SPCA1/2 type Ca^{2+}/Mn^{2+} pumps. Of these two, SERCA2 is enriched in the *cis*-Golgi, while SPCA1 is mainly present in the *trans*-Golgi (Vangheluwe et al. 2009). Unlike SERCAs, SPCAs are also engaged in Mn^{2+} transport (Vangheluwe et al. 2009; Wong et al. 2013). In addition to SPCAs, recent evidence suggests that TMEM165 mutations in patients cause a type II congenital disorder of glycosylation in humans by interfering with Mn^{2+} and Ca^{2+}/H^{+} transport (Dulary et al. 2017; Thines et al. 2018). Manganese is an essential trace metal and important co-factor needed for the catalytic activity of many inverting Golgi glycosyltransferases such as β(1,4)GalT1. The DXD motif typically present in these enzymes plays a key role in Mn^{2+}-mediated donor substrate (UDP-Gal) binding (Breton et al. 2005). Based upon the recent structure of the β(1,4)GalT1 homodimer (Harrus et al. 2018), Mn^{2+} appears to regulate transitions of the lid and the "Trp loop" that define the open (inactive) and closed (active) states of the enzyme. Accordingly, the Met340H mutant form of the enzyme that binds Mn^{2+} 25 times more avidly, blocks the β(1,4)GalT1 in the closed state, inactivates the enzyme, and prevents its ability to form homodimers.

7.4.3 Golgi Redox State

Reactive oxygen species and low oxygen tension (hypoxia) also contribute to Golgi glycosylation potential. Most often, their effects are mediated by hypoxia-inducible factors (HIF1–3) that regulate the expression of a number of N-glycosylation-associated genes, including MGATs (MGAT2, MGAT-3, and MGAT5a and 5b), fucosyltransferases (FuT1, 2 and 7), sialyltransferases (ST3Gal1 and ST6Gal1) as well as nucleotide sugar transporters for UDP-galactose, CMP-sialic acid and UDP-N-acetylglycosamine (Koike et al. 2004; Shirato et al. 2010; Belo et al. 2015). Based upon these observations, Taniguchi et al. (Taniguchi et al. 2016) introduced the term "Glyco-redox" to link altered glycosylation with oxidative stress generated by hypoxia or reactive oxygen species (ROS). Their close association may also contribute to neurodegenerative disorders such as Parkinson's disease, Alzheimer's disease, and amyotrophic lateral sclerosis (ALS). Hypoxia (or HIFs) may also induce cleavage of cell surface N-linked glycans and thereby affect cell–extracellular matrix interactions (Taniguchi et al. 2016; Eguchi et al. 2002, 2005). Oxidative stress and altered glycosylation have also been linked to high-fat diet, obesity, and the onset of type II diabetes mellitus (Ohtsubo and Marth 2006; Ohtsubo 2010; Ohtsubo et al. 2011). Marth and co-workers showed in their studies (Ohtsubo 2010; Ohtsubo et al. 2011) that high levels of free fatty acids inhibit the expression of MGAT4a, a glycosyltransferase needed for $\beta(1,4)$-GlcNAc branching of N-glycans as well as GLUT-2 glucose transporter in pancreatic β-cells. The $\beta(1,4)$-GlcNAc branch is normally required for cell surface localization of the glucose transporter, and thus for glucose transport into cells. Without the $\beta(1,4)$-GlcNAc branch, the GLUT-2 remains intracellular, leading to decreased glucose import, insulin export, and accumulation of glucose in the blood.

Recent evidence indicates that hypoxia can modulate N-glycosylation also in a HIF-independent manner via affecting the oxidative potential of the Golgi lumen (Hassinen et al. 2019). Surprisingly, in normoxic conditions, it is higher than that of the ER (the main site of disulfide bond formation in the cells). In hypoxic cells, however, Golgi oxidative potential equals that of the ER in normoxic cells. The cells also displayed less sialic acid in their cell surface N-glycans. Interestingly, this was shown to be associated with reduced formation of surface-exposed disulfide bonds in ST6Gal1 (and likely also in some other sialyltransferases including ST3Gal3), loss of its catalytic activity, and inability to interact with $\beta(1,4)$GalT1 (Hassinen et al. 2019). Therefore, the high oxidative potential in the Golgi lumen appears to be necessary for the catalytic activity of certain sialyltransferases. This "redox switch" guarantees that the ST6Gal1 remains inactive until it reaches the Golgi compartment where it is expected to function. Likewise, the $\beta(1,4)$GalT1 enzyme acquires full activity also in the acidic Golgi compartment after interacting with the ST6Gal1 sialyltransferase.

7.5 Concluding Remarks

N-glycosylation is a frequent and complex modification of proteins, and essential for both uni- and multicellular life. It regulates a plethora of cellular functions that range from protein folding, trafficking, sorting, localization, half-life, and signaling to proliferation, migration, and adhesion with its surroundings. Therefore, it is also not surprising that we currently know a vast number of human disorders that are caused by, or are associated with, altered *N*-glycosylation. While previous work has provided us a clear overall picture of the basic principles in *N*-glycan biosynthesis, there is a big gap in our understanding of the factors that underlie cell-, tissue-, or organism-specific glycosylation patterns and their dynamic variability that starts during embryonic development continuing thereafter throughout our lives. We currently know that factors such as pH, redox potential, and changes in ion fluxes mainly in the Golgi compartment fundamentally affect and regulate the functioning and activity of glycosyltransferases expressed in a cell. Yet, there are many questions that remain unanswered. For example, how and why cells have evolved in such a complex way to make their *N*-glycans, needing removal of some sugar residues and replacing them with others instead of adding the right sugar in the beginning? Perhaps there is an evolutionary reason, as yeasts (an early eukaryote) produce mainly high mannose type *N*-glycans which needed to be modified to different ones in order to provoke immune responses only against them and other pathogens, and thereby survive. And what is the reason (or cause) of producing *N*-glycans which differ between two identical protein molecules? Does it result from a "sloppy" machinery that is prone to mistakes, or is there some purpose or benefit behind? Increasing biodiversity perhaps? Or is it a mark of ongoing evolution and trials to find the best fit for changing conditions? Is it regulated, or random? List is endless.

Nevertheless, these examples emphasize the need to understand in much more detail how glycans are made, how their synthesis is regulated and to what extent. An important issue also to keep in mind when one aims to produce optimally glycosylated antibodies for therapeutic use is to realize that yeasts, other lower eukaryotes or bacteria might not be the best choices to be used as hosts, as glycosylation is not just a simple outcome of enzymes present. It requires also conditions that support their full activity and complex mutual interplay necessary for their efficient functioning. Finally, we infer that there is an urgent need for developing more effective glycoengineering tools to edit glycans at will and thereby improve physicochemical and pharmacological properties of glycoprotein-based therapeutic compounds.

Acknowledgments We wish to acknowledge all the previous group members of the "glycan biosynthesis" group in Oulu University, as without their output during the years writing of this review would not have been possible.

Compliance with Ethical Standards

Conflict of Interest The authors declare that they have no conflict of interest.

Ethical Approval This chapter does not contain any studies with human participants or animals performed by any of the authors.

References

Angata T, Varki A (2002) Chemical diversity in the sialic acids and related α-keto acids: an evolutionary perspective. Chem Rev 102(2):439–470

Banda K, Gregg C, Chow R, Varki N, Varki A (2012) Metabolism of vertebrate amino sugars with N-glycolyl groups. J Biol Chem 287(34):28852–28864

Belo A, van Vliet S, Maus A, Laan L, Nauta T, Koolwijk P et al (2015) Hypoxia inducible factor 1α down regulates cell surface expression of α1,2-fucosylated glycans in human pancreatic adenocarcinoma cells. FEBS Lett 589(18):2359–2366

Benyair R, Ogen-Shtern N, Lederkremer G (2015) Glycan regulation of ER-associated degradation through compartmentalization. Semin Cell Dev Biol 41:99–109

Breton C, Šnajdrová L, Jeanneau C, Koča J, Imberty A (2005) Structures and mechanisms of glycosyltransferases. Glycobiology 16(2):29R–37R

Caramelo J, Castro O, Alonso L, de Prat-Gay G, Parodi A (2003) UDP-Glc:glycoprotein glucosyltransferase recognizes structured and solvent accessible hydrophobic patches in molten globule-like folding intermediates. Proc Natl Acad Sci 100(1):86–91

Chanat E, Huttner W (1991) Milieu-induced, selective aggregation of regulated secretory proteins in the trans-Golgi network. J Cell Biol 115(6):1505–1519

Dell A, Galadari A, Sastre F, Hitchen P (2010) Similarities and differences in the glycosylation mechanisms in prokaryotes and eukaryotes. Int J Microbiol 2010:1–14

Dimitroff C (2019) I-branched carbohydrates as emerging effectors of malignant progression. Proc Natl Acad Sci 116(28):13729–13737

Dulary E, Potelle S, Legrand D, Foulquier F (2017) TMEM165 deficiencies in congenital disorders of glycosylation type II (CDG-II): clues and evidences for roles of the protein in Golgi functions and ion homeostasis. Tissue Cell 49(2):150–156

Dunphy W (1985) Attachment of terminal N-acetylglucosamine to asparagine-linked oligosaccharides occurs in central cisternae of the Golgi stack. Cell 40(2):463–472

Eguchi H, Ikeda Y, Koyota S, Honke K, Suzuki K, Gutteridge J et al (2002) Oxidative damage due to copper ion and hydrogen peroxide induces GlcNAc-specific cleavage of an Asn-linked oligosaccharide. J Biochem 131(3):477–484

Eguchi H, Ikeda Y, Ookawara T, Koyota S, Fujiwara N, Honke K et al (2005) Modification of oligosaccharides by reactive oxygen species decreases sialyl Lewis x-mediated cell adhesion. Glycobiology 15(11):1094–1101

Francisco R, Pascoal C, Marques-da-Silva D, Brasil S, Pimentel-Santos F, Altassan R et al (2020) New insights into immunological involvement in congenital disorders of glycosylation (CDG) from a people-centric approach. J Clin Med 9(7):2092

Freeze H, Chong J, Bamshad M, Ng B (2014) Solving glycosylation disorders: fundamental approaches reveal complicated pathways. Am J Hum Genet 94(2):161–175

Galli C, Bernasconi R, Soldà T, Calanca V, Molinari M (2011) Malectin participates in a backup glycoprotein quality control pathway in the mammalian ER. PLoS One 6(1):e16304

Gao X (2004) Physical interactions between the Alg1, Alg2, and Alg11 mannosyltransferases of the endoplasmic reticulum. Glycobiology 14(6):559–570

Grinna L, Robbins P (1979) Glycoprotein biosynthesis. Rat liver microsomal glucosidases which process oligosaccharides. J Biol Chem 254(18):8814–8818

Hanna M, Peotter J, Frankel E, Audhya A (2018) Membrane transport at an organelle interface in the early secretory pathway: take your coat off and stay a while. BioEssays 40(7):1800004

Harrus D, Khoder-Agha F, Peltoniemi M, Hassinen A, Ruddock L, Kellokumpu S et al (2018) The dimeric structure of wild-type human glycosyltransferase B4GalT1. PLoS One 13(10): e0205571

Hassinen A, Kellokumpu S (2014) Organizational interplay of Golgi N-glycosyltransferases involves organelle microenvironment-dependent transitions between enzyme homo- and heteromers. J Biol Chem 289(39):26937–26948

Hassinen A, Pujol F, Kokkonen N, Pieters C, Kihlström M, Korhonen K et al (2011) Functional organization of Golgi N- and O-glycosylation pathways involves pH-dependent complex formation that is impaired in cancer cells. J Biol Chem 286(44):38329–38340

Hassinen A, Khoder-Agha F, Khosrowabadi E, Mennerich D, Harrus D, Noel M et al (2019) A Golgi-associated redox switch regulates catalytic activation and cooperative functioning of ST6Gal-I with B4GalT-I. Redox Biol 24:101182

Helenius J, Ng D, Marolda C, Walter P, Valvano M, Aebi M (2002) Translocation of lipid-linked oligosaccharides across the ER membrane requires Rft1 protein. Nature 415(6870):447–450. References

Janssen M, Waanders E, Woudenberg J, Lefeber D, Drenth J (2010) Congenital disorders of glycosylation in hepatology: the example of polycystic liver disease. J Hepatol 52(3):432–440

Kasturi L, Eshleman J, Wunner W, Shakin-Eshleman S (1995) The hydroxy amino acid in an Asn-X-Ser/Thr sequon can Influence N-linked core glycosylation efficiency and the level of expression of a cell surface glycoprotein. J Biol Chem 270(24):14756–14761

Kasturi L, Chen H, Shakin-Eshleman S (1997) Regulation of N-linked core glycosylation: use of a site-directed mutagenesis approach to identify Asn-Xaa-Ser/Thr sequons that are poor oligosaccharide acceptors. Biochem J 323(2):415–419

Kelleher D, Banerjee S, Cura A, Samuelson J, Gilmore R (2007) Dolichol-linked oligosaccharide selection by the oligosaccharyltransferase in protist and fungal organisms. J Cell Biol 177 (1):29–37

Kellokumpu S (2019) Golgi pH, ion and redox homeostasis: how much do they really matter? Front Cell Dev Biol 7:1–15

Kellokumpu S, Hassinen A, Glumoff T (2016) Glycosyltransferase complexes in eukaryotes: long-known, prevalent but still unrecognized. Cell Mol Life Sci 73(2):305–325

Khoder-Agha F, Sosicka P, Escriva Conde M, Hassinen A, Glumoff T, Olczak M, Kellokumpu S (2019a) N-acetylglucosaminyltransferases and nucleotide sugar transporters form multi-enzyme–multi-transporter assemblies in Golgi membranes in vivo. Cell Mol Life Sci 76:1821–1832

Khoder-Agha F, Harrus D, Brysbaert G, Lensink M, Harduin-Lepers A, Glumoff T et al (2019b) Assembly of B4GALT1/ST6GAL1 heteromers in the Golgi membranes involves lateral interactions via highly charged surface domains. J Biol Chem 294(39):14383–14393

Khosrowabadi E, Kellokumpu S (2020) Golgi pH and ion homeostasis in health and disease. Rev Physiol Biochem Pharmacol. https://doi.org/10.1007/112_2020_49

Kiss J, Rougon G (1997) Cell biology of polysialic acid. Curr Opin Neurobiol 7(5):640–646

Kizuka Y, Taniguchi N (2018) Neural functions of bisecting GlcNAc. Glycoconj J 35(4):345–351

Klein A, Roussel P (1998) O-acetylation of sialic acids. Biochimie 80(1):49–57

Kobata A, Matsuoka Y, Kuroki M, Yamashita K (1995) The sugar chain structures of carcinoembryonic antigens and related normal antigens. Pure Appl Chem 67(10):1689–1698

Koike T, Kimura N, Miyazaki K, Yabuta T, Kumamoto K, Takenoshita S et al (2004) Hypoxia induces adhesion molecules on cancer cells: a missing link between Warburg effect and induction of selectin-ligand carbohydrates. Proc Natl Acad Sci 101(21):8132–8137

Kornfeld R, Kornfeld S (1985) Assembly of asparagine-linked oligosaccharides. Annu Rev Biochem 54(1):631–664

Lee L, Lin C, Fanayan S, Packer N, Thaysen-Andersen M (2014) Differential site accessibility mechanistically explains subcellular-specific N-glycosylation determinants. Front Immunol 5:404

Lu J, Takahashi T, Ohoka A, Nakajima K, Hashimoto R, Miura N et al (2012) Alg14 organizes the formation of a multiglycosyltransferase complex involved in initiation of lipid-linked oligosaccharide biosynthesis. Glycobiology 22(4):504–516

Lu H, Fermaintt C, Cherepanova N, Gilmore R, Yan N, Lehrman M (2018) Mammalian STT3A/B oligosaccharyltransferases segregate N-glycosylation at the translocon from lipid-linked oligosaccharide hydrolysis. Proc Natl Acad Sci 115(38):9557–9562

Medus M, Gomez G, Zacchi L, Couto P, Labriola C, Labanda M et al (2017) N-glycosylation triggers a dual selection pressure in eukaryotic secretory proteins. Sci Rep 7(1):8788

Medzihradszky K, Kaasik K, Chalkley R (2015) Tissue-specific glycosylation at the glycopeptide level. Mol Cell Proteomics 14(8):2103–2110

Nakano M, Mishra S, Tokoro Y, Sato K, Nakajima K, Yamaguchi Y et al (2019) Bisecting GlcNAc is a general suppressor of terminal modification of N-glycan. Mol Cell Proteomics 18 (10):2044–2057

Nawa D, Shimada O, Kawasaki N, Matsumoto N, Yamamoto K (2007) Stable interaction of the cargo receptor VIP36 with molecular chaperone BiP. Glycobiology 17(9):913–921

Nilsson T, Au C, Bergeron J (2009) Sorting out glycosylation enzymes in the Golgi apparatus. FEBS Lett 583(23):3764–3769

Noffz C, Keppler-Ross S, Dean N (2009) Hetero-oligomeric interactions between early glycosyltransferases of the dolichol cycle. Glycobiology 19(5):472–478

Oh D (2015) Glyco-engineering strategies for the development of therapeutic enzymes with improved efficacy for the treatment of lysosomal storage diseases. BMB Rep 48(8):438–444

Ohtsubo K (2010) Targeted genetic inactivation of N-acetylglucosaminyltransferase-IVa impairs insulin secretion from pancreatic β cells and evokes type 2 diabetes. Methods Enzymol 479:205–222

Ohtsubo K, Marth J (2006) Glycosylation in cellular mechanisms of health and disease. Cell 126 (5):855–867

Ohtsubo K, Chen M, Olefsky J, Marth J (2011) Pathway to diabetes through attenuation of pancreatic beta cell glycosylation and glucose transport. Nat Med 17(9):1067–1075

Paroutis P, Touret N, Grinstein S (2004) The pH of the secretory pathway: measurement, determinants, and regulation. Physiology 19(4):207–215

Peotter J, Kasberg W, Pustova I, Audhya A (2019) COPII-mediated trafficking at the ER/ERGIC interface. Traffic 20(7):491–503

Pizzo P, Lissandron V, Pozzan T (2010) The trans-Golgi compartment. Commun Integr Biol 3 (5):462–464

Qian Y, Lee I, Lee W, Qian M, Kudo M, Canfield W et al (2010) Functions of the α, β, and γ subunits of UDP-GlcNAc:lysosomal enzymen-acetylglucosamine-1-phosphotransferase. J Biol Chem 285(5):3360–3370

Qin S, Hu D, Matsumoto K, Takeda K, Matsumoto N, Yamaguchi Y et al (2012) Malectin forms a complex with ribophorin I for enhanced association with misfolded glycoproteins. J Biol Chem 287(45):38080–38089

Rabouille C, Hui N, Hunte F, Kieckbusch R, Berger E, Warren G et al (1995) Mapping the distribution of Golgi enzymes involved in the construction of complex oligosaccharides. Cell Sci 108(Pt 4):1617–1627

Rao R, Wollenweber B (2010) Do N-glycoproteins have preference for specific sequons? Bioinformation 5(5):208–212

Reily C, Stewart TJ, Renfrow MB, Novak J (2019) Glycosylation in health and disease. Nat Rev Nephrol 15:346–366

Rivinoja A, Kokkonen N, Kellokumpu I, Kellokumpu S (2006) Elevated Golgi pH in breast and colorectal cancer cells correlates with the expression of oncofetal carbohydrate T-antigen. J Cell Physiol 208(1):167–174

Rivinoja A, Hassinen A, Kokkonen N, Kauppila A, Kellokumpu S (2009) Elevated Golgi pH impairs terminal N-glycosylation by inducing mislocalization of Golgi glycosyltransferases. J Cell Physiol 220(1):144–154

Ruddock L, Molinari M (2006) N-glycan processing in ER quality control. J Cell Sci 119 (21):4373–4380

Schallus T, Jaeckh C, Fehér K, Palma A, Liu Y, Simpson J et al (2008) Malectin: a novel carbohydrate-binding protein of the endoplasmic reticulum and a candidate player in the early steps of protein N-glycosylation. Mol Biol Cell 19(8):3404–3414

Schallus T, Fehér K, Sternberg U, Rybin V, Muhle-Goll C (2010) Analysis of the specific interactions between the lectin domain of malectin and diglucosides. Glycobiology 20 (8):1010–1020

Schulz BL (2012) Beyond the sequon: sites of N-glycosylation. In: Petrescu S (ed) Glycosylation. Intech, Rijeka, pp 21–40

Shirato K, Nakajima K, Korekane H, Takamatsu S, Gao C, Angata T et al (2010) Hypoxic regulation of glycosylation via the N-acetylglucosamine cycle. J Clin Biochem Nutr 48(1):20–25

Shrimal S, Ng B, Losfeld M, Gilmore R, Freeze H (2013a) Mutations in STT3A and STT3B cause two congenital disorders of glycosylation. Hum Mol Genet 22(22):4638–4645

Shrimal S, Trueman S, Gilmore R (2013b) Extreme C-terminal sites are post-translocationally glycosylated by the STT3B isoform of the OST. J Cell Biol 201(1):81–95

Spiro R (2002) Protein glycosylation: nature, distribution, enzymatic formation, and disease implications of glycopeptide bonds. Glycobiology 12(4):43R–56R

Stanley P (2016) N-linked glycans (N-glycans). In: Encyclopedia of cell biology. Elsevier Science, Waltham, MA, pp 339–346

Taniguchi N, Kizuka Y, Takamatsu S, Miyoshi E, Gao C, Suzuki K et al (2016) Glyco-redox, a link between oxidative stress and changes of glycans: lessons from research on glutathione, reactive oxygen and nitrogen species to glycobiology. Arch Biochem Biophys 595:72–80

Thines L, Deschamps A, Sengottaiyan P, Savel O, Stribny J, Morsomme P (2018) The yeast protein Gdt1p transports Mn2+ions and thereby regulates manganese homeostasis in the Golgi. J Biol Chem 293(21):8048–8055

Thompson A, Williams R, Hakki Z, Alonzi D, Wennekes T, Gloster T et al (2012) Structural and mechanistic insight into N-glycan processing by endo-α-mannosidase. Proc Natl Acad Sci 109(3):781–786

Van Baelen K, Dode L, Vanoevelen J, Callewaert G, De Smedt H, Missiaen L et al (2004) The Ca2+/Mn2+ pumps in the Golgi apparatus. Biochim Biophys Acta Mol Cell Res 1742(1–3):103–112

van de Bovenkamp F, Hafkenscheid L, Rispens T, Rombouts Y (2016) The emerging importance of IgG fab glycosylation in immunity. J Immunol 196(4):1435–1441

Vangheluwe P, Sepúlveda M, Missiaen L, Raeymaekers L, Wuytack F, Vanoevelen J (2009) Intracellular Ca2+- and Mn2+-transport ATPases. Chem Rev 109(10):4733–4759

Vanoevelen J, Dode L, Raeymaekers L, Wuytack F, Missiaen L (2007) Diseases involving the Golgi calcium pump. Subcell Biochem 45:385–404

Varki A, Gagneux P (2017) Essentials of glycobiology: biological functions of glycans, 3rd edn. Cold Spring Harbor Laboratory Press, Cold Spring Harbor, NY

Wang J, Gao W, Grimm R, Jiang S, Liang Y, Ye H et al (2017) A method to identify trace sulfated IgG N-glycans as biomarkers for rheumatoid arthritis. Nat Commun 8(1):631

Wong A, Capitanio P, Lissandron V, Bortolozzi M, Pozzan T, Pizzo P (2013) Heterogeneity of Ca2+ handling among and within Golgi compartments. J Mol Cell Biol 5(4):266–276

Xu M, Motabar O, Ferrer M, Marugan J, Zheng W, Ottinger E (2016) Disease models for the development of therapies for lysosomal storage diseases. Ann N Y Acad Sci 1371(1):15–29

Yang Q, Fu M, Gao H, Yamamoto K, Hu D, Qin S (2018) Subcellular distribution of endogenous malectin under rest and stress conditions is regulated by ribophorin I. Glycobiology 28(6):374–381

Zapun A, Petrescu S, Rudd P, Dwek R, Thomas D, Bergeron J (1997) Conformation-independent binding of monoglucosylated ribonuclease B to calnexin. Cell 88(1):29–38

Zheng C, Liu H, Yuan S, Zhou J, Zhang B (2010) Molecular basis of LMAN1 in coordinating LMAN1-MCFD2 cargo receptor formation and ER-to-Golgi transport of FV/FVIII. Blood 116(25):5698–5706

Zielinska D, Gnad F, Wiśniewski J, Mann M (2010) Precision mapping of an in vivo N-glycoproteome reveals rigid topological and sequence constraints. Cell 141(5):897–907

Zuber C, Spiro M, Guhl B, Spiro R, Roth J (2000) Golgi apparatus immunolocalization of endomannosidase suggests post-endoplasmic reticulum glucose trimming: implications for quality control. Mol Biol Cell 11(12):4227–4240

Chapter 8
Genetic Regulation of Immunoglobulin G Glycosylation

Azra Frkatovic, Olga O. Zaytseva, and Lucija Klaric

Contents

8.1	Introduction	260
8.2	Heritability of the Human IgG *N*-Glycome	261
8.3	Linkage Studies of Mouse *N*-glycome	264
8.4	Genome-Wide Association Studies of Human *N*-Glycome	265
	8.4.1 Genomic Loci Associated with IgG *N*-Glycosylation	268
	8.4.2 Suggestive Associations	276
	8.4.3 Functional Network of Loci Associated with IgG Glycosylation	278
8.5	Pleiotropy with Complex Traits and Diseases	280
8.6	Conclusions	281
References		282

Abstract Defining the genetic components that control glycosylation of the human immunoglobulin G (IgG) is an ongoing effort, which has so far been addressed by means of heritability, linkage and genome-wide association studies (GWAS). Unlike the synthesis of proteins, *N*-glycosylation biosynthesis is not a template-driven process, but rather a complex process regulated by both genetic and environmental factors. Current heritability studies have shown that while up to 75% of the variation in levels of some IgG glycan traits can be explained by genetics, some glycan traits are completely defined by environmental influences. Advances in both high-throughput genotyping and glycan quantification methods have enabled genome-wide association studies that are increasingly used to estimate associations of millions of single-nucleotide polymorphisms and glycosylation traits. Using this method, 18 genomic regions have so far been robustly associated with IgG *N*-glycosylation, discovering associations with genes encoding glycosyltransferases, but also transcription factors, co-factors, membrane transporters and other genes

---

A. Frkatovic · O. O. Zaytseva
Glycoscience Research Laboratory, Genos Ltd., Zagreb, Croatia

L. Klaric (✉)
MRC Human Genetics Unit, Institute of Genetics and Cancer, University of Edinburgh, Edinburgh, UK
e-mail: lklaric@ed.ac.uk

with no apparent role in IgG glycosylation. Further computational analyses have shown that IgG glycosylation is likely to be regulated through the expression of glycosyltransferases, but have also for the first time suggested which transcription factors are involved in the process. Moreover, it was also shown that IgG glycosylation and inflammatory diseases share common underlying causal genetic variants, suggesting that studying genetic regulation of IgG glycosylation helps not only to better understand this complex process but can also contribute to understanding why glycans are changed in disease. However, further studies are needed to unravel whether changes in IgG glycosylation are causing these diseases or the changes in the glycome are caused by the disease.

Keywords Immunoglobulin G · *N*-glycosylation · Genome-wide association study · Heritability

Abbreviations

CC	Collaborative Cross
CD4	Helper T-cells
CD8	Cytotoxic T-cells
CD14	Macrophages
CD15	Neutrophils
CD19	B-cells
DZ	Dizygotic
ER	Endoplasmic reticulum
GWAS	Genome-wide association study
HDL	High-density lipoprotein
IgG	Immunoglobulin G
LD	Linkage disequilibrium
MZ	Monozygotic
SNP	Single nucleotide polymorphism

8.1 Introduction

N-glycans attached to the immunoglobulin G (IgG) molecules influence the effector functions of IgG (Ferrara et al. 2011; Dekkers et al. 2017; Banda et al. 2008), reviewed in Gudelj et al. (2018), antibody half-life in the bloodstream, antigen-binding, auto-reactivity, and immune complex formation (van de Bovenkamp et al. 2016). Although *N*-glycan biosynthesis is, to a certain degree, stochastic and not template-driven, it is not a random assembly of all possible sugar residues. It is rather a complex but well-governed process that results in specific *N*-glycan profiles, characteristic for secreted IgG, produced by a specific cell type, namely, plasma cells. Defining the genetic component that controls this complex process is one of the essential aims of glycobiology, which will provide new insights in the mechanisms of a number of pathologies where IgG is involved, as well as improve our understanding of humoral immune functions in general.

The biosynthesis of the conserved *N*-glycan core begins in the endoplasmic reticulum and continues through *cis*- and medial Golgi, while the majority of features characteristic for IgG glycans, like fucosylation and antennary modifications, are created in the *trans*-Golgi (Varki et al. 2017). Regulation of *N*-glycosylation is a complex process, dependent on the expression and activity of glycosyltransferases (GTs) (Klaric et al. 2020; Nairn et al. 2008), competition between GTs for their substrates, GT transport and localisation within Golgi, membrane properties, pH and ionic composition of Golgi apparatus (Kellokumpu 2019), availability and trafficking of sugar donors, as well as on the properties of the glycosylated protein, such as amino acid sequence and folding [reviewed in Pothukuchi et al. (2019)]. Naturally, expression and activity of all protein components involved in the glycosylation process are also controlled at transcriptional and post-transcriptional levels. Therefore, a diverse and complex network of interactions taking place on all possible cellular levels defines the types of *N*-glycans produced by a cell and attached to specific protein cargos. The resulting structural spectrum and corresponding quantities of glycans produced by a certain cell type or found on a certain glycoprotein are usually referred to as *N*-glycome.

The *N*-glycome of human IgG is distinct from those of other antibody classes (Clerc et al. 2016), and mainly includes complex-type biantennary *N*-glycans, with or without core fucose and bisecting *N*-acetylglucosamine (GlcNAc), with almost no highly-branched structures (Pučić et al. 2011). Both or only one antenna can be decorated with galactose residues, on top of which an *N*-acetylneuraminic acid residue can be added. A small amount of high-mannose *N*-glycans is also present in the IgG *N*-glycome. IgG *N*-glycosylation is known to be perturbed in diseases, ageing (Gudelj et al. 2018) and pregnancy (Bondt et al. 2014). Thus, one may regard these changing glycan features as potential biomarkers of health and age (Gudelj et al. 2018; Krištić et al. 2014; Vanhooren et al. 2007, 2009; Vilaj et al. 2019). A study of global variation of IgG glycome in human populations (Stambuk et al. 2020) showed that IgG galactosylation correlates with socioeconomic development indicators of the country of residence, probably reflecting exposure to pathogens, lifestyle, healthcare quality, and, perhaps, genetic differences between human populations. Evidence suggests that complex interactions between environment and genetic sequence have a vast impact on glycan biosynthesis, resulting in immediate glycan change or creating lasting modifications that are maintained through epigenetic mechanisms (Zoldos et al. 2012).

8.2 Heritability of the Human IgG *N*-Glycome

The main measurable parameter that gives us information on the relative contributions of the genes and environment to the variation of a certain phenotype—for instance, IgG *N*-glycome composition—is heritability. Heritability, in a broad sense, is defined as the proportion of the observed phenotypic variance that can be attributed to variance of the genotype (Visscher et al. 2008). However, what the studies of heritability most commonly report, is the so-called "narrow-sense heritability", estimated from empirical data as a ratio of the variance of additive genetic effects to the total observed variance of a phenotype of interest (Visscher et al. 2008). Heritability estimates provide a better understanding of how exactly genes and environment shape a certain phenotype, allow to make meaningful comparisons between populations and are essential prerequisites for gene-mapping studies that can pinpoint genes responsible for the phenotype of interest and discover new biomarkers of disease. There are multiple technical approaches to heritability estimation, and the choice of a particular method depends on the available observations and the purpose of the study (Visscher et al. 2008). One of the earliest analyses of IgG *N*-glycome heritability was performed in a cohort of 906 individuals from a Croatian island of Vis using genealogical information (Pučić et al. 2011). In general, heritability of *N*-glycan levels varied from insignificant to over 50% in the case of sialylation-related traits (Fig. 8.1). This approach to estimating heritability, however, does not account for the possible contribution of a shared environment between related individuals, and therefore, some of the heritability estimates might be inflated.

A classic approach to study relative contributions of genetic and environmental factors to human traits is twin studies (Tan et al. 2015). Usually, this kind of

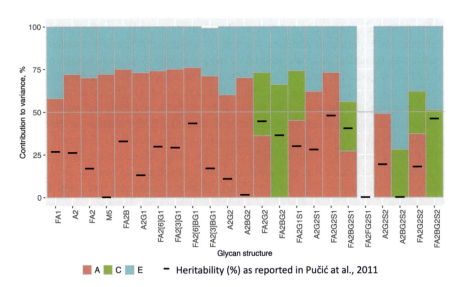

Fig. 8.1 Heritability estimates and partitioning of variance for IgG *N*-glycans based on data published in Pučić et al. (2011) and Menni et al. (2013). Variance components, as estimated by Menni et al using twins study approach: A, additive genetic variance (heritability); C, common environment variance; E, unique environment variance. X-axis: IgG glycosylation traits; y-axis: contributions of A, C, and E components to the total variance of glycosylation traits (%). Black lines represent heritability estimates by Pučić et al. based on pedigrees (genealogical information)

methodology employs a cohort of di- (DZ) and monozygotic (MZ) twins. MZ twins are genetically identical and DZ twins share approximately a half of their genes, so when a phenotype has significantly higher concordance in MZ twins, than in DZ, we can assume that this phenotype is largely influenced by the genes. At the same time, both DZ and MZ twins are exposed to similar environments, including uterine environment, parenting, education, family lifestyle, and quality of life, so the heritability estimates based on twin studies can also be somewhat inflated. However, when a phenotype has a similar concordance rate between MZ and DZ twins, it is supposed to be under the influence of the environmental factors shared by the twins. If a phenotype is largely discordant between both MZ and DZ twins, it is assumed to be controlled by environmental factors that are unique for an individual, for instance, personal lifestyle, life choices, accidents, etc. Modern methods of mathematical modelling using maximum likelihood approaches allow to decompose the variance observed in a quantitative trait measured in MZ and DZ twins into genetic, shared, and unique environmental components (Tan et al. 2015). Such an approach was applied to study the heritability of IgG N-glycosylation features (Menni et al. 2013). In this study, N-glycans were enzymatically released from the total pool of IgG isolated from plasma of 220 MZ and 310 DZ twin pairs and chromatographically separated into 24 peaks, each containing one major N-glycan structure. Heritability was estimated for relative abundances of 22 individual N-glycan structures found on IgG and for 54 derived traits that describe relative abundances of N-glycans that share structural features (e.g. all glycans with core fucose). For more than half of all N-glycan traits studied, 51 of the 76, at least 50% of the observed variance was explained by genetic factors. In total, only three individual glycans and four derived traits were not influenced by genetic factors at all. It is difficult to directly compare the results of this study with heritability estimates based on genealogy (Pučić et al. 2011). However, one can note that some of the sialylation traits, that were found to be highly heritable in the genealogical study, were not as much controlled by genetics according to the twin analysis and instead were under the considerable influence of the shared environment, which was not accounted for in Pučić et al. (2011). Relative abundances of most neutral glycan structures were found to be highly heritable (heritability >50%) and almost not affected by shared environment, while for individual sialylated structures the degree of estimated heritability varied from 0 to over 70% (Fig. 8.1).

Such discrepancies between heritability estimates of the same phenotype in different studies are not uncommon and are bound to happen when heritability is estimated with different methods in different populations (Tenesa and Haley 2013). One might ask a logical question—which estimate is then closer to the truth? Unfortunately, there is no clear answer to such a question. Estimated heritability of the same phenotypic trait can differ between populations because of the differences in frequencies of certain mutations, or because variability of a trait changes depending on the environment. An interaction between genes and environment is an important factor when the studied phenotype is concerned, so that in particular environments a phenotype may become more or less heritable (Visscher et al. 2008). It is widely accepted that even though it is hard to generalise heritability estimates obtained in a single study (Tenesa and Haley 2013; Moore and Shenk 2017), nevertheless, the estimates obtained in different studies give a researcher a general idea of how relatively important are nature and nurture for the studied trait. For instance, in the case of IgG *N*-glycan traits, the findings published in Pučić et al. (2011) and Menni et al. (2013) can provide the lower and the upper limits of possible heritability of *N*-glycan abundances in the IgG *N*-glycome.

However, these methods give no information on what kind of genes and how many of them drive the observed variability of the IgG *N*-glycome composition. This question can be answered by the means of genetic association studies that are specifically designed to test if a specific region of a chromosome or a specific mutation in the genome could be controlling a certain trait.

8.3 Linkage Studies of Mouse *N*-glycome

Genetic association studies are performed to find out if a certain genetic locus could be influencing a phenotype of interest. This kind of study uses the information on how common polymorphisms, such as point mutations, insertions, and deletions are distributed in the population. The linkage disequilibrium (LD) mapping approach uses the fact that some groups of genetic variants situated within the same recombination block on a chromosome are more likely to be inherited together. Such groups of alleles are called haplotypes. By testing in a population whether a qualitative phenotype or the value of a quantitative phenotype, is correlated with the presence of a certain known haplotype, linkage analysis can identify chromosomal regions harbouring the genes influencing the trait of interest. It is assumed, that even if a causal mutation is not genotyped in the studied population, it will be tagged by some common genetic variants in LD. Further inspection of the identified locus can reveal the truly causal variants influencing the studied phenotype.

A linkage study directed at IgG *N*-glycosylation regulation was performed in the Collaborative Cross (CC) mouse cohort, by quantifying IgG *N*-glycans in 589 mice of 95 inbred strains (Krištić et al. 2018). The CC resource was specifically developed for mapping genes that underlie complex phenotypes in mice, commonly used as model organisms (Morahan et al. 2008). It provides a population of inbred mouse strains, which are derived from crosses of eight founder strains. Thus, the genomes of the CC inbred strains result from recombination of the eight founder genomes, and each strain has a unique mosaic pattern of inheritance of the genetic variants originating from the founders. The position of chromosomal blocks inherited from any of the eight founders in each of the CC strains was defined by genotyping. By correlating abundances of certain *N*-glycans in the IgG *N*-glycomes with the presence of genome regions derived from certain founder strains one can tell which parts of the mouse genome are harbouring candidate genetic variants that affect IgG *N*-glycosylation. Krištić et al. identified at least five genetic regions in the mouse genome that are potentially involved in the regulation of 11 individual *N*-glycan structures. The sizes of the chromosomal blocks inherited from founder strains in this case spanned several megabases and could contain dozens of genes and their promoter regions, which made prioritisation of candidate genes an important task. Krištić et al. focused primarily on the genes in the loci of interest that contained missense mutations specific for the founder alleles that correlated with differing abundances of glycan structures. For instance, the levels of bisected *N*-glycans were associated with missense genetic variants in the *Mgat3* gene, encoding β-1,4-mannosyl-glycoprotein-4-β-*N*-acetylglucosaminyltransferase (GnT-III or MGAT3) (Taniguchi et al. 2014), an enzyme responsible for the addition of bisecting GlcNAc. In general, the percentage of bisected structures in the IgG *N*-glycome of most commonly used laboratory mouse strains is rather low, around 1–3% (de Haan et al. 2017; Zaytseva et al. 2018), so that in some studies it was not even reported (Blomme et al. 2011). However, CC mice that inherited an *Mgat3* allele from either of the two outbred CC founder strains, CAST or PWK, exhibited a significantly higher incidence of bisection in their IgG *N*-glycomes. Other proposed candidate genes, supposedly affecting digalactosylated and monosialylated glycans, were *Ighg1*, *Ighg2*, and *Ighg2c*, encoding the constant regions of immunoglobulin heavy gamma chains 1, 2b, and 2c, respectively. These genes are a part of a cluster of tightly linked genes, coding for constant regions of Ig heavy chains and are mainly inherited together, so it was not possible to attribute the differences in IgG *N*-glycosylation to any of these genes specifically. Studies of IgG subclass-specific *N*-glycomes, however, show that a relatively rare IgG1 allotype encoded by an *Ighg1* allele found in C57BL/6, CD1 and NOD mice is characterised by lower sialylation and galactosylation of *N*-glycans attached to the CH2 domain of IgG. This finding is in line with the studies showing that point mutations leading to amino acid substitutions in the heavy chain of IgG can result in alterations of IgG *N*-glycome (Lund et al. 1996; Rose et al. 2013).

8.4 Genome-Wide Association Studies of Human *N*-Glycome

Linkage analyses are a great tool for mapping which genomic loci have an effect on a phenotype, but suffer from a low mapping resolution—given the low number of recombination events per meiosis, associated regions covered many genes, requiring further analyses to pinpoint the causal gene. At the beginning of the twenty-first century, the Human Genome Project (Lander et al. 2001) released the first sequence of the human genome, which, together with the quantification of LD structure from the HapMap project (Belmont et al. 2005), facilitated the development of dense commercial Single Nucleotide Polymorphism (SNP) arrays. These arrays captured most of the common genetic variation and enabled affordable genotyping of hundreds of thousands of SNPs in thousands of individuals and were a start of the genome-wide association study era. Genome-wide association studies (GWAS) are high-resolution association scans across the whole genome, where each SNP in the genome is tested for association with a phenotype. As a result, these analyses "zoom in" from associations with several megabases long regions in the linkage studies, to individual SNPs. The availability of reliable high-throughput methods for glycan analysis has enabled GWAS to help further elucidate the genetic background of IgG glycosylation by testing the association of the glycan levels with SNPs measured in a large group of individuals.

The first GWAS of IgG *N*-glycome was performed by Lauc et al. (2013), testing for genome-wide associations with 77 IgG *N*-glycome traits measured by ultra-performance liquid chromatography (UPLC) in 2247 individuals of European descent. Nine genomic regions (also called loci) were found to be significant on the genome-wide level (association p-value $\leq 5 \times 10^{-8}$). Four of these regions contained genes encoding glycosyltransferases (*ST6GAL1*, *B4GALT1*, *MGAT3*, *FUT8*)—enzymes catalysing the addition of sugar units to the growing glycan chain. Although not having an apparent role in glycosylation, five additional loci were discovered containing the following genes: *IKZF1*, *IL6ST-ANKRD55*, *SUV420H1*, *ABCF2-SMARCD3* and *SMARCB1-DERL3*. Statistically significant associations of glycan traits with variants at glycosyltransferase loci gave the first indication that GWA study for IgG glycans could identify genes relevant for glycosylation, as well as identify novel genes that have a potentially important, yet unknown role in this process.

The next attempt to discover additional genomic loci associated with IgG glycosylation applied a multivariate approach to GWAS of 23 directly measured IgG *N*-glycosylation phenotypes measured by UPLC in the ORCADES (Orkney Complex Disease Study; $N = 1960$) (Shen et al. 2017). In contrast to the more commonly used univariate approach, where an SNP is associated with one phenotype at a time, multivariate approaches jointly analyse the association of multiple phenotypes with an SNP (Solovieff et al. 2013). This approach takes into account the correlation structure among glycan traits, which might be due to shared genetic regulation, and provides greater power to detect associations (Solovieff et al. 2013), resulting in the discovery of more associated loci compared to a univariate study of the same sample size. In Shen et al., individual *N*-glycan traits were grouped to describe certain features of IgG *N*-glycome, such as fucosylation, bisection, galactosylation, etc., and association tests were performed jointly using all glycan traits from a specific group. This resulted in replication of five loci previously identified in Lauc et al. (2013), but also in the discovery of five novel loci: *IGH*, *ELL2*, *HLA-B-C*, *FUT6-FUT3*, *AZI1, and TMEM105* (Shen et al. 2017). These newly discovered loci were not detected in the univariate analysis, further confirming and addressing the need for multivariate approaches in complex trait analysis.

In contrast to the GWA studies by Lauc et al. (2013) and Shen et al. (2017), the GWAS by Wahl et al. (2018) was performed on liquid chromatography-electrospray mass spectrometry (LC–ESI-MS) measurements in KORA F4 (Cooperative Health Research in the Region of Augsburg; $N = 1836$). LC-ESI-MS glycoprofiling provides glycan measurements specific for IgG subclasses. With the new data type, the study replicated six of the previously known loci and discovered one novel locus on chromosome 1, containing the *RUNX3* gene, but, more importantly, highlighted that the major difference between genetic regulation of *N*-glycomes of different IgG subclasses lies in the regulation of bisection and fucosylation.

The outcomes of these GWA studies indicated the polygenic nature of IgG *N*-glycosylation, where apart from glycosyltransferase genes that have a big effect on glycosylation, many other genes with smaller effects are also involved in the process. This, in turn, suggests that many associated genes cannot be detected in small sample sizes due to low statistical power. Therefore, a GWAS of 77 IgG *N*-glycan traits measured by UPLC was performed by combining summary statistics from four European cohorts with a total sample size of 8090 (Klaric et al. 2020). As the result of this genome-wide meta-analysis, 13 of the previously known loci were replicated and 14 novel loci were identified. Since genomic regions associated with glycan traits can span multiple genes, it is often a challenge to point to the actual causal genes involved in IgG glycosylation. While the previous studies focused mainly on the pre-existing knowledge on biological functions of the genes to suggest the most plausible candidate, this study applied several computational approaches to refine the list of IgG glycosylation-related genes. For obvious reasons, candidate genes that are known to be involved in protein glycosylation, such as enzymes directly involved in the process, were prioritised first. Next were genes with non-synonymous variants associated with IgG *N*-glycosylation since implied changes in amino acid sequence could potentially impact protein structure and function, thus influencing IgG glycosylation. Genetic variants can often affect multiple traits, a phenomenon known as pleiotropy. In the next prioritisation approach, the candidate variants were tested for pleiotropy with gene expression, i.e. if the same variant could be affecting both *N*-glycosylation of IgG and expression of any genes in the tissues relevant for IgG production or functions. Last but not least, candidate associations were subjected to pathway enrichment analysis (also referred to in this text as gene-set enrichment analyses). This type of gene prioritisation uses the information on which genes belong to certain biological pathways based on their co-expression in different cells and tissues or/and on well-known and characterised biological pathways. Then the method allows to test if the current list of genes is statistically significantly overrepresented in some specific pathways and decide if certain candidate genes are more relevant to the *N*-glycosylation of IgG than the others. When no other evidence was available for any of the genes in the locus, the gene closest to the strongest association signal was reported (also called positional mapping). The overall results of these prioritisation efforts can be seen in Fig. 8.2, while Table 8.1 lists gene associations for each glycan trait. Details for gene-prioritisation for each genomic locus, organised by chromosome, can be seen in Fig. 8.2 and paragraphs that follow.

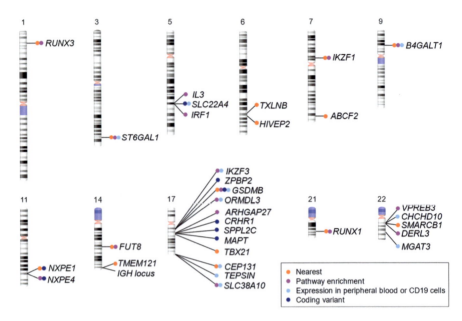

Fig. 8.2 Chromosomal location of currently known associations with IgG glycosylation across human genome. The annotated genes were prioritised based on the approaches labelled with coloured dots: being the nearest to the strongest association in the region (orange), pathway enrichment (purple), pleiotropy with gene expression in peripheral blood or CD19 B cells (light blue), and presence of missense mutations (dark blue)

8.4.1 Genomic Loci Associated with IgG N-Glycosylation

8.4.1.1 Chromosome 1

The locus on chromosome 1 identified in GWAS by Wahl et al. (2018) and later replicated in Klaric et al. (2020) contains a single gene *runt-related transcription factor 3* (*RUNX3*), which codes for a transcription factor with an important role in a range of developmental processes including haematopoiesis (Voon et al. 2015). Evidence suggests that *RUNX3* might play a role in B cell maturation (Whiteman and Farrell 2006) and T cell differentiation, particularly into CD4+ helper and CD8+ cytotoxic cells (Overgaard et al. 2015; Steinke et al. 2014). Previously, it was shown that IgG1 glycosylation depends on B cell stimuli, including T cell-derived cytokines and metabolites (Wang et al. 2011), hence Wahl et al. (2018) suggested that *RUNX3* could indirectly affect the glycosylation of antibodies by influencing T cell differentiation.

8 Genetic Regulation of Immunoglobulin G Glycosylation

Table 8.1 Detailed list of associations of IgG glycans and loci across human genome based on Klaric et al. (2020)

Genomic Locus	Prioritised Gene	A2*	FA2	FA2B	A2G1	FA2[6]G1	FA2[3]G1	FA2[6]BG1	FA2[3]BG1	A2G2	A2BG2	FA2G2	FA2BG2	FA2G1S1	A2G2S1	FA2G2S1	FA2BG2S1	FA2FG2S1	A2BG2S2	FA2G2S2	FA2BG2S2	FBG2* / FG2*	FG2S1/FG2+FG2S1+FG2S (I)
1:25226001-25345011	RUNX3																						
3:186712711-186739421	ST6GAL1									■	■		■		■	■			■	■	■		
5:131028218-131835599	IRF1; IL3; SLC22A4	■																					
6:139617560-139696003	TXLNB																						
6:143150223-143203591	HIVEP2													■					■				
7:50325717-50351683	IKZF1				■																		
7:150906463-150969535	ABCF2											■			■								
9:33041761-33186080	B4GALT1		■																				
11:114328627-114450529	NXPE1; NXPE4																					■	
14:65472801-66284081	FUT8				■																		
14:105866019-106002352	TMEM121; IGH	■									■		■										
17:37903731-38112190	ORMDL3; GSDMB; IKZF3; ZPBP2	■										■											
17:43463492-44690083	CRHR1; SPPL2C; MAPT; ARHGAP27							■	■														
17:45516683-45674272	TBX21									■													
17:79165171-79257880	SLC38A10; CEP131; TEPSIN		■	■				■	■				■										
21:36546759-36665202	RUNX1																						
22:24093789-24182500	SMARCB1; CHCHD10; VPREB3												■										
22:38774449-39860368	MGAT3														■	■				■			

8.4.1.2 Chromosome 3

The strongest association from all four conducted GWA studies was observed in the region on chromosome 3 harbouring *β-galactoside-α-2,6-sialyltransferase 1* (*ST6GAL1*) gene. It codes for a sialyltransferase involved in the addition of α2,6-linked sialic acid residues to Galβ1,4-GlcNAc structures in *N*-linked glycans of glycoproteins (Weinstein et al. 1982; Dall'Olio 2000), thus making it the most plausible candidate gene in the region.

8.4.1.3 Chromosome 5

One of the genomic regions on chromosome 5 associated with the percentage of agalactosylated (G0) glycans in total IgG glycans (Klaric et al. 2020) spans 800 kb and 11 genes: *SLC22A4, IRF1, IL3, SLC22A5, CSF2, FNIP1, P4HA2, P4HA2-AS1, MEIKIN, PDLIM4,* and *ACSL6*. In this locus, gene-set enrichment analysis prioritised *interferon regulatory factor 1* (*IRF1*) and *interleukin 3* (*IL3*) genes. *IRF1* gene codes for a transcriptional regulator of T cell differentiation and is involved in cell proliferation, DNA damage response and apoptosis, as well as the activation of genes in innate and adaptive immune response (Oshima et al. 2004; Fragale et al. 2008). *IL3* gene codes for interleukin 3, a cytokine with an important function in the differentiation and proliferation of haematopoietic and lymphoid cell lineages. Pleiotropy analysis between gene expression and IgG glycosylation also suggested *solute carrier family 22 member 4* (*SLC22A4*), gene encoding a sodium-ion dependent plasma membrane transporter of organic cations (Tokuhiro et al. 2003) as a candidate gene in this locus. In addition, glycosylation-associated variant in *SLC22A4* is a missense variant resulting in leucine to phenylalanine change (rs1050152:Leu>Phe) but was predicted to have no influence on the protein structure. However, obtained results and previous indications for functions of *SLC22A4*, *IRF1* and *IL3* genes could not provide sufficient evidence to prioritise one gene over the others in the given locus.

8.4.1.4 Chromosome 6

The first locus on chromosome 6 contains a single gene, *taxilin beta* (*TXLNB*). *TXLNB* gene is expressed in skeletal muscle and heart tissues and has a role in myogenesis (Sakane et al. 2016). No additional evidence other than positional mapping was found to explain and support the relationship of this gene with IgG glycosylation events.

Another significant association was found in the locus harbouring *human immunodeficiency virus type I enhancer-binding protein 2* (*HIVEP2*) gene, which encodes a transcription factor involved in the regulation of gene expression during brain development (Takagi et al. 2006). *HIVEP2* knock-out in mice has shown to upregulate the gene expression of nuclear factor kappa-light-chain-enhancer of activated B Cells (NF-κB) and cause inflammation in several brain areas, implicating a likely causal relationship between immune response and neurodevelopmental disorders (Takao et al. 2013; Choi et al. 2015).

8.4.1.5 Chromosome 7

A genomic region on chromosome 7 harbouring *IKAROS family zinc finger 1* (*IKZF1*) gene was among the five discovered loci by Lauc et al. (2013), with no previously implicated roles in protein glycosylation. The association with this region was later replicated both by Wahl et al. (2018) and Klaric et al. (2020). *IKZF1* gene codes for DNA-binding protein Ikaros, responsible for the regulation of lymphocyte differentiation and has been previously implicated to influence effector pathways in the humoral immune response by controlling class switching recombination (Sellars et al. 2009). As shown in Lauc et al. (2013), the *IKZF1* knock-out mice exhibit somewhat lower levels of core fucosylation and higher levels of glycans with bisecting GlcNAc, implying that *IKZF1* is potentially involved in the regulation of fucosylation by promoting the addition of bisecting GlcNAc (Lauc et al. 2013). Klaric et al. further hypothesised that *IKZF1* and *IKAROS family zinc finger 3* (*IKZF3*) genes play a role in regulation of expression of *fucosyltransferase 8* (*FUT8*) gene and performed a functional validation in a B cell derived lymphoblastoid cell line, MATAT6. Knock-down of *IKZF1* resulted in downregulated expression of *IKZF3* gene and significant up-regulation of *FUT8* expression, thus resulting in increased levels of fucosylated structures (Klaric et al. 2020). The experiment provided direct evidence of *IKZF1* role in regulation of *FUT8* gene expression and regulation of IgG glycosylation.

The region on chromosome 7 containing *SWI/SNF-related, matrix-associated, actin-dependent regulator of chromatin, subfamily D, member 3* (*SMARCD3*) and *ATP binding cassette subfamily F member 2* (*ABCF2*) genes was also indicated as genome-wide significant in GWAS by Lauc et al. (2013) and replicated in a subsequent study by Klaric et al. (2020). The *ABCF2* gene codes for the ATP-binding cassette, sub-family F, member 2 protein which has a potential function in transmembrane transport of molecules (Liu et al. 2016). *SMARCD3* is involved in the SWI/SNF chromatin remodelling complex which is likely to regulate transcription of genes by altering the chromatin structure and exposing genes regulatory regions to transcriptional machinery (Wang et al. 1996). Expression of *ABCF2* was detected in both peripheral blood and immune cells, as well as expression of *SMARCD3* in the CD15 neutrophil cells, but there was no evidence for pleiotropy with IgG *N*-glycosylation (Klaric et al. 2020).

8.4.1.6 Chromosome 9

The genomic locus identified on chromosome 9 contains *beta-1,4-galactosyltransferase 1* (B4GALT1), gene coding for an enzyme catalysing the transfer of galactose from UDP-galactose to the GlcNAc residue in the nonreducing end of N-linked glycans, resulting in the disaccharide moiety LacNAc with a β1–4-glycosidic linkage (Hennet 2002). In a study by Lauc et al., variants from this region were mainly associated with differences in sialylation and percentage of bisection rather than galactosylation measures (Lauc et al. 2013). However, associations with these traits were still considered plausible since galactosylation is a prerequisite for the addition of a sialic acid residue, further supported by the evidence that enzymes adding galactose and bisecting GlcNAc units compete for the same substrate (Fukuta et al. 2000). In Klaric et al., *B4GALT1* was linked to immune system processes and variants identified in the locus were pleiotropic with gene expression of *B4GALT1* in CD4 helper T cells and CD19 B cells (Klaric et al. 2020), suggesting that expression of this gene in relevant cells influences IgG glycosylation.

8.4.1.7 Chromosome 11

A region on chromosome 11 contains two genes encoding members of neurexophilin and PC-esterase domain family: *NXPE1* and *NXPE4*. The role of *NXPE4* was indicated in regulation of colorectal cancer development and progression, likely functioning as a tumour suppressor (Liu et al. 2019). Both genes contain a missense variant associated with IgG glycans but given the weak LD between the two SNPs, it might be an indication for two separate association signals, each affecting a different set of IgG glycan traits. While no evidence of pleiotropy with gene expression was found, *NXPE4* was prioritised for being a member of pathways enriched for IgG glycosylation genes (Klaric et al. 2020).

8.4.1.8 Chromosome 14

Associations with variants in the region on chromosome 14, harbouring *FUT8* gene were discovered and replicated in all four GWA studies (Klaric et al. 2020; Lauc et al. 2013; Shen et al. 2017; Wahl et al. 2018). *FUT8* gene codes for fucosyltransferase 8, an enzyme responsible for the transfer of fucose from GDP-fucose to the innermost GlcNAc residue of *N*-linked glycans, resulting in α1,6-fucose residue (Taniguchi et al. 2014). Previously described and well-understood function of *FUT8* explains its involvement in IgG glycosylation and clearly points to *FUT8* as the causal gene in the region. It is interesting to note that all associations with this locus are with afucosylated glycans (Table 8.1). While perhaps contra intuitive, this might suggest that availability of the substrate (afucosylated glycan) is a rate-limiting step in fucosylation from the perspective of genetic regulation.

The second region on chromosome 14, discovered both in the multivariate IgG *N*-glycome GWAS (Shen et al. 2017) and Klaric et al. (2020) contains several genes. *Transmembrane protein 121* (*TMEM121*) has no function described to date, while *immunoglobulin heavy locus* (*IGH*) spans genes encoding heavy chains of immunoglobulins, including IgG heavy chain genes (*IGHG*). Although the direct relationship between *IGH* locus and glycosylation is not yet understood, previous research showed that the amino-acid sequence of the constant region of the IgG heavy chain has an impact on the *N*-glycome composition (Lund et al. 1996; Rose et al. 2013). In addition, a QTL study of IgG *N*-glycome regulation in mice proposed genes coding for the constant region of IgG heavy chain as candidate genes influencing sialylation and galactosylation of IgG (Krištić et al. 2018).

8.4.1.9 Chromosome 17

The first locus on chromosome 17 harbours four genes: *ORMDL sphingolipid biosynthesis regulator 3* (*ORMDL3*), *gasdermin B* (*GSDMB*), *IKAROS family zinc finger 3* (*IKZF3*) and *zona pellucida binding protein 2* (*ZPBP2*) (Klaric et al. 2020). Both *GSDMB* and *ZPBP2* genes contain missense SNPs, rs2305479, resulting in a potentially damaging change from glycine to arginine in GSDMB, and rs11557467, resulting in a tolerated change from serine to isoleucine in ZPBP2. Pleiotropy between glycosylation and gene expression was observed for *GSDMB* and *ORMDL3 genes* in CD19 B cells, *ORMDL3* in CD4 helper and CD8 cytotoxic T cells and expression of *IKZF3* in peripheral blood. All three genes, *GSDMB*, *ORMDL3* and *IKZF3*, were also prioritised by the pathway enrichment analysis, making them all potential gene candidates in the region (Klaric et al. 2020). GSDMB is a member of a protein family likely having a role in pyroptosis (highly inflammatory form of programmed cell death), thereby triggering a strong inflammatory response (Ding et al. 2016). *ORMDL3* gene was shown to modulate T-lymphocyte activation via controlling the influx of calcium ions in the endoplasmic reticulum upon antigen binding (Carreras-Sureda et al. 2013). IKZF3 is a member of a family of haematopoietic-specific transcription factors that play a critical role in regulating B cell and T cell development (Rebollo and Schmitt 2003). In addition, *ORMDL3-GSDMB-IKZF3-ZPBP2* region has previously been identified as a susceptibility locus for childhood-onset asthma (Moffatt et al. 2007), inflammatory bowel disease (Anderson et al. 2011), type 1 diabetes (Saleh et al. 2011), and rheumatoid arthritis (Stahl et al. 2010), suggesting that future insights into the role of these genes in IgG glycosylation could also increase our understanding of the disease development and progression.

The second region on chromosome 17 spans 1.5 Mb and harbours 12 genes, for some of which evidence for prioritisation was found. Pathway enrichment analysis prioritised *Rho GTPase activating protein 27* (*ARHGAP27*) (Klaric et al. 2020), a gene that encodes a member of a protein family that activates Rho-type GTP metabolising enzymes and whose function is implicated in mitogen-activated protein kinase signalling (Julià et al. 2018). Three genes, *corticotropin-releasing hormone receptor 1* (*CRHR1*), *signal peptide peptidase like 2C* (*SPPL2C*) and *microtubule-associated protein tau* (*MAPT*), contain a non-synonymous variant associated with glycosylation (rs16940665, stop lost variant in *CRHR1*; rs12185233, potentially damaging variant resulting in a change from Arginine to Proline in *SPPL2C*; rs754512, stop gained variant in *MAPT*) (Klaric et al. 2020). *CRHR1* gene encodes a receptor that binds the corticotrophin-releasing hormone family, which plays various roles in stress response but also in inflammatory processes (Zhu et al. 2011). Intramembrane-cleaving aspartic protease encoded by the *SPPL2C* gene was shown to participate in vesicular transport and possibly cause retention of cargo proteins in the ER, thereby affecting cellular processes, such as protein glycosylation, through miss-localisation of glycan-modifying enzymes (Papadopoulou et al. 2019). *MAPT* gene codes for the tau protein which is expressed throughout the central nervous system and was previously implicated in neurodegeneration (Strang et al. 2019). Strong LD in the region and lack of evidence for pleiotropy with gene expression make it hard to point to a single gene candidate.

T-Box 21 (*TBX21*) gene is located closest to the strongest signal in the third region on chromosome 17 (Klaric et al. 2020). TBX21 is a member of the T-box family of transcription factors expressed in multiple cells of the adaptive and innate immune system, with an important role in development, survival, and activation of the cells in immune response (Lazarevic et al. 2013). The region contains expression quantitative trait loci (eQTL) weakly associated with the expression of *TBX21* in CD4 helper T cells (Klaric et al. 2020). Additional four genes can be found in the region: *EF-hand calcium-binding domain 13* (*EFCAB13*), *aminopeptidase puromycin sensitive* (*NPEPPS*), *karyopherin subunit beta 1* (*KPNB1*), and *TBK1 binding protein 1* (*TBKBP1*). However, insufficient evidence for their prioritisation positions *TBX21* as the top gene candidate for this locus.

Four genes, *NADH:ubiquinone oxidoreductase complex assembly factor 8* (*NDUFAF8*), *solute carrier family 38 member 10* (*SLC38A10*), *adaptor related protein complex 4 accessory protein* (*TEPSIN*) and *centrosomal protein 131* (*CEP131*), are found in the locus on chromosome 17 first discovered by Shen et al. (2017) and replicated in Klaric et al. (2020). Evidence for pleiotropy of gene expression and IgG glycosylation was found for the following: expression of *SLC38A10* in macrophages, expression of *TEPSIN* in neutrophils and peripheral blood, and expression of *CEP131* in peripheral blood (Klaric et al. 2020). *CEP131* gene is involved in the formation of cilia and flagellum but is also shown to be an important regulator of genome stability (Hall et al. 2013; Staples et al. 2012). *CEP131* gene is located closest to the strongest association in the region, while pathway enrichment analysis prioritised *SLC38A10* (Klaric et al. 2020). *SLC38A10* gene encodes a member of the solute carrier (SLC) family-38 of transporters with a suggested role in signalling pathways regulating protein synthesis (Tripathi et al. 2019). Given that none of the genes in the region have functions linking them to glycosylation or immunity in general, all of them still remain biologically plausible target genes for IgG *N*-glycosylation.

8.4.1.10 Chromosome 21

The region on chromosome 21 is located downstream from *runt-related transcription factor 1* (*RUNX1*) gene, which codes for a transcription factor important for embryonic development, tumorigenesis, and inflammatory response (Scheitz and Tumbar 2013) and with a crucial role in haematopoiesis (Okuda et al. 1996). Additionally, due to existing evidence of interaction with RUNX3 (Spender et al. 2005) and role in pathways enriched for IgG glycosylation, it is considered a biologically plausible candidate.

8.4.1.11 Chromosome 22

SWI/SNF related, matrix associated, actin-dependent regulator of chromatin, subfamily B, member 1 (*SMARCB1*) is one of six genes found in a locus on chromosome 22. SMARCB1 gene encodes a core subunit of the SWI/sucrose non-fermenting ATP-dependent chromatin remodelling complex (SWI/SNF), which plays a key role in the regulation of gene transcription (Kalimuthu and Chetty 2016). SMARCB1 also has an immune system-related function as its role was described in inhibition of antiviral activity, neurodevelopment, tumour formation, cell differentiation, and proliferation (Pottier et al. 2007). Another gene in the region, *derlin 3* (*DERL3*), codes for a component of endoplasmic reticulum-associated degradation for misfolded luminal glycoproteins (Oda et al. 2006). Expression of *coiled-coil-helix-coiled-coil-helix domain containing 10* (*CHCHD10*), mitochondrial protein gene in peripheral blood was shown to be pleiotropic with IgG glycosylation. *V-set pre-b cell surrogate light chain 3* (*VPREB3*) gene encodes a protein involved in B lymphocyte maturation (Melchers 2005) and was prioritised for being a member of pathways enriched for IgG glycosylation. Due to the implications for the interaction of SWI/SNF complex with *RUNX1* gene in controlling expression of the genes in haematopoietic cell line (Bakshi et al. 2010), *SMARCB1* together with *VPREB3*, *CHCHD10*, and *DERL3* genes remains a candidate gene for IgG N-glycosylation.

The second locus on chromosome 22 harbours three genes including *TGF-beta activated kinase 1* (*MAP3K7*) *binding protein 1* (*TAB1*), *synaptogyrin 1* (*SYNGR1*), and gene encoding the enzyme mannosyl (β-1,4-)-glycoprotein β-1,4-N-acetylglucosaminyltransferase (*MGAT3*) (Klaric et al. 2020). The *SYNGR1* gene encodes an integral membrane protein expressed throughout the nervous system and is likely involved in traffic through the plasma membrane (Stenius et al. 1995; Baumert et al. 1990). Expression of *SYNGR1* (in CD8 cytotoxic T cells and peripheral blood) and *MGAT3* (in CD19 B cells) is pleiotropic with IgG glycosylation (Klaric et al. 2020); however, the most plausible candidate gene in this region is *MGAT3* since the encoded N-acetylglucosaminyltransferase III enzyme is known to catalyse the addition of GlcNAc to the core β-mannose residue of N-glycans by β1,4-linkage (Taniguchi et al. 2014).

8.4.2 Suggestive Associations

Genome-wide association studies are a powerful tool for detecting genetic associations, but, as all statistical tests, suffer from false-positive associations. Given that in each GWA study millions of association tests are performed (for each variant in the genome), using the classical p-value threshold of 0.05 would just by chance result in tens of thousands of associations classified as significant, even when the null hypothesis of no association is true. To mitigate this problem, in a typical GWA study the significance threshold is set to 5×10^{-8}, correcting the "classical" 0.05 threshold for the number of independent tests performed, based on an assumption that there are roughly 1,000,000 independent SNPs in the genome (0.05/ 1,000,000 = 5×10^{-8}). This does not, however, protect fully from false-positive associations, it merely reduces their number. To further validate the findings from these studies, an additional step is usually performed—repeating the GWAS, on samples collected in cohorts that were not used in the discovery analysis (so-called replication GWAS). Since there is a delicate balance between having a discovery study with as high statistical power as possible, but still leaving some samples aside for replicating the findings, replication GWAS are often performed on several orders of magnitude fewer samples, resulting in a reduced power in replication study. In previous paragraphs, we reported only robust, replicated findings. However, since replication studies are usually underpowered, some of the non-replicated findings might still be true associations. For the interest of readers, we are reporting these findings in the following paragraphs, but stress that these should be considered only for exploratory purposes.

Variants in chromosome 5 region were located in *interleukin 6 signal transducer (IL6ST)* (Lauc et al. 2013), a gene which codes for signalling receptor subunit shared by multiple cytokines such as ciliary neurotrophic factor (CNTF), interleukin 6 (IL6), oncostatin M (OSM), and leukaemia inhibitory factor (LIF) (Rose-John 2018). Variants in *IL6ST* were previously linked to rheumatoid arthritis and multiple myeloma susceptibility, as well as components of metabolic syndrome (Stahl et al. 2010; Birmann et al. 2009; Gottardo et al. 2008). However, it must be stressed that this genomic locus was not significant in Klaric et al. (2020). While its biological function might seem relevant for IgG glycosylation, there is no robust evidence for the involvement of *IL6ST* in IgG glycosylation.

The second interval on chromosome 5, spanning *elongation factor for RNA polymerase II (ELL2)* gene was one of the newly discovered associations in multivariate GWAS (Shen et al. 2017) and then further confirmed, but not replicated, in the bigger univariate GWAS (Klaric et al. 2020). *ELL2* gene encodes a transcription elongation factor with a role in controlling expression of membrane-associated immunoglobulin by B cells and secretion of immunoglobulin by plasma cells (Santos et al. 2011; Benson et al. 2012). *ELL2* is involved in the regulation of the processing of mRNA transcribed from *IGH* locus, which contains genes coding for heavy chains of immunoglobulins (Martincic et al. 2009). Interestingly, *IGH* locus was also significantly associated with glycan phenotypes in the same study, indicating the potential biological link between *ELL2* and IgG glycosylation.

First discovered in the multivariate GWAS and rediscovered in the biggest IgG N-glycome GWAS, *HLA (human leukocyte antigen)* locus on chromosome 6 is an expected association due to the direct role of this gene-dense region in innate and adaptive immunity. It spans more than a hundred genes that have been associated with multiple autoimmune diseases (Shiina et al. 2009). However, due to the complexity of the region, with an extended range of high linkage disequilibrium covering many genes, it is challenging to point to specific variants and genes influencing IgG glycosylation.

A region on chromosome 7 harbours two genes: *diacylglycerol lipase beta (DAGLB)* and *KDEL endoplasmic reticulum protein retention receptor 2 (KDELR2)* (Klaric et al. 2020). *DAGLB* gene codes for a serine hydrolase enzyme involved in proinflammatory signalling in neuroinflammation, with enriched activity in macrophages. *KDELR2* encodes a member of the KDEL receptor family with a role in retrieving ER-resident proteins from the Golgi apparatus back to ER (Capitani and Sallese 2009). Proximity to the strongest glycan-SNP association in the region prioritised the *DAGLB* gene (Klaric et al. 2020) but given the lack of replication of the locus in other cohorts, its role in IgG glycosylation should be considered with caution.

Outer dense fiber of sperm tails 1 *(ODF1)* gene is found in the region on chromosome 8 (Klaric et al. 2020). *ODF1* gene encodes a protein located in outer dense fiber in mammalian sperm tail with a role in spermatogenesis, however, its exact function is not yet fully understood (Amaral et al. 2013). There is no current evidence linking it to the IgG glycosylation process or immune response in general.

A variant on chromosome 11 found in *SUV420H1-CHKA* locus was associated with FA1 glycans in the first IgG N-glycome GWAS (Lauc et al. 2013). Histone-Lysine N-methyltransferase *(SUV420H1)* gene encodes an enzyme that trimethylates lysine 20 of histone H4, thereby affecting the activity of various genes (Schotta et al. 2004) and, via epigenetic mechanisms, might be involved in proviral silencing in germline and somatic cells (Matsui et al. 2010). *Choline kinase alpha (CHKA)* gene is known to be involved in phospholipid biosynthesis and tumour cell growth (Aoyama et al. 2004). Existing evidence of strong associations between glycomics and lipidomics favours *CHKA* as a plausible candidate gene from this region on chromosome 11 since lipid environment is considered to affect the activity of glycosyltransferases (Igl et al. 2011). However, it must be stressed that this association was not confirmed in Klaric et al. (2020) and is therefore not considered robust.

A region on chromosome 16 contains five genes including *component of oligomeric Golgi complex 7 (COG7), Golgi associated, gamma adaptin ear containing, ARF binding protein 2 (GGA2), glutamyl-tRNA synthetase 2, mitochondrial (EARS2), NADH:ubiquinone oxidoreductase subunit AB1 (NDUFAB1),* and *ubiquitin family domain containing 1 (UBDF1)* genes (Klaric et al. 2020). *COG7* gene encodes a member of the Golgi-localised protein complex essential for proper protein N-glycosylation. Mutations in *COG7* were indicated in patients with congenital disorders of glycosylation (Wu et al. 2004). On the other hand, *GGA2* gene codes for a member of a protein family with a role in protein trafficking between the *trans*-Golgi network and the lysosome (Hirst et al. 2000). *GGA2* gene was prioritised by pathway enrichment analysis, while the expression of *COG7* in peripheral blood was pleiotropic with IgG glycosylation (Klaric et al. 2020). Given their localisation in the Golgi apparatus, where glycosylation mainly takes place, and prioritisation evidence, both genes remain plausible gene candidates in the region, but have not been replicated in other cohorts.

Mitochondrial genome maintenance exonuclease 1 (*MGME1*) *gene* is found in the locus on chromosome 20 (Klaric et al. 2020). *MGME1* encodes a mitochondrial nuclease which plays an important role in the maintenance and correct metabolism of mitochondrial DNA (El-Hattab et al. 2017). However, besides this gene being the closest to the IgG glycosylation associated SNPs (Klaric et al. 2020), there is no additional evidence for how *the MGME1* gene might affect IgG glycosylation.

Multivariate GWAS approach for IgG glycome resulted in the discovery of a region on chromosome 19 containing the *FUT3-5-6* gene cluster (Shen et al. 2017). The gene cluster codes for fucosyltransferases, enzymes involved in transfer of fucose from donor to acceptor molecules with a role in determining Lewis blood groups (Taniguchi et al. 2014). Furthermore, it was shown that these genes catalyse the addition of antennary fucose (Ma et al. 2006). *Fucosyltransferase 6* (*FUT6*) was previously associated with glycosylation of plasma proteins (Lauc et al. 2010; Sharapov et al. 2019) and was rediscovered in Klaric et al. (2020) as being associated with GP20, a structure that was recently reported to have antennary fucose (Russell et al. 2017), thereby suggesting the potential functional effect of this gene on IgG glycosylation.

Another associated region on chromosome 19 includes *regulatory factor X associated ankyrin-containing protein* (*RFXANK*) gene (Klaric et al. 2020), coding for a subunit of RFX complex, trimeric transcription factor interacting with MHCII promoters and activating the gene expression (Ting and Trowsdale 2002). According to the pleiotropy analysis, the same underlying variant in the locus influences IgG glycosylation and expression of the RFXANK gene in peripheral blood (Klaric et al. 2020). Together with its involvement in regulation of expression of immune system relevant genes, this evidence supports *RFXANK* as a gene candidate for IgG glycosylation, but as all other associations reported in this section, needs to be treated with caution for lack of replication.

8.4.3 Functional Network of Loci Associated with IgG Glycosylation

To understand mechanisms underlying glycosylation of IgG, the natural next step is to ask in which biological pathways are these genes enriched and how the non-glycosylation genes influence glycosylation of IgG.

To put these findings in a biological context, Klaric et al. (2020) applied Data-driven Expression Prioritised Integration for Complex Traits (Pers et al. 2015), a method that tests for enrichment of the list of genes in predefined pathways. The main advantage of this method is that pathways are not only based on known molecular and biochemical pathways, but also integrate information from gene co-expression, mouse gene knock-out studies and protein–protein interaction databases. While there were no statistically significantly enriched pathways below a strict significance threshold, a more relaxed threshold allowing for 20% of false positives uncovered many relevant pathways, with the strongest evidence for enrichment of IgG glycosylation genes in B cell differentiation, activation, and proliferation, abnormal immune system physiology, immunoglobulin production, production of molecular mediators of immune response, thymus hyperplasia, abnormal Peyer's patch morphology (germinal centres responsible for differentiation of B cells in intestine) (Spencer and Sollid 2016), and protein–protein interaction subnetworks of two transcription regulators, serum response factor (SRF) and inhibitor of DNA binding 2 (ID2) (Klaric et al. 2020). This method also provides an insight into tissue enrichment, highlighting that IgG glycosylation-related genes are mostly expressed in haemic and immune system tissues, primarily B-lymphocytes and antibody-producing cells (Klaric et al. 2020).

Genes with the strongest influence on IgG glycosylation were known glycosyltransferase enzymes from the known biological pathway, namely fucosyltransferase FUT8, sialyltransferase ST6GAL1, galactosyltransferase B4GALT1, and MGAT3, an enzyme that produces the bisecting GlcNAc. This finding was not surprising and, in a way, provides a proof of principle that genome-wide association studies can be used to pinpoint genes that regulate IgG glycosylation. The next step was to propose how the other, non-glycosylation genes influence IgG glycosylation through their direct or indirect effect on glycosyltransferases. As we have seen in the previous sections, associated SNPs within glycosyltransferase genes are non-coding and are likely to regulate their expression. With an assumption that a glycosyltransferase and any gene that has an influence on the enzyme (either by controlling its expression, controlling substrate availability, or a different mechanism) will have a similar effect on all glycan traits associated with that enzyme, Klaric et al. (2020) created a functional network of IgG glycosylation and proposed which genes are more likely to influence which glycosyltransferase.

The strongest evidence was found for regulation of expression of *MGAT3* by transcription factors RUNX1 and RUNX3, and chromatin remodelling protein SMARCB1. The network analysis revealed that SNPs from *SMARCB1*, *RUNX1*, *RUNX3* and *MGAT3* loci have strikingly similar effects on IgG glycosylation, suggesting that all of these genes together affect bisecting of IgG *N*-glycans. The strongest associated variant within the *MGAT3* locus is within a binding site of RUNX3 transcription factor. While there was no publicly available information on

RUNX1 binding sites at the time of the study, the two proteins, RUNX1 and RUNX3 are homologous and bind for the same DNA motif. In addition, there was evidence of SMARCB1 and RUNX1 interacting in immortalised T-lymphocytes (Bakshi et al. 2010). Altogether, this suggests that transcription factors RUNX1 and RUNX3, together with the chromatin remodelling protein SMARCB1 could be regulating transcription of *MGAT3*.

The second strongest evidence from the network analysis for non-glycosylation genes influencing expression of glycosyltransferases was found for regulation of expression of *FUT8* by transcription factors IKZF1 and IKZF3. A functional study by Klaric et al. (2020) has shown that knock-down of *IKZF1* in a B cell-derived lymphoblastoid cell line, MATAT6, down-regulates expression of *IKZF3* and upregulates *FUT8* expression, which results in increased IgG fucosylation.

Less evidence and less clear mechanistic insight were found for the remaining two glycosyltransferases, *ST6GAL1* and *B4GALT1*. HIVEP2 was suggested to regulate expression of galactosyltransferase *B4GALT1* and *ELL2* and *NXPE1-NXPE4* were suggested to regulate gene expression of sialyltransferase *ST6GAL1*. However, there is no clear mechanistic explanation for either of the two.

8.5 Pleiotropy with Complex Traits and Diseases

IgG glycans are long known to be changed in various diseases and physiological changes, from ageing to autoimmune, inflammatory, and infectious diseases to cancer, being both biomarkers of the disease and biomarkers of its progression [reviewed in Gudelj et al. (2018)]. Indeed, one of the earliest findings related to IgG glycosylation was that of aberrant galactosylation in patients with rheumatoid arthritis (Parekh et al. 1985), followed by extensive research into its effects in ageing (Vanhooren et al. 2007, 2010; Ruhaak et al. 2011; Krištić et al. 2014). While there is increasing evidence that glycans may be involved in every major disease (National Research Council 2012), the mechanisms behind these changes still remain unclear. Genome-wide association studies are also a useful tool in this regard—finding genetic variants that influence both glycosylation and disease risk can help understanding aberrant glycosylation in those diseases. Such variants that influence two or more traits are called pleiotropic. Klaric et al. (2020) investigated pleiotropy of IgG glycosylation-associated variants by exploring which other diseases the same variants were associated with and assessing whether both the disease risk and glycosylation are likely to be regulated by the same underlying causal variant. While there was evidence of shared variants with 83 different diseases, mostly affecting the immune and inflammatory system, but also digestive and neurological system, 3 IgG glycosylation loci and 8 diseases were pleiotropic, suggesting that the same underlying causal variant influences both the disease risk and IgG glycosylation. Namely, variants associated with IgG agalactosylation within the *IRF1-SLC22A4* were pleiotropic with risk for Crohn's disease, IgG agalactosylation and fucosylation associated variants from the *ORMDL3-GSDMB-IKZF3-ZPBP2* locus were pleiotropic with risk for inflammatory bowel disease, ulcerative colitis, asthma, rheumatoid arthritis, primary biliary cirrhosis and high-density lipoprotein (HDL) cholesterol level, and mono- and digalactosylated core fucosylated traits with bisecting GlcNAc-associated variants from the *CRHR1-SPPL2C-MAPT-ARHGAP27* locus were pleiotropic with Parkinson's disease. For the remaining 75 diseases, it was either impossible to assess pleiotropy for a lack of publicly available summary statistics data or the disease risk was more likely to be controlled by a different underlying causal variant (Klaric et al. 2020). It is important to note that while the two traits can be pleiotropic (regulated by the same underlying causal variant), these types of analyses do not infer the direction of the causality and do not clarify whether the change in glycan levels is causing the disease or the disease is causing the change in glycan levels.

Another point to consider is that, while two traits might be regulated by the same underlying variant, tissue context is also of importance—the same variant can have an effect on a different trait depending on the tissue. For example, the observed pleiotropy of IgG glycosylation and Crohn's disease does not necessarily imply causality—it is possible that the same variant is independently influencing glycosylation of IgG in B cells and Crohn's disease risk in T cells. To further elucidate such paradigms, more detailed analyses of causality, supported by functional follow-up are needed.

8.6 Conclusions

A key insight from genetic studies of IgG *N*-glycome is that genetic regulation of IgG glycosylation is a highly complex process and, apart from the main glycosylation enzymes, involves possibly hundreds of other genes which operate to collectively influence the *N*-glycosylation of IgG. Although each of the variants identified in genetic studies of IgG *N*-glycome makes a small contribution to the IgG glycosylation process, in aggregate they can still explain only a portion of variation in IgG glycans (0.5–21.9%) (Klaric et al. 2020), implicating that further studies, with an even higher number of samples and including variants at the rare allele frequency spectrum are needed. Studying genetic regulation of IgG glycosylation provides not only better insight into this complex process, but through discovering pleiotropic variants, SNPs associated with both glycosylation and complex disease, can also help understanding aberrant glycosylation in those diseases. However, further studies are needed to determine direction of the causality, assessing whether changes in IgG glycosylation are causing the disease or the disease is causing changes in IgG glycosylation.

Compliance with Ethical Standards

Funding The work of LK was supported by an RCUK Innovation Fellowship from the National Productivity Investment Fund (MR/R026408/1). The work of OOZ was supported by the Croatian National Centre of Research Excellence in Personalized Healthcare grant (#KK.01.1.1.01.0010). The work of AF was supported by European Union's Horizon 2020 research and innovation programme IMforFUTURE under H2020-MSCA-ITN grant agreement number 721815.

Conflict of Interest AF and OOZ are the employees of Genos Ltd., which specialises in high-throughput glycoanalysis. LK declares that she has no conflict of interest.

Ethical Approval All procedures performed in studies involving human participants were in accordance with the ethical standards of the institutional and/or national research committee and with the 1964 Helsinki declaration and its later amendments or comparable ethical standards.

Informed Consent Informed consent was obtained from all individual participants included in the studies.

References

Amaral A, Castillo J, Estanyol JM, Ballesca JL, Ramalho-Santos J, Oliva R (2013) Human sperm tail proteome suggests new endogenous metabolic pathways. Mol Cell Proteomics 12 (2):330–342

Anderson CA, Boucher G, Lees CW, Franke A, D'Amato M, Taylor KD et al (2011) Meta-analysis identifies 29 additional ulcerative colitis risk loci, increasing the number of confirmed associations to 47. Nat Publ Group 43:246–252

Aoyama C, Liao H, Ishidate K (2004) Structure and function of choline kinase isoforms in mammalian cells. Prog Lipid Res 43:266–281

Bakshi R, Hassan MQ, Pratap J, Lian JB, Montecino MA, van Wijnen AJ et al (2010) The human SWI/SNF complex associates with RUNX1 to control transcription of hematopoietic target genes. J Cell Physiol 225(2):569–576

Banda NK, Wood AK, Takahashi K, Levitt B, Rudd PM, Royle L et al (2008) Initiation of the alternative pathway of murine complement by immune complexes is dependent on N-glycans in IgG antibodies. Arthritis Rheum 58(10):3081–3089

Baumert M, Takei K, Hartinger J, Burger PM, Fischer von Mollard G, Maycox PR et al (1990) P29: a novel tyrosine-phosphorylated membrane protein present in small clear vesicles of neurons and endocrine cells. J Cell Biol 110(4):1285–1294

Belmont JW, Boudreau A, Leal SM, Hardenbol P, Pasternak S, Wheeler DA et al (2005) A haplotype map of the human genome. Nature 437(7063):1299–1320

Benson MJ, Äijö T, Chang X, Gagnon J, Pape UJ, Anantharaman V et al (2012) Heterogeneous nuclear ribonucleoprotein L-like (hnRNPLL) and elongation factor, RNA polymerase II, 2 (ELL2) are regulators of mRNA processing in plasma cells. Proc Natl Acad Sci U S A 109 (40):16252–16257

Birmann BM, Tamimi RM, Giovannucci E, Rosner B, Hunter DJ, Kraft P et al (2009) Insulin-like growth factor-1- and interleukin-6-related gene variation and risk of multiple myeloma. Cancer Epidemiol Biomarkers Prev 18(1):282

Blomme B, Van Steenkiste C, Grassi P, Haslam SM, Dell A, Callewaert N et al (2011) Alterations of serum protein N-glycosylation in two mouse models of chronic liver disease are hepatocyte and not B cell driven. Am J Physiol Gastrointest Liver Physiol 300(5):G833–GG42

Bondt A, Rombouts Y, Selman MHJ, Hensbergen PJ, Reiding KR, Hazes JMW et al (2014) Immunoglobulin G (IgG) fab glycosylation analysis using a new mass spectrometric high-throughput profiling method reveals pregnancy-associated changes. Mol Cell Proteomics 13 (11):3029–3039

Capitani M, Sallese M (2009) The KDEL receptor: new functions for an old protein. FEBS Lett 583 (23):3863–3871

Carreras-Sureda A, Cantero-Recasens G, Rubio-moscardo F, Kiefer K, Peinelt C, Niemeyer BA et al (2013) ORMDL3 modulates store-operated calcium entry and lymphocyte activation. Hum Mol Genet 22(3):519–530

Choi JK, Zhu A, Jenkins BG, Hattori S, Kil KE, Takagi T et al (2015) Combined behavioral studies and in vivo imaging of inflammatory response and expression of mGlu5 receptors in schnurri-2 knockout mice. Neurosci Lett 609:159–164

Clerc F, Reiding KR, Jansen BC, Kammeijer GSM, Bondt A, Wuhrer M (2016) Human plasma protein N-glycosylation. Glycoconj J 33(3):309–343

Dall'Olio F (2000) The sialyl-α2,6-lactosaminyl-structure: biosynthesis and functional role. Glycoconj J 17(10):669–676

de Haan N, Reiding KR, Krištić J, Hipgrave Ederveen AL, Lauc G, Wuhrer M (2017) The N-glycosylation of mouse immunoglobulin G (IgG)-fragment crystallizable differs between IgG subclasses and strains. Front Immunol 8:608

Dekkers G, Treffers L, Plomp R, Bentlage AEH, de Boer M, Koeleman CAM et al (2017) Decoding the human immunoglobulin G-glycan repertoire reveals a spectrum of Fc-receptor- and complement-mediated-effector activities. Front Immunol 8:877

Ding J, Wang K, Liu W, She Y, Sun Q, Shi J et al (2016) Pore-forming activity and structural autoinhibition of the gasdermin family. Nature 535:111–116

El-Hattab AW, Craigen WJ, Scaglia F (2017) Mitochondrial DNA maintenance defects. Elsevier B. V., Amsterdam, pp 1539–1555

Ferrara C, Grau S, Jager C, Sondermann P, Brunker P, Waldhauer I et al (2011) Unique carbohydrate-carbohydrate interactions are required for high affinity binding between FcgammaRIII and antibodies lacking core fucose. Proc Natl Acad Sci U S A 108(31):12669–12674

Fragale A, Gabriele L, Stellacci E, Borghi P, Perrotti E, Ilari R et al (2008) IFN regulatory factor-1 negatively regulates CD4+CD25+ regulatory T cell differentiation by repressing Foxp3 expression. J Immunol 181(3):1673–1682

Fukuta K, Abe R, Yokomatsu T, Omae F, Asanagi M, Makino T (2000) Control bisecting GlcNAc addition to N-linked sugar chains. J Biol Chem 275(31):23456–23461

Gottardo L, De Cosmo S, Zhang Y-Y, Powers C, Prudente S, Marescotti MC et al (2008) A polymorphism at the IL6ST (gp130) locus is associated with traits of the metabolic syndrome. Obesity 16(1):205–210

Gudelj I, Lauc G, Pezer M (2018) Immunoglobulin G glycosylation in aging and diseases. Cell Immunol 333(January):65–79

Hall EA, Keighren M, Ford MJ, Davey T, Jarman AP, Smith LB et al (2013) Acute versus chronic loss of mammalian Azi1/Cep131 results in distinct ciliary phenotypes. PLoS Genet 9(12):e1003928

Hennet T (2002) The galactosyltransferase family. Cell Mol Life Sci 59(7):1081–1095

Hirst J, Lui WWY, Bright NA, Totty N, Seaman MNJ, Robinson MS (2000) A family of proteins with γ-adaptin and VHS domains that facilitate trafficking between the trans-golgi network and the vacuole/lysosome. J Cell Biol 149(1):67–79

Igl W, Polaek O, Gornik O, Kneević A, Puić M, Novokmet M et al (2011) Glycomics meets lipidomics—associations of N-glycans with classical lipids, glycerophospholipids, and sphingolipids in three European populations. Mol BioSyst 7(6):1852–1862

Julià A, López-Longo FJ, Pérez Venegas JJ, Bonàs-Guarch S, Olivé À, Andreu JL et al (2018) Genome-wide association study meta-analysis identifies five new loci for systemic lupus erythematosus. Arthritis Res Ther 20(1):100

Kalimuthu SN, Chetty R (2016) Gene of the month: SMARCB1. J Clin Pathol 69(6):484–489

Kellokumpu S (2019) Golgi pH, ion and redox homeostasis: how much do they really matter? Front Cell Dev Biol 7:93

Klaric L, Tsepilov YA, Stanton CM, Mangino M, Sikka TT, Esko T et al (2020) Glycosylation of immunoglobulin G is regulated by a large network of genes pleiotropic with inflammatory diseases. Sci Adv 6(8):eaax0301

Krištić J, Vučković F, Menni C, Klarić L, Keser T, Beceheli I et al (2014) Glycans are a novel biomarker of chronological and biological ages. J Gerontol A Biol Sci Med Sci 69(7):779–789

Krištić J, Zaytseva OO, Ram R, Nguyen Q, Novokmet M, Vučković F et al (2018) Profiling and genetic control of the murine immunoglobulin G glycome. Nat Chem Biol 14(5):516–524

Lander ES, Linton LM, Birren B, Nusbaum C, Zody MC, Baldwin J et al (2001) Initial sequencing and analysis of the human genome. Nature 409(6822):860–921

Lauc G, Essafi A, Huffman JE, Hayward C, Knežević A, Kattla JJ et al (2010) Genomics meets glycomics-the first GWAS study of human N-glycome identifies HNF1α as a master regulator of plasma protein fucosylation. PLoS Genet 6(12):e1001256

Lauc G, Huffman JE, Pučić M, Zgaga L, Adamczyk B, Mužinić A et al (2013) Loci associated with N-glycosylation of human immunoglobulin G show pleiotropy with autoimmune diseases and haematological cancers. PLoS Genet 9(1):e1003225

Lazarevic V, Glimcher LH, Lord GM (2013) T-bet: a bridge between innate and adaptive immunity. Nat Rev Immunol 13:777–789

Liu X, Li S, Peng W, Feng S, Feng J, Mahboob S et al (2016) Genome-wide identification, characterization and Phylogenetic analysis of ATP-binding cassette (ABC) transporter genes in common carp (cyprinus carpio). PLoS One 11(4):e0153246

Liu YR, Hu Y, Zeng Y, Li ZX, Zhang HB, Deng JL et al (2019) Neurexophilin and PC-esterase domain family member 4 (NXPE4) and prostate androgen-regulated mucin-like protein 1 (PARM1) as prognostic biomarkers for colorectal cancer. J Cell Biochem 120(10):18041–18052

Lund J, Takahashi N, Pound JD, Goodall M, Jefferis R (1996) Multiple interactions of IgG with its core oligosaccharide can modulate recognition by complement and human Fc gamma receptor I and influence the synthesis of its oligosaccharide chains. J Immunol 157(11):4963

Ma B, Simala-Grant JL, Taylor DE (2006) Fucosylation in prokaryotes and eukaryotes. Oxford Academic, Oxford, pp 158–184

Martincic K, Alkan SA, Cheatle A, Borghesi L, Milcarek C (2009) Transcription elongation factor ELL2 directs immunoglobulin secretion in plasma cells by stimulating altered RNA processing. Nat Immunol 10(10):1102–1109

Matsui T, Leung D, Miyashita H, Maksakova IA, Miyachi H, Kimura H et al (2010) Proviral silencing in embryonic stem cells requires the histone methyltransferase ESET. Nature 464 (7290):927–931

Melchers F (2005) The pre-B-cell receptor: selector of fitting immunoglobulin heavy chains for the B-cell repertoire. Nat Rev Immunol 5:578–584

Menni C, Keser T, Mangino M, Bell JT, Erte I, Akmačić I et al (2013) Glycosylation of immunoglobulin G: role of genetic and epigenetic influences. PLoS One 8(12):6–13

Moffatt MF, Kabesch M, Liang L, Dixon AL, Strachan D, Heath S et al (2007) Genetic variants regulating ORMDL3 expression contribute to the risk of childhood asthma. Nature 448 (7152):470–473

Moore DS, Shenk D (2017) The heritability fallacy. Wiley Interdiscip Rev Cogn Sci 8(1–2):e1400

Morahan G, Balmer L, Monley D (2008) Establishment of "The Gene Mine": a resource for rapid identification of complex trait genes. Mamm Genome 19(6):390–393

Nairn AV, York WS, Harris K, Hall EM, Pierce JM, Moremen KW (2008) Regulation of glycan structures in animal tissues: transcript profiling of glycan-related genes. J Biol Chem 283 (25):17298–17313

National Research Council (2012) Transforming glycoscience: a roadmap for the future. The National Academies Press, Washington, DC

Oda Y, Okada T, Yoshida H, Kaufman RJ, Nagata K, Mori K (2006) Derlin-2 and Derlin-3 are regulated by the mammalian unfolded protein response and are required for ER-associated degradation. J Cell Biol 172(3):383–393

Okuda T, Van Deursen J, Hiebert SW, Grosveld G, Downing JR (1996) AML1, the target of multiple chromosomal translocations in human leukemia, is essential for normal fetal liver hematopoiesis. Cell 84(2):321–330

Oshima S, Nakamura T, Namiki S, Tsuchiya K, Okamoto R, Yokota T et al (2004) Interferon regulatory factor 1 (IRF-1) and IRF-2 distinctively up-regulate gene expression and production of interleukin-7 in human intestinal epithelial cells interferon regulatory factor 1 (IRF-1) and IRF-2 distinctively up-regulate gene expression. Mol Cell Biol 24(14):6298–6310

Overgaard NH, Jung J-W, Steptoe RJ, Wells JW (2015) CD4+/CD8+ double-positive T cells: more than just a developmental stage? J Leukoc Biol 97(1):31–38

Papadopoulou AA, Müller SA, Mentrup T, Shmueli MD, Niemeyer J, Haug-Kröper M et al (2019) Signal peptide peptidase-like 2c impairs vesicular transport and cleaves SNARE proteins. EMBO Rep 20(3):e46451

Parekh RB, Dwek RA, Sutton BJ, Fernandes DL, Leung A, Stanworth D et al (1985) Association of rheumatoid arthritis and primary osteoarthritis with changes in the glycosylation pattern of total serum IgG. Nature 316:452–457

Pers TH, Karjalainen JM, Chan Y, Westra H-J, Wood AR, Yang J et al (2015) Biological interpretation of genome-wide association studies using predicted gene functions. Nat Commun 6:5890

Pothukuchi P, Agliarulo I, Russo D, Rizzo R, Russo F, Parashuraman S (2019) Translation of genome to glycome: role of the Golgi apparatus. FEBS Lett 593(17):2390–2411

Pottier N, Cheok MH, Yang W, Assem M, Tracey L, Obenauer JC et al (2007) Expression of SMARCB1 modulates steroid sensitivity in human lymphoblastoid cells: identification of a promoter snp that alters PARP1 binding and SMARCB1 expression. Hum Mol Genet 16(19):2261–2271

Pučić M, Knežević A, Vidic J, Adamczyk B, Novokmet M, Polašek O et al (2011) High throughput isolation and glycosylation analysis of IgG-variability and heritability of the IgG glycome in three isolated human populations. Mol Cell Proteomics 10(10):M111.010090

Rebollo A, Schmitt C (2003) Ikaros, Aiolos and Helios: transcription regulators and lymphoid malignancies. Immunol Cell Biol 81(3):171–175

Rose RJ, van Berkel PHC, van den Bremer ETJ, Labrijn AF, Vink T, Schuurman J et al (2013) Mutation of Y407 in the CH3 domain dramatically alters glycosylation and structure of human IgG. mAbs 5(2):219–228

Rose-John S (2018) Interleukin-6 family cytokines. Cold Spring Harbor Perspect Biol 10(2): a028415

Ruhaak LR, Uh H-W, Beekman M, Hokke CH, Westendorp RGJ, Houwing-Duistermaat J et al (2011) Plasma protein N-glycan profiles are associated with calendar age, familial longevity and health. J Proteome Res 10(4):1667–1674

Russell AC, Šimurina M, Garcia MT, Novokmet M, Wang Y, Rudan I et al (2017) The N-glycosylation of immunoglobulin G as a novel biomarker of Parkinson's disease. Glycobiology 27(5):501–510

Sakane H, Makiyama T, Nogami S, Horii Y, Akasaki K, Shirataki H (2016) β-Taxilin participates in differentiation of C2C12 myoblasts into myotubes. Exp Cell Res 345(2):230–238

Saleh NM, Raj SM, Smyth DJ, Wallace C, Howson JMM, Bell L et al (2011) Genetic association analyses of atopic illness and proinflammatory cytokine genes with type 1 diabetes. Diabetes Metab Res Rev 27(8):838–843

Santos P, Arumemi F, Park KS, Borghesi L, Milcarek C (2011) Transcriptional and epigenetic regulation of B cell development. Immunol Res 50:105–112

Scheitz CJF, Tumbar T (2013) New insights into the role of Runx1 in epithelial stem cell biology and pathology. J Cell Biochem 114(5):985–993

Schotta G, Lachner M, Sarma K, Ebert A, Sengupta R, Reuter G et al (2004) A silencing pathway to induce H3-K9 and H4-K20 trimethylation at constitutive heterochromatin. Genes Dev 18 (11):1251–1262

Sellars M, Reina-San-Martin B, Kastner P, Chan S (2009) Ikaros controls isotype selection during immunoglobulin class switch recombination. J Exp Med 206(5):1073–1087

Sharapov SZ, Tsepilov YA, Klaric L, Mangino M, Thareja G, Shadrina AS et al (2019) Defining the genetic control of human blood plasma N-glycome using genome-wide association study. Hum Mol Genet 28(12):2062–2077

Shen X, Klarić L, Sharapov SZ, Mangino M, Ning Z, Wu D et al (2017) Multivariate discovery and replication of five novel loci associated with immunoglobulin G N-glycosylation. Nat Commun 8(1):447

Shiina T, Hosomichi K, Inoko H, Kulski JK (2009) The HLA genomic loci map: expression, interaction, diversity and disease. J Hum Genet 54(1):15–39

Solovieff N, Cotsapas C, Lee PH, Purcell SM, Smoller JW (2013) Pleiotropy in complex traits: challenges and strategies. Nat Rev Genet 14(7):483–495

Spencer J, Sollid LM (2016) The human intestinal B-cell response. Mucosal Immunol 9 (5):1113–1124

Spender LC, Whiteman HJ, Karstegl CE, Farrell PJ (2005) Transcriptional cross-regulation of RUNX1 by RUNX3 in human B cells. Oncogene 24(11):1873–1881

Stahl EA, Raychaudhuri S, Remmers EF, Xie G, Eyre S, Thomson BP et al (2010) Genome-wide association study meta-analysis identifies seven new rheumatoid arthritis risk loci. Nat Genet 42 (6):508–514

Stambuk J, Nakic N, Vuckovic F, Pucic-Bakovic M, Razdorov G, Trbojevic-Akmacic I et al (2020) Global variability of the human IgG glycome. Aging (Albany NY) 12(15):15222–15259

Staples CJ, Myers KN, Beveridge RDD, Patil AA, Lee AJX, Swanton C et al (2012) The centriolar satellite protein Cep131 is important for genome stability. J Cell Sci 125(20):4770–4779

Steinke FC, Yu S, Zhou X, He B, Yang W, Zhou B et al (2014) TCF-1 and LEF-1 act upstream of Th-POK to promote the CD4+ T cell fate and interact with Runx3 to silence Cd4 in CD8+ T cells. Nat Immunol 15(7):646–656

Stenius K, Janz R, Südhof TC, Jahn R (1995) Structure of synaptogyrin (p29) defines novel synaptic vesicle protein. J Cell Biol 131(6 II):1801–1809

Strang KH, Golde TE, Giasson BI (2019) MAPT mutations, tauopathy, and mechanisms of neurodegeneration. Lab Invest 99:912–928

Takagi T, Jin W, Taya K, Watanabe G, Mori K, Ishii S (2006) Schnurri-2 mutant mice are hypersensitive to stress and hyperactive. Brain Res 1108(1):88–97

Takao K, Kobayashi K, Hagihara H, Ohira K, Shoji H, Hattori S et al (2013) Deficiency of schnurri-2, an MHC enhancer binding protein, induces mild chronic inflammation in the brain and confers molecular, neuronal, and behavioral phenotypes related to schizophrenia. Neuropsychopharmacology 38(8):1409–1425

Tan Q, Christiansen L, von Bornemann Hjelmborg J, Christensen K (2015) Twin methodology in epigenetic studies. J Exp Biol 218(Pt 1):134–139

Taniguchi N, Honke K, Fukuda M, Narimatsu H, Yamaguchi Y, Angata T (2014) Handbook of glycosyltransferases and related genes. Springer, Tokyo

Tenesa A, Haley CS (2013) The heritability of human disease: estimation, uses and abuses. Nat Rev Genet 14(2):139–149

Ting JPY, Trowsdale J (2002) Genetic control of MHC class II expression. Cell 109:21–33

Tokuhiro S, Yamada R, Chang X, Suzuki A, Kochi Y, Sawada T et al (2003) An intronic SNP in a RUNX1 binding site of SLC22A4, encoding an organic cation transporter, is associated with rheumatoid arthritis. Nat Genet 35(4):341–348

Tripathi R, Hosseini K, Arapi V, Fredriksson R, Bagchi S (2019) SLC38A10 (SNAT10) is located in ER and Golgi compartments and has a role in regulating nascent protein synthesis. Int J Mol Sci 20(24):6265 van de Bovenkamp FS, Hafkenscheid L, Rispens T, Rombouts Y (2016) The emerging importance of IgG Fab glycosylation in immunity. J Immunol 196(4):1435–1441

Vanhooren V, Desmyter L, Liu X-E, Cardelli M, Franceschi C, Federico A et al (2007) N-glycomic changes in serum proteins during human aging. Rejuvenation Res 10(4):521–531

Vanhooren V, Liu XE, Franceschi C, Gao CF, Libert C, Contreras R et al (2009) N-glycan profiles as tools in diagnosis of hepatocellular carcinoma and prediction of healthy human ageing. Mech Ageing Dev 130(1–2):92–97

Vanhooren V, Dewaele S, Libert C, Engelborghs S, De Deyn PP, Toussaint O et al (2010) Serum N-glycan profile shift during human ageing. Exp Gerontol 45(10):738–743

Varki A, Cummings RD, Esko JD, Stanley P, Hart GW, Aebi M et al (2017) Essentials of glycobiology, 3rd edn. Cold Spring Harbor Laboratory Press, Cold Spring Harbor, NY

Vilaj M, Gudelj I, Trbojević-Akmačić I, Lauc G, Pezer M (2019) IgG glycans as a biomarker of biological age. In: Moskalev A (ed) Biomarkers of human aging. Springer International, Cham, pp 81–99

Visscher PM, Hill WG, Wray NR (2008) Heritability in the genomics era—concepts and misconceptions. Nat Rev Genet 9(4):255–266

Voon DC-C, Hor YT, Ito Y (2015) The RUNX complex: reaching beyond haematopoiesis into immunity. Immunology 146(4):523–536

Wahl A, van den Akker E, Klaric L, Stambuk J, Benedetti E, Plomp R et al (2018) Genome-wide association study on immunoglobulin G glycosylation patterns. Front Immunol 9(Feb):277

Wang W, Xue Y, Zhou S, Kuo A, Cairns BR, Crabtree GR (1996) Diversity and specialization of mammalian SWI/SNF complexes. Genes Dev 10(17):2117–2130

Wang J, Balog CIA, Stavenhagen K, Koeleman CAM, Scherer HU, Selman MH et al (2011) Fc-glycosylation of IgG1 is modulated by B-cell stimuli. Mol Cell Proteomics 10(5): M110.004655

Weinstein J, de Souza-e-Silva U, Paulson JC (1982) Purification of a Gal/31+4GlcNAc a2+6 sialyltransferase and a Gal/l1-,3(4)GlcNAc a2+3 sialyltransferase to homogeneity from rat liver. J Biol Chem 257:13835–13844

Whiteman HJ, Farrell PJ (2006) RUNX expression and function in human B cells. Crit Rev Eukaryot Gene Expr 16(1):31–44

Wu X, Steet RA, Bohorov O, Bakker J, Newell J, Krieger M et al (2004) Mutation of the COG complex subunit gene COG7 causes a lethal congenital disorder. Nat Med 10(5):518–523

Zaytseva OO, Jansen BC, Hanić M, Mrčela M, Razdorov G, Stojković R et al (2018) MIgGGly (mouse IgG glycosylation analysis)—a high-throughput method for studying Fc-linked IgG N-glycosylation in mice with nanoUPLC-ESI-MS. Sci Rep 8(1):13688

Zhu H, Wang J, Li J, Li S (2011) Corticotropin-releasing factor family and its receptors: pro-inflammatory or anti-inflammatory targets in the periphery? Inflamm Res 60(8):715–721

Zoldos V, Horvat T, Novokmet M, Cuenin C, Muzinic A, Pucic M et al (2012) Epigenetic silencing of HNF1A associates with changes in the composition of the human plasma N-glycome. Epigenetics 7(2):164–172

Chapter 9
Epigenetics of Immunoglobulin G Glycosylation

Marija Klasić and Vlatka Zoldoš

Contents

9.1	Introduction	290
9.2	Regulation of Glycosyltransferases by Transcription Factors	291
9.3	Epigenetic Regulation of IgG Glycosylation	291
9.4	The Role of miRNAs in Protein *N*-Glycosylation	296
9.5	Conclusions	297
References		298

Abstract Alternative glycosylation of immunoglobulin G (IgG) affects its effector functions during the immune response. IgG glycosylation is altered in many diseases, but also during a healthy life of an individual. Currently, there is limited knowledge of factors that alter IgG glycosylation in the healthy state and factors involved in specific IgG glycosylation patterns associated with pathophysiology. Genetic background plays an important role, but epigenetic mechanisms also contribute to the alteration of IgG glycosylation patterns in healthy life and in disease. It is known that the expression of many glycosyltransferases is regulated by DNA methylation and by microRNA (miRNA) molecules, but the involvement of other epigenetic mechanisms, such as histone modifications, in the regulation of glycosylation-related genes (glycogenes) is still poorly understood. Recent studies have identified several differentially methylated loci associated with IgG glycosylation, but the mechanisms involved in the formation of specific IgG glycosylation patterns remain poorly understood.

Keywords Epigenetic · Gene regulation · DNA methylation · Histone modifications · miRNA · IgG glycosylation

---

M. Klasić · V. Zoldoš (✉)
Division of Molecular Biology, Department of Biology, Faculty of Science, University of Zagreb, Zagreb, Croatia
e-mail: vzoldos@biol.pmf.hr

Abbreviations

5-aza-CR	5-azacytidine
5-aza-dC	5-aza-2′-deoxycytidine
ADCC	antibody-dependent cellular cytotoxicity
AFP	α-fetoprotein
CML	chronic myeloid leukemia
CRC	colorectal carcinoma
EWAS	epigenome-wide association study
GlcNAc	*N*-acetylglucosamine
GWAS	genome-wide association study
HCC	hepatocellular carcinoma
IBD	inflammatory bowel disease
IgG	immunoglobulin G
miRNA	microRNA
PBMC	peripheral blood mononuclear cell
TSA	Trichostatin A

9.1 Introduction

Alterations in immunoglobulin G glycosylation are reported in many (patho)-physiological conditions and during aging, but the mechanisms responsible for alternative glycosylation are still poorly understood (Gudelj et al. 2018). The formation of a specific glycosylation pattern is a complex multi-level process involving transcriptional, posttranscriptional, translational, and posttranslational regulation (Neelamegham and Mahal 2016). The importance of the genetic component for IgG glycosylation was discussed in detail in the previous chapter. However, less explored are the epigenetic mechanisms responsible for cell type-specific gene expression profiles and thus specific glycosylation. With the rapid development of sophisticated methods for studying epigenetic mechanisms, especially on the genome-wide level, we are now collecting more and more data with the aim to understand which genes are involved in protein glycosylation, and how they are regulated. Moreover, the combination of glycan data and expression/methylation data of glycogenes in various diseases helps us to understand how these changes might be functionally relevant in pathogenesis (Klasić et al. 2016, 2018; Vojta et al. 2016; Menni et al. 2013; Saldova et al. 2011; Zoldoš et al. 2012; Cummings and Pierce 2014). Nevertheless, our knowledge of how specific IgG glycosylation patterns occur in many diseases and whether these aberrant glycan patterns are a cause or consequence of the disease is not well understood.

9.2 Regulation of Glycosyltransferases by Transcription Factors

Numerous studies have identified transcription factors that control the expression of glycogenes in a cell-specific or tissue-specific manner (Guo and Pierce 2015). For example, the transcription factors E1AF and E2F1 are known to upregulate *B4GALT1*, a gene responsible for galactosylation, in mammalian cells (Zhu et al. 2005; Wei et al. 2010). *ST6GAL1*, the gene encoding a sialyltransferase, is upregulated in liver cells by the transcription factors HNF-1, DBP, Sp-1, and Oct-1 (Svensson et al. 1990; Taniguchi et al. 2000). A recent study by Klarić and colleagues suggests that the transcription factors RUNX1 and RUNX3, together with the chromatin remodeler SMARCB1, regulate the expression of the *MGAT3* gene, which encodes the glycosyltransferase responsible for the addition of bisecting *N*-acetylglucosamine (GlcNAc) to the core of a glycan structure (Klarić et al. 2020). On the other hand, the study by Xu et al. demonstrated that *MGAT3* is downregulated through Wnt/β-catenin signaling in a colon carcinoma cell line (Xu et al. 2011). In the large genome-wide association study (GWAS) of IgG glycosylation, Klarić et al. discovered a possible mechanism involved in core fucosylation of IgG molecules. They performed a knockdown of the transcription factor IKZF1, resulting in an upregulation of the *FUT8* gene and an increase of core fucosylated glycans on IgG secreted by a lymphoblastoid cell line (Klarić et al. 2020). Interestingly, another study showed that the transcription factor HNF1A downregulates *FUT8*, which is responsible for core fucosylation, in liver cells (Lauc et al. 2010). Knock-down of *HNF1A* using RNAi induced a decrease of transcriptional expression in *FUT* genes responsible for antennary fucosylation, and an increase in *FUT8* expression responsible for core fucosylation (Lauc et al. 2010). Overall, these results provide new evidence that the regulation of glycosylation might be tissue-specific and cell-specific. Therefore, further studies are needed to elucidate which transcription factors in B cells are relevant for the establishment of alternative IgG glycosylation during numerous pathological conditions and throughout life.

9.3 Epigenetic Regulation of IgG Glycosylation

In the mammalian genome, DNA methylation occurs predominantly at the fifth carbon of cytosine in CpG dinucleotides. It is established in mammalian early development and maintained during life. Nevertheless, the methylation pattern can change during a life due to environmental factors (Herceg 2007; Martin and Fry 2018). Moreover, DNA methylation patterns are ubiquitously altered in many complex diseases, with changes best characterized in various tumors (Bergman and Cedar 2013). For example, global hypomethylation and promoter hypermethylation occur in tumors, leading to genomic instability and dysregulation of many genes (Jones and Baylin 2007).

Histone modifications, such as acetylation, methylation, phosphorylation, and ubiquitylation, affect chromatin structure by merely being there (e.g., acetylation adds a negative charge to histones that repels negatively charged DNA, resulting in chromatin relaxation) or by recruiting chromatin remodeling proteins (Bannister and Kouzarides 2011). Numerous studies have shown dysregulation of histone modifications in various diseases, including cancer and autoimmune diseases (Audia and Campbell 2016; Araki and Mimura 2017). Among the affected genes, altered histone modifications have been found in several glycogenes, but regulation of glycogenes by histone modifications is less well understood than regulation by DNA methylation. For example, the expression of *ST6GalNAc6*, a gene encoding the α2→6 sialyltransferase, is downregulated in human colon cancer cells compared to normal epithelial cells. Experiments performed by Miyazaki and colleagues indicated that the *ST6GalNAc6* gene is probably controlled by histone deacetylation in human colon cancer cells, resulting in a decrease in the expression of disialyl Lewis (a) structures (Miyazaki et al. 2004). Another study showed downregulation of the *DTDST* gene in colon cancer cells, but its expression was restored after treatment with histone deacetylase inhibitors, indicating an important role of histone acetylation in the regulation of this gene (Yusa et al. 2010). In addition, several studies showed neural cell-specific regulation of the *Mgat5b* gene by histone acetylation (Kizuka et al. 2011, 2014).

The influence of DNA methylation and histone modifications on protein glycosylation began to be studied intensively about 10 years ago. First studies reported the deregulation of glycogenes via changes in promoter methylation. These first studies used cancer cell lines as a model. Cells were treated with DNA methylation and histone deacetylase inhibitors, such as 5-azacytidine (5-aza-CR), 5-aza-2′-deoxycytidine (5-aza-dC), sodium butyrate, Trichostatin A (TSA), and valproic acid. The treatments resulted in global DNA hypomethylation and global acetylation and consequently changes in glycosylation of membrane and secreted glycoproteins (Klasić et al. 2016; Saldova et al. 2011; Horvat et al. 2012, 2013; Anugraham et al. 2014; Kohler et al. 2016). Treatment of cells with these inhibitors induced deregulation of many glycogenes through direct DNA demethylation and histone acetylation of their promoters or indirectly through epigenetic changes of upstream regulatory factors and yet unknown indirect mechanisms (Fig. 9.1). For example, hypomethylation of the *MGAT3* promoter resulted in an increase of gene expression and change of glycan structures with bisecting GlcNAc on membrane glycoproteins in ovarian carcinoma cells (Kohler et al. 2016). On the other hand, *MGAT5* expression was increased after 5-aza-dC treatment probably due to the activation of specific transcription activator, leading to an increased level of branched glycans on secreted glycoproteins in ovarian carcinoma cells (Saldova et al. 2011). It was also shown that the induced changes in *N*-glycosylation of membrane proteins were reversible upon the recovery of the cells in a medium free of DNA methylation and histone deacetylation inhibitors, suggesting that glycan change, due to epigenetic mechanisms, is reversible (Horvat et al. 2013).

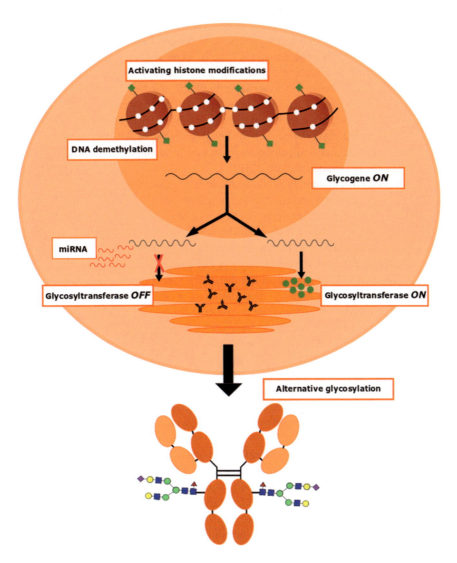

Fig. 9.1 Epigenetic mechanisms involved in alternative glycosylation of immunoglobulin G. Transcriptionally active chromatin leads to the expression of glycogenes. Glycogene mRNA can be processed and translated into a protein or inactivated posttranscriptionally by miRNA molecules, leading to a decrease in protein level. Differences in glycosyltransferase expression can lead to alternative glycosylation of immunoglobulin G. White circles—unmethylated DNA; green rectangles—activating modifications on histones

Epigenetic regulation of the glycogene *MGAT3* has been well characterized in several studies. Treatment of ovarian and hepatocellular carcinoma (HCC) cells with 5-aza-dC resulted in decreased methylation of *MGAT3* promoter and increased level of gene expression. Upregulation of *MGAT3* resulted in altered levels of glycans on total membrane and secreted glycoproteins. Treatment with 5-aza-dC induced an increase in *N*-glycans with bisecting GlcNAc on the cell membrane glycoproteins of ovarian carcinoma cells, whereas in HCC cells secreted glycoproteins showed less tetraantennary and core fucosylated structures (Klasić et al. 2016; Anugraham et al. 2014; Kohler et al. 2016). To further investigate the effect of DNA methylation on *MGAT3* expression and glycan phenotype, Josipović and colleagues used CRISPR/dCas9-DNMT3A molecular tool to specifically methylate a part of the CpG island of this gene. Targeted methylation led to downregulation of *MGAT3* in an ovarian adenocarcinoma cell line and consequently to a decrease of *N*-glycan structures with bisecting GlcNAc (Josipović et al. 2019).

A large study using publicly available methylation and expression data, combined with wet-lab experiments, revealed ten glycogenes that ubiquitously show aberrant expression patterns in different types of tumors and metastases through CpG methylation (Vojta et al. 2016). For some of them (*GALNT* and *MAN* genes), it was the first report that the change of expression occurs via aberrant promoter methylation. The importance of this study was a discovery of a new group of genes, glycogenes, that contribute to tumors as well as the discovery that aberrant expression of these genes due to aberrant promoter methylation is a way leading to characteristic glycosylation profiles often described in cancer (Stowell et al. 2015). Tumorigenesis and tumor progression are accompanied by significant changes in protein glycosylation. Tumor cells have an altered core of *N*-glycans and branching. For example, glycoproteins on the cell surface may be enriched with fucosylated and hypersialylated glycans during tumor transformation (Miyoshi et al. 2008; Dobie and Skropeta 2021). Changes can also be seen in glycoproteins secreted from cells, e.g., fucosylated form of α-fetoprotein (AFP) is present in the serum of patients with HCC and may serve as a biomarker that distinguishes HCC from other benign liver diseases (Moriya et al. 2013).

Recent GWASs showed that several transcription factors are also associated with changes in glycosylation. The transcription factor HNF1A, which is associated with fucose metabolism and recognized as a master regulator of plasma protein fucosylation, has been shown to be regulated by DNA methylation (Zoldoš et al. 2012). When regulatory CpG sites in the *HNF1A* regulatory region were manipulated using the TET1-CRISPR/dCas9 molecular tool, a decrease of complex glycan structures with core-fucose was reported in the ovarian adenocarcinoma cell line (Josipović et al. 2019).

Data on the influence of epigenetic mechanisms, such as DNA methylation, on the glycosylation of specific proteins, such as IgG antibody, is still very scarce. Glycans on immunoglobulin G have been shown to be stable in individuals during homeostasis but are displaying a great variability in a population (Pučić et al. 2011). Within an individual, IgG glycosylation pattern changes with lifestyle or aging, as well as in a particular disease (Krištić et al. 2014; Yu et al. 2016; Novokmet et al. 2014). Abundant changes in IgG glycosylation have been observed in many autoimmune diseases as well as in tumors (Gudelj et al. 2018; Klasić et al. 2018; Lauc et al. 2013, 2016). It has been shown that the level of agalactosylated glycans decreases in childhood and adolescence, and then increases with age (Parekh et al.

1988). In many autoimmune diseases (e.g., rheumatoid arthritis, inflammatory bowel disease (IBD), and systemic lupus erythematosus), there was a decrease in galactosylated and sialylated IgG glycans and an increase in fucosylated glycans and structures with bisecting GlcNAc. IgG glycosylation is altered in many tumors such as colorectal and hepatocellular carcinoma, lung cancer, multiple myeloma, and ovarian cancer [for a detailed review see Gudelj et al. (2018)]. Indeed, alternative IgG glycosylation is important for the effector function of this antibody and thus for an adequate immune response (Schwab and Nimmerjahn 2013). Therefore, it is extremely important to unravel the layers of regulation of IgG glycosylation and elucidate mechanisms beyond specific changes in disease. Many genes have been linked to IgG glycosylation through GWA studies. In addition to genes encoding glycosyltransferases and glycosydases, many transcription factors, receptors, signaling molecules, and chromatin remodelers have also been associated with IgG glycosylation (Klarić et al. 2020; Lauc et al. 2013; Shen et al. 2017; Benedetti et al. 2017; Wahl et al. 2018a). Epigenetic regulation of IgG glycosylation has just started to be elucidated. The first epigenome-wide association study (EWAS) addressing associations between whole-genome DNA methylation and IgG glycans was performed on the cohort of monozygotic and dizygotic twin pairs. The study identified differentially methylated positions in the *ANKRD11* and *SFRS10* loci associated with glycans with low heritability, and differentially methylated positions in the *NRN1L* and *QPCT* loci associated with IgG glycans with high heritability. These genes have not been previously reported to play a role in IgG glycosylation, but the results suggest that the IgG glycosylation pattern with low heritability may be epigenetically mediated (Menni et al. 2013). The second EWAS, performed on four large cohorts, discovered an association between IgG glycosylation and smoking through changes in the methylation of the *AHRR* gene. That gives rise to environmental influence on IgG glycosylation mediated by DNA methylation (Wahl et al. 2018b). Moreover, Krištić and colleagues found several significant associations between differentially methylated CpG site located 0.8–0.9 kb upstream of the *BACH2* transcription start site and IgG glycans (Krištić et al. 2014). *BACH2* methylation and expression were altered during aging (*BACH2* mRNA downregulation was reported in CD4+ T cells, CD8+ T cells, and CD19+ B cells in healthy people) and SNP in *BACH2* was previously associated with IgG galactosylation (Lauc et al. 2013; Hannum et al. 2013; Chi et al. 2019). The results from this study suggest that changes in IgG glycosylation during aging may be caused in part by changes in the *BACH2* gene expression, leading to an increased inflammation with age (Krištić et al. 2014).

To address the correlation between methylation in promoter regions of genes encoding glycosyltransferases and/or transcription factors associated with IgG glycosylation, Klasić and colleagues investigated possible correlations between *BACH2* and *MGAT3* methylation and IgG glycan profiles from patients with IBD. While *MGAT3* is a classical glycogene with a known function in the glycosylation network, *BACH2* encodes a transcription factor involved in B cell differentiation and maturation (Igarashi et al. 2014). However, its role in IgG glycosylation is still unknown. A study performed on several hundred patients with IBD, including Crohn's disease and ulcerative colitis, showed a decrease in *BACH2* and an increase in *MGAT3* promoter methylation in peripheral blood mononuclear cells (PBMCs) and CD19+ B cells separated from whole blood (Klasić et al. 2018). While *BACH2* methylation showed no significant correlation with glycan structures, the *MGAT3* promoter methylation correlated negatively with galactosylated and sialylated glycans, as well as with glycan structures with bisecting GlcNAc on IgG molecules in the same IBD patients. These results suggest a possible deregulation of *MGAT3* via DNA methylation in IBD patients leading to a specific IgG glycosylation pattern. Furthermore, the results suggest a not straightforward and complex role of transcription factors in the formation of specific IgG glycosylation patterns (Klasić et al. 2018).

9.4 The Role of miRNAs in Protein *N*-Glycosylation

As discussed above, glycogenes are partly regulated by transcription factors and epigenetic mechanisms, but a number of studies suggest that glycogene expression can in some cases only semi-quantitatively predict the corresponding enzyme activity and glycan expression. The relationship between glycogene expression and glycan structures is non-linear in most cases, suggesting another level of regulation of protein glycosylation, i.e., posttranscriptional regulation via miRNAs (Fig. 9.1) (Neelamegham and Mahal 2016; Thu and Mahal 2019). MiRNAs are small non-coding RNA molecules that act as posttranscriptional regulators of protein expression. They play an important role in developmental processes in eukaryotes and are dysregulated in many diseases (Ardekani and Naeini 2010). Recent studies have identified miRNAs as important regulators of glycosyltransferases expression, which are also highly dysregulated in different diseases. Several glycogenes associated with IgG glycosylation are known to be posttranscriptionally regulated by miRNAs. The *FUT8* gene, which encodes core fucosyltransferase, is regulated by four miRNAs in HCC, leading to dysregulation of protein expression and consequent changes in protein core fucosylation. *FUT8* is overexpressed in HCC cells and HCC tissue, leading to an increase in core fucosylated glycan structures on membrane and secreted glycoproteins. Overexpression of four miRNA molecules (miR-122, miR-34a, miR-26a, and miR-455-3p) targeting the *FUT8* transcript in HCC cells led to downregulation of *FUT8* and core fucosylation on secreted glycoproteins and suppression of HCC cells progression (Bernardi et al. 2013; Cheng et al. 2016). A similar effect of another miRNA molecule was observed in colorectal carcinoma (CRC) cells. Overexpression of miR-198 in CRC cells induced downregulation of FUT8 and inhibited cell proliferation, migration and invasion. These results were confirmed in vivo, where expression of miR-198 inhibited the growth and invasion of CRC in nude mice (Wang et al. 2014). Moreover, core fucosylation is important for IgG antibody function in antibody-dependent cellular cytotoxicity (ADCC) (Li et al. 2017; Pereira et al. 2018). Recent studies discovered altered core fucosylation on IgG molecules in patients with some autoimmune and alloimmune diseases as well as in tumors (Gudelj et al. 2018). Moreover, glycosyltransferases responsible for galactosylation, sialylation, and addition of bisecting *N*-acetylglucosamine (B4GALT1, ST6GAL1, and MGAT3) are also regulated by miRNAs at the posttranscriptional level in chronic myeloid leukemia (CML) and HCC cell lines (Liu et al. 2016; Han et al. 2018; Huang et al. 2018). While miRNAs have been shown to be important for the differentiation of B cells into IgG-producing plasma cells, current knowledge about miRNA regulation of IgG glycosylation patterns and the influence on its alternative glycosylation in different (patho)physiological states is insufficient (Vigorito et al. 2007). There is increasing evidence that miRNAs control the fine-tuning of glycosyltransferase expression, but our knowledge of the miRNA network for the assembly of specific glycan structures is still incomplete.

9.5 Conclusions

Changes in glycosylation of immunoglobulin G occur during the healthy life of an individual due to environmental factors, but also in many diseases due to homeostasis perturbation. However, the exact mechanisms leading to the formation of specific glycan patterns associated with a particular disease are still poorly understood. There is increasing evidence that both genetic and epigenetic components are involved in alternative IgG glycosylation. Moreover, several GWA studies of IgG glycosylation have found that not only glycosyltransferases and glycosidases are involved in IgG glycosylation, but also many transcription factors, receptors, signaling molecules, chromatin remodelers, etc. Therefore, recent studies used RNAi silencing or novel molecular tools such as CRISPR/dCa9 for activation and/or repression of gene transcriptional activity to elucidate the functional role of these GWAS hits in the regulation of IgG glycosylation. In addition, many studies have shown an epigenetic influence on alternative protein and IgG glycosylation observed in various diseases such as diabetes, chronic inflammatory diseases, and cancer. Many glycogenes as well as some genes encoding transcription factors are regulated by promoter methylation and epigenetic modifications have been shown to influence *N*-glycosylation. While histone modifications have been little studied in terms of protein glycosylation regulation, miRNA molecules appeared to fine-tune glycosyltransferase expression, and as a result, alter protein glycosylation. Several glycogenes associated with IgG glycosylation were shown to be posttranscriptionally regulated by miRNAs. Future challenges lie in uncovering a complex gene network involved in alternative IgG glycosylation ubiquitously involved in pathophysiological conditions, in order to understand molecular mechanisms beyond specific glycan patterns associated with a specific disease and to find new potential diagnostic, prognostic, and therapeutic strategies.

Ethics

Funding This study was funded by the European Structural and Investment Funds IRI (Grant# KK.01.2.1.01.0003), the Croatian National Centre of Research Excellence in Personalized Healthcare (Grant# KK.01.1.1.01.0010), the Center of Competence in Molecular Diagnostics (Grant# KK.01.2.2.03.0006) and the CRISPR/Cas9-CasMouse project (Grant# KK.01.1.1.04.0085).

Conflict of Interest Marija Klasić declares that she has no conflict of interest. Vlatka Zoldoš declares that she has no conflict of interest.

Ethical Approval This chapter does not contain any studies with human participants or animals performed by any of the authors.

References

Anugraham M, Jacob F, Nixdorf S, Everest-Dass AV, Heinzelmann-Schwarz V, Packer NH (2014) Specific glycosylation of membrane proteins in epithelial ovarian cancer cell lines: glycan structures reflect gene expression and DNA methylation status. Mol Cell Proteomics 13 (9):2213–2232

Araki Y, Mimura T (2017) The histone modification code in the pathogenesis of autoimmune diseases. Mediat Inflamm 2017:2608605

Ardekani AM, Naeini MM (2010) The role of microRNAs in human diseases. Avicenna J Med Biotechnol 2(4):161–179

Audia JE, Campbell RM (2016) Histone modifications and cancer. Cold Spring Harb Perspect Biol 8(4):1–32

Bannister AJ, Kouzarides T (2011) Regulation of chromatin by histone modifications. Cell Res 21 (3):381–395. Available from: http://www.nature.com/doifinder/10.1038/cr.2011.22

Benedetti E, Pučić-Baković M, Keser T, Wahl A, Hassinen A, Yang JY et al (2017) Network inference from glycoproteomics data reveals new reactions in the IgG glycosylation pathway. Nat Commun 8(1):1–15

Bergman Y, Cedar H (2013) DNA methylation dynamics in health and disease. Nat Struct Mol Biol 20(3):274–281

Bernardi C, Soffientini U, Piacente F, Tonetti MG (2013) Effects of microRNAs on fucosyltransferase 8 (FUT8) expression in hepatocarcinoma cells. PLoS One 8(10):6–11

Cheng L, Gao S, Song X, Dong W, Zhou H, Zhao L et al (2016) Comprehensive N-glycan profiles of hepatocellular carcinoma reveal association of fucosylation with tumor progression and regulation of FUT8 by microRNAs. Oncotarget 7(38):61199–61214

Chi VLD, Garaud S, De Silva P, Thibaud V, Stamatopoulos B, Berehad M et al (2019) Age-related changes in the BACH2 and PRDM1 genes in lymphocytes from healthy donors and chronic lymphocytic leukemia patients 11 Medical and Health Sciences 1107 Immunology 06 Biological Sciences 0604 Genetics. BMC Cancer 19(1):1–10

Cummings RD, Pierce JM (2014) The challenge and promise of glycomics. Chem Biol 21(1):1–15. Available from: https://doi.org/10.1016/j.chembiol.2013.12.010

Dobie C, Skropeta D (2021) Insights into the role of sialylation in cancer progression and metastasis. Br J Cancer 124(1):76–90. Available from: https://doi.org/10.1038/s41416-020-01126-7

Gudelj I, Lauc G, Pezer M (2018) Immunoglobulin G glycosylation in aging and diseases. Cell Immunol 333(July):65–79. Available from: https://doi.org/10.1016/j.cellimm.2018.07.009

Guo H, Pierce JM (2015) Transcriptional regulation of glycan expression. In: Taniguchi N, Endo T, Hart GW, Seeberger PH, Wong C-H (eds) Glycoscience: biology and medicine. Springer Japan, Tokyo, pp 1173–1180. Available from: https://doi.org/10.1007/978-4-431-54841-6_79

Han Y, Liu Y, Fu X, Zhang Q, Huang H, Zhang C et al (2018) miR-9 inhibits the metastatic ability of hepatocellular carcinoma via targeting beta galactoside alpha-2,6-sialyltransferase 1. J Physiol Biochem 74(3):491–501

Hannum G, Guinney J, Zhao L, Zhang L, Hughes G, Sadda SV et al (2013) Genome-wide methylation profiles reveal quantitative views of human aging rates. Mol Cell 49(2):359–367. Available from: https://doi.org/10.1016/j.molcel.2012.10.016

Herceg Z (2007) Epigenetics and cancer: towards an evaluation of the impact of environmental and dietary factors. Mutagenesis 22(2):91–103

Horvat T, Mužinić A, Barišić D, Bosnar MH, Zoldoš V (2012) Epigenetic modulation of the HeLa cell membrane N-glycome. Biochim Biophys Acta Gen Subj 1820(9):1412–1419

Horvat T, Deželjin M, Redžić I, Barišić D, Herak Bosnar M, Lauc G et al (2013) Reversibility of membrane N-glycome of HeLa cells upon treatment with epigenetic inhibitors. PLoS One 8 (1):1–9

Huang H, Liu Y, Yu P, Qu J, Guo Y, Li W et al (2018) MiR-23a transcriptional activated by Runx2 increases metastatic potential of mouse hepatoma cell via directly targeting Mgat3. Sci Rep 8 (1):1–11. Available from: https://doi.org/10.1038/s41598-018-25768-z

Igarashi K, Ochiai K, Itoh-Nakadai A, Muto A (2014) Orchestration of plasma cell differentiation by Bach2 and its gene regulatory network. Immunol Rev 261(1):116–125

Jones PA, Baylin SB (2007) The epigenomics of cancer. Cell 128(4):683–692

Josipović G, Tadić V, Klasić M, Zanki V, Bečeheli I, Chung F et al (2019) Antagonistic and synergistic epigenetic modulation using orthologous CRISPR/dCas9-based modular system. Nucleic Acids Res 47(18):9637–9657

Kizuka Y, Kitazume S, Yoshida M, Taniguchi N (2011) Brain-specific expression of N-acetylglucosaminyltransferase IX (GnT-IX) is regulated by epigenetic histone modifications. J Biol Chem 286(36):31875–31884

Kizuka Y, Kitazume S, Okahara K, Villagra A, Sotomayor EM, Taniguchi N (2014) Epigenetic regulation of a brain-specific glycosyltransferase N-acetylglucosaminyltransferase-IX (GnT-IX) by specific chromatin modifiers. J Biol Chem 289(16):11253–11261

Klarić L, Tsepilov YA, Stanton CM, Mangino M, Sikka TT, Esko T et al (2020) Glycosylation of immunoglobulin G is regulated by a large network of genes pleiotropic with inflammatory diseases. Sci Adv 6(8):eaax0301

Klasić M, Krištić J, Korać P, Horvat T, Markulin D, Vojta A et al (2016) DNA hypomethylation upregulates expression of the MGAT3 gene in HepG2 cells and leads to changes in N-glycosylation of secreted glycoproteins. Sci Rep 6:24363

Klasić M, Markulin D, Vojta A, Samaržija I, Biruš I, Dobrinić P et al (2018) Promoter methylation of the MGAT3 and BACH2 genes correlates with the composition of the immunoglobulin G glycome in inflammatory bowel disease. Clin Epigenetics 10(1):75

Kohler RS, Anugraham M, López MN, Xiao C, Schoetzau A, Hettich T et al (2016) Epigenetic activation of MGAT3 and corresponding bisecting GlcNAc shortens the survival of cancer patients. Oncotarget 7(32):51674–51686

Krištić J, Vučković F, Menni C, Klarić L, Keser T, Beceheli I et al (2014) Glycans are a novel biomarker of chronological and biological ages. Journals Gerontol Ser A Biol Sci Med Sci 69 (7):779–789

Lauc G, Essafi A, Huffman JE, Hayward C, Knežević A, Kattla JJ, et al. Genomics meets glycomics-the first GWAS study of human N-Glycome identifies HNF1α as a master regulator of plasma protein fucosylation. PLoS Genet. 2010 [cited 2014 Mar 26];6(12):e1001256. Available from: http://www.pubmedcentral.nih.gov/articlerender.fcgi?artid=3009678&tool=pmcentrez&rendertype=abstract

Lauc G, Huffman JE, Pučić M, Zgaga L, Adamczyk B, Mužinić A et al (2013) Loci associated with N-glycosylation of human immunoglobulin G show pleiotropy with autoimmune diseases and haematological cancers. PLOS Genet 9(1):e1003225. Available from: https://doi.org/10.1371/journal.pgen.1003225

Lauc G, Pezer M, Rudan I, Campbell H (2016) Mechanisms of disease: the human N-glycome. Biochim Biophys Acta Gen Subj 1860(8):1574–1582. Available from: https://doi.org/10.1016/j.bbagen.2015.10.016

Li T, DiLillo DJ, Bournazos S, Giddens JP, Ravetch JV, Wang LX (2017) Modulating IgG effector function by Fc glycan engineering. Proc Natl Acad Sci U S A 114(13):3485–3490

Liu YX, Wang L, Liu WJ, Zhang HT, Xue JH, Zhang ZW et al (2016) MIR-124-3p/B4GALT1 axis plays an important role in SOCS3-regulated growth and chemo-sensitivity of CML. J Hematol Oncol 9(1):1–12. Available from: https://doi.org/10.1186/s13045-016-0300-3

Martin EM, Fry RC (2018) Environmental influences on the epigenome: exposure- associated DNA methylation in human populations. Annu Rev Public Health 39:309–333

Menni C, Keser T, Mangino M, Bell JT, Erte I, Akmačić I et al (2013) Glycosylation of immunoglobulin G: role of genetic and epigenetic influences. PLoS One 8(12):6–13

Miyazaki K, Ohmori R, Izawa M, Koike T, Kumamoto K, Furukawa K et al (2004) Loss of disialyl Lewisa, the ligand for lymphocyte inhibitory receptor sialic acid-binding immunoglobulin-like lectin-7 (Siglec-7) associated with increased sialyl Lewisa expression on human colon cancers. Cancer Res 64(13):4498–4505

Miyoshi E, Moriwaki K, Nakagawa T (2008) Biological function of fucosylation in cancer biology. J Biochem 143(6):725–729

Moriya S, Morimoto M, Numata K, Nozaki A, Shimoyama Y, Kondo M et al (2013) Fucosylated fraction of alpha-fetoprotein as a serological marker of early hepatocellular carcinoma. Anticancer Res 33(3):997–1002

Neelamegham S, Mahal LK (2016) Multi-level regulation of cellular glycosylation: from genes to transcript to enzyme to structure. Curr Opin Struct Biol 40(Figure 1):145–152. Available from: https://doi.org/10.1016/j.sbi.2016.09.013

Novokmet M, Lukić E, Vučković F, Durić Ž, Keser T, Rajšl K et al (2014) Changes in IgG and total plasma protein glycomes in acute systemic inflammation. Sci Rep 4:1–10

Parekh R, Roitt I, Isenberg D, Dwek R, Rademacher T (1988) Age-related galactosylation of the N-linked oligosaccharides of human serum IgG. J Exp Med 167(5):1731–1736

Pereira NA, Chan KF, Lin PC, Song Z (2018) The "less-is-more" in therapeutic antibodies: Afucosylated anti-cancer antibodies with enhanced antibody-dependent cellular cytotoxicity. MAbs 10(5):693–711. Available from: https://doi.org/10.1080/19420862.2018.1466767

Pučić M, Knežević A, Vidič J, Adamczyk B, Novokmet M, Polašek O et al (2011) High throughput isolation and glycosylation analysis of IgG-variability and heritability of the IgG glycome in three isolated human populations. Mol Cell Proteomics 10(10):1–15

Saldova R, Dempsey E, Pérez-Garay M, Mariño K, Watson JA, Blanco-Fernández A et al (2011) 5-AZA-2′-deoxycytidine induced demethylation influences N-glycosylation of secreted glycoproteins in ovarian cancer. Epigenetics 6(11):1362–1372

Schwab I, Nimmerjahn F (2013) Intravenous immunoglobulin therapy: how does IgG modulate the immune system? Nat Rev Immunol 13(3):176–189. Available from: https://doi.org/10.1038/nri3401

Shen X, Klarić L, Sharapov S, Mangino M, Ning Z, Wu D et al (2017) Multivariate discovery and replication of five novel loci associated with immunoglobulin G N-glycosylation. Nat Commun 8(1):1–10

Stowell SR, Ju T, Cummings RD (2015) Protein glycosylation in cancer. Annu Rev Pathol 10:473–510. Available from: https://pubmed.ncbi.nlm.nih.gov/25621663

Svensson EC, Soreghan B, Paulson JC (1990) Organization of the beta-galactoside alpha 2,6-sialyltransferase gene. Evidence for the transcriptional regulation of terminal glycosylation. J Biol Chem 265(34):20863–20868. Available from: http://europepmc.org/abstract/MED/2249992

Taniguchi A, Hasegawa Y, Higai K, Matsumoto K (2000) Transcriptional regulation of human beta-galactoside alpha2, 6-sialyltransferase (hST6Gal I) gene during differentiation of the HL-60 cell line. Glycobiology 10(6):623–628

Thu CT, Mahal LK (2019) Sweet control: microRNA regulation of the glycome. Biochemistry 59 (34):3098–3110

Vigorito E, Perks KL, Abreu-Goodger C, Bunting S, Xiang Z, Kohlhaas S et al (2007) microRNA-155 regulates the generation of immunoglobulin class-switched plasma cells. Immunity 27 (6):847–859

Vojta A, Samaržija I, Bočkor L, Zoldoš V (2016) Glyco-genes change expression in cancer through aberrant methylation. Biochim Biophys Acta Gen Subj 1860(8):1776–1785

Wahl A, van den Akker E, Klaric L, Štambuk J, Benedetti E, Plomp R et al (2018a) Genome-wide association study on immunoglobulin G glycosylation patterns. Front Immunol 9(Feb):1–14

Wahl A, Kasela S, Carnero-Montoro E, van Iterson M, Štambuk J, Sharma S et al (2018b) IgG glycosylation and DNA methylation are interconnected with smoking. Biochim Biophys Acta Gen Subj 1862(3):637–648

Wang M, Wang J, Kong X, Chen H, Wang Y, Qin M et al (2014) MiR-198 represses tumor growth and metastasis in colorectal cancer by targeting fucosyl transferase 8. Sci Rep 4:1–10

Wei Y, Zhou F, Ge Y, Chen H, Cui C, Liu D et al (2010) Regulation of the $β1,4$-galactosyltransferase i promoter by E2F1. J Biochem 148(3):263–271

Xu Q, Akama R, Isaji T, Lu Y, Hashimoto H, Kariya Y et al (2011) Wnt/$β$-catenin signaling down-regulates N-acetylglucosaminyltransferase III expression: the implications of two mutually exclusive pathways for regulation. J Biol Chem 286(6):4310–4318

Yu X, Wang Y, Kristic J, Dong J, Chu X, Ge S et al (2016) Profiling IgG N-glycans as potential biomarker of chronological and biological ages: a community-based study in a Han Chinese population. Medicine (Baltimore) 95(28):1–10

Yusa A, Miyazaki K, Kimura N, Izawa M, Kannagi R (2010) Epigenetic silencing of the sulfate transporter gene DTDST induces sialyl Lewisx expression and accelerates proliferation of colon cancer cells. Cancer Res 70(10):4064–4073

Zhu X, Jiang J, Shen H, Wang H, Zong H, Li Z et al (2005) Elevated $β1,4$-galactosyltransferase I in highly metastatic human lung cancer cells: identification of E1AF as important transcription activator. J Biol Chem 280(13):12503–12516

Zoldoš V, Horvat T, Novokmet M, Cuenin C, Mužinić A, Pučić M et al (2012) Epigenetic silencing of HNF1A associates with changes in the composition of the human plasma N-glycome. Epigenetics 7(2):164–172. Available from: https://doi.org/10.4161/epi.7.2.18918

Chapter 10
Immunoglobulin G Glycosylation Changes in Aging and Other Inflammatory Conditions

Fabio Dall'Olio and Nadia Malagolini

Contents

10.1	Premise	305
10.2	IgG Glycosylation	305
10.3	Changes in IgG Asn_{297} Glycans Associated with Inflammatory Diseases	307
10.4	IgG Glycosylation as a Predictor of Disease Onset, Progression, and Therapy Response	310
	10.4.1 Prediction of Disease Onset	310
	10.4.2 Prediction of Progression and Therapy Response	311
10.5	IgG Glycosylation in Aging	311
10.6	The Inflammaging Is a Link Between Aging and Inflammation	313
10.7	How Altered IgG Glycosylation Drives Inflammation	314
	10.7.1 Activation of Complement Through the Lectin or the Classical Pathways	314
	10.7.2 Binding to Fcγ Receptors	314
	10.7.3 Binding on Lectin Receptors of Antigen-Presenting Cells: Role in the Intravenous Administration of High Doses IgG (IVIG)	318
	10.7.4 Anti IgG Autoantibodies	320
10.8	Molecular Bases of N-Glycosylation Changes	321
10.9	Conclusions	322
References		323

Abstract Among the multiple roles played by protein glycosylation, the fine regulation of biological interactions is one of the most important. The asparagine 297 (Asn_{297}) of IgG heavy chains is decorated by a diantennary glycan bearing a number of galactose and sialic acid residues on the branches ranging from 0 to 2. In addition, the structure can present core-linked fucose and/or a bisecting GlcNAc. In many inflammatory and autoimmune conditions, as well as in metabolic, cardiovascular, infectious, and neoplastic diseases, the IgG Asn_{297}-linked glycan becomes less

---

The gene names and enzyme names are according to HUGO nomenclature rules (https://www.genenames.org/)

F. Dall'Olio (✉) · N. Malagolini
Department of Experimental, Diagnostic and Specialty Medicine (DIMES), University of Bologna, Bologna, Italy
e-mail: fabio.dallolio@unibo.it

© The Author(s), under exclusive license to Springer Nature Switzerland AG 2021
M. Pezer (ed.), *Antibody Glycosylation*, Experientia Supplementum 112,
https://doi.org/10.1007/978-3-030-76912-3_10

sialylated and less galactosylated, leading to increased expression of glycans terminating with GlcNAc. These conditions alter also the presence of core-fucose and bisecting GlcNAc. Importantly, similar glycomic alterations are observed in aging. The common condition, shared by the above-mentioned pathological conditions and aging, is a low-grade, chronic, asymptomatic inflammatory state which, in the case of aging, is known as inflammaging. Glycomic alterations associated with inflammatory diseases often precede disease onset and follow remission. The aberrantly glycosylated IgG glycans associated with inflammation and aging can sustain inflammation through different mechanisms, fueling a vicious loop. These include complement activation, Fcγ receptor binding, binding to lectin receptors on antigen-presenting cells, and autoantibody reactivity. The complex molecular bases of the glycomic changes associated with inflammation and aging are still poorly understood.

Keywords Glycosylation in aging · Inflammaging · Intravenous immunoglobulin · Hypogalactosylated antibodies · Glycosyltransferases

Abbreviations

ACPA	Anti-citrullinated protein antibodies
ADCC	Antibody-dependent cell cytotoxicity
AGEs	Advanced glycation end-products
ANCA	Anti-neutrophil cytoplasmic antibodies
APC	Antigen-presenting cells
DAMPS	Danger-associated molecular patterns
DC-SIGN	Dendritic cell-specific ICAM-grabbing non-integrin
FcγR	Fcγ receptor
Fuc	Fucose
Gal	Galactose
GalNAc	*N*-acetylgalactosamine
Glc	Glucose
GlcNAc	*N*-acetylglucosamine
GWAS	Genome-wide association studies
IVIG	Intravenous immunoglobulin
MALDI-TOF-MS	Matrix-assisted laser desorption/ionization-time of flight-mass spectrometry
Man	Mannose
MBL	Mannose-binding lectin
NK	Natural killer cells
PAMPS	Pathogen-associated molecular patterns
RA	Rheumatoid arthritis
RF	Rheumatoid factors
SASP	Senescence-associated secretory phenotype

SNP	Single nucleotide polymorphysms
STAT6	Signal transducer and activator of transcription 6
TNF	Tumor necrosis factor

10.1 Premise

The title of this chapter refers to aging as one of the many conditions associated with inflammation. The aging-associated inflammation is referred to as inflammaging and is low-grade, chronic, and asymptomatic. A similar state of low-grade chronic inflammation is common to various pathological conditions, including metabolic and cardiovascular diseases and cancer.

10.2 IgG Glycosylation

The asparagine residue 297 (Asn_{297}) of the IgG heavy chains bears a diantennary complex type glycan whose basic structure can undergo some functionally important variations under physiological and pathological conditions. On the simplest structure depicted in Fig. 10.1 (boxed), an α1,6-linked Fuc residue can be added to the innermost GlcNAc residue. This modification, which is known as "core fucosylation" and is mediated by fucosyltransferase 8 (FUT8), can take place also on other more complex structures, as shown. Another modification that can take place on the simplest structure, as well as on more complex glycans, is the addition of a β1,4-linked GlcNAc residue to the "central" mannose residue. The enzyme responsible for this modification, which is known as "bisecting" GlcNAc, is β1,4 N-acetylglucosaminyltransferase 3 (MGAT3). The two branches of the glycans can be elongated by the addition of one or two β1,4-linked galactose residues and of one or two α2,6-linked sialic acid residues. The addition of galactose can be mediated by a family of β1,4-galactosyltransferases (B4GALTs), although B4GALT3 has been suggested to play a major role (Schwedler et al. 2018). On the other hand, the addition of α2,6-linked sialic acid is mediated by sialyltransferase ST6GAL1 (Dall'Olio 2000). Glycosylation is a stochastic process regulated mainly by the relative abundance of specific glycosyltransferases. This leads to the phenomenon known as microheterogeneity, which means that the sugar structures attached to a specific glycosylation site in a given glycoprotein display a certain degree of variability. For example, the core fucose, the bisecting GlcNAc, and the other modifications mentioned above can be present on some but not all molecules. It should be kept in mind that while the vast majority of plasma glycoproteins are originated by the liver, antibodies are generated and glycosylated by plasma cells. The position to which the two N-glycans are attached to the two heavy chains of IgG allows them to regulate the interaction with Fcγ receptors (FcγR) (Krapp et al. 2003; Barb and Prestegard 2011; Yamaguchi et al. 1760) and their effector functions (Raju

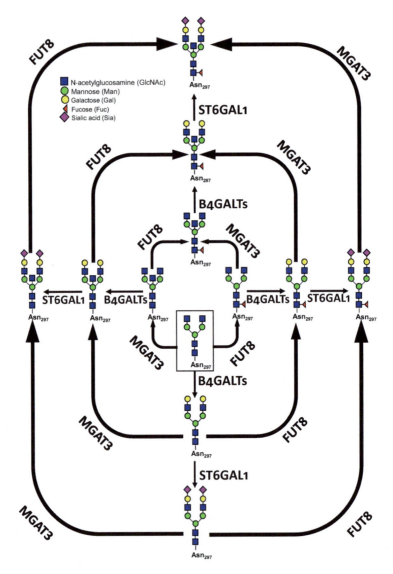

Fig. 10.1 Schematic representation of the different glycoforms which can be attached to Asn_{297} of the IgG heavy chains and their biosynthesis. The simplest structure (boxed) can be modified by the addition of "core fucose" by FUT8 or by addition of "bisecting GlcNAc," mediated by MGAT3. Otherwise, this structure can be β1,4-galactosylated by B4GALTs and successively α2,6-sialylated by ST6GAL1. Addition of core fucose and/or bisecting GlcNAc can take place also on more complex structures. Likewise, galactosylation and sialylation can take place also on core-fucosylated and/or bisected structures. The number of galactose and/or sialic acid residues on these structures ranges from 0 to 2. Owing to the phenomenon of microheterogeneity, all these structures can theoretically be linked to Asn_{297}. Their relative abundance is affected by multiple factors determined by physiological and pathological conditions, as well as the genetic background. Not all the enzymatic reactions can necessarily take place in any order, owing to the substrate specificity of the glycosyltransferase involved

2008). It is well known that glycosylation is a process associated with intracellular membranes. However, it has been shown that it can take place extracellularly, where plasmatic glycosyltransferases are able to glycosylate soluble glycoproteins utilizing sugar-nucleotides contained in platelets and exosomes as donor substrates (Wandall et al. 2012; Jones et al. 2012). In particular, antibodies can undergo extracellular sialylation, catalyzed by plasmatic sialyltransferase ST6GAL1 (Jones et al. 2012, 2016), which utilizes the CMP-sialic acid contained inside platelets as donor substrate (Lee et al. 2014). However, it is not clear which is the real relevance of extracellular glycosylation in shaping the glycosylation pattern of IgG. In fact, the level of $\alpha 2,6$ sialylation of IgG does not correlate with the level of plasmatic ST6GAL1 of healthy subjects (Catera et al. 2016). Moreover, extracellular sialylation plays a very limited role in the sialylation of the bulk of circulating IgG in resting mice (Schaffert et al. 2019) but becomes relevant as a consequence of inflammatory stimuli (Manhardt et al. 2017).

10.3 Changes in IgG Asn_{297} Glycans Associated with Inflammatory Diseases

More than 30 years ago, it was reported for the first time that in the autoimmune inflammatory conditions known as rheumatoid arthritis (RA) and in osteoarthritis Asn_{297}-linked glycans displayed a shift from fully galactosylated structures to GlcNAc-terminating forms (Parekh et al. 1985). Although other modifications of Asn_{297} glycans, such as increased bisecting GlcNAc and core fucosylation, were found to be associated with inflammatory conditions, the hypogalactosylation of the branches is certainly the most relevant (reviewed in (Yamaguchi and Barb 2020)).

Successively, the presence of agalactosylated N-linked chains was found to be associated with different diseases sharing an inflammatory condition (Table 10.1). Besides classical autoimmune/inflammatory conditions, hypogalactosylation of Asn_{297} IgG was found to be associated also with metabolic, cardiovascular and nervous system diseases, as well as with infectious diseases and several neoplasms. The observed changes are not identical among diseases and in some cases allow to distinguish the specific diseases (Axford et al. 2003; Pilkington et al. 1995a; Martin et al. 2001; Liu et al. 2019a, 2020a; Lemmers et al. 2017). In sera of RA patients, reduced galactosylation is detectable also in the N-linked chains of glycoproteins other than IgG (Nakagawa et al. 2007), although hypogalactosylation does not affect glycans attached to IgG light chains (Holland et al. 2006; Mimura et al. 2007; Youings et al. 1996; Bondt et al. 2016) nor to IgA (Field et al. 1994). These findings are not easy to reconcile. In fact, on the one hand, they suggest that inflammatory diseases alter glycosylation in both liver and lymphocytes. On the other hand, they suggest that altered glycosylation is restricted to a specific glycosylation site of IgG produced by B lymphocytes/plasma cells. Decreased galactosylation levels were also associated with mouse models of arthritis (Rook et al. 1991a; Yagev et al. 1993; Kuroda et al. 2001; Bodman et al. 1994; Endo et al. 1993).

Table 10.1 Human diseases associated with plasma hypogalactosylated IgG

Autoimmune/inflammatory diseases	References
Rheumatoid arthritis	Parekh et al. (1985), Axford et al. (2003), Pilkington et al. (1995a), Martin et al. (2001), Su et al. (2020)
Primary osteoarthritis	Parekh et al. (1985), Bond et al. (1997)
Juvenile onset rheumatoid arthritis	Bond et al. (1997), Parekh et al. (1988a), Sumar et al. (1991)
Juvenile idiopathic arthritis	Cheng et al. (2017)
Juvenile chronic arthritis	Flogel et al. (1998)
Psoriatic arthritis	Martin et al. (2001)
Anchylosing spondylitis	Martin et al. (2001), Liu et al. (2019b), Chou et al. (2010)
Spondyloarthropathy	Liu et al. (2020a), Leirisalo-Repo et al. (1999)
Erythema nodosum leprosum	Filley et al. (1989)
Takayasu's arteritis	Hernandez-Pando et al. (1994)
Sjogren's syndrome	Bond et al. (1997)
Systemic vasculitides associated with ANCA	Holland et al. (2002, 2006)
Systemic lupus erythematosus	Pilkington et al. (1995a, 1996a), Vuckovic et al. (2015)
Autoimmune hemolytic anemia	Sonneveld et al. (2017)
Systemic sclerosis	Liu et al. (2020b)
Lambert–Eaton myasthenic syndrome	Selman et al. (2011)
Myositis	Perdivara et al. (2011)
Myastenia gravis	Pilkington et al. (1995a)
Granulomatosis with polyangitis	Kemna et al. (2017)
Guillain–Barre syndrome	Fokkink et al. (2014)
Chronic inflammatory demyelinating polyneuropathy	Wong et al. (2016)
Crohn disease and ulcerative colitis	Bond et al. (1997), Dube et al. (1990), Shinzaki et al. (2008, 2013), Simurina et al. (2018), Theodoratou et al. (2014)
Pancreatitis	Chen et al. (2014)
Alzheimer disease	Lundstrom et al. (2014)
Multiple sclerosis	Wuhrer et al. (2015)
Chronic obstructive pulmonary disease (COPD)	Komaromy et al. (2020)
Metabolic and cardiovascular diseases	
Central adiposity	Liu et al. (2019c)
Ischemic heart disease (in RA)	Troelsen et al. (2007)
Dilated cardiomyopathy	Reinke et al. (2019)
Hypertension	Knezevic et al. (2010), Liu et al. (2018a), Gao et al. (2017)
Nonalcoholic fat liver disease (NAFLD)	Zhao et al. (2018)
Nonalcoholic steatohepatitis (NASH)	Blomme et al. (2011)
Obesity	Nikolac et al. (2014)

(continued)

Table 10.1 (continued)

Autoimmune/inflammatory diseases	References
Type 2 diabetes	Liu et al. (2019a), Lemmers et al. (2017)
Galactosemia	Coman et al. (2010)
Infectious diseases	
HBV hepatitis	Ho et al. (2014)
HBV-associated fibrosis	Gui et al. (2010), Ho et al. (2015)
HCV-associated fibrosis	Mehta et al. (2008)
Adult tuberculosis	Parekh et al. (1989), Rademacher et al. (1988), Liu et al. (2020c), Rook et al. (1994)
Childhood tuberculosis	Pilkington et al. (1996b)
Parasitic infections	de Jong et al. (2016)
Cancers	
Gastric	Kanoh et al. (2004a), Xin et al. (1995), Bones et al. (2011), Qin et al. (2019, 2020)
Prostate	Kanoh et al. (2004b)
Lung	Kanoh et al. (2004a, 2006), Chen et al. (2013)
Breast	Pierce et al. (2010)
Ovarian	Saldova et al. (2007)
Pancreatic	Chen et al. (2014), Shih et al. (2019)
Colon	Liu et al. (2019d, 2020d), Zou et al. (2020), Vuckovic et al. (2016)
Cholangiocarcinoma	Chang et al. (2018)
Neuroblastoma	Qin et al. (2018)
Hematological malignancies	de Haan et al. (2018a)

In many circumstances, IgG glycosylation closely follows disease remission or progression. For example, in HBV hepatitis, the level of agalactosylated IgG decreases after entecavir therapy (Ho et al. 2014). Some conditions are known to cause RA disease remission and a concomitant normalization of IgG Asn_{297} galactosylation. Examples include pregnancy (Rook et al. 1991b; Alavi et al. 2000; van de Geijn et al. 2009; Pekelharing et al. 1988; Bondt et al. 2013), fasting (Kjeldsen-Kragh et al. 1996), and treatment with anti-TNF (Van Beneden et al. 2009; Croce et al. 2007; Pasek et al. 2006; Collins et al. 2013) or anti-IL-6 (Mesko et al. 2012) antibodies. Pregnancy reduces the agalactosylation of IgG per se, independently on RA (Reiding et al. 2017; Bondt et al. 2014). Interestingly, the level of agalactosylated IgG is always lower in the fetus than in its mother (Williams et al. 1995; Kibe et al. 1996), suggesting a kind of selective transport. Nevertheless, agalactosylated antibodies are involved in the mother-to-child transmission of inflammatory conditions, such as systemic lupus erythematosus (Pilkington et al. 1996a) and myastenia gravis (Pilkington et al. 1995b).

Anti-neutrophil cytoplasmic antibodies (ANCAs) are a group of autoantibodies, mainly of the IgG type, against antigens in the cytoplasm of neutrophils, which are produced in various autoimmune inflammatory conditions, but in particular in systemic vasculitis. In some cases, the glycosylation of ANCA antibodies differs from that of total IgG, providing different information on the disease status. For example, in granulomatosis with polyangiitis, low galactosylation and sialyation in total IgG1 but not in anti-ANCA IgG1 predicts disease reactivation (Kemna et al. 2017). Moreover, information on disease status can be provided by glycosylation of antigen-specific ANCAs. For example, Fc glycosylation of total IgG was significantly reduced in patients with active ANCA-associated vasculitis. Clinical remission was associated with complete glycan normalization for proteinase3-ANCA patients but not for myeloperoxidase-ANCA patients (Lardinois et al. 2019).

IgG glycosylation is influenced also by socioeconomic conditions. In fact, galactosylation levels were generally lower in less affluent countries and in less urban communities, probably reflecting increased immune activation and consequent inflammation (de Jong et al. 2016).

10.4 IgG Glycosylation as a Predictor of Disease Onset, Progression, and Therapy Response

10.4.1 Prediction of Disease Onset

In some circumstances, IgG glycosylation changes anticipate the onset of disease. The changes are not limited to galactosylation, but involve core-fucosylation, bisecting GlcNAc, and sialylation. Examples of changes detectable before the onset of the disease are reported in Table 10.2. Noteworthy, some modifications, including core-fucosylation and sialylation could be different in different diseases.

Table 10.2 Asn_{297} IgG glycosylation changes as predictive factors of disease

Disease	Core fucosylation	Bisecting GlcNAc	Galactosylation	Sialylation	References
Rheumatoid arthritis			Lower		Ercan et al. (2010), Gudelj et al. (2018), Rombouts et al. (2015)
Granulomatosis with polyangiitis			Lower	Lower	Kemna et al. (2017)
Autoimmune thyroid diseases	Lower				Martin et al. (2020a, b)
Ischemic stroke		Higher	Lower	Lower	Liu et al. (2018b)
Cardiovascular diseases				Higher	Menni et al. (2018)
Liver disease	Higher	Higher	Lower		Zhao et al. (2018)
Plasma markers of inflammation	Higher		Lower	Lower	Plomp et al. (2017)

10.4.2 Prediction of Progression and Therapy Response

Good examples of therapy response prediction in autoimmune/inflammatory diseases are provided by the association between low galactosylation and poor response to methotrexate therapy in early RA (Lundstrom et al. 2017) and to TNF blockers in ankylosing spondylitis (Liu et al. 2019b).

Among infectious diseases, in HBV hepatitis, the presence of fully galactosylated IgG bearing core fucose and bisecting GlcNAc is associated with an attenuated liver inflammation. The response to treatment of these patients was unfavorable, probably because of the importance of the inflammatory response in antiviral defense (Ho et al. 2019). The crucial role played by core-fucosylation in regulating the progression of some infectious diseases is demonstrated by the following examples. In Dengue virus-infected patients producing afucosylated IgG1, infection frequently progresses toward hemorrhagic fever and shock because of the enhanced affinity for FcγRIIIA of afucosylated IgG (see below) which triggers platelet reduction (Wang and Ravetch 2019; Wang et al. 2017; Lok 2017). Moreover, in *Fut8(−/−)* mice (whose IgG is not core-fucosylated) infected with *Salmonella typhimurium*, the presence of bacteria colonizing the cecum was increased and the production of specific antibodies was decreased, compared with wild type mice (Zahid et al. 2020). A group of meningococcus-infected children develops sepsis. These patients display lower fucosylation and higher bisection of IgG1 than age-matched healthy controls (de Haan et al. 2018b). HIV infection is associated with an increased level of fucosylated glycans, which is associated with lower antibody-dependent cell-mediated cytotoxicity (ADCC) (Vadrevu et al. 2018).

Among cancers, in cholangiocarcinoma, the presence of fucosylated agalactosylated IgG1 induces metastasis and early recurrence. This is due to the activation by these antibodies of tumor-associated macrophages M2, which exert a tumor-promoting activity (Chang et al. 2018). After bone marrow transplantation, leukemia patients display a low level of IgG galactosylation, which is more similar to their pretransplantation profiles than to profiles of the donors (de Haan et al. 2018a). This suggests that hypogalactosylation is not directly linked to the presence of cancer.

10.5 IgG Glycosylation in Aging

Few years after the discovery of undergalactosylated IgG in serum of RA patients, a similar alteration in IgG of aging people was reported (Parekh et al. 1988b). Successively the hypogalactosylation of IgG in aging was confirmed by Yamada et al. (Yamada et al. 1997) and by many other investigations (reviewed in (Cobb 2020; de Haan et al. 2020)). Besides hypogalactosylation, also bisecting GlcNAc, shows age dependence (Yamada et al. 1997). A gender dependence of these modifications was suggested by a study showing the increase of agalactosylated structures only in females, while the increase of bisecting GlcNAc was confirmed in both genders (Shikata et al. 1998). Interestingly, the most prominent drop in the levels of galactosylated and sialylated glycoforms in females was observed around the age of 45–60 years when females usually enter menopause (Bakovic et al. 2013; Ercan et al. 2017). In the past years, different high-throughput techniques have allowed the detailed analysis of a very large number of plasma or serum specimens of various cohorts (Dall'Olio et al. 2013). Analysis of Italian and Belgian cohorts by a DNA sequencer used for the separation of N-glycans, revealed an increase of core-fucosylated, agalactosylated biantennary N-linked chains with or without a bisecting GlcNAc and a concomitant decrease of core-fucosylated di-galactosylated structures in people over 60 (Vanhooren et al. 2007, 2008, 2009). The very complex pattern of glycoforms generated by these techniques required focusing on those provided with the highest biological significance. The GlycoAge test is a good example of an index that combines the predictive potential of key glycan structures. This marker of aging is calculated as the Log of the ratio between a core-fucosylated diantennary N-glycan with two terminal GlcNAc residues and an identical glycan with two terminal galactose residues (Vanhooren et al. 2010). The dependence of the galactosylation pattern by the biological age, rather than by the calendar age, is demonstrated by the observation that the GlycoAge test of people affected by accelerated aging (progeroid) syndromes, such as Werner or Cockaine, was similar to that of centenarians, rather than to that of healthy age-matched controls (Vanhooren et al. 2007, 2010). Reduced galactosylation is associated also with Down syndrome, which presents a feature of accelerated aging (Borelli et al. 2015). Unexpectedly, in children the GlycoAge test was not lower than that of young adults (Catera et al. 2016), indicating that the reduced galactosylation starts with adulthood. Japanese semisupercentanarians (mean age 107 years) display, besides an increase of agalacto- and/or bisecting N-glycans, an increase of multibranched and highly sialylated N-glycans (Miura et al. 2015; Miura and Endo 1860). Analysis by MALDI-TOF-MS of subjects enrolled in the Leiden Longevity Study revealed that age-related glycosylation changes were sex-specific only below the age of 60, with young females having higher galactosylation than males, confirming previous results (Bakovic et al. 2013; Ercan et al. 2017).

Crucial questions on the relationship between IgG glycome and aging regard: (1) the value of glycosylation in prediction of longevity; (2) the relative contribution of genome and environment in shaping the glycome; and (3) the variability of the plasma glycome among healthy individuals and its stability over a short period of times. About the first point, changes of IgG glycome allow to explain up to 58% of variance in chronological age, more than telomere length (Kristic et al. 2014). Prediction of longevity has been attributed to low levels of agalactosylated forms containing a bisecting GlcNAc in people below the age of 60 (Ruhaak et al. 2010, 2011). About the second point, a strong indication about the relative contribution of genome and environment to a given trait can be obtained by the study of monozygotic and dizygotic twins. Through this approach, it has been established that about two-thirds of the plasma glycomic traits are genetically determined, while one-third is mainly controlled by the environment (Menni et al. 2013). Glycosyltransferase polymorphisms do not play a major role in shaping IgG glycome (Knezevic et al. 2009, 2010). About the third point, in the majority of individuals, the N-glycome displays a "normal" pattern, while it is different in a few outliers (Pucic et al. 2010). Over few days, the glycome of healthy individuals undergoes very little or no changes, while minor changes were observed over a 1-year-long period (Gornik et al. 2009). Although aging is associated with increased IgG agalactosylation, the transition between childhood and adolescence is associated with an opposite trend. In fact, core fucosylation and the level of agalactosylated glycans decreased while digalactosylated glycans increased with age, in both plasma and IgG glycome (Pucic et al. 2012).

Experimental systems confirm the age-associated glycomic changes observed in humans. In fact, decreased IgG galactosylation was observed in all seven mice strain examined between 2 and 8 months of age (Bodman et al. 1994). Moreover, caloric restriction, which in mice is associated with extended lifespan (Weindruch et al. 1988), reverts the increase of agalactosylated N-glycans (Vanhooren et al. 2011).

10.6 The Inflammaging Is a Link Between Aging and Inflammation

As previously suggested (Dall'Olio et al. 2013; Dall'Olio 2018), the age-associated inflammatory status known as inflammaging can be the link to explain the largely overlapping glycomic changes observed in inflammatory conditions and aging. The inflammaging is a chronic, low-grade, asymptomatic inflammatory status, which accompanies aging. This condition is multifactorial and depends on the long life exposure to pro-inflammatory stimuli of microbial, environmental, and endogenous origin (Franceschi 2007; Franceschi et al. 2007, 2018). Some of the stimuli putatively triggering inflammaging include (but are not limited to) the pro-inflammatory cytokines released by senescent cells displaying the "senescence associated secretory phenotype" (SASP) (Rodier and Campisi 2011), the danger-associated molecular patterns (DAMPS), released by necrotic or damaged cells, the pathogen-associated molecular patterns (PAMPS) associated with microorganisms and the advanced glycation end-products (AGEs) released by glycated proteins. Senescent cells are more numerous in elderly people, while DAMPS, PAMPS, and AGEs stimulate various receptors of the innate immune system, resulting in its chronic low-grade activation. Although the notion that the glycomic shift observed in aging is somehow related to inflammaging is highly plausible, observations in humans (Ruhaak et al. 2011) and in animal models (Vanhooren et al. 2011), support the possibility that the age-associated N-glycomic shift is regulated by metabolic pathways, independently of the inflammatory status.

10.7 How Altered IgG Glycosylation Drives Inflammation

Glycosylation at Asn_{297} provides a very good example of the fine-tuning of biological functions by glycans. Carbohydrate structures associated with aging and inflammation can sustain inflammation, triggering a vicious self-sustaining inflammatory loop (Dall'Olio et al. 2013). A seminal demonstration of the pathogenic effect of agalactosylated IgG was the demonstration that galactosidase-treated IgG exhibited increased ability to induce arthritis in healthy mice (Rademacher et al. 1994). The mechanisms which could causally link antibody glycosylation with inflammation include modulation of: (1) complement activation; (2) binding to Fcγ receptors; (3) binding to lectin receptors on antigen-presenting cells; (4) formation of anti-IgG autoantibodies. The effect of differential glycosylation on the different biological functions of IgG is summarized in Table 10.3 and depicted in Fig. 10.2.

10.7.1 Activation of Complement Through the Lectin or the Classical Pathways

It has been reported that the ability of hypogalactosylated IgG to activate complement was higher because of a stronger interaction with mannose-binding lectin (MBL), which is the first component of the lectin pathway of complement (Malhotra et al. 1995; Ezekowitz 1995) [reviewed in: (Rudd et al. 2001; Arnold et al. 2006)]. This notion was challenged by the fact that in RA patients deficient for MBL (Stanworth et al. 1998), as well as in MBL-deficient mice (Nimmerjahn et al. 2007), the pro-inflammatory activity of agalactosylated IgG was not impaired. Moreover, agalactosylated IgG from the plasma of patients affected by inflammatory bowel diseases failed to activate the lectin complement pathway (Nakajima et al. 2011). A study in a mouse model also excludes a role for differential galactosylation in complement activation through the lectin pathway (Ito et al. 2014). Regarding the activation of the classical complement pathway, it has been reported that high galactosylation (Peschke et al. 2017; Dekkers et al. 2017) and sialylation (Zabczynska et al. 2020) of IgG Asn_{297} increased binding and complement activation. This finding was unexpected, considering the pro-inflammatory role of complement, and remains to be clarified (Dekkers et al. 2018a).

10.7.2 Binding to Fcγ Receptors

The most relevant members of the IgG Fc receptors (FcγR) are the following: FcγRI (CD64), a high-affinity receptor expressed mainly on macrophages and neutrophils, very important for phagocytosis; FcγRIIIA (CD16), a low-affinity receptor expressed on NK and macrophages, very important for ADCC; FcγRIIB (CD32),

Table 10.3 Effect of glycosylation modifications on IgG function

		Core fucosylation	Bisecting GlcNAc	Galactosylation	Sialylation
Biological functions	Complement activation via lectin pathway			Down (Malhotra et al. 1995)	
	Complement activation via classical pathway	Unchanged (Niwa et al. 2005)	Unchanged (Dekkers et al. 2017)	Up (Peschke et al. 2017; Dekkers et al. 2017; Kiyoshi et al. 2018; Boyd et al. 1995; Hodoniczky et al. 2005)	Up (Dekkers et al. 2017; Zabczynska et al. 2020) Down (Quast et al. 2015)
	ADCC	Down (Dekkers et al. 2017; Niwa et al. 2005; Shields et al. 2002; Shinkawa et al. 2003; Satoh et al. 2006; Iida et al. 2006, 2009; Bruggeman et al. 2017; Peipp et al. 2008; Dekkers et al. 2018b; Patel et al. 2019; Falconer et al. 2018; Sakae et al. 2017; Harbison and Fadda 2020; Edri-Brami et al. 2019; Wirt et al. 2017; Tsukimura et al. 2017; Zhou et al. 2019)	Up (Umana et al. 1999; Davies et al. 2001; Hodoniczky et al. 2005)	Up (Kumpel et al. 1995; Thomann et al. 2015) Up (associated with hypofucosylation) (Dekkers et al. 2017; Kiyoshi et al. 2018)	Down[a] (Zabczynska et al. 2020; Li et al. 2017; Scallon et al. 2007) Unchanged (Thomann et al. 2015)
	Phagocytosis	Down (Shibata-Koyama et al. 2009)		Up (Kumpel et al. 1995)	
Receptor binding	Inhibitory FCγRIIB (CD32)			Up (Karsten et al. 2012)	Down (Tanigaki et al. 2018)
	High-affinity FcγRI (CD64)	Unchanged (Bruggeman et al. 2017)			Unchanged (Thomann et al. 2015)
	Low-affinity FcγRII (CD32)	Down (Subedi and Barb 2016)		Unchanged (Groenink et al. 1996) Up (Thomann et al. 2015)	Unchanged (Thomann et al. 2015)

(continued)

Table 10.3 (continued)

	Core fucosylation	Bisecting GlcNAc	Galactosylation	Sialylation
Low-affinity FcγRIII (CD16)	Down (Dekkers et al. 2017; Bruggeman et al. 2017; Wirt et al. 2017; Tsukimura et al. 2017; Subedi and Barb 2016)	Up (Davies et al. 2001) Unchanged (Dekkers et al. 2017)	Up (Adler et al. 1995; Thomann et al. 2015) Unchanged (Peschke et al. 2017; Subedi and Barb 2016)	Up (Adler et al. 1995) Unchanged (Dekkers et al. 2017; Thomann et al. 2015)
DC-SIGN		Down (Yabe et al. 2010)	Down (Yabe et al. 2010)	Up (Anthony et al. 2008a, 2011)

[a]Only if core-fucose is present

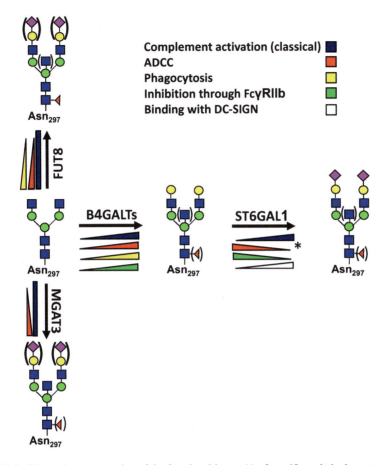

Fig. 10.2 Schematic representation of the functional impact(s) of specific carbohydrate structures on IgG Asn_{297}. The four modifications considered are: core fucosylation, bisecting GlcNAc, galactosylation, and sialylation. Where indicated in parenthesis, the structures can be present or not on that structure. The five IgG functions considered are: complement activation (classical pathway) (blue); ADCC (red); phagocytosis (yellow); immune inhibition through FcγRIIb (green); binding to DC-SIGN (white). Rectangles indicate that this function is not affected by that carbohydrate modification. Triangles indicate that the function is higher in the carbohydrate structure close to the base. *ADCC is reduced by sialylation only if core-fucosylation is present

a low-affinity receptor expressed by B cells, macrophages and dendritic cells which provides feedback inhibitory responses. Glycosylation of IgG at Asn_{297} plays a major role in regulating the binding of IgG to Fcγ receptors of the different classes (Nimmerjahn et al. 2007; Nimmerjahn and Ravetch 2006, 2008; Albert et al. 2008; Li et al. 2017; Karsten et al. 2012). A major role is played by core-fucosylation which inhibits ADCC, mainly mediated by low-affinity FcγRIII on natural killer (NK) cells (Dekkers et al. 2017; Niwa et al. 2005; Shields et al. 2002; Shinkawa et al. 2003; Satoh et al. 2006; Iida et al. 2006, 2009; Bruggeman et al. 2017) (Table 10.3).

However, low fucose enhances only NK-mediated ADCC, while neutrophil-mediated ADCC is enhanced by high fucose (Peipp et al. 2008), probably because it is mediated mainly by a different receptor. Glycosylation of FcγRs modulates the binding of IgG glycoforms (Hayes et al. 2014) because of interactions between the carbohydrate chains of the antibody and of the receptor (Ferrara et al. 2011). In particular, the core fucosylation of Asn_{297} of IgG hinders its interaction with Asn_{162} glycan of FcγRIII, lowering the strength of the IgG- FcγRIII receptor interaction (Dekkers et al. 2018b; Patel et al. 2019; Falconer et al. 2018; Sakae et al. 2017; Harbison and Fadda 2020). The finding that the glycan structures displayed by Asn_{162} of FcγRIII from NK cells of different individuals are highly variable and displaying very different ability to bind IgGs, provides an additional level of variability of immune functions among individuals (Patel et al. 2019).

Regarding the role of bisecting GlcNAc, it has been reported to increase binding of IgG to FcγRIII, enhancing ADCC (Umana et al. 1999; Davies et al. 2001) or to leave it unchanged (Dekkers et al. 2017). Galactosylation has been reported to be necessary for IgG binding to FcγRIII and ADCC (Adler et al. 1995; Kumpel et al. 1995; Kiyoshi et al. 2018), particularly when IgG is hypofucosylated (Dekkers et al. 2017). However, the binding with CD32 does not appear to be modulated by terminal galactose (Groenink et al. 1996). Fully galactosylated IgG that are produced during pregnancy cross the placenta better and provide the fetus and the newborn infant with antibodies optimized for NK ADCC (Pillai 2019). The IgG Fc glycosylation patterns of preterm infants display a more pro-inflammatory phenotype that is not observed in their mothers, suggesting an enrichment process. These IgGs may contribute to the increased risk for sustained inflammatory diseases in preterm infants (Twisselmann et al. 2018). The presence of α2,6-linked sialic acid has been reported to increase (Adler et al. 1995) or to leave unaffected (Dekkers et al. 2017) the binding to FcγRIII. Functionally, higher sialylation decreases ADCC (Zabczynska et al. 2020; Scallon et al. 2007). These apparently inconsistent results could be reconciled considering that sialylation exerts a relevant inhibition of ADCC only when core fucose is present (Li et al. 2017). The level of sialylation of antibodies elicited by vaccination toward influenza virus hemagglutinin positively affects the efficacy of the response because it induces the expansion of B lymphocytes carrying B cell receptors with higher affinity (Wang et al. 2015).

10.7.3 Binding on Lectin Receptors of Antigen-Presenting Cells: Role in the Intravenous Administration of High Doses IgG (IVIG)

Antigen-presenting cells (APC), in particular macrophages and dendritic cells, express on their surface various lectin receptors able to bind a myriad of carbohydrate ligands. Among the most relevant are the mannose-binding receptor and the C-type lectin DC-SIGN (Dendritic Cell-Specific Intercellular adhesion molecule-3-

Grabbing Non-integrin). The interaction between hypogalactosylated IgG with the mannose-binding receptor (Dong et al. 1999) and DC-SIGN (Yabe et al. 2010) has been reported to be increased.

An increased binding to DC-SIGN of α2,6-sialylated IgG has been reported as an essential step of the mechanisms underlying the anti-inflammatory effect of high doses IgG (IVIG) (Anthony et al. 2008a, 2011). Several autoimmune diseases can be beneficially, although transiently, treated with the intravenous administration of high doses IgG (IVIG) (Baerenwaldt et al. 2010). This treatment induces a functional silencing program similar to anergy in human B cells (Seite et al. 2014). Several lines of evidence indicate that the anti-inflammatory effect of IgG is attributable only to the minor α2,6-sialylated fraction of IgG (Kaneko et al. 2006; Anthony et al. 2008b), which allows interaction with the mouse lectin SIGN-R1 and its human orthologue DC-SIGN (Anthony et al. 2008a).

A model based on mouse experiments proposed that after binding of α2,6-sialylated antibodies to DC-SIGN, dendritic cells release IL-33 which induces the expansion of IL-4 producing basophils (Schwab et al. 2014a; Fillatreau 2014). The two Th2 cytokines upregulate the inhibitory FcγRIIb receptor in macrophages, resulting in the inhibition of inflammation (Anthony et al. 2011). As previously mentioned, glycosylation has been shown to take place extracellularly in some circumstances and could play a role in these mechanisms. In fact, IgG extracellularly glycosylated by soluble galactosyltransferases and ST6GAL1 are able to mimic the effects of IVIG through DC-SIGN, STAT6 signaling, and FcγRIIb (Pagan et al. 2018). In this process, the sialylation of pathogenic IgG was localized at the site of inflammation, because of the local availability of nucleotide-sugar donors released by platelets (Pagan et al. 2018). Thus, inflammation activates a self-inhibiting mechanism which, through the elevation of plasmatic ST6GAL1 (Lammers and Jamieson 1986) and the local presence of CMP-sialic acid donor leads to the extracellular α2,6-sialylation of Asn_{297}-linked IgG glycans, activating an inhibitory Fcγ receptor; this mechanism is exploited by IVIG. In addition, IVIG protects from experimental allergic bronchopulmonary aspergillosis via a sialylation-dependent mechanism, through reduced recruitment of eosinophils and inhibition of Th2 and Th17 responses, with an enhanced involvement of regulatory T cells and IL-10 (Bozza et al. 2019). Classical anaphylactic reactions are mediated by IgE. However, high allergen levels have the potential to mediate anaphylaxis through the binding of allergen-specific IgG to FcγRs on different immune cell types (Miyajima et al. 1997). This type of anaphylaxis can be inhibited by IgG, through a sialylation-dependent mechanism (Epp et al. 2018). However, this view on the mechanisms underlying IVIG is not unanimously accepted (Novokmet et al. 2014; Schwab et al. 2014b; Ballow 2014) and other mechanisms are probably operating. For example, a Fab-dependent but sialylation-independent mechanism through which IVIG exerts anti-inflammatory effects is based on stimulation of autophagy by dendritic cells and M1 macrophages (Das et al. 2020). The picture cannot be considered complete without mentioning recent papers questioning the proposed mechanism (Nagelkerke and Kuijpers 2014). In particular, it has been reported that IgGs do not interact with DC-SIGN, regardless of their glycosylation state (Temming et al. 2019; Yu et al.

2013), that basophils are not required (Campbell et al. 2014), that the amount of IL-33 released is not sufficient (Sharma et al. 2014) and that IgG sialylation is not necessary for T cells reciprocal modulation (Othy et al. 2014), phagocytosis by macrophages (Nagelkerke et al. 2014) or amelioration of murine immune thrombocytopenia (Leontyev et al. 2012).

An unusual mechanism has been claimed to play a role in cancer immunosuppression. Cancer cells could ectopically produce IgG which are abundantly sialylated. The binding of these antibodies to inhibitory sialic-acid receptors of the Siglec family would promote immunosuppression (Wang et al. 2019).

10.7.4 Anti IgG Autoantibodies

The presence of anti-IgG antibodies characterizes RA and other autoimmune diseases. These antibodies, which are often referred to as rheumatoid factors (RF), are in some cases directed against agalactosylated IgG (Mimura et al. 2004, 2005; Nishijima et al. 2001; Das et al. 2004; Maeno et al. 2004; Matsumoto et al. 2000; Soltys et al. 1994; Imafuku et al. 2003) (reviewed in (Goulabchand et al. 2014)). Some studies proposed that hypogalactosylation would promote the formation of aggregates of IgG and RF (Matsumoto et al. 2000; Leader et al. 1996), although this conclusion was not supported by others (Soltys et al. 1994; Imafuku et al. 2003; Falkenburg et al. 2017). Anti–citrullinated protein antibodies (ACPA) are highly specific autoantibodies for a subgroup of RA patients who have a severe erosive disease. An involvement of hypogalactosylated ACPA in disease pathogenesis is suggested by the following observations. (1) The concentration of hypogactosylated IgG1 ACPA is higher in the synovial fluid than in plasma of early RA patients (Scherer et al. 2010). (2) The level of IgG galactosylation was much lower in ACPA than in the rest of IgG molecules (Ercan et al. 2010). (3) The concentration of ACPA correlates with that of hypogalactosylated antibodies in RA patients (Schwedler et al. 2018). (4) In RA women with ACPA, the pregnancy-induced increase of IgG galactosylation affected ACPA and not total IgG (Bondt et al. 2018).

Sialylation of IgG plays an anti-inflammatory key role even through autoantibodies. This notion is supported by the following observations. In both human RA patients and mouse models of arthritis, the sialylation level of IgG is lower; consistently, sialylation of anti-collagen antibodies attenuates their arthritogenic activity (Ohmi et al. 2016). To explain the higher risk of developing RA by postmenopausal women, it has been proposed that, like in mice, estrogens activate ST6GAL1 in antibody-producing cells. Decrease in estrogens after menopause would lead to decreased α2,6 sialylation of IgG (Engdahl et al. 2018). In mouse models of lupus nephritis and RA, the presence of sialylated IgGs directed against self-antigen reduced inflammation through reduction of inflammatory T helper cells (Bartsch et al. 2018). In RA patients, desialylated, but not sialylated, immune complexes enhance osteoclastogenesis, while the Fc sialylation of IgG autoantibodies regulates the bone architecture (Harre et al. 2015). The anti-inflammatory role of sialylated IgG is confirmed also in glomerulonephritis (Otani et al. 2012).

10.8 Molecular Bases of N-Glycosylation Changes

A number of investigation report that the glycosylation changes associated with inflammatory diseases and aging are specific for Asn_{297} of IgG, while other studies report changes also in glycans linked to the light chains and in non IgG glycoproteins. Investigations on the enzymatic bases of IgG hypogalactosylation have utilized total lymphocyte populations or isolated B lymphocytes from human inflammatory conditions and mouse experimental systems. The studies summarized below are often conflicting and a conclusive evidence on the molecular bases of glycosylation changes of IgG is still lacking. The primary limit of these studies is in the use of B lymphocytes, rather than plasma cells, as an enzyme source. The first are more abundant and easy to obtain than the latter, but their biosynthetic apparatus is not necessarily representative of that of plasma cells, which are the main source of circulating antibodies. Lymphocytes from RA patients can produce in vitro hypogalactosylated IgG (Bodman et al. 1992), while an inverse relationship between hypogactosylated IgG and the level of B4GALTs has been reported by some studies (Axford et al. 1987, 1992; Alavi and Axford 1995). On the other hand, other studies reported a similar galactosyltransferase level in B lymphocytes of RA patients and healthy controls (Furukawa et al. 1990; Keusch et al. 1998; Delves et al. 1990; Jeddi et al. 1996). The importance of environmental factors in the regulation of IgG galactosylation is demonstrated by the fact that all-trans retinoic acid is able to modify specifically galactosylation of IgG1, without affecting glycosylation of other glycoproteins (Wang et al. 2011). Conflicting results have been obtained also from murine experimental models. In splenic lymphocytes of the arthritis-prone MLR lpr/lpr mice, the level of B4GALT1 transcript was found to be reduced (Jeddi et al. 1994, 1996). However, a lower galactosyltransferase activity was detected in peripheral, but not in splenic B lymphocytes in mouse models of arthritis (Axford et al. 1994; Alavi et al. 1998). B4GALT activity was found to be lower in hybridomas producing rheumatoid factors than in hybridomas secreting irrelevant antibodies (Axford et al. 1994).

As previously mentioned, the production of α2,6-sialylated IgG is a part of an anti-inflammatory response. In mice, these anti-inflammatory α2,6-sialylated IgG can be produced against T cell-independent antigens or against T cell-dependent antigens in a tolerogenic microenvironment through modulation of ST6GAL1 expression in plasma cells (Hess et al. 2013; Oefner et al. 2012). The increased level of bisecting GlcNAc associated with aging could be at least in part due to the fact that agalactosylated biantennary N-linked chains, which accumulate in aging, are a preferred substrate for MGAT3 (Narasimhan 1982), which is the only enzyme mediating the addition of bisecting GlcNAc (Ihara et al. 1993). On the other hand, promoter methylation of *MGAT3* gene correlates with the composition of the immunoglobulin G glycome in inflammatory bowel disease (Klasic et al. 2018).

Besides mechanistic studies, the bases of IgG glycome changes have been investigated through genomic and glycomic approaches. Genome-wide association studies (GWAS) indicated a link between given glycosidic traits and genetic loci associated with glycosyltransferases and with autoimmune and inflammatory diseases (Lauc et al. 2013). GWAS identified an association of agalactosylated sugar structures and single nucleotide polymorphisms (SNPs) in the loci *FUT8* and *ESR2*, which encodes estrogen receptor β, but not in *MGAT3* or *B4GALT1* (Lauc et al. 2009). Although the association between polymorphisms in glycosyltransferase genes and specific glycomic traits is rarely strong, polymorphisms in several transcription factors controlling key glycosyltransferases and inflammatory diseases have been identified. For example, *FUT8* has been shown to be associated with polymorphism in the *HNF1A* gene, encoding the transcription factor HNF1α (Sharapov et al. 2019; Lauc et al. 2010), and *IKZF1* (Klaric et al. 2020), while *MGAT3* is under the control of transcription factors RUNX1 and RUNX3 (Wahl et al. 2018).

10.9 Conclusions

After more than three decades of intensive investigation in the field, it is clear that a few glycomic traits of IgG are associated with inflammatory/autoimmune, metabolic, and infectious diseases, aging and cancer. Frequently, these changes precede disease onset and revert upon disease remission. The conditions associated with these IgG glycosylation changes share an inflammatory status which, in the case of aging, is referred to as "Inflammaging." Through different mechanisms, these aberrantly glycosylated IgG can fuel inflammation, triggering a vicious loop. Thus, it is very likely that the Asn_{297} glycosylation changes are intimately associated with the very complex mechanisms governing inflammation, even though the detailed molecular links between control of glycosylation in B lymphocytes and plasma cells and inflammation remain elusive. Considering the causal role of these glycans in sustaining or inhibiting inflammation, the full comprehension of their biosynthesis will provide a new tool for the manipulation of the inflammatory process.

Compliance with Ethical Standards

Conflict of Interest Author F.D declares that he has no conflict of interest. Author N.M. declares that she has no conflict of interest.

Ethical Approval This article does not contain any studies with human participants or animals performed by any of the authors.

References

Adler Y, Lamour A, Jamin C, Menez JF, Le Corre R, Shoenfeld Y, Youinou P (1995) Impaired binding capacity of asialyl and agalactosyl IgG to Fcγ receptors. Clin Exp Rheumatol 13:315–319

Alavi A, Axford J (1995) β1,4-galactosyltransferase variations in rheumatoid arthritis. Adv Exp Med Biol 376:185–192

Alavi A, Axford JS, Hay FC, Jones MG (1998) Tissue-specific galactosyltransferase abnormalities in an experimental model of rheumatoid arthritis. Ann Med Interne (Paris) 149:251–260

Alavi A, Arden N, Spector TD, Axford JS (2000) Immunoglobulin G glycosylation and clinical outcome in rheumatoid arthritis during pregnancy. J Rheumatol 27:1379–1385

Albert H, Collin M, Dudziak D, Ravetch JV, Nimmerjahn F (2008) *In vivo* enzymatic modulation of IgG glycosylation inhibits autoimmune disease in an IgG subclass-dependent manner. Proc Natl Acad Sci USA 105:15005–15009

Anthony RM, Wermeling F, Karlsson MC, Ravetch JV (2008a) Identification of a receptor required for the anti-inflammatory activity of IVIG. Proc Natl Acad Sci USA 105:19571–19578

Anthony RM, Nimmerjahn F, Ashline DJ, Reinhold VN, Paulson JC, Ravetch JV (2008b) Recapitulation of IVIG anti-inflammatory activity with a recombinant IgG Fc. Science 320:373–376

Anthony RM, Kobayashi T, Wermeling F, Ravetch JV (2011) Intravenous gammaglobulin suppresses inflammation through a novel TH2 pathway. Nature 475:110–113

Arnold JN, Dwek RA, Rudd PM, Sim RB (2006) Mannan binding lectin and its interaction with immunoglobulins in health and in disease. Immunol Lett 106:103–110

Axford JS, Mackenzie L, Lydyard PM, Hay FC, Isenberg DA, Roitt IM (1987) Reduced B-cell galactosyltransferase activity in rheumatoid arthritis. Lancet 2:1486–1488

Axford JS, Sumar N, Alavi A, Isenberg DA, Young A, Bodman KB, Roitt IM (1992) Changes in normal glycosylation mechanisms in autoimmune rheumatic disease. J Clin Invest 89:1021–1031

Axford JS, Alavi A, Bond A, Hay FC (1994) Differential B lymphocyte galactosyltransferase activity in the MRL mouse model of rheumatoid arthritis. Autoimmunity 17:157–163

Axford JS, Cunnane G, Fitzgerald O, Bland JM, Bresnihan B, Frears ER (2003) Rheumatic disease differentiation using immunoglobulin G sugar printing by high density electrophoresis. J Rheumatol 30:2540–2546

Baerenwaldt A, Biburger M, Nimmerjahn F (2010) Mechanisms of action of intravenous immunoglobulins. Expert Rev Clin Immunol 6:425–434

Bakovic MP, Selman MH, Hoffmann M, Rudan I, Campbell H, Deelder AM, Lauc G, Wuhrer M (2013) High-throughput IgG fc N-glycosylation profiling by mass spectrometry of glycopeptides. J Proteome Res 12:821–831

Ballow M (2014) Mechanisms of immune regulation by IVIG. Curr Opin Allergy Clin Immunol 14:509–515

Barb AW, Prestegard JH (2011) NMR analysis demonstrates immunoglobulin G N-glycans are accessible and dynamic. Nat Chem Biol 7:147–153

Bartsch YC, Rahmoller J, Mertes MMM, Eiglmeier S, Lorenz FKM, Stoehr AD, Braumann D, Lorenz AK, Winkler A, Lilienthal GM, Petry J, Hobusch J, Steinhaus M, Hess C, Holecska V, Schoen CT, Oefner CM, Leliavski A, Blanchard V, Ehlers M (2018) Sialylated autoantigen-reactive IgG antibodies attenuate disease development in autoimmune mouse models of lupus nephritis and rheumatoid arthritis. Front Immunol 9:1183. https://doi.org/10.3389/fimmu.2018.01183

Blomme B, Francque S, Trepo E, Libbrecht L, Vanderschaeghe D, Verrijken A, Pattyn P, Van Nieuwenhove Y, Van De PD, Geerts A, Colle I, Delanghe J, Moreno C, Van Gaal L, Callewaert N (2011) Van Vlierberghe. H. N-glycan based biomarker distinguishing non-alcoholic steatohepatitis from steatosis independently of fibrosis. Dig, Liver Dis

Bodman KB, Sumar N, Mackenzie LE, Isenberg DA, Hay FC, Roitt IM, Lydyard PM (1992) Lymphocytes from patients with rheumatoid arthritis produce agalactosylated IgG in vitro. Clin Exp Immunol 88:420–423

Bodman KB, Hutchings PR, Jeddi PA, Delves PJ, Rook GA, Sumar N, Roitt IM, Lydyard PM (1994) IgG glycosylation in autoimmune-prone strains of mice. Clin Exp Immunol 95:103–107

Bond A, Alavi A, Axford JS, Bourke BE, Bruckner FE, Kerr MA, Maxwell JD, Tweed KJ, Weldon MJ, Youinou P, Hay FC (1997) A detailed lectin analysis of IgG glycosylation, demonstrating disease specific changes in terminal galactose and N-acetylglucosamine. J Autoimmun 10:77–85

Bondt A, Selman MH, Deelder AM, Hazes JM, Willemsen SP, Wuhrer M, Dolhain RJ (2013) Association between galactosylation of immunoglobulin G and improvement of rheumatoid arthritis during pregnancy is independent of sialylation. J Proteome Res 12:4522–4531

Bondt A, Rombouts Y, Selman MH, Hensbergen PJ, Reiding KR, Hazes JM, Dolhain RJ, Wuhrer M (2014) Immunoglobulin G (IgG) Fab glycosylation analysis using a new mass spectrometric high-throughput profiling method reveals pregnancy-associated changes. Mol Cell Proteomics 13:3029–3039

Bondt A, Wuhrer M, Kuijper TM, Hazes JM, Dolhain RJ (2016) Fab glycosylation of immunoglobulin G does not associate with improvement of rheumatoid arthritis during pregnancy. Arthritis Res Ther 18:274. https://doi.org/10.1186/s13075-016-1172-1

Bondt A, Hafkenscheid L, Falck D, Kuijper TM, Rombouts Y, Hazes JMW, Wuhrer M, Dolhain RJEM (2018) ACPA IgG galactosylation associates with disease activity in pregnant patients with rheumatoid arthritis. Ann Rheum Dis 77:1130–1136. https://doi.org/10.1136/annrheumdis-2018-212946

Bones J, Byrne JC, O'Donoghue N, McManus C, Scaife C, Boissin H, Nastase A, Rudd PM (2011) Glycomic and glycoproteomic analysis of serum from patients with stomach cancer reveals potential markers arising from host defense response mechanisms. J Proteome Res 10:1246–1265

Borelli V, Vanhooren V, Lonardi E, Reiding KR, Capri M, Libert C, Garagnani P, Salvioli S, Franceschi C, Wuhrer M (2015) Plasma N-glycome signature of down syndrome. J Proteome Res 14:4232–4245. https://doi.org/10.1021/acs.jproteome.5b00356

Boyd PN, Lines AC, Patel AK (1995) The effect of the removal of sialic acid, galactose and total carbohydrate on the functional activity of Campath-1H. Mol Immunol 32:1311–1318

Bozza S, Kasermann F, Kaveri SV, Romani L, Bayry J (2019) Intravenous immunoglobulin protects from experimental allergic bronchopulmonary aspergillosis via a sialylation-dependent mechanism. Eur J Immunol 49:195–198. https://doi.org/10.1002/eji.201847774

Bruggeman CW, Dekkers G, Bentlage AEH, Treffers LW, Nagelkerke SQ, Lissenberg-Thunnissen S, Koeleman CAM, Wuhrer M, van den Berg TK, Rispens T, Vidarsson G, Kuijpers TW (2017) Enhanced effector functions due to antibody defucosylation depend on the effector cell Fcγ receptor profile. J Immunol 199:204–211. https://doi.org/10.4049/jimmunol.1700116

Campbell IK, Miescher S, Branch DR, Mott PJ, Lazarus AH, Han D, Maraskovsky E, Zuercher AW, Neschadim A, Leontyev D, McKenzie BS, Kasermann F (2014) Therapeutic effect of IVIG on inflammatory arthritis in mice is dependent on the Fc portion and independent of sialylation or basophils. J Immunol 192:5031–5038

Catera M, Borelli V, Malagolini N, Chiricolo M, Venturi G, Reis C, Osorio H, Abruzzo PM, Capri M, Monti D, Ostan R, Franceschi C, Dall'Olio F (2016) Identification of novel plasma glycosylation-associated markers of aging. Oncotarget. https://doi.org/10.18632/oncotarget.7059

Chang TT, Tsai HW, Ho CH (2018) Fucosyl-Agalactosyl IgG(1) induces cholangiocarcinoma metastasis and early recurrence by activating tumor-associated macrophage. Cancers (Basel) 10. https://doi.org/10.3390/cancers10110460

Chen G, Wang Y, Qin X, Li H, Guo Y, Wang Y, Liu H, Wang X, Song G, Li F, Li F, Guo S, Qiu L, Li Z (2013) Change in IgG1 Fc N-linked glycosylation in human lung cancer: age- and sex-related diagnostic potential. Electrophoresis 34:2407–2416

Chen G, Li H, Qiu L, Qin X, Liu H, Li Z (2014) Change of fucosylated IgG2 Fc-glycoforms in pancreatitis and pancreatic adenocarcinoma: a promising disease-classification model. Anal Bioanal Chem 406:267–273

Cheng HD, Stockmann H, Adamczyk B, McManus CA, Ercan A, Holm IA, Rudd PM, Ackerman ME, Nigrovic PA (2017) High-throughput characterization of the functional impact of IgG Fc glycan aberrancy in juvenile idiopathic arthritis. Glycobiology 27:1099–1108. https://doi.org/10.1093/glycob/cwx082

Chou CL, Wu MJ, Yu CL, Lu MC, Hsieh SC, Wu TH, Chou CT, Tsai CY (2010) Anti-agalactosyl IgG antibody in ankylosing spondylitis and psoriatic arthritis. Clin Rheumatol 29:875–881

Cobb BA (2020) The history of IgG glycosylation and where we are now. Glycobiology 30:202–213. https://doi.org/10.1093/glycob/cwz065

Collins ES, Galligan MC, Saldova R, Adamczyk B, Abrahams JL, Campbell MP, Ng CT, Veale DJ, Murphy TB, Rudd PM, Fitzgerald O (2013) Glycosylation status of serum in inflammatory arthritis in response to anti-TNF treatment. Rheumatology (Oxford) 52:1572–1582

Coman DJ, Murray DW, Byrne JC, Rudd PM, Bagaglia PM, Doran PD, Treacy EP (2010) Galactosemia, a single gene disorder with epigenetic consequences. Pediatr Res 67:286–292

Croce A, Firuzi O, Altieri F, Eufemi M, Agostino R, Priori R, Bombardieri M, Alessandri C, Valesini G, Saso L (2007) Effect of infliximab on the glycosylation of IgG of patients with rheumatoid arthritis. J Clin Lab Anal 21:303–314

Dall'Olio F, Vanhooren V, Chen CC, Slagboom PE, Wuhrer M, Franceschi C (2013) N-glycomic biomarkers of biological aging and longevity: a link with inflammaging. Ageing Res Rev 12:685–698

Dall'Olio F (2000) The sialyl-$\alpha 2,6$-lactosaminyl-structure: biosynthesis and functional role. Glycoconj J 17:669–676

Dall'Olio F (2018) Glycobiology of aging. Subcell Biochem 90:505–526. https://doi.org/10.1007/978-981-13-2835-0_17

Das H, Atsumi T, Fukushima Y, Shibuya H, Ito K, Yamada Y, Amasaki Y, Ichikawa K, Amengual O, Koike T (2004) Diagnostic value of antiagalactosyl IgG antibodies in rheumatoid arthritis. Clin Rheumatol 23:218–222

Das M, Karnam A, Stephen-Victor E, Gilardin L, Bhatt B, Kumar S, Rambabu N, Patil V, Lecerf M, Kasermann F, Bruneval P, Narayanaswamy BK, Benveniste O, Kaveri SV, Bayry J (2020) Intravenous immunoglobulin mediates anti-inflammatory effects in peripheral blood mononuclear cells by inducing autophagy. Cell Death Dis 11:50. https://doi.org/10.1038/s41419-020-2249-y

Davies J, Jiang L, Pan LZ, LaBarre MJ, Anderson D, Reff M (2001) Expression of GnTIII in a recombinant anti-CD20 CHO production cell line: expression of antibodies with altered glycoforms leads to an increase in ADCC through higher affinity for FC gamma RIII. Biotechnol Bioeng 74:288–294

de Haan N, van Tol MJD, Driessen GJ, Wuhrer M, Lankester AC (2018a) Immunoglobulin G fragment crystallizable glycosylation after hematopoietic stem cell transplantation is dissimilar to donor profiles. Front Immunol 9:1238. https://doi.org/10.3389/fimmu.2018.01238

de Haan N, Boeddha NP, Ekinci E, Reiding KR, Emonts M, Hazelzet JA, Wuhrer M, Driessen GJ (2018b) Differences in IgG fc glycosylation are associated with outcome of Pediatric meningococcal sepsis. MBio 9. https://doi.org/10.1128/mBio.00546-18

de Haan N, Falck D, Wuhrer M (2020) Monitoring of immunoglobulin N- and O-glycosylation in health and disease. Glycobiology 30:226–240. https://doi.org/10.1093/glycob/cwz048

de Jong SE, Selman MH, Adegnika AA, Amoah AS, van Riet E, Kruize YC, Raynes JG, Rodriguez A, Boakye D, von Mutius E, Knulst AC, Genuneit J, Cooper PJ, Hokke CH, Wuhrer M, Yazdanbakhsh M (2016) IgG1 Fc N-glycan galactosylation as a biomarker for immune activation. Sci Rep 6:28207. https://doi.org/10.1038/srep28207

Dekkers G, Treffers L, Plomp R, Bentlage AEH, de Boer M, Koeleman CAM, Lissenberg-Thunnissen SN, Visser R, Brouwer M, Mok JY, Matlung H, van den Berg TK, van Esch WJE, Kuijpers TW, Wouters D, Rispens T, Wuhrer M, Vidarsson G (2017) Decoding the human immunoglobulin G-glycan repertoire reveals a spectrum of Fc-receptor- and complement-mediated-effector activities. Front Immunol 8:877. https://doi.org/10.3389/fimmu.2017.00877

Dekkers G, Rispens T, Vidarsson G (2018a) Novel concepts of altered immunoglobulin G galactosylation in autoimmune diseases. Front Immunol 9:553. https://doi.org/10.3389/fimmu.2018.00553

Dekkers G, Bentlage AEH, Plomp R, Visser R, Koeleman CAM, Beentjes A, Mok JY, van Esch WJE, Wuhrer M, Rispens T, Vidarsson G (2018b) Conserved FcγR-glycan discriminates between fucosylated and afucosylated IgG in humans and mice. Mol Immunol 94:54–60. https://doi.org/10.1016/j.molimm.2017.12.006

Delves PJ, Lund T, Axford JS, Alavi-Sadrieh A, Lydyard PM, Mackenzie L, Smith MD, Kidd VJ (1990) Polymorphism and expression of the galactosyltransferase-associated protein kinase gene in normal individuals and galactosylation-defective rheumatoid arthritis patients. Arthritis Rheum 33:1655–1664

Dong X, Storkus WJ, Salter RD (1999) Binding and uptake of agalactosyl IgG by mannose receptor on macrophages and dendritic cells. J Immunol 163:5427–5434

Dube R, Rook GA, Steele J, Brealey R, Dwek R, Rademacher T, Lennard-Jones J (1990) Agalactosyl IgG in inflammatory bowel disease: correlation with C-reactive protein. Gut 31:431–434

Edri-Brami M, Rosental B, Hayoun D, Welt M, Rosen H, Wirguin I, Nefussy B, Drory VE, Porgador A, Lichtenstein RG (2012) Glycans in sera of amyotrophic lateral sclerosis patients and their role in killing neuronal cells. PLoS One 7:e35772

Endo T, Iwakura Y, Kobata A (1993) Structural changes in the N-linked sugar chains of serum immunoglobulin G of HTLV-I transgenic mice. Biochem Biophys Res Commun 192:1004–1010

Engdahl C, Bondt A, Harre U, Raufer J, Pfeifle R, Camponeschi A, Wuhrer M, Seeling M, Martensson IL, Nimmerjahn F, Kronke G, Scherer HU, Forsblad-d'Elia H, Schett G (2018) Estrogen induces St6gal1 expression and increases IgG sialylation in mice and patients with rheumatoid arthritis: a potential explanation for the increased risk of rheumatoid arthritis in postmenopausal women. Arthritis Res Ther 20:84. https://doi.org/10.1186/s13075-018-1586-z

Epp A, Hobusch J, Bartsch YC, Petry J, Lilienthal GM, Koeleman CAM, Eschweiler S, Mobs C, Hall A, Morris SC, Braumann D, Engellenner C, Bitterling J, Rahmoller J, Leliavski A, Thurmann R, Collin M, Moremen KW, Strait RT, Blanchard V, Petersen A, Gemoll T, Habermann JK, Petersen F, Nandy A, Kahlert H, Hertl M, Wuhrer M, Pfutzner W, Jappe U, Finkelman FD, Ehlers M (2018) Sialylation of IgG antibodies inhibits IgG-mediated allergic reactions. J Allergy Clin Immunol 141:399–402. https://doi.org/10.1016/j.jaci.2017.06.021

Ercan A, Cui J, Chatterton DE, Deane KD, Hazen MM, Brintnell W, O'Donnell CI, Derber LA, Weinblatt ME, Shadick NA, Bell DA, Cairns E, Solomon DH, Holers VM, Rudd PM, Lee DM (2010) Aberrant IgG galactosylation precedes disease onset, correlates with disease activity, and is prevalent in autoantibodies in rheumatoid arthritis. Arthritis Rheum 62:2239–2248

Ercan A, Kohrt WM, Cui J, Deane KD, Pezer M, Yu EW, Hausmann JS, Campbell H, Kaiser UB, Rudd PM, Lauc G, Wilson JF, Finkelstein JS, Nigrovic PA (2017) Estrogens regulate glycosylation of IgG in women and men. JCI Insight 2:e89703. https://doi.org/10.1172/jci.insight.89703

Ezekowitz RA (1995) Agalactosyl IgG and mannose-binding proteins: biochemical nicety or pathophysiological paradigm? Nat Med 1:207–208

Falconer DJ, Subedi GP, Marcella AM, Barb AW (2018) Antibody fucosylation lowers the FcγRIIIa/CD16a affinity by limiting the conformations sampled by the N162-glycan. ACS Chem Biol 13:2179–2189. https://doi.org/10.1021/acschembio.8b00342

Falkenburg WJJ, Kempers AC, Dekkers G, Ooijevaar-de HP, Bentlage AEH, Vidarsson G, van Schaardenburg D, Toes REM, Scherer HU, Rispens T (2017) Rheumatoid factors do not preferentially bind to ACPA-IgG or IgG with altered galactosylation. Rheumatology (Oxford) 56:2025–2030. https://doi.org/10.1093/rheumatology/kex284

Ferrara C, Grau S, Jager C, Sondermann P, Brunker P, Waldhauer I, Hennig M, Ruf A, Rufer AC, Stihle M, Umana P, Benz J (2011) Unique carbohydrate-carbohydrate interactions are required for high affinity binding between FcγRIII and antibodies lacking core fucose. Proc Natl Acad Sci USA 108:12669–12674

Field MC, Amatayakul-Chantler S, Rademacher TW, Rudd PM, Dwek RA (1994) Structural analysis of the N-glycans from human immunoglobulin A1: comparison of normal human serum immunoglobulin A1 with that isolated from patients with rheumatoid arthritis. Biochem J 299(Pt 1):261–275

Fillatreau S (2014) Sweetened antibodies against humoral autoimmunity: sialylated antibodies are required for IVIg-mediated therapy. Eur J Immunol 44:1276–1280

Filley E, Andreoli A, Steele J, Waters M, Wagner D, Nelson D, Tung K, Rademacher T, Dwek R, Rook GA (1989) A transient rise in agalactosyl IgG correlating with free interleukin 2 receptors, during episodes of erythema nodosum leprosum. Clin Exp Immunol 76:343–347

Flogel M, Lauc G, Gornik I, Macek B (1998) Fucosylation and galactosylation of IgG heavy chains differ between acute and remission phases of juvenile chronic arthritis. Clin Chem Lab Med 36:99–102

Fokkink WJ, Selman MH, Dortland JR, Durmus B, Kuitwaard K, Huizinga R, van Rijs W, Tio-Gillen AP, van Doorn PA, Deelder AM, Wuhrer M, Jacobs BC (2014) IgG Fc N-glycosylation in Guillain-Barre syndrome treated with immunoglobulins. J Proteome Res 13:1722–1730

Franceschi C (2007) Inflammaging as a major characteristic of old people: can it be prevented or cured? Nutr Rev 65:S173–S176

Franceschi C, Capri M, Monti D, Giunta S, Olivieri F, Sevini F, Panourgia MP, Invidia L, Celani L, Scurti M, Cevenini E, Castellani GC, Salvioli S (2007) Inflammaging and anti-inflammaging: a systemic perspective on aging and longevity emerged from studies in humans. Mech Ageing Dev 128:92–105

Franceschi C, Garagnani P, Parini P, Giuliani C, Santoro A (2018) Inflammaging: a new immune-metabolic viewpoint for age-related diseases. Nat Rev Endocrinol 14:576–590. https://doi.org/10.1038/s41574-018-0059-4

Furukawa K, Matsuta K, Takeuchi F, Kosuge E, Miyamoto T, Kobata A (1990) Kinetic study of a galactosyltransferase in the B cells of patients with rheumatoid arthritis. Int Immunol 2:105–112

Gao Q, Dolikun M, Stambuk J, Wang H, Zhao F, Yiliham N, Wang Y, Trbojevic-Akmacic I, Zhang J, Fang H, Sun Y, Peng H, Zhao Z, Liu D, Liu J, Li Q, Sun Q, Wu L, Lauc G, Wang W, Song M (2017) Immunoglobulin G N-glycans as potential postgenomic biomarkers for hypertension in the Kazakh population. OMICS 21:380–389. https://doi.org/10.1089/omi.2017.0044

Gornik O, Wagner J, Pucic M, Knezevic A, Redzic I, Lauc G (2009) Stability of N-glycan profiles in human plasma. Glycobiology 19:1547–1553

Goulabchand R, Vincent T, Batteux F, Eliaou JF, Guilpain P (2014) Impact of autoantibody glycosylation in autoimmune diseases. Autoimmun Rev 13:742–750

Groenink J, Spijker J, van den Herik-Oudijk IE, Boeije L, Rook G, Aarden L, Smeenk R, van de Winkel JG, van den Broek MF (1996) On the interaction between agalactosyl IgG and Fc gamma receptors. Eur J Immunol 26:1404–1407

Gudelj I, Salo PP, Trbojevic-Akmacic I, Albers M, Primorac D, Perola M, Lauc G (2018) Low galactosylation of IgG associates with higher risk for future diagnosis of rheumatoid arthritis during 10 years of follow-up. Biochim Biophys Acta Mol basis Dis 1864:2034–2039. https://doi.org/10.1016/j.bbadis.2018.03.018

Gui HL, Gao CF, Wang H, Liu XE, Xie Q, Dewaele S, Wang L, Zhuang H, Contreras R, Libert C, Chen C (2010) Altered serum N-glycomics in chronic hepatitis B patients. Liver Int 30:259–267

Harbison A, Fadda E (2020) An atomistic perspective on antibody-dependent cellular cytotoxicity quenching by core-fucosylation of IgG1 Fc N-glycans from enhanced sampling molecular dynamics. Glycobiology 30:407–414. https://doi.org/10.1093/glycob/cwz101

Harre U, Lang SC, Pfeifle R, Rombouts Y, Fruhbeisser S, Amara K, Bang H, Lux A, Koeleman CA, Baum W, Dietel K, Grohn F, Malmstrom V, Klareskog L, Kronke G, Kocijan R, Nimmerjahn F,

Toes RE, Herrmann M, Scherer HU, Schett G (2015) Glycosylation of immunoglobulin G determines osteoclast differentiation and bone loss. Nat Commun 6:6651. https://doi.org/10.1038/ncomms7651

Hayes JM, Frostell A, Cosgrave EF, Struwe WB, Potter O, Davey GP, Karlsson R, Anneren C, Rudd PM (2014) Fc γ receptor glycosylation modulates the binding of IgG glycoforms: a requirement for stable antibody interactions. J Proteome Res 13:5471–5485

Hernandez-Pando R, Reyes P, Espitia C, Wang Y, Rook G, Mancilla R (1994) Raised agalactosyl IgG and antimycobacterial humoral immunity in Takayasu's arteritis. J Rheumatol 21:1870–1876

Hess C, Winkler A, Lorenz AK, Holecska V, Blanchard V, Eiglmeier S, Schoen AL, Bitterling J, Stoehr AD, Petzold D, Schommartz T, Mertes MM, Schoen CT, Tiburzy B, Herrmann A, Kohl J, Manz RA, Madaio MP, Berger M, Wardemann H, Ehlers M (2013) T cell-independent B cell activation induces immunosuppressive sialylated IgG antibodies. J Clin Invest 123:3788–3796

Ho CH, Chien RN, Cheng PN, Liu CK, Su CS, Wu IC, Liu WC, Chen SH, Chang TT (2014) Association of serum IgG N-glycome and transforming growth factor-β1 with hepatitis B virus e antigen seroconversion during entecavir therapy. Antivir Res 111:121–128

Ho CH, Chien RN, Cheng PN, Liu JH, Liu CK, Su CS, Wu IC, Li IC, Tsai HW, Wu SL, Liu WC, Chen SH, Chang TT (2015) Aberrant serum immunoglobulin G glycosylation in chronic hepatitis B is associated with histological liver damage and reversible by antiviral therapy. J Infect Dis 211:115–124

Ho CH, Chen SH, Tsai HW, Wu IC, Chang TT (2019) Fully galactosyl-fucosyl-bisected IgG1 reduces anti-HBV efficacy and liver histological improvement. Antivir Res 163:1–10. https://doi.org/10.1016/j.antiviral.2018.12.021

Hodoniczky J, Zheng YZ, James DC (2005) Control of recombinant monoclonal antibody effector functions by Fc N-glycan remodeling in vitro. Biotechnol Prog 21:1644–1652

Holland M, Takada K, Okumoto T, Takahashi N, Kato K, Adu D, Ben Smith A, Harper L, Savage CO, Jefferis R (2002) Hypogalactosylation of serum IgG in patients with ANCA-associated systemic vasculitis. Clin Exp Immunol 129:183–190

Holland M, Yagi H, Takahashi N, Kato K, Savage CO, Goodall DM, Jefferis R (2006/1760) Differential glycosylation of polyclonal IgG, IgG-Fc and IgG-Fab isolated from the sera of patients with ANCA-associated systemic vasculitis. Biochim Biophys Acta 669–677

Ihara Y, Nishikawa A, Tohma T, Soejima H, Niikawa N, Taniguchi N (1993) cDNA cloning, expression, and chromosomal localization of human N-acetylglucosaminyltransferase III (GnT-III). J Biochem (Tokyo) 113:692–698

Iida S, Misaka H, Inoue M, Shibata M, Nakano R, Yamane-Ohnuki N, Wakitani M, Yano K, Shitara K, Satoh M (2006) Nonfucosylated therapeutic IgG1 antibody can evade the inhibitory effect of serum immunoglobulin G on antibody-dependent cellular cytotoxicity through its high binding to FcγRIIIa. Clin Cancer Res 12:2879–2887

Iida S, Kuni-Kamochi R, Mori K, Misaka H, Inoue M, Okazaki A, Shitara K, Satoh M (2009) Two mechanisms of the enhanced antibody-dependent cellular cytotoxicity (ADCC) efficacy of non-fucosylated therapeutic antibodies in human blood. BMC Cancer 9:58

Imafuku Y, Yoshida H, Yamada Y (2003) Reactivity of agalactosyl IgG with rheumatoid factor. Clin Chim Acta 334:217–223

Ito K, Furukawa J, Yamada K, Tran NL, Shinohara Y, Izui S (2014) Lack of galactosylation enhances the pathogenic activity of IgG1 but not IgG2a anti-erythrocyte autoantibodies. J Immunol 192:581–588

Jeddi PA, Lund T, Bodman KB, Sumar N, Lydyard PM, Pouncey L, Heath LS, Kidd VJ, Delves PJ (1994) Reduced galactosyltransferase mRNA levels are associated with the agalactosyl IgG found in arthritis-prone MRL-lpr/lpr strain mice. Immunology 83:484–488

Jeddi PA, Bodman-Smith KB, Lund T, Lydyard PM, Mengle-Gaw L, Isenberg DA, Youinou P, Delves PJ (1996) Agalactosyl IgG and β-1,4-galactosyltransferase gene expression in rheumatoid arthritis patients and in the arthritis-prone MRL lpr/lpr mouse. Immunology 87:654–659

Jones MB, Nasirikenari M, Lugade AA, Thanavala Y, Lau JT (2012) Anti-inflammatory IgG production requires functional P1 promoter in beta-Galactoside α2,6-Sialyltransferase 1 (ST6Gal-1) gene. J Biol Chem 287:15365–15370

Jones MB, Oswald DM, Joshi S, Whiteheart SW, Orlando R, Cobb BA (2016) B-cell-independent sialylation of IgG. Proc Natl Acad Sci USA 113:7207–7212. https://doi.org/10.1073/pnas.1523968113

Kaneko Y, Nimmerjahn F, Ravetch JV (2006) Anti-inflammatory activity of immunoglobulin G resulting from Fc sialylation. Science 313:670–673

Kanoh Y, Mashiko T, Danbara M, Takayama Y, Ohtani S, Imasaki T, Abe T, Akahoshi T (2004a) Analysis of the oligosaccharide chain of human serum immunoglobulin G in patients with localized or metastatic cancer. Oncology 66:365–370

Kanoh Y, Mashiko T, Danbara M, Takayama Y, Ohtani S, Egawa S, Baba S, Akahoshi T (2004b) Changes in serum IgG oligosaccharide chains with prostate cancer progression. Anticancer Res 24:3135–3139

Kanoh Y, Ohara T, Mashiko T, Abe T, Masuda N, Akahoshi T (2006) Relationship between N-linked oligosaccharide chains of human serum immunoglobulin G and serum tumor markers with non-small cell lung cancer progression. Anticancer Res 26:4293–4297

Karsten CM, Pandey MK, Figge J, Kilchenstein R, Taylor PR, Rosas M, McDonald JU, Orr SJ, Berger M, Petzold D, Blanchard V, Winkler A, Hess C, Reid DM, Majoul IV, Strait RT, Harris NL, Kohl G, Wex E, Ludwig R, Zillikens D, Nimmerjahn F, Finkelman FD, Brown GD, Ehlers M, Kohl J (2012) Anti-inflammatory activity of IgG1 mediated by Fc galactosylation and association of FcγRIIB and dectin-1. Nat Med 18:1401–1406

Kemna MJ, Plomp R, van Paassen P, Koeleman CA, Jansen BC, Damoiseaux JG, Cohen Tervaert JW, Wuhrer M (2017) Galactosylation and sialylation levels of IgG predict relapse in patients with PR3-ANCA associated Vasculitis. EBioMedicine 17:108–118. https://doi.org/10.1016/j.ebiom.2017.01.033

Keusch J, Lydyard PM, Berger EG, Delves PJ (1998) B lymphocyte galactosyltransferase protein levels in normal individuals and in patients with rheumatoid arthritis. Glycoconj J 15:1093–1097

Kibe T, Fujimoto S, Ishida C, Togari H, Wada Y, Okada S, Nakagawa H, Tsukamoto Y, Takahashi N (1996) Glycosylation and placental transport of immunoglobulin G. J Clin Biochem Nutr 21:57–63

Kiyoshi M, Caaveiro JMM, Tada M, Tamura H, Tanaka T, Terao Y, Morante K, Harazono A, Hashii N, Shibata H, Kuroda D, Nagatoishi S, Oe S, Ide T, Tsumoto K, Ishii-Watabe A (2018) Assessing the heterogeneity of the Fc-glycan of a therapeutic antibody using an engineered FcγReceptor IIIa-immobilized column. Sci Rep 8:3955. https://doi.org/10.1038/s41598-018-22199-8

Kjeldsen-Kragh J, Sumar N, Bodman-Smith K, Brostoff J (1996) Changes in glycosylation of IgG during fasting in patients with rheumatoid arthritis. Br J Rheumatol 35:117–119

Klaric L, Tsepilov YA, Stanton CM, Mangino M, Sikka TT, Esko T, Pakhomov E, Salo P, Deelen J, McGurnaghan SJ, Keser T, Vuckovic F, Ugrina I, Kristic J, Gudelj I, Stambuk J, Plomp R, Pucic-Bakovic M, Pavic T, Vilaj M, Trbojevic-Akmacic I, Drake C, Dobrinic P, Mlinarec J, Jelusic B, Richmond A, Timofeeva M, Grishchenko AK, Dmitrieva J, Bermingham ML, Sharapov SZ, Farrington SM, Theodoratou E, Uh HW, Beekman M, Slagboom EP, Louis E, Georges M, Wuhrer M, Colhoun HM, Dunlop MG, Perola M, Fischer K, Polasek O, Campbell H, Rudan I, Wilson JF, Zoldos V, Vitart V, Spector T, Aulchenko YS, Lauc G, Hayward C (2020) Glycosylation of immunoglobulin G is regulated by a large network of genes pleiotropic with inflammatory diseases. Sci Adv 6:eaax0301. https://doi.org/10.1126/sciadv.aax0301

Klasic M, Markulin D, Vojta A, Samarzija I, Birus I, Dobrinic P, Ventham NT, Trbojevic-Akmacic I, Simurina M, Stambuk J, Razdorov G, Kennedy NA, Satsangi J, Dias AM, Pinho S, Annese V, Latiano A, D'Inca R, Lauc G, Zoldos V (2018) Promoter methylation of the MGAT3 and BACH2 genes correlates with the composition of the immunoglobulin G glycome in inflammatory bowel disease. Clin Epigenetics 10:75. https://doi.org/10.1186/s13148-018-0507-y

Knezevic A, Polasek O, Gornik O, Rudan I, Campbell H, Hayward C, Wright A, Kolcic I, O'Donoghue N, Bones J, Rudd PM, Lauc G (2009) Variability, heritability and environmental determinants of human plasma N-glycome. J Proteome Res 8:694–701

Knezevic A, Gornik O, Polasek O, Pucic M, Redzic I, Novokmet M, Rudd PM, Wright AF, Campbell H, Rudan I, Lauc G (2010) Effects of aging, body mass index, plasma lipid profiles, and smoking on human plasma N-glycans. Glycobiology 20:959–969

Komaromy A, Reider B, Jarvas G, Guttman A (2020) Glycoprotein biomarkers and analysis in chronic obstructive pulmonary disease and lung cancer with special focus on serum immunoglobulin G. Clin Chim Acta 506:204–213. https://doi.org/10.1016/j.cca.2020.03.041

Krapp S, Mimura Y, Jefferis R, Huber R, Sondermann P (2003) Structural analysis of human IgG-Fc glycoforms reveals a correlation between glycosylation and structural integrity. J Mol Biol 325:979–989

Kristic J, Vuckovic F, Menni C, Klaric L, Keser T, Beceheli I, Pucic-Bakovic M, Novokmet M, Mangino M, Thaqi K, Rudan P, Novokmet N, Sarac J, Missoni S, Kolcic I, Polasek O, Rudan I, Campbell H, Hayward C, Aulchenko Y, Valdes A, Wilson JF, Gornik O, Primorac D, Zoldos V, Spector T, Lauc G (2014) Glycans are a novel biomarker of chronological and biological ages. J Gerontol A Biol Sci Med Sci 69:779–789

Kumpel BM, Wang Y, Griffiths HL, Hadley AG, Rook GA (1995) The biological activity of human monoclonal IgG anti-D is reduced by β-galactosidase treatment. Hum Antibodies Hybridomas 6:82–88

Kuroda Y, Nakata M, Hirose S, Shirai T, Iwamoto M, Izui S, Kojima N, Mizuochi T (2001) Abnormal IgG galactosylation in MRL-lpr/lpr mice: pathogenic role in the development of arthritis. Pathol Int 51:909–915

Lammers G, Jamieson JC (1986) Studies on the effect of experimental inflammation on sialyltransferase in the mouse and Guinea pig. Comp Biochem Physiol B 84:181–187

Lardinois OM, Deterding LJ, Hess JJ, Poulton CJ, Henderson CD, Jennette JC, Nachman PH, Falk RJ (2019) Immunoglobulins G from patients with ANCA-associated vasculitis are atypically glycosylated in both the Fc and Fab regions and the relation to disease activity. PLoS One 14: e0213215. https://doi.org/10.1371/journal.pone.0213215

Lauc G, Huffman J, Hayward C, Knezevic A, Polasek O, Gornik O, Vitart V, Kolcic I, Biloglav Z, Zgaga L, Hastie ND, Wright AF, Campbell H, Rudd PM, Rudan I (2009) Genome-wide association study identifies *FUT8* and *ESR2* as co-regulators of a bi-antennary N-linked glycan A2 (GlcNAc2Man3GlcNAc2) in human plasma proteins. Nat Preccedings

Lauc G, Essafi A, Huffman JE, Hayward C, Knezevic A, Kattla JJ, Polasek O, Gornik O, Vitart V, Abrahams JL, Pucic M, Novokmet M, Redzic I, Campbell S, Wild SH, Borovecki F, Wang W, Kolcic I, Zgaga L, Gyllensten U, Wilson JF, Wright AF, Hastie ND, Campbell H, Rudd PM, Rudan I (2010) Genomics meets glycomics-the first GWAS study of human N-glycome identifies HNF1α as a master regulator of plasma protein fucosylation. PLoS Genet 6:e1001256

Lauc G, Huffman JE, Pucic M, Zgaga L, Adamczyk B, Muzinic A, Novokmet M, Polasek O, Gornik O, Kristic J, Keser T, Vitart V, Scheijen B, Uh HW, Molokhia M, Patrick AL, McKeigue P, Kolcic I, Lukic IK, Swann O, van Leeuwen FN, Ruhaak LR, Houwing-Duistermaat JJ, Slagboom PE, Beekman M, de Craen AJ, Deelder AM, Zeng Q, Wang W, Hastie ND, Gyllensten U, Wilson JF, Wuhrer M, Wright AF, Rudd PM, Hayward C, Aulchenko Y, Campbell H, Rudan I (2013) Loci associated with N-glycosylation of human immunoglobulin g show pleiotropy with autoimmune diseases and haematological cancers. PLoS Genet 9:e1003225

Leader KA, Lastra GC, Kirwan JR, Elson CJ (1996) Agalactosyl IgG in aggregates from the rheumatoid joint. Br J Rheumatol 35:335–341

Lee MM, Nasirikenari M, Manhardt CT, Ashline DJ, Hanneman AJ, Reinhold VN, Lau JT (2014) Platelets support extracellular sialylation by supplying the sugar donor substrate. J Biol Chem 289:8742–8748

Leirisalo-Repo M, Hernandez-Munoz HE, Rook GA (1999) Agalactosyl IgG is elevated in patients with active spondyloarthropathy. Rheumatol Int 18:171–176

Lemmers RFH, Vilaj M, Urda D, Agakov F, Simurina M, Klaric L, Rudan I, Campbell H, Hayward C, Wilson JF, Lieverse AG, Gornik O, Sijbrands EJG, Lauc G, van Hoek M (2017) IgG glycan patterns are associated with type 2 diabetes in independent European populations. Biochim Biophys Acta 1861:2240–2249. https://doi.org/10.1016/j.bbagen.2017.06.020

Leontyev D, Katsman Y, Ma XZ, Miescher S, Kasermann F, Branch DR (2012) (1799–1805) Sialylation-independent mechanism involved in the amelioration of murine immune thrombocytopenia using intravenous gammaglobulin. Transfusion 52

Li T, DiLillo DJ, Bournazos S, Giddens JP, Ravetch JV, Wang LX (2017) Modulating IgG effector function by Fc glycan engineering. Proc Natl Acad Sci USA 114:3485–3490. https://doi.org/10.1073/pnas.1702173114

Liu JN, Dolikun M, Stambuk J, Trbojevic-Akmacic I, Zhang J, Wang H, Zheng DQ, Zhang XY, Peng HL, Zhao ZY, Liu D, Sun Y, Sun Q, Li QH, Zhang JX, Sun M, Cao WJ, Momcilovic A, Razdorov G, Wu LJ, Russell A, Wang YX, Song MS, Lauc G, Wang W (2018a) The association between subclass-specific IgG Fc N-glycosylation profiles and hypertension in the Uygur, Kazak, Kirgiz, and Tajik populations. J Hum Hypertens 32:555–563. https://doi.org/10.1038/s41371-018-0071-0

Liu D, Zhao Z, Wang A, Ge S, Wang H, Zhang X, Sun Q, Cao W, Sun M, Wu L, Song M, Zhou Y, Wang W, Wang Y (2018b) Ischemic stroke is associated with the pro-inflammatory potential of N-glycosylated immunoglobulin G. J Neuroinflammation 15:123. https://doi.org/10.1186/s12974-018-1161-1

Liu J, Dolikun M, Stambuk J, Trbojevic-Akmacic I, Zhang J, Zhang J, Wang H, Meng X, Razdorov G, Menon D, Zheng D, Wu L, Wang Y, Song M, Lauc G, Wang W (2019a) Glycomics for type 2 diabetes biomarker discovery: promise of immunoglobulin G subclass-specific fragment crystallizable N-glycosylation in the Uyghur population. OMICS 23:640–648. https://doi.org/10.1089/omi.2019.0052

Liu J, Zhu Q, Han J, Zhang H, Li Y, Ma Y, He D, Gu J, Zhou X, Reveille JD, Jin L, Zou H, Ren S, Wang J (2019b) IgG Galactosylation status combined with MYOM2-rs2294066 precisely predicts anti-TNF response in ankylosing spondylitis. Mol Med 25:25. https://doi.org/10.1186/s10020-019-0093-2

Liu D, Li Q, Dong J, Li D, Xu X, Xing W, Zhang X, Cao W, Hou H, Wang H, Song M, Tao L, Kang X, Meng Q, Wang W, Guo X, Wang Y (2019c) The association between normal BMI with central adiposity and proinflammatory potential immunoglobulin G N-glycosylation. Diabetes Metab Syndr Obes 12:2373–2385. https://doi.org/10.2147/DMSO.S216318

Liu S, Huang Z, Zhang Q, Fu Y, Cheng L, Liu BF, Liu X (2019d) Profiling of isomer-specific IgG N-glycosylation in cohort of Chinese colorectal cancer patients. Biochim Biophys Acta Gen Subj 1864:129510. https://doi.org/10.1016/j.bbagen.2019.129510

Liu J, Zhu Q, Han J, Zhang H, Li Y, Ma Y, Ji H, He D, Gu J, Zhou X, Reveille JD, Jin L, Zou H, Ren S, Wang J (2020a) The IgG galactosylation ratio is higher in spondyloarthritis patients and associated with the MRI score. Clin Rheumatol 39:2317–2323. https://doi.org/10.1007/s10067-020-04998-5

Liu Q, Lin J, Han J, Zhang Y, Lu J, Tu W, Zhao Y, Guo G, Chu H, Pu W, Liu J, Ma Y, Chen X, Zhang R, Gu J, Zou H, Jin L, Wu W, Ren S, Wang J (2020b) Immunoglobulin G galactosylation levels are decreased in systemic sclerosis patients and differ according to disease subclassification. Scand J Rheumatol 49:146–153. https://doi.org/10.1080/03009742.2019.1641615

Liu P, Ren S, Xie Y, Liu C, Qin W, Zhou Y, Zhang M, Yang Q, Chen XC, Liu T, Yao Q, Xiao Z, Gu J, Zhang XL (2020c) Quantitative analysis of serum-based IgG agalactosylation for tuberculosis auxiliary diagnosis. Glycobiology 30:746–759. https://doi.org/10.1093/glycob/cwaa021

Liu S, Fu Y, Huang Z, Liu Y, Liu BF, Cheng L, Liu X (2020d) A comprehensive analysis of subclass-specific IgG glycosylation in colorectal cancer progression by nanoLC-MS/MS. Analyst. https://doi.org/10.1039/d0an00369g

Lok SM (2017) Unsweetened IgG is bad for dengue patients. Cell Host Microbe 21:312–314. https://doi.org/10.1016/j.chom.2017.02.011

Lundstrom SL, Yang H, Lyutvinskiy Y, Rutishauser D, Herukka SK, Soininen H, Zubarev RA (2014) Blood plasma IgG Fc glycans are significantly altered in Alzheimer's disease and progressive mild cognitive impairment. J Alzheimers Dis 38:567–579

Lundstrom SL, Hensvold AH, Rutishauser D, Klareskog L, Ytterberg AJ, Zubarev RA, Catrina AI (2017) IgG Fc galactosylation predicts response to methotrexate in early rheumatoid arthritis. Arthritis Res Ther 19:182. https://doi.org/10.1186/s13075-017-1389-7

Maeno N, Takei S, Fujikawa S, Yamada Y, Imanaka H, Hokonohara M, Kawano Y, Oda H (2004) Antiagalactosyl IgG antibodies in juvenile idiopathic arthritis, juvenile onset Sjogren's syndrome, and healthy children. J Rheumatol 31:1211–1217

Malhotra R, Wormald MR, Rudd PM, Fischer PB, Dwek RA, Sim RB (1995) Glycosylation changes of IgG associated with rheumatoid arthritis can activate complement via the mannose-binding protein. Nat Med 1:237–243

Manhardt CT, Punch PR, Dougher CWL, Lau JTY (2017) Extrinsic sialylation is dynamically regulated by systemic triggers in vivo. J Biol Chem 292:13514–13520. https://doi.org/10.1074/jbc.C117.795138

Martin K, Talukder R, Hay FC, Axford JS (2001) Characterization of changes in IgG associated oligosaccharide profiles in rheumatoid arthritis, psoriatic arthritis, and ankylosing spondylitis using fluorophore linked carbohydrate electrophoresis. J Rheumatol 28:1531–1536

Martin TC, Ilieva KM, Visconti A, Beaumont M, Kiddle SJ, Dobson RJB, Mangino M, Lim EM, Pezer M, Steves CJ, Bell JT, Wilson SG, Lauc G, Roederer M, Walsh JP, Spector TD, Karagiannis SN (2020a) Dysregulated antibody, natural killer cell and immune mediator profiles in autoimmune thyroid diseases. Cell 9:665. https://doi.org/10.3390/cells9030665

Martin TC, Simurina M, Zabczynska M, Martinic KM, Rydlewska M, Pezer M, Kozlowska K, Burri A, Vilaj M, Turek-Jabrocka R, Krnjajic-Tadijanovic M, Trofimiuk-Muldner M, Ugrina I, Litynska A, Hubalewska-Dydejczyk A, Trbojevic-Akmacic I, Lin EM, Walsh JP, Pochec E, Spector TD, Wilson SG, Lauc G (2020b) Decreased immunoglobulin G core fucosylation, a player in antibody-dependent cell-mediated cytotoxicity, is associated with autoimmune thyroid diseases. Mol Cell Proteomics 19:774–792. https://doi.org/10.1074/mcp.RA119.001860

Matsumoto A, Shikata K, Takeuchi F, Kojima N, Mizuochi T (2000) Autoantibody activity of IgG rheumatoid factor increases with decreasing levels of galactosylation and sialylation. J Biochem (Tokyo) 128:621–628

Mehta AS, Long RE, Comunale MA, Wang M, Rodemich L, Krakover J, Philip R, Marrero JA, Dwek RA, Block TM (2008) Increased levels of galactose-deficient anti-Gal immunoglobulin G in the sera of hepatitis C virus-infected individuals with fibrosis and cirrhosis. J Virol 82:1259–1270

Menni C, Keser T, Mangino M, Bell JT, Erte I, Akmacic I, Vuckovic F, Pucic BM, Gornik O, McCarthy MI, Zoldos V, Spector TD, Lauc G, Valdes AM (2013) Glycosylation of immunoglobulin G: role of genetic and epigenetic influences. PLoS One 8:e82558

Menni C, Gudelj I, Macdonald-Dunlop E, Mangino M, Zierer J, Besic E, Joshi PK, Trbojevic-Akmacic I, Chowienczyk PJ, Spector TD, Wilson JF, Lauc G, Valdes AM (2018) Glycosylation profile of immunoglobulin G is cross-sectionally associated with cardiovascular disease risk score and subclinical atherosclerosis in two independent cohorts. Circ Res 122:1555–1564. https://doi.org/10.1161/CIRCRESAHA.117.312174

Mesko B, Poliska S, Szamosi S, Szekanecz Z, Podani J, Varadi C, Guttman A, Nagy L (2012) Peripheral blood gene expression and IgG glycosylation profiles as markers of tocilizumab treatment in rheumatoid arthritis. J Rheumatol 39:916–928

Mimura Y, Ihn H, Jinnin M, Asano Y, Yamane K, Yazawa N, Tamaki K (2004) Rheumatoid factor isotypes and anti-agalactosyl IgG antibodies in systemic sclerosis. Br J Dermatol 151:803–808

Mimura Y, Ihn H, Jinnin M, Asano Y, Yamane K, Yazawa N, Tamaki K (2005) Serum levels of anti-agalactosyl IgG antibodies in mixed connective tissue disease. Br J Dermatol 152:806–807

Mimura Y, Ashton PR, Takahashi N, Harvey DJ, Jefferis R (2007) Contrasting glycosylation profiles between Fab and Fc of a human IgG protein studied by electrospray ionization mass spectrometry. J Immunol Methods 326:116–126

Miura Y, Endo T (1860) Glycomics and glycoproteomics focused on aging and age-related diseases - Glycans as a potential biomarker for physiological alterations. Biochim Biophys Acta 2016:1608–1614. https://doi.org/10.1016/j.bbagen.2016.01.013

Miura Y, Hashii N, Tsumoto H, Takakura D, Ohta Y, Abe Y, Arai Y, Kawasaki N, Hirose N, Endo T (2015) Change in N-glycosylation of plasma proteins in Japanese semisupercentenarians. PLoS One 10:e0142645. https://doi.org/10.1371/journal.pone.0142645

Miyajima I, Dombrowicz D, Martin TR, Ravetch JV, Kinet JP, Galli (1997) Systemic anaphylaxis in the mouse can be mediated largely through IgG1 and FcγRIII. Assessment of the cardiopulmonary changes, mast cell degranulation, and death associated with active or IgE- or IgG1-dependent passive anaphylaxis. J Clin Invest 99:901–914. https://doi.org/10.1172/JCI119255

Nagelkerke SQ, Kuijpers TW (2014) Immunomodulation by IVIg and the role of Fc-γ receptors: classic mechanisms of action after all? Front Immunol 5:674

Nagelkerke SQ, Dekkers G, Kustiawan I, van de Bovenkamp FS, Geissler J, Plomp R, Wuhrer M, Vidarsson G, Rispens T, van den Berg TK, Kuijpers TW (2014) Inhibition of FcγR-mediated phagocytosis by IVIg is independent of IgG-Fc sialylation and FcγRIIb in human macrophages. Blood 124:3709–3718

Nakagawa H, Hato M, Takegawa Y, Deguchi K, Ito H, Takahata M, Iwasaki N, Minami A, Nishimura S (2007) Detection of altered N-glycan profiles in whole serum from rheumatoid arthritis patients. J Chromatogr B Analyt Technol Biomed Life Sci 853:133–137

Nakajima S, Iijima H, Shinzaki S, Egawa S, Inoue T, Mukai A, Hayashi Y, Kondo J, Akasaka T, Nishida T, Kanto T, Morii E, Mizushima T, Miyoshi E, Tsujii M, Hayashi N (2011) Functional analysis of agalactosyl IgG in inflammatory bowel disease patients. Inflamm Bowel Dis 17:927–936

Narasimhan S (1982) Control of glycoprotein synthesis. UDP-GlcNAc:glycopeptide β4-N-acetylglucosaminyltransferase III, an enzyme in hen oviduct which adds GlcNAc in β1-4 linkage to the β-linked mannose of the trimannosyl core of N-glycosyl oligosaccharides. J Biol Chem 257:10235–10242

Nikolac PM, Pucic BM, Kristic J, Novokmet M, Huffman JE, Vitart V, Hayward C, Rudan I, Wilson JF, Campbell H, Polasek O, Lauc G, Pivac N (2014) The association between galactosylation of immunoglobulin G and body mass index. Prog Neuro-Psychopharmacol Biol Psychiatry 48C:20–25

Nimmerjahn F, Ravetch JV (2006) Fcγ receptors: old friends and new family members. Immunity 24:19–28

Nimmerjahn F, Ravetch JV (2008) Fcγ receptors as regulators of immune responses. Nat Rev Immunol 8:34–47

Nimmerjahn F, Anthony RM, Ravetch JV (2007) Agalactosylated IgG antibodies depend on cellular Fc receptors for *in vivo* activity. Proc Natl Acad Sci USA 104:8433–8437

Nishijima C, Sato S, Takehara K (2001) Anti-agalactosyl IgG antibodies in sera from patients with systemic sclerosis. J Rheumatol 28:1847–1851

Niwa R, Natsume A, Uehara A, Wakitani M, Iida S, Uchida K, Satoh M, Shitara K (2005) IgG subclass-independent improvement of antibody-dependent cellular cytotoxicity by fucose removal from Asn297-linked oligosaccharides. J Immunol Methods 306:151–160

Novokmet M, Lukic E, Vuckovic F, Ethuric Z, Keser T, Rajsl K, Remondini D, Castellani G, Gasparovic H, Gornik O, Lauc G (2014) Changes in IgG and total plasma protein glycomes in acute systemic inflammation. Sci Rep 4:4347

Oefner CM, Winkler A, Hess C, Lorenz AK, Holecska V, Huxdorf M, Schommartz T, Petzold D, Bitterling J, Schoen AL, Stoehr AD, Vu VD, Darcan-Nikolaisen Y, Blanchard V, Schmudde I, Laumonnier Y, Strover HA, Hegazy AN, Eiglmeier S, Schoen CT, Mertes MM, Loddenkemper C, Lohning M, Konig P, Petersen A, Luger EO, Collin M, Kohl J, Hutoff A, Hamelmann E, Berger M, Wardemann H, Ehlers M (2012) Tolerance induction with T cell-dependent protein antigens induces regulatory sialylated IgGs. J Allergy Clin Immunol 129:1647–1655

Ohmi Y, Ise W, Harazono A, Takakura D, Fukuyama H, Baba Y, Narazaki M, Shoda H, Takahashi N, Ohkawa Y, Ji S, Sugiyama F, Fujio K, Kumanogoh A, Yamamoto K, Kawasaki N, Kurosaki T, Takahashi Y, Furukawa K (2016) Sialylation converts arthritogenic IgG into inhibitors of collagen-induced arthritis. Nat Commun 7:11205. https://doi.org/10.1038/ncomms11205

Otani M, Kuroki A, Kikuchi S, Kihara M, Nakata J, Ito K, Furukawa J, Shinohara Y, Izui S (2012) Sialylation determines the nephritogenicity of IgG3 cryoglobulins. J Am Soc Nephrol 23:1869–1878

Othy S, Topcu S, Saha C, Kothapalli P, Lacroix-Desmazes S, Kasermann F, Miescher S, Bayry J, Kaveri SV (2014) Sialylation may be dispensable for reciprocal modulation of helper T cells by intravenous immunoglobulin. Eur J Immunol 44:2059–2063

Pagan JD, Kitaoka M, Anthony RM (2018) Engineered sialylation of pathogenic antibodies in vivo attenuates autoimmune disease. Cell 172:564–577. https://doi.org/10.1016/j.cell.2017.11.041

Parekh RB, Dwek RA, Sutton BJ, Fernandes DL, Leung A, Stanworth D, Rademacher TW, Mizuochi T, Taniguchi T, Matsuta K, Takeuki F, Nagano Y, Miyamoto T, Kobata A (1985) Association of rheumatoid arthritis and primary osteoarthritis with changes in the glycosylation pattern of total serum IgG. Nature 316:452–457

Parekh RB, Roitt IM, Isenberg DA, Dwek RA, Ansell BM, Rademacher TW (1988a) Galactosylation of IgG associated oligosaccharides: reduction in patients with adult and juvenile onset rheumatoid arthritis and relation to disease activity. Lancet 1:966–969

Parekh R, Roitt I, Isenberg D, Dwek R, Rademacher T (1988b) Age-related galactosylation of the N-linked oligosaccharides of human serum IgG. J Exp Med 167:1731–1736

Parekh R, Isenberg D, Rook G, Roitt I, Dwek R, Rademacher T (1989) A comparative analysis of disease-associated changes in the galactosylation of serum IgG. J Autoimmun 2:101–114

Pasek M, Duk M, Podbielska M, Sokolik R, Szechinski J, Lisowska E, Krotkiewski H (2006) Galactosylation of IgG from rheumatoid arthritis (RA) patients—changes during therapy. Glycoconj J 23:463–471

Patel KR, Nott JD, Barb AW (2019) Primary human natural killer cells retain proinflammatory IgG1 at the cell surface and express CD16a glycoforms with donor-dependent variability. Mol Cell Proteomics 18:2178–2190. https://doi.org/10.1074/mcp.RA119.001607

Peipp M, Lammerts van Bueren JJ, Schneider-Merck T, Bleeker WW, Dechant M, Beyer T, Repp R, van Berkel PH, Vink T, van de Winkel JG, Parren PW, Valerius T (2008) Antibody fucosylation differentially impacts cytotoxicity mediated by NK and PMN effector cells. Blood 112:2390–2399

Pekelharing JM, Hepp E, Kamerling JP, Gerwig GJ, Leijnse B (1988) Alterations in carbohydrate composition of serum IgG from patients with rheumatoid arthritis and from pregnant women. Ann Rheum Dis 47:91–95

Perdivara I, Peddada SD, Miller FW, Tomer KB, Deterding LJ (2011) Mass spectrometric determination of IgG subclass-specific glycosylation profiles in siblings discordant for myositis syndromes. J Proteome Res 10:2969–2978

Peschke B, Keller CW, Weber P, Quast I, Lunemann JD (2017) Fc-galactosylation of human immunoglobulin gamma isotypes improves C1q binding and enhances complement-dependent cytotoxicity. Front Immunol 8:646. https://doi.org/10.3389/fimmu.2017.00646

Pierce A, Saldova R, Abd Hamid UM, Abrahams JL, McDermott EW, Evoy D, Duffy MJ, Rudd PM (2010) Levels of specific glycans significantly distinguish lymph node-positive from lymph node-negative breast cancer patients. Glycobiology 20:1283–1288

Pilkington C, Yeung E, Isenberg D, Lefvert AK, Rook GA (1995a) Agalactosyl IgG and antibody specificity in rheumatoid arthritis, tuberculosis, systemic lupus erythematosus and myasthenia gravis. Autoimmunity 22:107–111

Pilkington C, Lefvert AK, Rook GA (1995b) Neonatal myasthenia gravis and the role of agalactosyl IgG. Autoimmunity 21:131–135

Pilkington C, Taylor PV, Silverman E, Isenberg DA, Costello AM, Rook GA (1996a) Agalactosyl IgG and materno-fetal transmission of autoimmune neonatal lupus. Rheumatol Int 16:89–94

Pilkington C, Basaran M, Barlan I, Costello AM, Rook GA (1996b) Raised levels of agalactosyl IgG in childhood tuberculosis. Trans R Soc Trop Med Hyg 90:167–168

Pillai S (2019) Sugar mommy. Sci Immunol 4. https://doi.org/10.1126/sciimmunol.aaz2439

Plomp R, Ruhaak LR, Uh HW, Reiding KR, Selman M, Houwing-Duistermaat JJ, Slagboom PE, Beekman M, Wuhrer M (2017) Subclass-specific IgG glycosylation is associated with markers of inflammation and metabolic health. Sci Rep 7:12325. https://doi.org/10.1038/s41598-017-12495-0

Pucic M, Pinto S, Novokmet M, Knezevic A, Gornik O, Polasek O, Vlahovicek K, Wang W, Rudd PM, Wright AF, Campbell H, Rudan I, Lauc G (2010) Common aberrations from the normal human plasma N-glycan profile. Glycobiology 20:970–975

Pucic M, Muzinic A, Novokmet M, Skledar M, Pivac N, Lauc G, Gornik O (2012) Changes in plasma and IgG N-glycome during childhood and adolescence. Glycobiology 22:975–982

Qin W, Pei H, Qin R, Zhao R, Han J, Zhang Z, Dong K, Ren S, Gu J (2018) Alteration of serum IgG galactosylation as a potential biomarker for diagnosis of neuroblastoma. J Cancer 9:906–913. https://doi.org/10.7150/jca.22014

Qin R, Yang Y, Qin W, Han J, Chen H, Zhao J, Zhao R, Li C, Gu Y, Pan Y, Wang X, Ren S, Sun Y, Gu J (2019) The value of serum immunoglobulin G glycome in the preoperative discrimination of peritoneal metastasis from advanced gastric cancer. J Cancer 10:2811–2821. https://doi.org/10.7150/jca.31380

Qin R, Yang Y, Chen H, Qin W, Han J, Gu Y, Pan Y, Cheng X, Zhao J, Wang X, Ren S, Sun Y, Gu J (2020) Prediction of neoadjuvant chemotherapeutic efficacy in patients with locally advanced gastric cancer by serum IgG glycomics profiling. Clin Proteomics 17:4. https://doi.org/10.1186/s12014-020-9267-8

Quast I, Keller CW, Maurer MA, Giddens JP, Tackenberg B, Wang LX, Munz C, Nimmerjahn F, Dalakas MC, Lunemann JD (2015) Sialylation of IgG Fc domain impairs complement-dependent cytotoxicity. J Clin Invest 125:4160–4170. https://doi.org/10.1172/JCI82695

Rademacher TW, Parekh RB, Dwek RA, Isenberg D, Rook G, Axford JS, Roitt I (1988) The role of IgG glycoforms in the pathogenesis of rheumatoid arthritis. Springer Semin Immunopath 10:231–249

Rademacher TW, Williams P, Dwek RA (1994) Agalactosyl glycoforms of IgG autoantibodies are pathogenic. Proc Natl Acad Sci USA 91:6123–6127

Raju TS (2008) Terminal sugars of Fc glycans influence antibody effector functions of IgGs. Curr Opin Immunol 20:471–478

Reiding KR, Vreeker GCM, Bondt A, Bladergroen MR, Hazes JMW, van der Burgt YEM, Wuhrer M, Dolhain RJEM (2017) Serum protein N-glycosylation changes with rheumatoid arthritis disease activity during and after pregnancy. Front Med (Lausanne) 4:241. https://doi.org/10.3389/fmed.2017.00241

Reinke Y, Konemann S, Chamling B, Gross S, Weitmann K, Hoffmann W, Klingel K, Nauck M, Fielitz J, Dorr M, Felix SB (2019) Sugars make the difference – glycosylation of cardiodepressant antibodies regulates their activity in dilated cardiomyopathy. Int J Cardiol. https://doi.org/10.1016/j.ijcard.2019.04.025

Rodier F, Campisi J (2011) Four faces of cellular senescence. J Cell Biol 192:547–556

Rombouts Y, Ewing E, van de Stadt LA, Selman MH, Trouw LA, Deelder AM, Huizinga TW, Wuhrer M, van Schaardenburg D, Toes RE, Scherer HU (2015) Anti-citrullinated protein antibodies acquire a pro-inflammatory Fc glycosylation phenotype prior to the onset of rheumatoid arthritis. Ann Rheum Dis 74:234–241

Rook G, Thompson S, Buckley M, Elson C, Brealey R, Lambert C, White T, Rademacher T (1991a) The role of oil and agalactosyl IgG in the induction of arthritis in rodent models. Eur J Immunol 21:1027–1032

Rook GA, Steele J, Brealey R, Whyte A, Isenberg D, Sumar N, Nelson JL, Bodman KB, Young A, Roitt IM, Williams P, Scragg IG, Edge CI, Arkwright PD, Ashford D, Wormald MR, Rudd PM,

Redman CWG, Dwek R, Rademacher T (1991b) Changes in IgG glycoform levels are associated with remission of arthritis during pregnancy. J Autoimmun 4:779–794

Rook GA, Onyebujoh P, Wilkins E, Ly HM, Al Attiyah R, Bahr G, Corrah T, Hernandez H, Stanford JL (1994) A longitudinal study of per cent agalactosyl IgG in tuberculosis patients receiving chemotherapy, with or without immunotherapy. Immunology 81:149–154

Rudd PM, Elliott T, Cresswell P, Wilson IA, Dwek RA (2001) Glycosylation and the immune system. Science 291:2370–2376

Ruhaak LR, Uh HW, Beekman M, Koeleman CA, Hokke CH, Westendorp RG, Wuhrer M, Houwing-Duistermaat JJ, Slagboom PE, Deelder AM (2010) Decreased levels of bisecting GlcNAc glycoforms of IgG are associated with human longevity. PLoS One 5:e12566

Ruhaak LR, Uh HW, Beekman M, Hokke CH, Westendorp RG, Houwing-Duistermaat J, Wuhrer M, Deelder AM, Slagboom PE (2011) Plasma protein N-glycan profiles are associated with calendar age, familial longevity and health. J Proteome Res 10:1667–1674

Sakae Y, Satoh T, Yagi H, Yanaka S, Yamaguchi T, Isoda Y, Iida S, Okamoto Y, Kato K (2017) Conformational effects of N-glycan core fucosylation of immunoglobulin G Fc region on its interaction with Fcγ receptor IIIa. Sci Rep 7:13780. https://doi.org/10.1038/s41598-017-13845-8

Saldova R, Royle L, Radcliffe CM, Abd Hamid UM, Evans R, Arnold JN, Banks RE, Hutson R, Harvey DJ, Antrobus R, Petrescu SM, Dwek RA, Rudd PM (2007) Ovarian cancer is associated with changes in glycosylation in both acute-phase proteins and IgG. Glycobiology 17:1344–1356

Satoh M, Iida S, Shitara K (2006) Non-fucosylated therapeutic antibodies as next-generation therapeutic antibodies. Expert Opin Biol Ther 6:1161–1173

Scallon BJ, Tam SH, McCarthy SG, Cai AN, Raju TS (2007) Higher levels of sialylated Fc glycans in immunoglobulin G molecules can adversely impact functionality. Mol Immunol 44:1524–1534

Schaffert A, Hanic M, Novokmet M, Zaytseva O, Kristic J, Lux A, Nitschke L, Peipp M, Pezer M, Hennig R, Rapp E, Lauc G, Nimmerjahn F (2019) Minimal B cell extrinsic IgG glycan modifications of pro- and anti-inflammatory IgG preparations in vivo. Front Immunol 10:3024. https://doi.org/10.3389/fimmu.2019.03024

Scherer HU, van der Woude D, Ioan-Facsinay A, el Bannoudi H, Trouw LA, Wang J, Haupl T, Burmester GR, Deelder AM, Huizinga TW, Wuhrer M, Toes RE (2010) Glycan profiling of anti-citrullinated protein antibodies isolated from human serum and synovial fluid. Arthritis Rheum 62:1620–1629

Schwab I, Mihai S, Seeling M, Kasperkiewicz M, Ludwig RJ, Nimmerjahn F (2014a) Broad requirement for terminal sialic acid residues and FcγRIIB for the preventive and therapeutic activity of intravenous immunoglobulins in vivo. Eur J Immunol 44:1444–1453

Schwab I, Lux A, Nimmerjahn F (2014b) Reply to – IVIG pluripotency and the concept of Fc-sialylation: challenges to the scientist. Nat Rev Immunol 14:349

Schwedler C, Haupl T, Kalus U, Blanchard V, Burmester GR, Poddubnyy D, Hoppe B (2018) Hypogalactosylation of immunoglobulin G in rheumatoid arthritis: relationship to HLA-DRB1 shared epitope, anticitrullinated protein antibodies, rheumatoid factor, and correlation with inflammatory activity. Arthritis Res Ther 20:44. https://doi.org/10.1186/s13075-018-1540-0

Seite JF, Goutsmedt C, Youinou P, Pers JO, Hillion S (2014) Intravenous immunoglobulin induces a functional silencing program similar to anergy in human B cells. J Allergy Clin Immunol 133:181–188

Selman MH, Niks EH, Titulaer MJ, Verschuuren JJ, Wuhrer M, Deelder AM (2011) IgG Fc N-glycosylation changes in Lambert-Eaton Myasthenic syndrome and myasthenia gravis. J Proteome Res 10:143–152

Sharapov SZ, Tsepilov YA, Klaric L, Mangino M, Thareja G, Shadrina AS, Simurina M, Dagostino C, Dmitrieva J, Vilaj M, Vuckovic F, Pavic T, Stambuk J, Trbojevic-Akmacic I, Kristic J, Simunovic J, Momcilovic A, Campbell H, Doherty M, Dunlop MG, Farrington SM, Pucic-Bakovic M, Gieger C, Allegri M, Louis E, Georges M, Suhre K, Spector T, Williams

FMK, Lauc G, Aulchenko YS (2019) Defining the genetic control of human blood plasma N-glycome using genome-wide association study. Hum Mol Genet 28:2062, 5374524–2077. https://doi.org/10.1093/hmg/ddz054

Sharma M, Schoindre Y, Hegde P, Saha C, Maddur MS, Stephen-Victor E, Gilardin L, Lecerf M, Bruneval P, Mouthon L, Benveniste O, Kaveri SV, Bayry J (2014) Intravenous immunoglobulin-induced IL-33 is insufficient to mediate basophil expansion in autoimmune patients. Sci Rep 4:5672

Shibata-Koyama M, Iida S, Misaka H, Mori K, Yano K, Shitara K, Satoh M (2009) Nonfucosylated rituximab potentiates human neutrophil phagocytosis through its high binding for FcγRIIIb and MHC class II expression on the phagocytotic neutrophils. Exp Hematol 37:309–321

Shields RL, Lai J, Keck R, O'Connell LY, Hong K, Meng YG, Weikert SH, Presta LG (2002) Lack of fucose on human IgG1 N-linked oligosaccharide improves binding to human Fcγ RIII and antibody-dependent cellular toxicity. J Biol Chem 277:26733–26740

Shih HC, Chang MC, Chen CH, Tsai IL, Wang SY, Kuo YP, Chen CH, Chang YT (2019) High accuracy differentiating autoimmune pancreatitis from pancreatic ductal adenocarcinoma by immunoglobulin G glycosylation. Clin Proteomics 16:1. https://doi.org/10.1186/s12014-018-9221-1

Shikata K, Yasuda T, Takeuchi F, Konishi T, Nakata M, Mizuochi T (1998) Structural changes in the oligosaccharide moiety of human IgG with aging. Glycoconj J 15:683–689

Shinkawa T, Nakamura K, Yamane N, Shoji-Hosaka E, Kanda Y, Sakurada M, Uchida K, Anazawa H, Satoh M, Yamasaki M, Hanai N, Shitara K (2003) The absence of fucose but not the presence of galactose or bisecting N-acetylglucosamine of human IgG1 complex-type oligosaccharides shows the critical role of enhancing antibody-dependent cellular cytotoxicity. J Biol Chem 278:3466–3473

Shinzaki S, Iijima H, Nakagawa T, Egawa S, Nakajima S, Ishii S, Irie T, Kakiuchi Y, Nishida T, Yasumaru M, Kanto T, Tsujii M, Tsuji S, Mizushima T, Yoshihara H, Kondo A, Miyoshi E, Hayashi N (2008) IgG oligosaccharide alterations are a novel diagnostic marker for disease activity and the clinical course of inflammatory bowel disease. Am J Gastroenterol 103:1173–1181

Shinzaki S, Kuroki E, Iijima H, Tatsunaka N, Ishii M, Fujii H, Kamada Y, Kobayashi T, Shibukawa N, Inoue T, Tsujii M, Takeishi S, Mizushima T, Ogata A, Naka T, Plevy SE, Takehara T, Miyoshi E (2013) Lectin-based immunoassay for aberrant IgG glycosylation as the biomarker for Crohn's disease. Inflamm Bowel Dis 19:321–331

Simurina M, de Haan N, Vuckovic F, Kennedy NA, Stambuk J, Falck D, Trbojevic-Akmacic I, Clerc F, Razdorov G, Khon A, Latiano A, D'Inca R, Danese S, Targan S, Landers C, Dubinsky M, McGovern DPB, Annese V, Wuhrer M, Lauc G (2018) Glycosylation of immunoglobulin G associates with clinical features of inflammatory bowel diseases. Gastroenterology. https://doi.org/10.1053/j.gastro.2018.01.002

Soltys AJ, Hay FC, Bond A, Axford JS, Jones MG, Randen I, Thompson KM, Natvig JB (1994) The binding of synovial tissue-derived human monoclonal immunoglobulin M rheumatoid factor to immunoglobulin G preparations of differing galactose content. Scand J Immunol 40:135–143

Sonneveld ME, de Haas M, Koeleman C, de Haan N, Zeerleder SS, Ligthart PC, Wuhrer M, van der Schoot CE, Vidarsson G (2017) Patients with IgG1-anti-red blood cell autoantibodies show aberrant Fc-glycosylation. Sci Rep 7:8187. https://doi.org/10.1038/s41598-017-08654-y

Stanworth SJ, Donn RP, Hassall A, Dawes P, Ollier W, Snowden N (1998) Absence of an association between mannose-binding lectin polymorphism and rheumatoid arthritis. Br J Rheumatol 37:186–188

Su Z, Xie Q, Wang Y, Li Y (2020) Abberant immunoglobulin G glycosylation in rheumatoid arthritis by LTQ-ESI-MS. Int J Mol Sci 21. https://doi.org/10.3390/ijms21062045

Subedi GP, Barb AW (2016) The immunoglobulin G1 N-glycan composition affects binding to each low affinity Fcγ receptor. MAbs 1–13. https://doi.org/10.1080/19420862.2016.1218586

Sumar N, Isenberg DA, Bodman KB, Soltys A, Young A, Leak AM, Round J, Hay FC, Roitt IM (1991) Reduction in IgG galactose in juvenile and adult onset rheumatoid arthritis measured by a lectin binding method and its relation to rheumatoid factor. Ann Rheum Dis 50:607–610

Tanigaki K, Sacharidou A, Peng J, Chambliss KL, Yuhanna IS, Ghosh D, Ahmed M, Szalai AJ, Vongpatanasin W, Mattrey RF, Chen Q, Azadi P, Lingvay I, Botto M, Holland WL, Kohler JJ, Sirsi SR, Hoyt K, Shaul PW, Mineo C (2018) Hyposialylated IgG activates endothelial IgG receptor FcγRIIB to promote obesity-induced insulin resistance. J Clin Invest 128:309–322. https://doi.org/10.1172/JCI89333

Temming AR, Dekkers G, van de Bovenkamp FS, Plomp HR, Bentlage AEH, Szittner Z, Derksen NIL, Wuhrer M, Rispens T, Vidarsson G (2019) Human DC-SIGN and CD23 do not interact with human IgG. Sci Rep 9:9995. https://doi.org/10.1038/s41598-019-46484-2

Theodoratou E, Campbell H, Ventham NT, Kolarich D, Pucic-Bakovic M, Zoldos V, Fernandes D, Pemberton IK, Rudan I, Kennedy NA, Wuhrer M, Nimmo E, Annese V, McGovern DP, Satsangi J, Lauc G (2014) The role of glycosylation in IBD. Nat Rev Gastroenterol Hepatol 11:588–600

Thomann M, Schlothauer T, Dashivets T, Malik S, Avenal C, Bulau P, Ruger P, Reusch D (2015) In vitro glycoengineering of IgG1 and its effect on Fc receptor binding and ADCC activity. PLoS One 10:e0134949. https://doi.org/10.1371/journal.pone.0134949

Troelsen LN, Garred P, Madsen HO, Jacobsen S (2007) Genetically determined high serum levels of mannose-binding lectin and agalactosyl IgG are associated with ischemic heart disease in rheumatoid arthritis. Arthritis Rheum 56:21–29

Tsukimura W, Kurogochi M, Mori M, Osumi K, Matsuda A, Takegawa K, Furukawa K, Shirai T (2017) Preparation and biological activities of anti-HER2 monoclonal antibodies with fully core-fucosylated homogeneous bi-antennary complex-type glycans. Biosci Biotechnol Biochem 81:2353–2359. https://doi.org/10.1080/09168451.2017.1394813

Twisselmann N, Bartsch YC, Pagel J, Wieg C, Hartz A, Ehlers M, Hartel C (2018) IgG fc glycosylation patterns of preterm infants differ with gestational age. Front Immunol 9:3166. https://doi.org/10.3389/fimmu.2018.03166

Umana P, Jean-Mairet J, Moudry R, Amstutz H, Bailey JE (1999) Engineered glycoforms of an antineuroblastoma IgG1 with optimized antibody-dependent cellular cytotoxic activity. Nat Biotechnol 17:176–180

Vadrevu SK, Trbojevic-Akmacic I, Kossenkov AV, Colomb F, Giron LB, Anzurez A, Lynn K, Mounzer K, Landay AL, Kaplan RC, Papasavvas E, Montaner LJ, Lauc G, Abdel-Mohsen M (2018) Frontline science: plasma and immunoglobulin G galactosylation associate with HIV persistence during antiretroviral therapy. J Leukoc Biol. https://doi.org/10.1002/JLB.3HI1217-500R

Van Beneden K, Coppieters K, Laroy W, De Keyser F, Hoffman IE, Van den BF, Vander CB, Drennan M, Jacques P, Rottiers P, Verbruggen G, Contreras R, Callewaert N, Elewaut D (2009) Reversible changes in serum immunoglobulin galactosylation during the immune response and treatment of inflammatory autoimmune arthritis. Ann Rheum Dis 68:1360–1365

van de Geijn FE, Wuhrer M, Selman MH, Willemsen SP, de Man YA, Deelder AM, Hazes JM, Dolhain RJ (2009) Immunoglobulin G galactosylation and sialylation are associated with pregnancy-induced improvement of rheumatoid arthritis and the postpartum flare: results from a large prospective cohort study. Arthritis Res Ther 11:R193

Vanhooren V, Desmyter L, Liu XE, Cardelli M, Franceschi C, Federico A, Libert C, Laroy W, Dewaele S, Contreras R, Chen C (2007) N-glycomic changes in serum proteins during human aging. Rejuvenation Res 10:521–531a

Vanhooren V, Laroy W, Libert C, Chen C (2008) N-glycan profiling in the study of human aging. Biogerontology 9:351–356

Vanhooren V, Liu XE, Franceschi C, Gao CF, Libert C, Contreras R, Chen C (2009) N-glycan profiles as tools in diagnosis of hepatocellular carcinoma and prediction of healthy human ageing. Mech Ageing Dev 130:92–97

Vanhooren V, Dewaele S, Libert C, Engelborghs S, De Deyn PP, Toussaint O, Debacq-Chainiaux F, Poulain M, Glupczynski Y, Franceschi C, Jaspers K, van der Pluijm I, Hoeijmakers J, Chen CC (2010) Serum N-glycan profile shift during human ageing. Exp Gerontol 45:738–743

Vanhooren V, Dewaele S, Kuro O, Taniguchi N, Dolle L, van Grunsven LA, Makrantonaki E, Zouboulis CC, Chen CC, Libert C (2011) Alteration in N-glycomics during mouse aging: a role for FUT8. Aging Cell 10:1056–1066

Vuckovic F, Kristic J, Gudelj I, Teruel AM, Keser T, Pezer M, Pucic-Bakovic M, Stambuk J, Trbojevic-Akmacic I, Barrios C, Pavic T, Menni C, Wang Y, Zhou Y, Cui L, Song H, Zeng Q, Guo X, Pons-Estel BA, McKeigue P, Patrick AL, Gornik O, Spector TD, Harjacek M, Alarcon-Riquelme M, Molokhia M, Wang W, Lauc G (2015) Systemic lupus erythematosus associates with the decreased immunosuppressive potential of the IgG glycome. Arthritis Rheumatol 67:2978–2989. https://doi.org/10.1002/art.39273

Vuckovic F, Theodoratou E, Thaci K, Timofeeva M, Vojta A, Stambuk J, Pucic-Bakovic M, Derek L, Servis D, Rudd P, Wennerstrom A, Aulchenko Y, Farrington S, Perola M, Dunlop M, Campbell H, Lauc G (2016) IgG glycome in colorectal cancer. Clin Cancer Res 22:3078–3086. https://doi.org/10.1158/1078-0432.CCR-15-1867

Wahl A, van den Akker E, Klaric L, Stambuk J, Benedetti E, Plomp R, Razdorov G, Trbojevic-Akmacic I, Deelen J, van H D, Slagboom PE, Vuckovic F, Grallert H, Krumsiek J, Strauch K, Peters A, Meitinger T, Hayward C, Wuhrer M, Beekman M, Lauc G, Gieger C (2018) Genome-wide association study on immunoglobulin G glycosylation patterns. Front Immunol 9:277. https://doi.org/10.3389/fimmu.2018.00277

Wandall HH, Rumjantseva V, Sorensen AL, Patel-Hett S, Josefsson EC, Bennett EP, Italiano JE Jr, Clausen H, Hartwig JH, Hoffmeister KM (2012) The origin and function of platelet glycosyltransferases. Blood 120:626–635

Wang TT, Ravetch JV (2019) Functional diversification of IgGs through Fc glycosylation. J Clin Invest 129:3492–3498. https://doi.org/10.1172/JCI130029

Wang J, Balog CI, Stavenhagen K, Koeleman CA, Scherer HU, Selman MH, Deelder AM, Huizinga TW, Toes RE, Wuhrer M (2011) Fc-glycosylation of IgG1 is modulated by B-cell stimuli. Mol Cell Proteomics 10:M110

Wang TT, Maamary J, Tan GS, Bournazos S, Davis CW, Krammer F, Schlesinger SJ, Palese P, Ahmed R, Ravetch JV (2015) Anti-HA glycoforms drive B cell affinity selection and determine influenza vaccine efficacy. Cell 162:160–169. https://doi.org/10.1016/j.cell.2015.06.026

Wang TT, Sewatanon J, Memoli MJ, Wrammert J, Bournazos S, Bhaumik SK, Pinsky BA, Chokephaibulkit K, Onlamoon N, Pattanapanyasat K, Taubenberger JK, Ahmed R, Ravetch JV (2017) IgG antibodies to dengue enhanced for FcγRIIIA binding determine disease severity. Science 355:395–398. https://doi.org/10.1126/science.aai8128

Wang Z, Geng Z, Shao W, Liu E, Zhang J, Tang J, Wang P, Sun X, Xiao L, Xu W, Zhang Y, Cui H, Zhang L, Yang X, Chang X, Qiu X (2019) Cancer-derived sialylated IgG promotes tumor immune escape by binding to Siglecs on effector T cells. Cell Mol Immunol. https://doi.org/10.1038/s41423-019-0327-9

Weindruch R, Naylor PH, Goldstein AL, Walford RL (1988) Influences of aging and dietary restriction on serum thymosin α 1 levels in mice. J Gerontol 43:B40–B42

Williams PJ, Arkwright PD, Rudd P, Scragg IG, Edge CJ, Wormald MR, Rademacher TW (1995) Short communication: selective placental transport of maternal IgG to the fetus. Placenta 16:749–756

Wirt T, Rosskopf S, Rosner T, Eichholz KM, Kahrs A, Lutz S, Kretschmer A, Valerius T, Klausz K, Otte A, Gramatzki M, Peipp M, Kellner C (2017) An fc double-engineered CD20 antibody with enhanced ability to trigger complement-dependent cytotoxicity and antibody-dependent cell-mediated cytotoxicity. Transfus Med Hemother 44:292–300. https://doi.org/10.1159/000479978

Wong AH, Fukami Y, Sudo M, Kokubun N, Hamada S, Yuki N (2016) Sialylated IgG-Fc: a novel biomarker of chronic inflammatory demyelinating polyneuropathy. J Neurol Neurosurg Psychiatry 87:275–279. https://doi.org/10.1136/jnnp-2014-309964

Wuhrer M, Selman MH, McDonnell LA, Kumpfel T, Derfuss T, Khademi M, Olsson T, Hohlfeld R, Meinl E, Krumbholz M (2015) Pro-inflammatory pattern of IgG1 Fc glycosylation in multiple sclerosis cerebrospinal fluid. J Neuroinflammation 12:235. https://doi.org/10.1186/s12974-015-0450-1

Xin Y, Lasker JM, Lieber CS (1995) Serum carbohydrate-deficient transferrin: mechanism of increase after chronic alcohol intake. Hepatology 22:1462–1468

Yabe R, Tateno H, Hirabayashi J (2010) Frontal affinity chromatography analysis of constructs of DC-SIGN, DC-SIGNR and LSECtin extend evidence for affinity to agalactosylated N-glycans. FEBS J 277:4010–4026

Yagev H, Frenkel A, Cohen IR, Friedman A (1993) Adjuvant arthritis is associated with changes in the glycosylation of serum IgG1 and IgG2b. Clin Exp Immunol 94:452–458

Yamada E, Tsukamoto Y, Sasaki R, Yagyu K, Takahashi N (1997) Structural changes of immunoglobulin G oligosaccharides with age in healthy human serum. Glycoconj J 14:401–405

Yamaguchi Y, Barb AW (2020) A synopsis of recent developments defining how N-glycosylation impacts immunoglobulin G structure and function. Glycobiology 30:214–225. https://doi.org/10.1093/glycob/cwz068

Yamaguchi Y, Nishimura M, Nagano M, Yagi H, Sasakawa H, Uchida K, Shitara K, Kato K (1760) Glycoform-dependent conformational alteration of the Fc region of human immunoglobulin G1 as revealed by NMR spectroscopy. Biochim Biophys Acta 2006:693–700

Youings A, Chang SC, Dwek RA, Scragg IG (1996) Site-specific glycosylation of human immunoglobulin G is altered in four rheumatoid arthritis patients. Biochem J 314(Pt 2):621–630

Yu X, Vasiljevic S, Mitchell DA, Crispin M, Scanlan CN (2013) Dissecting the molecular mechanism of IVIg therapy: the interaction between serum IgG and DC-SIGN is independent of antibody glycoform or Fc domain. J Mol Biol 425:1253–1258

Zabczynska M, Polak K, Kozlowska K, Sokolowski G, Pochec E (2020) The contribution of IgG glycosylation to antibody-dependent cell-mediated cytotoxicity (ADCC) and complement-dependent cytotoxicity (CDC) in Hashimoto's thyroiditis: an in vitro model of thyroid autoimmunity. Biomol Ther 10:171. https://doi.org/10.3390/biom10020171

Zahid D, Zhang N, Fang H, Gu J, Li M, Li W (2020) Loss of core fucosylation suppressed the humoral immune response in *Salmonella typhimurium* infected mice. J Microbiol Immunol Infect 30034–30037. https://doi.org/10.1016/j.jmii.2020.02.006

Zhao ZY, Liu D, Cao WJ, Sun M, Song MS, Wang W, Wang YX (2018) Association between IgG N-glycans and nonalcoholic fatty liver disease in Han Chinese. Biomed Environ Sci 31:454–458. https://doi.org/10.3967/bes2018.059

Zhou J, Gao H, Xie W, Li Y (2019) FcγR-binding affinity of monoclonal murine IgG1s carrying different N-linked Fc oligosaccharides. Biochem Biophys Res Commun 520:8–13. https://doi.org/10.1016/j.bbrc.2019.09.068

Zou Y, Hu J, Jie J, Lai J, Li M, Liu Z, Zou X (2020) Comprehensive analysis of human IgG Fc N-glycopeptides and construction of a screening model for colorectal cancer. J Proteome 213:103616. https://doi.org/10.1016/j.jprot.2019.103616

Chapter 11
Estrogen-Driven Changes in Immunoglobulin G Fc Glycosylation

Kaitlyn A. Lagattuta and Peter A. Nigrovic

Contents

11.1	Introduction	343
11.2	IgG Fc Glycan Changes with Age, Sex, and Pregnancy	344
11.3	Estrogens as the First Confirmed Modulators of the Human IgG Fc Glycome	345
11.4	Implications of Estrogen-Induced Fc Glycans Changes on IgG Function	348
11.5	Implications of Estrogen-Induced Fc Glycan Changes for Pregnancy	349
11.6	In Search of Mechanism	351
	11.6.1 Possibility 1: Transcriptional Activation of B4GALT Genes by Regulatory Complexes Containing ERα/ERβ	352
	11.6.2 Possibility 2: Modulation of β4 Galactosyltransferase Enzymatic Activity by Estradiol	352
	11.6.3 Possibility 3: Secondary Cytokine Signaling	353
	11.6.4 Possibility 4: Post-Secretory Glycan Modification	354
11.7	Future Directions	355
References		356

Abstract Glycosylation within the immunoglobulin G (IgG) Fc region modulates its ability to engage complement and Fc receptors, affording the opportunity to fine-tune effector functions. Mechanisms regulating IgG Fc glycans remain poorly understood. Changes accompanying menarche, menopause, and pregnancy have long implicated hormonal factors. Intervention studies now confirm that estrogens enhance IgG Fc galactosylation, in females and also in males, defining the first pathway modulating Fc glycans and thereby a new link between sex and immunity. This mechanism may participate in fetal-maternal immunity, antibody-mediated inflammation, and other aspects of age- and sex-specific immune function. Here

---

K. A. Lagattuta
Harvard-MIT MD-PhD Program, Harvard Medical School, Boston, MA, USA

P. A. Nigrovic (✉)
Division of Immunology, Boston Children's Hospital, Boston, MA, USA

Division of Rheumatology, Inflammation, and Immunity, Brigham and Women's Hospital, Harvard Medical School, Boston, MA, USA
e-mail: peter.nigrovic@childrens.harvard.edu

© The Author(s), under exclusive license to Springer Nature Switzerland AG 2021
M. Pezer (ed.), *Antibody Glycosylation*, Experientia Supplementum 112,
https://doi.org/10.1007/978-3-030-76912-3_11

we review the changes affecting the IgG Fc glycome from childhood through old age, the evidence establishing a role for estrogens, and research directions to uncover associated mechanisms that may inform therapeutic intervention.

Keywords Estrogen · Antibody · Glycan · Pregnancy · Autoimmunity · Sex

Abbreviations

ADCC	Antibody-dependent cellular toxicity
β4Gal-T1	β-1,4-galactosyltransferase 1
BCR	B cell receptor
DNA	Deoxyribonucleic acid
E1	Estrone
E2	17(beta)-estradiol
E3	Estriol
ERα	Estrogen receptor α
ERβ	Estrogen receptor β
ERE	Estrogen response element
ERRγ	Transcription factor estrogen-related receptor gamma
Fab region	Antigen-binding fragment
Fc region	Crystallizable fragment
FcRn	Neonatal Fc receptor
G0	Agalactosylated
G0F	Agalactosylated, fucosylated, non-bisected
G0FB	Agalactosylated, fucosylated, bisected
G1	Monogalactosylated
G2	Digalactosylated
GALE	UDP-galactose 4′-epimerase
GALK	Galactokinase
GALT	Galactose-1-phosphate uridyltransferase
GlcNAc	N-acetyl-glucosamine
GnRH	Gonadotropin-releasing hormone
GWAS	Genome-wide association study
HT-SELEX	High-throughput systematic evolution of ligands by exponential enrichment
Ig	Immunoglobulin
IL-6	Interleukin 6
IL-21	Interleukin 21
JIA	Juvenile idiopathic arthritis
MANOVA	Multivariate analysis of variance
mRNA	Messenger ribonucleic acid
MS	Mass spectrometry
N-glycan	Oligosaccharide attached to the nitrogen atom of an asparagine residue

NK cells	Natural killer cells
RA	Rheumatoid arthritis
SLE	Systemic lupus erythematosus
SNP	Single nucleotide polymorphism
St6Gal-T1	ST6 β-galactoside α-2,6-sialyltransferase 1
UDP	Uridine diphosphate
UMP^{2-}	Uridine monophosphate
UPLC	Ultra-performance liquid chromatography

11.1 Introduction

The "omics" revolution has unveiled the staggering diversity and specificity of the immune system at an ever-higher resolution. The breadth of receptor diversity achieved by stochastic DNA rearrangement and somatic hypermutation pertains to antigen recognition, helping to define which immune cells and molecular programs are called to respond to an environmental challenge. However, the immune system must also carefully regulate the downstream pathways engaged by antigen-specific mechanisms to ensure that the immune response remains appropriate to the nature and extent of the threat. Balancing the risks of infection, autoimmunity, and autoinflammation, the immune system responds with the "right cells," and "right severity," at the "right place" and "right time"—an intersection we might call *situationally appropriate reactivity*.

It is reasonable to suspect that what is "right" might change with context and over the course of a lifetime. One moment of the unique challenge is pregnancy, during which a female must maintain her own immune surveillance while simultaneously tolerating and also defending an evolving foreign body within her. Here, situationally appropriate reactivity likely requires a recalibration of immune mechanisms.

One well-known immunomodulating factor is the antibody Fc region, serving to direct humoral responses to specific cell types, effector mechanisms, and compartments. Distinct Fc heavy chains define antibody classes: at a first approximation, IgM for early responses and complement activation, IgA for mucosal immunity, IgE for mast cell immunity, and IgG for most other functions, specified further through subclasses IgG1-IgG4. The characteristics of each of these distinct protein backbones are further refined by post-translational variation: glycosylation within the Fc region.

Position 297 of each IgG heavy chain holds an asparagine (N) residue decorated with a biantennary ("2-armed") glycan varying in length from 7 to 13 sugar residues. The core of this N-glycan is comprised of 4*N*-acetyl-glucosamines (GlcNAc) and 3 mannose molecules. The GlcNAc closest to the core is usually but not invariably fucosylated, and an additional GlcNAc bisects the two glycan arms in ~10% of IgG Fc glycans. The core GlcNAc terminating each of the two arms can receive a galactose, defining glycan families termed G0, G1, or G2 (for a-, mono-, and di-galactosylation, respectively). Each galactose can in turn be extended by sialylation, giving rise to glycans bearing zero, one, or two sialic acids. Together with minor species, there are over 60 different glycoforms an IgG Fc glycan could assume, approximately 30 of which occur with appreciable frequency.

Glycoform variation in the Fc region, as compared to the Fab region, is of particular interest because it impacts core antibody effector functions. Contained within the "cage" formed by the two heavy chains, glycans maintain the quaternary structure required for interaction with complement and Fc receptors (Sondermann et al. 2000; Barb and Prestegard 2011; Subedi and Barb 2015). Without Fc glycans, IgG cannot interact with complement or most Fc receptors (Nose and Wigzell 1983). However, defining structure-function correlations for individual IgG Fc glycans has proven complex. The four IgG subclasses differ in heavy chain sequence, and lessons learned with one form may not apply to another. Murine and human IgG isotypes, as well as Fc receptors, differ, complicating the determination of glycan-dependent antibody functions in vivo. IgG changes conformation when bound to the antigen, raising the possibility that studies using free monomeric antibody may not reflect IgG "in action." The two glycans within an IgG Fc region can be different, yielding pair effects that may have effector consequences distinct from those observed in homogeneously glycosylated IgG. Finally, IgG Fc glycosylation is a "zero sum game": an increase in the abundance of one glycoform necessarily comes at the expense of one or more others, leading to highly complex changes in the effector capacity of the IgG pool taken as a whole. Therefore, while it seems clear that changes in Fc glycans have consequences for humoral immune function, predicting and even measuring these effects has not proven straightforward.

In this review, we summarize variation in IgG Fc glycoforms over the lifespan, with a focus on the evidence that now establishes an unequivocal role for estrogens in these changes. We explore potential physiologic and pathophysiologic implications for this phenomenon, including in pregnancy and inflammatory diseases. Finally, we discuss candidate mechanisms by which estrogens could alter Fc glycans. Though much remains unknown, we suggest ways in which advancing techniques in glycobiology and next-generation sequencing will further deepen our understanding of developmental immunology.

11.2 IgG Fc Glycan Changes with Age, Sex, and Pregnancy

The relative abundance of glycan species within the IgG Fc region varies with age (Parekh et al. 1988a; Yamada et al. 1997). The largest adult series, assessing 5117 adults aged 18–95 across four populations, found a significant association with age for the large majority of IgG glycans (Kristic et al. 2014). The most consistent changes affect galactosylation, with the abundance of G0 forms increasing and of G2 forms decreasing progressively with age. Similar charges have been observed in other cohorts, with a decrease in sialylation accompanying the reduction in IgG Fc galactosylation (Yu et al. 2016; Bakovic et al. 2013). Pediatric series also reveal age-dependent variation. Though the G1 fraction remains relatively stable at ~40–45% of all glycans throughout adulthood, this fraction is reduced to ~35% in the first few years of life, a period during which G2 glycoforms are somewhat more abundant (Parekh et al. 1988a; Shikata et al. 1998; Cheng et al. 2020). After these first few years, galactosylation rises modestly with age through childhood (the reverse of the adult trend) and fucosylation declines slightly (Cheng et al. 2020; Pucic et al. 2012; de Haan et al. 2016; Pezer et al. 2016; van Erp et al. 2020). In both adults and children, age-associated changes are sufficiently regular to enable approximate estimation of age from IgG glycans alone (Kristic et al. 2014; Cheng et al. 2020).

A consistent finding across many of these population studies is a shift in IgG glycans in females coincident with periods of hormonal transition (Yamada et al. 1997; Kristic et al. 2014; Bakovic et al. 2013; Shikata et al. 1998; Pucic et al. 2012; Knezevic et al. 2009; Chen et al. 2012; Ercan et al. 2017).

In particular, females of reproductive age demonstrate a greater proportion of galactosylated forms than males of the same age, with these sex-based trajectories converging rapidly around age 50, the typical age of menopause (Kristic et al. 2014; Bakovic et al. 2013; Ercan et al. 2017). A reciprocal shift is observed around the age of menarche, with females beginning to exhibit greater IgG galactosylation than males around age 12 years (Cheng et al. 2020; Pucic et al. 2012). Although females and males both undergo a so-called "mini-puberty" in infancy and early childhood, characterized by activation of the hypothalamic–pituitary–gonadal axis shortly after birth, no associated variation in IgG galactosylation is observed, potentially because sex hormone levels remain low (Cheng et al. 2020; Lanciotti et al. 2018; Bidlingmaier et al. 1973).

Changes accompanying major hormonal inflection points are especially intriguing in light of the similar IgG Fc glycan shifts observed during pregnancy. First reported in 1991, multiple studies have now confirmed that IgG Fc galactosylation and sialylation increase during pregnancy, peaking during the third trimester (Rook et al. 1991; Wuhrer et al. 2007; van de Geijn et al. 2009; Bondt et al. 2014). Many physiological adaptations coincide with these changes, among which is a nearly 20-fold rise in circulating estrogens (Abbassi-Ghanavati et al. 2009). These observations, along with the evidence suggesting an increase in galactosylation at menarche and a decrease at menopause, are consistent with the possibility that estrogens promote IgG Fc glycosylation.

11.3 Estrogens as the First Confirmed Modulators of the Human IgG Fc Glycome

In order of increasing potency, the three endogenous human estrogens are estriol (E3), estrone (E1), and 17(beta)-estradiol (E2). Estriol is produced almost exclusively during pregnancy by the placenta and is suspected to influence uteroplacental blood flow and consequent cerebral development during fetal life. Estradiol is the major circulating estrogen from puberty to menopause, including in pregnancy. Estrone exists in equilibrium with estradiol via 17β-hydroxysteroid dehydrogenase and gradually becomes the only remaining estrogen as the ovaries cease to produce estradiol at menopause (Tepperman and Tepperman 1987). These molecules differ only by one hydroxyl or ketone moiety and bind the same estrogen receptors with affinities that correspond to their potencies (Fig. 11.1). Consequently, with the exception of subtle differences not further considered here, they exert highly similar biological effects and are collectively referred to as "estrogens."

To test whether female sex hormones directly regulate IgG Fc glycosylation, Ercan et al. quantitated the abundance of IgG-associated glycoforms in two populations of adult blood donors, confirming an increase in the abundance of G0 glycans in females at age 50 years, a difference that was further sharpened in 189 females in whom premenopausal vs. postmenopausal status was known by history (Ercan et al. 2017). The glycan shift was noted primarily as an increase in

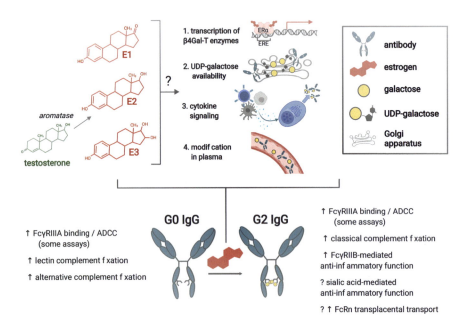

Fig. 11.1 IgG Fc glycan regulation by estrogens: potential mechanisms and functional implications. Estrogen occurs in three forms, estrone (E1), estradiol (E2), and estriol (E3), the latter produced principally from the placenta during pregnancy. Testosterone modulates IgG Fc glycans via interconversion to estradiol. Shown are four candidate mechanisms, not mutually exclusive, that may contribute to the effect of estrogens to increase the abundance of G2 IgG Fc glycans at the expense of G0 IgG Fc glycans. Potential implications of this shift for IgG effector functions are shown; note that enhanced FcγRIIIA binding is reported for both G0 and G2 glycans (please see text for discussion of function). *β4Gal-T* β4 galactosyltransferase, *ERα* estrogen receptor alpha, *ERE* estrogen response element, *UDP* uridine diphosphate, *ADCC* antibody-dependent cellular cytotoxicity, *FcRn* neonatal Fc receptor

the most abundant fucosylated non-bisected glycoform (G0F) but was much more modest in the bisected form G0FB.

To establish a causal role for estrogens, these investigators took advantage of several cohorts of human subjects undergoing hormonal manipulation. Van Pelt et al. randomized 119 healthy postmenopausal females to receive conjugated estrogens (a combination of estradiol, estrone, and equine estrogens), an estradiol mimetic (raloxifene), or placebo during a six-month exercise intervention (Van Pelt et al. 2014). Ercan et al. analyzed a random subset of these patients ($N = 58$) and found that, over an 18-month span, patients in the placebo arm exhibited a stable or slightly increased proportion of G0F IgG glycans, as expected given the known effect of age; by contrast, each estrogen arm exhibited a decrease of approximately 10% (Ercan et al. 2017). Shea et al. used leuprolide (an analog of gonadotropin-releasing hormone, GnRH) to suppress ovarian estrogen production in 70 healthy premenopausal females, with or without add-back transdermal estradiol for 5 months; both groups were then observed to the recovery of spontaneous menses (Shea et al. 2015). Analyzing a random subset ($N = 21$) of these patients, Ercan et al. found that the placebo group exhibited an immediate increase in IgG G0F glycans, a change not observed in the estrogen group. Both groups returned to pre-study G0F levels upon recovery from leuprolide (Ercan et al. 2017). Further analysis of samples from these premenopausal females confirmed that leuprolide changed the overall IgG glycan profile to resemble that of older females, but this change was fully prevented by estradiol (Juric et al. 2020). These studies establish a causal role for estrogens in the regulation of IgG Fc glycosylation in females, consistent with the direction and magnitude of the effects of menarche and menopause observed in population studies.

In males, a small amount of testosterone is routinely converted to estradiol by aromatase enzymes. To assess the effect of testosterone and estrogen on IgG glycans in males, Ercan et al. examined 40 patients from a cohort of males undergoing hormonal manipulation (Finkelstein et al. 2013). These subjects received the GnRH agonist goserelin acetate to suppress endogenous testosterone production, followed by replacement transdermal testosterone, placebo, or transdermal testosterone plus anastrozole (an aromatase inhibitor that blocks the conversion of testosterone to estradiol). Over a period of 12 weeks, males receiving testosterone exhibited no change in IgG glycans, whereas placebo subjects exhibited a significant increase in G0F glycans. Critically, patients are given testosterone plus anastrozole demonstrated a G0F increment identical to placebo, establishing that estrogens drive IgG Fc galactosylation also in males, albeit to a lesser degree than females because estrogen levels are substantially lower. By contrast, testosterone has no direct effect on IgG Fc glycosylation.

In a concordant study, Engdahl et al. characterized IgG Fc glycans in a cohort of 49 postmenopausal female rheumatoid arthritis (RA) patients randomized to hormone replacement therapy (in most patients, estradiol with the progestin norethindrone acetate) or no treatment. Patients receiving hormone supplementation demonstrated an increase in IgG Fc galactosylation as well as sialyation that inversely correlated with disease activity (Engdahl et al. 2018). It was not established whether the increase in sialylated forms was a direct effect of hormone supplementation or instead mediated by the increase in galactosylation, a prerequisite to sialylation. The authors found that mRNA for ST6 β-galactoside α-2,6-sialyltransferase 1 (St6Gal-T1, encoded by *St6gal1*), an enzyme that adds sialic acid to IgG glycans, increased in response to estradiol in murine plasmablasts both in vitro and in vivo (mRNA and protein), and in human ex vivo-induced plasmablasts (mRNA only). Of note, mRNA for an enzyme that adds galactose to IgG glycans, β-1,4-galactosyltransferase 1 (β4Gal-T1), trended similarly but failed to achieve statistical significance in the sample size taken due to larger variance (Engdahl et al. 2018). Since galactosylated forms of IgG indeed increased in response to estradiol, it is unknown whether *St6gal1* transcription was induced directly by estrogen receptor activation or by some other regulator responsive to the increased galactosylated IgG substrate availability. Thus, these data are consistent with a direct effect of estrogens on IgG Fc sialylation but do not yet permit a firm conclusion.

With respect to galactosylation, these studies confirm that estrogens modulate IgG Fc glycans in humans, in females as well as males. Estrogens are thus the first and to date only mediator confirmed to have such an effect in vivo, driving the increase in IgG Fc galactosylation that accompanies menarche and disappears with menopause. Estrogens are likely also a major contributor to IgG glycan changes in pregnancy, though this connection has not been established experimentally. Testosterone plays no evident estrogen-independent role in the control of IgG Fc glycans. Effects of hormones such as progesterone and prolactin, or other factors associated with menarche, menopause, and pregnancy, remain to be determined.

11.4 Implications of Estrogen-Induced Fc Glycans Changes on IgG Function

Extrapolating from IgG Fc glycosylation changes to effector function is complex, as noted above. G0 glycoforms are typically regarded as rendering IgG pro-inflammatory, whereas G2 glycoforms and terminal sialic acid (obligatorily attached to galactose) are considered anti-inflammatory. Yet published findings conflict. G0 IgG activate complement via the mannose-binding lectin pathway and the alternative pathway (Malhotra et al. 1995; Arnold et al. 2006; Banda et al. 2008). Enzymatic removal of galactose enhances binding to FcγRIIIA and thus antibody-dependent cellular cytotoxicity (ADCC) (Ackerman et al. 2013). Conversely, G2 IgG engages immunoregulatory mechanisms by binding FcγRIIB, while terminal sialic acid is implicated in a range of anti-inflammatory and immunoregulatory functions (Karsten et al. 2012; Wang and Ravetch 2019). However, IgG Fc galactosylation can also be shown to enhance FcγRIIIA binding, ADCC, and complement fixation via the classical pathway (C1q), while effects of sialylation are not invariably observed (Dekkers et al. 2017; de Taeye et al. 2019; Jennewein et al. 2019). Thus, no straightforward answer is yet available to the question of whether estrogen makes the overall pool of IgG more or less poised to promote inflammation.

The suggestion that G0 IgG is "net" pro-inflammatory comes principally from studies in inflammatory arthritis. Patients with RA and juvenile idiopathic arthritis (JIA) have long been recognized to exhibit an excess of G0 IgG, even preceding disease onset, and glycan aberrancy correlate with disease activity (Parekh et al. 1985, 1988b; Ercan et al. 2010, 2012a, b; Rombouts et al. 2015). G0 IgG appears especially pathogenic in murine models of antibody-mediated arthritis, while sialylation confers protection (Engdahl et al. 2018; Rademacher et al. 1994; Pfeifle et al. 2017). Human RA often improves transiently with pregnancy, only to flare after parturition, events that correlate with changes in IgG Fc galactosylation (Rook et al. 1991; van de Geijn et al. 2009; Bondt et al. 2013; Reiding et al. 2017). However, although RA incidence peaks in postmenopausal females, males also demonstrate a late peak, and hormone replacement therapy has no consistent effect on disease incidence or activity (Walitt et al. 2008; Hall et al. 1994). Of note, RA patients develop serologic evidence of inflammation well before clinical findings, and IgG Fc galactosylation varies with disease activity, rendering it very challenging to distinguish whether the relationship of IgG Fc glycan changes with disease activity represents one of cause, effect, or both (Ercan et al. 2012b; Deane et al. 2010).

More generally, females appear less susceptible to infection and more prone to autoimmunity, including autoantibody-driven conditions such as RA and systemic lupus erythematosus (SLE), compared to males (vom Steeg and Klein 2016; Libert et al. 2010). These differences integrate a host of immune and non-immune sexual dimorphisms, but would be unexpected if estrogens simply diminished IgG potency. Extrapolating from these observations, it is likely that the overall effects of estrogens on IgG potency will not be reflected accurately in a simple stronger/weaker dichotomy, but will likely vary with physiological context.

11.5 Implications of Estrogen-Induced Fc Glycan Changes for Pregnancy

The highest circulating estrogen concentrations in normal human physiology are attained during pregnancy. Elevated levels of estradiol and estriol are sustained for months, raising the possibility that pregnancy could be an important physiological context for estrogen-induced changes in IgG Fc glycans. A shift toward anti-inflammatory IgG Fc glycoforms, for example, could be an immune adaptation that helps maintain tolerance to the fetus, although no studies have yet tested this hypothesis.

One intriguing possibility is that estrogens regulate antibody transfer across the placenta. The central molecular mechanism mediating this transfer is the neonatal Fc receptor (FcRn) (Roopenian and Akilesh 2007; Martinez et al. 2018). FcRn has a high affinity for the Fc portion of IgG at acidic pH (<6.5), but not at physiological pH (Roopenian and Akilesh 2007; Rodewald 1976). Syncytiotrophoblasts employ FcRn to capture IgG antibodies from maternal circulation in acidic endosomes and then deliver these vesicles to fetal endothelial cells, where physiological pH promotes dissociation of IgG from FcRn. IgG1 and IgG4 are transported with somewhat greater efficiency than IgG2 and IgG3 (Malek et al. 1996; Stapleton et al. 2011; Einarsdottir et al. 2014a, b). Much remains unknown about this important process, including whether antibodies against different targets equilibrate evenly across the placenta (Jennewein et al. 2019; Martinez et al. 2018, 2019; Fu et al. 2016).

As noted above, maternal IgG becomes more galactosylated as pregnancy progresses, likely due at least in part to estrogens. Intriguingly, comparison of total IgG from umbilical cord blood to maternal blood shows a further skew away from G0 and toward G2 glycoforms, suggesting selection for antibodies based on their glycosylation (termed a "placental sieve") (Jennewein et al. 2019; Williams et al. 1995; Kibe et al. 1996; Jansen et al. 2016). Relative to pre-term infants, antibodies in term infants are significantly more galactosylated, perhaps reflecting the cumulative effects of the sieve through gestation (Twisselmann et al. 2018). Proposed mechanisms underlying this skew include preferential affinity of G2 IgG for FcRn and possible engagement of the Fc receptors FcγRIIIa or FcγRIIb, also expressed at the maternal-fetal interface (Martinez et al. 2018, 2019; Kameda et al. 1991; Takizawa et al. 2005; Ishikawa et al. 2015; Mishima et al. 2007). These G2 antibodies may be especially well-suited to activate neonatal NK cells, providing enhanced immune defense to the fetus (Jennewein et al. 2019). In this manner, estrogen-driven enhancement of maternal IgG Fc galactosylation may set the stage for the provision of particularly protective antibodies for the fetus.

However, these observations remain controversial. The extent of G2 skewing is subtle, suggesting that any "sieve" is of low efficiency. IgG recognizing different antigens may exhibit differential glycan enrichment or lack enrichment altogether (Martinez et al. 2019). Further, when IgG Fc glycans are assessed in an isotype-specific manner, glycan-specific enrichment is largely absent, suggesting that the whole-IgG observations could reflect at least in part the known selectivity of FcRn for IgG1 (Malek et al. 1996; Stapleton et al. 2011; Einarsdottir et al. 2014a, b). In many experimental systems, Fc glycosylation has no impact on FcRn binding, and mice engineered to express human FcRn exhibit no differential IgG transfer to the fetus as a function of IgG Fc glycans (Liu et al. 2011; Souders et al. 2015; Cymer et al. 2017; Borghi et al. 2020). Thus, the extent to which estrogen-driven changes in IgG Fc galactosylation facilitate transplacental transfer remains uncertain. Indeed, the G2 skew of maternal IgG Fc glycans driven by estrogens in pregnancy renders a placental G2 "sieve" largely redundant. Rather, enhanced FcγRIIIa engagement by G2 IgG could still promote NK cell-mediated protection of the fetus and neonate, the role proposed for the sieve, providing an appealing teleological explanation for why estrogens regulate IgG Fc glycans at all (Dekkers et al. 2017; Jennewein et al. 2019; Kimura et al. 2000).

11.6 In Search of Mechanism

The molecular processes that link estradiol to IgG Fc glycan variation remain unknown. Presumably, the mechanisms responsible for these shifts must begin with an estrogen receptor, include a β4 galactosyltransferase, and potentially involve a sialyltransferase. There are seven β4 galactosyltransferase enzymes in humans (β4Gal-T (Sondermann et al. 2000; Barb and Prestegard 2011; Subedi and Barb 2015; Nose and Wigzell 1983; Parekh et al. 1988a; Yamada et al. 1997; Kristic et al. 2014), encoded by *B4GALT* (Sondermann et al. 2000; Barb and Prestegard 2011; Subedi and Barb 2015; Nose and Wigzell 1983; Parekh et al. 1988a; Yamada et al. 1997; Kristic et al. 2014)) with varying cellular localizations and substrate specificities. β4Gal-T1 has been shown to induce large changes in IgG Fc galactosylation in human cell lines, but the contribution of the other six galactosyltransferases has yet to be resolved (Kimura et al. 2000).

IgG N-glycan modification occurs predominantly in the Golgi apparatus. B cells express both estrogen receptors, ERα and ERβ, through many developmental stages (Kanda and Tamaki 1999) and this expression has been accompanied by estrogen responsivity in various forms, including increased production of IgM and IgG, reduced B cell receptor (BCR) signaling, and resistance to apoptosis (Kanda and Tamaki 1999; Grimaldi et al. 2002; Hill et al. 2011). Residing typically within the nucleus, ERα and ERβ act as transcription factors to change the expression of genes whose promoters are enriched for estrogen response elements (EREs). The genes of three enzymes that add galactose to glycan structures (β-1,4-galactosyltransferase 1, 3, and 7) contain an ERE in their promoter, but binding to ERα or Erβ has yet to be confirmed experimentally (Bourdeau et al. 2004).

In the absence of evidence for direct transcriptional regulation by estrogen, four distinct non-mutually-exclusive mechanisms exist whereby estrogen could enhance IgG Fc galactosylation (Sondermann et al. 2000). ERα and ERβ coordinate with some other transcription factor to increase the expression of a β4 galactosyltransferase in antibody-secreting B cells (Barb and Prestegard 2011); estrogens modulate the activity of a β4 galactosyltransferase via changes in substrate availability or product stability (Subedi and Barb 2015); changes in B cell galactosylation activity are mediated by cytokine signaling from other cells sensitive to estrogens (Nose and Wigzell 1983); or the effect of estrogens is independent of B cells altogether and is instead related to the plasma activity of β4 galactosyltransferases or differential lifespan of IgG Fc glycoforms in the plasma (Fig. 11.1).

11.6.1 Possibility 1: Transcriptional Activation of B4GALT Genes by Regulatory Complexes Containing ERα/ERβ

Some experimental and computational evidence has been gathered for the possibility that ERα and ERβ coordinate with some other transcription factor to increase the expression of a *B4GALT* gene. Multiple studies have found that ERα and ERβ form a complex with Sp1 (Porter et al. 1997; Saville et al. 2000; Bartella et al. 2012; Barreto-Andrade et al. 2018), a transcriptional activator of several *B4GALT* genes in certain lineages (Sato and Furukawa 2004; Jiang et al. 2006; Shen et al. 2007). Undirected high-throughput assays to detect protein–DNA interactions, on the other hand, suggest that the transcription factor ERRγ (estrogen-related receptor gamma, encoded by *ESRRG*, an orphan nuclear receptor named solely for sequence homology to the nuclear estrogen receptor) may mediate the connection between ERα and *B4GALT1*. HT-SELEX (high-throughput systematic evolution of ligands by exponential enrichment) of 984 human transcription factors reported interactions between ERRγ and the promoter of *B4GALT1*, as well as an interaction between *ESRRG* and ERα (Jolma et al. 2013). Integration of various high-throughput protein–DNA interactions in cell lines validated the ERRγ-*B4GALT1* interaction, further confirmed by homology-based molecular dynamics models (Kulakovskiy et al. 2013; Pujato et al. 2014). Direct experimental validation by antibody-secreting cells is needed to confirm this signal further and establish its role in estrogen-driven glycan changes.

11.6.2 Possibility 2: Modulation of β4 Galactosyltransferase Enzymatic Activity by Estradiol

The frequency of galactosylated glycans reflects not only the expression levels of *B4GALT* genes but also the availability of the galactose substrate. This substrate, uridine diphosphate (UDP)-galactose, is formed in the cytosol when galactose-1-phosphate uridyltransferase (GALT) exchanges the phosphate on a galactose molecule phosphorylated by galactokinase (GALK) for UDP (Slepak et al. 2005). In the cytosol, UDP-galactose 4′-epimerase (GALE) maintains a balance between UDP-galactose and UDP-glucose through reversible epimerization (Seo et al. 2019; Broussard et al. 2020). UDP-galactose translocator enzymes then deliver UDP-galactose to the Golgi lumen against a severe UDP-galactose concentration gradient, anti-porting uridine monophosphate (UMP^{2-}) to leverage the acidity of the cytosol relative to the Golgi compartment (Parker and Newstead 2019). Thus, substrate availability for the β4 galactosyltransferase enzymes is dependent on the activity of GALK, GALT, GALE, and UDP-galactose transporters, as well as the relative concentrations of UMP^{2-} and H^+ between the Golgi apparatus and the cytosol. Evidence for ERα/β-driven changes in these factors is limited, since the hypothesis that estrogens modulate the availability of substrate for galactosyltransferases has not been specifically investigated. Homology-based molecular dynamics models have suggested that *SLC35E3*, an orphan translocator with marked sequence homology to Golgi membrane nucleotide sugar transporters, is a direct target of ERα (Pujato et al. 2014). More intriguingly, high-throughput DNA–protein interaction assays have reported that ERγ binds the promoters for *GALT* and *GALE* in addition to *B4GALT1*, furthering the evidence that it will be fruitful to investigate this transcription factor (Jolma et al. 2013; Kulakovskiy et al. 2013). In RA, some studies but not others have found reduced activity of β4 galactosyltransferases (Furukawa et al. 1990; Axford et al. 1987; Jeddi et al. 1996). Of note, IgG Fc galactosylation is a reversible reaction, susceptible to hydrolysis of galactose by the enzyme β-galactosidase. Elevated activity of this enzyme has been reported in RA; regulation by estrogen has not been explored (Su et al. 2020).

11.6.3 Possibility 3: Secondary Cytokine Signaling

Many cells express ERα and ERβ, raising the possibility that the effect of estrogens on IgG is mediated indirectly via mediators produced by other lineages. For example, in cultured B cells, IgG Fc galactosylation is augmented by interleukin 21 (IL-21) and decreased by all-trans retinoic acid (Wang et al. 2011). The enormous number of potentially relevant cell–cell interactions render this question well-suited for genome-wide association studies (GWAS), which employ inter-individual genetic variation to probe many biological hypotheses in parallel. Lauc et al. were the first to study the frequency of IgG glycoforms as quantitative traits, querying for associated variation with 296,619 single-nucleotide polymorphisms (SNPs) across the genome (Lauc et al. 2013). In the discovery cohort (four European populations, $N = 2247$), the authors observed nine genome-wide significant associations between genetic loci and IgG Fc glycan traits. For galactosylation, these loci included a region containing *B4GALT1* and a second region containing the genes *ANKRD55* and *IL6ST*. While the function of *ANKRD55* is undefined, *IL6ST* encodes a subunit of several cytokine receptors, including the interleukin 6 (IL-6) receptor. SNPs in the *ANKRD55-IL6ST* locus exhibited opposite effects on the frequency of G0 and G2 IgG glycans, consistent with an underlying change in the net galactosylation rate. Though the lead *ANKRD55-IL6ST* SNP failed to replicate in a second cohort ($N = 1848$) with respect to G2 frequency ($p = 0.6$) and reached only nominal significance ($p = 0.01$) for G0 frequency, this could potentially be an artifact of population stratification or differential glycoform ascertainment in ultra-performance liquid chromatography (UPLC) compared to mass spectrometry (MS).

Several GWAS have since extended the approach put forward by Lauc et al., each replicating the expected association with *B4GALT1* (Shen et al. 2017; Wahl et al. 2018; Klaric et al. 2020). Shen et al. reanalyzed a subset of the individuals in Lauc et al., but in recognition of the inherent correlation structure between glycoform frequencies, used multivariate analysis of variance (MANOVA) to uncover SNPs explaining variance in the overall distribution of IgG glycoforms (Shen et al. 2017).

Despite this promising methodological approach, these analyses were not suited to find specific galactosylation regulators because glycoform variations were modeled *within* each galactosylation state rather than between them (the distribution of G0F, G0N, and G0, for example, rather than the distribution of G0F, G1F, and G2F). Wahl et al. examined MS data in a separate cohort ($N = 1823$) and noted several associations between *RUNX3*, a transcription factor frequently silenced in cancer, and galactosylation and bisection glycoforms from IgG2 (Wahl et al. 2018; Bae and Choi 2004). With the largest IgG glycome GWAS to date ($N = 8090$), Klarić et al. clarified that the *RUNX3* glycan phenotype effect size distribution resembled that of enzymes known to induce bisection rather than enzymes known to induce galactosylation (Klaric et al. 2020). This study also uncovered and replicated an association between the frequency of digalactosylated glycoforms and *HIVEP2*, encoding a zinc finger transcription factor that has been implicated in T cell activation (Nomura et al. 1991). It has been noted that the expression of this transcription factor is reduced in breast tumor samples, but this appears to be independent of estrogen receptor expression, and we are unaware of any other *HIVEP2* connections to estrogen biology (Fujii et al. 2005). While Klarić et al. replicated the *ANKRD55-IL6ST* signal with nearly genome-wide significance ($P = 7.22 \times 10^{-6}$), Wahl et al. did not, perhaps due to their conservative Bonferroni correction that assumed 50 independent tests (given the unique correlation structure of glycome phenotypes, robust statistical inference requires multivariate models or permutation simulations to uncover the true number of independent tests).

The evidence for a role of IL-6 in multiple cohorts is intriguing, given that estrogen negatively regulates IL-6 signaling in many cells, including those in the B cell lineage (Pottratz et al. 1994; Stein and Yang 1995; Kurebayashi et al. 1997; Liu et al. 2005; Canellada et al. 2008). These observations raise the possibility that estrogens modulate IgG Fc galactosylation in part through interference with IL-6 production or signaling, a pathway exploited by therapeutic blockade of IL-6 in RA (Emery et al. 2008). Whether IL-6 in fact modulates IgG Fc galactosylation, and which other mediators also participate in this regulation, are outstanding questions that should be interrogated experimentally.

11.6.4 Possibility 4: Post-Secretory Glycan Modification

A final possibility is that estrogen regulation of IgG Fc glycans bypasses B cells entirely. Considering the IgG lifespan in three phases—generation, circulation, and degradation—brings to light the possibility that estrogens might modify IgG Fc glycans after release from B cells. Recent studies have shown that enzymes capable of modifying IgG Fc galactose and sialic acid content are present and active in the plasma (Wandall et al. 2012; Jones et al. 2016; Catera et al. 2016). ST6Gal-T1, for example, is released by the liver as an acute phase reactant (Dalziel et al. 1999). Conditional knock-out mouse models demonstrate that IgG sialylation can occur even when expression of *St6gal1* by B cells has been ablated, though follow-up work suggests that such effects are modest (Jones et al. 2016; Schaffert et al. 2019). Whether estrogens modulate post-secretory changes in IgG glycans remains unknown, although some evidence implicates estrogens in the activation of platelets, which are known to release soluble β4Gal-T1 along with sufficient levels of sugar nucleotides to galactosylate exogenous substances (Maccarrone et al. 2002; Moro et al. 2005). We are unaware of any data implicating estrogens in antibody degradation and whether it may preferentially do so for certain glycoforms.

11.7 Future Directions

The promotion of IgG Fc galactosylation by estrogens represents a new link between sex and immunity, adding a contextual layer that brings us closer to understanding situationally appropriate immune reactivity over the lifespan (Fig. 11.1). The physiological consequences of this estrogen effect remain undefined. The especially marked changes in IgG Fc glycans during pregnancy, presumably related at least in part to high and sustained levels of estrogens, favor a role in the protection of the fetus and/or mother. The mechanisms mediating estrogen-driven IgG galactosylation remain unknown, requiring experiments to clarify whether the estrogen effect is mediated directly by B cells expressing estrogen receptors or via other pathways. Convergent studies highlight potential roles for Sp1, ERRγ, and IL-6 signaling, justifying targeted experimental follow-up.

It remains unknown whether the effect of estrogens on glycans is unique to the IgG Fc region or occurs more broadly among N-glycosylated proteins, including in other types of antibody. Menarche- and menopause-related changes in whole-plasma glycans are comparable, if somewhat less marked than those found in IgG (Pucic et al. 2012; Knezevic et al. 2009, 2010). Since whole-plasma samples contain IgG-derived glycans, it is difficult to determine if such changes arise from any proteins beyond IgG. Nonetheless, glycoform shifts with age, in pregnancy, and in arthritis and other autoimmune diseases highlight the importance of IgG Fc glycans as an axis of immunoregulation (Cheng et al. 2020). Ultimately, decoding the pathways that define the IgG Fc glycome will provide a much higher resolution view of immune development and attendant opportunities for therapeutic intervention.

Compliance with Ethical Standards

Funding KAL is supported by Award Number T32GM007753 from the National Institute of General Medical Sciences. PAN is funded by NIAMS awards 2R01 AR065538, R01 AR075906, R01 AR073201, P30 AR070253, R21 AR076630 and NHLBI award R21 HL150575; the Fundación Bechara; the Samara Jan Turkel Clinical Center for Pediatric Autoimmune Disease at Boston Children's Hospital; and the Arbuckle Family Fund for Arthritis Research.

Conflict of Interest KAL and PAN declare no related conflict of interest.

Ethical Approval This article does not contain any studies with human participants performed by any of the authors.

References

Abbassi-Ghanavati M, Greer LG, Cunningham FG (2009) Pregnancy and laboratory studies: a reference table for clinicians. Obstet Gynecol 114(6):1326–1331
Ackerman ME, Crispin M, Yu X et al (2013) Natural variation in Fc glycosylation of HIV-specific antibodies impacts antiviral activity. J Clin Invest 123(5):2183–2192
Arnold JN, Dwek RA, Rudd PM, Sim RB (2006) Mannan binding lectin and its interaction with immunoglobulins in health and in disease. Immunol Lett 106(2):103–110
Axford JS, Mackenzie L, Lydyard PM, Hay FC, Isenberg DA, Roitt IM (1987) Reduced B-cell galactosyltransferase activity in rheumatoid arthritis. Lancet 2(8574):1486–1488
Bae SC, Choi JK (2004) Tumor suppressor activity of RUNX3. Oncogene 23(24):4336–4340
Bakovic MP, Selman MH, Hoffmann M et al (2013) High-throughput IgG Fc N-glycosylation profiling by mass spectrometry of glycopeptides. J Proteome Res 12(2):821–831
Banda NK, Wood AK, Takahashi K et al (2008) Initiation of the alternative pathway of murine complement by immune complexes is dependent on N-glycans in IgG antibodies. Arthritis Rheum 58(10):3081–3089
Barb AW, Prestegard JH (2011) NMR analysis demonstrates immunoglobulin G N-glycans are accessible and dynamic. Nat Chem Biol 7(3):147–153
Barreto-Andrade JN, de Fatima LA, Campello RS, Guedes JAC, de Freitas HS, Machado M (2018) Estrogen receptor 1 (ESR1) enhances Slc2a4/GLUT4 expression by a SP1 cooperative mechanism. Int J Med Sci 15(12):1320–1328
Bartella V, Rizza P, Barone I et al (2012) Estrogen receptor beta binds Sp1 and recruits a corepressor complex to the estrogen receptor alpha gene promoter. Breast Cancer Res Treat 134(2):569–581
Bidlingmaier F, Wagner-Barnack M, Butenandt O, Knorr D (1973) Plasma estrogens in childhood and puberty under physiologic and pathologic conditions. Pediatr Res 7(11):901–907
Bondt A, Selman MH, Deelder AM et al (2013) Association between galactosylation of immunoglobulin G and improvement of rheumatoid arthritis during pregnancy is independent of sialylation. J Proteome Res 12(10):4522–4531
Bondt A, Rombouts Y, Selman MH et al (2014) Immunoglobulin G (IgG) fab glycosylation analysis using a new mass spectrometric high-throughput profiling method reveals pregnancy-associated changes. Mol Cell Proteomics 13(11):3029–3039
Borghi S, Bournazos S, Thulin NK et al (2020) FcRn, but not FcgammaRs, drives maternal-fetal transplacental transport of human IgG antibodies. Proc Natl Acad Sci U S A 117(23):12943–12951
Bourdeau V, Deschenes J, Metivier R et al (2004) Genome-wide identification of high-affinity estrogen response elements in human and mouse. Mol Endocrinol 18(6):1411–1427

Broussard A, Florwick A, Desbiens C et al (2020) Human UDP-galactose 4'-epimerase (GALE) is required for cell-surface glycome structure and function. J Biol Chem 295(5):1225–1239

Canellada A, Alvarez I, Berod L, Gentile T (2008) Estrogen and progesterone regulate the IL-6 signal transduction pathway in antibody secreting cells. J Steroid Biochem Mol Biol 111 (3–5):255–261

Catera M, Borelli V, Malagolini N et al (2016) Identification of novel plasma glycosylation-associated markers of aging. Oncotarget 7(7):7455–7468

Chen G, Wang Y, Qiu L et al (2012) Human IgG Fc-glycosylation profiling reveals associations with age, sex, female sex hormones and thyroid cancer. J Proteome 75(10):2824–2834

Cheng HD, Tirosh I, de Haan N et al (2020) IgG Fc glycosylation as an axis of humoral immunity in childhood. J Allergy Clin Immunol 145(2):710–713. e719

Cymer F, Schlothauer T, Knaupp A, Beck H (2017) Evaluation of an FcRn affinity chromatographic method for IgG1-type antibodies and evaluation of IgG variants. Bioanalysis 9 (17):1305–1317

Dalziel M, Lemaire S, Ewing J, Kobayashi L, Lau JT (1999) Hepatic acute phase induction of murine beta-galactoside alpha 2,6 sialyltransferase (ST6Gal I) is IL-6 dependent and mediated by elevation of exon H-containing class of transcripts. Glycobiology 9(10):1003–1008

de Haan N, Reiding KR, Driessen G, van der Burg M, Wuhrer M (2016) Changes in healthy human IgG Fc-glycosylation after birth and during early childhood. J Proteome Res 15(6):1853–1861

de Taeye SW, Rispens T, Vidarsson G (2019) The ligands for human IgG and their effector functions. Antibodies (Basel) 8(2)

Deane KD, O'Donnell CI, Hueber W et al (2010) The number of elevated cytokines and chemokines in preclinical seropositive rheumatoid arthritis predicts time to diagnosis in an age-dependent manner. Arthritis Rheum 62(11):3161–3172

Dekkers G, Treffers L, Plomp R et al (2017) Decoding the human immunoglobulin G-glycan repertoire reveals a spectrum of Fc-receptor- and complement-mediated-effector activities. Front Immunol 8:877

Einarsdottir H, Ji Y, Visser R et al (2014a) H435-containing immunoglobulin G3 allotypes are transported efficiently across the human placenta: implications for alloantibody-mediated diseases of the newborn. Transfusion 54(3):665–671

Einarsdottir HK, Stapleton NM, Scherjon S et al (2014b) On the perplexingly low rate of transport of IgG2 across the human placenta. PLoS One 9(9):e108319

Emery P, Keystone E, Tony HP et al (2008) IL-6 receptor inhibition with tocilizumab improves treatment outcomes in patients with rheumatoid arthritis refractory to anti-tumour necrosis factor biologicals: results from a 24-week multicentre randomised placebo-controlled trial. Ann Rheum Dis 67(11):1516–1523

Engdahl C, Bondt A, Harre U et al (2018) Estrogen induces St6gal1 expression and increases IgG sialylation in mice and patients with rheumatoid arthritis: a potential explanation for the increased risk of rheumatoid arthritis in postmenopausal women. Arthritis Res Ther 20(1):84

Ercan A, Cui J, Chatterton DE et al (2010) Aberrant IgG galactosylation precedes disease onset, correlates with disease activity, and is prevalent in autoantibodies in rheumatoid arthritis. Arthritis Rheum 62(8):2239–2248

Ercan A, Barnes MG, Hazen M et al (2012a) Multiple juvenile idiopathic arthritis subtypes demonstrate pro-inflammatory IgG glycosylation. Arthritis Rheum 64(9):3025–3033

Ercan A, Cui J, Hazen MM et al (2012b) Hypogalactosylation of serum N-glycans fails to predict clinical response to methotrexate and TNF inhibition in rheumatoid arthritis. Arthritis Res Ther 14(2):R43

Ercan A, Kohrt WM, Cui J et al (2017) Estrogens regulate glycosylation of IgG in women and men. JCI insight 2(4):e89703

Finkelstein JS, Lee H, Burnett-Bowie SA et al (2013) Gonadal steroids and body composition, strength, and sexual function in men. N Engl J Med 369(11):1011–1022

Fu C, Lu L, Wu H et al (2016) Placental antibody transfer efficiency and maternal levels: specific for measles, coxsackievirus A16, enterovirus 71, poliomyelitis I-III and HIV-1 antibodies. Sci Rep 6:38874

Fujii H, Gabrielson E, Takagaki T, Ohtsuji M, Ohtsuji N, Hino O (2005) Frequent down-regulation of HIVEP2 in human breast cancer. Breast Cancer Res Treat 91(2):103–112

Furukawa K, Matsuta K, Takeuchi F, Kosuge H, Miyamoto T, Kobata A (1990) Kinetic study of a galactosyltransferase in the B cells of patients with rheumatoid arthritis. Int Immunol 2(1):105–112

Grimaldi CM, Cleary J, Dagtas AS, Moussai D, Diamond B (2002) Estrogen alters thresholds for B cell apoptosis and activation. J Clin Invest 109(12):1625–1633

Hall GM, Daniels M, Huskisson EC, Spector TD (1994) A randomised controlled trial of the effect of hormone replacement therapy on disease activity in postmenopausal rheumatoid arthritis. Ann Rheum Dis 53(2):112–116

Hill L, Jeganathan V, Chinnasamy P, Grimaldi C, Diamond B (2011) Differential roles of estrogen receptors alpha and beta in control of B-cell maturation and selection. Mol Med 17(3–4):211–220

Ishikawa T, Takizawa T, Iwaki J et al (2015) Fc gamma receptor IIb participates in maternal IgG trafficking of human placental endothelial cells. Int J Mol Med 35(5):1273–1289

Jansen BC, Bondt A, Reiding KR, Scherjon SA, Vidarsson G, Wuhrer M (2016) MALDI-TOF-MS reveals differential N-linked plasma- and IgG-glycosylation profiles between mothers and their newborns. Sci Rep 6:34001

Jeddi PA, Bodman-Smith KB, Lund T et al (1996) Agalactosyl IgG and beta-1,4-galactosyltransferase gene expression in rheumatoid arthritis patients and in the arthritis-prone MRL lpr/lpr mouse. Immunology 87(4):654–659

Jennewein MF, Goldfarb I, Dolatshahi S et al (2019) Fc glycan-mediated regulation of placental antibody transfer. Cell 178(1):202–215. e214

Jiang J, Shen J, Wu T et al (2006) Down-regulation of beta1,4-galactosyltransferase V is a critical part of etoposide-induced apoptotic process and could be mediated by decreasing Sp1 levels in human glioma cells. Glycobiology 16(11):1045–1051

Jolma A, Yan J, Whitington T et al (2013) DNA-binding specificities of human transcription factors. Cell 152(1–2):327–339

Jones MB, Oswald DM, Joshi S, Whiteheart SW, Orlando R, Cobb BA (2016) B-cell-independent sialylation of IgG. Proc Natl Acad Sci U S A 113(26):7207–7212

Juric J, Kohrt WM, Kifer D et al (2020) Effects of estradiol on biological age measured using the glycan age index. Aging (Albany NY) 12(19):19756–19765

Kameda T, Koyama M, Matsuzaki N, Taniguchi T, Saji F, Tanizawa O (1991) Localization of three subtypes of Fc gamma receptors in human placenta by immunohistochemical analysis. Placenta 12(1):15–26

Kanda N, Tamaki K (1999) Estrogen enhances immunoglobulin production by human PBMCs. J Allergy Clin Immunol 103(2 Pt 1):282–288

Karsten CM, Pandey MK, Figge J et al (2012) Anti-inflammatory activity of IgG1 mediated by Fc galactosylation and association of FcgammaRIIB and dectin-1. Nat Med 18(9):1401–1406

Kibe T, Fujimoto S, Ishida C et al (1996) Glycosylation and placental transport of immunoglobulin G. J Clin Biochem Butr 21(1):57–63

Kimura S, Numaguchi M, Kaizu T, Kim D, Takagi Y, Gomi K (2000) High galactosylation of oligosaccharides in umbilical cord blood IgG, and its relationship to placental function. Clin Chim Acta 299(1–2):169–177

Klaric L, Tsepilov YA, Stanton CM et al (2020) Glycosylation of immunoglobulin G is regulated by a large network of genes pleiotropic with inflammatory diseases. Sci Adv 6(8):eaax0301

Knezevic A, Polasek O, Gornik O et al (2009) Variability, heritability and environmental determinants of human plasma N-glycome. J Proteome Res 8(2):694–701

Knezevic A, Gornik O, Polasek O et al (2010) Effects of aging, body mass index, plasma lipid profiles, and smoking on human plasma N-glycans. Glycobiology 20(8):959–969

Kristic J, Vuckovic F, Menni C et al (2014) Glycans are a novel biomarker of chronological and biological ages. J Gerontol A Biol Sci Med Sci 69(7):779–789

Kulakovskiy IV, Medvedeva YA, Schaefer U et al (2013) HOCOMOCO: a comprehensive collection of human transcription factor binding sites models. Nucleic Acids Res 41(Database issue):D195–D202

Kurebayashi S, Miyashita Y, Hirose T, Kasayama S, Akira S, Kishimoto T (1997) Characterization of mechanisms of interleukin-6 gene repression by estrogen receptor. J Steroid Biochem Mol Biol 60(1–2):11–17

Lanciotti L, Cofini M, Leonardi A, Penta L, Esposito S (2018) Up-To-Date review about minipuberty and overview on hypothalamic-pituitary-gonadal axis activation in fetal and neonatal life. Front Endocrinol (Lausanne) 9:410

Lauc G, Huffman JE, Pucic M et al (2013) Loci associated with N-glycosylation of human immunoglobulin G show pleiotropy with autoimmune diseases and haematological cancers. PLoS Genet 9(1):e1003225

Libert C, Dejager L, Pinheiro I (2010) The X chromosome in immune functions: when a chromosome makes the difference. Nat Rev Immunol 10(8):594–604

Liu H, Liu K, Bodenner DL (2005) Estrogen receptor inhibits interleukin-6 gene expression by disruption of nuclear factor kappaB transactivation. Cytokine 31(4):251–257

Liu L, Stadheim A, Hamuro L et al (2011) Pharmacokinetics of IgG1 monoclonal antibodies produced in humanized Pichia pastoris with specific glycoforms: a comparative study with CHO produced materials. Biologicals 39(4):205–210

Maccarrone M, Bari M, Battista N, Finazzi-Agro A (2002) Estrogen stimulates arachidonoylethanolamide release from human endothelial cells and platelet activation. Blood 100(12):4040–4048

Malek A, Sager R, Kuhn P, Nicolaides KH, Schneider H (1996) Evolution of maternofetal transport of immunoglobulins during human pregnancy. Am J Reprod Immunol 36(5):248–255

Malhotra R, Wormald MR, Rudd PM, Fischer PB, Dwek RA, Sim RB (1995) Glycosylation changes of IgG associated with rheumatoid arthritis can activate complement via the mannose-binding protein. Nat Med 1(3):237–243

Martinez DR, Fouda GG, Peng X, Ackerman ME, Permar SR (2018) Noncanonical placental Fc receptors: what is their role in modulating transplacental transfer of maternal IgG? PLoS Pathog 14(8):e1007161

Martinez DR, Fong Y, Li SH et al (2019) Fc characteristics mediate selective placental transfer of IgG in HIV-infected women. Cell 178(1):190–201. e111

Mishima T, Kurasawa G, Ishikawa G et al (2007) Endothelial expression of Fc gamma receptor IIb in the full-term human placenta. Placenta 28(2–3):170–174

Moro L, Reineri S, Piranda D et al (2005) Nongenomic effects of 17beta-estradiol in human platelets: potentiation of thrombin-induced aggregation through estrogen receptor beta and Src kinase. Blood 105(1):115–121

Nomura N, Zhao MJ, Nagase T et al (1991) HIV-EP2, a new member of the gene family encoding the human immunodeficiency virus type 1 enhancer-binding protein. Comparison with HIV-EP1/PRDII-BF1/MBP-1. J Biol Chem 266(13):8590–8594

Nose M, Wigzell H (1983) Biological significance of carbohydrate chains on monoclonal antibodies. Proc Natl Acad Sci USA 80(21):6632–6636

Parekh RB, Dwek RA, Sutton BJ et al (1985) Association of rheumatoid arthritis and primary osteoarthritis with changes in the glycosylation pattern of total serum IgG. Nature 316(6027):452–457

Parekh R, Roitt I, Isenberg D, Dwek R, Rademacher T (1988a) Age-related galactosylation of the N-linked oligosaccharides of human serum IgG. J Exp Med 167(5):1731–1736

Parekh RB, Roitt IM, Isenberg DA, Dwek RA, Ansell BM, Rademacher TW (1988b) Galactosylation of IgG associated oligosaccharides: reduction in patients with adult and juvenile onset rheumatoid arthritis and relation to disease activity. Lancet 1(8592):966–969

Parker JL, Newstead S (2019) Gateway to the Golgi: molecular mechanisms of nucleotide sugar transporters. Curr Opin Struct Biol 57:127–134

Pezer M, Stambuk J, Perica M et al (2016) Effects of allergic diseases and age on the composition of serum IgG glycome in children. Sci Rep 6:33198

Pfeifle R, Rothe T, Ipseiz N et al (2017) Regulation of autoantibody activity by the IL-23-TH17 axis determines the onset of autoimmune disease. Nat Immunol 18(1):104–113

Porter W, Saville B, Hoivik D, Safe S (1997) Functional synergy between the transcription factor Sp1 and the estrogen receptor. Mol Endocrinol 11(11):1569–1580

Pottratz ST, Bellido T, Mocharla H, Crabb D, Manolagas SC (1994) 17 beta-estradiol inhibits expression of human interleukin-6 promoter-reporter constructs by a receptor-dependent mechanism. J Clin Invest 93(3):944–950

Pucic M, Muzinic A, Novokmet M et al (2012) Changes in plasma and IgG N-glycome during childhood and adolescence. Glycobiology 22(7):975–982

Pujato M, Kieken F, Skiles AA, Tapinos N, Fiser A (2014) Prediction of DNA binding motifs from 3D models of transcription factors; identifying TLX3 regulated genes. Nucleic Acids Res 42 (22):13500–13512

Rademacher TW, Williams P, Dwek RA (1994) Agalactosyl glycoforms of IgG autoantibodies are pathogenic. Proc Natl Acad Sci USA 91(13):6123–6127

Reiding KR, Vreeker GCM, Bondt A et al (2017) Serum protein N-glycosylation changes with rheumatoid arthritis disease activity during and after pregnancy. Front Med (Lausanne) 4:241

Rodewald R (1976) pH-dependent binding of immunoglobulins to intestinal cells of the neonatal rat. J Cell Biol 71(2):666–669

Rombouts Y, Ewing E, van de Stadt LA et al (2015) Anti-citrullinated protein antibodies acquire a pro-inflammatory fc glycosylation phenotype prior to the onset of rheumatoid arthritis. Ann Rheum Dis 74(1):234–241

Rook GA, Steele J, Brealey R et al (1991) Changes in IgG glycoform levels are associated with remission of arthritis during pregnancy. J Autoimmun 4(5):779–794

Roopenian DC, Akilesh S (2007) FcRn: the neonatal Fc receptor comes of age. Nat Rev Immunol 7 (9):715–725

Sato T, Furukawa K (2004) Transcriptional regulation of the human beta-1,4-galactosyltransferase V gene in cancer cells: essential role of transcription factor Sp1. J Biol Chem 279 (38):39574–39583

Saville B, Wormke M, Wang F et al (2000) Ligand-, cell-, and estrogen receptor subtype (alpha/beta)-dependent activation at GC-rich (Sp1) promoter elements. J Biol Chem 275(8):5379–5387

Schaffert A, Hanic M, Novokmet M et al (2019) Minimal B cell extrinsic IgG glycan modifications of pro- and anti-inflammatory IgG preparations in vivo. Front Immunol 10:3024

Seo A, Gulsuner S, Pierce S et al (2019) Inherited thrombocytopenia associated with mutation of UDP-galactose-4-epimerase (GALE). Hum Mol Genet 28(1):133–142

Shea KL, Gavin KM, Melanson EL et al (2015) Body composition and bone mineral density after ovarian hormone suppression with or without estradiol treatment. Menopause 22 (10):1045–1052

Shen J, Jiang J, Wei Y et al (2007) Two specific inhibitors of the phosphatidylinositol 3-kinase LY294002 and wortmannin up-regulate beta1,4-galactosyltransferase I and thus sensitize SMMC-7721 human hepatocarcinoma cells to cycloheximide-induced apoptosis. Mol Cell Biochem 304(1–2):361–367

Shen X, Klaric L, Sharapov S et al (2017) Multivariate discovery and replication of five novel loci associated with immunoglobulin G N-glycosylation. Nat Commun 8(1):447

Shikata K, Yasuda T, Takeuchi F, Konishi T, Nakata M, Mizuochi T (1998) Structural changes in the oligosaccharide moiety of human IgG with aging. Glycoconj J 15(7):683–689

Slepak T, Tang M, Addo F, Lai K (2005) Intracellular galactose-1-phosphate accumulation leads to environmental stress response in yeast model. Mol Genet Metab 86(3):360–371

Sondermann P, Huber R, Oosthuizen V, Jacob U (2000) The 3.2-A crystal structure of the human IgG1 Fc fragment-Fc gammaRIII complex. Nature 406(6793):267–273

Souders CA, Nelson SC, Wang Y, Crowley AR, Klempner MS, Thomas W (2015) A novel in vitro assay to predict neonatal Fc receptor-mediated human IgG half-life. mAbs 7(5):912–921

Stapleton NM, Andersen JT, Stemerding AM et al (2011) Competition for FcRn-mediated transport gives rise to short half-life of human IgG3 and offers therapeutic potential. Nat Commun 2:599

Stein B, Yang MX (1995) Repression of the interleukin-6 promoter by estrogen receptor is mediated by NF-kappa B and C/EBP beta. Mol Cell Biol 15(9):4971–4979

Su Z, Gao J, Xie Q, Wang Y, Li Y (2020) Possible role of beta-galactosidase in rheumatoid arthritis. Mod Rheumatol 30(4):671–680

Subedi GP, Barb AW (2015) The structural role of antibody N-glycosylation in receptor interactions. Structure 23(9):1573–1583

Takizawa T, Anderson CL, Robinson JM (2005) A novel Fc gamma R-defined, IgG-containing organelle in placental endothelium. J Immunol 175(4):2331–2339

Tepperman J, Tepperman HM (1987) Metabolic and endocine physiology, 5th edn. Year Book Medical, Chicago

Twisselmann N, Bartsch YC, Pagel J et al (2018) IgG Fc glycosylation patterns of preterm infants differ with gestational age. Front Immunol 9:3166

van de Geijn FE, Wuhrer M, Selman MH et al (2009) Immunoglobulin G galactosylation and sialylation are associated with pregnancy-induced improvement of rheumatoid arthritis and the postpartum flare: results from a large prospective cohort study. Arthritis Res Ther 11(6):R193

van Erp EA, Lakerveld AJ, de Graaf E et al (2020) Natural killer cell activation by respiratory syncytial virus-specific antibodies is decreased in infants with severe respiratory infections and correlates with Fc-glycosylation. Clin Transl Immunol 9(2):e1112

Van Pelt RE, Gozansky WS, Wolfe P et al (2014) Estrogen or raloxifene during postmenopausal weight loss: adiposity and cardiometabolic outcomes. Obesity (Silver Spring MD) 22(4):1024–1031

vom Steeg LG, Klein SL (2016) SeXX matters in infectious disease pathogenesis. PLoS Pathog 12(2):e1005374

Wahl A, van den Akker E, Klaric L et al (2018) Genome-wide association study on immunoglobulin G glycosylation patterns. Front Immunol 9:277

Walitt B, Pettinger M, Weinstein A et al (2008) Effects of postmenopausal hormone therapy on rheumatoid arthritis: the women's health initiative randomized controlled trials. Arthritis Rheum 59(3):302–310

Wandall HH, Rumjantseva V, Sorensen AL et al (2012) The origin and function of platelet glycosyltransferases. Blood 120(3):626–635

Wang TT, Ravetch JV (2019) Functional diversification of IgGs through fc glycosylation. J Clin Invest 129(9):3492–3498

Wang J, Balog CI, Stavenhagen K et al (2011) Fc-glycosylation of IgG1 is modulated by B-cell stimuli. Mol Cell Proteomics 10(5):M110 004655

Williams PJ, Arkwright PD, Rudd P et al (1995) Short communication: selective placental transport of maternal IgG to the fetus. Placenta 16(8):749–756

Wuhrer M, Stam JC, van de Geijn FE et al (2007) Glycosylation profiling of immunoglobulin G (IgG) subclasses from human serum. Proteomics 7(22):4070–4081

Yamada E, Tsukamoto Y, Sasaki R, Yagyu K, Takahashi N (1997) Structural changes of immunoglobulin G oligosaccharides with age in healthy human serum. Glycoconj J 14(3):401–405

Yu X, Wang Y, Kristic J et al (2016) Profiling IgG N-glycans as potential biomarker of chronological and biological ages: a community-based study in a Han Chinese population. Medicine (Baltimore) 95(28):e4112

Part III
Effector Functions and Diseases

Chapter 12
Sweet Rules: Linking Glycosylation to Antibody Function

Falk Nimmerjahn and Anja Werner

Contents

12.1	Introduction	367
12.2	Regulation of Antibody Glycosylation	369
12.3	Impact of Glycosylation on IgG Activity	373
	12.3.1 Impact of Sialic Acid on IgG Dependent Effector Functions	375
	12.3.2 Impact of Galactose on IgG Dependent Effector Functions	376
	12.3.3 Impact of Fucose on IgG Dependent Effector Functions	377
	12.3.4 Impact of IgG Glycosylation on Antibody Half-Life	377
	12.3.5 IgG Sialylation and the Anti-inflammatory Activity of IgG	378
12.4	Impact of Glycosylation on IgA Function	380
12.5	Impact of Glycosylation on IgE Function	382
12.6	Summary and Outlook	384
References		384

Abstract Antibodies produced upon infections with pathogenic microorganisms are essential for clearing primary infections and for providing the host with long-lasting immunity. Moreover, antibodies have become the most widely used platform for developing novel therapies against cancer and autoimmunity, requiring an in-depth understanding of how antibodies mediate their activity in vivo and which factors modulate pro- or anti-inflammatory antibody activities. Since the discovery that select residues present in the sugar domain attached to the immunoglobulin G (IgG) fragment crystallizable (Fc) region can modulate both, pro- and anti-inflammatory effector functions, a wealth of studies has focused on understanding how IgG glycosylation is regulated and how this knowledge can be used to optimize

---

F. Nimmerjahn (✉)
Chair of Genetics, Department of Biology, Institute of Genetics, University of Erlangen-Nürnberg, Erlangen, Germany

Medical Immunology Campus Erlangen, Erlangen, Germany
e-mail: falk.nimmerjahn@fau.de

A. Werner
Chair of Genetics, Department of Biology, Institute of Genetics, University of Erlangen-Nürnberg, Erlangen, Germany

© The Author(s), under exclusive license to Springer Nature Switzerland AG 2021
M. Pezer (ed.), *Antibody Glycosylation*, Experientia Supplementum 112,
https://doi.org/10.1007/978-3-030-76912-3_12

therapeutic antibody activity. With the introduction of glycoengineered afucosylated antibodies in cancer therapy and the initiation of clinical testing of highly sialylated anti-inflammatory antibodies the proof-of-concept that understanding antibody glycosylation can lead to clinical innovation has been provided. The focus of this review is to summarize recent insights into how antibody glycosylation is regulated in vivo and how select sugar residues impact IgG function.

Keywords B cells · Antibody glycosylation · Therapeutic antibodies · Fc-receptors

Abbreviations

ADCC	Antibody-dependent cell-mediated cytotoxicity
Asp	Asparagine
B4GALT1	β-1,4-galactosyltransferase 1
BCR	B cell receptor
CD	Cluster of differentiation
CDC	Complement-dependent cytotoxicity
CMP-SA	Cytidine monophospho-sialic acid
ER	Endoplasmic reticulum
Fab	Fragment antigen binding
Fc	Fragment crystallizable
FcαR	Fc-alpha receptor
FcεR	Fc-epsilon receptor
FcγR	Fc gamma receptor
FcRn	Neonatal Fc receptor
FNAIT	Fetal or neonatal alloimmune thrombocytopenia
G0	Agalactosylated
G1	Monogalactosylated
G2	Digalactosylated
GlcNAc	N-acetylglucosamine
IFNγ	Interferon gamma
Ig	Immunoglobulin
IL	Interleukin
ITAM	Immunoreceptor tyrosine-based activation motif
ITIM	Immunoreceptor tyrosine-based inhibitory motif
IVIg	Intravenous immunoglobulin G
JC	Joining chain
K_A	Association constant
Man	Mannose
MBL	Mannose-binding lectin
MHCII	Major histocompatibility complex II
N	Asparagine
NK cell	Natural killer cell

PC	Plasma cell
PNGaseF	Peptide-*N*4-(*N*-acetyl-beta-glucosaminyl) asparagine amidase
RA	Rheumatoid arthritis
SC	Secretory component
ST6Gal1	β-galactoside-α2,6-sialyltransferase 1
STAT6	Signal transducer and activator of transcription 6
TCR	T cell receptor
TD	T cell-dependent
Tfh cell	T follicular helper cell
Th cell	T helper cell
TI	T cell-independent
TNFα	Tumor necrosis factor-alpha
Treg	Regulatory T cell

12.1 Introduction

With the introduction of second-generation glycoengineered therapeutic antibodies optimized for Fc receptor binding into clinical application, the field of antibody glycosylation has shifted from a mere descriptive, method oriented research area to a highly translational field with proven clinical impact (Li et al. 2017b; Yu et al. 2017). More recently, highly sialylated immunomodulatory IgG preparations have successfully passed first clinical test phases, emphasizing that a detailed understanding of the molecular and cellular pathways modulating antibody glycosylation can drive clinical innovation (Arroyo et al. 2019; Seeling et al. 2017). All human and mouse antibody isotypes are characterized by specific sugar moieties attached to the antibody fragment crystallizable (Fc) domain (Fig. 12.1). While IgM, IgA, and IgE isotypes have several N-linked glycosylation sites in the Fc fragment, immunoglobulin G (IgG) is characterized by only one sugar moiety attached to Asn297 of each CH2 domain (Arnold et al. 2007). In addition to N-linked sugar moieties, IgA and certain other immunoglobulin isotypes and subclasses may also contain highly complex O-linked sugar residues (Arnold et al. 2007). Although the existence of these sugar residues has been known for decades, a more in-depth understanding of the effects these post-translational modifications may have on antibody activity have only been obtained during the last ten to fifteen years. Due to the highly complex nature of the multiple sugar domains present in IgM, IgA, and IgE, an initial focus of the research was to study the single sugar domain present in each of the two IgG heavy chains. Despite the seemingly simple structure of this branched sugar domain, containing a heptameric core structure of *N*-acetylglucosamine (GlcNAc) and mannose (Man) residues, the variable addition of terminal galactose and sialic acid residues in combination with branched fucose and GlcNAc residues can create a high level of variability. Indeed, more than 30 different IgG glycosylation variants may be generated in principle, when anticipating that both sugar moieties within one

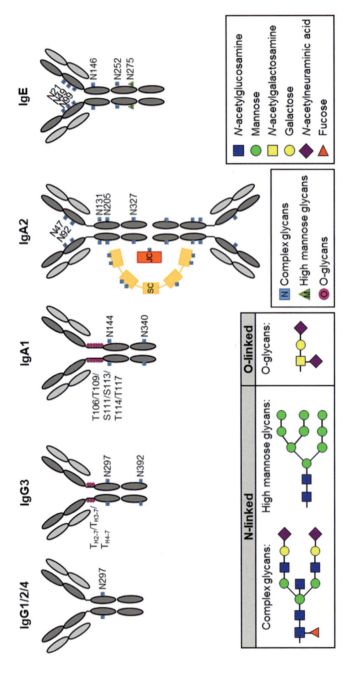

Fig. 12.1 Glycosylation sites and sugar moieties of antibody isotypes. Schematic representation of the glycosylation sites in IgG, IgA, and IgE molecules. The glycosylation sites for IgG are indicated by their amino acid number according to Arnold et al. (2007) and Plomp et al. (2016), while the other isotypes follow UniProt numbering. Each immunoglobulin consists of two heavy (dark gray) and two light chains (light gray), with IgA1 forming dimers complexed with the joining chain (JC; orange) and the secretory component (SC; yellow). N-linked glycosylation sites are divided into complex and high mannose glycans depicted by a blue square and a green triangle, respectively. Furthermore, O-linked glycans are illustrated by a purple circle. Schematic representative glycan structures for N-linked complex and high mannose glycans, as well as O-linked glycans, are depicted

IgG molecule are the same (Arnold et al. 2007; Kobata 2008; Raju 2008); if the individual IgG heavy chains contain different sugar structures, theoretically more than 500 IgG glycosylation variants may be generated (Holland et al. 2006; Masuda et al. 2000; Saphire et al. 2002). Potentially limiting the overall complexity, recent reports suggest that individual IgG subclasses may have distinct glycosylation patterns (de Haan et al. 2017; Kao et al. 2017; Wuhrer et al. 2007). In addition to the Fc-linked sugar moiety, some IgG antibodies may also have a sugar domain attached to the antibody variable domain (Fragment antigen binding, Fab). The glycosylation acceptor site for such sugar domains is not encoded in the germline but rather generated de novo during the process of antibody hypermutation (van de Bovenkamp et al. 2018). Indeed, it was suggested that Fab-linked sugar moieties play an important role in antigen recognition. In contrast to Fc-linked sugar domains, Fab-associated sugar structures are characterized by a higher level of fully processed sugar moieties, rich in terminal galactose and sialic acid residues. In this review, we will focus on the function of Fc-linked sugar structures, and the interested reader is directed to the excellent recent review article on Fab glycosylation (van de Bovenkamp et al. 2018).

12.2 Regulation of Antibody Glycosylation

Glycan biosynthesis takes place in the endoplasmic reticulum (ER) and the Golgi apparatus of B cells (van Kooyk et al. 2013). After an initial synthesis of high mannose N-glycans in the ER, proteins are transferred to the Golgi, where the trimming and addition of further sugar moieties are catalyzed by different glycosyltransferases, which are differentially expressed within the Golgi (Boune et al. 2020; van Kooyk et al. 2013). Their activity can be dependent on, for example, transcription factors (Klaric et al. 2020; Rajput et al. 1996), availability of sugar donors (Milewski et al. 1991) or cytokines, like tumor necrosis factor-alpha (TNFα) (Garcia-Vallejo et al. 2005; Gringhuis et al. 2005). Thus, the ultimate glycan structure of glycoproteins like IgG is largely determined by the selective activity of glycosidases and glycosyltransferases as well as by individual metabolic conditions within the cell, enabling to adopt IgG glycosylation to altered environmental conditions (Wang et al. 2011). In addition to B cell-intrinsic antibody glycosylation, it was recently proposed that IgG sialylation might occur independently of the secretory pathway allowing a B cell-independent modification of IgG glycosylation (Dougher et al. 2017; Jones et al. 2016; Lee et al. 2014; Manhardt et al. 2017). With respect to the use of therapeutic antibodies, this would pose a major problem, as such B cell-independent modifications of IgG glycosylation occurring upon IgG injection may alter IgG function in vivo. Mechanistically, it was suggested that soluble sialyltransferases may be generated by proteolytic cleavage of the membrane-associated enzymes in liver hepatocytes or platelets. Upon release into the circulation, CMP-sialic acid (CMP-SA) precursors may become conjugated to existing IgG sugar structures extracellularly (Lee et al. 2014; Manhardt et al. 2017; Wandall et al.

2012). As the appropriate sugar donors are not believed to be present in the circulation at sufficient amounts, however, it was proposed that degranulating platelets may provide CMP-SA, allowing for a B cell-extrinsic IgG glycosylation. Indeed, thrombin-activated platelets were shown to efficiently substitute for synthetic CMP-SA in providing sugar donors for extracellular β-galactoside-α2,6-sialyltransferase 1 (ST6Gal1)-mediated IgG sialylation (Lee et al. 2014). However, recent studies suggest that the efficiency of this B cell-extrinsic IgG sialylation pathway compared to B cell-intrinsic IgG sialylation may be very low. Thus, no major increase in serum IgG sialylation was detectable in mice with a B cell-specific knockout of ST6Gal1 (Ohmi et al. 2016). Furthermore, the injection of desialylated IgG into B cell-deficient mice (lacking B cell-intrinsic but not extrinsic IgG sialylation pathways) revealed that B cell-extrinsic IgG glycosylation occurs at very low levels and predominantly in the more easily accessible Fab fragment of the IgG molecule (Schaffert et al. 2019).

The general idea that antibody glycosylation may impact its function comes from the finding that antibody glycosylation is not stable, but can change upon immune stimulation. It has been firmly established, for example, that during microbial infection, vaccination, autoimmune inflammation, aging, and pregnancy characteristic changes in serum IgG glycosylation affecting most dominantly terminal sugar residues can occur (Bartsch et al. 2020; de Haan et al. 2016; Jansen et al. 2016; Kaneko et al. 2006; Kao et al. 2017; Kristic et al. 2014; Mahan et al. 2016; Mehta et al. 2008; Moore et al. 2005; Parekh et al. 1985, 1988; Pezer et al. 2016; Pucic et al. 2012; Selman et al. 2012; Twisselmann et al. 2018; Vadrevu et al. 2018). Further along these lines, it was demonstrated that socio-economic factors might impact antibody glycosylation (Mahan et al. 2016; Stambuk et al. 2020). For example, it was shown that the serum IgG glycome in individuals living in developing countries, like Africa, was characterized by glycan traits with high levels of G0 glycan forms lacking terminal galactose and sialic acid residues. In contrast, in industrial countries having well-developed health systems and sufficient access to water and food, IgG glycan traits with higher levels of galactosylated and sialylated glycan species were observed (Mahan et al. 2016; Stambuk et al. 2020). In addition to the more frequent alterations in antibody sialylation and galactosylation, IgG core fucosylation, which is highly abundant in IgG sugar domains, can be strongly reduced on platelet specific alloantibodies in a disease called fetal or neonatal alloimmune thrombocytopenia (FNAIT) (Kapur et al. 2014; Sonneveld et al. 2016). Moreover, low levels of IgG fucosylation were detected on Dengue virus-specific antibodies and correlated with more severe infections (Wang et al. 2017).

In mice, vaccination studies have provided valuable insights into how different pathways of B cell activation can impact IgG glycan composition. While immunization with T cell-independent (TI) antigens or T cell-dependent (TD) antigens without adjuvants mostly led to the formation of sialylated IgG molecules, stimulation with TD antigens under inflammatory conditions induced IgG antibodies with low levels of sialic acid residues (Hess et al. 2013; Oefner et al. 2012) (Fig. 12.2a). Whereas many studies were investigating global changes in IgG glycosylation, more recent reports investigated how individual IgG subclass glycosylation was affected

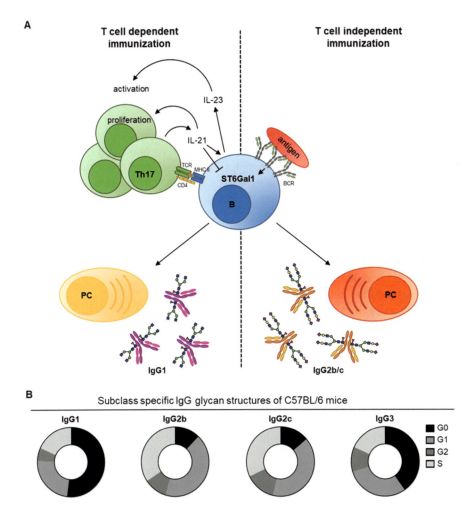

Fig. 12.2 Impact of immunization and IgG subclass on antibody glycosylation. (**a**) *Left:* Interaction between the T cell receptor (TCR) on T helper 17 cells (Th17) and major histocompatibility complex II (MHC II) on B cells (B) under pro-inflammatory conditions leads to the release of Interleukin 21 (IL-21) by Th17 cells which in turn induces Th17 proliferation and IL-23 production of B cells leading to T cell activation. Furthermore, T cell-dependent stimulation and IL-21 inhibit the β-galactoside alpha-2,6-sialyltransferase 1 (ST6Gal1), inducing the generation of non-sialylated IgG1 antibodies by plasma cells (PC). *Right:* Crosslinking of the B cell receptor (BCR) by an antigen without adjuvants induces ST6Gal1 activation leading to the production of sialylated IgG antibodies. (**b**) IgG subclass-specific Fc glycosylation patterns in C57BL/6 mice. Depicted is the proportion of agalactosylated (G0), monogalactosylated (G1), digalactosylated (G2), and sialylated glycan species of the different IgG subclasses IgG1, IgG2b, IgG2c, and IgG3

upon vaccination. Interestingly, IgG subclass-specific changes were observed after stimulation with TI and TD antigens (Kao et al. 2017). Of note, IgG subclass-specific differences in glycosylation have already been noted under steady-state conditions (de Haan et al. 2017; Kao et al. 2017). While murine IgG1 molecules were largely present in the G0 glycoform, IgG2a/c and IgG2b associated sugar moieties contained mono- or digalactosylated glycan structures with additional terminal sialic acid residues (Fig. 12.2b) (Kao et al. 2017). While it is difficult to assess the steady-state IgG subclass glycosylation in humans due to the constant exposure to immune stimulants (infections, vaccination), it was suggested that human IgG1 has higher levels of galactose containing sugar structures than IgG2 and IgG4 (Plomp et al. 2017). In addition, serum IgG3 seemed to contain more sialylated sugar domains compared to the IgG1 subclass (Sonneveld et al. 2018). As human IgG3 (and IgG1) functionally behaves very equal to mouse IgG2a/b/c subclasses, this result would be in line with the mouse IgG subclass glycosylation data. These results suggest that either IgG subclass structure or genetic factors linking IgG class switch to glycosylation play a major role in determining the precise composition of IgG subclass glycosylation.

Moreover, there is accumulating evidence that the type of T helper cell response—namely Th1, Th2, or Th17—can impose a distinct IgG glycosylation pattern. For example, B cells treated with Interleukin-21 (IL-21) and CpG significantly increased the generation of galactosylated and sialylated N-glycan species while decreasing bisecting GlcNAc (Wang et al. 2011). However, IL-21 in the context of Th17 cells and under pathogenic conditions seems to have the opposite effect on IgG glycosylation (Hess et al. 2013; Pfeifle et al. 2017). Thus, in a mouse model of rheumatoid arthritis (RA) IL-21 triggered the downregulation of ST6Gal1 expression in newly developing plasma blasts and plasma cells via IL-23-activated Th17 cells (Pfeifle et al. 2017). The IL-23-Th17 axis, which has already been associated with the pathogenesis of RA before (Lubberts 2015), seems to be responsible to control the intrinsic inflammatory activity of autoantibodies by shifting IgG glycosylation to pro-inflammatory sugar structures, triggering the clinical onset of autoimmune arthritis (Pfeifle et al. 2017). Moreover, IL-21 may act in an autocrine loop specifically supporting the proliferation and differentiation of Th17 cells (Korn et al. 2007). Furthermore, IL-21 may induce IL-23 production by B cells, further enhancing the autoimmune process by a positive feedback mechanism (Lee et al. 2015). The pivotal role of T follicular helper 17 (Tfh17) cells in modulating IgG sialylation was confirmed recently by showing that Tfh17 induction by IL-6 and IL-23 leads to a downregulation of ST6Gal1 allowing the production of desialylated pro-inflammatory IgG molecules (Bartsch et al. 2020). A similar inhibition of ST6Gal1 expression in germinal center B cells was also noted for water-in-oil adjuvants, which trigger IFNγ-producing Tfh1 cells (Riteau et al. 2016). Of note, not only IgG glycosylation, but also Fc-gamma receptor (FcγR) expression and IgG class-switching can be regulated via cytokines, creating complex regulatory networks, which need to be fine-tuned to ensure optimal antibody activity.

12.3 Impact of Glycosylation on IgG Activity

Before we discuss the impact of individual sugar residues on IgG activity in more detail, it is helpful to provide a short overview of the family of Fcγ-receptors (FcγRs). Fcγ-receptors bind the IgG Fc fragment near the hinge-proximal region in a 1:1 complex and include several activating FcγRs (FcγRI, FcγRIII, FcγRIV in mice and FcγRIA, FcγRIIA, FcγRIIC, FcγRIIIA, FcγRIIIB in humans) and one inhibitory FcγR (FcγRIIB) (Sondermann et al. 2000). Activating FcγRs are associated with a dimer of the common FcRγ chain, which contains an immunoreceptor tyrosine-based activation motif (ITAM), while the inhibitory FcγRIIB transmits its signals via an immunoreceptor tyrosine-based inhibitory motif (ITIM) present in its cytoplasmic region (Fig. 12.3a) (Brandsma et al. 2016; Nimmerjahn and Ravetch 2008). As activatory and inhibitory type I FcγRs are usually co-expressed on the same cell, a balanced immune response can be triggered. Apart from FcγR expression level, the different affinities of individual IgG subclasses to select FcγRs to modulate the magnitude of downstream effector pathways. For example, the mouse and human high-affinity receptor FcγRI binds IgG2a (in mice) or IgG1 and IgG3 (in humans) with an affinity of 10^7–10^8 M^{-1} while most of the other FcγRs have a 100–1000-fold lower affinity for IgG subclasses (Fig. 12.3a). Furthermore, mouse IgG1 and IgG2a/c or IgG2b bind differentially to activating and inhibitory FcγRs, suggesting that IgG antibodies are subject to a subclass-specific negative regulation via the inhibitory FcγRIIB (Nimmerjahn and Ravetch 2005). For example, IgG1 binds with an approximately tenfold higher affinity to the inhibitory FcγRIIB compared to the binding to its activating receptor, FcγRIII. In contrast, IgG2a/c or IgG2b has a higher affinity for activating FcγRs, FcγRI and FcγRIV, resulting in a lower threshold for cell activation (Bruhns and Jonsson 2015; Nimmerjahn and Ravetch 2005; Takai 2005). To account for these differences in affinity and use them as a predictive tool for IgG subclass activity in vivo, the concept of the A/I ratio, in which the affinity of an individual subclass for the activating receptor is divided by the affinity for the inhibitory FcγRIIB, has been developed and was validated by bioinformatic modeling more recently (Nimmerjahn and Ravetch 2005; Robinett et al. 2018).

As mentioned before, changes in IgG glycosylation can alter IgG affinity (and hence the A/I ratio), allowing to use of this concept for assessing the potential impact of alterations in IgG affinity on IgG activity in vivo. In general, IgG glycosylation is critical to allow functional binding of the IgG-Fc domain to Fc receptors (Arnold et al. 2007; Nimmerjahn and Ravetch 2006, 2008; Walker et al. 1989). Crystal structural analysis revealed that the glycosylated horseshoe-shaped IgG Fc fragment can occur in an open or closed conformation while aglycosylated human Fc fragments seemed to predominantly exist in the closed IgG-Fc conformation, prohibiting functional Fc receptor binding (Ahmed et al. 2014; Borrok et al. 2012; Jefferis and Lund 1997; Krapp et al. 2003). The open conformation is probably supported by hydrophobic interactions between the two mannose residues in the α1,3-branch of the two oligosaccharides (Krapp et al. 2003; Raju 2008). Interestingly, individual

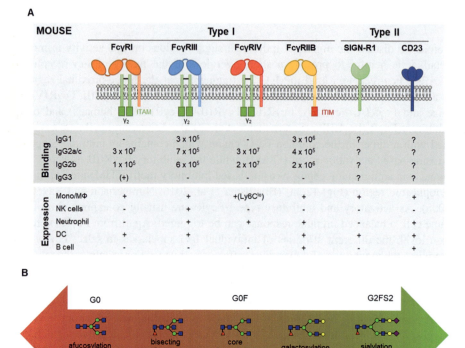

Fig. 12.3 IgG binding to FcγRs and impact of glycosylation on IgG activity. (**a**) *Upper part:* Schematic representations of murine Type I and Type II FcRs with associated gamma chain (γ2) and immunoreceptor tyrosine-based inhibitory or activation motif (ITIM or ITAM). *Middle part:* Binding affinities of different murine IgG subclasses to individual Type I and Type II Fc receptors given as K_A (M^{-1}). -: no binding, (+): possible binding, ?: no reported affinity values. *Lower part:* Expression profile of individual FcRs on different immune cells. +: expression, -: no expression, +(Ly6C^{lo}): expression only on Ly6C^{lo} monocytes. (**b**) Impact of different IgG glycosylation patterns on antibody activity. Depending on the Fc glycan composition IgG molecules exhibit either more pro- or anti-inflammatory effector functions. Representative glycan structures associated with different IgG activities are depicted

IgG subclasses seem to have a differential requirement for the length of the N297-linked sugar moiety. While a single GlcNAc residue was sufficient to maintain the activity of mouse IgG2a and human IgG1 subclasses, trimming down the Fc-linked sugar moiety to one residue largely abrogated the activity of mouse IgG1 and IgG2b subclasses (Kao et al. 2015). Moreover, the four murine IgG subclasses (IgG1, IgG2b, IgG2a/c, and IgG3) are differentially glycosylated with 50 % of IgG1 being present in the G0 form, while IgG2b and IgG2a/c molecules show an increased proportion of galactosylated and sialylated glycan species (Fig. 12.2b), creating a complex situation when trying to decipher the contribution of individual sugar moieties on IgG function. Yet another factor that is rarely considered when discussing the impact of IgG glycosylation on IgG function is the significant impact of higher order immune complexes, the natural ligand of the low-affinity FcγRs, may have. Thus, even aglycosylated IgG immune complexes demonstrated residual binding to the high-affinity FcγRI (Lux et al. 2013).

12.3.1 Impact of Sialic Acid on IgG Dependent Effector Functions

With respect to terminal sialic acid residues, several groups observed that high levels of IgG sialylation reduce antibody-dependent effector functions, such as antibody-dependent cell-mediated cytotoxicity (ADCC), phagocytosis, and complement-dependent cytotoxicity (CDC) (Kaneko et al. 2006; Li et al. 2017a; Quast et al. 2015; Scallon et al. 2007). Mechanistically, it was suggested that a lower level of binding of sialylated IgG variants to mouse and human activating FcγRs expressed on natural killer (NK) cells or myeloid cells may explain this reduced in vivo activity. In line with such a model, a change in IgG-Fc structure towards a more closed conformation was observed when high levels of terminal sialic acid residues were present in the IgG Fc-sugar structure, possibly explaining the reduction in binding to FcγRs (Ahmed et al. 2014; Sondermann et al. 2013). However, other studies found no major impact of sialylation on IgG structure (Crispin et al. 2013) and either no (Thomann et al. 2015) or even an increased activity of highly sialylated IgG on ADCC or CDC activity (Lin et al. 2015). Whether different methods to prepare highly sialylated IgG, the specific recombinant antibody used, or the assay systems employed to assess IgG activity underlie these inconsistent results will have to be addressed in future studies.

Of note, it was also demonstrated that different IgG subclasses were affected differentially by sialylation. Thus, mouse IgG1 antibodies showed a stronger reduction in activity compared to mouse IgG2b, further emphasizing that, especially with respect to changes of individual sugar residues, each IgG subclasses needs to be considered individually (Nimmerjahn and Ravetch 2005). A plausible explanation for these IgG subclass-specific effects is the varying baseline affinity of different IgG subclasses for different activating FcγRs. Whereas IgG1 shows a weak binding to its activating FcγR, FcγRIII, and hence may be impacted stronger by even small reductions in affinity, IgG2b has a roughly tenfold higher affinity for its major activating FcγR, FcγRIV, and thus may be more resistant to sialic acid-dependent affinity modulations (Kaneko et al. 2006). Further along these lines, IgG antibodies mediate their activity in vivo as higher order immune complexes, which may further mitigate sialylation-dependent reductions in affinity for individual FcγRs via avidity effects. Indeed, even aglycosylated IgG has some residual binding to select activating FcγRs as an immune complex (Lux et al. 2013). In addition, Scallon and colleagues observed for select antibodies that changes in antibody affinity to its target antigen may occur upon hypersialylation, suggesting that under certain conditions not only the Fc but also the Fab fragment of the IgG molecule may be affected by sialylation (Scallon et al. 2007). In summary, more detailed further studies will be necessary to assess the impact of IgG sialylation on FcγR binding, taking into account IgG subclasses and individual FcγRs. Apart from a potential impact of highly sialylated IgG molecules to classical FcγR binding, terminal sialic acid residues were also suggested to confer an active anti-inflammatory activity to IgG (Fig. 12.3b), which will be discussed later in this review.

12.3.2 Impact of Galactose on IgG Dependent Effector Functions

With respect to IgG galactosylation, there is the long-standing observation that IgG lacking terminal galactose residues (G0 glycoform) is associated with high disease activity (Arnold et al. 2007). Apart from establishing IgG-G0 forms as a valid biomarker of active disease, it was suggested that IgG glycoforms lacking galactose residues might also have a higher pro-inflammatory activity (Fig. 12.3b) due to a better binding to activating FcγRs or the classical complement pathway. Initial evidence along these lines was provided by data demonstrating that IgG-G0 glycoforms may allow a better binding to mannose binding lectin (MBL) and hence may trigger the MBL pathway of complement activation (Malhotra et al. 1995). A later study performed in MBL deficient mice, however, demonstrated that IgG-G0 glycoforms did not lose in vivo activity and continued to operate via binding to classical FcγRs, arguing against a major contribution of the complement pathway to the activity of this IgG glycoform at least in mouse model systems (Nimmerjahn et al. 2007). In further contrast to the expectation that agalactosylated IgG glycoforms have a higher pro-inflammatory activity, more recent studies suggest that the presence of galactose residues improves IgG binding to both, FcγRs and the classical complement pathway. For example, an increased binding of galactosylated human IgG1 to FcγRIIA and FcγRIIIA was noted, which translated to an increased ADCC activity (Dashivets et al. 2015; Subedi and Barb 2016; Thomann et al. 2015); and high levels of IgG1 galactosylation also improved binding to C1q and complement-dependent cytotoxicity at least in vitro (Peschke et al. 2017). Complicating the situation even further, it was noted that highly galactosylated IgG1 molecules, present as immune complexes, are associated with an enhanced anti-inflammatory activity (Karsten et al. 2012). Thus, injection of galactosylated IgG1 immune complexes interfered with autoantibody activity in several mouse models of autoimmunity by promoting the association between FcγRIIB and Dectin-1, further suggesting that galactose residues increase IgG binding to FcγRs (Karsten et al. 2012). As discussed for terminal sialic acid residues, more studies with individual mouse and human IgG subclasses will be critical to resolve this issue. At present, however, there seems to be no link between the increased abundance of agalactosylated IgG species in patients with infections or inflammation and a higher activity of this IgG glycosylation variant.

12.3.3 Impact of Fucose on IgG Dependent Effector Functions

The most convincing data that an altered IgG glycosylation translates into increased IgG activity exists for the penultimate fucose residue. While initial studies also pointed out that IgG1 molecules bearing bisecting *N*-acetylglucosamine residues have a higher ADCC activity (Davies et al. 2001; Umana et al. 1999), later studies demonstrated that this effect was explained due to the concomitant lack of fucose residues in these IgG preparations (Dekkers et al. 2017; Shields et al. 2002; Shinkawa et al. 2003). Non-fucosylated antibodies of all IgG subclasses bind with higher affinity to the activating FcγRIIIA, substantially increasing ADCC activity and independently of IgG subclass (Niwa et al. 2005a). Of note, antigen binding and CDC seem not to be altered by removal of core fucose (Chung et al. 2012; Niwa et al. 2005a). Therefore, generating afucosylated IgG has become a general strategy to selectively enhance therapeutic antibody binding to FcγRIIIA (Niwa et al. 2005b; Satoh et al. 2006). Mechanistically, crystal structural analysis revealed that the penultimate fucose residue of IgG comes in close contact with an N-linked sugar moiety present in FcγRIIIA, hindering a tight protein-protein interaction and thus resulting in a lower-affine interaction (Ferrara et al. 2006, 2011). It is interesting to note that this IgG glycosylation variant selectively modulates the binding to FcγRIIIA (CD16), while it does not affect IgG binding to other activating or the inhibitory FcγRIIB.

12.3.4 Impact of IgG Glycosylation on Antibody Half-Life

In general, IgG molecules exhibit quite long half-lives due to their binding to the neonatal Fc receptor (FcRn) (Roopenian and Akilesh 2007). This receptor is broadly expressed by hematopoietic and parenchymal cells and binds IgG molecules at the CH2–CH3 interface of the Fc. It protects IgG molecules from rapid degradation in lysosomes by a specific pH-dependent recycling process (Roopenian and Akilesh 2007), therefore increasing antibody half-lives. In contrast to the essential role of IgG glycosylation for binding to classical FcγRs, FcRn binding to IgG occurs in the IgG CH3 domain and thus far from the IgG-linked sugar domain. Indeed, the sugar moiety of IgG is not involved in FcRn binding as aglycosylated IgG can still bind FcRn and displays a normal half-life. Nonetheless, some IgG glycosylation variants were shown to have an altered in vivo half-life. For example, antibodies with terminal sialic acid residues were suggested to have a slightly longer half-life than non-sialylated antibodies (Bas et al. 2019; Raju et al. 2001). Conversely, glycoproteins containing high mannose structures exhibit a reduced serum half-life due to their ability to binding to the mannose receptor (Jones et al. 2007; Kanda et al. 2007; Raju et al. 2001; Wright et al. 2000). However, another study investigating mannose enriched IgG preparations found no significant differences in half-life between complex-type and high mannose type glycans (Millward et al. 2008). Thus, while IgG glycosylation is not critical for FcRn binding, some sugar moieties may slightly impact IgG half-life in vivo.

12.3.5 IgG Sialylation and the Anti-inflammatory Activity of IgG

Since the discovery that the infusion of high doses of intravenous immunoglobulin G (IVIg) consisting of pooled serum IgG preparations from thousands of donors can suppress a wide variety of chronic inflammatory and autoimmune diseases, scientists have tried to identify the mechanism underlying this activity (Bayry et al. 2011; Schwab and Nimmerjahn 2013). As this review focuses on IgG glycosylation, we will not cover all potential molecular and cellular pathways underlying IVIg activity. The interested reader is directed to several excellent reviews covering this topic in greater detail (Bayry et al. 2011; Bussel 2006; Schwab and Nimmerjahn 2013). With respect to a potential role of IgG glycosylation in the immunomodulatory activity of IgG, the previously mentioned observation that serum IgG antibodies have reduced levels of galactosylation and sialylation during active inflammation suggested that these terminal sugar moieties might play an active role in suppressing inflammation independently of classical FcγRs. Evidence along these lines was first provided by Kaneko and colleagues in 2006, demonstrating that in animal models, IVIg lost its activity if the entire IgG sugar moiety or more selectively α2,6-linked sialic acid residues were removed (Kaneko et al. 2006). These results were further validated in a variety of other classical and humanized mouse model systems (Bozza et al. 2019; Fiebiger et al. 2015; Massoud et al. 2014; Schwab et al. 2012, 2014, 2015). Of note, several groups also noted sialic acid independent IVIg activities, suggesting that both sialic acid-dependent and -independent pathways may underly IVIg activity (Campbell et al. 2014; Issekutz et al. 2015; Leontyev et al. 2012; Othy et al. 2014). More importantly, from a clinical perspective, enriching IVIg for terminal sialic acid residues resulted in increased immunosuppressive activity, opening a new avenue for potentially enhancing IVIg activity (Anthony et al. 2008a; Kaneko et al. 2006). In an effort to replace hypersialylated IVIg infusion, soluble variants of the galactosyltransferase B4GALT1 and ST6Gal1 were generated to allow an in vivo galactosylation and sialylation of autoantibodies. Indeed, co-injection of these enzymes induced increased IgG galactosylation and sialylation at the site of inflammation, leading to an amelioration of autoimmune pathology (Pagan et al. 2018). Of note, activated platelets present at the site of autoantibody activity were demonstrated to release sugar donor molecules required for autoantibody galactosylation and sialylation in vivo (Pagan et al. 2018).

Based on this conceptual framework, highly pure tetrasialylated IVIg preparations were generated and shown to have a superior anti-inflammatory activity in various pre-clinical model systems (Washburn et al. 2015). More recently, tetrasialylated IVIg preparations successfully passed first clinical trials in patients with Immune thrombocytopenia, suggesting that the basic concepts identified in pre-clinical mouse model systems translate to the human situation (Arroyo et al. 2019). Once approved, the data obtained in various human autoimmune disease patient cohorts will have to show how broad the clinical benefit of hypersialylated IVIg preparations is.

With respect to the molecular pathways underlying the anti-inflammatory activity of hypersialylated IgG, no unifying model has emerged yet. Starting with several studies by Anthony and colleagues, several other groups have noted that C- or I-type lectin receptors, such as DC-SIGN (murine homolog: SIGNR1), DCIR, CD22, and CD23 seem to be involved in the anti-inflammatory and immunomodulatory activity of sialylated IgG in different model systems in vivo (Anthony et al. 2008b; Fiebiger et al. 2015; Massoud et al. 2014; Pagan et al. 2018; Schwab et al. 2012; Seite et al. 2010; Wang et al. 2015). Extending these observations, it was suggested that hypersialylated IgG may directly bind to these so-called type II FcRs and induce an upregulation of FcγRIIB on innate immune effector cells via IL-33 and IL-4-producing basophils (Anthony et al. 2011; Fiebiger et al. 2015; Pincetic et al. 2014; Sondermann et al. 2013). Mechanistically, the increased expression of FcγRIIB on neutrophils and myeloid cells increases the threshold for activation via autoantibody immune complexes, thereby mitigating pro-inflammatory effector pathways and inducing resolution of inflammation. Confirming this observation, the injection of IL-4 alone was sufficient to ameliorate autoantibody-dependent inflammation via IL-4R and STAT6 signaling in myeloid cells (Anthony et al. 2011; Wermeling et al. 2013). Apart from the upregulation of FcγRIIB on innate immune effector cells, this immunomodulatory pathway was also responsible for the expansion of regulatory T cells (Treg), which play a critical role in suppressing T cell-dependent autoimmune diseases such as multiple sclerosis (Fiebiger et al. 2015). Importantly both, an upregulation of FcγRIIB and an expansion of Tregs were observed in human patients receiving IVIg therapy, validating the results obtained in mouse model systems (Ephrem et al. 2008; Schwab et al. 2015; Tackenberg et al. 2009). In addition to acting on innate immune cells, sialylated IgG molecules were shown to feedback on B cells via CD23. Binding to CD23 induced an upregulation of the inhibitory FcγRIIB on germinal center B cells, leading to a higher threshold for B cell activation and generation of higher affine antibody responses (Wang et al. 2015). With respect to autoimmune diseases, this pathway may block the continuous production of autoantibodies by newly developing autoreactive B cells in the germinal center and via induction of apoptosis in autoantibody-producing plasma cells, and hence explain some of the long-lasting effects of IVIg therapy.

However, several publications challenged the concept that IgG sialylation allows a direct binding of IgG to type II FcRs via altering IgG structure (Temming et al.

2019; Yu et al. 2013). Thus, these receptors may still be involved in the immunomodulatory pathway of IVIg activity but not represent the initial triggering receptor setting the whole anti-inflammatory cascade into motion. As different C-type lectin receptors seemed to be involved depending on the organ affected by the disease, a better understanding of local cell types and receptors participating in IVIg activity would be very helpful. In summary, more studies will be necessary to resolve the exact pathways underlying the sialic acid-dependent anti-inflammatory activities of IVIg in different disease entities. The promising first results of hypersialylated IVIg in human clinical trials may help to resolve these issues in the near future by allowing to study immunological effects triggered via hypersialylated IVIg preparations in human patient populations.

12.4 Impact of Glycosylation on IgA Function

Unlike the well-established connection between IgG glycosylation and modulation of IgG effector function, the impact of IgA glycosylation on IgA effector functions is less well understood (reviewed in (Arnold et al. 2005, 2007); schematic representation see Fig. 12.1). This is partly due to the fact that mice and humans significantly differ with respect to this antibody isotype. Thus, mice only have one IgA subclass and lack an orthologue to the human FcαRI (CD89). IgA is the most abundant immunoglobulin on inner body surfaces and plays an essential role in mucosal homeostasis of the gastrointestinal, respiratory, and genitourinary tract (Kerr 1990). Besides its transport to mucosal tissues by plasma cells residing in the lamina propria, it is also the second most abundant immunoglobulin in the serum (Hansen et al. 2019; Kerr 1990). Secretory IgA and serum IgA differ with respect to secretory IgA being present as a dimer in a complex with the joining chain and the secretory component, and serum IgA being present predominantly in a monomeric form (Fig. 12.1) (Kerr 1990). The two IgA subclasses, IgA1 and IgA2, can be further subdivided into two IgA2 allotypic variants, named IgA2m(1) and IgA2m(2) (Woof and Kerr 2006). IgA1 and IgA2 differ by 13 amino acids found in the hinge region of IgA1, which is highly O-glycosylated in IgA1 molecules (van Egmond et al. 2001). IgA1 predominates in the serum, whereas IgA1 and IgA2 are more evenly distributed on mucosal surfaces (van Egmond et al. 2001). Apart from simple pathogen neutralization, more recent studies suggest that IgA may also actively participate in the regulation of immune responses (Hansen et al. 2019). The major effector cells expressing CD89 are monocytes, macrophages, neutrophils, eosinophils, and monocyte-derived dendritic cells, which also express FcγRs. In further similarity to activating FcγRs, CD89 also associates with the common FcRγ-chain for signal transduction (Hansen et al. 2019; van Egmond et al. 2001) and is also characterized by a low affinity (K_a approximately 10^6 M^{-1}), resulting in a rapid dissociation of monomeric IgA (Wines et al. 1999). Binding of IgA as an immune complex can trigger a variety of effector responses, like the release of cytokines and chemokines, degranulation, phagocytosis, or ADCC (Bakema and van Egmond 2011). Of note, several studies have shown that IgA may contribute to the pathology of autoimmune diseases, like rheumatoid arthritis, celiac disease or inflammatory bowel disease, suggesting that IgA can be a driver of inflammation (reviewed in (Hansen et al. 2019)). In contrast, others have noted that serum IgA may exert anti-inflammatory effects and protect against autoimmune responses (Lecocq et al. 2013; Pilette et al. 2010; Rossato et al. 2015). Thus, in line with IgG, IgA molecules may have a dual functionality promoting both, pro- and anti-inflammatory immune responses, raising the question to what extent IgA glycosylation may impact these different effector functions.

In contrast to the single sugar moiety in the IgG Fc domain, which is largely buried in the hydrophobic pocket between the two IgG heavy chains, the different sugar moieties of IgA are exposed on the protein surface. Compared to IgG, glycosylation of IgA is more complex and differs between IgA1 and IgA2 subclasses. While IgA1 molecules contain two conserved N-glycosylation sites (N144 in the Cα2 and N340 in the Cα3 domain) and several sites of O-glycosylation in the extended hinge region of each IgA heavy chain, IgA2 has four conserved N- (N47 in the Cα1, N131 and N205 in the Cα2, and N327 in the Cα3 domain) but no O-glycosylation sites (Fig. 12.1) (de Haan et al. 2020). Whereas O-glycans consist of one core *N*-acetylgalactosamine with β1,3-linked galactose residues, which can also be sialylated (Fig. 12.1) (Deshpande et al. 2010; Franc et al. 2013; Novak et al. 2013), the majority of N-glycosylation sites of IgA are of the complex-type and vary between IgA1 and IgA2. Creating further complexity, IgA glycosylation in the serum differs from glycosylation of secretory IgA (Plomp et al. 2018; Steffen et al. 2020). For example, almost 90 % of serum IgA is either mono- or di-sialylated while secretory IgA has only 13 % sialylated glycan structures (Arnold et al. 2007; Plomp et al. 2018). Furthermore, secretory IgA is dominated by glycan structures terminating in galactose (23%) and *N*-acetylglucosamine (52%) and contains 75% bisecting *N*-acetylglucosamine and 50% core fucose residues. With 29% bisecting *N*-acetylglucosamine and 32% fucose residues, these branching sugar residues are present at a much lower abundance in serum IgA1. Moreover, secretory IgA dimers are covalently linked with the joining chain and the secretory component, which also carries complex N-glycosylation sites (Arnold et al. 2007; Deshpande et al. 2010). Although IgA molecules have no C1q binding site, it was shown that secretory IgA contains exposed mannose and *N*-acetylglucosamine residues, which are masked by the secretory component but can be unmasked upon interaction with bacteria allowing complement activation via mannose-binding lectin, promoting pathogen opsonization and phagocytosis (Royle et al. 2003). More recent publications revealed that the composition of individual N-linked sugar moieties within IgA might differ (Chandler et al. 2019; Steffen et al. 2020). For example, the N144 and N131 associated sugar moieties in IgA1 and IgA2, respectively, were virtually non-fucosylated, whereas all other N-linked sugar structures carried fucose residues. While all N-linked sugar structures in both, IgA1 and IgA2 were highly sialylated, the N340 (IgA1) and N327 (IgA2) sugar moieties showed mostly di-sialylated sugar residues, whereas the predominant glycoform in all other N-linked sugar structures was mono-sialylated. Of note, a slightly higher level of di-sialylated sugar structures was noted on IgA1 molecules, which had a lower capacity to activate neutrophils and cause pro-inflammatory effector functions (Steffen et al. 2020). Interestingly, the global removal of sialic acid residues seemed to specifically enhance the pro-inflammatory activity of IgA1, suggesting that sialic acid residues specifically regulated IgA1 activity (Steffen et al. 2020). In line with a model in which IgA sialylation regulates binding to CD89, Basset and colleagues noted an enhanced binding of neuraminidase treated and reduced binding of hypersialylated IgA to CD89 expressing cells (Basset et al. 1999). In stark contrast to IgG, a similar effect was noted for PNGaseF treated IgA1, in which all of the N-linked sugar domains are removed, indicating that in general N-linked sugar domains are not essential for or may even limit IgA binding to CD89 (Gomes et al. 2008; Mattu et al. 1998; Oortwijn et al. 2007; Steffen et al. 2020). With respect to IgA half-life, it is important to note that not FcRn but rather the hepatic asialo receptor is involved in IgA removal from the serum. As this receptor specifically recognizes asialylated proteins, this may lead to different half-lives of sialylated (longer half-life) compared to non-sialylated (shorter half-life) IgA molecules in vivo. Combined with the generally low serum half-life of IgA this creates a complex scenario when trying to distinguish the effects of sialylation on IgA half-life and CD89 dependent effector functions. Although much more work is necessary to fully understand the impact of the individual sugar moieties on IgA function, the most recent evidence suggests that similar to IgG, sialylation may modulate IgA effector functions.

12.5 Impact of Glycosylation on IgE Function

IgE is the least abundant immunoglobulin in the serum and is critical for defending the host against infections with parasites. In addition, the pathogenic role of IgE as a trigger of allergic responses is a major clinical problem, necessitating an in-depth understanding of IgE function (Gould and Sutton 2008; Gounni et al. 1994). In the body, IgE is bound to its high-affinity receptor, FcεRI (K_a approximately 10^{10} M^{-10} (Young et al. 1995)) on mast cells and basophils, where it triggers the release of inflammatory mediators, like histamine, tumor necrosis factor (TNF), and prostaglandin D2 upon antigen recognition (Galli and Tsai 2012; Gould and Sutton 2008). Furthermore, CD23 (also known as FcεRII) has been identified as low-affinity receptor for IgE, which is expressed on T and B cells, follicular dendritic cells, and epithelial cells (Armitage et al. 1989; Rieber et al. 1993; Yu et al. 2003). CD23 can occur either in a soluble or in a membrane-bound form and is important for the regulation of IgE homeostasis (Shade et al. 2019) and the elimination of intracellular pathogens (Vouldoukis et al. 1995). IgE molecules consist of four Cε heavy chain domains with a rigid hinge region (Arnold et al. 2005; Oettgen 2016) and are the most heavily glycosylated immunoglobulins (Dorrington and Bennich 1978) (Fig. 12.1). It has seven N-glycosylation sites in the IgE heavy chain, with N383 being non-glycosylated, N394 carrying only oligomannose structures, and the remaining sites being occupied by complex glycans (Baenziger et al. 1974a, b;

Dorrington and Bennich 1978; Shade et al. 2020). Although it has been described that glycan removal leads to reduced FcεRI binding (Bjorklund et al. 1999; Nettleton and Kochan 1995), others found almost no impact of IgE glycosylation on its receptor binding capacity (Basu et al. 1993; Vercelli et al. 1989). Moreover, several early studies tried to analyze the complex glycan structure of IgE (Baenziger et al. 1974a, b; Dorrington and Bennich 1978; Nettleton and Kochan 1995). However, as the level of IgE is very low those studies were usually performed on myeloma-derived IgE or did not distinguish between individual N-glycosylation sites making it difficult to draw conclusions about site-specific glycosylation patterns within IgE. The first insights into site-specific IgE glycosylation was provided by Plomp and colleagues, who also compared serum IgE glycosylation between myeloma patients, healthy and hyperimmune donors (Plomp et al. 2014). This study demonstrated that IgE from myeloma patients was characterized by decreased levels of bisecting *N*-acetylglucosamines, which was further associated with tumor progression (Balog et al. 2012; Miwa et al. 2012). More recently, a detailed site-specific IgE glycan analysis demonstrated that the N-linked sites at N140, N168, N218, and N265 (N21, N49, N99 and N146 according to UniProt numbering) are almost completely mono- or di-sialylated while only 30% of the N371 (N252 according to UniProt numbering) were glycosylated; no sugar moiety was present at the N383 (N264 according to UniProt numbering) site, consistent with previous results (Arnold et al. 2004; Baenziger et al. 1974a, b; Dorrington and Bennich 1978; Plomp et al. 2014; Shade et al. 2020; Wu et al. 2016). With respect to the impact of the different N-linked glycosylation sites on binding to the high-affinity FcεRI, several studies suggest a prominent role for the N394-linked (N275 according to UniProt numbering) high mannose structure, which is conserved in IgE molecules of all mammalian species (Bjorklund et al. 1999; Shade et al. 2019). A selective enzymatic removal of the N394 sugar moiety altered the secondary structure of IgE prevented binding to FcεRI on mast cells and attenuated IgE-mediated anaphylaxis in vivo (Shade et al. 2015). In addition, some minor effects of the other N-linked complex sugar structures have been identified (Shade et al. 2019). While removal of the complex glycans at position N265 and/or N371 minimally affected FcεRI binding, mutation of the N-linked sites in the Fab region showed slightly reduced IgE-mediated mast cell degranulation, suggesting a possible interaction of IgE Fab glycans with its antigen (Shade et al. 2015, 2019). Moreover, galectins, a class of proteins that bind specifically to β-galactoside sugars and are involved in many physiological functions such as inflammation, immune responses, cell migration, autophagy, and signaling (Johannes et al. 2018), were shown to bind IgE molecules, suggesting that complex IgE glycans might be involved in IgE effector functions (Shade et al. 2019). High-affinity binding of IgE glycans to galectins can regulate IgE activity and have an anti-allergic effect by preventing IgE-antigen complex formation. By comparing IgE glycosylation between atopic and non-atopic individuals, reduced amounts of bisecting *N*-acetylglucosamines and terminal galactose residues but increased levels of sialic acids were detected in allergen-specific IgE. Of note, removal of sialic acid residues attenuated IgE-dependent effector cell degranulation and anaphylaxis (Shade et al. 2020).

Thus, in contrast to IgG and IgA isotypes, where terminal sialic acid residues conferred a reduced pro-inflammatory activity, terminal sialic acid residues were critical for IgE-dependent effector functions. This finding would be in line with the Th2 cytokine-dependent immune responses triggered by parasite infections, which may change the B cell-intrinsic glycosylation machinery towards a high sialylation state.

12.6 Summary and Outlook

In summary, research over the last decade has firmly established that glycosylation plays an active role in modulating the activity of different antibody isotypes in vivo. While the multiple and complex sugar moieties present in IgA and IgE create a way more complex scenario compared to IgG, several studies have started to show that either select sugar moieties or individual sugar residues such as sialic acids seem to play a crucial role in fine-tuning antibody activity. With respect to the pathways determining how a plasma cell glycosylates different antibody isotypes, much more studies are necessary to obtain a clear picture of the underlying factors involved in modulating antibody glycosylation in vivo. The observation that certain cytokines or organ environments in which the plasma cell resides may impact antibody glycosylation provide a first roadmap for further investigations. Finally, the encouraging data from the clinical use of glycoengineered antibodies underscores the translational impact this field of research already has and surely will have in the near future.

Acknowledgment This manuscript was supported by funding from the German Research Foundation to F.N. (DFG FOR 2886, FOR 2953, NI711/9-1, and CRC1181-A07).

Compliance with Ethical Standards

Conflict of Interest Falk Nimmerjahn declares that he has no conflict of interest. Anja Werner declares that she has no conflict of interest.

Ethical Approval This article does not contain any studies with human participants or animals performed by any of the authors.

References

Ahmed AA, Giddens J, Pincetic A, Lomino JV, Ravetch JV, Wang LX, Bjorkman PJ (2014) Structural characterization of anti-inflammatory immunoglobulin G Fc proteins. J Mol Biol 426:3166–3179

Anthony RM, Nimmerjahn F, Ashline DJ, Reinhold VN, Paulson JC, Ravetch JV (2008a) Recapitulation of IVIG anti-inflammatory activity with a recombinant IgG Fc. Science 320:373–376

Anthony RM, Wermeling F, Karlsson MC, Ravetch JV (2008b) Identification of a receptor required for the anti-inflammatory activity of IVIG. Proc Natl Acad Sci USA 105:19571–19578

Anthony RM, Kobayashi T, Wermeling F, Ravetch JV (2011) Intravenous gammaglobulin suppresses inflammation through a novel T(H)2 pathway. Nature 475:110–113

Armitage RJ, Goff LK, Beverley PC (1989) Expression and functional role of CD23 on T cells. Eur J Immunol 19:31–35

Arnold JN, Radcliffe CM, Wormald MR, Royle L, Harvey DJ, Crispin M, Dwek RA, Sim RB, Rudd PM (2004) The glycosylation of human serum IgD and IgE and the accessibility of identified oligomannose structures for interaction with mannan-binding lectin. J Immunol 173:6831–6840

Arnold JN, Royle L, Dwek RA, Rudd PM, Sim RB (2005) Human immunoglobulin glycosylation and the lectin pathway of complement activation. Adv Exp Med Biol 564:27–43

Arnold JN, Wormald MR, Sim RB, Rudd PM, Dwek RA (2007) The impact of glycosylation on the biological function and structure of human immunoglobulins. Annu Rev Immunol 25:21–50

Arroyo S, Tiessen RG, Denney WS, Jin J, van Lersel MP, Zeitz H, Manning AM, Schipperus MR, Bussel JB (2019) Hyper-sialylated IgG M254, an innovative therapeutic candidate, evaluated in healthy volunteers and in patients with immune thrombocytopenia purpura: safety, tolerability, pharmacokinetics, and pharmacodynamics. Blood 134

Baenziger J, Kornfeld S, Kochwa S (1974a) Structure of the carbohydrate units of IgE immunoglobulin. I. Over-all composition, glycopeptide isolation, and structure of the high mannose oligosaccharide unit. J Biol Chem 249:1889–1896

Baenziger J, Kornfeld S, Kochwa S (1974b) Structure of the carbohydrate units of IgE immunoglobulin. II. Sequence of the sialic acid-containing glycopeptides. J Biol Chem 249:1897–1903

Bakema JE, van Egmond M (2011) The human immunoglobulin A Fc receptor FcalphaRI: a multifaceted regulator of mucosal immunity. Mucosal Immunol 4:612–624

Balog CI, Stavenhagen K, Fung WL, Koeleman CA, McDonnell LA, Verhoeven A, Mesker WE, Tollenaar RA, Deelder AM, Wuhrer M (2012) N-glycosylation of colorectal cancer tissues: a liquid chromatography and mass spectrometry-based investigation. Mol Cell Proteomics 11:571–585

Bartsch YC, Eschweiler S, Leliavski A, Lunding HB, Wagt S, Petry J, Lilienthal GM, Rahmoller J, de Haan N, Holscher A et al (2020) IgG Fc sialylation is regulated during the germinal center reaction following immunization with different adjuvants. J Allergy Clin Immunol 146:652–666 e611

Bas M, Terrier A, Jacque E, Dehenne A, Pochet-Beghin V, Beghin C, Dezetter AS, Dupont G, Engrand A, Beaufils B et al (2019) Fc sialylation prolongs serum half-life of therapeutic antibodies. J Immunol 202:1582–1594

Basset C, Devauchelle V, Durand V, Jamin C, Pennec YL, Youinou P, Dueymes M (1999) Glycosylation of immunoglobulin A influences its receptor binding. Scand J Immunol 50:572–579

Basu M, Hakimi J, Dharm E, Kondas JA, Tsien WH, Pilson RS, Lin P, Gilfillan A, Haring P, Braswell EH et al (1993) Purification and characterization of human recombinant IgE-Fc fragments that bind to the human high affinity IgE receptor. J Biol Chem 268:13118–13127

Bayry J, Negi VS, Kaveri SV (2011) Intravenous immunoglobulin therapy in rheumatic diseases. Nat Rev Rheumatol 7:349–359

Bjorklund JE, Karlsson T, Magnusson CG (1999) N-glycosylation influences epitope expression and receptor binding structures in human IgE. Mol Immunol 36:213–221

Borrok MJ, Jung ST, Kang TH, Monzingo AF, Georgiou G (2012) Revisiting the role of glycosylation in the structure of human IgG Fc. ACS Chem Biol 7:1596–1602

Boune S, Hu P, Epstein AL, Khawli LA (2020) Principles of N-linked glycosylation variations of IgG-based therapeutics: pharmacokinetic and functional considerations. Antibodies (Basel) 9

Bozza S, Kasermann F, Kaveri SV, Romani L, Bayry J (2019) Intravenous immunoglobulin protects from experimental allergic bronchopulmonary aspergillosis via a sialylation-dependent mechanism. Eur J Immunol 49:195–198

Brandsma AM, Hogarth PM, Nimmerjahn F, Leusen JH (2016) Clarifying the confusion between cytokine and Fc receptor "common gamma chain". Immunity 45:225–226

Bruhns P, Jonsson F (2015) Mouse and human FcR effector functions. Immunol Rev 268:25–51

Bussel J (2006) Treatment of immune thrombocytopenic purpura in adults. Semin Hematol 43:S3-10; discussion S18-19

Campbell IK, Miescher S, Branch DR, Mott PJ, Lazarus AH, Han D, Maraskovsky E, Zuercher AW, Neschadim A, Leontyev D et al (2014) Therapeutic effect of IVIG on inflammatory arthritis in mice is dependent on the Fc portion and independent of sialylation or basophils. J Immunol 192:5031–5038

Chandler KB, Mehta N, Leon DR, Suscovich TJ, Alter G, Costello CE (2019) Multi-isotype glycoproteomic characterization of serum antibody heavy chains reveals isotype- and subclass-specific N-glycosylation profiles. Mol Cell Proteomics 18:686–703

Chung S, Quarmby V, Gao X, Ying Y, Lin L, Reed C, Fong C, Lau W, Qiu ZJ, Shen A et al (2012) Quantitative evaluation of fucose reducing effects in a humanized antibody on Fcgamma receptor binding and antibody-dependent cell-mediated cytotoxicity activities. MAbs 4:326–340

Crispin M, Yu X, Bowden TA (2013) Crystal structure of sialylated IgG Fc: implications for the mechanism of intravenous immunoglobulin therapy. Proc Natl Acad Sci USA 110:E3544–E3546

Dashivets T, Thomann M, Rueger P, Knaupp A, Buchner J, Schlothauer T (2015) Multi-angle effector function analysis of human monoclonal IgG glycovariants. PLoS One 10:e0143520

Davies J, Jiang L, Pan LZ, LaBarre MJ, Anderson D, Reff M (2001) Expression of GnTIII in a recombinant anti-CD20 CHO production cell line: Expression of antibodies with altered glycoforms leads to an increase in ADCC through higher affinity for FC gamma RIII. Biotechnol Bioeng 74:288–294

de Haan N, Reiding KR, Driessen G, van der Burg M, Wuhrer M (2016) Changes in healthy human IgG Fc-glycosylation after birth and during early childhood. J Proteome Res 15:1853–1861

de Haan N, Reiding KR, Kristic J, Hipgrave Ederveen AL, Lauc G, Wuhrer M (2017) The N-glycosylation of mouse immunoglobulin G (IgG)-fragment crystallizable differs between IgG subclasses and strains. Front Immunol 8:608

de Haan N, Falck D, Wuhrer M (2020) Monitoring of immunoglobulin N- and O-glycosylation in health and disease. Glycobiology 30:226–240

Dekkers G, Treffers L, Plomp R, Bentlage AEH, de Boer M, Koeleman CAM, Lissenberg-Thunnissen SN, Visser R, Brouwer M, Mok JY et al (2017) Decoding the human immunoglobulin G-glycan repertoire reveals a spectrum of Fc-receptor- and complement-mediated-effector activities. Front Immunol 8:877

Deshpande N, Jensen PH, Packer NH, Kolarich D (2010) GlycoSpectrumScan: fishing glycopeptides from MS spectra of protease digests of human colostrum sIgA. J Proteome Res 9:1063–1075

Dorrington KJ, Bennich HH (1978) Structure-function relationships in human immunoglobulin E. Immunol Rev 41:3–25

Dougher CWL, Buffone A Jr, Nemeth MJ, Nasirikenari M, Irons EE, Bogner PN, Lau JTY (2017) The blood-borne sialyltransferase ST6Gal-1 is a negative systemic regulator of granulopoiesis. J Leukoc Biol 102:507–516

Ephrem A, Chamat S, Miquel C, Fisson S, Mouthon L, Caligiuri G, Delignat S, Elluru S, Bayry J, Lacroix-Desmazes S et al (2008) Expansion of CD4+CD25+ regulatory T cells by intravenous immunoglobulin: a critical factor in controlling experimental autoimmune encephalomyelitis. Blood 111:715–722

Ferrara C, Stuart F, Sondermann P, Brunker P, Umana P (2006) The carbohydrate at FcgammaRIIIa Asn-162. An element required for high affinity binding to non-fucosylated IgG glycoforms. J Biol Chem 281:5032–5036

Ferrara C, Grau S, Jager C, Sondermann P, Brunker P, Waldhauer I, Hennig M, Ruf A, Rufer AC, Stihle M et al (2011) Unique carbohydrate-carbohydrate interactions are required for high affinity binding between FcgammaRIII and antibodies lacking core fucose. Proc Natl Acad Sci USA 108:12669–12674

Fiebiger BM, Maamary J, Pincetic A, Ravetch JV (2015) Protection in antibody- and T cell-mediated autoimmune diseases by antiinflammatory IgG Fcs requires type II FcRs. Proc Natl Acad Sci USA 112:E2385–E2394

Franc V, Rehulka P, Raus M, Stulik J, Novak J, Renfrow MB, Sebela M (2013) Elucidating heterogeneity of IgA1 hinge-region O-glycosylation by use of MALDI-TOF/TOF mass spectrometry: role of cysteine alkylation during sample processing. J Proteomics 92:299–312

Galli SJ, Tsai M (2012) IgE and mast cells in allergic disease. Nat Med 18:693–704

Garcia-Vallejo JJ, van Dijk W, van Die I, Gringhuis SI (2005) Tumor necrosis factor-alpha up-regulates the expression of beta1,4-galactosyltransferase I in primary human endothelial cells by mRNA stabilization. J Biol Chem 280:12676–12682

Gomes MM, Wall SB, Takahashi K, Novak J, Renfrow MB, Herr AB (2008) Analysis of IgA1 N-glycosylation and its contribution to FcalphaRI binding. Biochemistry 47:11285–11299

Gould HJ, Sutton BJ (2008) IgE in allergy and asthma today. Nat Rev Immunol 8:205–217

Gounni AS, Lamkhioued B, Ochiai K, Tanaka Y, Delaporte E, Capron A, Kinet JP, Capron M (1994) High-affinity IgE receptor on eosinophils is involved in defence against parasites. Nature 367:183–186

Gringhuis SI, Garcia-Vallejo JJ, van Het Hof B, van Dijk W (2005) Convergent actions of I kappa B kinase beta and protein kinase C delta modulate mRNA stability through phosphorylation of 14-3-3 beta complexed with tristetraprolin. Mol Cell Biol 25:6454–6463

Hansen IS, Baeten DLP, den Dunnen J (2019) The inflammatory function of human IgA. Cell Mol Life Sci 76:1041–1055

Hess C, Winkler A, Lorenz AK, Holecska V, Blanchard V, Eiglmeier S, Schoen AL, Bitterling J, Stoehr AD, Petzold D et al (2013) T cell-independent B cell activation induces immunosuppressive sialylated IgG antibodies. J Clin Invest 123:3788–3796

Holland M, Yagi H, Takahashi N, Kato K, Savage CO, Goodall DM, Jefferis R (2006) Differential glycosylation of polyclonal IgG, IgG-Fc and IgG-Fab isolated from the sera of patients with ANCA-associated systemic vasculitis. Biochim Biophys Acta 1760:669–677

Issekutz AC, Rowter D, Miescher S, Kasermann F (2015) Intravenous IgG (IVIG) and subcutaneous IgG (SCIG) preparations have comparable inhibitory effect on T cell activation, which is not dependent on IgG sialylation, monocytes or B cells. Clin Immunol

Jansen BC, Bondt A, Reiding KR, Lonardi E, de Jong CJ, Falck D, Kammeijer GS, Dolhain RJ, Rombouts Y, Wuhrer M (2016) Pregnancy-associated serum N-glycome changes studied by high-throughput MALDI-TOF-MS. Sci Rep 6:23296

Jefferis R, Lund J (1997) Glycosylation of antibody molecules: structural and functional significance. Chem Immunol 65:111–128

Johannes L, Jacob R, Leffler H (2018) Galectins at a glance. J Cell Sci 131

Jones AJ, Papac DI, Chin EH, Keck R, Baughman SA, Lin YS, Kneer J, Battersby JE (2007) Selective clearance of glycoforms of a complex glycoprotein pharmaceutical caused by terminal N-acetylglucosamine is similar in humans and cynomolgus monkeys. Glycobiology 17:529–540

Jones MB, Oswald DM, Joshi S, Whiteheart SW, Orlando R, Cobb BA (2016) B-cell-independent sialylation of IgG. Proc Natl Acad Sci USA 113:7207–7212

Kanda Y, Yamada T, Mori K, Okazaki A, Inoue M, Kitajima-Miyama K, Kuni-Kamochi R, Nakano R, Yano K, Kakita S et al (2007) Comparison of biological activity among nonfucosylated therapeutic IgG1 antibodies with three different N-linked Fc oligosaccharides: the high-mannose, hybrid, and complex types. Glycobiology 17:104–118

Kaneko Y, Nimmerjahn F, Ravetch JV (2006) Anti-inflammatory activity of immunoglobulin G resulting from Fc sialylation. Science 313:670–673

Kao D, Danzer H, Collin M, Gross A, Eichler J, Stambuk J, Lauc G, Lux A, Nimmerjahn F (2015) A monosaccharide residue is sufficient to maintain mouse and human IgG subclass activity and directs IgG effector functions to cellular Fc receptors. Cell Rep 13:2376–2385

Kao D, Lux A, Schaffert A, Lang R, Altmann F, Nimmerjahn F (2017) IgG subclass and vaccination stimulus determine changes in antigen specific antibody glycosylation in mice. Eur J Immunol 47:2070–2079

Kapur R, Kustiawan I, Vestrheim A, Koeleman CA, Visser R, Einarsdottir HK, Porcelijn L, Jackson D, Kumpel B, Deelder AM et al (2014) A prominent lack of IgG1-Fc fucosylation of platelet alloantibodies in pregnancy. Blood 123:471–480

Karsten CM, Pandey MK, Figge J, Kilchenstein R, Taylor PR, Rosas M, McDonald JU, Orr SJ, Berger M, Petzold D et al (2012) Anti-inflammatory activity of IgG1 mediated by Fc galactosylation and association of FcgammaRIIB and dectin-1. Nat Med 18:1401–1406

Kerr MA (1990) The structure and function of human IgA. Biochem J 271:285–296

Klaric L, Tsepilov YA, Stanton CM, Mangino M, Sikka TT, Esko T, Pakhomov E, Salo P, Deelen J, McGurnaghan SJ et al (2020) Glycosylation of immunoglobulin G is regulated by a large network of genes pleiotropic with inflammatory diseases. Sci Adv 6:eaax0301

Kobata A (2008) The N-linked sugar chains of human immunoglobulin G: their unique pattern, and their functional roles. Biochim Biophys Acta 1780:472–478

Korn T, Bettelli E, Gao W, Awasthi A, Jager A, Strom TB, Oukka M, Kuchroo VK (2007) IL-21 initiates an alternative pathway to induce proinflammatory T(H)17 cells. Nature 448:484–487

Krapp S, Mimura Y, Jefferis R, Huber R, Sondermann P (2003) Structural analysis of human IgG-Fc glycoforms reveals a correlation between glycosylation and structural integrity. J Mol Biol 325:979–989

Kristic J, Vuckovic F, Menni C, Klaric L, Keser T, Beceheli I, Pucic-Bakovic M, Novokmet M, Mangino M, Thaqi K et al (2014) Glycans are a novel biomarker of chronological and biological ages. J Gerontol A Biol Sci Med Sci 69:779–789

Lecocq M, Detry B, Guisset A, Pilette C (2013) FcalphaRI-mediated inhibition of IL-12 production and priming by IFN-gamma of human monocytes and dendritic cells. J Immunol 190:2362–2371

Lee MM, Nasirikenari M, Manhardt CT, Ashline DJ, Hanneman AJ, Reinhold VN, Lau JT (2014) Platelets support extracellular sialylation by supplying the sugar donor substrate. J Biol Chem 289:8742–8748

Lee Y, Mitsdoerffer M, Xiao S, Gu G, Sobel RA, Kuchroo VK (2015) IL-21R signaling is critical for induction of spontaneous experimental autoimmune encephalomyelitis. J Clin Invest 125:4011–4020

Leontyev D, Katsman Y, Ma XZ, Miescher S, Kasermann F, Branch DR (2012) Sialylation-independent mechanism involved in the amelioration of murine immune thrombocytopenia using intravenous gammaglobulin. Transfusion 52:1799–1805

Li T, DiLillo DJ, Bournazos S, Giddens JP, Ravetch JV, Wang LX (2017a) Modulating IgG effector function by Fc glycan engineering. Proc Natl Acad Sci USA 114:3485–3490

Li W, Zhu Z, Chen W, Feng Y, Dimitrov DS (2017b) Crystallizable fragment glycoengineering for therapeutic antibodies development. Front Immunol 8:1554

Lin CW, Tsai MH, Li ST, Tsai TI, Chu KC, Liu YC, Lai MY, Wu CY, Tseng YC, Shivatare SS et al (2015) A common glycan structure on immunoglobulin G for enhancement of effector functions. Proc Natl Acad Sci USA 112:10611–10616

Lubberts E (2015) The IL-23-IL-17 axis in inflammatory arthritis. Nat Rev Rheumatol 11:415–429

Lux A, Yu X, Scanlan CN, Nimmerjahn F (2013) Impact of immune complex size and glycosylation on IgG binding to human FcgammaRs. J Immunol 190:4315–4323

Mahan AE, Jennewein MF, Suscovich T, Dionne K, Tedesco J, Chung AW, Streeck H, Pau M, Schuitemaker H, Francis D et al (2016) Antigen-specific antibody glycosylation is regulated via vaccination. PLoS Pathog 12:e1005456

Malhotra R, Wormald MR, Rudd PM, Fischer PB, Dwek RA, Sim RB (1995) Glycosylation changes of IgG associated with rheumatoid arthritis can activate complement via the mannose-binding protein. Nat Med 1:237–243

Manhardt CT, Punch PR, Dougher CWL, Lau JTY (2017) Extrinsic sialylation is dynamically regulated by systemic triggers in vivo. J Biol Chem 292:13514–13520

Massoud AH, Yona M, Xue D, Chouiali F, Alturaihi H, Ablona A, Mourad W, Piccirillo CA, Mazer BD (2014) Dendritic cell immunoreceptor: a novel receptor for intravenous immunoglobulin mediates induction of regulatory T cells. J Allergy Clin Immunol 133:853–863 e855

Masuda K, Yamaguchi Y, Kato K, Takahashi N, Shimada I, Arata Y (2000) Pairing of oligosaccharides in the Fc region of immunoglobulin G. FEBS Lett 473:349–357

Mattu TS, Pleass RJ, Willis AC, Kilian M, Wormald MR, Lellouch AC, Rudd PM, Woof JM, Dwek RA (1998) The glycosylation and structure of human serum IgA1, Fab, and Fc regions and the role of N-glycosylation on Fcalpha receptor interactions. J Biol Chem 273:2260–2272

Mehta AS, Long RE, Comunale MA, Wang M, Rodemich L, Krakover J, Philip R, Marrero JA, Dwek RA, Block TM (2008) Increased levels of galactose-deficient anti-Gal immunoglobulin G in the sera of hepatitis C virus-infected individuals with fibrosis and cirrhosis. J Virol 82:1259–1270

Milewski S, Mignini F, Borowski E (1991) Synergistic action of nikkomycin X/Z with azole antifungals on Candida albicans. J Gen Microbiol 137:2155–2161

Millward TA, Heitzmann M, Bill K, Langle U, Schumacher P, Forrer K (2008) Effect of constant and variable domain glycosylation on pharmacokinetics of therapeutic antibodies in mice. Biologicals 36:41–47

Miwa HE, Song Y, Alvarez R, Cummings RD, Stanley P (2012) The bisecting GlcNAc in cell growth control and tumor progression. Glycoconj J 29:609–618

Moore JS, Wu X, Kulhavy R, Tomana M, Novak J, Moldoveanu Z, Brown R, Goepfert PA, Mestecky J (2005) Increased levels of galactose-deficient IgG in sera of HIV-1-infected individuals. AIDS 19:381–389

Nettleton MY, Kochan JP (1995) Role of glycosylation sites in the IgE Fc molecule. Int Arch Allergy Immunol 107:328–329

Nimmerjahn F, Ravetch JV (2005) Divergent immunoglobulin g subclass activity through selective Fc receptor binding. Science 310:1510–1512

Nimmerjahn F, Ravetch JV (2006) Fcgamma receptors: old friends and new family members. Immunity 24:19–28

Nimmerjahn F, Ravetch JV (2008) Fcgamma receptors as regulators of immune responses. Nat Rev Immunol 8:34–47

Nimmerjahn F, Anthony RM, Ravetch JV (2007) Agalactosylated IgG antibodies depend on cellular Fc receptors for in vivo activity. Proc Natl Acad Sci USA 104:8433–8437

Niwa R, Natsume A, Uehara A, Wakitani M, Iida S, Uchida K, Satoh M, Shitara K (2005a) IgG subclass-independent improvement of antibody-dependent cellular cytotoxicity by fucose removal from Asn297-linked oligosaccharides. J Immunol Methods 306:151–160

Niwa R, Sakurada M, Kobayashi Y, Uehara A, Matsushima K, Ueda R, Nakamura K, Shitara K (2005b) Enhanced natural killer cell binding and activation by low-fucose IgG1 antibody results in potent antibody-dependent cellular cytotoxicity induction at lower antigen density. Clin Cancer Res 11:2327–2336

Novak J, Renfrow MB, Gharavi AG, Julian BA (2013) Pathogenesis of immunoglobulin A nephropathy. Curr Opin Nephrol Hypertens 22:287–294

Oefner CM, Winkler A, Hess C, Lorenz AK, Holecska V, Huxdorf M, Schommartz T, Petzold D, Bitterling J, Schoen AL et al (2012) Tolerance induction with T cell-dependent protein antigens induces regulatory sialylated IgGs. J Allergy Clin Immunol 129:1647–1655 e1613

Oettgen HC (2016) Fifty years later: Emerging functions of IgE antibodies in host defense, immune regulation, and allergic diseases. J Allergy Clin Immunol 137:1631–1645

Ohmi Y, Ise W, Harazono A, Takakura D, Fukuyama H, Baba Y, Narazaki M, Shoda H, Takahashi N, Ohkawa Y et al (2016) Sialylation converts arthritogenic IgG into inhibitors of collagen-induced arthritis. Nat Commun 7:11205

Oortwijn BD, Roos A, van der Boog PJ, Klar-Mohamad N, van Remoortere A, Deelder AM, Daha MR, van Kooten C (2007) Monomeric and polymeric IgA show a similar association with the myeloid FcalphaRI/CD89. Mol Immunol 44:966–973

Othy S, Topcu S, Saha C, Kothapalli P, Lacroix-Desmazes S, Kasermann F, Miescher S, Bayry J, Kaveri SV (2014) Sialylation may be dispensable for reciprocal modulation of helper T cells by intravenous immunoglobulin. Eur J Immunol 44:2059–2063

Pagan JD, Kitaoka M, Anthony RM (2018) Engineered sialylation of pathogenic antibodies in vivo attenuates autoimmune disease. Cell 172:564–577 e513

Parekh RB, Dwek RA, Sutton BJ, Fernandes DL, Leung A, Stanworth D, Rademacher TW, Mizuochi T, Taniguchi T, Matsuta K et al (1985) Association of rheumatoid arthritis and primary osteoarthritis with changes in the glycosylation pattern of total serum IgG. Nature 316:452–457

Parekh R, Roitt I, Isenberg D, Dwek R, Rademacher T (1988) Age-related galactosylation of the N-linked oligosaccharides of human serum IgG. J Exp Med 167:1731–1736

Peschke B, Keller CW, Weber P, Quast I, Lunemann JD (2017) Fc-galactosylation of human immunoglobulin gamma isotypes improves C1q binding and enhances complement-dependent cytotoxicity. Front Immunol 8:646

Pezer M, Stambuk J, Perica M, Razdorov G, Banic I, Vuckovic F, Gospic AM, Ugrina I, Vecenaj A, Bakovic MP et al (2016) Effects of allergic diseases and age on the composition of serum IgG glycome in children. Sci Rep 6:33198

Pfeifle R, Rothe T, Ipseiz N, Scherer HU, Culemann S, Harre U, Ackermann JA, Seefried M, Kleyer A, Uderhardt S et al (2017) Regulation of autoantibody activity by the IL-23-TH17 axis determines the onset of autoimmune disease. Nat Immunol 18:104–113

Pilette C, Detry B, Guisset A, Gabriels J, Sibille Y (2010) Induction of interleukin-10 expression through Fcalpha receptor in human monocytes and monocyte-derived dendritic cells: role of p38 MAPKinase. Immunol Cell Biol 88:486–493

Pincetic A, Bournazos S, DiLillo DJ, Maamary J, Wang TT, Dahan R, Fiebiger BM, Ravetch JV (2014) Type I and type II Fc receptors regulate innate and adaptive immunity. Nat Immunol 15:707–716

Plomp R, Hensbergen PJ, Rombouts Y, Zauner G, Dragan I, Koeleman CA, Deelder AM, Wuhrer M (2014) Site-specific N-glycosylation analysis of human immunoglobulin e. J Proteome Res 13:536–546

Plomp R, Bondt A, de Haan N, Rombouts Y, Wuhrer M (2016) Recent advances in clinical glycoproteomics of immunoglobulins (Igs). Mol Cell Proteomics 15:2217–2228

Plomp R, Ruhaak LR, Uh HW, Reiding KR, Selman M, Houwing-Duistermaat JJ, Slagboom PE, Beekman M, Wuhrer M (2017) Subclass-specific IgG glycosylation is associated with markers of inflammation and metabolic health. Sci Rep 7:12325

Plomp R, de Haan N, Bondt A, Murli J, Dotz V, Wuhrer M (2018) Comparative glycomics of immunoglobulin A and G from saliva and plasma reveals biomarker potential. Front Immunol 9:2436

Pucic M, Muzinic A, Novokmet M, Skledar M, Pivac N, Lauc G, Gornik O (2012) Changes in plasma and IgG N-glycome during childhood and adolescence. Glycobiology 22:975–982

Quast I, Keller CW, Maurer MA, Giddens JP, Tackenberg B, Wang LX, Munz C, Nimmerjahn F, Dalakas MC, Lunemann JD (2015) Sialylation of IgG Fc domain impairs complement-dependent cytotoxicity. J Clin Invest 125:4160–4170

Rajput B, Shaper NL, Shaper JH (1996) Transcriptional regulation of murine beta1,4-galactosyltransferase in somatic cells. Analysis of a gene that serves both a housekeeping and a mammary gland-specific function. J Biol Chem 271:5131–5142

Raju TS (2008) Terminal sugars of Fc glycans influence antibody effector functions of IgGs. Curr Opin Immunol 20:471–478

Raju TS, Briggs JB, Chamow SM, Winkler ME, Jones AJ (2001) Glycoengineering of therapeutic glycoproteins: in vitro galactosylation and sialylation of glycoproteins with terminal N-acetylglucosamine and galactose residues. Biochemistry 40:8868–8876

Rieber EP, Rank G, Kohler I, Krauss S (1993) Membrane expression of Fc epsilon RII/CD23 and release of soluble CD23 by follicular dendritic cells. Adv Exp Med Biol 329:393–398

Riteau N, Radtke AJ, Shenderov K, Mittereder L, Oland SD, Hieny S, Jankovic D, Sher A (2016) Water-in-oil-only adjuvants selectively promote T follicular helper cell polarization through a type I IFN and IL-6-dependent pathway. J Immunol 197:3884–3893

Robinett RA, Guan N, Lux A, Biburger M, Nimmerjahn F, Meyer AS (2018) Dissecting FcgammaR regulation through a multivalent binding model. Cell Syst 7:41–48 e45

Roopenian DC, Akilesh S (2007) FcRn: the neonatal Fc receptor comes of age. Nat Rev Immunol 7:715–725

Rossato E, Ben Mkaddem S, Kanamaru Y, Hurtado-Nedelec M, Hayem G, Descatoire V, Vonarburg C, Miescher S, Zuercher AW, Monteiro RC (2015) Reversal of arthritis by human monomeric IgA through the receptor-mediated SH2 domain-containing phosphatase 1 inhibitory pathway. Arthritis Rheumatol 67:1766–1777

Royle L, Roos A, Harvey DJ, Wormald MR, van Gijlswijk-Janssen D, el Redwan RM, Wilson IA, Daha MR, Dwek RA, Rudd PM (2003) Secretory IgA N- and O-glycans provide a link between the innate and adaptive immune systems. J Biol Chem 278:20140–20153

Saphire EO, Stanfield RL, Crispin MD, Parren PW, Rudd PM, Dwek RA, Burton DR, Wilson IA (2002) Contrasting IgG structures reveal extreme asymmetry and flexibility. J Mol Biol 319:9–18

Satoh M, Iida S, Shitara K (2006) Non-fucosylated therapeutic antibodies as next-generation therapeutic antibodies. Expert Opin Biol Ther 6:1161–1173

Scallon BJ, Tam SH, McCarthy SG, Cai AN, Raju TS (2007) Higher levels of sialylated Fc glycans in immunoglobulin G molecules can adversely impact functionality. Mol Immunol 44:1524–1534

Schaffert A, Hanic M, Novokmet M, Zaytseva O, Kristic J, Lux A, Nitschke L, Peipp M, Pezer M, Hennig R et al (2019) Minimal B cell extrinsic IgG glycan modifications of pro- and anti-inflammatory IgG preparations in vivo. Front Immunol 10:3024

Schwab I, Nimmerjahn F (2013) Intravenous immunoglobulin therapy: how does IgG modulate the immune system? Nat Rev Immunol 13:176–189

Schwab I, Biburger M, Kronke G, Schett G, Nimmerjahn F (2012) IVIg-mediated amelioration of ITP in mice is dependent on sialic acid and SIGNR1. Eur J Immunol 42:826–830

Schwab I, Mihai S, Seeling M, Kasperkiewicz M, Ludwig R, Nimmerjahn F (2014) Broad requirement for terminal sialic acid residues and FcgRIIB for the preventive and therapeutic activity of intravenous immunoglobulins in vivo. Eur J Immunol

Schwab I, Lux A, Nimmerjahn F (2015) Pathways responsible for human autoantibody and therapeutic intravenous IgG activity in humanized mice. Cell Rep 13:610–620

Seeling M, Bruckner C, Nimmerjahn F (2017) Differential antibody glycosylation in autoimmunity: sweet biomarker or modulator of disease activity? Nat Rev Rheumatol 13:621–630

Seite JF, Cornec D, Renaudineau Y, Youinou P, Mageed RA, Hillion S (2010) IVIg modulates BCR signaling through CD22 and promotes apoptosis in mature human B lymphocytes. Blood 116:1698–1704

Selman MH, de Jong SE, Soonawala D, Kroon FP, Adegnika AA, Deelder AM, Hokke CH, Yazdanbakhsh M, Wuhrer M (2012) Changes in antigen-specific IgG1 Fc N-glycosylation upon influenza and tetanus vaccination. Mol Cell Proteomics 11(M111):014563

Shade KT, Platzer B, Washburn N, Mani V, Bartsch YC, Conroy M, Pagan JD, Bosques C, Mempel TR, Fiebiger E, Anthony RM (2015) A single glycan on IgE is indispensable for initiation of anaphylaxis. J Exp Med 212:457–467

Shade KT, Conroy ME, Anthony RM (2019) IgE Glycosylation in Health and Disease. Curr Top Microbiol Immunol 423:77–93

Shade KC, Conroy ME, Washburn N, Kitaoka M, Huynh DJ, Laprise E, Patil SU, Shreffler WG, Anthony RM (2020) Sialylation of immunoglobulin E is a determinant of allergic pathogenicity. Nature 582:265–270

Shields RL, Lai J, Keck R, O'Connell LY, Hong K, Meng YG, Weikert SH, Presta LG (2002) Lack of fucose on human IgG1 N-linked oligosaccharide improves binding to human Fcgamma RIII and antibody-dependent cellular toxicity. J Biol Chem 277:26733–26740

Shinkawa T, Nakamura K, Yamane N, Shoji-Hosaka E, Kanda Y, Sakurada M, Uchida K, Anazawa H, Satoh M, Yamasaki M et al (2003) The absence of fucose but not the presence of galactose or bisecting N-acetylglucosamine of human IgG1 complex-type oligosaccharides shows the critical role of enhancing antibody-dependent cellular cytotoxicity. J Biol Chem 278:3466–3473

Sondermann P, Huber R, Oosthuizen V, Jacob U (2000) The 3.2-A crystal structure of the human IgG1 Fc fragment-Fc gammaRIII complex. Nature 406:267–273

Sondermann P, Pincetic A, Maamary J, Lammens K, Ravetch JV (2013) General mechanism for modulating immunoglobulin effector function. Proc Natl Acad Sci USA 110:9868–9872

Sonneveld ME, Natunen S, Sainio S, Koeleman CA, Holst S, Dekkers G, Koelewijn J, Partanen J, van der Schoot CE, Wuhrer M, Vidarsson G (2016) Glycosylation pattern of anti-platelet IgG is stable during pregnancy and predicts clinical outcome in alloimmune thrombocytopenia. Br J Haematol 174:310–320

Sonneveld ME, Koeleman CAM, Plomp HR, Wuhrer M, van der Schoot CE, Vidarsson G (2018) Fc-glycosylation in human IgG1 and IgG3 is similar for both total and anti-red-blood cell anti-K antibodies. Front Immunol 9:129

Stambuk J, Nakic N, Vuckovic F, Pucic-Bakovic M, Razdorov G, Trbojevic-Akmacic I, Novokmet M, Keser T, Vilaj M, Stambuk T et al (2020) Global variability of the human IgG glycome. Aging (Albany NY) 12:15222–15259

Steffen U, Koeleman CA, Sokolova MV, Bang H, Kleyer A, Rech J, Unterweger H, Schicht M, Garreis F, Hahn J et al (2020) IgA subclasses have different effector functions associated with distinct glycosylation profiles. Nat Commun 11:120

Subedi GP, Barb AW (2016) The immunoglobulin G1 N-glycan composition affects binding to each low affinity Fc gamma receptor. MAbs 8:1512–1524

Tackenberg B, Jelcic I, Baerenwaldt A, Oertel WH, Sommer N, Nimmerjahn F, Lunemann JD (2009) Impaired inhibitory Fcgamma receptor IIB expression on B cells in chronic inflammatory demyelinating polyneuropathy. Proc Natl Acad Sci USA 106:4788–4792

Takai T (2005) Fc receptors and their role in immune regulation and autoimmunity. J Clin Immunol 25:1–18

Temming AR, Dekkers G, van de Bovenkamp FS, Plomp HR, Bentlage AEH, Szittner Z, Derksen NIL, Wuhrer M, Rispens T, Vidarsson G (2019) Human DC-SIGN and CD23 do not interact with human IgG. Sci Rep 9:9995

Thomann M, Schlothauer T, Dashivets T, Malik S, Avenal C, Bulau P, Ruger P, Reusch D (2015) In vitro glycoengineering of IgG1 and its effect on Fc receptor binding and ADCC activity. PLoS One 10:e0134949

Twisselmann N, Bartsch YC, Pagel J, Wieg C, Hartz A, Ehlers M, Hartel C (2018) IgG Fc glycosylation patterns of preterm infants differ with gestational age. Front Immunol 9:3166

Umana P, Jean-Mairet J, Moudry R, Amstutz H, Bailey JE (1999) Engineered glycoforms of an antineuroblastoma IgG1 with optimized antibody-dependent cellular cytotoxic activity. Nat Biotechnol 17:176–180

Vadrevu SK, Trbojevic-Akmacic I, Kossenkov AV, Colomb F, Giron LB, Anzurez A, Lynn K, Mounzer K, Landay AL, Kaplan RC et al (2018) Frontline Science: Plasma and immunoglobulin G galactosylation associate with HIV persistence during antiretroviral therapy. J Leukoc Biol 104:461–471

van de Bovenkamp FS, Derksen NIL, Ooijevaar-de Heer P, van Schie KA, Kruithof S, Berkowska MA, van der Schoot CE, van der Burg M, Gils A et al (2018) Adaptive antibody diversification through N-linked glycosylation of the immunoglobulin variable region. Proc Natl Acad Sci USA 115:1901–1906

van Egmond M, Damen CA, van Spriel AB, Vidarsson G, van Garderen E, van de Winkel JG (2001) IgA and the IgA Fc receptor. Trends Immunol 22:205–211

van Kooyk Y, Kalay H, Garcia-Vallejo JJ (2013) Analytical tools for the study of cellular glycosylation in the immune system. Front Immunol 4:451

Vercelli D, Helm B, Marsh P, Padlan E, Geha RS, Gould H (1989) The B-cell binding site on human immunoglobulin E. Nature 338:649–651

Vouldoukis I, Riveros-Moreno V, Dugas B, Ouaaz F, Becherel P, Debre P, Moncada S, Mossalayi MD (1995) The killing of Leishmania major by human macrophages is mediated by nitric oxide induced after ligation of the Fc epsilon RII/CD23 surface antigen. Proc Natl Acad Sci USA 92:7804–7808

Walker MR, Lund J, Thompson KM, Jefferis R (1989) Aglycosylation of human IgG1 and IgG3 monoclonal antibodies can eliminate recognition by human cells expressing Fc gamma RI and/or Fc gamma RII receptors. Biochem J 259:347–353

Wandall HH, Rumjantseva V, Sorensen AL, Patel-Hett S, Josefsson EC, Bennett EP, Italiano JE Jr, Clausen H, Hartwig JH, Hoffmeister KM (2012) The origin and function of platelet glycosyltransferases. Blood 120:626–635

Wang J, Balog CI, Stavenhagen K, Koeleman CA, Scherer HU, Selman MH, Deelder AM, Huizinga TW, Toes RE, Wuhrer M (2011) Fc-glycosylation of IgG1 is modulated by B-cell stimuli. Mol Cell Proteomics 10(M110):004655

Wang TT, Maamary J, Tan GS, Bournazos S, Davis CW, Krammer F, Schlesinger SJ, Palese P, Ahmed R, Ravetch JV (2015) Anti-HA glycoforms drive B cell affinity selection and determine influenza vaccine efficacy. Cell 162:160–169

Wang TT, Sewatanon J, Memoli MJ, Wrammert J, Bournazos S, Bhaumik SK, Pinsky BA, Chokephaibulkit K, Onlamoon N, Pattanapanyasat K et al (2017) IgG antibodies to dengue enhanced for FcgammaRIIIA binding determine disease severity. Science 355:395–398

Washburn N, Schwab I, Ortiz D, Bhatnagar N, Lansing JC, Medeiros A, Tyler S, Mekala D, Cochran E, Sarvaiya H et al (2015) Controlled tetra-Fc sialylation of IVIg results in a drug candidate with consistent enhanced anti-inflammatory activity. Proc Natl Acad Sci USA 112: E1297–E1306

Wermeling F, Anthony RM, Brombacher F, Ravetch JV (2013) Acute inflammation primes myeloid effector cells for anti-inflammatory STAT6 signaling. Proc Natl Acad Sci USA 110:13487–13491

Wines BD, Hulett MD, Jamieson GP, Trist HM, Spratt JM, Hogarth PM (1999) Identification of residues in the first domain of human Fc alpha receptor essential for interaction with IgA. J Immunol 162:2146–2153

Woof JM, Kerr MA (2006) The function of immunoglobulin A in immunity. J Pathol 208:270–282

Wright A, Sato Y, Okada T, Chang K, Endo T, Morrison S (2000) In vivo trafficking and catabolism of IgG1 antibodies with Fc associated carbohydrates of differing structure. Glycobiology 10:1347–1355

Wu G, Hitchen PG, Panico M, North SJ, Barbouche MR, Binet D, Morris HR, Dell A, Haslam SM (2016) Glycoproteomic studies of IgE from a novel hyper IgE syndrome linked to PGM3 mutation. Glycoconj J 33:447–456

Wuhrer M, Stam JC, van de Geijn FE, Koeleman CA, Verrips CT, Dolhain RJ, Hokke CH, Deelder AM (2007) Glycosylation profiling of immunoglobulin G (IgG) subclasses from human serum. Proteomics 7:4070–4081

Young RJ, Owens RJ, Mackay GA, Chan CM, Shi J, Hide M, Francis DM, Henry AJ, Sutton BJ, Gould HJ (1995) Secretion of recombinant human IgE-Fc by mammalian cells and biological activity of glycosylation site mutants. Protein Eng 8:193–199

Yu LC, Montagnac G, Yang PC, Conrad DH, Benmerah A, Perdue MH (2003) Intestinal epithelial CD23 mediates enhanced antigen transport in allergy: evidence for novel splice forms. Am J Physiol Gastrointest Liver Physiol 285:G223–G234

Yu X, Vasiljevic S, Mitchell DA, Crispin M, Scanlan CN (2013) Dissecting the molecular mechanism of IVIg therapy: the interaction between serum IgG and DC-SIGN is independent of antibody glycoform or Fc domain. J Mol Biol 425:1253–1258

Yu X, Marshall MJE, Cragg MS, Crispin M (2017) Improving antibody-based cancer therapeutics through glycan engineering. BioDrugs 31:151–166

Chapter 13
Immunoglobulin G Glycosylation in Diseases

Marija Pezer

Contents

13.1	Introduction	396
13.2	IgG Glycans are an Integral Structural and Functional Part of the Molecule	396
13.3	IgG Glycans Affect IgG Functions	397
	13.3.1 Fc Glycans	397
	13.3.2 Fab Glycans	399
13.4	Regulation of IgG Glycosylation	399
13.5	Common IgG Glycosylation Pattern in Inflammatory Diseases and Aging	400
13.6	Role of Skewed IgG Glycosylation in Diseases	401
13.7	Perspectives for IgG Glycosylation in Precision Medicine	409
13.8	Conclusions	410
References		411

Abstract Changes in immunoglobulin G (IgG) glycosylation pattern have been observed in a vast array of auto- and alloimmune, infectious, cardiometabolic, malignant, and other diseases. This chapter contains an updated catalog of over 140 studies within which IgG glycosylation analysis was performed in a disease setting. Since the composition of IgG glycans is known to modulate its effector functions, it is suggested that a changed IgG glycosylation pattern in patients might be involved in disease development and progression, representing a predisposition and/or a functional effector in disease pathology. In contrast to the glycopattern of bulk serum IgG, which likely relates to the systemic inflammatory background, the glycosylation profile of antigen-specific IgG probably plays a direct role in disease pathology in several infectious and allo- and autoimmune antibody-dependent diseases. Depending on the specifics of any given disease, IgG glycosylation readout might therefore in the future be developed into a useful clinical biomarker or a supplementary to currently used biomarkers.

Keywords IgG glycosylation · Differential glycosylation · Disease · Biomarker

M. Pezer (✉)
Glycoscience Research Laboratory, Genos Ltd., Zagreb, Croatia
e-mail: mpezer@genos.hr

Abbreviations

ACPA	Anti-citrullinated protein antibody
ADCC	Antibody-dependent cell-mediated cytotoxicity
Asn	Asparagine
CH2	Constant heavy 2
Fab	Fragment antigen binding
Fc	Fragment crystallizable
FcγRs	Fcγ receptors
FNAIT	Fetal and neonatal alloimmune thrombocytopenia
GlcNac	*N*-acetylglucosamine
HFD	High-fat diet
HDFN	Haemolytic disease of the fetus and newborn
IgG	Immunoglobulin G
IVIg	Intravenous immunoglobulin
RA	Rheumatoid arthritis

13.1 Introduction

Since the first reports on glycans attached to immunoglobulin G (IgG) in the 1970s (Ciccimarra et al. 1976; Williams et al. 1973; Koide et al. 1977; Hymes et al. 1979) and the seminal papers by Parekh and al. on the association of a changed IgG glycome composition with a diseased status and aging, (Parekh et al. 1985, 1988) IgG glycans are today universally recognized as modulators of IgG activity (Yamaguchi and Barb 2020). The importance of IgG glycome composition is implied in various physiological and pathological states. IgG glycans are discussed as potential contributors to disease development and progression, as well as a diagnostic, prognostic, and follow-up biomarker. This chapter is a brief update and extension of our comprehensive review on IgG glycosylation in aging and diseases published 3 years ago (Gudelj et al. 2018a), with a focus on the potential functionality of the skewed IgG glycosylation pattern. The table presents the updated list of publications that examined IgG glycosylation in various diseased states.

13.2 IgG Glycans are an Integral Structural and Functional Part of the Molecule

IgG glycans represent about 15% of the molecule's weight (Arnold et al. 2007). Each IgG molecule contains an N-glycan covalently attached to the conserved asparagine (Asn) at position 297 within the Fc region on each of the two heavy chains (Shade and Anthony 2013). In addition, 15–20% IgG molecules contain an

N-glycan within the Fab region, attached to the asparagine within an N-glycosylation sequon formed by somatic hypermutation during affinity maturation (Dunn-Walters et al. 2000; van de Bovenkamp et al. 2016).

Fc N-glycans are placed in the cavity between the CH2 domains of the two opposing heavy chains (Pincetic et al. 2014; Deisenhofer et al. 1976) and are important for the molecule's structural integrity, stability, and serum-half life (Boune et al. 2020; Cymer et al. 2018). They are also involved in the modulation of IgG effector functions, by affecting the molecule's affinity toward its ligands and receptors: type I and type II Fc receptors, C1q complement component, mannan-binding lectin, etc. (Pincetic et al. 2014; Peschke et al. 2017; Malhotra et al. 1995; Dekkers et al. 2017). Although markedly less explored than Fc glycans, Fab glycans are also reported to affect IgG's biological properties and effector functions, such as half-life, stability, solubility, and antigen-binding (van de Bovenkamp et al. 2016, 2018a; Wu et al. 2010; Wright et al. 1991; Higel et al. 2016; Liu 2015, 2018).

13.3 IgG Glycans Affect IgG Functions

The composition of both Fab and Fc glycans has been confirmed to influence IgG functionality and activity. Since this has been described in detail in Chap. 12, the main findings are only briefly summarized here as a reminder for the reader.

13.3.1 Fc Glycans

Due to the positioning of the Fc N-glycan at the Asn-297, structural differences of the N-glycans attached to the Fc region influence the affinity to the IgG ligands and receptors that interact with IgG at the CH2 domain and the CH2-CH3 domain interface (Dekkers et al. 2017; Reusch and Tejada 2015; Li et al. 2017; Wada et al. 2019; Vidarsson et al. 2014).

Core-Fucosylation Contrary to most other plasma proteins, over 90% of all Fc glycans are core-fucosylated (fucosylated glycans, F) (van de Bovenkamp et al. 2016; Štambuk et al. 2020; Baković et al. 2013; Clerc et al. 2016). The lack of core fucose significantly increases the IgG's affinity for the Fcγ receptor III (FcγRIII), both A and B, enhancing the FcγRIII-mediated effector functions, particularly the antibody-dependent cell-mediated cytotoxicity (ADCC) (Dekkers et al. 2017; Shields et al. 2002; Shinkawa et al. 2003). This prominent effect of alternative Fc glycosylation on the IgG function found its application in the industrial production of therapeutic monoclonal antibodies (Garber 2018).

Bisection Up to 10% of all IgG Fc glycans are bisected, i.e., contain a bisecting N-acetylglucosamine (GlcNAc) (bisected glycans, B) (van de Bovenkamp et al. 2016). Since the presence of GlcNAc and core fucose, to a degree, preclude each other during glycan synthesis (Benedetti et al. 2017; Schuster et al. 2005; Ferrara et al. 2006), the increase in binding affinity fo FcγRIII sometimes associated with bisected glycans (Umaña et al. 1999; Davies et al. 2001; Lifely et al. 1995) cannot be easily uncoupled from the same effect observed for core-fucosylated IgG glycans (Shinkawa et al. 2003).

Galactosylation Galactosylation is the IgG glycosylation trait with the most pronounced inter-individual variation (Huhn et al. 2009; Gornik et al. 2012). On average, about 35% of IgG Fc glycans contain no terminal galactose residues (agalactosylated glycans, G0), about 35% contain one (monogalactosylated glycans, G1), and about 15% contain two terminal galactoses (digalactosylated glycans, G2) (Baković et al. 2013; Huffman et al. 2014). Terminal galactoses modulate IgG inflammatory potential by affecting binding affinities to complement components and FcγRs. Agalactosylated Fc glycans are considered to act pro-inflammatory by activating the complement through the alternative pathway (Banda et al. 2008), and the lectin pathway after binding to the mannose-binding lectin (Malhotra et al. 1995; Ji et al. 2002; Arnold et al. 2006). Galactosylation was also held responsible for the anti-inflammatory activity of immune complexes by binding to the inhibitory FcγRIIB (Karsten et al. 2012). However, Fc galactosylation has also been reported to enhance complement-dependent cytotoxicity (CDC) through the classical complement pathway by increasing the IgG's affinity for the C1q complement component (Peschke et al. 2017; Boyd et al. 1995; Hodoniczky et al. 2005). Likewise, by increasing the affinity of IgG for FcγRs, it enhances the downstream processes mediated by FcγRs, in particular ADCC (Dekkers et al. 2017; Kumpel et al. 1994, 1995; Houde et al. 2010; Subedi and Barb 2016). We should therefore not rush to proclaim terminal IgG galactosylation simply "anti-inflammatory," before considering the entire context and the nature and extent of IgG involvement in the process we are investigating.

Sialylation On average, 10–15% of IgG Fc glycans carry a single terminal sialic acid (monosialylated glycans, S1) or two sialic acids (disialylated glycans, S2) (Baković et al. 2013; Huffman et al. 2014). Similar to terminal galactosylation, sialylation is most often discussed as a modulator of IgG functions regarding inflammation (Böhm et al. 2014).

The importance of sialylation became evident when the presence of the sialylated Fc fraction was discovered indispensable for the anti-inflammatory activity of the intravenous immunoglobulin (IVIg) preparation in a K/BxN serum-transfer mouse model of RA (Kaneko et al. 2006). Mouse studies on several antibody-dependent autoimmune disease models helped elucidate the mechanistic pathway for its activity, starting with the binding of the sialylated Fc fraction to specific ICAM-3 grabbing non-integrin-related 1 (SIGN-R1) on the surface of splenic macrophages and ending in enhanced FcγRIIB expression on the effector macrophages (Kaneko et al. 2006; Schwab and Nimmerjahn 2013; Anthony et al. 2008, 2011; Schwab et al. 2012, 2014; Washburn et al. 2015; Galeotti et al. 2017; Fiebiger et al. 2015). However, this finding did not hold in several other in vitro and in vivo models, nor human studies (Galeotti et al. 2017; Guhr et al. 2011; Leontyev et al. 2012; Campbell et al. 2014; Temming et al. 2019). This confirms the well-established notion that the IVIg mode of action is complex and tightly connected with the corresponding immune context.

Depending on the sialylation status of the Fc glycan, the Fc domain is suggested to adopt either an "open" or a "closed" conformation, for sialylated and asialylated glycans, respectively. The "open" conformation favors binding to the type I FcγRs, whereas the "closed" conformation favors the binding of type II FcRs (Pincetic et al. 2014; Sondermann et al. 2013). Terminal sialylation is thus proposed to act as a switch between two distinct immunological effector functions.

To summarize—agalactosylated, asialylated, and bisected IgG molecules are often simply described as "pro-inflammatory," and terminally galactosylated and sialylated IgG molecules as "anti-inflammatory," while afucosylated IgG has an augmented capacity to trigger ADCC through enhanced FcγRIIIA binding. We should, however, always bear in mind that this generalization is a simplification, and exercise caution when considering its implications.

13.3.2 Fab Glycans

As expected, Fab glycans are mostly reported to affect antigen-binding (Wright et al. 1991; Coloma et al. 1999; Schneider et al. 2015; Wallick et al. 1988; Tachibana et al. 1997; Leibiger et al. 1999; Khurana et al. 1997; Man Sung Co et al. 1993; Fujimura et al. 2000; Van De Bovenkamp et al. 2018b). Besides the obvious, they are also suggested to influence IgG aggregation and precipitation (Courtois et al. 2016), immune complex formation (Gutierrez et al. 2006), and have a role in the IVIg mode of action (Käsermann et al. 2012; Wiedeman et al. 2013; Massoud et al. 2014; Séïté et al. 2010, 2014).

13.4 Regulation of IgG Glycosylation

IgG glycosylation is a complex trait, influenced by both, genetics (Menni et al. 2013; Pučić et al. 2011; Klarić et al. 2020) and the environment (Štambuk et al. 2020; Yu et al. 2016; Krištić et al. 2014; De Jong et al. 2016). More precisely, the compound IgG glycosylation pattern seems to be, to different degrees, modulated by IgG aminoacid sequence (Lund et al. 1996; Zaitseva et al. 2018; Yu et al. 2013), the intra- and extracellular milieu affecting the glycosylation machinery (Ohtsubo and Marth 2006; Oefner et al. 2012; Bartsch et al. 2020; Hess et al. 2013; Canellada et al. 2002; Gutiérrez et al. 2001; Miranda et al. 2005; Wang et al. 2011; Pfeifle et al. 2017; Liu et al. 2014; Fan et al. 2015), and environmental factors (Novokmet et al. 2014; Greto et al. 2020; Ercan et al. 2017; Engdahl et al. 2018; Klasić et al. 2018; Tijardović et al. 2019; Sarin et al. 2019; Peng et al. 2019). Solving the outstanding question of IgG glycosylation regulation would likely bring us one step closer to understanding the possible functionality of changes in IgG glycan composition in different physiological and pathological states.

13.5 Common IgG Glycosylation Pattern in Inflammatory Diseases and Aging

Advances in the development of high-throughput glycomic and glycoproteomic analyses (Huhn et al. 2009; Mariño et al. 2010; Trbojević-Akmačić et al. 2016, 2017) have enabled a significant number of large-scale epidemiological studies examining total IgG glycosylation pattern in diseased *vs.* healthy control subjects (Štambuk et al. 2020; Singh et al. 2020; Lemmers et al. 2017; Menni et al. 2018; Šimurina et al. 2018; Theodoratou et al. 2016; Wahl et al. 2018). In many of the diseases that were studied, a similar pattern emerged: diseased subjects often exhibited a decreased abundance of galactosylated, sialylated, and—occasionally—an increased abundance of bisected bulk IgG glycans when compared to healthy controls (Fig. 13.1). In addition, the trend was often associated with disease severity and reverted to baseline values upon successful application of therapy. Interestingly, the same pattern that was observed in diseases with an inflammatory component was also evident in aging subjects (Fig. 13.1) (Gudelj et al. 2018a; Lauc 2016). This "pro-inflammatory IgG glycome composition" is likely associated with

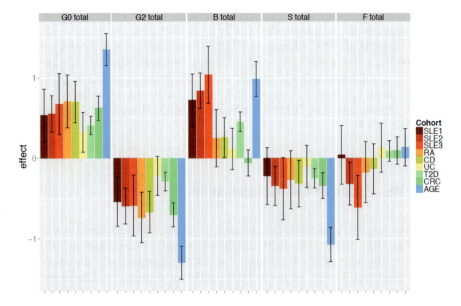

Fig. 13.1 General patterns of IgG glycosylation changes are similar in several diseases and aging. The effect (shown on the *y*-axis) is shown as the difference between means of case and control populations (for aging, population over *vs.* population under 50 years of age), expressed in standard deviations. The whiskers represent the 95% confidence interval. The reference populations for disease cohorts are age- and sex-matched healthy controls. *SLE* systemic lupus erythematosus, *RA* rheumatoid arthritis, *CD* Crohn's disease, *UC* ulcerous colitis, *T2D* type 2 diabetes mellitus, *CRC* colorectal carcinoma. *G0* agalactosylated glycans, *G2* digalactosylated glycans, *B* bisected glycans, *S* sialylated glycans, *F* core-fucosylated glycans. Reused with permission from Lauc (2016)

the common background inflammatory component of the studied diseases. In some cases, it might reflect a predisposition toward disease development, or/and even be involved as an effector of inflammation. Additionally, it might represent a consequence of environmental exposure to antigens through a lifetime or unhealthy lifestyle choices.

Indeed, the composition of IgG glycome was reported to associate with many physiological and biochemical traits, as well as with traits correlated to inflammation and poor metabolic health (Gudelj et al. 2018a) and the expected lifespan (Štambuk et al. 2020). IgG glycome was thus suggested to be a biomarker of general immune activation (De Jong et al. 2016), while we propose total IgG glycoprofile can be positioned as a read-out of a general state of health, i.e. biological age (Vilaj et al. 2019).

13.6 Role of Skewed IgG Glycosylation in Diseases

When we take into account the complexity of the IgG's multiple roles in our immune system, it is no wonder there is no single common interpretation of the altered IgG glycopattern across the wide spectrum of diseases (Table 13.1). The multiple pleiotropic loci, i.e. shared associations of IgG glycome composition and autoimmune, inflammatory, and other diseases (Klarić et al. 2020; Lauc et al. 2013), as well as the changes in IgG glycopattern preceding disease development—such as in the case of RA (Gudelj et al. 2018b) and cardiovascular diseases (Menni et al. 2018)—suggest that a skewed bulk serum IgG glycoprofile might reflect a disease risk or predisposition. This predisposition can manifest through an inherited (Klarić et al. 2020; Lauc et al. 2013) or acquired propensity for inflammation modulation (Franceschi et al. 2018).

In most other cases, when it comes to glycosylation of the bulk serum IgG, the role of a shifted glycosylation pattern is not clear. As already mentioned, decreased galactosylation and sialylation often accompany diseases that involve an inflammatory immune response. The evidence that would enable us to unambiguously determine whether the "pro-inflammatory" IgG glycoforms represent one of many drivers of disease pathology or merely byproducts of the inflamed immune system is still lacking. The current consensus is that total IgG glycopattern is likely relevant in the general modulation of the immune activation threshold.

In some cases, however, a clear link/evidence for the functionality of IgG glycans is provided. A mouse study investigating the link between obesity and the development of hypertension resulted in a very intriguing observation. Hyposialylated IgG from mice in which obesity was induced by a high-fat diet (HFD) induced an elevated blood pressure when transferred to IgG-deficient mice. Moreover, supplementing HFD-feed mice with a sialic acid precursor, N-acetyl-D-mannosamine (ManNAc), resulted in the restoration of the baseline level of IgG sialylation and protected them from obesity-induced hypertension development (Peng et al. 2019). This finding thus demonstrated the functional role of IgG glycans

Table 13.1 Diseases exhibiting an altered serum IgG glycosylation profile

	↓	↑
G	**Inflammatory diseases and states** Takayasu's arteritis (Hernandez-Pando et al. 1994) Adult periodontal disease (Novak et al. 2005) Nonalcoholic steatohepatitis (Vanderschaeghe et al. 2018) IgG4-related disease (Culver et al. 2019) Primary sclerosing cholangitis (Culver et al. 2019) **Autoimmune diseases** Rheumatoid arthritis—total (Parekh et al. 1985, 1988; Bond et al. 1996; Van de Geijn et al. 2009; Young et al. 1991; Axford et al. 1992; Engdahl et al. 2018; Gudelj et al. 2018b; Gindzienska-Sieskiewicz et al. 2007; Tomana et al. 1988; Bodman-Smith et al. 1996; Gińdzieńska-Sieśkiewicz et al. 2016; Pekelharing et al. 1988; Pilkington et al. 1995), ACPA (Ercan et al. 2010; Rombouts et al. 2015; Bond et al. 1997, 2018), RF (Matsumoto et al. 2000) Osteoarthritis (Parekh et al. 1985; Bond et al. 1997) Juvenile onset rheumatoid arthritis (Parekh et al. 1988; Flögel et al. 1998; Sumar et al. 1991; Ercan et al. 2012; Cheng et al. 2017) Systemic lupus erythematosus (Tomana et al. 1988, 1992; Pilkington et al. 1995; Bond et al. 1997; Vučković et al. 2015) Lupus nephritis (Bhargava et al. 2021) Inflammatory bowel disease: Crohn's disease and ulcerative colitis (Šimurina et al. 2018; Tomana et al. 1988; Bond et al. 1997; Dubé et al. 1990; Go et al. 1994; Shinzaki et al. 2008; Nakajima et al. 2011; Trbojevic Akmacic et al. 2015; Parekh et al. 1989; Miyoshi et al. 2016) Sjögren's syndrome (Bond et al. 1996, 1997) Neonatal lupus (Pilkington et al. 1996a) Spondyloarthropathy (Bond et al. 1997; Leirisalo-Repo et al. 1999) ANCA-associated vasculitis—total (Holland et al. 2002, 2006; Espy et al. 2011; Kemna et al. 2017; Wuhrer et al. 2015), ANCA (Kemna et al. 2017; Wuhrer et al. 2015) Coeliac disease (Cremata et al. 2003) Lambert–Eaton myasthenic syndrome (Selman et al. 2011) Myasthenia gravis (Selman et al. 2011) Myositis (Perdivara et al. 2011) Guillain–Barré syndrome (Fokkink et al. 2014a; b)	**Alloimmune diseases** Fetal or neonatal alloimmune thrombocytopenia—anti-HPA (Sonneveld et al. 2016; Wuhrer et al. 2009) Hemolytic disease of the fetus and newborn—anti-c, anti-E (Sonneveld et al. 2017a) **Cancers** Thyroid cancer (Chen et al. 2012) Multiple myeloma (Mittermayr et al. 2017) Mammary gland hyperplasia (Meng et al. 2020) **Infectious diseases** Measles—anti-measles (Larsen et al. 2020) Mumps—anti-mumps (Larsen et al. 2020) Parvovirus-B19 infection—anti-B19 (Larsen et al. 2020) COVID-19—anti-S (Larsen et al. 2020), anti-N (Larsen et al. 2020) RSV infection—anti-RSV (van Erp et al. 2020) Tuberculosis—antigen-specific (Lu et al. 2020) **Other diseases** Parkinson's disease (Russell et al. 2017)

(continued)

Table 13.1 (continued)

↓	↑
Poor glycemic control and impaired renal function in type I diabetes (Bermingham et al. 2018)	
Autoimmune hemolytic anemia—total and anti-RBC (Sonneveld et al. 2017b)	
Membranous nephropathy (Haddad et al. 2021)	
Alloimmune diseases	
Hemolytic disease of the fetus and newborn—anti-K (Sonneveld et al. 2018)	
Cancers	
Multiple myeloma (Nishiura et al. 1990; Aurer et al. 2007)	
Bone disease in multiple myeloma (Westhrin et al. 2020)	
Ovarian cancer—total (Gerçel-Taylor et al. 2001; Saldova et al. 2007; Alley et al. 2012; Qian et al. 2013; Ruhaak et al. 2016), tumor-reactive (Gerçel-Taylor et al. 2001)	
Prostate cancer (Kanoh et al. 2004a, 2008, 2009; Kazuno et al. 2016)	
Non-small cell cancer (Kanoh et al. 2006)	
Gastric cancer (Kanoh et al. 2004b, 2008; Bones et al. 2010, 2011; Kodar et al. 2012)	
Lung cancer (Kanoh et al. 2004b, 2008; Chen et al. 2013)	
Colorectal carcinoma (Theodoratou et al. 2016; Vučković et al. 2016)	
Breast cancer (Kawaguchi-Sakita et al. 2016)	
Malignant hematological diseases[a] (de Haan et al. 2018a)	
Infectious diseases	
Leprosy—Erythema nodosum leprosum (Filley et al. 1989)	
Tuberculosis (Rook et al. 1994; Pilkington et al. 1995, 1996b; Parekh et al. 1989; Filley et al. 1989; Rademacher et al. 1988; Lu et al. 2016)	
Infective endocarditis (Bond et al. 1997)	
HIV infection—total (Ackerman et al. 2013; Moore et al. 2005; Muenchhoff et al. 2020), anti-HIV (Ackerman et al. 2013; Larsen et al. 2020)	
Hepatitis C: liver fibrosis, cirrhosis—anti-Gal (Mehta et al. 2008)	
Hepatitis B: chronic infection (Ho et al. 2015); liver cirrhosis – total (Ho et al. 2015), anti-Gal (Mehta et al. 2008)	
Visceral leishmaniasis (Gardinassi et al. 2014)	
CMV infection—anti-CMV (Larsen et al. 2020)	

(continued)

Table 13.1 (continued)

	↓	↑
	COVID-19 [139] **Other diseases** Castleman's disease (Nakao et al. 1991) Galactosaemia (Coss et al. 2012; Knerr et al. 2015; Maratha et al. 2016; Stockmann et al. 2016; Coman et al. 2010; Coss et al. 2014) Alzheimer's disease (Lundström et al. 2014) Asthma? (De Jong et al. 2016; Pezer et al. 2016) Chronic kidney disease (Barrios et al. 2016) Hypertension (Wang et al. 2016; Gao et al. 2017) Type II diabetes (Lemmers et al. 2017; Li et al. 2019) Nonalcoholic fatty liver disease (Zhao et al. 2018) Ischemic stroke (Liu et al. 2018) Hyperuricemia (Hou et al. 2019) Diabetic retinopathy (Wu et al. 2021)	
S	**Inflammatory diseases and conditions** Primary sclerosing cholangitis (Culver et al. 2019) **Autoimmune diseases** Rheumatoid arthritis—total (Parekh et al. 1985; Engdahl et al. 2018; Gudelj et al. 2018b; Gińdzieńska-Sieśkiewicz et al. 2016), ACPA (Scherer et al. 2010), RF (Matsumoto et al. 2000) Osteoarthritis (Parekh et al. 1985) ANCA-associated vasculitis—total (Espy et al. 2011; Kemna et al. 2017; Wuhrer et al. 2015), ANCA (Kemna et al. 2017; Wuhrer et al. 2015) Systemic lupus erythematosus—total (Vučković et al. 2015; Chen et al. 2015), ANA (Magorivska et al. 2016) Inflammatory bowel disease: Crohn's disease (Trbojevic Akmacic et al. 2015) Juvenile onset rheumatoid arthritis (Cheng et al. 2017) Antiphospholipid syndrome (Fickentscher et al. 2015) Autoimmune hemolytic anemia—? (Sonneveld et al. 2017b) **Alloimmune diseases** Hemolytic disease of the fetus and new-born—anti-K (Sonneveld et al. 2018) **Infectious diseases** Visceral leishmaniasis (Gardinassi et al. 2014)	**Autoimmune diseases** Autoimmune hemolytic anemia—anti-RBC (Sonneveld et al. 2017b) **Alloimmune diseases** Fetal or neonatal alloimmune thrombocytopenia—anti-HPA (Sonneveld et al. 2016; Wuhrer et al. 2009) **Cancers** Multiple myeloma (Aurer et al. 2007; Fleming et al. 1998) Thyroid cancer (Chen et al. 2012) Lung cancer (Ruhaak et al. 2013) **Infectious diseases** Parvovirus-B19 infection—anti-B19 (Larsen et al. 2020) COVID-19—anti-S (Larsen et al. 2020), anti-N (Larsen et al. 2020) Recurrent respiratory infections (Cheng et al. 2020) RSV infection—anti-RSV (van Erp et al. 2020) HIV infection—anti-HIV (Muenchhoff et al. 2020) Tuberculosis—antigen-specific (Lu et al. 2020)

(continued)

Table 13.1 (continued)

	↓	↑
	Tuberculosis (Lu et al. 2016) HIV infection—total (Vadrevu et al. 2018), anti-HIV (Larsen et al. 2020) Meningococcal sepsis (de Haan et al. 2018b) CMV infection — anti-CMV (Larsen et al. 2020) COVID-19 [139] **Cancers** Ovarian cancer (Saldova et al. 2007) Colorectal carcinoma (Theodoratou et al. 2016; Vučković et al. 2016) Malignant hematological diseases[a] (de Haan et al. 2018a) Monoclonal gammopathy of undetermined significance (Bosseboeuf et al. 2017) Multiple myeloma (Bosseboeuf et al. 2017) Bone disease in multiple myeloma (Westhrin et al. 2020) **Other diseases** Alzheimer's disease (Lundström et al. 2014) Chronic kidney disease (Barrios et al. 2016) Type II diabetes (Lemmers et al. 2017) Hypertension (Peng et al. 2019; Gao et al. 2017) Parkinson's disease (Russell et al. 2017) Ischemic stroke (Liu et al. 2018) Hyperuricemia (Hou et al. 2019) Dementia (Zhang et al. 2021)	
F	**Inflammatory diseases and conditions** Inflammation severity (Novokmet et al. 2014) Low back pain (Freidin et al. 2016) **Autoimmune diseases** Systemic lupus erythematosus? (Vučković et al. 2015; Sjöwall et al. 2015) ANCA-associated vasculitis—ANCA (Kemna et al. 2017) Inflammatory bowel disease: ulcerative colitis (Šimurina et al. 2018) Autoimmune thyroid diseases (Martin et al. 2020) Multiple sclerosis (Cvetko et al. 2020) **Alloimmune diseases** Fetal or neonatal alloimmune thrombocytopenia—anti-HPA (Kapur et al. 2014b; Sonneveld et al. 2016; Wuhrer et al. 2009) Hemolytic disease of the fetus and newborn—anti-D (Kapur et al. 2014a), anti-c, anti-E, anti-K (Sonneveld et al. 2017a, 2018) **Infectious diseases**	**Autoimmune diseases** Juvenile onset rheumatoid arthritis (Flögel et al. 1998) Rheumatoid arthritis—total (Gińdzieńska-Sieśkiewicz et al. 2016; Gornik et al. 1999), ACPA (Rombouts et al. 2015) Systemic lupus erythematosus? (Vučković et al. 2015; Sjöwall et al. 2015) ANCA-associated vasculitis (Kemna et al. 2017) Inflammatory bowel disease: Crohn's disease (Šimurina et al. 2018) **Infectious diseases** Visceral leishmaniasis (Gardinassi et al. 2014) Tuberculosis (Lu et al. 2016) HIV infection—total (Vadrevu et al. 2018), anti-HIV (Muenchhoff et al. 2020) **Cancers** Hepatocellular carcinoma (Comunale et al. 2006) Multiple myeloma (Westhrin et al. 2020)

(continued)

Table 13.1 (continued)

	↓	↑
	Dengue fever progressing to dengue hemorrhagic fever or dengue shock syndrome—anti-ENV, anti-NS1, anti-HA (Wang et al. 2017) Meningococcal sepsis (de Haan et al. 2018b) HIV infection—anti-HIV (Ackerman et al. 2013; Larsen et al. 2020) CMV infection—anti-CMV (Larsen et al. 2020) Mumps—anti-mumps (Larsen et al. 2020) COVID-19—anti-S (Larsen et al. 2020), anti-N (Larsen et al. 2020) anti-RBD (Chakraborty et al. 2021) Tuberculosis—antigen-specific (Lu et al. 2020) **Cancers** Multiple myeloma (Mittermayr et al. 2017) Malignant hematological diseases[a] (de Haan et al. 2018a) **Other diseases** Dementia (Zhang et al. 2021) Kidney function decline in type II diabetes (Singh et al. 2020) Non-malignant hematological diseases[b] (de Haan et al. 2018a)	**Other diseases** Galactosaemia (Maratha et al. 2016) Hypertension (Gao et al. 2017) Nonalcoholic fatty liver disease (Zhao et al. 2018)
B	**Inflammatory diseases and conditions** IgG4-related disease (Culver et al. 2019) **Autoimmune diseases** Osteoarthritis (Bond et al. 1997) ANCA-associated vasculitis—total (Kemna et al. 2017; Wuhrer et al. 2015), ANCA (Wuhrer et al. 2015) Autoimmune hemolytic anemia—anti-RBC (Sonneveld et al. 2017b) **Alloimmune diseases** Hemolytic disease of the fetus and newborn—anti-c (Sonneveld et al. 2017a) **Infectious diseases** Visceral leishmaniasis (Gardinassi et al. 2014) HIV infection—anti-HIV (Larsen et al. 2020; Muenchhoff et al. 2020) COVID-19—total (Larsen et al. 2020; Petrović et al. 2020), anti-S (Larsen et al. 2020), anti-N (Larsen et al. 2020) **Cancers** Thyroid cancer (Chen et al. 2012) **Other diseases** Hypertension (Wang et al. 2016; Gao et al.	**Inflammatory diseases and conditions** Low back pain (Freidin et al. 2016) Primary sclerosing cholangitis (Culver et al. 2019) COPD (Pavić et al. 2018) **Autoimmune diseases** Rheumatoid arthritis (Pekelharing et al. 1988; Bond et al. 1996, 1997) Juvenile onset rheumatoid arthritis (Bond et al. 1996, 1997) Inflammatory bowel disease: Crohn's disease and ulcerative colitis (Bond et al. 1997) Lambert–Eaton myasthenic syndrome (Selman et al. 2011) Systemic lupus erythematosus (Vučković et al. 2015) Lupus nephritis (Bhargava et al. 2021) **Alloimmune diseases** Hemolytic disease of the fetus and newborn—anti-K (Sonneveld et al. 2018) **Infectious diseases** Infective endocarditis (Bond et al. 1997) Meningococcal sepsis (de Haan et al. 2018b) CMV infection—anti-CMV (Larsen et al.

(continued)

Table 13.1 (continued)

	↓	↑
	2017) Galactosaemia (Maratha et al. 2016)	2020) Mumps—anti-mumps (Larsen et al. 2020) Parvovirus-B19 infection—anti-B19 (Larsen et al. 2020) Recurrent respiratory infections (Cheng et al. 2020) Tuberculosis—antigen-specific (Lu et al. 2020) **Cancers** Colorectal carcinoma (Theodoratou et al. 2016) Malignant hematological diseases[a] (de Haan et al. 2018a) **Other diseases** Chronic kidney disease (Barrios et al. 2016) Type II diabetes (Lemmers et al. 2017) Nonalcoholic fatty liver disease (Zhao et al. 2018) Ischemic stroke (Liu et al. 2018) Kidney function decline in type II diabetes (Singh et al. 2020) Dementia (Zhang et al. 2021)
H	**Inflammatory diseases and conditions** IgG4-related disease (Culver et al. 2019) **Infectious diseases** Meningococcal sepsis (de Haan et al. 2018b) **Cancers** Malignant hematological diseases[a] (de Haan et al. 2018a)	
M		**Autoimmune diseases** Multiple sclerosis (Cvetko et al. 2020)

"Down" arrow (↓) refers to a decreased and "up" arrow (↑) to an increased proportion of the corresponding IgG glycosylation trait (as calculated in the corresponding publication) in patients suffering from the disease compared to healthy controls and/or in association with disease activity and severity. In the case of antigen-specific IgG, the arrows refer to the comparison between antigen-specific and total IgG and/or to the association with disease activity and severity. Due to the complexity of IgG glycosylation in a disease setting, the associations shown here are simplified and do not reflect the particulars, such as IgG subclass and clonality, IgG region (total *vs.* Fab *vs.* Fc), analytical methodology, calculation of derived glycosylation traits, subject demographics, clinical parameters, etc. For details, readers are advised to consult the original publications.
G galactosylated, *S* sialylated, *F* core-fucosylated, *B* bisected, *H* hybrid, *M* high-mannose glycans. *ACPA* anti-citrullinated protein antibody, *ANA* anti-nuclear antibody, *ANCA* anti-neutrophil cytoplasmic antibody, *CMV* cytomegalovirus, *COPD* chronic obstructive pulmonary disease, *COVID-19* corona virus disease 2019, *ENV* envelope protein, *HA* hemagglutinin, *HPA* human platelet antigen, *N* nucleocapsid protein, *NS1* non-structural protein 1, *RBC* red blood cell, *RBD* receptor binding domain, *RSV* respiratory syncytial virus, *S* spike protein. Modified (updated) from our previous review (Gudelj et al. 2018a)—an open-access article, available under the terms of the Creative Commons Attribution License (CC BY): https://creativecommons.org/licenses/by/4.0/
[a]Malignant hematological diseases: acute lymphoblastic leukemia, myelodysplastic syndrome/ acute myeloblastic leukemia, acute myeloblastic leukemia
[b]Non-malignant hematological diseases: thalassemia, Fanconi anemia, sickle cell disease, severe aplastic anemia, progressive bone marrow failure, neutropenia congenita, Glanzmann thrombasthenia, hemophagocytic lymphohistiocytosis, X-linked lymphoproliferative disease

in the development of hypertension. Interestingly, the same treatment restored IgG sialylation and reduced tumor load and bone loss in a mouse model of myeloma (Westhrin et al. 2020).

On the level of total serum IgG, increased level of glycosylation of the Fab region observed in some malignant diseases (Zhu et al. 2002, 2003; Radcliffe et al. 2007; Coelho et al. 2010; McCann et al. 2008) is proposed to contribute to disease development and progression by enhancing tumor cell persistence and expansion (Coelho et al. 2010; Amin et al. 2015).

Glycosylation changes on antigen-specific IgG are more likely to be directly involved in disease pathology in case of antibody-mediated auto- or alloimmune diseases or defense from pathogens in case of infectious diseases. The role of differential IgG glycosylation in these cases corresponds to the specifics of a particular disease and the molecular mechanisms underlying its pathology.

In addition to the change in total IgG, multiple infectious diseases are characterized by a distinct glycosylation pattern of relevant antigen-specific IgG in comparison to total IgG (Table 13.1). This implies a distinct regulation of IgG glycosylation, depending on both the disease and the antigen (Ackerman et al. 2013), even within a single individual (Mahan et al. 2016). This supports the notion that IgG glycome relevance should be interpreted in the disease-specific functional context.

One of the rare instances where the role of IgG glycosylation is mechanistically explained is once more linked to the enhanced affinity of afucosylated IgG molecules for FcγRIIIA. In the case of dengue fever, occasionally a secondary, heterologous dengue infection results in severe dengue hemorrhagic fever and dengue shock syndrome. This is attributed to antibody-dependent enhancement (ADE) of the disease by cross-reactivity of afucosylated anti-dengue IgG with platelet antigens, resulting in platelet depletion (Wang et al. 2017). Additionally, the enhanced binding of afucosylated IgG to FcγRIIA and FcγRIIIA promotes the FcγR-mediated viral entry and signaling in cells bearing these receptors on their surface, primarily monocytes and macrophages, resulting in infection progression (Thulin et al. 2020).

A similar relevance for afucosylated antigen-specific IgG is observed in COVID-19 patients. Anti-SARS-CoV-2 IgG with a higher core-fucosylation level is associated with unaided clearance of the infection (Larsen et al. 2020). By contrast, critically ill patients display lower levels of fucosylated anti-SARS-CoV-2 IgG (Larsen et al. 2020; Chakraborty et al. 2021). Furthermore, in in vitro studies afucosylated anti-S/-RBD antibodies were shown to induce enhanced natural killer (NK) cell degranulation (Chakraborty et al. 2021) and elevated production of pro-inflammatory cytokines by primary monocytes and alveolar macrophages, which is likely the background of the severe disease phenotype associated with this glycoprofile in vivo (Larsen et al. 2020; Chakraborty et al. 2021; Hoepel et al. 2020).

Similarly, afucosylated antigen-specific IgG in fetal and neonatal alloimmune thrombocytopenia (FNAIT) and hemolytic disease of the fetus and newborn (HDFN) are thought to contribute, again through enhanced FcγRIIIA-mediated mechanisms, namely phagocytosis and ADCC, to the more severe disease phenotype (Kapur et al. 2014a, b; Sonneveld et al. 2016, 2017a).

In lupus nephritis, a serious complication of SLE, the presence of core fucose was shown to induce upregulated calcium/calmodulin kinase IV expression in podocytes, leading to podocyte injury and limited nephrin synthesis. In the same experimental setting, the presence of terminal galactoses acted protectively (Bhargava et al. 2021).

An interesting recent finding on the importance of Fab glycans emerged in the most explored disease in the context of IgG glycosylation. In RA, a high percentage of anti-citrullinated protein antibody (ACPA) is additionally glycosylated at the Fab region (Rombouts et al. 2016; Hafkenscheid et al. 2017), a feature distinguishing RA patients from ACPA$^+$ but healthy subjects (Kissel et al. 2019; Hafkenscheid et al. 2019). This suggests Fab glycosylation of ACPA might be mechanistically involved in RA development (Rombouts et al. 2016).

13.7 Perspectives for IgG Glycosylation in Precision Medicine

A skewed IgG glycoprofile in comparison to the personal baseline value (requiring longitudinal monitoring) or in comparison to ethnicity-, age-, and sex-matched subjects (requiring a population baseline cohort) in a cross-sectional experimental design, might indicate an increased risk for disease development (Gudelj et al. 2018b), or disease progression (Gudelj et al. 2018a). However, since the alterations in bulk serum IgG glycome composition are not disease-specific, they cannot be used as a stand-alone diagnostic marker. A total IgG glycoprofile of the composition significantly removed from the baseline can instead be used as an indication of a necessity for an examination by an expert clinician.

In case of an established diagnosis, bulk IgG glycome might serve as a predictor of disease progression—e.g., decreased IgG2/3 galactosylation in patients progressing from undifferentiated to rheumatoid arthritis (Sénard et al. 2021). Similarly, IgG glycome is proposed to bear potential for a useful add-on tool for monitoring functional disease progression and response to therapy (Parekh et al. 1988; Kanoh et al. 2004a, 2008; Váradi et al. 2015; Collins et al. 2013; Van Zeben et al. 1994; Rook et al. 1994; Pasek et al. 2006; Gindzienska-Sieskiewicz et al. 2007; Croce et al. 2007; Ercan et al. 2010).

The relevance and biomarker potential of IgG glycome analysis is more evident in some cases of antigen-specific IgG. For instance, due to the increased level of ACPA Fab glycosylation in individuals at risk for RA development, IgG glycome analysis might in the future provide the currently missing understanding (and biomarker) for the first determining pathogenic event leading to disease development (Rombouts et al. 2015; Scherer et al. 2010). Furthermore, as already mentioned, in several diseases a particular antigen-specific IgG glycopattern is associated with a risk for the severe phenotype (Kapur et al. 2014a, b; Sonneveld et al. 2016; Sonneveld et al. 2017a). Similarly, following the mechanistical explanation for the role of afucosylated anti-dengue IgG described in the previous section, afucosylated maternal anti-dengue IgG is proposed to denote a susceptibility to symptomatic dengue infection in infants (Thulin et al. 2020). The knowledge that a particular glycan profile of antigen-specific IgG, including post-vaccination status for some infectious diseases, is related to the risk of developing (the severe form of) a disease might in the future enable or aid the stratification of patients at risk and timely preventive action.

Another sought-after biomarker type is the one enabling patient stratification aiming at improved differential (sub-)diagnosis and subsequent selection of appropriate therapeutic measures. Differential IgG glycosylation was also suggested as a possibility for such applications. Indeed, the IgG sialylation level predicted response to therapy in Kawasaki disease (Ogata et al. 2013), and the galactosylation level response to anti-tumor necrosis factor (TNF) therapy in RA and Crohn's disease (Váradi et al. 2015), and response to methotrexate therapy in RA (Lundström et al. 2017). Having the means to distinguish non-responders before the very initiation of long and expensive therapeutic treatments is truly an exciting prospect.

In summary, there are multiple possibilities for IgG glycosylation to enter the arena of clinical disease management. Currently, all of the possible applications mentioned here are still at the level of basic research and further studies are necessary to validate the initial findings and propel the IgG glycome analysis to the status of a full-fledged clinical biomarker.

13.8 Conclusions

IgG glycans can modulate virtually all of its numerous effector roles, the specifics depending on the disease and immune context. The associations of multiple IgG glycosylation traits with an immense array of heterogeneous diseases and their different stages imply that there is no single pathway connecting IgG glycome composition and disease development and progression.

Many inflammatory, autoimmune, infectious, cardiometabolic, and neoplastic diseases share a common IgG glycosylation profile of bulk (total) serum IgG, also characteristic for aging and often described as "pro-inflammatory": a decreased level of galactosylated and sialylated glycans, and (sometimes) an increased level of bisected IgG glycans. This pattern is presumably associated with an inflammatory disease component as a part or consequence of disease pathology, or environmental events, such as antigen exposure. It might be mechanistically involved in disease advancement through modulation of inflammation, and, in some cases, manifest before the occurrence of symptoms, thus representing disease predisposition or mark the risk for disease development or progression.

When it comes to a distinct glycosylation profile of antigen-specific versus total serum IgG, IgG glycans are more likely to be directly involved in disease pathogenesis and progression through disease-specific effector mechanisms. This is often the case with afucosylated IgG glycans enhancing the affinity of IgG toward FcγRIIIA.

The read-out of IgG glycosylation has a potential for an (add-on) biomarker helping improve current algorithms for disease prediction and diagnosis, patient stratification, monitoring of disease progression, and response to therapy.

Compliance with Ethical Standards

Funding This work was supported by the European Structural and Investment Funds CEKOM (Grant# KK.01.2.2.03.0006).

Conflict of Interest MP is an employee of Genos Ltd.—a private company that specializes in high-throughput glycomic analysis and has several patents in the field, and of Genos Glycoscience Ltd.—a spin-off of Genos Ltd. that commercializes its scientific discoveries.

Ethical Approval This article does not contain any studies with human participants.

References

Ackerman ME, Crispin M, Yu X, Baruah K, Boesch AW, Harvey DJ, Dugast AS, Heizen EL, Ercan A, Choi I, Streeck H, Nigrovic PA, Bailey-Kellogg C, Scanlan C, Alter G (2013) Natural variation in Fc glycosylation of HIV-specific antibodies impacts antiviral activity. J Clin Invest 123:2183–2192. https://doi.org/10.1172/JCI65708

Alley WR, Vasseur JA, Goetz JA, Svoboda M, Mann BF, Matei DE, Menning N, Hussein A, Mechref Y, Novotny MV (2012) N-linked glycan structures and their expressions change in the blood sera of ovarian cancer patients. J Proteome Res 11:2282–2300. https://doi.org/10.1021/pr201070k

Amin R, Mourcin F, Uhel F, Pangault C, Ruminy P, Dupré L, Guirriec M, Marchand T, Fest T, Lamy T, Tarte K (2015) DC-SIGN-expressing macrophages trigger activation of mannosylated IgM B-cell receptor in follicular lymphoma. Blood 126:1911–1920. https://doi.org/10.1182/blood-2015-04-640912

Anthony RM, Nimmerjahn F, Ashline DJ, Reinhold VN, Paulson JC, Ravetch JV (2008) Recapitulation of IVIG anti-inflammatory activity with a recombinant IgG fc. Science (80-) 320:373–376. https://doi.org/10.1126/science.1154315

Anthony RM, Kobayashi T, Wermeling F, Ravetch JV (2011) Intravenous gammaglobulin suppresses inflammation through a novel T H 2 pathway. Nature 475:110–114. https://doi.org/10.1038/nature10134

Arnold JN, Dwek RA, Rudd PM, Sim RB (2006) Mannan binding lectin and its interaction with immunoglobulins in health and in disease. Immunol Lett 106:103–110. https://doi.org/10.1016/j.imlet.2006.05.007

Arnold JN, Wormald MR, Sim RB, Rudd PM, Dwek RA (2007) The impact of glycosylation on the biological function and structure of human immunoglobulins. Annu Rev Immunol 25:21–50. https://doi.org/10.1146/annurev.immunol.25.022106.141702

Aurer I, Lauc G, Dumić J, Rendić D, Matišić D, Miloš M, Heffer-Lauc M, Flogel M, Labar B (2007) Aberrant glycosylation of IgG heavy chain in multiple myeloma. Coll Antropol 31:247–251

Axford JS, Sumar N, Alavi A, Isenberg DA, Young A, Bodman KB, Roitt IM (1992) Changes in normal glycosylation mechanisms in autoimmune rheumatic disease. J Clin Invest 89:1021–1031. https://doi.org/10.1172/JCI115643

Baković MP, Selman MHJ, Hoffmann M, Rudan I, Campbell H, Deelder AM, Lauc G, Wuhrer M (2013) High-throughput IgG Fc N-glycosylation profiling by mass spectrometry of glycopeptides. J Proteome Res 12:821–831. https://doi.org/10.1021/pr300887z

Banda NK, Wood AK, Takahashi K, Levitt B, Rudd PM, Royle L, Abrahams JL, Stahl GL, Holers VM, Arend WP (2008) Initiation of the alternative pathway of murine complement by immune complexes is dependent on N-glycans in IgG antibodies. Arthritis Rheum 58:3081–3089. https://doi.org/10.1002/art.23865

Barrios C, Zierer J, Gudelj I, Stambuk J, Ugrina I, Rodríguez E, Soler MJ, Pavic T, Simurina M, Keser T, Pucic-Bakovic M, Mangino M, Pascual J, Spector TD, Lauc G, Menni C (2016) Glycosylation profile of IgG in moderate kidney dysfunction. J Am Soc Nephrol 27:933–941. https://doi.org/10.1681/ASN.2015010109

Bartsch YC, Eschweiler S, Leliavski A, Lunding HB, Wagt S, Petry J, Lilienthal GM, Rahmöller J, de Haan N, Hölscher A, Erapaneedi R, Giannou AD, Aly L, Sato R, de Neef LA, Winkler A, Braumann D, Hobusch J, Kuhnigk K, Krémer V, Steinhaus M, Blanchard V, Gemoll T, Habermann JK, Collin M, Salinas G, Manz RA, Fukuyama H, Korn T, Waisman A, Yogev N, Huber S, Rabe B, Rose-John S, Busch H, Berberich-Siebelt F, Hölscher C, Wuhrer M, Ehlers M (2020) IgG Fc sialylation is regulated during the germinal center reaction following immunization with different adjuvants. J Allergy Clin Immunol 146:652–666.e11. https://doi.org/10.1016/j.jaci.2020.04.059

Benedetti E, Pučić-Baković M, Keser T, Wahl A, Hassinen A, Yang JY, Liu L, Trbojević-Akmačić I, Razdorov G, Štambuk J, Klarić L, Ugrina I, Selman MHJ, Wuhrer M, Rudan I, Polasek O, Hayward C, Grallert H, Strauch K, Peters A, Meitinger T, Gieger C, Vilaj M, Boons GJ, Moremen KW, Ovchinnikova T, Bovin N, Kellokumpu S, Theis FJ, Lauc G, Krumsiek J (2017) Network inference from glycoproteomics data reveals new reactions in the IgG glycosylation pathway. Nat Commun 8:1483. https://doi.org/10.1038/s41467-017-01525-0

Bermingham ML, Colombo M, McGurnaghan SJ, Blackbourn LAK, Vučković F, Baković MP, Trbojević-Akmačić I, Lauc G, Agakov F, Agakova AS, Hayward C, Klarić L, Palmer CNA, Petrie JR, Chalmers J, Collier A, Green F, Lindsay RS, Macrury S, McKnight JA, Patrick AW, Thekkepat S, Gornik O, McKeigue PM, Colhoun HM (2018) N-glycan profile and kidney disease in type 1 diabetes. Diabetes Care 41:79–87. https://doi.org/10.2337/dc17-1042

Bhargava R, Lehoux S, Maeda K, Tsokos MG, Krishfield S, Ellezian LY, Pollak M, Stillman IE, Cummings RD, Tsokos GC (2021) Aberrantly glycosylated IgG elicits pathogenic signaling in podocytes and signifies lupus nephritis. JCI Insight. https://doi.org/10.1172/jci.insight.147789

Bodman-Smith K, Sumar N, Sinclair H, Roitt I, Isenberg D, Young A (1996) Agalactosyl IgG [gal (o)]-an analysis of its clinical utility in the long-term follow-up of patients with rheumatoid arthritis. Br J Rheumatol 35:1063–1066. https://doi.org/10.1093/rheumatology/35.11.1063

Böhm S, Kao D, Nimmerjahn F (2014) Sweet and sour: the role of glycosylation for the anti-inflammatory activity of immunoglobulin G. Curr Top Microbiol Immunol 382:393–417. https://doi.org/10.1007/978-3-319-07911-0-18

Bond A, Alavi A, Axford JS, Youinou P, Hay FC (1996) The relationship between exposed galactose and N-acetylglucosamine residues on IgG in rheumatoid arthritis (RA), juvenile chronic arthritis (JCA) and Sjogren's syndrome (SS). Clin Exp Immunol 105:99–103. https://doi.org/10.1046/j.1365-2249.1996.d01-741.x

Bond A, Alavi A, Axford JS, Bourke BE, Bruckner FE, Kerr MA, Maxwell JD, Tweed KJ, Weldon MJ, Youinou P, Hay FC (1997) A detailed lectin analysis of IgG glycosylation, demonstrating disease specific changes in terminal galactose and N-acetylglucosamine. J Autoimmun 10:77–85. https://doi.org/10.1006/jaut.1996.0104

Bondt A, Hafkenscheid L, Falck D, Kuijper TM, Rombouts Y, Hazes JMW, Wuhrer M, Dolhain RJEM (2018) ACPA IgG galactosylation associates with disease activity in pregnant patients with rheumatoid arthritis. Ann Rheum Dis 77:1130–1136. https://doi.org/10.1136/annrheumdis-2018-212946

Bones J, Mittermayr S, O'Donoghue N, Guttman A, Rudd PM (2010) Ultra performance liquid chromatographic profiling of serum N-glycans for fast and efficient identification of cancer associated alterations in glycosylation. Anal Chem 82:10208–10215. https://doi.org/10.1021/ac102860w

Bones J, Byrne JC, Odonoghue N, McManus C, Scaife C, Boissin H, Nastase A, Rudd PM (2011) Glycomic and glycoproteomic analysis of serum from patients with stomach cancer reveals potential markers arising from host defense response mechanisms. J Proteome Res 10:1246–1265. https://doi.org/10.1021/pr101036b

Bosseboeuf A, Allain-Maillet S, Mennesson N, Tallet A, Rossi C, Garderet L, Caillot D, Moreau P, Piver E, Girodon F, Perreault H, Brouard S, Nicot A, Bigot-Corbe E, Hermouet S, Harb J (2017) Pro-inflammatory state in monoclonal gammopathy of undetermined significance and in multiple myeloma is characterized by low sialylation of pathogen-specific and other monoclonal immunoglobulins. Front Immunol 8. https://doi.org/10.3389/fimmu.2017.01347

Boune S, Hu P, Epstein AL, Khawli LA (2020) Principles of N-linked glycosylation variations of IgG-based therapeutics: pharmacokinetic and functional considerations. Antibodies 9:22. https://doi.org/10.3390/antib9020022

Boyd PN, Lines AC, Patel AK (1995) The effect of the removal of sialic acid, galactose and total carbohydrate on the functional activity of Campath-1H. Mol Immunol 32:1311–1318. https://doi.org/10.1016/0161-5890(95)00118-2

Campbell IK, Miescher S, Branch DR, Mott PJ, Lazarus AH, Han D, Maraskovsky E, Zuercher AW, Neschadim A, Leontyev D, McKenzie BS, Käsermann F (2014) Therapeutic effect of IVIG on inflammatory arthritis in mice is dependent on the Fc portion and independent of sialylation or basophils. J Immunol 192:5031–5038. https://doi.org/10.4049/jimmunol.1301611

Canellada A, Blois S, Gentile T, Margni Idehu RA (2002) In vitro modulation of protective antibody responses by estrogen, progesterone and interleukin-6. Am J Reprod Immunol 48:334–343. https://doi.org/10.1034/j.1600-0897.2002.01141.x

Chakraborty S, Gonzalez J, Edwards K, Mallajosyula V, Buzzanco AS, Sherwood R, Buffone C, Kathale N, Providenza S, Xie MM, Andrews JR, Blish CA, Singh U, Dugan H, Wilson PC, Pham TD, Boyd SD, Nadeau KC, Pinsky BA, Zhang S, Memoli MJ, Taubenberger JK, Morales T, Schapiro JM, Tan GS, Jagannathan P, Wang TT (2021) Proinflammatory IgG Fc structures in patients with severe COVID-19. Nat Immunol 22:67–73. https://doi.org/10.1038/s41590-020-00828-7

Chen G, Wang Y, Qiu L, Qin X, Liu H, Wang X, Wang Y, Song G, Li F, Guo Y, Li F, Guo S, Li Z (2012) Human IgG Fc-glycosylation profiling reveals associations with age, sex, female sex hormones and thyroid cancer. J Proteome 75:2824–2834. https://doi.org/10.1016/j.jprot.2012.02.001

Chen G, Wang Y, Qin X, Li H, Guo Y, Wang Y, Liu H, Wang X, Song G, Li F, Li F, Guo S, Qiu L, Li Z (2013) Change in IgG1 Fc N-linked glycosylation in human lung cancer: age- and sex-related diagnostic potential. Electrophoresis 34:2407–2416. https://doi.org/10.1002/elps.201200455

Chen XX, Chen YQ, Ye S (2015) Measuring decreased serum IgG sialylation: a novel clinical biomarker of lupus. Lupus 24:948–954. https://doi.org/10.1177/0961203315570686

Cheng HD, Stöckmann H, Adamczyk B, McManus CA, Ercan A, Holm IA, Rudd PM, Ackerman ME, Nigrovic PA (2017) High-throughput characterization of the functional impact of IgG Fc glycan aberrancy in juvenile idiopathic arthritis. Glycobiology 27:1099–1108. https://doi.org/10.1093/glycob/cwx082

Cheng HD, Tirosh I, de Haan N, Stöckmann H, Adamczyk B, McManus CA, O'Flaherty R, Greville G, Saldova R, Bonilla FA, Notarangelo LD, Driessen GJ, Holm IA, Rudd PM, Wuhrer M, Ackerman ME, Nigrovic PA (2020) IgG Fc glycosylation as an axis of humoral immunity in childhood. J Allergy Clin Immunol 145:710–713.e9. https://doi.org/10.1016/j.jaci.2019.10.012

Ciccimarra F, Rosen FS, Schneeberger E, Merler E (1976) Failure of heavy chain glycosylation of IgG in some patients with common, variable agammaglobulinemia. J Clin Invest 57:1386–1390. https://doi.org/10.1172/JCI108407

Clerc F, Reiding KR, Jansen BC, Kammeijer GSM, Bondt A, Wuhrer M (2016) Human plasma protein N-glycosylation. Glycoconj J 33:309–343. https://doi.org/10.1007/s10719-015-9626-2

Coelho V, Krysov S, Ghaemmaghami AM, Emara M, Potter KN, Johnson P, Packham G, Martinez-Pomares L, Stevenson FK (2010) Glycosylation of surface Ig creates a functional bridge between human follicular lymphoma and microenvironmental lectins. Proc Natl Acad Sci USA 107:18587–18592. https://doi.org/10.1073/pnas.1009388107

Collins ES, Galligan MC, Saldova R, Adamczyk B, Abrahams JL, Campbell MP, Ng CT, Veale DJ, Murphy TB, Rudd PM, FitzGerald O (2013) Glycosylation status of serum in inflammatory arthritis in response to anti-TNF treatment. Rheumatol (United Kingdom) 52:1572–1582. https://doi.org/10.1093/rheumatology/ket189

Coloma MJ, Trinh RK, Martinez AR, Morrison SL (1999) Position effects of variable region carbohydrate on the affinity and in vivo behavior of an anti-(1-->6) dextran antibody. J Immunol 162:2162–2170. http://www.ncbi.nlm.nih.gov/pubmed/9973491

Coman DJ, Murray DW, Byrne JC, Rudd PM, Bagaglia PM, Doran PD, Treacy EP (2010) Galactosemia, a single gene disorder with epigenetic consequences. Pediatr Res 67:286–292. https://doi.org/10.1203/PDR.0b013e3181cbd542

Comunale MA, Lowman M, Long RE, Krakover J, Philip R, Seeholzer S, Evans AA, Hann HWL, Block TM, Mehta AS (2006) Proteomic analysis of serum associated fucosylated glycoproteins in the development of primary hepatocellular carcinoma. J Proteome Res 5:308–315. https://doi.org/10.1021/pr050328x

Coss KP, Byrne JC, Coman DJ, Adamczyk B, Abrahams JL, Saldova R, Brown AY, Walsh O, Hendroff U, Carolan C, Rudd PM, Treacy EP (2012) IgG N-glycans as potential biomarkers for determining galactose tolerance in classical galactosaemia. Mol Genet Metab 105:212–220. https://doi.org/10.1016/j.ymgme.2011.10.018

Coss KP, Hawkes CP, Adamczyk B, Stöckmann H, Crushell E, Saldova R, Knerr I, Rubio-Gozalbo ME, Monavari AA, Rudd PM, Treacy EP (2014) N-glycan abnormalities in children with galactosemia. J Proteome Res 13:385–394. https://doi.org/10.1021/pr4008305

Courtois F, Agrawal NJ, Lauer TM, Trout BL (2016) Rational design of therapeutic mAbs against aggregation through protein engineering and incorporation of glycosylation motifs applied to bevacizumab. MAbs 8:99–112. https://doi.org/10.1080/19420862.2015.1112477

Cremata JA, Sorell L, Montesino R, García R, Mata M, Cabrera G, Galvan JA, García G, Valdés R, Garrote JA (2003) Hypogalactosylation of serum IgG in patients with coeliac disease. Clin Exp Immunol 133:422–429. https://doi.org/10.1046/j.1365-2249.2003.02220.x

Croce A, Firuzi O, Altieri F, Eufemi M, Agostino R, Priori R, Bombardieri M, Alessandri C, Valesini G, Saso L (2007) Effect of infliximab on the glycosylation of IgG of patients with rheumatoid arthritis. J Clin Lab Anal 21:303–314. https://doi.org/10.1002/jcla.20191

Culver EL, van de Bovenkamp FS, Derksen NIL, Koers J, Cargill T, Barnes E, de Neef LA, Koeleman CAM, Aalberse RC, Wuhrer M, Rispens T (2019) Unique patterns of glycosylation in immunoglobulin subclass G4-related disease and primary sclerosing cholangitis. J Gastroenterol Hepatol 34:1878–1886. https://doi.org/10.1111/jgh.14512

Cvetko A, Kifer D, Gornik O, Klarić L, Visser E, Lauc G, Wilson JF, Štambuk T (2020) Glycosylation alterations in multiple sclerosis show increased proinflammatory potential. Biomedicine 8:410. https://doi.org/10.3390/biomedicines8100410

Cymer F, Beck H, Rohde A, Reusch D (2018) Therapeutic monoclonal antibody N-glycosylation – structure, function and therapeutic potential. Biologicals 52:1–11. https://doi.org/10.1016/j.biologicals.2017.11.001

Davies J, Jiang L, Pan LZ, Labarre MJ, Anderson D, Reff M (2001) Expression of GnTIII in a recombinant anti-CD20 CHO production cell line: expression of antibodies with altered glycoforms leads to an increase in ADCC through higher affinity for FcγRIII. Biotechnol Bioeng 74:288–294. https://doi.org/10.1002/bit.1119

de Haan N, van Tol MJD, Driessen GJ, Wuhrer M, Lankester AC (2018a) Immunoglobulin G fragment crystallizable glycosylation after hematopoietic stem cell transplantation is dissimilar to donor profiles. Front Immunol 9:1. https://doi.org/10.3389/fimmu.2018.01238

de Haan N, Boeddha NP, Ekinci E, Reiding KR, Emonts M, Hazelzet JA, Wuhrer M, Driessen GJ (2018b) Differences in IgG Fc glycosylation are associated with outcome of pediatric meningococcal sepsis. MBio 9. https://doi.org/10.1128/mBio.00546-18

De Jong SE, Selman MHJ, Adegnika AA, Amoah AS, Van Riet E, Kruize YCM, Raynes JG, Rodriguez A, Boakye D, Von Mutius E, Knulst AC, Genuneit J, Cooper PJ, Hokke CH, Wuhrer M, Yazdanbakhsh M (2016) IgG1 Fc N-glycan galactosylation as a biomarker for immune activation. Sci Rep 6:28207. https://doi.org/10.1038/srep28207

Deisenhofer J, Colman PM, Epp O, Huber R (1976) Kristallographische Studien an einem Fc-fragment, II. Ein vollständiges Modell nach einer Fourier-Synthese bei 3.5Å Auflösung, Hoppe. Seylers Z Physiol Chem 357:1421–1434. https://doi.org/10.1515/bchm2.1976.357.2.1421

Dekkers G, Treffers L, Plomp R, Bentlage AEH, de Boer M, Koeleman CAM, Lissenberg-Thunnissen SN, Visser R, Brouwer M, Mok JY, Matlung H, van den Berg TK, van Esch WJE, Kuijpers TW, Wouters D, Rispens T, Wuhrer M, Vidarsson G (2017) Decoding the human immunoglobulin G-glycan repertoire reveals a spectrum of Fc-receptor- and complement-mediated-effector activities. Front Immunol 8:877. https://doi.org/10.3389/fimmu.2017.00877

Dubé R, Rook GAW, Steele J, Brealey R, Dwek R, Rademacher T, Lennard-Jones J (1990) Agalactosyl IgG in inflammatory bowel disease: correlation with C-reactive protein. Gut 31:431–434. https://doi.org/10.1136/gut.31.4.431

Dunn-Walters D, Boursier L, Spencer J (2000) Effect of somatic hypermutation on potential N-glycosylation sites in human immunoglobulin heavy chain variable regions. Mol Immunol 37:107–113. https://doi.org/10.1016/S0161-5890(00)00038-9

Engdahl C, Bondt A, Harre U, Raufer J, Pfeifle R, Camponeschi A, Wuhrer M, Seeling M, Mårtensson IL, Nimmerjahn F, Krönke G, Scherer HU, Forsblad-d'Elia H, Schett G (2018) Estrogen induces St6gall expression and increases IgG sialylation in mice and patients with rheumatoid arthritis: a potential explanation for the increased risk of rheumatoid arthritis in postmenopausal women. Arthritis Res Ther 20. https://doi.org/10.1186/s13075-018-1586-z

Ercan A, Cui J, Chatterton DEW, Deane KD, Hazen MM, Brintnell W, O'Donnell CI, Derber LA, Weinblatt ME, Shadick NA, Bell DA, Cairns E, Solomon DH, Holers VM, Rudd PM, Lee DM (2010) Aberrant IgG galactosylation precedes disease onset, correlates with disease activity, and is prevalent in autoantibodies in rheumatoid arthritis. Arthritis Rheum 62:2239–2248. https://doi.org/10.1002/art.27533

Ercan A, Barnes MG, Hazen M, Tory H, Henderson L, Dedeoglu F, Fuhlbrigge RC, Grom A, Holm IA, Kellogg M, Kim S, Adamczyk B, Rudd PM, Son MB, Sundel RP, Foell D, Glass DN, Thompson SD, Nigrovic PA (2012) Multiple juvenile idiopathic arthritis subtypes demonstrate proinflammatory IgG glycosylation. Arthritis Rheum 64:3025–3033. https://doi.org/10.1002/art.34507

Ercan A, Kohrt WM, Cui J, Deane KD, Pezer M, Yu EW, Hausmann JS, Campbell H, Kaiser UB, Rudd PM, Lauc G, Wilson JF, Finkelstein JS, Nigrovic PA (2017) Estrogens regulate glycosylation of IgG in women and men. JCI Insight 2:e89703. https://doi.org/10.1172/jci.insight.89703

Espy C, Morelle W, Kavian N, Grange P, Goulvestre C, Viallon V, Chéreau C, Pagnoux C, Michalski JC, Guillevin L, Weill B, Batteux F, Guilpain P (2011) Sialylation levels of antiproteinase 3 antibodies are associated with the activity of granulomatosis with polyangiitis (Wegener's). Arthritis Rheum 63:2105–2115. https://doi.org/10.1002/art.30362

Fan Y, Jimenez Del Val I, Müller C, Wagtberg Sen J, Rasmussen SK, Kontoravdi C, Weilguny D, Andersen MR (2015) Amino acid and glucose metabolism in fed-batch CHO cell culture affects antibody production and glycosylation. Biotechnol Bioeng 112:521–535. https://doi.org/10.1002/bit.25450

Ferrara C, Brünker P, Suter T, Moser S, Püntener U, Umaña P (2006) Modulation of therapeutic antibody effector functions by glycosylation engineering: influence of golgi enzyme localization domain and co-expression of heterologous β1, 4-N-acetylglucosaminyltransferase III and Golgi α-mannosidase II. Biotechnol Bioeng 93:851–861. https://doi.org/10.1002/bit.20777

Fickentscher C, Magorivska I, Janko C, Biermann M, Bilyy R, Nalli C, Tincani A, Medeghini V, Meini A, Nimmerjahn F, Schett G, Muñoz LE, Andreoli L, Herrmann M (2015) The pathogenicity of anti-β 2GP1-IgG autoantibodies depends on Fc glycosylation. J Immunol Res 2015:638129. https://doi.org/10.1155/2015/638129

Fiebiger BM, Maamary J, Pincetic A, Ravetch JV (2015) Protection in antibody- and T cell-mediated autoimmune diseases by antiinflammatory IgG Fcs requires type II FcRs. Proc Natl Acad Sci U S A 112:E2385–E2394. https://doi.org/10.1073/pnas.1505292112

Filley E, Andreoli A, Steele J, Waters M, Wagner D, Nelson D, Tung K, Rademacher T, Dwek R, Rook GAW (1989) A transient rise in agalactosyl IgG correlating with free interleukin 2 receptors, during episodes of erythema nodosum leprosum. Clin Exp Immunol 76:343–347. http://www.ncbi.nlm.nih.gov/pubmed/2787714

Fleming SC, Smith S, Knowles D, Skillen A, Self CH (1998) Increased sialylation of oligosaccharides on IgG paraproteins – a potential new tumour marker in multiple myeloma. J Clin Pathol 51:825–830. https://doi.org/10.1136/jcp.51.11.825

Flögel M, Lauc G, Gornik I, Maček B (1998) Fucosylation and galactosylation of IgG heavy chains differ between acute and remission phases of juvenile chronic arthritis. Clin Chem Lab Med 36:99–102. https://doi.org/10.1515/CCLM.1998.018

Fokkink WJR, Selman MHJ, Dortland JR, Durmuş B, Kuitwaard K, Huizinga R, Van Rijs W, Tio-Gillen AP, Van Doorn PA, Deelder AM, Wuhrer M, Jacobs BC (2014a) IgG Fc N-glycosylation in Guillain-Barré syndrome treated with immunoglobulins. J Proteome Res 13:1722–1730. https://doi.org/10.1021/pr401213z

Fokkink WJR, Selman MHC, Wuhrer M, Jacobs BC (2014b) Immunoglobulin G Fc N-glycosylation in Guillain-Barré syndrome treated with intravenous immunoglobulin. Clin Exp Immunol 178:105–107. https://doi.org/10.1111/cei.12530

Franceschi C, Garagnani P, Parini P, Giuliani C, Santoro A (2018) Inflammaging: a new immune–metabolic viewpoint for age-related diseases. Nat Rev Endocrinol 14:576–590. https://doi.org/10.1038/s41574-018-0059-4

Freidin MB, Keser T, Gudelj I, Stambuk J, Vucenovic D, Allegri M, Pavic T, Simurina M, Fabiane SM, Lauc G, Williams FMK (2016) The association between low Back pain and composition of IgG glycome. Sci Rep 6:26815. https://doi.org/10.1038/srep26815

Fujimura Y, Tachibana H, Eto N, Yamada K (2000) Antigen binding of an ovomucoid-specific antibody is affected by a carbohydrate chain located on the light chain variable region. Biosci Biotechnol Biochem 64:2298–2305. https://doi.org/10.1271/bbb.64.2298

Galeotti C, Kaveri SV, Bayry J (2017) IVIG-mediated effector functions in autoimmune and inflammatory diseases. Int Immunol 29:491–498. https://doi.org/10.1093/intimm/dxx039

Gao Q, Dolikun M, Štambuk J, Wang H, Zhao F, Yiliham N, Wang Y, Trbojević-Akmačić I, Zhang J, Fang H, Sun Y, Peng H, Zhao Z, Liu D, Liu J, Li Q, Sun Q, Wu L, Lauc G, Wang W, Song M (2017) Immunoglobulin G N-glycans as potential postgenomic biomarkers for hypertension in the Kazakh population. Omi A J Integr Biol 21:380–389. https://doi.org/10.1089/omi.2017.0044

Garber K (2018) No added sugar: antibody makers find an upside to 'no fucose'. Nat Biotechnol 36:1025–1027. https://doi.org/10.1038/nbt1118-1025

Gardinassi LG, Dotz V, Ederveen AH, De Almeida RP, Costa CHN, Costa DL, De Jesus AR, Mayboroda OA, Garcia GR, Wuhrer M, de Miranda Santos IKF (2014) Clinical severity of visceral leishmaniasis is associated with changes in immunoglobulin G Fc N-glycosylation. MBio 5:e01844. https://doi.org/10.1128/mBio.01844-14

Gerçel-Taylor Ç, Bazzett LB, Taylor DD (2001) Presence of aberrant tumor-reactive immunoglobulins in the circulation of patients with ovarian cancer. Gynecol Oncol 81:71–76. https://doi.org/10.1006/gyno.2000.6102

Gindzienska-Sieskiewicz E, Klimiuk PA, Kisiel DG, Gindzienski A, Sierakowski S (2007) The changes in monosaccharide composition of immunoglobulin G in the course of rheumatoid arthritis. Clin Rheumatol 26:685–690. https://doi.org/10.1007/s10067-006-0370-7

Gińdzieńska-Sieśkiewicz E, Radziejewska I, Domysławska I, Klimiuk PA, Sulik A, Rojewska J, Gabryel-Porowska H, Sierakowski S (2016) Changes of glycosylation of IgG in rheumatoid arthritis patients treated with methotrexate. Adv Med Sci 61:193–197. https://doi.org/10.1016/j.advms.2015.12.009

Go MF, Schrohenloher RE, Tomana M (1994) Deficient galactosylation of serum IgG in inflammatory bowel disease: correlation with disease activity. J Clin Gastroenterol 18:86–87. https://doi.org/10.1097/00004836-199401000-00021

Gornik I, Maravić G, Dumić J, Flögel M, Lauc G (1999) Fucosylation of IgG heavy chains is increased in rheumatoid arthritis. Clin Biochem 32:605–608. https://doi.org/10.1016/S0009-9120(99)00060-0

Gornik O, Pavić T, Lauc G (2012) Alternative glycosylation modulates function of IgG and other proteins - implications on evolution and disease. Biochim Biophys Acta - Gen Subj 1820:1318–1326. https://doi.org/10.1016/j.bbagen.2011.12.004

Greto V, Cvetko A, Štambuk T, Dempster N, Kifer D, Deriš H, Cindrić A, Vučković F, Falchi M, Gillies R, Tomlinson J, Gornik O, Sgromo B, Spector T, Menni C, Geremia A, Arancibia-Cárcamo C, Lauc G (2020) Extensive weight loss can reduce immune age by altering IgG N-glycosylation. MedRxiv. 2020.04.24.20077867. https://doi.org/10.1101/2020.04.24.20077867

Gudelj I, Lauc G, Pezer M (2018a) Immunoglobulin G glycosylation in aging and diseases. Cell Immunol 333:65–79. https://doi.org/10.1016/j.cellimm.2018.07.009

Gudelj I, Salo PP, Trbojević-Akmačić I, Albers M, Primorac D, Perola M, Lauc G (2018b) Low galactosylation of IgG associates with higher risk for future diagnosis of rheumatoid arthritis during 10 years of follow-up. Biochim Biophys Acta Mol basis Dis 1864:2034–2039. https://doi.org/10.1016/j.bbadis.2018.03.018

Guhr T, Bloem J, Derksen NIL, Wuhrer M, Koenderman AHL, Aalberse RC, Rispens T (2011) Enrichment of sialylated IgG by lectin fractionation does not enhance the efficacy of immunoglobulin G in a murine model of immune thrombocytopenia. PLoS One 6:e21246. https://doi.org/10.1371/journal.pone.0021246

Gutiérrez G, Malan Borel I, Margni RA (2001) The placental regulatory factor involved in the asymmetric IgG antibody synthesis responds to IL-6 features. J Reprod Immunol 49:21–32. https://doi.org/10.1016/S0165-0378(00)00074-7

Gutierrez G, Gentile T, Miranda S, Margni RA (2006) Asymmetric antibodies: a protective arm in pregnancy. Chem Immunol Allergy 89:158–168. https://doi.org/10.1159/000087964

Haddad G, Lorenzen JM, Ma H, de Haan N, Seeger H, Zaghrini C, Brandt S, Kölling M, Wegmann U, Kiss B, Pál G, Gál P, Wüthrich RP, Wuhrer M, Beck LH, Salant DJ, Lambeau G, Kistler AD (2021) Altered glycosylation of IgG4 promotes lectin complement pathway activation in anti-PLA2R1-associated membranous nephropathy. J Clin Invest 131. https://doi.org/10.1172/JCI140453

Hafkenscheid L, Bondt A, Scherer HU, Huizinga TWJ, Wuhrer M, Toes REM, Rombouts Y (2017) Structural analysis of variable domain glycosylation of anti-citrullinated protein antibodies in rheumatoid arthritis reveals the presence of highly sialylated glycans. Mol Cell Proteomics 16:278–287. https://doi.org/10.1074/mcp.M116.062919

Hafkenscheid L, de Moel E, Smolik I, Tanner S, Meng X, Jansen BC, Bondt A, Wuhrer M, Huizinga TWJ, Toes REM, El-Gabalawy H, Scherer HU (2019) N-linked glycans in the variable domain of IgG anti–citrullinated protein antibodies predict the development of rheumatoid arthritis. Arthritis Rheumatol 71:1626–1633. https://doi.org/10.1002/art.40920

Hernandez-Pando R, Reyes P, Espitia C, Wang Y, Rook G, Mancilla R (1994) Raised agalactosyl IgG and antimycobacterial humoral immunity in Takayasu's arteritis. J Rheumatol 21:1870–1876. http://www.ncbi.nlm.nih.gov/pubmed/7837153

Hess C, Winkler A, Lorenz AK, Holecska V, Blanchard V, Eiglmeier S, Schoen AL, Bitterling J, Stoehr AD, Petzold D, Schommartz T, Mertes MMM, Schoen CT, Tiburzy B, Herrmann A, Köhl J, Manz RA, Madaio MP, Berger M, Wardemann H, Ehlers M (2013) T cell-independent B cell activation induces immunosuppressive sialylated IgG antibodies. J Clin Invest 123:3788–3796. https://doi.org/10.1172/JCI65938

Higel F, Seidl A, Sörgel F, Friess W (2016) N-glycosylation heterogeneity and the influence on structure, function and pharmacokinetics of monoclonal antibodies and Fc fusion proteins. Eur J Pharm Biopharm 100:94–100. https://doi.org/10.1016/j.ejpb.2016.01.005

Ho CH, Chien RN, Cheng PN, Liu JH, Liu CK, Su CS, Wu IC, Li IC, Tsai HW, Wu SL, Liu WC, Chen SH, Chang TT (2015) Aberrant serum immunoglobulin G glycosylation in chronic hepatitis B is associated with histological liver damage and reversible by antiviral therapy. J Infect Dis 211:115–124. https://doi.org/10.1093/infdis/jiu388

Hodoniczky J, Yuan ZZ, James DC (2005) Control of recombinant monoclonal antibody effector functions by Fc N-glycan remodeling in vitro. Biotechnol Prog 21:1644–1652. https://doi.org/10.1021/bp050228w

Hoepel W, Chen H-J, Allahverdiyeva S, Manz X, Aman J, Bonta P, Brouwer P, de Taeye S, Caniels T, van der Straten K, Golebski K, Griffith G, Jonkers R, Larsen M, Linty F, Neele A, Nouta J, van Baarle F, van Drunen C, Vlaar A, de Bree G, Sanders R, Willemsen L, Wuhrer M, Bogaard HJ, van Gils M, Vidarsson G, de Winther M, den Dunnen J (2020) Anti-SARS-CoV-2 IgG from severely ill COVID-19 patients promotes macrophage hyper-inflammatory responses. BioRxiv. 2020.07.13.190140. https://doi.org/10.1101/2020.07.13.190140

Holland M, Takada K, Okumoto T, Takahashi N, Kato K, Adu D, Ben-Smith A, Harper L, Savage COS, Jefferis R (2002) Hypogalactosylation of serum IgG in patients with ANCA-associated systemic vasculitis. Clin Exp Immunol 129:183–190. https://doi.org/10.1046/j.1365-2249.2002.01864.x

Holland M, Yagi H, Takahashi N, Kato K, Savage COS, Goodall DM, Jefferis R (2006) Differential glycosylation of polyclonal IgG, IgG-Fc and IgG-Fab isolated from the sera of patients with ANCA-associated systemic vasculitis. Biochim Biophys Acta - Gen Subj 1760:669–677. https://doi.org/10.1016/j.bbagen.2005.11.021

Hou H, Xu X, Sun F, Zhang X, Dong H, Wang L, Ge S, An K, Sun Q, Li Y, Cao W, Song M, Hu S, Xing W, Wang W, Li D, Wang Y (2019) Hyperuricemia is associated with immunoglobulin G N-glycosylation: a community-based study of glycan biomarkers. Omi A J Integr Biol 23:660–667. https://doi.org/10.1089/omi.2019.0004

Houde D, Peng Y, Berkowitz SA, Engen JR (2010) Post-translational modifications differentially affect IgG1 conformation and receptor binding. Mol Cell Proteomics 9:1716–1728. https://doi.org/10.1074/mcp.M900540-MCP200

Huffman JE, Pučić-Baković M, Klarić L, Hennig R, Selman MHJ, Vučković F, Novokmet M, Krištić J, Borowiak M, Muth T, Polašek O, Razdorov G, Gornik O, Plomp R, Theodoratou E, Wright AF, Rudan I, Hayward C, Campbell H, Deelder AM, Reichl U, Aulchenko YS, Rapp E, Wuhrer M, Lauc G (2014) Comparative performance of four methods for high-throughput glycosylation analysis of immunoglobulin G in genetic and epidemiological research. Mol Cell Proteomics 13:1598–1610. https://doi.org/10.1074/mcp.M113.037465

Huhn C, Selman MHJ, Ruhaak LR, Deelder AM, Wuhrer M (2009) IgG glycosylation analysis. Proteomics 9:882–913. https://doi.org/10.1002/pmic.200800715

Hymes AJ, Mullinax GL, Mullinax F (1979) Immunoglobulin carbohydrate requirement for formation of an IgG-IgG complex. J Biol Chem 254:3148–3151. https://doi.org/10.1016/S0021-9258(18)50733-X

Ji H, Ohmura K, Mahmood U, Lee DM, Hofhuis FMA, Boackle SA, Takahashi K, Holers VM, Walport M, Gerard C, Ezekowitz A, Carroll MC, Brenner M, Weissleder R, Verbeek JS, Duchatelle V, Degott C, Benoist C, Mathis D (2002) Arthritis critically dependent on innate immune system players. Immunity 16:157–168. https://doi.org/10.1016/S1074-7613(02)00275-3

Kaneko Y, Nimmerjahn F, Ravetch JV (2006) Anti-inflammatory activity of immunoglobulin G resulting from Fc sialylation. Science (80-) 313:670–673. https://doi.org/10.1126/science. 1129594

Kanoh Y, Mashiko T, Danbara M, Takayama Y, Ohtani S, Egawa S, Baba S, Akahoshi T (2004a) Changes in serum IgG oligosaccharide chains with prostate cancer progression. Anticancer Res 24:3135–3139. http://www.ncbi.nlm.nih.gov/pubmed/15510601

Kanoh Y, Mashiko T, Danbara M, Takayama Y, Ohtani S, Imasaki T, Abe T, Akahoshi T (2004b) Analysis of the oligosaccharide chain of human serum immunoglobulin G in patients with localized or metastatic cancer. Oncology 66:365–370. https://doi.org/10.1159/000079484

Kanoh Y, Ohara T, Mashiko T, Abe T, Masuda N, Akahoshi T (2006) Relationship between N-linked oligosaccharide chains of human serum immunoglobulin G and serum tumor markers with non-small cell lung cancer progression. Anticancer Res 26:4293–4297

Kanoh Y, Ohara T, Tadano T, Kanoh M, Akahoshi T (2008) Changes to N-linked oligosaccharide chains of human serum immunoglobulin G and matrix metalloproteinase-2 with cancer progression. Anticancer Res 28:715–720. http://www.ncbi.nlm.nih.gov/pubmed/18507012

Kanoh Y, Egawa S, Baba S, Akahoshi T (2009) Associations of IgG N-linked oligosaccharide chains and proteases in sera of prostate cancer patients with and without α2-macroglobulin deficiency. J Clin Lab Anal 23:125–131. https://doi.org/10.1002/jcla.20302

Kapur R, Della Valle L, Sonneveld M, Hipgrave Ederveen A, Visser R, Ligthart P, de Haas M, Wuhrer M, van der Schoot CE, Vidarsson G (2014a) Low anti-RhD IgG-Fc-fucosylation in pregnancy: a new variable predicting severity in haemolytic disease of the fetus and newborn. Br J Haematol 166:936–945. https://doi.org/10.1111/bjh.12965

Kapur R, Kustiawan I, Vestrheim A, Koeleman CAM, Visser R, Einarsdottir HK, Porcelijn L, Jackson D, Kumpel B, Deelder AM, Blank D, Skogen B, Killie MK, Michaelsen TE, De Haas M, Rispens T, Van Der Schoot CE, Wuhrer M, Vidarsson G (2014b) A prominent lack of IgG1-fc fucosylation of platelet alloantibodies in pregnancy. Blood 123:471–480. https://doi.org/10.1182/blood-2013-09-527978

Karsten CM, Pandey MK, Figge J, Kilchenstein R, Taylor PR, Rosas M, McDonald JU, Orr SJ, Berger M, Petzold D, Blanchard V, Winkler A, Hess C, Reid DM, Majoul IV, Strait RT, Harris NL, Köhl G, Wex E, Ludwig R, Zillikens D, Nimmerjahn F, Finkelman FD, Brown GD, Ehlers M, Köhl J (2012) Anti-inflammatory activity of IgG1 mediated by Fc galactosylation and association of FcγRIIB and dectin-1. Nat Med 18:1401–1406. https://doi.org/10.1038/nm. 2862

Käsermann F, Boerema DJ, Rüegsegger M, Hofmann A, Wymann S, Zuercher AW, Miescher S (2012) Analysis and functional consequences of increased fab-sialylation of intravenous immunoglobulin (IVIG) after lectin fractionation. PLoS One 7:37243. https://doi.org/10.1371/journal. pone.0037243

Kawaguchi-Sakita N, Kaneshiro-Nakagawa K, Kawashima M, Sugimoto M, Tokiwa M, Suzuki E, Kajihara S, Fujita Y, Iwamoto S, Tanaka K, Toi M (2016) Serum immunoglobulin G fc region N-glycosylation profiling by matrix-assisted laser desorption/ionization mass spectrometry can distinguish breast cancer patients from cancer-free controls. Biochem Biophys Res Commun 469:1140–1145. https://doi.org/10.1016/j.bbrc.2015.12.114

Kazuno S, Furukawa JI, Shinohara Y, Murayama K, Fujime M, Ueno T, Fujimura T (2016) Glycosylation status of serum immunoglobulin G in patients with prostate diseases. Cancer Med 5:1137–1146. https://doi.org/10.1002/cam4.662

Kemna MJ, Plomp R, van Paassen P, Koeleman CAM, Jansen BC, Damoiseaux JGMC, Cohen Tervaert JW, Wuhrer M (2017) Galactosylation and sialylation levels of IgG predict relapse in patients with PR3-ANCA associated vasculitis. EBioMedicine 17:108–118. https://doi.org/10. 1016/j.ebiom.2017.01.033

Khurana S, Raghunathan V, Salunke DM (1997) The variable domain glycosylation in a monoclonal antibody specific to GnRH modulates antigen binding. Biochem Biophys Res Commun 234:465–469. https://doi.org/10.1006/bbrc.1997.5929

Kissel T, Van Schie KA, Hafkenscheid L, Lundquist A, Kokkonen H, Wuhrer M, Huizinga TWJ, Scherer HU, Toes R, Rantapää-Dahlqvist S (2019) On the presence of HLA-SE alleles and ACPA-IgG variable domain glycosylation in the phase preceding the development of rheumatoid arthritis. Ann Rheum Dis 78:1616–1620. https://doi.org/10.1136/annrheumdis-2019-215698

Klarić L, Tsepilov YA, Stanton CM, Mangino M, Sikka TT, Esko T, Pakhomov E, Salo P, Deelen J, McGurnaghan SJ, Keser T, Vučković F, Ugrina I, Krištić J, Gudelj I, Štambuk J, Plomp R, Pučić-Baković M, Pavić T, Vilaj M, Trbojević-Akmačić I, Drake C, Dobrinić P, Mlinarec J, Jelušić B, Richmond A, Timofeeva M, Grishchenko AK, Dmitrieva J, Bermingham ML, Sharapov SZ, Farrington SM, Theodoratou E, Uh HW, Beekman M, Slagboom EP, Louis E, Georges M, Wuhrer M, Colhoun HM, Dunlop MG, Perola M, Fischer K, Polasek O, Campbell H, Rudan I, Wilson JF, Zoldoš V, Vitart V, Spector T, Aulchenko YS, Lauc G, Hayward C (2020) Glycosylation of immunoglobulin G is regulated by a large network of genes pleiotropic with inflammatory diseases. Sci Adv 6:eaax0301. https://doi.org/10.1126/sciadv.aax0301

Klasić M, Markulin D, Vojta A, Samaržija I, Biruš I, Dobrinić P, Ventham NT, Trbojević-Akmačić I, Šimurina M, Štambuk J, Razdorov G, Kennedy NA, Satsangi J, Dias AM, Pinho S, Annese V, Latiano A, D'Inca R, Lauc G, Zoldoš V (2018) Promoter methylation of the MGAT3 and BACH2 genes correlates with the composition of the immunoglobulin G glycome in inflammatory bowel disease. Clin Epigenetics 10:75. https://doi.org/10.1186/s13148-018-0507-y

Knerr I, Coss KP, Kratzsch J, Crushell E, Clark A, Doran P, Shin Y, Stöckmann H, Rudd PM, Treacy E (2015) Effects of temporary low-dose galactose supplements in children aged 5-12 y with classical galactosemia: a pilot study. Pediatr Res 78:272–279. https://doi.org/10.1038/pr.2015.107

Kodar K, Stadlmann J, Klaamas K, Sergeyev B, Kurtenkov O (2012) Immunoglobulin G Fc N-glycan profiling in patients with gastric cancer by LC-ESI-MS: relation to tumor progression and survival. Glycoconj J 29:57–66. https://doi.org/10.1007/s10719-011-9364-z

Koide N, Nose M, Muramatsu T (1977) Recognition of IgG by Fc receptor and complement: effects of glycosidase digestion. Biochem Biophys Res Commun 75:838–844. https://doi.org/10.1016/0006-291X(77)91458-9

Krištić J, Vučković F, Menni C, Klarić L, Keser T, Beceheli I, Pučić-Baković M, Novokmet M, Mangino M, Thaqi K, Rudan P, Novokmet N, Šarac J, Missoni S, Kolčić I, Polašek O, Rudan I, Campbell H, Hayward C, Aulchenko Y, Valdes A, Wilson JF, Gornik O, Primorac D, Zoldoš V, Spector T, Lauc G (2014) Glycans are a novel biomarker of chronological and biological ages. Journals Gerontol - Ser A Biol Sci Med Sci 69:779–789. https://doi.org/10.1093/gerona/glt190

Kumpel BM, Rademacher TW, Rook GAW, Williams PJ, Wilson IBH (1994) Galactosylation of human IgG monoclonal anti-D produced by EBV-transformed B-lympboblastoid cell lines is dependent on culture method and affects Fc receptor-mediated functional activity. Hum Antibodies Hybridomas 5:143–151

Kumpel BM, Wang Y, Griffiths HL, Hadley AG, Rook GAW (1995) The biological activity of human monoclonal IgG anti-D is reduced by β-galactosidase treatment. Hum Antibodies 6:82–88. https://doi.org/10.3233/HAB-1995-6301

Larsen MD, de Graaf EL, Sonneveld ME, Plomp HR, Nouta J, Hoepel W, Chen H-J, Linty F, Visser R, Brinkhaus M, Šuštić T, de Taeye SW, Bentlage AEH, Toivonen S, Koeleman CAM, Sainio S, Kootstra NA, Brouwer PJM, Geyer CE, Derksen NIL, Wolbink G, de Winther M, Sanders RW, van Gils MJ, de Bruin S, Vlaar APJ, Rispens T, den Dunnen J, Zaaijer HL, Wuhrer M, van der Schoot CE, Vidarsson G (2020) Afucosylated IgG characterizes enveloped viral responses and correlates with COVID-19 severity. Science 80. https://doi.org/10.1126/science.abc8378

Lauc G (2016) Precision medicine that transcends genomics: glycans as integrators of genes and environment. Biochim Biophys Acta - Gen Subj 1860:1571–1573. https://doi.org/10.1016/j.bbagen.2016.05.001

Lauc G, Huffman JE, Pučić M, Zgaga L, Adamczyk B, Mužinić A, Novokmet M, Polašek O, Gornik O, Krištić J, Keser T, Vitart V, Scheijen B, Uh HW, Molokhia M, Patrick AL, McKeigue P, Kolčić I, Lukić IK, Swann O, van Leeuwen FN, Ruhaak LR, Houwing-Duistermaat JJ, Slagboom PE, Beekman M, de Craen AJM, Deelder AM, Zeng Q, Wang W, Hastie ND, Gyllensten U, Wilson JF, Wuhrer M, Wright AF, Rudd PM, Hayward C, Aulchenko Y, Campbell H, Rudan I (2013) Loci associated with N-glycosylation of human immunoglobulin G show pleiotropy with autoimmune diseases and haematological cancers. PLoS Genet 9:e1003225. https://doi.org/10.1371/journal.pgen.1003225

Leibiger H, Wüstner D, Stigler RD, Marx U (1999) Variable domain-linked oligosaccharides of a human monoclonal IgG: structure and influence on antigen binding. Biochem J 338:529–538. https://doi.org/10.1042/0264-6021:3380529

Leirisalo-Repo M, Hernandez-Munoz HE, Rook GAW (1999) Agalactosyl IgG is elevated in patients with active spondyloarthropathy. Rheumatol Int 18:171–176. https://doi.org/10.1007/s002960050080

Lemmers RFH, Vilaj M, Urda D, Agakov F, Šimurina M, Klaric L, Rudan I, Campbell H, Hayward C, Wilson JF, Lieverse AG, Gornik O, Sijbrands EJG, Lauc G, van Hoek M (2017) IgG glycan patterns are associated with type 2 diabetes in independent European populations. Biochim Biophys Acta - Gen Subj 1861:2240–2249. https://doi.org/10.1016/j.bbagen.2017.06.020

Leontyev D, Katsman Y, Ma XZ, Miescher S, Käsermann F, Branch DR (2012) Sialylation-independent mechanism involved in the amelioration of murine immune thrombocytopenia using intravenous gammaglobulin. Transfusion 52:1799–1805. https://doi.org/10.1111/j.1537-2995.2011.03517.x

Li T, DiLillo DJ, Bournazos S, Giddens JP, Ravetch JV, Wang LX (2017) Modulating IgG effector function by Fc glycan engineering. Proc Natl Acad Sci USA 114:3485–3490. https://doi.org/10.1073/pnas.1702173114

Li X, Wang H, Russell A, Cao W, Wang X, Ge S, Zheng Y, Guo Z, Hou H, Song M, Yu X, Wang Y, Hunter M, Roberts P, Lauc G, Wang W (2019) Type 2 diabetes mellitus is associated with the immunoglobulin G N-glycome through putative proinflammatory mechanisms in an Australian population. Omi A J Integr Biol 23:631–639. https://doi.org/10.1089/omi.2019.0075

Lifely MR, Hale C, Boyce S, Keen MJ, Phillips J (1995) Glycosylation and biological activity of CAMPATH-1H expressed in different cell lines and grown under different culture conditions. Glycobiology 5:813–822. https://doi.org/10.1093/glycob/5.8.813

Liu L (2015) Antibody glycosylation and its impact on the pharmacokinetics and pharmacodynamics of monoclonal antibodies and Fc-fusion proteins. J Pharm Sci 104:1866–1884. https://doi.org/10.1002/jps.24444

Liu L (2018) Pharmacokinetics of monoclonal antibodies and Fc-fusion proteins. Protein Cell 9:15–32. https://doi.org/10.1007/s13238-017-0408-4

Liu B, Spearman M, Doering J, Lattová E, Perreault H, Butler M (2014) The availability of glucose to CHO cells affects the intracellular lipid-linked oligosaccharide distribution, site occupancy and the N-glycosylation profile of a monoclonal antibody. J Biotechnol 170:17–27. https://doi.org/10.1016/j.jbiotec.2013.11.007

Liu D, Zhao Z, Wang A, Ge S, Wang H, Zhang X, Sun Q, Cao W, Sun M, Wu L, Song M, Zhou Y, Wang W, Wang Y (2018) Ischemic stroke is associated with the pro-inflammatory potential of N-glycosylated immunoglobulin G. J Neuroinflammation 15. https://doi.org/10.1186/s12974-018-1161-1

Lu LL, Chung AW, Rosebrock TR, Ghebremichael M, Yu WH, Grace PS, Schoen MK, Tafesse F, Martin C, Leung V, Mahan AE, Sips M, Kumar MP, Tedesco J, Robinson H, Tkachenko E, Draghi M, Freedberg KJ, Streeck H, Suscovich TJ, Lauffenburger DA, Restrepo BI, Day C, Fortune SM, Alter G (2016) A functional role for antibodies in tuberculosis. Cell 167:433–443. e14. https://doi.org/10.1016/j.cell.2016.08.072

Lu LL, Das J, Grace PS, Fortune SM, Restrepo BI, Alter G (2020) Antibody Fc glycosylation discriminates between latent and active tuberculosis. J Infect Dis 222:2093–2102. https://doi.org/10.1093/infdis/jiz643

Lund J, Takahashi N, Pound JD, Goodall M, Jefferis R (1996) Multiple interactions of IgG with its core oligosaccharide can modulate recognition by complement and human Fc gamma receptor I and influence the synthesis of its oligosaccharide chains. J Immunol 157:4963–4969. http://www.ncbi.nlm.nih.gov/pubmed/8943402

Lundström SL, Yang H, Lyutvinskiy Y, Rutishauser D, Herukka SK, Soininen H, Zubarev RA (2014) Blood plasma IgG Fc glycans are significantly altered in Alzheimer's disease and progressive mild cognitive impairment. J Alzheimers Dis 38:567–579. https://doi.org/10.3233/JAD-131088

Lundström SL, Hensvold AH, Rutishauser D, Klareskog L, Ytterberg AJ, Zubarev RA, Catrina AI (2017) IgG fc galactosylation predicts response to methotrexate in early rheumatoid arthritis. Arthritis Res Ther 19:182. https://doi.org/10.1186/s13075-017-1389-7

Magorivska I, Muñoz LE, Janko C, Dumych T, Rech J, Schett G, Nimmerjahn F, Bilyy R, Herrmann M (2016) Sialylation of anti-histone immunoglobulin G autoantibodies determines their capabilities to participate in the clearance of late apoptotic cells. Clin Exp Immunol 184:110–117. https://doi.org/10.1111/cei.12744

Mahan AE, Jennewein MF, Suscovich T, Dionne K, Tedesco J, Chung AW, Streeck H, Pau M, Schuitemaker H, Francis D, Fast P, Laufer D, Walker BD, Baden L, Barouch DH, Alter G (2016) Antigen-specific antibody glycosylation is regulated via vaccination. PLoS Pathog 12: e1005456. https://doi.org/10.1371/journal.ppat.1005456

Malhotra R, Wormald MR, Rudd PM, Fischer PB, Dwek RA, Sim RB (1995) Glycosylation changes of IgG associated with rheumatooid arthritis can activate complement via the mannose-binding protein. Nat Med 1:237–243. https://doi.org/10.1038/nm0395-237

Man Sung Co, Scheinberg DA, Avdalovic NM, Mcgraw K, Vasquez M, Caron PC, Queen C (1993) Genetically engineered deglycosylation of the variable domain increases the affinity of an anti-CD33 monoclonal antibody. Mol Immunol 30:1361–1367. https://doi.org/10.1016/0161-5890(93)90097-U

Maratha A, Stockmann H, Coss KP, Estela Rubio-Gozalbo M, Knerr I, Fitzgibbon M, McVeigh TP, Foley P, Moss C, Colhoun HO, Van Erven B, Stephens K, Doran P, Rudd P, Treacy E (2016) Classical galactosaemia: novel insights in IgG N-glycosylation and N-glycan biosynthesis. Eur J Hum Genet 24:976–984. https://doi.org/10.1038/ejhg.2015.254

Mariño K, Bones J, Kattla JJ, Rudd PM (2010) A systematic approach to protein glycosylation analysis: a path through the maze. Nat Chem Biol 6:713–723. https://doi.org/10.1038/nchembio.437

Martin TC, Šimurina M, Zabczynska M, Kavur M, Rydlewska M, Pezer M, Kozłowska K, Burri A, Vilaj M, Turek-Jabrocka R, Krnjajić-Tadijanović M, Trofimiuk-Müldner M, Ugrina I, Lityn A, Hubalewska-Dydejczyk A, Trbojevic-Akmacic I, Lim EM, Walsh JP, Pocheć E, Spector TD, Wilson SG, Lauc G (2020) Decreased immunoglobulin G core fucosylation, a player in antibody-dependent cell-mediated cytotoxicity, is associated with autoimmune thyroid diseases. Mol Cell Proteomics 19:774–792. https://doi.org/10.1074/mcp.RA119.001860

Massoud AH, Yona M, Xue D, Chouiali F, Alturaihi H, Ablona A, Mourad W, Piccirillo CA, Mazer BD (2014) Dendritic cell immunoreceptor: a novel receptor for intravenous immunoglobulin mediates induction of regulatory T cells. J Allergy Clin Immunol 133:853–863. e5. https://doi.org/10.1016/j.jaci.2013.09.029

Matsumoto A, Shikata K, Takeuchi F, Kojima N, Mizuochi T (2000) Autoantibody activity of IgG rheumatoid factor increases with decreasing levels of galactosylation and sialylation. J Biochem 128:621–628. https://doi.org/10.1093/oxfordjournals.jbchem.a022794

McCann KJ, Ottensmeier CH, Callard A, Radcliffe CM, Harvey DJ, Dwek RA, Rudd PM, Sutton BJ, Hobby P, Stevenson FK (2008) Remarkable selective glycosylation of the immunoglobulin variable region in follicular lymphoma. Mol Immunol 45:1567–1572. https://doi.org/10.1016/j.molimm.2007.10.009

Mehta AS, Long RE, Comunale MA, Wang M, Rodemich L, Krakover J, Philip R, Marrero JA, Dwek RA, Block TM (2008) Increased levels of galactose-deficient anti-gal immunoglobulin G in the sera of hepatitis C virus-infected individuals with fibrosis and cirrhosis. J Virol 82:1259–1270. https://doi.org/10.1128/jvi.01600-07

Meng Z, Li C, Ding G, Cao W, Xu X, Heng Y, Deng Y, Li Y, Zhang X, Li D, Wang W, Wang Y, Xing W, Hou H (2020) Glycomics: immunoglobulin GN-glycosylation associated with mammary gland hyperplasia in women. Omi A J Integr Biol 24:551–558. https://doi.org/10.1089/omi.2020.0091

Menni C, Keser T, Mangino M, Bell JT, Erte I, Akmačić I, Vučković F, Baković MP, Gornik O, McCarthy MI, Zoldoš V, Spector TD, Lauc G, Valdes AM (2013) Glycosylation of immunoglobulin G: role of genetic and epigenetic influences. PLoS One 8:e82558. https://doi.org/10.1371/journal.pone.0082558

Menni C, Gudelj I, MacDonald-Dunlop E, Mangino M, Zierer J, Bešić E, Joshi PK, Trbojević-Akmačić I, Chowienczyk PJ, Spector TD, Wilson JF, Lauc G, Valdes AM (2018) Glycosylation profile of immunoglobulin G is cross-sectionally associated with cardiovascular disease risk score and subclinical atherosclerosis in two independent cohorts. Circ Res 122:1555–1564. https://doi.org/10.1161/CIRCRESAHA.117.312174

Miranda S, Canellada A, Gentile T, Margni R (2005) Interleukin-6 and dexamethasone modulate in vitro asymmetric antibody synthesis and UDP-Glc glycoprotein glycosyltransferase activity. J Reprod Immunol 66:141–150. https://doi.org/10.1016/j.jri.2005.04.001

Mittermayr S, Lê GN, Clarke C, Millán Martín S, Larkin AM, O'Gorman P, Bones J (2017) Polyclonal immunoglobulin G N-glycosylation in the pathogenesis of plasma cell disorders. J Proteome Res 16:748–762. https://doi.org/10.1021/acs.jproteome.6b00768

Miyoshi E, Shinzaki S, Fujii H, Iijima H, Kamada Y, Takehara T (2016) Role of aberrant IgG glycosylation in the pathogenesis of inflammatory bowel disease. Proteomics – Clin Appl 10:384–390. https://doi.org/10.1002/prca.201500089

Moore JS, Wu X, Kulhavy R, Tomana M, Novak J, Moldoveanu Z, Brown R, Goepfert PA, Mestecky J (2005) Increased levels of galactose-deficient IgG in sera of HIV-1-infected individuals. AIDS 19:381–389. https://doi.org/10.1097/01.aids.0000161767.21405.68

Muenchhoff M, Chung AW, Roider J, Dugast A-S, Richardson S, Kløverpris H, Leslie A, Ndung'u T, Moore P, Alter G, Goulder PJR (2020) Distinct immunoglobulin Fc glycosylation patterns are associated with disease nonprogression and broadly neutralizing antibody responses in children with HIV infection. MSphere 5. https://doi.org/10.1128/msphere.00880-20

Nakajima S, Iijima H, Shinzaki S, Egawa S, Inoue T, Mukai A, Hayashi Y, Kondo J, Akasaka T, Nishida T, Kanto T, Morii T, Mizushima T, Miyoshi E, Tsujii M, Hayashi N (2011) Functional analysis of agalactosyl IgG in inflammatory bowel disease patients. Inflamm Bowel Dis 17:927–936. https://doi.org/10.1002/ibd.21459

Nakao H, Nishikawa A, Nishiura T, Kanayama Y, Tarui S, Taniguchi N (1991) Hypogalactosylation of immunoglobulin G sugar chains and elevated serum interleukin 6 in Castleman's disease. Clin Chim Acta 197:221–228. https://doi.org/10.1016/0009-8981(91)90142-Y

Nishiura T, Kanayama Y, Tomiyama Y, Iida M, Karasuno T, Nakao H, Yonezawa T, Tarui S, Fujii S, Nishikawa A, Taniguchi N (1990) Carbohydrate analysis of immunoglobulin G myeloma proteins by lectin and high performance liquid chromatography: role of glycosyltransferases in the structures. Cancer Res 50:5345–5350. http://www.ncbi.nlm.nih.gov/pubmed/2386941

Novak J, Tomana M, Shah GR, Brown R, Mestecky J (2005) Heterogeneity of IgG glycosylation in adult periodontal disease. J Dent Res 84:897–901. https://doi.org/10.1177/154405910508401005

Novokmet M, Lukić E, Vučković F, Durić Ž, Keser T, Rajšl K, Remondini D, Castellani G, Gašparović H, Gornik O, Lauc G (2014) Changes in IgG and total plasma protein glycomes in acute systemic inflammation. Sci Rep 4. https://doi.org/10.1038/srep04347

Oefner CM, Winkler A, Hess C, Lorenz AK, Holecska V, Huxdorf M, Schommartz T, Petzold D, Bitterling J, Schoen AL, Stoehr AD, Vu Van D, Darcan-Nikolaisen Y, Blanchard V, Schmudde I, Laumonnier Y, Ströver HA, Hegazy AN, Eiglmeier S, Schoen CT, Mertes MMM, Loddenkemper C, Löhning M, König P, Petersen A, Luger EO, Collin M, Köhl J, Hutloff A, Hamelmann E, Berger M, Wardemann H, Ehlers M (2012) Tolerance induction with T cell-dependent protein antigens induces regulatory sialylated IgGs. J Allergy Clin Immunol 129:1647–1655.e13. https://doi.org/10.1016/j.jaci.2012.02.037

Ogata S, Shimizu C, Franco A, Touma R, Kanegaye JT, Choudhury BP, Naidu NN, Kanda Y, Hoang LT, Hibberd ML, Tremoulet AH, Varki A, Burns JC (2013) Treatment response in Kawasaki disease is associated with sialylation levels of endogenous but not therapeutic intravenous immunoglobulin G. PLoS One 8:e81448. https://doi.org/10.1371/journal.pone.0081448

Ohtsubo K, Marth JD (2006) Glycosylation in cellular mechanisms of health and disease. Cell 126:855–867. https://doi.org/10.1016/j.cell.2006.08.019

Parekh RB, Dwek RA, Sutton BJ, Fernandes DL, Leung A, Stanworth D, Rademacher TW, Mizuochi T, Taniguchi T, Matsuta K, Takeuchi F, Nagano Y, Miyamoto T, Kobata A (1985) Association of rheumatoid arthritis and primary osteoarthritis with changes in the glycosylation pattern of total serum IgG. Nature 316:452–457. https://doi.org/10.1038/316452a0

Parekh RB, Isenberg DA, Ansell BM, Roitt IM, Dwek RA, Rademacher TW (1988) Galactosylation of IgG associated oligosaccharides: reduction in patients with adult and juvenile onset rheumatoid arthritis and relation to disease activity. Lancet 331:966–969. https://doi.org/10.1016/S0140-6736(88)91781-3

Parekh R, Isenberg D, Rook G, Roitt I, Dwek R, Rademacher T (1989) A comparative analysis of disease-associated changes in the galactosylation of serum IgG. J Autoimmun 2:101–114. https://doi.org/10.1016/0896-8411(89)90148-0

Pasek M, Duk M, Podbielska M, Sokolik R, Szechiński J, Lisowska E, Krotkiewski H (2006) Galactosylation of IgG from rheumatoid arthritis (RA) patients – changes during therapy. Glycoconj J 23:463–471. https://doi.org/10.1007/s10719-006-5409-0

Pavić T, Dilber D, Kifer D, Selak N, Keser T, Ljubičić CDS, Vukić Dugac A, Lauc G, Rumora L, Gornik O (2018) N-glycosylation patterns of plasma proteins and immunoglobulin G in chronic obstructive pulmonary disease 11 medical and health sciences 1102 cardiorespiratory medicine and haematology. J Transl Med 16:323. https://doi.org/10.1186/s12967-018-1695-0

Pekelharing JM, Hepp E, Kamerling JP, Gerwig GJ, Leijnse B (1988) Alterations in carbohydrate composition of serum IgG from patients with rheumatoid arthritis and from pregnant women. Ann Rheum Dis 47:91–95. https://doi.org/10.1136/ard.47.2.91

Peng J, Vongpatanasin W, Sacharidou A, Kifer D, Yuhanna IS, Banerjee S, Tanigaki K, Polasek O, Chu H, Sundgren NC, Rohatgi A, Chambliss KL, Lauc G, Mineo C, Shaul PW (2019) Supplementation with the sialic acid precursor N-acetyl-D-Mannosamine breaks the link between obesity and hypertension. Circulation 140:2005–2018. https://doi.org/10.1161/CIRCULATIONAHA.119.043490

Perdivara I, Peddada SD, Miller FW, Tomer KB, Deterding LJ (2011) Mass spectrometric determination of IgG subclass-specific glycosylation profiles in siblings discordant for myositis syndromes. J Proteome Res 10:2969–2978. https://doi.org/10.1021/pr200397h

Peschke B, Keller CW, Weber P, Quast I, Lünemann JD (2017) Fc-galactosylation of human immunoglobulin gamma isotypes improves C1q binding and enhances complement-dependent cytotoxicity. Front Immunol 8:646. https://doi.org/10.3389/fimmu.2017.00646

Petrović T, Alves I, Bugada D, Pascual J, Vučković F, Skelin A, Gaifem J, Villar-Garcia J, Vicente MM, Fernandes Â, Dias AM, Kurolt I-C, Markotić A, Primorac D, Soares A, Malheiro L, Trbojević-Akmačić I, Abreu M, Castro RSE, Bettinelli S, Callegaro A, Arosio M, Sangiorgio L, Lorini LF, Castells X, Horcajada JP, Pinho SS, Allegri M, Barrios C, Lauc G (2020) Composition of the immunoglobulin G glycome associates with the severity of COVID-19. Glycobiology. https://doi.org/10.1093/glycob/cwaa102

Pezer M, Stambuk J, Perica M, Razdorov G, Banic I, Vuckovic F, Gospic AM, Ugrina I, Vecenaj A, Bakovic MP, Lokas SB, Zivkovic J, Plavec D, Devereux G, Turkalj M, Lauc G (2016) Effects of allergic diseases and age on the composition of serum IgG glycome in children. Sci Rep 6:33198. https://doi.org/10.1038/srep33198

Pfeifle R, Rothe T, Ipseiz N, Scherer HU, Culemann S, Harre U, Ackermann JA, Seefried M, Kleyer A, Uderhardt S, Haugg B, Hueber AJ, Daum P, Heidkamp GF, Ge C, Böhm S, Lux A, Schuh W, Magorivska I, Nandakumar KS, Lönnblom E, Becker C, Dudziak D, Wuhrer M, Rombouts Y, Koeleman CA, Toes R, Winkler TH, Holmdahl R, Herrmann M, Blüml S, Nimmerjahn F, Schett G, Krönke G (2017) Regulation of autoantibody activity by the IL-23-T H 17 axis determines the onset of autoimmune disease. Nat Immunol 18:104–113. https://doi.org/10.1038/ni.3579

Pilkington C, Yeung E, Isenberg D, Lefvert AK, Rook GAW (1995) Agalactosyl IgG and antibody specificity in rheumatoid arthritis, tuberculosis, systemic lupus erythematosus and myasthenia gravis. Autoimmunity 22:107–111. https://doi.org/10.3109/08916939508995306

Pilkington C, Taylor PV, Silverman E, Isenberg DA, Costello AMDL, Rook GAW (1996a) Agalactosyl IgG and materno-fetal transmission of autoimmune neonatal lupus. Rheumatol Int 16:89–94. https://doi.org/10.1007/BF01409979

Pilkington C, Basaran M, Barlan I, Costello AMDL, Rook GAW (1996b) Raised levels of agalactosyl IgG in childhood tuberculosis. Trans R Soc Trop Med Hyg 90:167–168. https://doi.org/10.1016/S0035-9203(96)90124-8

Pincetic A, Bournazos S, Dilillo DJ, Maamary J, Wang TT, Dahan R, Fiebiger BM, Ravetch JV (2014) Type i and type II Fc receptors regulate innate and adaptive immunity. Nat Immunol 15:707–716. https://doi.org/10.1038/ni.2939

Pučić M, Knežević A, Vidič J, Adamczyk B, Novokmet M, Polašek O, Gornik O, Šupraha-Goreta S, Wormald MR, Redžic I, Campbell H, Wright A, Hastie ND, Wilson JF, Rudan I, Wuhrer M, Rudd PM, Josić D, Lauc G (2011) High throughput isolation and glycosylation analysis of IgG-variability and heritability of the IgG glycome in three isolated human populations. Mol Cell Proteomics 10:M111.010090–M111.010090. https://doi.org/10.1074/mcp.M111.010090

Qian Y, Wang Y, Zhang X, Zhou L, Zhang Z, Xu J, Ruan Y, Ren S, Xu C, Gu J (2013) Quantitative analysis of serum IgG galactosylation assists differential diagnosis of ovarian cancer. J Proteome Res 12:4046–4055. https://doi.org/10.1021/pr4003992

Radcliffe CM, Arnold JN, Suter DM, Wormald MR, Harvey DJ, Royle L, Mimura Y, Kimura Y, Sim RB, Inogès S, Rodriguez-Calvillo M, Zabalegui N, De Cerio ALD, Potter KN, Mockridge CI, Dwek RA, Bendandi M, Rudd PM, Stevenson FK (2007) Human follicular lymphoma cells contain oligomannose glycans in the antigen-binding site of the B-cell receptor. J Biol Chem 282:7405–7415. https://doi.org/10.1074/jbc.M602690200

Rademacher TW, Parekh RB, Dwek RA, Isenberg D, Rook G, Axford JS, Roitt I (1988) The role of IgG glycoforms in the pathogenesis of rheumatoid arthritis. Springer Semin Immunopathol 10:231–249. https://doi.org/10.1007/BF01857227

Reusch D, Tejada ML (2015) Fc glycans of therapeutic antibodies as critical quality attributes. Glycobiology 25:1325–1334. https://doi.org/10.1093/glycob/cwv065

Rombouts Y, Ewing E, Van De Stadt LA, Selman MHJ, Trouw LA, Deelder AM, Huizinga TWJ, Wuhrer M, Van Schaardenburg D, Toes REM, Scherer HU (2015) Anti-citrullinated protein antibodies acquire a pro-inflammatory Fc glycosylation phenotype prior to the onset of rheumatoid arthritis. Ann Rheum Dis 74:234–241. https://doi.org/10.1136/annrheumdis-2013-203565

Rombouts Y, Willemze A, Van Beers JJBC, Shi J, Kerkman PF, Van Toorn L, Janssen GMC, Zaldumbide A, Hoeben RC, Pruijn GJM, Deelder AM, Wolbink G, Rispens T, Van Veelen PA, Huizinga TWJ, Wuhrer M, Trouw LA, Scherer HU, Toes REM (2016) Extensive glycosylation of ACPA-IgG variable domains modulates binding to citrullinated antigens in rheumatoid arthritis. Ann Rheum Dis 75:578–585. https://doi.org/10.1136/annrheumdis-2014-206598

Rook GAW, Onyebujoh P, Wilkins E, Ly HM, Al Attiyah R, Bahr G, Corrah T, Hernandez H, Stanford JL (1994) A longitudinal study of per cent agalactosyl IgG in tuberculosis patients receiving chemotherapy, with or without immunotherapy. Immunology 81:149–154

Williams RC, Osterland CK, Margherita S, Tokuda S, Messner RP (1973) Studies of biologic and serologic activities of rabbit IgG antibody depleted of carbohydrate residues. J Immunol 111:1690–1698

Yamaguchi Y, Barb AW (2020) A synopsis of recent developments defining how N-glycosylation impacts immunoglobulin G structure and function. Glycobiology 30:214–225. https://doi.org/10.1093/glycob/cwz068

Shade K-T, Anthony R (2013) Antibody glycosylation and inflammation. Antibodies 2:392–414. https://doi.org/10.3390/antib2030392

van de Bovenkamp FS, Hafkenscheid L, Rispens T, Rombouts Y (2016) The emerging importance of IgG Fab glycosylation in immunity. J Immunol 196:1435–1441. https://doi.org/10.4049/jimmunol.1502136

van de Bovenkamp FS, Derksen NIL, van Breemen MJ, de Taeye SW, Ooijevaar-de Heer P, Sanders RW, Rispens T (2018a) Variable domain N-linked glycans acquired during antigen-specific immune responses can contribute to immunoglobulin G antibody stability. Front Immunol 9:740. https://doi.org/10.3389/fimmu.2018.00740

Wu SJ, Luo J, O'Neil KT, Kang J, Lacy ER, Canziani G, Baker A, Huang M, Tang QM, Raju TS, Jacobs SA, Teplyakov A, Gilliland GL, Feng Y (2010) Structure-based engineering of a monoclonal antibody for improved solubility. Protein Eng Des Sel 23:643–651. https://doi.org/10.1093/protein/gzq037

Wright A, Tao MH, Kabat EA, Morrison SL (1991) Antibody variable region glycosylation: position effects on antigen binding and carbohydrate structure. EMBO J 10:2717–2723. https://doi.org/10.1002/j.1460-2075.1991.tb07819.x

Wada R, Matsui M, Kawasaki N (2019) Influence of N-glycosylation on effector functions and thermal stability of glycoengineered IgG1 monoclonal antibody with homogeneous glycoforms. MAbs 11:350–372. https://doi.org/10.1080/19420862.2018.1551044

Vidarsson G, Dekkers G, Rispens T (2014) IgG subclasses and allotypes: from structure to effector functions. Front Immunol 5:520. https://doi.org/10.3389/fimmu.2014.00520

Štambuk J, Nakić N, Vučković F, Pučić-Baković M, Razdorov G, Trbojević-Akmačić I, Novokmet M, Keser T, Vilaj M, Štambuk T, Gudelj I, Šimurina M, Song M, Wang H, Salihović MP, Campbell H, Rudan I, Kolčić I, Eller LA, McKeigue P, Robb ML, Halfvarson J, Kurtoglu M, Annese V, Škarić-Jurić T, Molokhia M, Polašek O, Hayward C, Kibuuka H, Thaqi K, Primorac D, Gieger C, Nitayaphan S, Spector T, Wang Y, Tillin T, Chaturvedi N, Wilson JF, Schanfield M, Filipenko M, Wang W, Lauc G (2020) Global variability of the human IgG glycome. Aging (Albany NY) 12:1–13. https://doi.org/10.18632/AGING.103884

Shields RL, Lai J, Keck R, O'Connell LY, Hong K, Gloria Meng Y, Weikert SHA, Presta LG (2002) Lack of fucose on human IgG1 N-linked oligosaccharide improves binding to human FcγRIII and antibody-dependent cellular toxicity. J Biol Chem 277:26733–26740. https://doi.org/10.1074/jbc.M202069200

Shinkawa T, Nakamura K, Yamane N, Shoji-Hosaka E, Kanda Y, Sakurada M, Uchida K, Anazawa H, Satoh M, Yamasaki M, Hanai N, Shitara K (2003) The absence of fucose but not the presence of galactose or bisecting N-acetylglucosamine of human IgG1 complex-type oligosaccharides shows the critical role of enhancing antibody-dependent cellular cytotoxicity. J Biol Chem 278:3466–3473. https://doi.org/10.1074/jbc.M210665200

Schuster M, Umana P, Ferrara C, Brünker P, Gerdes C, Waxenecker G, Wiederkum S, Schwager C, Loibner H, Himmler G, Mudde GC (2005) Improved effector functions of a therapeutic monoclonal Lewis Y-specific antibody by glycoform engineering. Cancer Res 65:7934–7941. https://doi.org/10.1158/0008-5472.CAN-04-4212

Umaña P, Jean-Mairet J, Moudry R, Amstutz H, Bailey JE (1999) Engineered glycoforms of an antineuroblastoma IgG1 with optimized antibody-dependent cellular cytotoxic activity. Nat Biotechnol 17:176–180. https://doi.org/10.1038/6179

Subedi GP, Barb AW (2016) The immunoglobulin G1 N-glycan composition affects binding to each low affinity Fc γ receptor. MAbs 8:1512–1524. https://doi.org/10.1080/19420862.2016.1218586

Schwab I, Nimmerjahn F (2013) Intravenous immunoglobulin therapy: how does IgG modulate the immune system? Nat Rev Immunol 13:176–189. https://doi.org/10.1038/nri3401

Schwab I, Biburger M, Krönke G, Schett G, Nimmerjahn F (2012) IVIg-mediated amelioration of ITP in mice is dependent on sialic acid and SIGNR1. Eur J Immunol 42:826–830. https://doi.org/10.1002/eji.201142260

Washburn N, Schwabb I, Ortiz D, Bhatnagar N, Lansing JC, Medeiros A, Tyler S, Mekala D, Cochran E, Sarvaiya H, Garofalo K, Meccariello R, Meador JW, Rutitzky L, Schultes BC, Ling L, Avery W, Nimmerjahn F, Manning AM, Kaundinya GV, Bosques CJ (2015) Controlled tetra-fc sialylation of IVIg results in a drug candidate with consistent enhanced anti-inflammatory activity. Proc Natl Acad Sci USA 112:E1297–E1306. https://doi.org/10.1073/pnas.1422481112

Schwab I, Mihai S, Seeling M, Kasperkiewicz M, Ludwig RJ, Nimmerjahn F (2014) Broad requirement for terminal sialic acid residues and FcγRIIB for the preventive and therapeutic activity of intravenous immunoglobulins in vivo. Eur J Immunol 44:1444–1453. https://doi.org/10.1002/eji.201344230

Temming AR, Dekkers G, van de Bovenkamp FS, Plomp HR, Bentlage AEH, Szittner Z, Derksen NIL, Wuhrer M, Rispens T, Vidarsson G (2019) Human DC-SIGN and CD23 do not interact with human IgG. Sci Rep 9:1–10. https://doi.org/10.1038/s41598-019-46484-2

Sondermann P, Pincetic A, Maamary J, Lammens K, Ravetch JV (2013) General mechanism for modulating immunoglobulin effector function. Proc Natl Acad Sci USA 110:9868–9872. https://doi.org/10.1073/pnas.1307864110

Schneider D, Von Minden MD, Alkhatib A, Setz C, Van Bergen CAM, Benkißer-Petersen M, Wilhelm I, Villringer S, Krysov S, Packham G, Zirlik K, Römer W, Buske C, Stevenson FK, Veelken H, Jumaa H (2015) Lectins from opportunistic bacteria interact with acquired variable-region glycans of surface immunoglobulin in follicular lymphoma. Blood 125:3287–3296. https://doi.org/10.1182/blood-2014-11-609404

Wallick SC, Kabat EA, Morrison SL (1988) Glycosylation of a VH residue of a monoclonal antibody against α(1→6) dextran increases its affinity for antigen. J Exp Med 168:1099–1109. https://doi.org/10.1084/jem.168.3.1099

Tachibana H, Kim JY, Shirahata S (1997) Building high affinity human antibodies by altering the glycosylation on the light chain variable region in N-acetylglucosamine-supplemented hybridoma cultures. Cytotechnology 23:151–159. https://doi.org/10.1023/a:1007980032042

Van De Bovenkamp FS, Derksen NIL, Ooijevaar-de Heer P, Van Schie KA, Kruithof S, Berkowska MA, Ellen van der Schoot C, IJspeert H, Van Der Burg M, Gils A, Hafkenscheid L, Toes REM, Rombouts Y, Plomp R, Wuhrer M, Marieke van Ham S, Vidarsson G, Rispens T (2018b) Adaptive antibody diversification through N-linked glycosylation of the immunoglobulin variable region. Proc Natl Acad Sci U S A 115:1901–1906. https://doi.org/10.1073/pnas.1711720115

Wiedeman AE, Santer DM, Yan W, Miescher S, Käsermann F, Elkon KB (2013) Contrasting mechanisms of interferon-α inhibition by intravenous immunoglobulin after induction by immune complexes versus toll-like receptor agonists. Arthritis Rheum 65:2713–2723. https://doi.org/10.1002/art.38082

Séïté JF, Cornec D, Renaudineau Y, Youinou P, Mageed RA, Hillion S (2010) IVIg modulates BCR signaling through CD22 and promotes apoptosis in mature human B lymphocytes. Blood 116:1698–1704. https://doi.org/10.1182/blood-2009-12-261461

Séïté JF, Goutsmedt C, Youinou P, Pers JO, Hillion S (2014) Intravenous immunoglobulin induces a functional silencing program similar to anergy in human B cells. J Allergy Clin Immunol 133. https://doi.org/10.1016/j.jaci.2013.08.042

Yu X, Wang Y, Kristic J, Dong J, Chu X, Ge S, Wang H, Fang H, Gao Q, Liu D, Zhao Z, Peng H, Bakovic MP, Wu L, Song M, Rudan I, Campbell H, Lauc G, Wang W (2016) Profiling IgG

N-glycans as potential biomarker of chronological and biological ages: a community-based study in a Han Chinese population. Med (United States) 95:e4112. https://doi.org/10.1097/MD.0000000000004112

Zaitseva O, Krištić J, Jansen B, Stojković R, Pezer M, Morahan G, Lauc G (2018) Fc-linked N-glycosylation of murine IgG1 variants. In: 2nd GlycoCom 1st Hum. Glycome Project Meeting, pp 62–63

Yu X, Baruah K, Harvey DJ, Vasiljevic S, Alonzi DS, Song BD, Higgins MK, Bowden TA, Scanlan CN, Crispin M (2013) Engineering hydrophobic protein-carbohydrate interactions to fine-tune monoclonal antibodies. J Am Chem Soc 135:9723–9732. https://doi.org/10.1021/ja4014375

Wang J, Balog CIA, Stavenhagen K, Koeleman CAM, Scherer HU, Selman MHJ, Deelder AM, Huizinga TWJ, Toes REM, Wuhrer M (2011) Fc-glycosylation of IgG1 is modulated by B-cell stimuli. Mol Cell Proteomics 10:M110 004655. https://doi.org/10.1074/mcp.M110.004655

Tijardović M, Marijančević D, Bok D, Kifer D, Lauc G, Gornik O, Keser T (2019) Intense physical exercise induces an anti-inflammatory change in IgG N-glycosylation profile. Front Physiol 10:1522. https://doi.org/10.3389/fphys.2019.01522

Sarin HV, Gudelj I, Honkanen J, Ihalainen JK, Vuorela A, Lee JH, Jin Z, Terwilliger JD, Isola V, Ahtiainen JP, Häkkinen K, Jurić J, Lauc G, Kristiansson K, Hulmi JJ, Perola M (2019) Molecular pathways mediating immunosuppression in response to prolonged intensive physical training, low-energy availability, and intensive weight loss. Front Immunol 10:907. https://doi.org/10.3389/fimmu.2019.00907

Trbojević-Akmačić I, Vilaj M, Lauc G (2016) High-throughput analysis of immunoglobulin G glycosylation. Expert Rev Proteomics 13:523–534. https://doi.org/10.1080/14789450.2016.1174584

Trbojević-Akmačić I, Ugrina I, Lauc G (2017) Comparative analysis and validation of different steps in glycomics studies. Methods Enzymol 586:37–55. https://doi.org/10.1016/bs.mie.2016.09.027

Singh SS, Heijmans R, Meulen CKE, Lieverse AG, Gornik O, Sijbrands EJG, Lauc G, Van Hoek M (2020) Association of the IgG N-glycome with the course of kidney function in type 2 diabetes. BMJ Open Diabetes Res Care 8:e001026. https://doi.org/10.1136/bmjdrc-2019-001026

Šimurina M, de Haan N, Vučković F, Kennedy NA, Štambuk J, Falck D, Trbojević-Akmačić I, Clerc F, Razdorov G, Khon A, Latiano A, D'Incà R, Danese S, Targan S, Landers C, Dubinsky M, Campbell H, Zoldoš V, Permberton IK, Kolarich D, Fernandes DL, Theorodorou E, Merrick V, Spencer DI, Gardner RA, Doran R, Shubhakar A, Boyapati R, Rudan I, Lionetti P, Krištić J, Novokmet M, Pučić-Baković M, Gornik O, Andriulli A, Cantoro L, Sturniolo G, Fiorino G, Manetti N, Arnott ID, Noble CL, Lees CW, Shand AG, Ho GT, Dunlop MG, Murphy L, Gibson J, Evenden L, Wrobel N, Gilchrist T, Fawkes A, Kammeijer GSM, Vojta A, Samaržija I, Markulin D, Klasić M, Dobrinić P, Aulchenko Y, van den Heuve T, Jonkers D, Pierik M, McGovern DPB, Annese V, Wuhrer M, Lauc G (2018) Glycosylation of immunoglobulin G associates with clinical features of inflammatory bowel diseases. Gastroenterology 154:1320–1333.e10. https://doi.org/10.1053/j.gastro.2018.01.002

Theodoratou E, Thaçi K, Agakov F, Timofeeva MN, Stambuk J, Pueìc-Bakovic M, Vuèkovic F, Orchard P, Agakova A, Din FVN, Brown E, Rudd PM, Farrington SM, Dunlop MG, Campbell H, Lauc G (2016) Glycosylation of plasma IgG in colorectal cancer prognosis. Sci Rep 6:28098. https://doi.org/10.1038/srep28098

Wahl A, van den Akker E, Klaric L, Štambuk J, Benedetti E, Plomp R, Razdorov G, Trbojević-Akmačić I, Deelen J, van Heemst D, Eline Slagboom P, Vučković F, Grallert H, Krumsiek J, Strauch K, Peters A, Meitinger T, Hayward C, Wuhrer M, Beekman M, Lauc G, Gieger C (2018) Genome-wide association study on immunoglobulin G glycosylation patterns. Front Immunol 9:277. https://doi.org/10.3389/fimmu.2018.00277

Vilaj M, Gudelj I, Trbojević-Akmačić I, Lauc G, Pezer M (2019) IgG glycans as a biomarker of biological age. In: Moskalev A (ed) Biomarkers hum aging. Springer, New York, pp 81–99. https://doi.org/10.1007/978-3-030-24970-0_7

Westhrin M, Kovcic V, Zhang Z, Moen SH, Vikene Nedal TM, Bondt A, Holst S, Misund K, Buene G, Sundan A, Waage A, Slørdahl TS, Wuhrer M, Standal T (2020) Monoclonal immunoglobulins promote bone loss in multiple myeloma. Blood 136:2656–2666. https://doi.org/10.1182/BLOOD.2020006045

Zhu D, McCarthy H, Ottensmeier CH, Johnson P, Hamblin TJ, Stevenson FK (2002) Acquisition of potential N-glycosylation sites in the immunoglobulin variable region by somatic mutation is a distinctive feature of follicular lymphoma. Blood 99:2562–2568. https://doi.org/10.1182/blood.V99.7.2562

Zhu D, Ottensmeier CH, Du MQ, McCarthy H, Stevenson FK (2003) Incidence of potential glycosylation sites in immunoglobulin variable regions distinguishes between subsets of Burkitt's lymphoma and mucosa-associated lymphoid tissue lymphoma. Br J Haematol 120:217–222. https://doi.org/10.1046/j.1365-2141.2003.04064.x

Wang TT, Sewatanon J, Memoli MJ, Wrammert J, Bournazos S, Bhaumik SK, Pinsky BA, Chokephaibulkit K, Onlamoon N, Pattanapanyasat K, Taubenberger JK, Ahmed R, Ravetch JV (2017) IgG antibodies to dengue enhanced by FcγRIIIA binding determine disease severity. Science 355:395–398. https://doi.org/10.1126/science.aai8128

Thulin NK, Brewer RC, Sherwood R, Bournazos S, Edwards KG, Ramadoss NS, Taubenberger JK, Memoli M, Gentles AJ, Jagannathan P, Zhang S, Libraty DH, Wang TT (2020) Maternal anti-dengue igg fucosylation predicts susceptibility to dengue disease in infants. Cell Rep 31. https://doi.org/10.1101/565259. Accessed 24 Apr 2021

Sonneveld ME, Natunen S, Sainio S, Koeleman CAM, Holst S, Dekkers G, Koelewijn J, Partanen J, van der Schoot CE, Wuhrer M, Vidarsson G (2016) Glycosylation pattern of anti-platelet IgG is stable during pregnancy and predicts clinical outcome in alloimmune thrombocytopenia. Br J Haematol 174:310–320. https://doi.org/10.1111/bjh.14053

Sonneveld ME, Koelewijn J, de Haas M, Admiraal J, Plomp R, Koeleman CAM, Hipgrave Ederveen AL, Ligthart P, Wuhrer M, van der Schoot CE, Vidarsson G (2017a) Antigen specificity determines anti-red blood cell IgG-Fc alloantibody glycosylation and thereby severity of haemolytic disease of the fetus and newborn. Br J Haematol 176:651–660. https://doi.org/10.1111/bjh.14438

Sénard T, Flouri I, Vučković F, Papadaki G, Goutakoli P, Pučić Baković M, Pezer M, Bertsias G, Lauc G, Sidiropoulos P (2021) Baseline IgG-Fc N-glycosylation profile is associated with long-term outcome in a cohort of early inflammatory arthritis patients (Submitted)

Van Zeben D, Rook GAW, Hazes JMW, Zwinderman AH, Zhang Y, Ghelani S, Rademacher TW, Breedveld FC (1994) Early agalactosylation of IGG is associated with a more progressive disease course in patients with rheumatoid arthritis: results of a follow-up study. Rheumatology 33:36–43. https://doi.org/10.1093/rheumatology/33.1.36

Váradi C, Holló Z, Póliska S, Nagy L, Szekanecz Z, Váncsa A, Palatka K, Guttman A (2015) Combination of IgG N-glycomics and corresponding transcriptomics data to identify anti-TNF-α treatment responders in inflammatory diseases. Electrophoresis 36:1330–1335. https://doi.org/10.1002/elps.201400575

Scherer HU, Van Der Woude D, Ioan-Facsinay A, El Bannoudi H, Trouw LA, Wang J, Häupl T, Burmester GR, Deelder AM, Huizinga TWJ, Wuhrer M, Toes REM (2010) Glycan profiling of anti-citrullinated protein antibodies isolated from human serum and synovial fluid. Arthritis Rheum 62:1620–1629. https://doi.org/10.1002/art.27414

Vanderschaeghe D, Meuris L, Raes T, Grootaert H, Van Hecke A, Verhelst X, Van de Velde F, Lapauw B, Van Vlierberghe H, Callewaert N (2018) Endoglycosidase s enables a highly simplified clinical chemistry procedure for direct assessment of serum IgG u. Mol Cell Proteomics 17:2508–2517. https://doi.org/10.1074/mcp.TIR118.000740

Tomana M, Schrohenloher RE, Koopman WJ, Alarcän GS, Paul WA (1988) Abnormal glycosylation of serum IgG from patients with chronic inflammatory diseases. Arthritis Rheum 31:333–338. https://doi.org/10.1002/art.1780310304

Van de Geijn FE, Wuhrer M, Selman MHJ, Willemsen SP, de Man YA, Deelder AM, Hazes JMW, Dolhain RJEM (2009) Immunoglobulin G galactosylation and sialylation are associated with pregnancy-induced improvement of rheumatoid arthritis and the postpartum flare: results from a large prospective cohort study. Arthritis Res Ther 11. https://doi.org/10.1186/ar2892

Young A, Sumar N, Bodman K, Goyal S, Sinclair H, Roitt I, Isenberg D (1991) Agalactosyl IgG: An aid to differential diagnosis in early synovitis. Arthritis Rheum 34:1425–1429. https://doi.org/10.1002/art.1780341113

Sumar N, Isenberg DA, Bodman KB, Soltys A, Young A, Leak AM, Round J, Hay FC, Roitt IM (1991) Reduction in IgG galactose in juvenile and adult onset rheumatoid arthritis measured by a lectin binding method and its relation to rheumatoid factor. Ann Rheum Dis 50:607–610. https://doi.org/10.1136/ard.50.9.607

Tomana M, Schrohenloher RE, Reveille JD, Arnett FC, Koopman WJ (1992) Abnormal galactosylation of serum IgG in patients with systemic lupus erythematosus and members of families with high frequency of autoimmune diseases. Rheumatol Int 12:191–194. https://doi.org/10.1007/BF00302151

Vučković F, Krištić J, Gudelj I, Teruel M, Keser T, Pezer M, Pučić-Baković M, Štambuk J, Trbojević-Akmačić I, Barrios C, Pavïc T, Menni C, Wang Y, Zhou Y, Cui L, Song H, Zeng Q, Guo X, Pons-Estel BA, McKeigue P, Patrick AL, Gornik O, Spector TD, Harjaček M, Alarcon-Riquelme M, Molokhia M, Wang W, Lauc G (2015) Association of systemic lupus erythematosus with decreased immunosuppressive potential of the IgG glycome. Arthritis Rheumatol 67:2978–2989. https://doi.org/10.1002/art.39273

Shinzaki S, Iijima H, Nakagawa T, Egawa S, Nakajima S, Ishii S, Irie T, Kakiuchi Y, Nishida T, Yasumaru M, Kanto T, Tsujii M, Tsuji S, Mizushima T, Yoshihara H, Kondo A, Miyoshi E, Hayashi N (2008) IgG oligosaccharide alterations are a novel diagnostic marker for disease activity and the clinical course of inflammatory bowel disease. Am J Gastroenterol 103:1173–1181. https://doi.org/10.1111/j.1572-0241.2007.01699.x

Trbojevic Akmacic I, Ventham NT, Theodoratou E, Vučković F, Kennedy NA, Krištić J, Nimmo ER, Kalla R, Drummond H, Štambuk J, Dunlop MG, Novokmet M, Aulchenko Y, Gornik O, Campbell H, Pučić Baković M, Satsangi J, Lauc G (2015) Inflammatory bowel disease associates with proinflammatory potential of the immunoglobulin G glycome. Inflamm Bowel Dis 21:1237–1247. https://doi.org/10.1097/MIB.0000000000000372

Wuhrer M, Stavenhagen K, Koeleman CAM, Selman MHJ, Harper L, Jacobs BC, Savage COS, Jefferis R, Deelder AM, Morgan M (2015) Skewed Fc glycosylation profiles of anti-proteinase 3 immunoglobulin G1 autoantibodies from granulomatosis with polyangiitis patients show low levels of bisection, galactosylation, and sialylation. J Proteome Res 14:1657–1665. https://doi.org/10.1021/pr500780a

Selman MHJ, Niks EH, Titulaer MJ, Verschuuren JJGM, Wuhrer M, Deelder AM (2011) IgG Fc N-glycosylation changes in Lambert-Eaton myasthenic syndrome and myasthenia gravis. J Proteome Res 10:143–152. https://doi.org/10.1021/pr1004373

Sonneveld ME, De Haas M, Koeleman C, De Haan N, Zeerleder SS, Ligthart PC, Wuhrer M, Van Der Schoot CE, Vidarsson G (2017b) Patients with IgG1-anti-red blood cell autoantibodies show aberrant Fc-glycosylation. Sci Rep 7:8187. https://doi.org/10.1038/s41598-017-08654-y

Sonneveld ME, Koeleman CAM, Plomp HR, Wuhrer M, van der Schoot CE, Vidarsson G (2018) Fc-glycosylation in human IgG1 and IgG3 is similar for both total and anti-red-blood cell anti-k antibodies. Front Immunol 9. https://doi.org/10.3389/fimmu.2018.00129

Saldova R, Royle L, Radcliffe CM, Hamid UMA, Evans R, Arnold JN, Banks RE, Hutson R, Harvey DJ, Antrobus R, Petrescu SM, Dwek RA, Rudd PM (2007) Ovarian cancer is associated with changes in glycosylation in both acute-phase proteins and IgG. Glycobiology 17:1344–1356. https://doi.org/10.1093/glycob/cwm100

Ruhaak LR, Kim K, Stroble C, Taylor SL, Hong Q, Miyamoto S, Lebrilla CB, Leiserowitz G (2016) Protein-specific differential glycosylation of immunoglobulins in serum of ovarian cancer patients. J Proteome Res 15:1002–1010. https://doi.org/10.1021/acs.jproteome.5b01071

Vučković F, Theodoratou E, Thaçi K, Timofeeva M, Vojta A, Štambuk J, Pučić-Baković M, Rudd PM, Derek L, Servis D, Wennerström A, Farrington SM, Perola M, Aulchenko Y, Dunlop MG,

Campbell H, Lauc G (2016) IgG glycome in colorectal cancer. Clin Cancer Res 22:3078–3086. https://doi.org/10.1158/1078-0432.CCR-15-1867

Stockmann H, Coss KP, Rubio-Gozalbo ME, Knerr I, Fitzgibbon M, Maratha A, Wilson J, Rudd P, Treacy EP (2016) IgG N-glycosylation galactose incorporation ratios for the monitoring of classical galactosaemia. JIMD Rep 27:47–53. https://doi.org/10.1007/8904_2015_490

Wang Y, Klarić L, Yu X, Thaqi K, Dong J, Novokmet M, Wilson J, Polasek O, Liu Y, Krištić J, Ge S, Pučić-Baković M, Wu L, Zhou Y, Ugrina I, Song M, Zhang J, Guo X, Zeng Q, Rudan I, Campbell H, Aulchenko Y, Lauc G, Wang W (2016) The association between glycosylation of immunoglobulin G and hypertension. Med (United States) 95:e3379. https://doi.org/10.1097/MD.0000000000003379

Zhao ZY, Liu D, Cao WJ, Sun M, Song MS, Wang W, Wang YX (2018) Association between IgG N-glycans and nonalcoholic fatty liver disease in Han Chinese. Biomed Environ Sci 31:454–458. https://doi.org/10.3967/bes2018.059

Wu Z, Pan H, Liu D, Zhou D, Tao L, Zhang J, Wang X, Li X, Wang Y, Wang W, Guo X (2021) Variation of IgG N-linked glycosylation profile in diabetic retinopathy. J Diabetes. https://doi.org/10.1111/1753-0407.13160

Wuhrer M, Porcelijn L, Kapur R, Koeleman CAM, Deelder AM, De Haas M, Vidarsson G (2009) Regulated glycosylation patterns of IgG during alloimmune responses against human platelet antigens. J Proteome Res 8:450–456. https://doi.org/10.1021/pr800651j

van Erp EA, Lakerveld AJ, de Graaf E, Larsen MD, Schepp RM, Hipgrave Ederveen AL, Ahout IML, de Haan CAM, Wuhrer M, Luytjes W, Ferwerda G, Vidarsson G, van Kasteren PB (2020) Natural killer cell activation by respiratory syncytial virus-specific antibodies is decreased in infants with severe respiratory infections and correlates with Fc-glycosylation. Clin Transl Immunol 9. https://doi.org/10.1002/cti2.1112

Russell AC, Šimurina M, Garcia MT, Novokmet M, Wang Y, Rudan I, Campbell H, Lauc G, Thomas MG, Wang W (2017) The N-glycosylation of immunoglobulin G as a novel biomarker of Parkinson's disease. Glycobiology 27:501–510. https://doi.org/10.1093/glycob/cwx022

Vadrevu SK, Trbojevic-Akmacic I, Kossenkov AV, Colomb F, Giron LB, Anzurez A, Lynn K, Mounzer K, Landay AL, Kaplan RC, Papasavvas E, Montaner LJ, Lauc G, Abdel-Mohsen M (2018) Frontline science: plasma and immunoglobulin G galactosylation associate with HIV persistence during antiretroviral therapy. J Leukoc Biol 104:461–471. https://doi.org/10.1002/JLB.3HI1217-500R

Zhang X, Yuan H, Lyu J, Meng X, Tian Q, Li Y, Zhang J, Xu X, Su J, Hou H, Li D, Sun B, Wang W, Wang Y (2021) Association of dementia with immunoglobulin G N-glycans in a Chinese Han population. Npj Aging Mech Dis 7. https://doi.org/10.1038/s41514-021-00055-w

Ruhaak LR, Nguyen UT, Stroble C, Taylor SL, Taguchi A, Hanash SM, Lebrilla CB, Kim K, Miyamoto S (2013) Enrichment strategies in glycomics-based lung cancer biomarker development. Proteomics – Clin Appl 7:664–676. https://doi.org/10.1002/prca.201200131

Sjöwall C, Zapf J, Von Löhneysen S, Magorivska I, Biermann M, Janko C, Winkler S, Bilyy R, Schett G, Herrmann M, Muñoz LE (2015) Altered glycosylation of complexed native IgG molecules is associated with disease activity of systemic lupus erythematosus. Lupus 24:569–581. https://doi.org/10.1177/0961203314558861

Chapter 14
Immunoglobulin A Glycosylation and Its Role in Disease

Alyssa L. Hansen, Colin Reily, Jan Novak, and Matthew B. Renfrow

Contents

14.1	Introduction	435
14.2	IgA Subtypes and Overall Structure	436
14.3	IgA Glycosylation Sites	438
14.4	Molecular Forms of IgA and Their Distributions (Monomeric and Polymeric IgA, SIgA, J Chain, Secretory Component)	440
14.5	IgA1 O-Glycosylation Pathways	441
14.6	Assessing Heterogeneity: Analysis of Glycan-Attachment Sites and Range of Heterogeneity	444
14.7	Observed IgA N-Glycosylation Heterogeneity	448
	14.7.1 IgA1/IgA2 Asn340/327 N-Glycosylation	449
	14.7.2 IgA1/IgA2 Asn144/131 N-Glycosylation	450
	14.7.3 IgA2 N-Glycosylation	451
	14.7.4 Joining Chain (J Chain) N-Glycosylation	451
	14.7.5 Secretory Component N-Glycosylation	452
14.8	IgA Receptors and Role of Glycans	452
14.9	Observed IgA1 O-Glycosylation Heterogeneity	454
14.10	IgA-Related Diseases	457
	14.10.1 IgA Nephropathy	457
	14.10.2 IgA Vasculitis with Nephritis (Formerly Known as Henoch–Schönlein Purpura Nephritis)	458
	14.10.3 Crohn's Disease	459
	14.10.4 Sjögren's Syndrome	459
	14.10.5 Systemic Lupus Erythematosus	459
	14.10.6 Rheumatoid Arthritis	460

A. L. Hansen · M. B. Renfrow (✉)
Department of Biochemistry and Molecular Genetics, University of Alabama at Birmingham, Birmingham, AL, USA
e-mail: renfrow@uab.edu

C. Reily
Departments of Medicine and Microbiology, University of Alabama at Birmingham, Birmingham, AL, USA

J. Novak (✉)
Department of Microbiology, University of Alabama at Birmingham, Birmingham, AL, USA
e-mail: jannovak@uab.edu

© The Author(s), under exclusive license to Springer Nature Switzerland AG 2021
M. Pezer (ed.), *Antibody Glycosylation*, Experientia Supplementum 112,
https://doi.org/10.1007/978-3-030-76912-3_14

	14.10.7	IgA Myeloma	460
	14.10.8	Celiac Disease	460
	14.10.9	Complement and IgA	461
14.11	Infectious Diseases		461
	14.11.1	Bacterial IgA-Specific Proteases	461
	14.11.2	Bacterial Glycosidases and Metabolic Utilization of IgA Glycans	462
	14.11.3	IgA as a Potential Therapeutic Antibody	463
14.12	Therapeutic IgA Antibodies		464
14.13	Conclusions		465
References			466

Abstract Human IgA is comprised of two subclasses, IgA1 and IgA2. Monomeric IgA (mIgA), polymeric IgA (pIgA), and secretory IgA (SIgA) are the main molecular forms of IgA. The production of IgA rivals all other immunoglobulin isotypes. The large quantities of IgA reflect the fundamental roles it plays in immune defense, protecting vulnerable mucosal surfaces against invading pathogens. SIgA dominates mucosal surfaces, whereas IgA in circulation is predominately monomeric. All forms of IgA are glycosylated, and the glycans significantly influence its various roles, including antigen binding and the antibody effector functions, mediated by the Fab and Fc portions, respectively. In contrast to its protective role, the aberrant glycosylation of IgA1 has been implicated in the pathogenesis of autoimmune diseases, such as IgA nephropathy (IgAN) and IgA vasculitis with nephritis (IgAVN). Furthermore, detailed characterization of IgA glycosylation, including its diverse range of heterogeneity, is of emerging interest. We provide an overview of the glycosylation observed for each subclass and molecular form of IgA as well as the range of heterogeneity for each site of glycosylation. In many ways, the role of IgA glycosylation is in its early stages of being elucidated. This chapter provides an overview of the current knowledge and research directions.

Keywords IgA glycosylation · IgA receptors · IgA1 · IgA2 · N-glycosylation · O-glycosylation · Mass spectrometry

Abbreviations

C1GalT1	Core 1 β1,3-galactosyltransferase
dIgA	Dimeric IgA
EBV	Epstein–Barr virus
ECD	Electron capture dissociation
ETD	Electron transfer dissociation
FcRL4	Fc receptor-like 4
Gal	Galactose
GalNAc	N-acetylgalactosamine
GalNAc-T	GalNAc-transferase
Gd-IgA1	Galactose-deficient IgA1

HC	Heavy chain
HR	Hinge region
Ig	Immunoglobulin
IgAN	IgA nephropathy
IgAV	IgA vasculitis
IgAVN	IgA vasculitis with nephritis
LC	Light chain
MBL	Mannose-binding lectin
mIgA	Monomeric IgA
MS	Mass spectrometry
pIgA	Polymeric IgA
RA	Rheumatoid arthritis
SA	Sialic acid
SC	Secretory component
SIgA	Secretory IgA
SLE	Systemic lupus erythematosus
TG2	Transglutaminase 2
V	Variable

14.1 Introduction

Immunoglobulin A (IgA) is one of the five primary immunoglobulins and its production in humans is greater than that of all other immunoglobulins. Humans expend a considerable amount of energy to produce IgA, as ~70 mg of IgA per kilogram of body weight is synthesized daily (Mestecky et al. 1986; Conley and Delacroix 1987). IgA is secreted by IgA-producing cells in two main molecular forms: monomeric IgA (mIgA) and dimeric IgA (dIgA), the latter having a joining chain (J chain) to bind two mIgA molecules. Furthermore, secretory IgA (SIgA) is the main form of IgA found on mucosal surfaces. SIgA has an additional protein chain attached, a secretory component, which is derived from polymeric immunoglobulin (Ig) receptor during transcytosis of dIgA through mucosal epithelial cells. IgA is the second most abundant Ig in the circulation, predominantly as mIgA, (~2 mg/mL), after IgG (~12 mg/mL); however, IgA is the most abundant antibody in external secretions (tears, saliva, colostrum, milk, nasal fluid, gallbladder bile, and intestinal fluid) of mucosal surfaces, secreted locally as SIgA (Jackson et al. 2005). The abundance of IgA highlights the important roles it plays in immune defense processes. Specifically, mucosal surfaces (i.e., respiratory, gastrointestinal, and genitourinary tracts) are often exposed to invading pathogens and IgA acts to protect such vulnerable areas from infection. Equipped with unique structural attributes of its heavy chain and its ability to form monomeric, dimeric, and higher polymeric forms (i.e., three or more mIgA molecules associated with J chain), both circulatory IgA and secretory forms of IgA are capable of neutralizing and removing pathogens by activating innate and acquired immune functions—blocking pathogens by antigen-specific and nonspecific binding (Russell 2007; Renegar and Small 1994; Phalipon et al. 2002).

Serum IgA is produced predominantly by plasma cells in the bone marrow. It is primarily a monomeric protein with a quaternary structure consisting of two heavy chains (HC) and two light chains (LC) linked by disulfide bonds. In contrast, mucosal IgA, specifically dIgA linked together by the J chain, is produced close to the epithelium by plasma cells in the lamina propria of mucosal surfaces (Koshland 1985). This IgA becomes SIgA when, during transcytosis, the secretory component (SC) is attached. Although most IgA-producing cells are localized in the gastrointestinal system, SIgA is also one of the major antibodies in human tears, saliva, milk, and colostrum (Underdown and Mestecky 1994). Maternal milk is essential for the passive immunity of infants (Newburg et al. 2005; Hanson 1998; Peterson et al. 2013), protecting babies from a variety of infections that otherwise may lead to sepsis and meningitis (Schroten et al. 1998). The diversity of IgA post-translational modifications, due to variable glycosylation, contributes to the effector functions of IgA.

Regardless of the source of IgA, heavy chains of all molecular forms of IgA, as well as the J chain and SC, are glycosylated. In general, glycosylation is the result of a non-template, multi-enzymatic multistep process that ultimately results in a range of glycan heterogeneity. Thus, the post-translationally modified IgA proteoforms are cell type- and lineage-specific. Overall, glycans associated with Igs have been shown to influence antibody functions. Specifically, glycosylation patterns differentially affect the effector functions of Igs (Lin et al. 2015; Li et al. 2017) depending on the type, branching, and modifications of *N*-glycans and/or the terminal sugars of *N*- and/or *O*-glycans, which include galactose and sialic acid (Steffen et al. 2020). In fact, Ig glycosylation can determine whether an antibody glycoform is pro-inflammatory or anti-inflammatory (Li et al. 2017; Pagan et al. 2018). The multiple biological roles of IgA glycans include glycan-mediated antigen-nonspecific binding to bacteria and viruses. The complexity is enhanced by the diverse glycan structures that create a range of heterogeneity.

Herein, we will focus on the glycosylation of human IgA. Specifically, we discuss the location and observed structures at individual amino-acid sites and/or individual clusters of sites. We include a summary of the assessment of native populations of IgA glycans, the range of glycans observed at each site, and how changes in the glycans at a given site have been connected to biological activity and/or diseases.

14.2 IgA Subtypes and Overall Structure

Human IgA exists in two subclasses: IgA1 and IgA2 (occurring in two allotypes—IgA2m(1) and IgA2m(2)) (Fig. 14.1). Each Ig comprises two HCs and two LCs, each with one variable (V) and one (LC) or three (HC) constant domains. Each domain is approximately 110–130 amino acids, averaging 12–13 kDa. Functionally, the

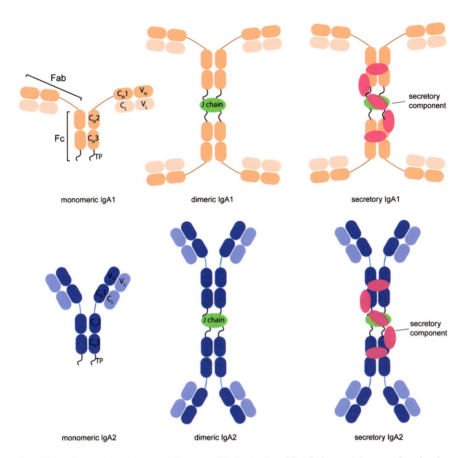

Fig. 14.1 Human IgA. Schematic diagram of IgA1 (top) and IgA2 (bottom) in several molecular forms (monomeric, dimeric, and secretory IgA). Although secretory IgA is depicted as a dimer, larger polymers such as trimers and tetramers can be formed. Heavy chains (V_H, C_H1, C_H2, and C_H3) are shown in orange (IgA1) and blue (IgA2), and light chains (V_L and C_L) are shown in light orange (IgA1) and light blue (IgA2). The tailpiece (TP) is portion of the C_H3 domain that is depicted as an extension. The J chain (green) is present in dimeric and secretory IgA, the secretory component (pink) is a component of secretory IgA

variable regions of HC and LC are responsible for antigen binding, while the constant domains are important for structure/function and defining/modifying the effector functions. Each of the two identical Fab regions of IgA consists of VH and VL and CH1 and CL, with CL being either kappa or lambda. HCs differ within their constant regions encoded by the Cα genes, Cα1 and 2. Each chain begins at its N-terminus with the variable region (VH), followed by the constant region, CH1, 2, and 3. CH1 is connected by the hinge region (HR) to the Fc region, consisting of CH2 and 3. Interchain disulfide bridges connect HC and LC of the Fab region and the two HC CH2 domains of the Fc region. Pairings between the collapsed β-barrel domains of adjacent chains (VH with VL, CH1 with CL, and both CH3 domains) are

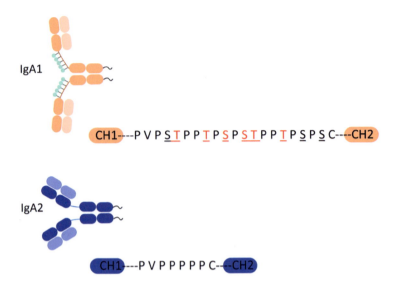

Fig. 14.2 IgA1 and IgA2 hinge regions. Comparison of amino-acid sequences of human IgA1 (top) and IgA2 (bottom) hinge regions. Human IgA1 has nine Ser (S) and Thr (T) amino-acid residues (underlined) in the hinge region segment (between constant regions C_H1 and C_H2 of the heavy chains). Usually, three to six clustered *O*-glycans are attached per hinge region (shown in red). IgA2 hinge region is shorter compared to that of IgA1, does not have Ser and Thr residues and, thus, IgA2 does not have *O*-glycans

mediated through non-covalent interactions, primarily hydrogen bonds and van der Waals contacts. A conserved 18-amino-acid extension of the CH3 heavy chain C-terminus is known as the tailpiece. This feature is unique for IgA and IgM and plays a critical role in dimer and pentamer formation, respectively, though binding with the J chain (Atkin et al. 1996; Yoo et al. 1999).

The HR between the Fab and Fc portions is another feature distinguishing each IgA isotype. IgA1, unlike IgA2, has an HR segment consisting of two octapeptide repeats rich in Ser, Thr, and Pro residues with nine potential *O*-glycosylation sites (Fig. 14.2). Usually, three to six of these sites are *O*-glycosylated (Novak et al. 2012, 2018). In contrast, the IgA2 HR lacks potential *O*-glycosylation sites. Unlike in IgG, IgA HR amino-acid sequence allows for greater flexibility, particularly for IgA1 with its extended sequence. Although this structural characteristic may enhance antigen binding by IgA, it also provides a target for bacterial IgA-specific proteases.

14.3 IgA Glycosylation Sites

The *N*-glycans at individual sites create a range of heterogeneity for both IgA isotypes, as they do for most Igs. Both IgA subclasses carry *N*-linked oligosaccharides. IgA1 and IgA2 have two and four conserved sites of *N*-glycosylation, respectively, all located in constant regions of the heavy chains. N-glycosylation significantly contributes to the total molecular mass of human IgA, accounting for 6–7% of the mass of IgA1, and 8–10% of the mass of IgA2 (Tomana et al. 1976). The IgA subclasses differ from other Igs in both the attachment sites and the overall positions of the N-linked glycans and the proximal disulfide bridges. IgA1 and IgA2 possess two similarly located conserved sites of N-linked glycosylation. The first site is at Asn144 (IgA1) and Asn131 (IgA2) in the CH2 domain of the heavy chain (the N-glycosylation sites herein are indicated by the residue number based on UniProt numbering, IgA1: P01876; IgA2: P01877) (UniProt Consortium 2017). The glycopeptide resulting from trypsin proteolytic digestion, as is conventional for analysis by mass spectrometry (MS), has an identical amino-acid sequence for both isotypes. Thus, if the two isotypes are not separated initially, these two N-glycosylation sites are observed as a mixture in standard MS glycosylation analysis (Steffen et al. 2020; Plomp et al. 2018). This is also the case for the second conserved site at Asn340 (IgA1) and Asn327 (IgA2) in the tailpiece. IgA2 has two additional sites of N-glycosylation, Asn47 located in the CH1 domain and in the CH2 domain at N205. All adjacent domains (VH with VL, CH1 with CL, and CH3 with CH3) are paired in close proximity to each other, except for the neighboring chains of the CH2 domains, which are not closely aligned. Such non-pairing is a feature also observed in IgG (CH2) and IgE (CH3). The potential solvent exposure, a consequence of the distance between the two CH2 domains, seems to be limited by the presence of N-glycans (located at Asn144 in IgA1 and Asn131 and Asn205 in IgA2) which shield the outer surface of the domain. The CH2 domain of IgA1 and IgA2 is stabilized by interchain disulfide bonds formed between three or four cysteine residues (Cys241, Cys242, Cys299, and Cys301) in the upper region of the domain. This feature is in contrast to IgG which instead has several disulfide connections located in the HR (2 HR disulfide bonds for IgG1 and IgG4, 4 for IgG2, and 11 for IgG3) (Liu and May 2012).

The tailpiece extension of the CH3 domain, of both IgA1 and IgA2, has a conserved N-glycan located at Asn340 and Asn327, respectively. The tailpiece, similar to the one found in IgM, contains a cysteine residue responsible for polymerization with the J chain. Previous studies have indicated that the tailpiece N-glycan of IgA plays an important role in J chain incorporation (Atkin et al. 1996; Sørensen et al. 2000). In plant-produced IgA proteins, the tailpiece is incompletely glycosylated and this deficiency may be the reason for the observed inefficient dimer formation (Göritzer et al. 2017, 2020; Westerhof et al. 2015; Castilho et al. 2018). Recently, it has been demonstrated that the tailpiece of IgA provides an innate line of defense against viruses, with the N-glycan mediating such activity (Maurer et al. 2018).

For IgA1, further glycosylation diversity arises through the clustered O-glycans attached to the HR. The glycosylated HR is a unique feature to Igs, shared by IgA1 and IgD, and in some forms of IgG3. In IgA1, these clustered O-linked glycans are composed of core 1 glycans or terminal or sialylated N-acetylgalactosamine (GalNAc), often attached to Thr225, Thr228, Ser230, Ser232, Thr233, and/or Thr236 (amino-acid numbering is based on conventionally used nomenclature for

IgA1 HR). Each Ser/Thr residue can be modified by a GalNAc residue that can be further extended by the addition of galactose (Gal) through a β1,3 glycosidic bond. Up to two sialic acid (SA) residues can be added, one attached to the GalNAc through an α2,6-linkage and the other to Gal by an α2,3-linkage.

14.4 Molecular Forms of IgA and Their Distributions (Monomeric and Polymeric IgA, SIgA, J Chain, Secretory Component)

Unlike most other Ig isotypes, IgA exists in multiple molecular forms (Fig. 14.1). IgA in the circulation is produced mainly in the bone marrow and to a lesser extent in the spleen and lymph nodes. The predominant form of serum IgA is monomeric. IgA destined for mucosal surfaces is produced locally by plasma cells in a polymeric form (pIgA), predominantly dimeric. The ~16-kDa J chain is an additional polypeptide that connects together two or more monomers of IgA via the Fc region tailpiece through disulfide bonds, forming pIgA (Fig. 14.1). Two of the J chain's eight cysteine residues are involved in covalent binding to IgAs tailpiece and the additional six cysteine residues form intramolecular disulfide bridges (Cys12-Cys100, Cys17-Cys91, and Cys108-Cys133) (Bastian et al. 1992). The J chain has a single N-linked glycan attached at Asn48. pIgA is formed after glycosylation occurs. Glycosylation is required for dimer assembly, as drastically reduced dimer formation was observed when Asn340 was substituted with alanine to prevent the attachment of N-glycans to the tailpiece (Atkin et al. 1996).

SIgA is a multi-polypeptide complex produced in mucosal tissues from pIgA, produced by local plasma cells, undergoing transcytosis through mucosal epithelial cells (Norderhaug et al. 1999). SIgA consists of an SC covalently attached to pIgA (i.e., IgA plus J chain). SIgA is mainly dimeric, although higher oligomers, including trimers and tetramers, are also present. The ~80-kDa SC is comprised of five immunoglobulin-like domains. To form SIgA, a polymeric immunoglobulin receptor-mediated pathway transports pIgA by transcytosis into the mucosal secretions. On reaching the mucosal surface, the polymeric immunoglobulin receptor (pIgR) that binds IgA is cleaved to release SIgA with bound SC which is a proteolytic cleavage product of the pIgR. SC is highly glycosylated by seven N-glycans, which contribute up to 25% of its molecular mass (Norderhaug et al. 1999). The SC glycans have several functions, in addition to protecting the SC and the SIgA from proteases (Crottet and Corthésy 1998), they can interact with adhesins and lectins. SC has been shown to bind to a range of bacteria via its glycans (Schroten et al. 1998; Borén et al. 1993; Wold et al. 1990), thereby inhibiting attachment and the subsequent infection of epithelial surfaces. The SC glycans are also involved in the localization of SIgA by anchoring the SIgA to the mucus lining the epithelial surface through its carbohydrate residues.

Circulating IgA is predominantly monomeric and predominantly of IgA1 subclass (~84% IgA1 and ~16% IgA2) (Mestecky et al. 1986), with a small proportion of pIgA (~5–8%) (Delacroix et al. 1982). Dimeric SIgA represents the dominant IgA at almost all mucosal surfaces, although the levels and subclass distributions among the different compartments of the mucosal immune system can vary considerably. For example, in saliva, 90–95% of total IgA is SIgA with a subclass distribution of ~65% IgA1 and ~35% IgA2. This is different for hepatic bile which has a much lower abundance of SIgA (~65%) and a subclass distribution of ~75% IgA1 and ~25% IgA2 (Delacroix et al. 1982). However, even the same compartment, sampled via different methods or times, can provide varying levels of isotype distribution and molecular forms. Furthermore, these distinct differences reflect the dominant Ig source (local production in mucosal tissue versus plasma), the expression of isotype-specific receptors that transport Igs, and the effects of specific regulatory mechanisms (e.g., cytokines and hormones) that influence Ig distribution.

14.5 IgA1 *O*-Glycosylation Pathways

The IgA1 HR with clustered sites of mucin-like *O*-glycosylation represents a unique acceptor site for a post-translational modification process that differs from most other Igs (only IgD has clustered *O*-glycans in its HR). Of the nine potential *O*-glycosylation sites in the HR of human IgA1, usually three to six are *O*-glycosylated (Baenziger and Kornfeld 1974; Mattu et al. 1998; Novak et al. 2000; Renfrow et al. 2007; Takahashi et al. 2010; Hiki et al. 1998; Iwase et al. 1996a; Royle et al. 2003; Franc et al. 2013; Wada et al. 2010a), although glycoforms with up to seven *O*-glycans have been observed in serum IgA1(Pouria et al. 2004). These clustered *O*-glycans of serum IgA1 are usually core 1 glycan (Field et al. 1989; Iwase et al. 1996b). The GalNAc-Gal disaccharide may be sialylated on GalNAc, Gal, or both sugars (Field et al. 1989; Takahashi et al. 2012). *N*-acetylneuraminic acid can be attached to GalNAc through an α2,6-linkage or to Gal by an α2,3-linkage. In addition to the core 1 glycans, some glycans may remain without Gal, i.e., as terminal GalNAc or sialylated GalNAc (Ohyama et al. 2020a, b). Unlike in the process of *N*-glycan biosynthesis, there is no control system for *O*-glycan biosynthesis, and *O*-glycosylation is also not required for folding or export of IgA1 (Gala and Morrison 2002).

Biosynthesis of IgA1 *O*-glycans occurs in the Golgi apparatus of IgA1-producing cells. As for other proteins with clustered *O*-glycans, it is a stepwise process that is not template-driven but rather controlled by expression, activity, and localization of different glycosyltransferases catalyzing the sequential addition of monosaccharides to the acceptor (glyco)protein during its transition through the Golgi apparatus (Reily et al. 2019). The clustered nature of the potential sites of *O*-glycosylation in the IgA1 HR creates a unique amino-acid synthesis template where each step of monosaccharide addition changes the template and has implications for the subsequent steps of monosaccharide addition at adjacent sites. This effect occurs due to

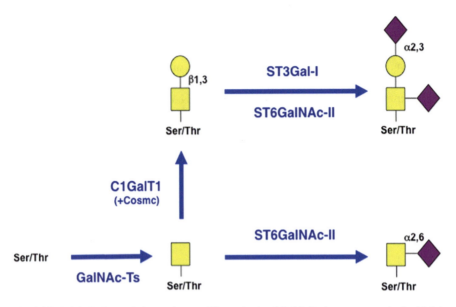

Fig. 14.3 IgA1 *O*-glycosylation pathways. Biosynthesis of IgA1 *O*-glycans occurs in the Golgi apparatus of IgA1-producing cells. The stepwise process begins with the attachment of GalNAc to some of the Ser/Thr residues in the hinge region catalyzed by GalNAc-transferases (GalNAc-Ts). The attached GalNAc residues can be then modified by addition of Gal, mediated by core 1 β1,3-galactosyltransferase (C1GalT1). Production of the active C1GalT1 enzyme depends on its chaperone (C1GalT1C1, Cosmc). The core 1 structures (GalNAc-Gal) of IgA1 can be further modified by sialyltransferases that attach sialic acid to Gal (mediated by an ST3Gal enzyme, e.g., ST3Gal-I) and/or GalNAc residues. Sialylation of GalNAc is mediated by ST6GalNAc-II, as the usual ST6GalNAc-I is not expressed in IgA1-producing cells. Conversely, if terminal GalNAc is sialylated by ST6GalNAc-II, this structure cannot be further modified

some of the glycosyltransferases having glycan-recognizing lectin domains that strongly influence their activity (Stewart et al. 2019, 2021). Also, as more *O*-glycan chains are added and extended in the confined space of the HR, there is a steric hindrance that precludes or inhibits further addition or extension of the clustered *O*-glycans. Still, these physical constraints on the HR create a consistent fidelity of the range and distribution of *O*-glycans that are attached. This is similar to the clustered *O*-glycosylation sites of mucins that have multiple amino-acid tandem repeats as opposed to the two found in IgA1.

IgA1 *O*-glycosylation begins with the attachment of GalNAc to some of the Ser/Thr residues in the HR. In humans, some members of a family of 20 enzymes, GalNAc-transferases (GalNAc-Ts), are involved in this process (Daniel et al. 2020) (Fig. 14.3). It has been proposed that GalNAc-T2 is the main GalNAc-T enzyme responsible for the initiation of *O*-glycosylation of IgA1 (Iwasaki et al. 2003), although other GalNAc-Ts may contribute to the process, including GalNAc-T1, -T11, and -T14 (Stewart et al. 2021; Daniel et al. 2020; Wandall et al. 2007).

GalNAc-Ts determine not only the sites of GalNAc attachment but also the final glycan density, as there are three to six sites per IgA1 HR that are typically glycosylated. Most GalNAc-T enzymes have two domains, a catalytic domain and lectin domain. Much of our understanding of GalNAc-T enzyme activities comes from studies of purified enzymes, synthetic peptides, glycopeptides, and glycoprotein substrates (Stewart et al. 2019, 2021; de Las et al. 2017). Data from multiple studies point to distinct and complementary roles of catalytic and lectin domains of GalNAc-Ts in the biosynthesis of clustered O-glycans of IgA1. Each monosaccharide addition can generate multiple isomers due to the selection of a specific attachment site. Whereas the selection of the first site is driven by the catalytic domain of the GalNAc-T enzyme, the subsequent site selection is substantially impacted by the lectin domain. The glycan density, i.e., the number of GalNAc residues per HR glycopeptide, depends on the expression and activity of GalNAc-Ts isoenzymes that can act on IgA1 and on the selection of the initiation and follow-up sites in the HR. Despite the variability of this step of O-glycan biosynthesis, most IgA1 HR glycoforms have three to six O-glycans.

The attached GalNAc residues can be modified by addition of Gal, mediated by core 1 β1,3-galactosyltransferase (C1GalT1) (Stewart et al. 2021; Ju et al. 2002, 2006; Aryal et al. 2012). The production of the active enzyme depends on its chaperone (C1GalT1C1, Cosmc) (Ju and Cummings 2002) (Fig. 14.3).

The core 1 structures of IgA1 may be further modified by sialyltransferases that attach sialic acid to Gal (mediated by a ST3Gal enzyme) and/or GalNAc residues (Stewart et al. 2021; Takahashi et al. 2014). Sialylation of GalNAc is mediated by ST6GalNAc-II, as the usual ST6GalNAc-I is not expressed in IgA1-producing cells (Takahashi et al. 2014; Raska et al. 2007; Stuchlova Horynova et al. 2015; Suzuki et al. 2008) (Fig. 14.3). Notably, sialylation of GalNAc on IgA1 by ST6GalNAc-II prevents subsequent galactosylation (Stewart et al. 2021; Stuchlova Horynova et al. 2015; Suzuki et al. 2014). Furthermore, studies with purified enzymes ST6GalNAc-II and ST3Gal-I and partially Gal-deficient IgA1 substrate revealed that prior sialylation by either enzyme influences the activity of the second enzyme (Stewart et al. 2021). These data suggest that the extent of sialylation of the clustered IgA1 HR segment is not only a net result of the enzyme activity but also involves steric hindrances of the clustered O-glycans (Novak et al. 2018; Stewart et al. 2021).

Additional knowledge about IgA1 O-glycan biosynthesis was obtained from genetic and genomic studies as well as studies of Epstein–Barr virus (EBV)-immortalized IgA1-secreting cells derived from the cells in human peripheral blood (Suzuki et al. 2008). EBV-immortalized IgA1-secreting cell lines derived from healthy individuals and patients with IgA nephropathy (IgAN) provided a new tool for comparative studies of IgA1 O-glycosylation in health and disease states. These cell lines produce IgA1 that mimics glycosylation of serum IgA1 of the respective donors (Suzuki et al. 2008). As detailed elsewhere, patients with IgAN have IgA1 in the glomerular immunodeposits and in the circulation enriched for glycoforms with some O-glycans deficient in Gal (Novak et al. 2018). Studies of EBV-immortalized IgA1-secreting cell lines from healthy individuals and patients IgAN revealed that dysregulation of expression and activity of several key enzymes is associated with elevated production of Gal-deficient IgA1 (Suzuki et al. 2008, 2014). Specifically, reduced expression/activity of C1GalT1 and its chaperone

Cosmc and elevated expression/activity of ST6GalNAc-II in the cells from IgAN patients are associated with reduced Gal content in the secreted IgA1 (Suzuki et al. 2008).

Studies of familial and sporadic IgAN cohorts revealed heritability of serum levels of Gal-deficient IgA1 (Gharavi et al. 2008; Hastings et al. 2010; Kiryluk et al. 2011). Genome-wide association studies (GWAS) revealed single nucleotide polymorphisms (SNPs) in the noncoding region of C1GALT1, the gene encoding the C1GalT1 galactosyltransferase, associated with serum levels of Gal-deficient IgA1 (Kiryluk et al. 2017; Gale et al. 2017). One of these studies also found associations between SNPs in the noncoding region of C1GALT1C1 (Cosmc), the gene encoding the C1GalT1-specific chaperone, and serum levels of Gal-deficient IgA1 (Kiryluk et al. 2017). siRNA knock-down experiments using immortalized IgA1-secreting cell lines further validated these findings (Kiryluk et al. 2017).

Furthermore, the genetically co-determined Gal content can be further influenced by some cytokines and growth factors, such as interleukin 6 (IL-6), IL-4, and leukemia inhibitory factor (LIF) (Suzuki et al. 2014; Yamada et al. 2010, 2020). This cytokine-mediated overproduction of Gal-deficient IgA1 is due to further dysregulation of expression/activity of specific enzymes (C1GalT1, Cosmc, ST6GalNAc-II) (Suzuki et al. 2014). This process, uniquely enhanced in the IgA1-producing cells from patients with IgAN, is associated with the enhanced and prolonged cytokine signaling, likely due to an aberrant regulation of cellular signaling in JAK-STAT pathways that are engaged by IL-6 or LIF in the cells derived from IgAN patients (Yamada et al. 2017, 2020).

These observed genetic and genomic differences provide points of investigation as to how the final fidelity of IgA1 clustered *O*-glycans can be altered and lead to differences in observed distributions of IgA1 glycosylated proteoforms (IgA1 *O*-glycoforms).

14.6 Assessing Heterogeneity: Analysis of Glycan-Attachment Sites and Range of Heterogeneity

Analysis of IgA1 and IgA2 glycosylation has proven technically challenging due to the multiple molecular forms of IgA, the overall heterogeneity of *N*- and *O*-glycosylation, and the nature of the IgA1 HR where *O*-glycans are attached. Historically, several strategies have been used, including monosaccharide compositional analysis by gas–liquid chromatography (Renfrow et al. 2005) or high-performance anion-exchange chromatography with pulsed-amperometric detection (HPEAC-PAD), Edman sequencing (Baenziger and Kornfeld 1974; Mattu et al. 1998), glycan-specific lectin blotting and ELISA (Moldoveanu et al. 2007), and more recently liquid chromatography-mass spectrometry (LC-MS) (Takahashi et al. 2010). For IgA1 *O*-glycosylation, lectin ELISA has become a standard means to allow high-throughput analyses in a quantitative manner for serum-based studies (Moldoveanu et al. 2007; Gomes et al. 2010; Moore et al. 2007); however, limitations for several of these methodologies include lack of individual glycoform specificity and the inability to provide information on the sites of attachment and heterogeneity in the context of amino-acid sequence. As a result, much of the standard analysis of IgA glycosylation has moved toward LC-MS.

MS-based techniques utilized to profile glycans and address glycan heterogeneity have become the standard tool for the structural characterization of carbohydrates. Over the last two decades, our understanding of the structures and diversity of glycans in biological samples has increased dramatically. This is the result of the significant improvements in the sensitivity and variety of mass spectrometry approaches. With the help of chromatographic separation, glycan derivatization to improve ionization, and tandem MS fragmentation techniques, oligosaccharide structures and sites of attachment can be elucidated (de Haan et al. 2019). Due to the increased use of IgG as a therapeutic, the majority of analytical development regarding glycosylation has focused on exploring *N*-glycan structures and heterogeneity. While the enzymatic release of *N*-glycans is still a common tool for global *N*-glycan analysis (glycomics), for Ig *N*-glycosylation, amino-acid site-specific analysis has become standard as well as MS analysis of intact proteins to assess the composition and distribution of *N*-glycan heterogeneity (de Haan et al. 2019). Such methodologies that have become standard for the analysis of IgG *N*-glycans are easily applicable to IgA *N*-glycans. As we will discuss more below, that trend has occurred but comparatively there are considerably more studies of *N*-glycans of IgG than IgA. For analysis of IgA *N*-glycans, as discussed above, nearly all sources of IgA in the body have some level of mixture of isotypes and molecular forms. Thus, many of the currently reported analyses of IgA glycosylation are mixtures of IgA produced from multiple sources, however, some groups have begun to make distinctions between various molecular forms.

The features of IgA1 clustered *O*-glycans create a unique analytical challenge. The *O*-glycans are often referred to as "mucin-like" given they mimic *O*-glycans found on the heavily *O*-glycosylated mucin family of proteins that line most epithelial surfaces. Mucins have a large number of tandem repeats that are *O*-glycosylated. The IgA1 HR has only two tandem repeats making it somewhat more amenable to heterogeneity analysis. The *O*-glycans of IgA1 have been implicated in the pathogenesis of IgAN (Tomana et al. 1997, 1999; Novak et al. 2005), and the closely related IgA vasculitis with nephritis (IgAVN) (Allen et al. 1998; Levinsky and Barratt 1979). The goal of identifying differences in IgA1 HR *O*-glycan heterogeneity and composition between patients and normal healthy individuals has driven the development of methodologies. This includes determining the range of IgA1 *O*-glycoforms present and the sites of *O*-glycan attachment defined in the context of adjacent sites as well as the heterogeneity at each individual site (Novak et al. 2018).

As stated, investigators in the IgAN field drove the initial progress in assessing IgA1 *O*-glycosylation. In 1996, Iwase et al. identified two IgA1 HR glycopeptides containing four or five *O*-glycan chains by use of MALDI-TOF mass spectrometry

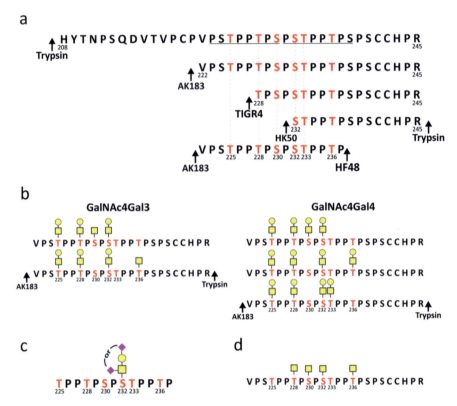

Fig. 14.4 Heterogeneity of IgA1 hinge-region O-glycosylation. (**a**) Amino-acid sequence of the hinge region of human IgA1 showing fragments cleaved by IgA-specific proteases (see Fig. 14.6 legend). The six commonly utilized O-glycosylation sites are highlighted in red and marked by the numbers below Ser/Thr residues for the AK183-HF48 fragment. The two tandem repeats are underlined in the trypsin fragment (top). (**b**) Examples of two identified IgA1 HR O-glycopeptide positional isomers, GalNAc4Gal3 and GalNAc4Gal4 produced by trypsin + AK183 IgA-specific protease. (**c**) Illustration of the potential for sialic acid to be attached to Gal and/or GalNAc residues creating isomers at a single site. (**d**) Four initial sites of GalNAc addition observed by in vitro GalNAc-T2 reactions (Stewart et al. 2019)

(Iwase et al. 1996a). These trypsin-released O-glycopeptides were confirmed as the IgA1 HR by sequential glycosidase treatment (neuraminidase, galactosidase, and N-acetylgalactosaminidase) and the spectra showed a shift in the peak distribution toward lower masses as carbohydrate residues of the O-glycan chains were removed. This promising beginning led to several reports of IgA1 O-glycopeptides isolated from serum (Iwase et al. 1998, 1999) and pooled sera of patients with IgAN (Odani et al. 2000), IgA1 isolated from pooled renal biopsies (Hiki et al. 2001), and tonsillar IgA1 (Horie et al. 2003). In 2000, Novak et al. reported the use of IgA-specific proteases that released distinct IgA1 HR fragments (Fig. 14.4) and provided novel means of generating IgA1 HR O-glycopeptides for analysis by mass spectrometry and lectins (Novak et al. 2000). This preparation was unique because one of the

IgA-specific proteases (Haemophilus HK50) cleaved between two sites of *O*-glycan attachment (cleavage after Pro231 between Ser230 and Ser232) and therefore provided a general localization of specific *O*-glycan chains N- or C-terminal to the cleavage site.

In terms of IgA1 *O*-glycan site localization, two groups initially identified the predominant sites of *O*-glycan attachment to include Thr225, Thr228, Ser230, Ser232, and Thr236 by use of modified N-terminal sequencing methods (Baenziger and Kornfeld 1974; Mattu et al. 1998). However, these reports analyzed the IgA1 population as a whole with no separation of individual IgA1 *O*-glycopeptides. Later, the introduction of electron radical fragmentation (electron capture dissociation, ECD, and electron transfer dissociation, ETD) in tandem MS lead to the direct assessment of individual IgA1 HR *O*-glycopeptide ions released by trypsin and other proteases, including IgA-specific proteases (Fig. 14.4a). A series of reports by the Renfrow group demonstrated the direct assessment of sites of *O*-glycan attachment for several IgA1 *O*-glycopeptide forms ranging from three to five glycans and having one or two sites comprised of only terminal GalNAc (i.e., galactose-deficient sites-containing IgA1; Gd-IgA1). In the 2012 paper, Takahashi et al. also reported the presence of several IgA1 *O*-glycopeptides that were comprised of isomeric mixtures of equal *O*-glycan chain compositions that were attached at alternate amino-acid sites in the IgA1 HR (Fig. 14.4b) (Takahashi et al. 2012). They also identified a predominant sixth site of *O*-glycan attachment at Thr233. The three attachment sites at Ser230, Thr236, and Thr233 were identified as the predominant sites with galactose-deficient IgA1 (Gd-IgA1) glycans in the IgA1 myeloma protein that was studied as well as two samples of serum IgA1 isolated from healthy individuals. Interestingly, it was the combination of the IgA-specific proteases (Fig. 14.4) plus the novel tandem MS fragmentation that allowed elucidation of many of the details of *O*-glycan attachment that are now considered standard in our understanding of what the heterogeneity of IgA1 *O*-glycans includes. Concurrent with these studies, a study comparing methodologies for assessing *O*-glycans was reported in 2010 using an IgA1 myeloma protein as the reference standard analyzed by fifteen different laboratories (Wada et al. 2010a). While none of the participating laboratories included amino-acid site-specific analysis of IgA1 myeloma reference standard, the study demonstrated that the range of IgA1 *O*-glycopeptides observed was consistent across MS platforms and that there was also a level of consistency in the relative abundance of individual IgA1 *O*-glycopeptides. Since these studies, there has been a steady increase in the use of MS and LC-MS methodologies to assess the distribution of IgA1 hinge-region *O*-glycosylated forms.

One area that remains a challenge is the assessment of IgA1 *O*-glycan sialylation. IgA1 *O*-glycans can be α2,6 sialylated or α2,3 sialylated at any given position. With anywhere from three to six *O*-glycans per hinge region, sialylation significantly increases the number of isomeric mixtures in a population of IgA1 *O*-glycosylated forms (Fig. 14.4c). Takahashi et al. demonstrated the potential for sialic acid isomers at a single site (Thr232). As such, many of the LC-MS analyses have used sialidases to remove the sialic acid residues and reduce the complexity of the IgA1 *O*-glycopeptide mixtures (Takahashi et al. 2010, 2012; Moldoveanu et al. 2007).

The goal of understanding the difference in IgA *O*-glycans in patients with IgAN vs. healthy individuals has led to the development of lectin-based quantitative ELISA tests (Moldoveanu et al. 2007; Zhao et al. 2012). Lectins recognize specific glycan structures and in the case of IgAN, several lectins specific for terminal GalNAc have been utilized in ELISA to provide an IgA1-specific assessment of the extent of Gal-deficient IgA1 in different cohorts. This is done by using IgA from serum or tissue samples and then probing for lectin reactivity (using lectins from *Helix pomatia* agglutinin, HPA or *Helix aspersa* agglutinin, HAA) compared to a reference standard. Through this analysis, investigators have been able to establish the range of Gd-IgA1 levels (based on standardized lectin reactivity) that exists in patients with IgAN, even to the point of utilizing Gd-IgA1 as a marker for IgAN disease progression (Zhao et al. 2012; Maixnerova et al. 2019; Reily et al. 2018). It is possible that LC-MS methodologies may at some point be able to provide a similar output, but currently, these lectin-based ELISA tests have been successfully implemented for larger-cohort studies.

14.7 Observed IgA *N*-Glycosylation Heterogeneity

The characterization and analysis of IgA *N*-glycosylation is a systematic process of localizing the composition and range of glycans linked to asparagine at the glycosylation sites. Assessing Ig *N*-glycosylation has become a common methodology in recent years due to the advent of therapeutic antibodies. This is more the case for IgG, but the principles of systematic mapping and assessment of the range of *N*-glycans at a given site by the use of mass spectrometry have become a tool utilized in the analysis of Igs. For example, these techniques can assess how glycosylation of IgG is affected in different forms, sources, and in responses to specific antigens/vaccines. In the case of IgA, there is the added complexity of assessing the J chain and SC sites of *N*-glycosylation as well as the *O*-glycosylated IgA1 HR (discussed below). To date, comprehensive studies have been performed for mIgA and SIgA from plasma, saliva, and colostrum samples and, in a few cases, these studies have included separation of IgA isotypes. Several different structures of *N*-glycans have been observed, contributing to heterogeneity for both IgA subclasses (Fig. 14.5). Glycan heterogeneity for mIgA and all potential SIgA *N*-glycosylation sites, including the seven on the SC, two on IgA1, four on IgA2, and one on the J chain, are detailed below. Overall, several sources identified IgA *N*-glycans to be mainly of the biantennary complex type, with various levels of galactosylation, sialylation, bisecting GlcNAc, and fucosylation (Steffen et al. 2020; Plomp et al. 2018; Goonatilleke et al. 2019; Bondt et al. 2016, 2017). The following text provides details of what has been observed to date, but continued experimental analysis of IgA glycan heterogeneity will ultimately aid in understanding the complex biological roles of IgA and SIgA (including SC and J chain) in health and disease.

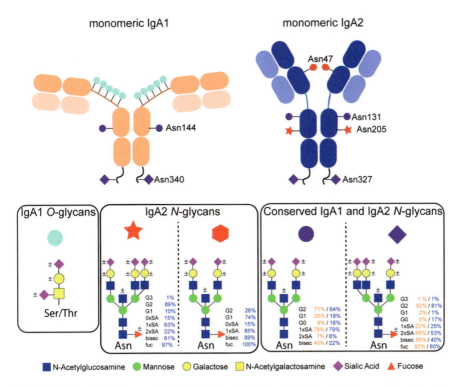

Fig. 14.5 Sites of IgA glycosylation. Illustration of IgA1 and IgA2 and their respective glycosylation sites. IgA1 has three to six observed sites of *O*-glycosylation (shown in cyan) and two IgA1 *N*-glycosylation sites (Asn144 and Asn340, shown in purple). IgA2 has four *N*-glycosylation sites (Asn47 and Asn205, shown in red; and Asn131 and Asn327, shown in purple). The sites in purple are conserved in both IgA subclasses. The outlined area contain representations of the possible *O*- and *N*-glycan structures reported at each site (Steffen et al. 2020; Plomp et al. 2018; Goonatilleke et al. 2019; Bondt et al. 2016, 2017; Deshpande et al. 2010; Huang et al. 2015; Gomes et al. 2008). The ± symbol indicates the variable presence of each monosaccharide. The relative abundance values (%) are as reported by Steffen et al., who performed site-specific MS quantification of glycans from isolated serum IgA1 and IgA2 (Steffen et al. 2020). Values in orange correspond to IgA1 and values in blue correspond to IgA2

14.7.1 IgA1/IgA2 Asn340/327 N-Glycosylation

The *N*-glycosylation sites Asn340 and Asn327 on the tailpiece of IgA1 and IgA2, respectively, are located on tryptic peptides common to both IgA subclasses. Therefore, unless IgA1 and IgA2 are separated (often via jacalin agarose that binds only IgA1), MS analysis will not distinguish between subclasses and information on site-specific glycosylation is gathered for total IgA. At Asn340/327, two different tryptic peptide sequences are commonly identified, the expected LAGKPTHVNVSVVMAEVDGTCY and the truncated LAGKPTHVNVSVVMAEVDGTC. The abundance of each peptide seems to depend on the origin of IgA, specifically IgA from plasma displays a higher abundance of the truncated peptide while IgA derived from saliva includes higher amounts of the peptide with the C-terminal tyrosine (Plomp et al. 2018). The glycans at this location are primarily biantennary complex-type; however, a low abundance of triantennary glycans (~5%) on the truncated peptide has been reported (Bondt et al. 2016). Additionally, high-mannose and hybrid glycans have been detected in low amounts on the non-truncated peptide (<5% IgA1; <20% IgA2) (Steffen et al. 2020). The relative abundance of bisection, galactosylation, sialylation, and fucosylation are similar for both the non-truncated and truncated peptide. Over 89% galactosylation, sialylation, and fucosylation has been observed for the N-glycans of IgA1. Sialylation of IgA1 non-truncated peptide was slightly higher compared to the truncated peptide (89% and 95%, respectively) (Bondt et al. 2016). When observed and reported, the non-truncated peptide on IgA2 displayed slightly lower bisection (<60%) compared to IgA1 (<40%). Fucosylation was also lower in IgA2 (80% IgA2 versus 95% IgA1) (Bondt et al. 2016). In SIgA versus mIgA, Plomp et al. (2018) noted lower fucosylation (1.4-times), galactosylation (2-times), and sialylation (5-times) and higher content of bisecting GlcNac (1.7-times) in salivary compared to plasma IgA. High-mannose glycans were significantly higher in salivary IgA.

14.7.2 IgA1/IgA2 Asn144/131 N-Glycosylation

IgA1 Asn144 and IgA2 Asn131 also share the same tryptic amino-acid sequence LSLHRPALEDLLLGSEANLTCTLTGLR, located in the CH2 domain. Analysis has shown that this site is primarily composed of complex-type N-glycans, and Plomp et al. (2018) identified approximately 20% high mannose-type glycans and up to 5% hybrid-type/mono-antennary glycans. Steffen et al. (2020) isolated IgA1 and IgA2, specifically identifying non-complex glycan abundance of each subclass at 5% and 20%, respectively. For IgA1, complete galactosylation and approximately 60% sialylation was observed. There was variation in the reported bisecting glycans of Asn144 on IgA1. For example, Bondt et al. reported approximately 25% bisecting glycans which increased to approximately 28% during pregnancy (Bondt et al. 2016). Other studies found it to be higher—between 40 and 50% for IgA1 (Steffen et al. 2020; Plomp et al. 2018). Furthermore, it has been well documented that glycans at Asn144/131 are almost entirely afucosylated (<1%), in contrast to Asn340/327 (Plomp et al. 2018; Bondt et al. 2016). Notable differences in plasma versus salivary IgA were seen for high-mannose glycans, bisecting GlcNAc, galactose, and sialylation (18-times, 1.5-times, 3-times, and 5-times higher in plasma IgA, respectively). This is an interesting finding, as a higher abundance of unprocessed glycans usually reflects rapid processing through the cellular machinery, differential expression of glycosylation enzymes, or restricted access to a specific glycosylation site.

14.7.3 IgA2 N-Glycosylation

For the glycosylation sites Asn47 in the CH1 domain and Asn205 in the CH2 domain, unique to IgA2, *N*-glycans were observed by Plomp et al. on the peptides SESGQNVTAR and TPLTANITK, respectively. The glycans at these sites were mostly complex biantennary (>98% of total abundance) (Plomp et al. 2018). For plasma IgA2, Steffen et al. observed that both glycan sites were mainly monosialylated, had a high level of bisecting GlcNAc, and were almost always fucosylated (Steffen et al. 2020). Differences in galactosylation were observed as Asn47 was mainly monogalactosylated while the antennae of Asn205 were often fully galactosylated. Overall, two- to fivefold differences were observed in the relative abundances of the different glycan types present on plasma vs. salivary IgA2. Specifically, salivary *N*-glycans at both Asn47 and Asn205 showed a higher degree of bisecting GlcNAc and a lower degree of galactosylation, sialylation, and sialylation per galactose compared to plasma *N*-glycans (Steffen et al. 2020; Plomp et al. 2018; Bondt et al. 2016).

14.7.4 Joining Chain (J Chain) N-Glycosylation

The single *N*-glycosylation site Asn71 on the J chain has been observed as two tryptic peptides: $EN_{71}ISDPTSPLR$ (JC Asn71) and the miscleaved $IIVPLNNREN_{71}ISDPTSPLR$ (JC Asn71m). Plomp et al. (2018) determined that this site contained between 20 and 50% monoantennary and hybrid-type glycans, which is significantly higher than the IgA constant-domain *N*-glycosylation sites. However, differences in glycans in salivary and plasma IgA J chain are consistent with observations for the IgA1 and IgA2 heavy chain *N*-glycosylation sites, namely a higher bisection (3.2 times higher) and lower galactosylation (1.1 times lower) and sialylation (1.7 times lower) in the saliva-derived samples. Surprisingly, the two glycopeptides, JC Asn71 and JC Asn71m, exhibited different glycoprofiles. The miscleaved glycopeptides were more abundant in fucosylation as compared to the expected tryptic glycopeptides (3.2-times higher in plasma and 2.9-times higher in saliva) (Plomp et al. 2018). This observation was previously observed of colostrum, as 10% tryptic and 67% miscleaved glycopeptides were fucosylated (Deshpande et al. 2010). The range of glycan heterogeneity at this site is enhanced as the tryptic glycopeptide exhibits partial or full sialylation, which is uncommon for the miscleaved glycopeptide.

14.7.5 Secretory Component N-Glycosylation

Seven *N*-glycosylation sites are located on six SC tryptic glycopeptides. *N*-glycans at Asn135, Asn186, Asn421, and Asn469 were determined to be complex-type and biantennary with antennae fully galactosylated and partially sialylated (Plomp et al. 2018; Deshpande et al. 2010). However, other reports showed that sialylation was uncommon for Asn135, Asn186, Asn469, and Asn499 and highest for Asn421 and that Asn135 was abundantly tetraantennary (Huang et al. 2015). Mono-antennary species were identified on Asn499 and the observed glycoforms carried zero to five fucose residues, contributing to both the core and antennary fucosylation. On Asn135, Asn469, and Asn499 between 1 and 4% bisection was observed. All glycosylation sites contained at least five galactose residues and one to three fucose residues.

Glycans Asn83 and Asn90 are located on the same tryptic peptide, complicating site-specific glycosylation analysis. This can be remedied by employing additional proteases. Furthermore, the joint glycan composition $H_{10}N_8F_{2-8}S_{0-3}$ at Asn83 and Asn90, reported by Plomp et al., indicates a similar composition to other SC glycans (Plomp et al. 2018).

14.8 IgA Receptors and Role of Glycans

As detailed above, IgA subclasses are quite diverse in the origin, molecular form, and site-specific glycosylation profiles. The isotype, glycosylation, and molecular form of IgA can impact interactions with various types of IgA receptors. In humans, these receptors include Fc receptors (Fcα-receptor I [FcαRI; CD89] and Fcα/μ receptor [CD351]), polymeric immunoglobulin receptor, transferrin receptor (CD71), lectins (e.g., asialoglycoprotein receptor on hepatocytes), cell-surface galactosyltransferase (e.g., β1,4-galactosyltransferase), and Fc receptor-like 4 (FcRL4, CD307d) (de Sousa-Pereira and Woof 2019; Monteiro and Van De Winkel 2003; Breedveld and van Egmond 2019; Cho et al. 2006; Aleyd et al. 2015; Honda et al. 2016; Yang et al. 2013; Tomana et al. 1993). FcαRI is specific for IgA1 and IgA2 and FcRL4 binds polymeric IgA and IgA found in immune complexes, whereas other receptors can bind other ligands in addition to IgA. Below, we discuss several IgA receptors, focusing on details with respect to glycosylation and molecular forms of IgA.

FcαRI is expressed by myeloid cells, such as monocytes, neutrophils, and some subsets of macrophages and dendritic cells (van Egmond et al. 2001). Serum IgA, predominantly monomeric IgA1, can bind to FcαRI (CD89) (Monteiro et al. 1990; Herr et al. 2003). This receptor plays the main role in the IgA-mediated clearance of pathogens and cancer cells (Woof and Kerr 2006; de Tymowski et al. 2019; Brandsma et al. 2019; Hansen et al. 2018, 2019; Heemskerk and van Egmond 2018). Both the ligand (IgA1 and IgA2) and the receptor are glycosylated (Mattu et al. 1998; Royle et al. 2003; de Sousa-Pereira and Woof 2019; Aleyd et al. 2015). Whereas IgA N-glycans affect IgA thermal stability but not receptor binding (Gomes et al. 2008; Göritzer et al. 2019), N-glycans of FcαRI significantly modulate binding affinity to IgA (Göritzer et al. 2019). Furthermore, the binding of IgA1 to FcαRI induces long-range conformational changes in IgA1, propagating up to the O-glycosylated HR (Posgai et al. 2018).

FcαRI does not contain any signaling motifs in its cytoplasmic tail and FcαRI-mediated signaling depends on the associated Fc receptor γ chain and its immunoreceptor tyrosine-based activation motifs. FcαRI-mediated activation can induce degranulation, phagocytosis, chemotaxis, and antibody-dependent cellular cytotoxicity. Furthermore, when IgA-FcαRI signaling is combined with activation through pattern recognition receptors (e.g., Toll-like receptors), cytokine production can be induced in antigen-presenting cells. This process is essential for controlling inflammation and inducing both innate and adaptive immune responses. Depending on whether IgA binds to FcαRI in a soluble or aggregated form, it can induce immunosuppressive or pro-inflammatory responses. Soluble forms of IgA in the circulation, monomers and dimers, have low affinity for FcαRI and bind only transiently, thus mediating inhibitory signaling under homeostatic conditions (Hansen et al. 2019). Similar inhibitory effects can be exerted by peptidomimetics and such approaches may be useful to prevent undesirable inflammatory conditions triggered by abnormal IgA-containing immune complexes, such as in IgA-mediated blistering skin diseases (Breedveld and van Egmond 2019; Heineke et al. 2017; Ben Mkaddem et al. 2019).

A recent study underlined differences in the effector functions of human serum IgA1 and IgA2 on myeloid cells. IgA proteins were heat-aggregated or immobilized to mimic immune complexes. Under these conditions, IgA2 was more effective than IgA1 in the induction of pro-inflammatory responses in neutrophils and macrophages (Steffen et al. 2020). However, these differences disappeared after the enzymatic removal of sialic acid or all N-glycans from IgA1. Thus, IgA effector functions depend on the subclass and glycosylation, as IgA1 and IgA2 have similar but different glycosylation profiles (IgA2 has 2 additional sites of N-glycosylation). Notably, IgA1 contains more sialic acid than IgA2, thus explaining the nature of distinct IgA subclass activities.

Asialoglycoprotein receptor on hepatocytes mediates clearance of IgA (and other asialoglycoproteins) from the circulation (Stockert et al. 1982; Baenziger and Maynard 1980; Baenziger and Fiete 1980; Rifai et al. 2000; Tomana et al. 1988). The binding of IgA is based on recognition of terminal Gal or GalNAc residues by the lectin receptor whereas sialic acid prevents this binding. The IgA molecules internalized by the asialoglycoprotein receptor are degraded and excreted into the bile. This catabolic pathway is thought to explain the relatively short half-life of IgA in the circulation (~5 days).

Polymeric immunoglobulin receptor mediates the specific transport of polymeric immunoglobulins (J-chain-containing polymeric IgA and IgM) across the mucosal epithelium into the secretions. The transport of dimeric IgA across the epithelium (transcytosis) begins with its binding to the polymeric immunoglobulin receptor at the basolateral surface of the epithelial cell, which is followed by internalization and transport via vesicular compartments to the apical surface of the cell. During the process, the polymeric immunoglobulin receptor is cleaved to release the SC part of the receptor. Secretory component is attached through a disulfide bridge to dimeric IgA, forming secretory IgA released at the apical surface of mucosal epithelium. Notably, SIgA1 exhibits different *O*-glycosylation compared to circulatory IgA1. Circulatory IgA1 contains core 1 *O*-glycans (i.e., GalNAc-Gal that may be sialylated) whereas secretory IgA1 has also core 2 *O*-glycans with extended branches (Royle et al. 2003).

The transferrin receptor (CD71) binds IgA1, but not IgA2, and the IgA1 binding is inhibitable by transferrin (Moura et al. 2001). CD71 binds polymeric, but not monomeric IgA1, and the binding is dependent on glycosylation, namely *O*-glycosylation (Moura et al. 2004). CD71 is thought to participate in the binding of pathogenic IgA1-containing immune complexes by mesangial cells in IgAN (Moura et al. 2004).

FcRL4, expressed by a subset of memory B cells in the epithelia, recognizes polymeric IgA with J chain and heat-aggregated IgA but not secretory IgA (Wilson et al. 2012; Liu et al. 2020). FcRL4 is an inhibitory receptor for IgA. In addition to the four extracellular C2-type Ig domains and a transmembrane domain, it has a cytoplasmic domain that contains three immune-receptor tyrosine-based inhibitory motifs. FcRL4 is thought to regulate B cell responses to mucosal commensal antigens (Liu et al. 2018) and FcRL4-positive B cells have significantly increased usage of the IgA isotype (Amara et al. 2017). FcRL4-positive B cells are also enriched in the joints of patients with rheumatoid arthritis and it is thought that the production of cytokines with bone remodeling activity contributes to the disease pathology (Amara et al. 2017). It is currently not known whether glycosylation of IgA impacts recognition by FcRL4.

These several examples illustrate how glycosylation can impact interactions of different molecular forms of IgA with various types of IgA receptors. Consequently, these glycan-dependent interactions impact the effector function of specific IgA glycoproteoforms.

14.9 Observed IgA1 *O*-Glycosylation Heterogeneity

Based on the involvement of IgA1 in the pathogenesis of IgAN and other similar diseases, considerable progress has been made in the analysis of *O*-glycosylation of serum/plasma IgA1. Much of the current knowledge of IgA1 heterogeneity is based on the analysis of IgA1 myeloma proteins that have been extensively used for the development and comparisons of methodologies discussed above (Renfrow et al. 2005, 2007; Takahashi et al. 2010, 2012). These proteins have also served as model proteins for in vitro and in vivo experiments to understand the pathogenesis of IgAN (Takahashi et al. 2014; Novak et al. 2007, 2011a, b, 2015; Yanagihara et al. 2012; Knoppova et al. 2016; Suzuki et al. 2009, 2019; Rizk et al. 2019; Moldoveanu et al.

2021). Additionally, as the IgA1 HR has only two amino-acid tandem repeats rather than multiple repeats found in mucins, it has been used in many studies as a template for the understanding of clustered O-glycan synthesis by glycosyltransferases (Wandall et al. 2007; Ten Hagen et al. 2003; Fritz et al. 2004). This has also contributed to the understanding of what the final IgA1 O-glycan heterogeneity is and how it could be altered in a disease such as IgAN.

Nomenclature for IgA1 O-glycosylation forms core 1 O-glycans is based on glycan composition starting with the first monosaccharide attached, GalNAc, followed by Gal and then sialic acid (SA). A designation of GalNAc4Gal4 implies four O-glycan chains per heavy-chain HR comprised of GalNAc + Gal disaccharides. A designation of GalNAc5Gal4 implies five O-glycan chains comprised of four disaccharides and a single GalNAc monosaccharide. When sialic acid residues are assigned, such as GalNAc4Gal3SA1 or GalNAc5Gal5SA2, there is no assumption of where or how the SA is linked. IgA1 with at least one GalNAc without Gal in the HR O-glycans is often referred to as Gd-IgA1. Among the observed Gd-IgA1 glycoforms, there are variants with one to three Gal-deficient sites per HR. However, it should be noted that these forms are seen in IgA1 from healthy individuals as well as patients with IgAN. Still, most patients with IgAN exhibit elevated circulatory levels of Gd-IgA1.

Early studies of IgA1 from patients with IgAN and healthy controls observed the HR with three to six O-glycans (Baenziger and Kornfeld 1974; Mattu et al. 1998; Takahashi et al. 2010, 2012). This distribution has been consistent across many analyses by MS of serum and IgA1 myeloma proteins. There have been some examples of IgA1 myeloma proteins with as few as one to three O-glycans per HR, but these appear to be outside the norm (Renfrow et al. 2007). Consistently, the distribution of O-glycosylated forms appears to center around those with four (GalNAc4Gal4) and five (GalNAc5Gal4) O-glycans (Novak et al. 2000; Takahashi et al. 2010; Iwase et al. 1996a; Wada et al. 2010a; Renfrow et al. 2005). The 2010 comparison study (Wada et al. 2010a) across fifteen different labs confirmed that there was a clear distinction in the abundance of glycoforms no matter what MS technology was used. MS studies of site-specific protein and peptide glycosylation heterogeneity have made use of relative quantitation to reflect the consistent pattern and abundance of glycoform distributions that are observed. Experimentally manipulated changes in abundance are consistent and reflect that the ionization is driven by the peptide portion of the molecule (Stewart et al. 2019). Several groups have demonstrated this for both N- and O-glycosylation (Steffen et al. 2020; Plomp et al. 2018; Goonatilleke et al. 2019; Deshpande et al. 2010; Huang et al. 2015) as well as in the analysis of therapeutic glycoproteins (Yang et al. 2016). When IgA1 populations are analyzed with sialidase pretreatment, the distribution is dominated by GalNAc4Gal4 and GalNAc5Gal4 (usually >50% of total combined) followed by a second tier including GalNac4Gal3, GalNAc5Gal3, GalNAc5Gal5, and GalNAc3Gal3 (usually 30–40% of total combined). The remaining distribution is comprised of GalNAc6Gal4, GalNAc6Gal3, GalNac6Gal5, GalNAc3Gal2, GalNAc4Gal2, and GalNAc5Gal2 (each representing <5% of total) (Takahashi et al. 2012; Ohyama et al. 2020b). As discussed more below, most of these identified

IgA1 *O*-glycopeptides are mixtures of amino-acid positional isomers (i.e., glycans attached at variable sites) further complicating the complete assignment of the total distribution of clustered *O*-glycan heterogeneity (Fig. 14.4b).

When viewed in the context of the IgA1 structure, the distribution of *O*-glycosylated forms is a result of the number of available Ser and Thr in the two tandem repeat amino-acid sequence, the enzymatic activity of the glycosyltransferases, and the three-dimensional constraints of the clustered amino-acid sites and flanking cysteines that are involved in disulfide bonds as part of the overall IgA1 quaternary structure. A recent analysis of in vitro reactions of GalNAc-T2 enzyme with IgA1 HR acceptor peptide identified a fast-rate phase of clustered GalNAc addition. For the GalNAc-T2 clustered activity on the native IgA1 HR, the fast phase ends after three GalNAc residues are added, matching the lower end of the predominant IgA1 *O*-glycosylated forms observed in serum IgA1 (Stewart et al. 2019). Interestingly, the natural occurrence of proteoforms with up to six *O*-glycan chains observed for serum IgA1 clustered *O*-glycans is less than what is observed for in vitro synthesis reactions on IgA1 HR peptide substrates where seven and eight additions of GalNAc have been reported (Stewart et al. 2019, 2021). This is where the overall structure of the IgA1 heavy chain and the flanking cysteine residues likely play a role in limiting the final *O*-glycan density. It is also possible that the concurrent addition of Gal and SA to the existing GalNAc residues limits the final heterogeneity as well.

As mentioned above, within this range of three to six *O*-glycans per HR, it is also the presence of amino-acid positional isomers that were initially identified in serum IgA1 (Takahashi et al. 2012). These isomeric mixtures occur from the outset of the synthesis process as demonstrated by in vitro GalNAc-T2 reactions (Stewart et al. 2019). Instead of a strictly ordered process of preferred amino-acid site addition, Stewart et al. demonstrated that there were four alternative initial sites of GalNAc addition at Thr228, Thr236, Ser230, and Ser232 with Thr228 and Thr236 being favored over the other two (Fig. 14.4d) (Stewart et al. 2019). Each initial site of GalNAc addition led to a unique combination of possible second sites of GalNAc addition. While the sites of GalNAc addition were all consistent with those seen with serum IgA1, these experiments revealed multiple GalNAc-addition pathways and explained the biosynthetic origin of isomeric mixtures of IgA1 *O*-glycosylated forms. Additional types of isomers can then occur due to variable sites of the addition of Gal and SA residues. Despite these variable *O*-glycosylation pathways, there is a consistent fidelity of the final IgA1 *O*-glycosylation distribution. For example, Ser230, Thr236, and Thr233 are the sites that show the predominant mixtures of isomers at the disaccharide and monosaccharide levels (Takahashi et al. 2012). These same sites are the predominant sites of Gal deficiency in serum IgA1. A recent study of serum IgA1 used enzymatic removal of GalNAc-Gal disaccharides (Ohyama et al. 2020b). This process resulted in consolidated mixtures of Gd-IgA1 forms with one, two, or three Gal-deficient sites (Ohyama et al. 2020b). These results corroborated Ser230, Thr236, and Thr233 as the predominant sites with Gal-deficient glycans but also indicated low levels of Thr228 and Thr232 with a single GalNAc residue as well. This study thus correlates with the data from in vitro studies of GalNAc-T2 that revealed multiple isomer possibilities. Since there are other GalNAc-Ts that have demonstrated activity on the IgA1 HR, it is possible that the distribution of IgA1 glyco-isomers is affected by expression patterns of GalNAc-Ts in specific IgA1-producing cells.

O-glycan heterogeneity of IgA1 can be impacted by additional constraints for the addition of Gal and SA. Based on the predominance of GalNAc4Gal4 and GalNAc5Gal4 HR glycoforms, it is logical to conclude the addition of up to four Gal residues is feasible for the C1GalT1 enzyme. The abundance of GalNAc5Gal5 HR glycoforms varies across some reports (~5–10%). This could reflect steric hindrances affecting the addition of a fifth Gal addition and/or a competition with SA addition occurring at the same time. The presence of SA on adjacent O-glycan chains has been demonstrated to inhibit C1GalT1 addition of Gal (Stewart et al. 2021; Takahashi et al. 2014; Suzuki et al. 2014). A similar inhibition has been demonstrated for SA addition by ST6GalNAc2 with prior addition of SA by ST3Gal1 and vice versa (Stewart et al. 2021; Suzuki et al. 2014). Interestingly, reports on O-glycan heterogeneity of serum IgA1 and IgA1 myeloma proteins have not found more than five SA residues for single IgA1 HR O-glycoforms despite there being two potential sites of addition on each glycan of core 1 disaccharide. This could be either the result of the steric hindrance of the clustered O-glycans or an artifact of analyzing IgA1 HR by positive-ion mode MS where SA residues add a negative charge. To date, it is unclear if a full accounting of the range of SA in IgA1 O-glycans has been accomplished.

In summary, while the generalized IgA1 O-glycan heterogeneity is consistent with three to six O-glycans per HR, there are underlying isomer mixtures that are likely cell-specific. Additionally, there is ample evidence from lectin ELISA studies that this complex heterogeneity is altered or shifted, reflecting elevated serum levels of Gd-IgA1 in patients with IgAN. The origin of these differences in IgA1 O-glycan heterogeneity is an active area of research.

14.10 IgA-Related Diseases

14.10.1 IgA Nephropathy

IgA nephropathy (IgAN) is an autoimmune disease characterized by the glomerular deposition of immune complexes containing galactose-deficient IgA1 (Gd-IgA1) (Knoppova et al. 2016; Rizk et al. 2019). This altered glycosylation results in the increased presence of terminal GalNAc or sialylated-GalNAc in the IgA1 HR. Gd-IgA1 in the glomerular immunodeposits is usually co-deposited with complement C3 and IgG autoantibodies specific to the abnormal glycosylation on the HR of Gd-IgA1. IgAN patients often have elevated levels of Gd-IgA1 and the corresponding IgG autoantibodies in the circulation, leading to the formation of Gd-IgA1-IgG circulating immune complexes. Circulatory levels of Gd-IgA1 are predictive of both disease progression and recurrence after transplantation (Berthoux et al. 2017). The current hypothesis on the pathobiology of IgAN highlights a four-hit mechanism, where elevated levels of Gd-IgA1 in the circulation (1) coupled with the production of anti-Gd-IgA1 IgG autoantibodies (2) leads to the formation of circulating immune complexes (3) some of which deposit in the glomeruli (4) to induce renal injury (Novak et al. 2008).

The galactose deficiency in IgAN affects HR glycans of the IgA1 and, thus, represents a change in O-glycosylation (Fig. 14.3). Gd-IgA1 in the circulation is predominantly in the polymeric (dimeric) form, although monomeric IgA1 is the predominant molecular form in the circulation; mechanisms leading to circulation of this molecular-form-specific effect are not known. In addition to genetically determined serum levels of Gd-IgA1, several studies show that some cytokines can elicit increased Gd-IgA1 production in IgA1-producing cells from IgAN patients. This effect is mediated by reduced expression of the galactosyltransferase encoded by the *C1GALT1* gene, C1GalT1, an enzyme that is responsible for addition of galactose to the GalNAc residues in the HR of IgA1 (Suzuki et al. 2014; Yamada et al. 2017, 2020). Two GWAS studies have found association between IgAN and SNPs in the *C1GALT1* locus, as well as its chaperone protein Cosmc encoded by *C1GALT1C1* (Kiryluk et al. 2017).

The glycosylation of Gd-IgA1 in IgAN exhibits microheterogeneity patterns, both at the level of site attachments and specific glycan composition at each site (Novak et al. 2018; Franc et al. 2013; Ohyama et al. 2020a; Moore et al. 2007). The implication for this variability in glycosylation motifs suggests a semi-stochastic process based on enzyme activities and locations in the Golgi apparatus. Further research is needed to ascertain details of the mechanisms affecting the levels of Gd-IgA1 in the circulation of IgAN patients.

14.10.2 IgA Vasculitis with Nephritis (Formerly Known as Henoch–Schönlein Purpura Nephritis)

IgA vasculitis (IgAV) with nephritis (IgAVN) exhibits kidney-pathology features similar to IgAN, including IgA1 immunodeposits in the mesangium (Selewski et al. 2018). Notable, some patients with IgAVN progress to IgAN. As in patients with IgAN, IgAVN patients have elevated circulating levels of Gd-IgA1 (Kiryluk et al. 2011; Suzuki et al. 2019; Lau et al. 2007; Nakazawa et al. 2019; Pillebout et al. 2017) that form pathogenic immune complexes (Suzuki et al. 2019; Novak et al. 2007). The Gd-IgA1 immune complexes in IgAV typically deposit in the small vessels, leading to systemic vasculitis. In patients with IgAV, only a small fraction develops nephritis, 4–6 weeks from the disease onset, with mesangial proliferation found upon renal biopsy (Lau et al. 2010; Boyd and Barratt 2011). These biopsies show deposition of Gd-IgA1 in a fashion similar to IgAN (Zhao et al. 2020; Sugiyama et al. 2020). This, along with genetic studies showing heritability of circulatory Gd-IgA1 levels in patients with IgAN and IgAVN, indicates a close relationship between IgAVN and IgAN (Kiryluk et al. 2011; Suzuki et al. 2018; Hastings et al. 2021).

14.10.3 Crohn's Disease

Crohn's disease is a subset of inflammatory bowel diseases, which affects approximately 1% of the US population, and is characterized by inflammation of the gastrointestinal tract (Hanauer 2006). Serological predictors of this disease have been difficult to determine, but recent work on IgA1 HR glycoforms found that decreased number of GalNAc residues per HR was associated with progression vs. recovery in Crohn's patients. Additionally, IgA1 with reduced content of *O*-glycans was found in the inflamed sections of intestinal biopsies from Crohn's patients (Inoue et al. 2012).

14.10.4 Sjögren's Syndrome

Sjögren's syndrome is an autoimmune disease that can affect the tear and salivary glands, with a minority of patients developing other rheumatoid complications, such as systemic lupus erythematosus and rheumatoid arthritis. Patients with Sjögren's syndrome have increased sialic acid content and decreased galactose content on both IgA1 and IgA2 *N*-glycans (Dueymes 1995). Additionally, Sjögren's syndrome patients also have elevated serum levels of IgA1, and follow-up analysis showed that IgA1 was predominantly oversialylated in *N*-glycans (Basset et al. 2000; Levy et al. 1994).

14.10.5 Systemic Lupus Erythematosus

Systemic lupus erythematosus (SLE) is an autoimmune disease where the immune system targets multiple tissues and organs, such as the brain, lung, kidneys, joints, and vasculature. Most SLE patients have a positive anti-nuclear antibody test, but additional tests for anti-ds DNA antibodies, anti-Smith antibodies, and anti-U1RP antibodies can be more specific (Mummert et al. 2018; Olsen et al. 2017). Elevated circulating levels of IgA, 4–6-times higher than in healthy individuals, and abnormal glycosylation of IgA were observed in SLE patients. IgA isolated from female SLE patients showed decreased levels of unbisected biantennary, and tri- and tetraantennary oligosaccharides. Additionally, decreased galactosylation of the HR of IgA1 was found in the same female SLE patients (Matei and Matei 2000).

14.10.6 Rheumatoid Arthritis

The clinical manifestations of rheumatoid arthritis (RA) are swelling and pain in the joints, bone, and cartilage due to inflammation from inappropriate targeting by the immune system. The standard test for RA is a blood analysis of IgG/IgA/IgM antibodies against IgG (RF; rheumatoid factor) and ACPA (anti-citrullinated protein antibodies) (Westra et al. 2021; Kurowska et al. 2017). Mass spectrometry analysis of IgA1 isolated from the circulation of RA patients showed decreased content of GalNAc in the HR of IgA1 (Wada et al. 2010b). Additionally, both IgG and IgA from synovial fluid showed differential galactosylation and sialylation at early and late stages of RA. Terminal $\alpha2,6$ sialic acid on IgA was lower in early RA compared to advanced RA, while terminal Gal on IgA was lower at early stages of RA and normalized later in the disease (Kratz et al. 2010).

14.10.7 IgA Myeloma

IgA myeloma is a clonal expansion of some IgA-producing cells that undergo genetic mutations leading to uncontrolled proliferation and increased serum levels of IgA (ranging from <30 g/L to >30 g/L) as the disease develops and progresses. Patients with IgA myeloma can develop multiple complications affecting the kidneys and bone and can exhibit various hematological pathologies. Production of IgA from clonal expansion in myeloma was found to include IgA glycoforms that were hyposialylated, which can affect FcαR1 binding (Basset et al. 1999; Bosseboeuf et al. 2020). Studies of the HR *O*-glycans from different IgA1 myeloma proteins have shown ranges of *O*-glycans outside the norm of three to six glycans observed in normal human serum (Renfrow et al. 2007). IgA1 deposition in the kidneys of some patients with IgA myeloma has been found concomitantly with under-galactosylated HRs, similar to the abnormality common to IgAN (Zickerman et al. 2000).

14.10.8 Celiac Disease

Celiac disease is an inflammatory intestinal condition brought on by antibodies targeting the transglutaminase 2 (TG2) protein and gluten-derived TG2-deamidated gliadin peptides. In patients with celiac disease, circulating IgA that targets TG2 has been found to be under-galactosylated, in addition to nonTG2 targeting IgA in circulation (Lindfors et al. 2011; Abbad et al. 2020). CD89 may be involved (Papista et al. 2015) and the interaction of multiple proteins, including CD71, was proposed (Lebreton et al. 2012; Papista et al. 2012).

14.10.9 Complement and IgA

IgA has been considered an anti-inflammatory immunoglobulin (Monteiro 2014; Diana et al. 2013; Ben Mkaddem et al. 2013). Unlike IgG, IgA-mediated activation of complement is not well understood, but it is generally accepted that IgA does not activate the classical complement pathway. However, IgA is thought to be able to activate alternative and lectin pathways of complement, under some circumstances. For example, the less abundant monomeric IgA2 was found to bind mannose-binding lectin (MBL) in ELISA. Additionally, when IgA1 or IgA2, both monomeric and polymeric, were treated with a galactosidase to remove galactose, their binding to MBL increased by over 10-fold (Terai et al. 2006). This would presumably be due to reducing the heterogeneity complexity of N-glycans in favor of high mannose glycans such as those observed at Asn340/327 sites. Additionally, IgA1 N-glycosylation is critical for binding C3, as evidenced by using point-mutation deletions in critical Asn sites in the Fc region (Chuang and Morrison 1997). This observation was further validated by inhibiting N-glycosylation of IgA using tunicamycin and testing for C3 binding (Zhang and Lachmann 1994). Interestingly, sialic acid content was also found to be critical for both IgA1 and IgA2 to fix C3b (Nikolova et al. 1994). In IgAN, where Gd-IgA1 is found in the polymeric form, the polymeric IgA1 is found to be able to bind MBL, resulting in C4 mesangial deposition (Oortwijn et al. 2006; Roos et al. 2001). Of note, complement activation resulting from abnormal IgA1 glycosylation can be found indirectly via anti-Gd-IgA1 IgG autoantibodies in IgAN, where the IgG presumably activates C3 when bound in an immune-complex with Gd-IgA1 (Novak et al. 2015; Knoppova et al. 2016; Rizk et al. 2019; Maillard et al. 2015).

14.11 Infectious Diseases

Some bacteria that can cause human infections utilize tools that negatively affect the defense mechanisms of the host. These tools include IgA-specific proteases and various glycosidases. Whereas the former act exclusively on IgA, the latter affect multiple glycoproteins at the mucosal surfaces. Below, we focus on several aspects of these bacterial virulence factors most relevant to IgA structural integrity and function.

14.11.1 Bacterial IgA-Specific Proteases

Multiple bacterial species that colonize human mucosal surfaces in the oral cavity, digestive tract, respiratory tract, and genital tract produce IgA-specific proteases. These bacteria include streptococci (e.g., *Streptococcus pneumoniae, S. sanguinis,*

Fig. 14.6 Bacterial IgA-specific proteases target the hinge region of IgA1. Amino-acid sequence of the hinge region of human IgA1. Six commonly utilized *O*-glycosylation sites are marked by the numbers below Ser/Thr residues. Numbers 1–5 with arrows above the amino-acid sequence mark peptide bonds that are cleaved by IgA-specific proteases produced by the following species and strains shown as examples: 1, *Clostridium ramosum* AK183; 2, *Streptococcus pneumoniae* TIGR4; 3, *Haemophilus influenzae* HK50; 4, *Neisseria gonorrhoeae* HF13; and 5, *Neisseria gonorrhoeae* HF48

S. mitis, and *S. oralis*), *Prevotella* and *Capnocytophaga* species and several *Neisseria*, *Clostridium*, and *Haemophilus* species (e.g., *Clostridium ramosum*, *Neisseria gonorrhoeae*, *N. meningitidis*, *Haemophilus influenzae*) (Kilian and Holmgren 1981; Kilian et al. 1979, 1983; Senior et al. 2000; Frandsen et al. 1987, 1995, 1997; Reinholdt et al. 1990; Poulsen et al. 1989, 1996; Lomholt et al. 1992; Lomholt and Kilian 1994; Reinholdt and Kilian 1997; Kosowska et al. 2002; Wang et al. 2020; Chi et al. 2017; Murphy et al. 2015; Janoff et al. 2014; Johnson et al. 2009; Wani et al. 1996).

The bacterial IgA-specific proteases are post-proline endopeptidases that include metalloproteases, serine proteases, and cysteine proteases. These proteases recognize and cleave HR of IgA1, except the protease from *Clostridium ramosum* that can cleave both IgA1 and IgA2, with variable site-specific preferences (Senior et al. 2000; Kosowska et al. 2002; Batten et al. 2003; Mistry and Stockley 2011; Senior and Woof 2005). Figure 14.6 shows examples of commonly recognized sites in IgA1 HR by several representative IgA-specific proteases.

It is thought that bacterial IgA-specific proteases represent a virulence factor for many pathogenic bacteria. Cleavage of IgA1 breaks the Fc fragment from its antigen-binding Fab fragments. This process thus reduces the protective effects of IgA1, such as inhibition of bacterial adherence and antibody complement-dependent killing of bacteria by phagocytic cells. In vitro studies of these proteases have demonstrated that the clustered IgA1 HR *O*-glycans can inhibit cleavage of some proteases indicating a shielding effect of this susceptible region of IgA1 by the *O*-glycans. Various studies examined and confirmed these concepts and searched for IgA-specific protease inhibitors (Wang et al. 2020; Janoff et al. 2014; Weiser et al. 2003; Garner et al. 2013).

14.11.2 Bacterial Glycosidases and Metabolic Utilization of IgA Glycans

Various bacteria occupying mucosal surfaces produce hydrolytic enzymes, such as endo- and exo-glycosidases, able to degrade *O*- and *N*-glycans of mucosal glycoproteins, including IgA. For example, in bacterial vaginosis, a common polymicrobial imbalance of the vaginal flora, IgA in the secretions is subjected to stepwise deglycosylation and subsequently enhanced proteolysis. These modifications are thought to compromise the ability of IgA to neutralize and eliminate pathogens (Lewis et al. 2012; Robinson et al. 2019; Govinden et al. 2018; Moncla et al. 2016; Smith and Ravel 2017). *Streptococcus pneumoniae* is a major airways pathogen that can produce multiple exo- and endoglycosidases to degrade *O*- and *N*-glycans of IgA and other glycoproteins (King et al. 2006; Marion et al. 2009; Syed et al. 2019; Robb et al. 2017; Blanchette et al. 2016; Kahya et al. 2017). For *S. pneumoniae* colonization of the upper airways, neuraminidase NanA is essential (Brittan et al. 2012). In some instances, a synergy between *S. pneumoniae* and influenza A sialidases can impact nasal colonization and middle ear infection by pneumococci (Wren et al. 2017). Conversely, IgA glycans, specifically sialic acid of the tail *N*-glycans, can inhibit influenza A and other enveloped viruses that use sialic acid as a receptor (Maurer et al. 2018). Another example of pathogenic bacteria utilizing glycosidases includes periodontitis and the associated bacteria, such as *Porphyromonas gingivalis* known to target host sialylated glycoproteins for immune dysregulation (Sudhakara et al. 2019). In the gastrointestinal system, IgA has multiple functions and glycans of secretory IgA1 (SIgA) are known to mediate the binding of SIgA to the microbiota. In fact, it has been proposed that IgA1 *N*- and *O*-glycans provide a link between innate and adaptive immune systems that sustain intestinal homeostasis (Royle et al. 2003; Reily et al. 2019; Mathias and Corthésy 2011a; Pabst and Slack 2020; Gupta et al. 2019; Nakajima et al. 2018; Corthésy 2013; Mathias and Corthésy 2011b). The net result of releasing the glycans is not only immune dysregulation and immune evasion but also a metabolic advantage. Some bacteria can metabolize the released glycans to support their growth (Perman and Modler 1982; Garbe and Collin 2012; Sjögren and Collin 2014).

14.11.3 IgA as a Potential Therapeutic Antibody

As outlined earlier, functions of human IgA subclasses, IgA1 and IgA2, are impacted by multiple aspects, including glycosylation, that directly and indirectly regulate IgA effector functions. For example, IgA1, but not IgA2, has *O*-glycans in addition to *N*-glycans and is more sialylated at the *N*-glycans. Consequently, IgA2 can induce pro-inflammatory responses in neutrophils and macrophages more effectively than IgA1 (Steffen et al. 2020). Conversely, polymeric IgA1, but not polymeric IgA2 or monomeric versions of either subclass, interacts with transferrin receptor CD71, apparently through IgA1 HR *O*-glycans (Moura et al. 2004). Furthermore, IgA glycans, specifically Gal and GalNAc, impact catabolism of IgA in the liver, a process that depends on the asialoglycoprotein receptor on hepatocytes and contributes to the short half-life of IgA in the circulation.

These aspects are at the forefront of recent research due to a growing interest in using IgA as a therapeutic antibody. The therapeutic applications of IgA may include systemic and mucosal delivery. In either setting, a better understanding is needed about the impact of producing-cell-specific glycosylation on IgA effector functions, receptor interactions, and pharmacokinetics. Biotechnological systems for the production of recombinant IgA, as well as other recombinant Igs, can include plants (Göritzer et al. 2017, 2019, 2020; Strasser et al. 2021; Kallolimath et al. 2016, 2020; Montero-Morales et al. 2019; Montero-Morales and Steinkellner 2018; Kallolimath and Steinkellner 2015). A better understanding of the differential glycosylation of various molecular forms and subclasses of IgA is needed to inform design and production of recombinant therapeutic IgA for future preclinical and clinical trials.

14.12 Therapeutic IgA Antibodies

Monoclonal antibody therapeutics have gained popularity in the medical field to treat a large set of pathologies due to their intrinsic specificities to target antigens. These therapies have largely been using IgG, although some research has been performed in vivo and in vitro to assess the feasibility of IgA therapeutics. IgA can be both pro-inflammatory and anti-inflammatory depending on the structure of its glycans which can regulate its binding to the FcαR1 and FcαR2 receptors. For example, treatment with IgA monoclonal antibodies against human epidermal growth factor receptor 2 (HER2) and epidermal growth factor receptor (EGFR) enhanced antitumor activity in FcαR1 transgenic mice (Boross et al. 2013; Meyer et al. 2016). This is thought to be done via activation of neutrophils and macrophages, which can be activated through IgA binding to FcαR1 and activating the immunoreceptor tyrosine-based activation motifs (ITAMs) pathway (Heineke and van Egmond 2017).

Targeting FcαR1 could prove to be an important clinical modality for IgA monoclonal treatments. However, a number of hurdles need to be overcome for IgA therapies to work. IgA is cleared more rapidly than IgG from the circulation in part due to asialoglycoprotein receptor expression in the liver. This receptor can bind multiple glycosylation sites on the IgA molecule, which can be inhibited by glycoengineering IgA to have increased terminal sialic acids (Rifai et al. 2000; Boross et al. 2013; Meyer et al. 2016). Interestingly, decreasing sialic acid content of IgA increases its pro-inflammatory capabilities (Steffen et al. 2020), leading to a trade-off between circulatory residence time (bioavailability) and the desired effector function of monoclonal IgA generated.

Production of properly folded and glycosylated IgA proteins can be difficult in mammalian systems, especially tuning the glycoengineering component to create the needed structures. This is in part due to the many different forms of *N*- and *O*-glycans that can be produced by mammalian cells, which makes it difficult to generate consistent glycoforms on a complex background (Montero-Morales and Steinkellner 2018). A more understudied way of generating consistent glycosylated proteins is to use modified plant cells that have undergone genetic modifications to specific glycosyltransferases involved in *N*- and *O*-glycosylation pathways. This approach was recently used successfully to generate IgG monoclonal therapies against the Ebola virus outbreak in Africa (Qiu et al. 2014; Davey et al. 2016). The generation of mammalian-like *N*-glycosylation-competent plants for IgG antibodies was first done in the *Nicotiana benthamiana* species, using genetic manipulations to inactivate the endogenous β1,2-xylosyltransferase (XylT) and α1,3-fucosyltransferase (FucT) genes (ΔXT/FT). This produced a biologically competent anti-HIV monoclonal antibody, with the same glycosylation as those expressed in Chinese hamster ovary (CHO) cell lines (Strasser et al. 2008). These ΔXT/FT plants have now been deployed to produce normally *N*- and *O*-glycosylated IgA1. While the whole IgA1 protein was not reproduced in the ΔXT/FT variant, the researchers were able to overcome hydroxyproline formation at the HR by overexpressing GalNAc-T2. This produced a variant of the core-1 motif-containing galactose, and only modest sialylation (Dicker et al. 2016). These studies that generate the technology for robust manufacturing of glycosylation competent IgA will be able to open up a larger sector for IgA therapeutics.

14.13 Conclusions

Understanding IgA glycosylation and its role in disease is still a developing field of study. However, with the help of MS, and recent technological advances, substantial progress has been made. Although still in its early stages, it is becoming more feasible to evaluate and decipher glycan structures and range of heterogeneities, especially when the different IgA isotypes and molecular forms are separated. Our knowledge of the role of IgA glycosylation in various diseases is currently based on observational studies. The one exception is the role that IgA1 *O*-glycans have been shown to play in the pathogenesis of IgA nephropathy. As the role of IgA in the immune response is further delineated from other immunoglobulins, the role of the glycans, especially those unique to the various IgA isotypes and molecular forms will be better understood. This will likely lead to IgA being more frequently used as a therapeutic to target specific disease states and/or locations within the human body.

Acknowledgments The authors express their appreciation to the colleagues and collaborators as well as to all volunteers participating in the cited studies. The authors apologize to their colleagues in the field whose work is not adequately discussed or cited in this review due to the space limitations.

Compliance with Ethical Standards

Funding The authors of this study were supported in part by NIH grants AR069516, DK106341, CD122194, GM098539, DK078244, and DK082753 and by a gift from the IGA Nephropathy Foundation of America.

Conflict of Interest MBR and JN are co-founders and co-owners of and consultants for Reliant Glycosciences, LLC. MBR and JN are co-inventors on US patent application 14/318,082 (assigned to UAB Research Foundation). ALH and CR declare no conflict of interest.

Ethical Approval Not applicable, as this is a review article and, thus, there are no human subjects recruited for this study or any animals used.

Informed Consent Not applicable, as this is a review article and, thus, there are no human subjects recruited for this study.

References

Abbad L, Monteiro RC, Berthelot L (2020) Food antigens and Transglutaminase 2 in IgA nephropathy: molecular links between gut and kidney. Mol Immunol 121:1–6

Aleyd E, Heineke MH, van Egmond M (2015) The era of the immunoglobulin A Fc receptor FcαRI; its function and potential as target in disease. Immunol Rev 268(1):123–138

Allen AC, Willis FR, Beattie TJ, Feehally J (1998) Abnormal IgA glycosylation in Henoch-Schönlein purpura restricted to patients with clinical nephritis. Nephrol Dial Transplant 13(4):930–934

Amara K, Clay E, Yeo L et al (2017) B cells expressing the IgA receptor FcRL4 participate in the autoimmune response in patients with rheumatoid arthritis. J Autoimmun 81:34–43

Aryal RP, Ju T, Cummings RD (2012) Tight complex formation between Cosmc chaperone and its specific client non-native T-synthase leads to enzyme activity and client-driven dissociation. J Biol Chem 287(19):15317–15329

Atkin JD, Pleass RJ, Owens RJ, Woof JM (1996) Mutagenesis of the human IgA1 heavy chain tailpiece that prevents dimer assembly. J Immunol 157(1):156–159

Baenziger JU, Fiete D (1980) Galactose and N-acetylgalactosamine-specific endocytosis of glycopeptides by isolated rat hepatocytes. Cell 22(2 Pt 2):611–620

Baenziger J, Kornfeld S (1974) Structure of the carbohydrate units of IgA1 immunoglobulin. II. Structure of the O-glycosidically linked oligosaccharide units. J Biol Chem 249(22):7270–7281

Baenziger JU, Maynard Y (1980) Human hepatic lectin. Physiochemical properties and specificity. J Biol Chem 255(10):4607–4613

Basset C, Devauchelle V, Durand V et al (1999) Glycosylation of immunoglobulin A influences its receptor binding. Scand J Immunol 50(6):572–579

Basset C, Durand V, Jamin C et al (2000) Increased N-linked glycosylation leading to oversialylation of monomeric immunoglobulin A1 from patients with Sjögren's syndrome. Scand J Immunol 51(3):300–306

Bastian A, Kratzin H, Eckart K, Hilschmann N (1992) Intra- and interchain disulfide bridges of the human J chain in secretory immunoglobulin A. Biol Chem Hoppe Seyler 373(12):1255–1263

Batten MR, Senior BW, Kilian M, Woof JM (2003) Amino acid sequence requirements in the hinge of human immunoglobulin A1 (IgA1) for cleavage by streptococcal IgA1 proteases. Infect Immun 71(3):1462–1469

Ben Mkaddem S, Rossato E, Heming N, Monteiro RC (2013) Anti-inflammatory role of the IgA Fc receptor (CD89): from autoimmunity to therapeutic perspectives. Autoimmun Rev 12(6):666–669

Ben Mkaddem S, Benhamou M, Monteiro RC (2019) Understanding Fc receptor involvement in inflammatory diseases: from mechanisms to new therapeutic tools. Front Immunol 10:811

Berthoux F, Suzuki H, Mohey H et al (2017) Prognostic value of serum biomarkers of autoimmunity for recurrence of IgA nephropathy after kidney transplantation. J Am Soc Nephrol 28(6):1943–1950

Blanchette KA, Shenoy AT, Milner J 2nd et al (2016) Neuraminidase A-exposed galactose promotes *Streptococcus pneumoniae* biofilm formation during colonization. Infect Immun 84(10):2922–2932

Bondt A, Nicolardi S, Jansen BC et al (2016) Longitudinal monitoring of immunoglobulin A glycosylation during pregnancy by simultaneous MALDI-FTICR-MS analysis of N- and O-glycopeptides. Sci Rep 6:27955

Bondt A, Nicolardi S, Jansen BC et al (2017) IgA N- and O-glycosylation profiling reveals no association with the pregnancy-related improvement in rheumatoid arthritis. Arthritis Res Ther 19(1):160

Borén T, Falk P, Roth KA, Larson G, Normark S (1993) Attachment of *Helicobacter pylori* to human gastric epithelium mediated by blood group antigens. Science 262(5141):1892–1895

Boross P, Lohse S, Nederend M et al (2013) IgA EGFR antibodies mediate tumour killing in vivo. EMBO Mol Med 5(8):1213–1226

Bosseboeuf A, Seillier C, Mennesson N et al (2020) Analysis of the targets and glycosylation of monoclonal IgAs from MGUS and myeloma patients. Front Immunol 11:854

Boyd JK, Barratt J (2011) Inherited IgA glycosylation pattern in IgA nephropathy and HSP nephritis: where do we go next? Kidney Int 80(1):8–10

Brandsma AM, Bondza S, Evers M et al (2019) Potent Fc receptor signaling by IgA leads to superior killing of cancer cells by neutrophils compared to IgG. Front Immunol 10:704

Breedveld A, van Egmond M (2019) IgA and FcαRI: pathological roles and therapeutic opportunities. Front Immunol 10:553

Brittan JL, Buckeridge TJ, Finn A, Kadioglu A, Jenkinson HF (2012) Pneumococcal neuraminidase A: an essential upper airway colonization factor for *Streptococcus pneumoniae*. Mol Oral Microbiol 27(4):270–283

Castilho A, Beihammer G, Pfeiffer C et al (2018) An oligosaccharyltransferase from Leishmania major increases the N-glycan occupancy on recombinant glycoproteins produced in *Nicotiana benthamiana*. Plant Biotechnol J 16(10):1700–1709

Chi YC, Rahkola JT, Kendrick AA et al (2017) *Streptococcus pneumoniae* IgA1 protease: a metalloprotease that can catalyze in a split manner in vitro. Protein Sci 26(3):600–610

Cho Y, Usui K, Honda S, Tahara-Hanaoka S, Shibuya K, Shibuya A (2006) Molecular characteristics of IgA and IgM Fc binding to the Fcalpha/muR. Biochem Biophys Res Commun 345(1):474–478

Chuang PD, Morrison SL (1997) Elimination of N-linked glycosylation sites from the human IgA1 constant region: effects on structure and function. J Immunol 158(2):724–732

Conley ME, Delacroix DL (1987) Intravascular and mucosal immunoglobulin A: two separate but related systems of immune defense? Ann Intern Med 106(6):892–899

Corthésy B (2013) Multi-faceted functions of secretory IgA at mucosal surfaces. Front Immunol 4:185

Crottet P, Corthésy B (1998) Secretory component delays the conversion of secretory IgA into antigen-binding competent F(ab')2: a possible implication for mucosal defense. J Immunol 161(10):5445–5453

Daniel EJP, Las Rivas M, Lira-Navarrete E et al (2020) Ser and Thr acceptor preferences of the GalNAc-Ts vary among isoenzymes to modulate mucin-type O-glycosylation. Glycobiology 30(11):910–922

Davey RT Jr, Dodd L, Proschan MA et al (2016) A randomized, controlled trial of ZMapp for Ebola virus infection. N Engl J Med 375(15):1448–1456

de Haan N, Falck D, Wuhrer M (2019) Monitoring of immunoglobulin N- and O-glycosylation in health and disease. Glycobiology 30(4):226–240

de Las RM, Lira-Navarrete E, Daniel EJP et al (2017) The interdomain flexible linker of the polypeptide GalNAc transferases dictates their long-range glycosylation preferences. Nat Commun 8(1):1959

de Sousa-Pereira P, Woof JM (2019) IgA: structure, function, and developability. Antibodies (Basel) 8(4)

de Tymowski C, Heming N, Correia MDT et al (2019) CD89 is a potent innate receptor for bacteria and mediates host protection from sepsis. Cell Rep 27(3):762–775.e765

Delacroix DL, Dive C, Rambaud JC, Vaerman JP (1982) IgA subclasses in various secretions and in serum. Immunology 47(2):383–385

Deshpande N, Jensen PH, Packer NH, Kolarich D (2010) GlycoSpectrumScan: fishing glycopeptides from MS spectra of protease digests of human colostrum sIgA. J Proteome Res 9(2):1063–1075

Diana J, Moura IC, Vaugier C et al (2013) Secretory IgA induces tolerogenic dendritic cells through SIGNR1 dampening autoimmunity in mice. J Immunol 191(5):2335–2343

Dicker M, Tschofen M, Maresch D et al (2016) Transient glyco-engineering to produce recombinant IgA1 with defined N- and O-glycans in plants. Front Plant Sci 7:18

Dueymes M, Bendaoud B, Pennec YL, Youinou P (1995) IgA glycosylation abnormalities in the serum of patients with primary Sjögren's syndrome. Clin Exp Rheumatol 13(2):247–250

Field MC, Dwek RA, Edge CJ, Rademacher TW (1989) O-linked oligosaccharides from human serum immunoglobulin A1. Biochem Soc Trans 17(6):1034–1035

Franc V, Řehulka P, Raus M et al (2013) Elucidating heterogeneity of IgA1 hinge-region O-glycosylation by use of MALDI-TOF/TOF mass spectrometry: role of cysteine alkylation during sample processing. J Proteome 92:299–312

Frandsen EV, Reinholdt J, Kilian M (1987) Enzymatic and antigenic characterization of immunoglobulin A1 proteases from Bacteroides and Capnocytophaga spp. Infect Immun 55(3):631–638

Frandsen EV, Reinholdt J, Kjeldsen M, Kilian M (1995) In vivo cleavage of immunoglobulin A1 by immunoglobulin A1 proteases from Prevotella and Capnocytophaga species. Oral Microbiol Immunol 10(5):291–296

Frandsen EV, Kjeldsen M, Kilian M (1997) Inhibition of Prevotella and Capnocytophaga immunoglobulin A1 proteases by human serum. Clin Diagn Lab Immunol 4(4):458–464

Fritz TA, Hurley JH, Trinh LB, Shiloach J, Tabak LA (2004) The beginnings of mucin biosynthesis: the crystal structure of UDP-GalNAc:polypeptide alpha-N-acetylgalactosaminyltransferase-T1. Proc Natl Acad Sci USA 101(43):15307–15312

Gala FA, Morrison SL (2002) The role of constant region carbohydrate in the assembly and secretion of human IgD and IgA1. J Biol Chem 277(32):29005–29011

Gale DP, Molyneux K, Wimbury D et al (2017) Galactosylation of IgA1 is associated with common variation in C1GALT1. J Am Soc Nephrol 28(7):2158–2166

Garbe J, Collin M (2012) Bacterial hydrolysis of host glycoproteins – powerful protein modification and efficient nutrient acquisition. J Innate Immun 4(2):121–131

Garner AL, Fullagar JL, Day JA, Cohen SM, Janda KD (2013) Development of a high-throughput screen and its use in the discovery of *Streptococcus pneumoniae* immunoglobulin A1 protease inhibitors. J Am Chem Soc 135(27):10014–10017

Gharavi AG, Moldoveanu Z, Wyatt RJ et al (2008) Aberrant IgA1 glycosylation is inherited in familial and sporadic IgA nephropathy. J Am Soc Nephrol 19(5):1008–1014

Gomes MM, Wall SB, Takahashi K, Novak J, Renfrow MB, Herr AB (2008) Analysis of IgA1 N-glycosylation and its contribution to FcαRI binding. Biochemistry 47(43):11285–11299

Gomes MM, Suzuki H, Brooks MT et al (2010) Recognition of galactose-deficient O-glycans in the hinge region of IgA1 by N-acetylgalactosamine-specific snail lectins: a comparative binding study. Biochemistry 49(27):5671–5682

Goonatilleke E, Smilowitz JT, Mariño KV, German BJ, Lebrilla CB, Barboza M (2019) Immunoglobulin A N-glycosylation presents important body fluid-specific variations in lactating mothers. Mol Cell Proteomics 18(11):2165–2177

Göritzer K, Maresch D, Altmann F, Obinger C, Strasser R (2017) Exploring site-specific N-glycosylation of HEK293 and plant-produced human IgA isotypes. J Proteome Res 16(7):2560–2570

Göritzer K, Turupcu A, Maresch D et al (2019) Distinct Fcα receptor N-glycans modulate the binding affinity to immunoglobulin A (IgA) antibodies. J Biol Chem 294(38):13995–14008

Göritzer K, Goet I, Duric S et al (2020) Efficient N-glycosylation of the heavy chain tailpiece promotes the formation of plant-produced dimeric IgA. Front Chem 8:346

Govinden G, Parker JL, Naylor KL, Frey AM, Anumba DOC, Stafford GP (2018) Inhibition of sialidase activity and cellular invasion by the bacterial vaginosis pathogen Gardnerella vaginalis. Arch Microbiol 200(7):1129–1133

Gupta S, Basu S, Bal V, Rath S, George A (2019) Gut IgA abundance in adult life is a major determinant of resistance to dextran sodium sulfate-colitis and can compensate for the effects of inadequate maternal IgA received by neonates. Immunology 158(1):19–34

Hanauer SB (2006) Inflammatory bowel disease: epidemiology, pathogenesis, and therapeutic opportunities. Inflamm Bowel Dis 12(Suppl 1):S3–S9

Hansen IS, Krabbendam L, Bernink JH et al (2018) FcαRI co-stimulation converts human intestinal CD103(+) dendritic cells into pro-inflammatory cells through glycolytic reprogramming. Nat Commun 9(1):863

Hansen IS, Baeten DLP, den Dunnen J (2019) The inflammatory function of human IgA. Cell Mol Life Sci 76(6):1041–1055

Hanson LA (1998) Breastfeeding provides passive and likely long-lasting active immunity. Ann Allergy Asthma Immunol 81(6):523–533; quiz 533–524, 537

Hastings MC, Moldoveanu Z, Julian BA et al (2010) Galactose-deficient IgA1 in African Americans with IgA nephropathy: serum levels and heritability. Clin J Am Soc Nephrol 5(11):2069–2074

Hastings MC, Rizk DV, Kiryluk K et al (2021) IgA vasculitis with nephritis: update of pathogenesis with clinical implications. Pediatr Nephrol

Heemskerk N, van Egmond M (2018) Monoclonal antibody-mediated killing of tumour cells by neutrophils. Eur J Clin Invest 48(Suppl 2):e12962

Heineke MH, van Egmond M (2017) Immunoglobulin A: magic bullet or Trojan horse? Eur J Clin Investig 47(2):184–192

Heineke MH, van der Steen LPE, Korthouwer RM et al (2017) Peptide mimetics of immunoglobulin A (IgA) and FcαRI block IgA-induced human neutrophil activation and migration. Eur J Immunol 47(10):1835–1845

Herr AB, Ballister ER, Bjorkman PJ (2003) Insights into IgA-mediated immune responses from the crystal structures of human FcalphaRI and its complex with IgA1-Fc. Nature 423(6940):614–620

Hiki Y, Tanaka A, Kokubo T et al (1998) Analyses of IgA1 hinge glycopeptides in IgA nephropathy by matrix-assisted laser desorption/ionization time-of-flight mass spectrometry. J Am Soc Nephrol 9(4):577–582

Hiki Y, Odani H, Takahashi M et al (2001) Mass spectrometry proves under-O-glycosylation of glomerular IgA1 in IgA nephropathy. Kidney Int 59(3):1077–1085

Honda S, Sato K, Totsuka N et al (2016) Marginal zone B cells exacerbate endotoxic shock via interleukin-6 secretion induced by Fcα/μR-coupled TLR4 signalling. Nat Commun 7:11498

Horie A, Hiki Y, Odani H et al (2003) IgA1 molecules produced by tonsillar lymphocytes are under-O-glycosylated in IgA nephropathy. Am J Kidney Dis 42(3):486–496

Huang J, Guerrero A, Parker E et al (2015) Site-specific glycosylation of secretory immunoglobulin A from human colostrum. J Proteome Res 14(3):1335–1349

Inoue T, Iijima H, Tajiri M et al (2012) Deficiency of N-acetylgalactosamine in O-linked oligosaccharides of IgA is a novel biologic marker for Crohn's disease. Inflamm Bowel Dis 18(9):1723–1734

Iwasaki H, Zhang Y, Tachibana K et al (2003) Initiation of O-glycan synthesis in IgA1 hinge region is determined by a single enzyme, UDP-N-acetyl-alpha-D-galactosamine:polypeptide N-acetylgalactosaminyltransferase 2. J Biol Chem 278(8):5613–5621

Iwase H, Tanaka A, Hiki Y et al (1996a) Estimation of the number of O-linked oligosaccharides per heavy chain of human serum IgA1 by matrix-assisted laser desorption ionization time-of-flight mass spectrometry (MALDI-TOFMS) analysis of the hinge glycopeptide. J Biochem 120 (2):393–397

Iwase H, Tanaka A, Hiki Y et al (1996b) Abundance of Gal beta 1,3GalNAc in O-linked oligosaccharide on hinge region of polymerized IgA1 and heat-aggregated IgA1 from normal human serum. J Biochem 120(1):92–97

Iwase H, Tanaka A, Hiki Y et al (1998) Application of matrix-assisted laser desorption ionization time-of-flight mass spectrometry to the analysis of glycopeptide-containing multiple O-linked oligosaccharides. J Chromatogr B Biomed Sci Appl 709(1):145–149

Iwase H, Tanaka A, Hiki Y et al (1999) Aggregated human serum immunoglobulin A1 induced by neuraminidase treatment had a lower number of O-linked sugar chains on the hinge portion. J Chromatogr B Biomed Sci Appl 724(1):1–7

Jackson S, Mestecky J, Moldoveanu Z, Spearman PW (2005) Appendix I – Collection and processing of human mucosal secretions. In: Mestecky J, Lamm ME, McGhee JR, Bienenstock J, Mayer L, Strober W (eds) Mucosal immunology, 3rd edn. Academic, Burlington, pp 1829–1839

Janoff EN, Rubins JB, Fasching C et al (2014) Pneumococcal IgA1 protease subverts specific protection by human IgA1. Mucosal Immunol 7(2):249–256

Johnson TA, Qiu J, Plaut AG, Holyoak T (2009) Active-site gating regulates substrate selectivity in a chymotrypsin-like serine protease the structure of *Haemophilus influenzae* immunoglobulin A1 protease. J Mol Biol 389(3):559–574

Ju T, Cummings RD (2002) A unique molecular chaperone Cosmc required for activity of the mammalian core 1 beta 3-galactosyltransferase. Proc Natl Acad Sci U S A 99(26):16613–16618

Ju T, Brewer K, D'Souza A, Cummings RD, Canfield WM (2002) Cloning and expression of human core 1 beta1,3-galactosyltransferase. J Biol Chem 277(1):178–186

Ju T, Zheng Q, Cummings RD (2006) Identification of core 1 O-glycan T-synthase from *Caenorhabditis elegans*. Glycobiology 16(10):947–958

Kahya HF, Andrew PW, Yesilkaya H (2017) Deacetylation of sialic acid by esterases potentiates pneumococcal neuraminidase activity for mucin utilization, colonization and virulence. PLoS Pathog 13(3):e1006263

Kallolimath S, Steinkellner H (2015) Glycosylation of plant produced human antibodies. Hum Antibodies 23(3–4):45–48

Kallolimath S, Castilho A, Strasser R et al (2016) Engineering of complex protein sialylation in plants. Proc Natl Acad Sci USA 113(34):9498–9503

Kallolimath S, Hackl T, Gahn R et al (2020) Expression profiling and glycan engineering of IgG subclass 1-4 in *Nicotiana benthamiana*. Front Bioeng Biotechnol 8:825

Kilian M, Holmgren K (1981) Ecology and nature of immunoglobulin A1 protease-producing streptococci in the human oral cavity and pharynx. Infect Immun 31(3):868–873

Kilian M, Mestecky J, Schrohenloher RE (1979) Pathogenic species of the genus Haemophilus and *Streptococcus pneumoniae* produce immunoglobulin A1 protease. Infect Immun 26(1):143–149

Kilian M, Thomsen B, Petersen TE, Bleeg HS (1983) Occurrence and nature of bacterial IgA proteases. Ann N Y Acad Sci 409:612–624

King SJ, Hippe KR, Weiser JN (2006) Deglycosylation of human glycoconjugates by the sequential activities of exoglycosidases expressed by *Streptococcus pneumoniae*. Mol Microbiol 59 (3):961–974

Kiryluk K, Moldoveanu Z, Sanders JT et al (2011) Aberrant glycosylation of IgA1 is inherited in both pediatric IgA nephropathy and Henoch-Schönlein purpura nephritis. Kidney Int 80 (1):79–87

Kiryluk K, Li Y, Moldoveanu Z et al (2017) GWAS for serum galactose-deficient IgA1 implicates critical genes of the O-glycosylation pathway. PLoS Genet 13(2):e1006609

Knoppova B, Reily C, Maillard N et al (2016) The origin and activities of IgA1-containing immune complexes in IgA nephropathy. Front Immunol 7:117

Koshland ME (1985) The coming of age of the immunoglobulin J chain. Annu Rev Immunol 3:425–453

Kosowska K, Reinholdt J, Rasmussen LK et al (2002) The Clostridium ramosum IgA proteinase represents a novel type of metalloendopeptidase. J Biol Chem 277(14):11987–11994

Kratz EM, Borysewicz K, Katnik-Prastowska I (2010) Terminal monosaccharide screening of synovial immunoglobulins G and A for the early detection of rheumatoid arthritis. Rheumatol Int 30(10):1285–1292

Kurowska W, Kuca-Warnawin EH, Radzikowska A, Maśliński W (2017) The role of anti-citrullinated protein antibodies (ACPA) in the pathogenesis of rheumatoid arthritis. Cent Eur J Immunol 42(4):390–398

Lau KK, Wyatt RJ, Moldoveanu Z et al (2007) Serum levels of galactose-deficient IgA in children with IgA nephropathy and Henoch-Schönlein purpura. Pediatr Nephrol 22(12):2067–2072

Lau KK, Suzuki H, Novak J, Wyatt RJ (2010) Pathogenesis of Henoch-Schönlein purpura nephritis. Pediatr Nephrol 25(1):19–26

Lebreton C, Ménard S, Abed J et al (2012) Interactions among secretory immunoglobulin A, CD71, and transglutaminase-2 affect permeability of intestinal epithelial cells to gliadin peptides. Gastroenterology 143(3):698–707.e694

Levinsky RJ, Barratt TM (1979) IgA immune complexes in Henoch-Schönlein purpura. Lancet 2 (8152):1100–1103

Levy Y, Dueymes M, Pennec YL, Shoenfeld Y, Youinou P (1994) IgA in Sjögren's syndrome. Clin Exp Rheumatol 12(5):543–551

Lewis WG, Robinson LS, Perry J et al (2012) Hydrolysis of secreted sialoglycoprotein immunoglobulin A (IgA) in ex vivo and biochemical models of bacterial vaginosis. J Biol Chem 287 (3):2079–2089

Li W, Zhu Z, Chen W, Feng Y, Dimitrov DS (2017) Crystallizable fragment glycoengineering for therapeutic antibodies development. Front Immunol 8:1554–1554

Lin C-W, Tsai M-H, Li S-T et al (2015) A common glycan structure on immunoglobulin G for enhancement of effector functions. Proc Natl Acad Sci USA 112(34):10611–10616

Lindfors K, Suzuki H, Novak J et al (2011) Galactosylation of serum IgA1 O-glycans in celiac disease. J Clin Immunol 31(1):74–79

Liu H, May K (2012) Disulfide bond structures of IgG molecules: structural variations, chemical modifications and possible impacts to stability and biological function. mAbs 4(1):17–23

Liu Y, McDaniel JR, Khan S et al (2018) Antibodies encoded by FCRL4-bearing memory B cells preferentially recognize commensal microbial antigens. J Immunol 200(12):3962–3969

Liu Y, Goroshko S, Leung LYT et al (2020) FCRL4 is an Fc receptor for systemic IgA, but not mucosal secretory IgA. J Immunol 205(2):533–538

Lomholt H, Kilian M (1994) Antigenic relationships among immunoglobulin A1 proteases from Haemophilus, Neisseria, and Streptococcus species. Infect Immun 62(8):3178–3183

Lomholt H, Poulsen K, Caugant DA, Kilian M (1992) Molecular polymorphism and epidemiology of Neisseria meningitidis immunoglobulin A1 proteases. Proc Natl Acad Sci USA 89 (6):2120–2124

Maillard N, Wyatt RJ, Julian BA et al (2015) Current understanding of the role of complement in IgA nephropathy. J Am Soc Nephrol 26(7):1503–1512

Maixnerova D, Ling C, Hall S et al (2019) Galactose-deficient IgA1 and the corresponding IgG autoantibodies predict IgA nephropathy progression. PLoS One 14(2):e0212254

Marion C, Limoli DH, Bobulsky GS, Abraham JL, Burnaugh AM, King SJ (2009) Identification of a pneumococcal glycosidase that modifies O-linked glycans. Infect Immun 77(4):1389–1396

Matei L, Matei I (2000) Lectin-binding profile of serum IgA in women suffering from systemic autoimmune rheumatic disorders. Rom J Intern Med 38–39:73–82

Mathias A, Corthésy B (2011a) N-Glycans on secretory component: mediators of the interaction between secretory IgA and gram-positive commensals sustaining intestinal homeostasis. Gut Microbes 2(5):287–293

Mathias A, Corthésy B (2011b) Recognition of gram-positive intestinal bacteria by hybridoma- and colostrum-derived secretory immunoglobulin A is mediated by carbohydrates. J Biol Chem 286 (19):17239–17247

Mattu TS, Pleass RJ, Willis AC et al (1998) The glycosylation and structure of human serum IgA1, Fab, and Fc regions and the role of N-glycosylation on Fcα receptor interactions. J Biol Chem 273(4):2260–2272

Maurer MA, Meyer L, Bianchi M et al (2018) Glycosylation of human IgA directly inhibits influenza A and other sialic-acid-binding viruses. Cell Rep 23(1):90–99

Mestecky J, Russell MW, Jackson S, Brown TA (1986) The human IgA system: a reassessment. Clin Immunol Immunopathol 40(1):105–114

Meyer S, Nederend M, Jansen JH et al (2016) Improved in vivo anti-tumor effects of IgA-Her2 antibodies through half-life extension and serum exposure enhancement by FcRn targeting. MAbs 8(1):87–98

Mistry DV, Stockley RA (2011) The cleavage specificity of an IgA1 protease from *Haemophilus influenzae*. Virulence 2(2):103–110

Moldoveanu Z, Wyatt RJ, Lee JY et al (2007) Patients with IgA nephropathy have increased serum galactose-deficient IgA1 levels. Kidney Int 71(11):1148–1154

Moldoveanu Z, Suzuki H, Reily C et al (2021) Experimental evidence of pathogenic role of IgG autoantibodies in IgA nephropathy. J Autoimmun 118:102593

Moncla BJ, Chappell CA, Debo BM, Meyn LA (2016) The effects of hormones and vaginal microflora on the glycome of the female genital tract: cervical-vaginal fluid. PLoS One 11(7): e0158687

Monteiro RC (2014) Immunoglobulin A as an anti-inflammatory agent. Clin Exp Immunol 178 (Suppl 1):108–110

Monteiro RC, Van De Winkel JG (2003) IgA Fc receptors. Annu Rev Immunol 21:177–204

Monteiro RC, Kubagawa H, Cooper MD (1990) Cellular distribution, regulation, and biochemical nature of an Fc alpha receptor in humans. J Exp Med 171(3):597–613

Montero-Morales L, Steinkellner H (2018) Advanced plant-based glycan engineering. Front Bioeng Biotechnol 6:81

Montero-Morales L, Maresch D, Crescioli S et al (2019) In planta glycan engineering and functional activities of IgE antibodies. Front Bioeng Biotechnol 7:242

Moore JS, Kulhavy R, Tomana M et al (2007) Reactivities of N-acetylgalactosamine-specific lectins with human IgA1 proteins. Mol Immunol 44(10):2598–2604

Moura IC, Centelles MN, Arcos-Fajardo M et al (2001) Identification of the transferrin receptor as a novel immunoglobulin (Ig)A1 receptor and its enhanced expression on mesangial cells in IgA nephropathy. J Exp Med 194(4):417–425

Moura IC, Arcos-Fajardo M, Sadaka C et al (2004) Glycosylation and size of IgA1 are essential for interaction with mesangial transferrin receptor in IgA nephropathy. J Am Soc Nephrol 15 (3):622–634

Mummert E, Fritzler MJ, Sjöwall C, Bentow C, Mahler M (2018) The clinical utility of anti-double-stranded DNA antibodies and the challenges of their determination. J Immunol Methods 459:11–19

Murphy TF, Kirkham C, Jones MM, Sethi S, Kong Y, Pettigrew MM (2015) Expression of IgA proteases by *Haemophilus influenzae* in the respiratory tract of adults with chronic obstructive pulmonary disease. J Infect Dis 212(11):1798–1805

Nakajima A, Vogelzang A, Maruya M et al (2018) IgA regulates the composition and metabolic function of gut microbiota by promoting symbiosis between bacteria. J Exp Med 215 (8):2019–2034

Nakazawa S, Imamura R, Kawamura M et al (2019) Evaluation of IgA1 O-glycosylation in Henoch-Schönlein purpura nephritis using mass spectrometry. Transplant Proc 51(5):1481–1487

Newburg DS, Ruiz-Palacios GM, Morrow AL (2005) Human milk glycans protect infants against enteric pathogens. Annu Rev Nutr 25:37–58

Nikolova EB, Tomana M, Russell MW (1994) The role of the carbohydrate chains in complement (C3) fixation by solid-phase-bound human IgA. Immunology 82(2):321–327

Norderhaug IN, Johansen FE, Schjerven H, Brandtzaeg P (1999) Regulation of the formation and external transport of secretory immunoglobulins. Crit Rev Immunol 19(5–6):481–508

Novak J, Tomana M, Kilian M et al (2000) Heterogeneity of O-glycosylation in the hinge region of human IgA1. Mol Immunol 37(17):1047–1056

Novak J, Tomana M, Matousovic K et al (2005) IgA1-containing immune complexes in IgA nephropathy differentially affect proliferation of mesangial cells. Kidney Int 67(2):504–513

Novak J, Moldoveanu Z, Renfrow MB et al (2007) IgA nephropathy and Henoch-Schoenlein purpura nephritis: aberrant glycosylation of IgA1, formation of IgA1-containing immune complexes, and activation of mesangial cells. Contrib Nephrol 157:134–138

Novak J, Julian BA, Tomana M, Mestecky J (2008) IgA glycosylation and IgA immune complexes in the pathogenesis of IgA nephropathy. Semin Nephrol 28(1):78–87

Novak J, Raskova Kafkova L, Suzuki H et al (2011a) IgA1 immune complexes from pediatric patients with IgA nephropathy activate cultured human mesangial cells. Nephrol Dial Transplant 26(11):3451–3457

Novak J, Moldoveanu Z, Julian BA et al (2011b) Aberrant glycosylation of IgA1 and anti-glycan antibodies in IgA nephropathy: role of mucosal immune system. Adv Otorhinolaryngol 72:60–63

Novak J, Julian BA, Mestecky J, Renfrow MB (2012) Glycosylation of IgA1 and pathogenesis of IgA nephropathy. Semin Immunopathol 34(3):365–382

Novak J, Rizk D, Takahashi K et al (2015) New insights into the pathogenesis of IgA nephropathy. Kidney Dis 1(1):8–18

Novak J, Barratt J, Julian BA, Renfrow MB (2018) Aberrant glycosylation of the IgA1 molecule in IgA nephropathy. Semin Nephrol 38(5):461–476

Odani H, Hiki Y, Takahashi M et al (2000) Direct evidence for decreased sialylation and galactosylation of human serum IgA1 Fc O-glycosylated hinge peptides in IgA nephropathy by mass spectrometry. Biochem Biophys Res Commun 271(1):268–274

Ohyama Y, Nakajima K, Renfrow MB, Novak J, Takahashi K (2020a) Mass spectrometry for the identification and analysis of highly complex glycosylation of therapeutic or pathogenic proteins. Expert Rev Proteomics 17(4):275–296

Ohyama Y, Yamaguchi H, Nakajima K et al (2020b) Analysis of O-glycoforms of the IgA1 hinge region by sequential deglycosylation. Sci Rep 10(1):671

Olsen NJ, Choi MY, Fritzler MJ (2017) Emerging technologies in autoantibody testing for rheumatic diseases. Arthritis Res Ther 19(1):172

Oortwijn BD, Roos A, Royle L et al (2006) Differential glycosylation of polymeric and monomeric IgA: a possible role in glomerular inflammation in IgA nephropathy. J Am Soc Nephrol 17(12):3529–3539

Pabst O, Slack E (2020) IgA and the intestinal microbiota: the importance of being specific. Mucosal Immunol 13(1):12–21

Pagan JD, Kitaoka M, Anthony RM (2018) Engineered sialylation of pathogenic antibodies in vivo attenuates autoimmune disease. Cell 172(3):564–577.e513

Papista C, Gerakopoulos V, Kourelis A et al (2012) Gluten induces coeliac-like disease in sensitised mice involving IgA, CD71 and transglutaminase 2 interactions that are prevented by probiotics. Lab Investig 92(4):625–635

Papista C, Lechner S, Ben Mkaddem S et al (2015) Gluten exacerbates IgA nephropathy in humanized mice through gliadin-CD89 interaction. Kidney Int 88(2):276–285

Perman JA, Modler S (1982) Glycoproteins as substrates for production of hydrogen and methane by colonic bacterial flora. Gastroenterology 83(2):388–393

Peterson R, Cheah WY, Grinyer J, Packer N (2013) Glycoconjugates in human milk: protecting infants from disease. Glycobiology 23(12):1425–1438

Phalipon A, Cardona A, Kraehenbuhl JP, Edelman L, Sansonetti PJ, Corthésy B (2002) Secretory component: a new role in secretory IgA-mediated immune exclusion in vivo. Immunity 17(1):107–115

Pillebout E, Jamin A, Ayari H et al (2017) Biomarkers of IgA vasculitis nephritis in children. PLoS One 12(11):e0188718

Plomp R, de Haan N, Bondt A, Murli J, Dotz V, Wuhrer M (2018) Comparative glycomics of immunoglobulin A and G from saliva and plasma reveals biomarker potential. Front Immunol 9:2436

Posgai MT, Tonddast-Navaei S, Jayasinghe M, Ibrahim GM, Stan G, Herr AB (2018) FcαRI binding at the IgA1 C(H)2-C(H)3 interface induces long-range conformational changes that are transmitted to the hinge region. Proc Natl Acad Sci USA 115(38):E8882–e8891

Poulsen K, Brandt J, Hjorth JP, Thøgersen HC, Kilian M (1989) Cloning and sequencing of the immunoglobulin A1 protease gene (iga) of *Haemophilus influenzae* serotype b. Infect Immun 57(10):3097–3105

Poulsen K, Reinholdt J, Kilian M (1996) Characterization of the *Streptococcus pneumoniae* immunoglobulin A1 protease gene (IgA) and its translation product. Infect Immun 64(10):3957–3966

Pouria S, Corran PH, Smith AC et al (2004) Glycoform composition profiling of O-glycopeptides derived from human serum IgA1 by matrix-assisted laser desorption ionization-time of flight-mass spectrometry. Anal Biochem 330(2):257–263

Qiu X, Wong G, Audet J et al (2014) Reversion of advanced Ebola virus disease in nonhuman primates with ZMapp. Nature 514(7520):47–53

Raska M, Moldoveanu Z, Suzuki H et al (2007) Identification and characterization of CMP-NeuAc: GalNAc-IgA1 alpha2,6-sialyltransferase in IgA1-producing cells. J Mol Biol 369(1):69–78

Reily C, Rizk DV, Julian BA, Novak J (2018) Assay for galactose-deficient IgA1 enables mechanistic studies with primary cells from IgA nephropathy patients. BioTechniques 65(2):71–77

Reily C, Stewart TJ, Renfrow MB, Novak J (2019) Glycosylation in health and disease. Nat Rev Nephrol 15(6):346–366

Reinholdt J, Kilian M (1997) Comparative analysis of immunoglobulin A1 protease activity among bacteria representing different genera, species, and strains. Infect Immun 65(11):4452–4459

Reinholdt J, Tomana M, Mortensen SB, Kilian M (1990) Molecular aspects of immunoglobulin A1 degradation by oral streptococci. Infect Immun 58(5):1186–1194

Renegar KB, Small PA Jr (1994) Passive immunization: systemic and mucosal. Handb Mucosal Immunol:347–356

Renfrow MB, Cooper HJ, Tomana M et al (2005) Determination of aberrant O-glycosylation in the IgA1 hinge region by electron capture dissociation fourier transform-ion cyclotron resonance mass spectrometry. J Biol Chem 280(19):19136–19145

Renfrow MB, Mackay CL, Chalmers MJ et al (2007) Analysis of O-glycan heterogeneity in IgA1 myeloma proteins by Fourier transform ion cyclotron resonance mass spectrometry: implications for IgA nephropathy. Anal Bioanal Chem 389(5):1397–1407

Rifai A, Fadden K, Morrison SL, Chintalacharuvu KR (2000) The N-glycans determine the differential blood clearance and hepatic uptake of human immunoglobulin (Ig)A1 and IgA2 isotypes. J Exp Med 191(12):2171–2182

Rizk DV, Saha MK, Hall S et al (2019) Glomerular immunodeposits of patients with IgA nephropathy are enriched for IgG autoantibodies specific for galactose-deficient IgA1. J Am Soc Nephrol 30(10):2017–2026

Robb M, Hobbs JK, Woodiga SA et al (2017) Molecular characterization of N-glycan degradation and transport in *Streptococcus pneumoniae* and its contribution to virulence. PLoS Pathog 13(1):e1006090

Robinson LS, Schwebke J, Lewis WG, Lewis AL (2019) Identification and characterization of NanH2 and NanH3, enzymes responsible for sialidase activity in the vaginal bacterium *Gardnerella vaginalis*. J Biol Chem 294(14):5230–5245

Roos A, Bouwman LH, van Gijlswijk-Janssen DJ, Faber-Krol MC, Stahl GL, Daha MR (2001) Human IgA activates the complement system via the mannan-binding lectin pathway. J Immunol 167(5):2861–2868

Royle L, Roos A, Harvey DJ et al (2003) Secretory IgA N- and O-glycans provide a link between the innate and adaptive immune systems*. J Biol Chem 278(22):20140–20153

Russell MW (2007) Biological functions of IgA. Mucosal Immune Defense: Immunoglobulin A:144–172

Schroten H, Stapper C, Plogmann R, Köhler H, Hacker J, Hanisch FG (1998) Fab-independent antiadhesion effects of secretory immunoglobulin A on S-fimbriated *Escherichia coli* are mediated by sialyloligosaccharides. Infect Immun 66(8):3971–3973

Selewski DT, Ambruzs JM, Appel GB et al (2018) Clinical characteristics and treatment patterns of children and adults with IgA nephropathy or IgA vasculitis: findings from the CureGN study. Kidney Int Rep 3(6):1373–1384

Senior BW, Woof JM (2005) Effect of mutations in the human immunoglobulin A1 (IgA1) hinge on its susceptibility to cleavage by diverse bacterial IgA1 proteases. Infect Immun 73 (3):1515–1522

Senior BW, Dunlop JI, Batten MR, Kilian M, Woof JM (2000) Cleavage of a recombinant human immunoglobulin A2 (IgA2)-IgA1 hybrid antibody by certain bacterial IgA1 proteases. Infect Immun 68(2):463–469

Sjögren J, Collin M (2014) Bacterial glycosidases in pathogenesis and glycoengineering. Future Microbiol 9(9):1039–1051

Smith SB, Ravel J (2017) The vaginal microbiota, host defence and reproductive physiology. J Physiol 595(2):451–463

Sørensen V, Rasmussen IB, Sundvold V, Michaelsen TE, Sandlie I (2000) Structural requirements for incorporation of J chain into human IgM and IgA. Int Immunol 12(1):19–27

Steffen U, Koeleman CA, Sokolova MV et al (2020) IgA subclasses have different effector functions associated with distinct glycosylation profiles. Nat Commun 11(1):120

Stewart TJ, Takahashi K, Whitaker RH et al (2019) IgA1 hinge-region clustered glycan fidelity is established early during semi-ordered glycosylation by GalNAc-T2. Glycobiology 29 (7):543–556

Stewart TJ, Takahashi K, Xu N et al (2021) Quantitative assessment of successive carbohydrate additions to the clustered O-glycosylation sites of IgA1 by glycosyltransferases. Glycobiology 31(5):540–556

Stockert RJ, Kressner MS, Collins JC, Sternlieb I, Morell AG (1982) IgA interaction with the asialoglycoprotein receptor. Proc Natl Acad Sci USA 79(20):6229–6231

Strasser R, Stadlmann J, Schähs M et al (2008) Generation of glyco-engineered *Nicotiana benthamiana* for the production of monoclonal antibodies with a homogeneous human-like N-glycan structure. Plant Biotechnol J 6(4):392–402

Strasser R, Seifert G, Doblin MS et al (2021) Cracking the "Sugar Code": a snapshot of N- and O-glycosylation pathways and functions in plants cells. Front Plant Sci 12:640919

Stuchlova Horynova M, Vrablikova A, Stewart TJ et al (2015) N-acetylgalactosaminide α2,6-sialyltransferase II is a candidate enzyme for sialylation of galactose-deficient IgA1, the key autoantigen in IgA nephropathy. Nephrol Dial Transplant 30(2):234–238

Sudhakara P, Sellamuthu I, Aruni AW (2019) Bacterial sialoglycosidases in virulence and pathogenesis. Pathogens 8(1)

Sugiyama M, Wada Y, Kanazawa N et al (2020) A cross-sectional analysis of clinicopathologic similarities and differences between Henoch-Schönlein purpura nephritis and IgA nephropathy. PLoS One 15(4):e0232194

Suzuki H, Moldoveanu Z, Hall S et al (2008) IgA1-secreting cell lines from patients with IgA nephropathy produce aberrantly glycosylated IgA1. J Clin Invest 118(2):629–639

Suzuki H, Fan R, Zhang Z et al (2009) Aberrantly glycosylated IgA1 in IgA nephropathy patients is recognized by IgG antibodies with restricted heterogeneity. J Clin Invest 119(6):1668–1677

Suzuki H, Raska M, Yamada K et al (2014) Cytokines alter IgA1 O-glycosylation by dysregulating C1GalT1 and ST6GalNAc-II enzymes. J Biol Chem 289(8):5330–5339

Suzuki H, Yasutake J, Makita Y et al (2018) IgA nephropathy and IgA vasculitis with nephritis have a shared feature involving galactose-deficient IgA1-oriented pathogenesis. Kidney Int 93 (3):700–705

Suzuki H, Moldoveanu Z, Julian BA, Wyatt RJ, Novak J (2019) Autoantibodies specific for galactose-deficient IgA1 in IgA vasculitis with nephritis. Kidney Int Rep 4(12):1717–1724

Syed S, Hakala P, Singh AK et al (2019) Role of pneumococcal NanA neuraminidase activity in peripheral blood. Front Cell Infect Microbiol 9:218

Takahashi K, Wall SB, Suzuki H et al (2010) Clustered O-glycans of IgA1: defining macro- and microheterogeneity by use of electron capture/transfer dissociation. Mol Cell Proteomics 9 (11):2545–2557

Takahashi K, Smith AD, Poulsen K et al (2012) Naturally occurring structural isomers in serum IgA1 o-glycosylation. J Proteome Res 11(2):692–702

Takahashi K, Raska M, Stuchlova Horynova M et al (2014) Enzymatic sialylation of IgA1 O-glycans: implications for studies of IgA nephropathy. PLoS One 9(2):e99026

Ten Hagen KG, Fritz TA, Tabak LA (2003) All in the family: the UDP-GalNAc:polypeptide N-acetylgalactosaminyltransferases. Glycobiology 13(1):1r-16r

Terai I, Kobayashi K, Vaerman JP, Mafune N (2006) Degalactosylated and/or denatured IgA, but not native IgA in any form, bind to mannose-binding lectin. J Immunol 177(3):1737–1745

Tomana M, Niedermeier W, Mestecky J, Skvaril F (1976) The differences in carbohydrate composition between the subclasses of IgA immunoglobulins. Immunochemistry 13 (4):325–328

Tomana M, Kulhavy R, Mestecky J (1988) Receptor-mediated binding and uptake of immuno- globulin A by human liver. Gastroenterology 94(3):762–770

Tomana M, Zikan J, Moldoveanu Z, Kulhavy R, Bennett JC, Mestecky J (1993) Interactions of cell- surface galactosyltransferase with immunoglobulins. Mol Immunol 30(3):265–275

Tomana M, Matousovic K, Julian BA, Radl J, Konecny K, Mestecky J (1997) Galactose-deficient IgA1 in sera of IgA nephropathy patients is present in complexes with IgG. Kidney Int 52 (2):509–516

Tomana M, Novak J, Julian BA, Matousovic K, Konecny K, Mestecky J (1999) Circulating immune complexes in IgA nephropathy consist of IgA1 with galactose-deficient hinge region and antiglycan antibodies. J Clin Invest 104(1):73–81

Underdown BJ, Mestecky J (1994) Mucosal immunoglobulins. In: Ogra PL, Mestecky J, Lamm ME, Strober W, McGhee JR, Bienenstock J (eds) Handbook of mucosal immunology. Aca- demic, Boston, pp 79–97

UniProt Consortium (2017) UniProt: the universal protein knowledgebase. Nucleic Acids Res 45 (D1):D158–d169

van Egmond M, Damen CA, van Spriel AB, Vidarsson G, van Garderen E, van de Winkel JG (2001) IgA and the IgA Fc receptor. Trends Immunol 22(4):205–211

Wada Y, Dell A, Haslam SM et al (2010a) Comparison of methods for profiling O-glycosylation: human proteome organisation human disease glycomics/proteome initiative multi-institutional study of IgA1. Mol Cell Proteomics 9(4):719–727

Wada Y, Tajiri M, Ohshima S (2010b) Quantitation of saccharide compositions of O-glycans by mass spectrometry of glycopeptides and its application to rheumatoid arthritis. J Proteome Res 9 (3):1367–1373

Wandall HH, Irazoqui F, Tarp MA et al (2007) The lectin domains of polypeptide GalNAc- transferases exhibit carbohydrate-binding specificity for GalNAc: lectin binding to GalNAc- glycopeptide substrates is required for high density GalNAc-O-glycosylation. Glycobiology 17 (4):374–387

Wang Z, Rahkola J, Redzic JS et al (2020) Mechanism and inhibition of *Streptococcus pneumoniae* IgA1 protease. Nat Commun 11(1):6063

Yoo EM, Coloma MJ, Trinh KR et al (1999) Structural requirements for polymeric immunoglobulin assembly and association with J chain. J Biol Chem 274(47):33771–33777

Westerhof LB, Wilbers RH, van Raaij DR et al (2015) Transient expression of secretory IgA in planta is optimal using a multi-gene vector and may be further enhanced by improving joining chain incorporation. Front Plant Sci 6:1200

Wold AE, Mestecky J, Tomana M et al (1990) Secretory immunoglobulin A carries oligosaccharide receptors for *Escherichia coli* type 1 fimbrial lectin. Infect Immun 58(9):3073–3077

Yamada K, Kobayashi N, Ikeda T et al (2010) Down-regulation of core 1 beta1,3-galactosyltransferase and Cosmc by Th2 cytokine alters O-glycosylation of IgA1. Nephrol Dial Transplant 25(12):3890–3897

Yamada K, Huang ZQ, Raska M et al (2020) Leukemia inhibitory factor signaling enhances production of galactose-deficient IgA1 in IgA nephropathy. Kidney Dis (Basel) 6(3):168–180

Yamada K, Huang ZQ, Raska M et al (2017) Inhibition of STAT3 signaling reduces IgA1 autoantigen production in IgA nephropathy. Kidney Int Rep 2(6):1194–1207

Zhao N, Hou P, Lv J et al (2012) The level of galactose-deficient IgA1 in the sera of patients with IgA nephropathy is associated with disease progression. Kidney Int 82(7):790–796

Yang X, Zhao Q, Zhu L, Zhang W (2013) The three complementarity-determining region-like loops in the second extracellular domain of human Fc alpha/mu receptor contribute to its binding of IgA and IgM. Immunobiology 218(5):798–809

Woof JM, Kerr MA (2006) The function of immunoglobulin A in immunity. J Pathol 208(2):270–282

Wilson TJ, Fuchs A, Colonna M (2012) Cutting edge: human FcRL4 and FcRL5 are receptors for IgA and IgG. J Immunol 188(10):4741–4745

Yanagihara T, Brown R, Hall S et al (2012) In vitro-generated immune complexes containing galactose-deficient IgA1 stimulate proliferation of mesangial cells. Results Immunol 2:166–172

Yang Y, Liu F, Franc V, Halim LA, Schellekens H, Heck AJR (2016) Hybrid mass spectrometry approaches in glycoprotein analysis and their usage in scoring biosimilarity. Nat Commun 7(1):13397

Zhao L, Peng L, Yang D et al (2020) Immunostaining of galactose-deficient IgA1 by KM55 is not specific for immunoglobulin A nephropathy. Clin Immunol 217:108483

Westra J, Brouwer E, Raveling-Eelsing E et al (2021) Arthritis autoantibodies in individuals without rheumatoid arthritis: follow-up data from a Dutch population-based cohort (Lifelines). Rheumatology (Oxford) 60(2):658–666

Zickerman AM, Allen AC, Talwar V et al (2000) IgA myeloma presenting as Henoch-Schönlein purpura with nephritis. Am J Kidney Dis 36(3):E19

Zhang W, Lachmann PJ (1994) Glycosylation of IgA is required for optimal activation of the alternative complement pathway by immune complexes. Immunology 81(1):137–141

Wani JH, Gilbert JV, Plaut AG, Weiser JN (1996) Identification, cloning, and sequencing of the immunoglobulin A1 protease gene of *Streptococcus pneumoniae*. Infect Immun 64(10):3967–3974

Weiser JN, Bae D, Fasching C, Scamurra RW, Ratner AJ, Janoff EN (2003) Antibody-enhanced pneumococcal adherence requires IgA1 protease. Proc Natl Acad Sci USA 100(7):4215–4220

Wren JT, Blevins LK, Pang B et al (2017) Pneumococcal Neuraminidase A (NanA) promotes biofilm formation and synergizes with influenza A virus in nasal colonization and middle ear infection. Infect Immun 85(4)

Part IV
Applications

Chapter 15
Importance and Monitoring of Therapeutic Immunoglobulin G Glycosylation

Yusuke Mimura, Radka Saldova, Yuka Mimura-Kimura, Pauline M. Rudd, and Roy Jefferis

Contents

15.1	Introduction	483
15.2	Influence of Fc Glycosylation on the Structure and Function of IgG-Fc	485
	15.2.1 Fucosylation of IgG-Fc	488
	15.2.2 Galactosylation of IgG-Fc	490
	15.2.3 Sialylation of IgG-Fc	491
15.3	Impact of Glycosylation of mAbs and Fc-Fusion Proteins on Pharmacokinetics	492
15.4	Glycosylation of Total IgG and Antigen-Specific IgG in Autoimmunity	494
15.5	Glycoengineering	496
	15.5.1 Chemoenzymatic Glycoengineering	496
	15.5.2 Antibody-Drug Conjugates Via Glycosylation	499
	15.5.3 Glycoengineering of Antibodies for Treatment of Cancers, Inflammation, and Infectious Diseases	500
15.6	Conclusion	503
References		504

Y. Mimura (✉) · Y. Mimura-Kimura
Department of Clinical Research, National Hospital Organization Yamaguchi Ube Medical Center, Ube, Japan
e-mail: mimura.yusuke.qy@mail.hosp.go.jp

R. Saldova
NIBRT GlycoScience Group, National Institute for Bioprocessing Research and Training, Mount Merrion, Blackrock, Dublin, Ireland

UCD School of Medicine, College of Health and Agricultural Science, University College Dublin, Belfield, Dublin, Ireland

P. M. Rudd
NIBRT GlycoScience Group, National Institute for Bioprocessing Research and Training, Mount Merrion, Blackrock, Dublin, Ireland

Bioprocessing Technology Institute, Singapore, Singapore

R. Jefferis
Institute of Immunology and Immunotherapy, College of Medical and Dental Sciences, University of Birmingham, Birmingham, UK

© The Author(s), under exclusive license to Springer Nature Switzerland AG 2021
M. Pezer (ed.), *Antibody Glycosylation*, Experientia Supplementum 112,
https://doi.org/10.1007/978-3-030-76912-3_15

Abstract The complex diantennary-type oligosaccharides at Asn297 residues of the IgG heavy chains have a profound impact on the safety and efficacy of therapeutic IgG monoclonal antibodies (mAbs). Fc glycosylation of a mAb is an established critical quality attribute (CQA), and its oligosaccharide profile is required to be thoroughly characterized by state-of-the-art analytical methods. The Fc oligosaccharides are highly heterogeneous, and the differentially glycosylated species (glycoforms) of IgG express unique biological activities. Glycoengineering is a promising approach for the production of selected mAb glycoforms with improved effector functions, and non- and low-fucosylated mAbs exhibiting enhanced antibody-dependent cellular cytotoxicity activity have been approved or are under clinical evaluation for treatment of cancers, autoimmune/chronic inflammatory diseases, and infection. Recently, the chemoenzymatic glycoengineering method that allows for the transfer of structurally defined oligosaccharides to Asn-linked GlcNAc residues with glycosynthase has been developed for remodeling of IgG-Fc oligosaccharides with high efficiency and flexibility. Additionally, various glycoengineering methods have been developed that utilize the Fc oligosaccharides of IgG as reaction handles to conjugate cytotoxic agents by "click chemistry", providing new routes to the design of antibody-drug conjugates (ADCs) with tightly controlled drug-antibody ratios (DARs) and homogeneity. This review focuses on current understanding of the biological relevance of individual IgG glycoforms and advances in the development of next-generation antibody therapeutics with improved efficacy and safety through glycoengineering.

Keywords Critical quality attribute · Drug-antibody conjugate · Endoglycosidase · Glycoengineering · Glycoform · Glycoprotein · Oligosaccharide · Recombinant antibody therapeutics

Abbreviations

2-AB	2-aminobenzamide
ACPA	Anti-citrullinated protein antibody
ADC	Antibody-drug conjugate
ADCC	Antibody-dependent cellular cytotoxicity
ADCP	Antibody-dependent cellular phagocytosis
CDC	Complement-dependent cytotoxicity
CHO	Chinese hamster ovary
CQA	Critical quality attribute
DAR	Drug-antibody ratio
DBCO	Dibenzoazacyclooctyne
ENGase	Endoglycosidase
FcγR	Receptor for Fc portion of IgG
FcRn	Neonatal Fc receptor
HILIC	Hydrophilic interaction liquid chromatography

IVIG	Intravenous immunoglobulin
PR3-ANCA	Proteinase 3-anti-neutrophil cytoplasmic antibody
RhIG	Anti-RhD IgG
RSV	Respiratory syncytial virus
sFcγRIIIa	Soluble form of FcγRIIIa
SG-Ox	Sialoglycan-oxazoline
SGP	Sialylglycopeptide

15.1 Introduction

Glycosylation of a glycoprotein is a complex and extensive posttranslational modification that influences protein conformation, stability, solubility, pharmacokinetics, biological activities, and immunogenicity (Mimura and Jefferis 2021; Rudd and Dwek 1997; Varki et al. 2017). The oligosaccharides attached at Asn297 residues of IgG-Fc are heterogeneous due to variable addition and processing of outer-arm sugar residues (sialic acid, galactose, and bisecting GlcNAc) and fucose onto the core diantennary heptasaccharide ($GlcNAc_2Man_3GlcNAc_2$, designated G0) (Arnold et al. 2006; Jefferis 2016, 2017c; Mimura and Jefferis 2021) (Fig. 15.1) and subject to alteration dependent on the expression hosts and culture conditions. The differentially glycosylated species (glycoforms) of IgG-Fc express unique biological activities, modulating antibody effector functions including antibody-dependent cellular cytotoxicity (ADCC) and complement-dependent cytotoxicity (CDC), and may be potentially immunogenic when bearing nonhuman oligosaccharides including α(1–3)-galactose and/or N-glycolylneuraminic acid (Cobb 2020; Mimura et al. 2018; de Taeye et al. 2019). Due to the profound impact of mAb glycoform on biological activity and clinical outcome, the glycoform profile is a CQA for each mAb and is characterized by state-of-the-art analytical methods to demonstrate consistency and comparability between batches, and biosimilarity between biosimilar and innovator products (Jefferis 2017a, b; Beck and Liu 2019; De Leoz et al. 2020) (Fig. 15.1).

Whether elevated serum levels of agalactosylation (G0) of IgG-Fc in certain chronic inflammatory diseases are associated with the disease activity remains a long-standing question. Site-specific glycosylation analyses of IgG-derived glycopeptides have revealed differences in galactosylation, sialylation, fucosylation, and bisection of glycoform profiles between autoantibody (antigen-specific IgG) and total (bulk, nonspecific) IgG (Kemna et al. 2017; Wuhrer et al. 2015). A new role of Fc glycosylation of total IgG has been proposed in which increased galactosylation/sialylation of total IgG results in diminished FcγR activation status in immune effector cells through increased FcγR occupancy (Dekkers et al. 2018), which provides insights into the generation of anti-inflammatory antibody therapeutics via glycoengineering.

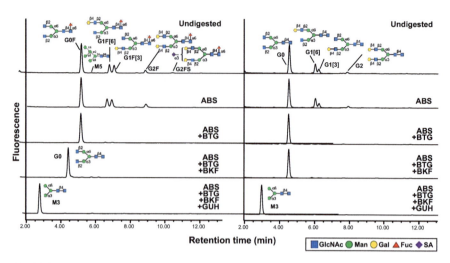

Fig. 15.1 Oligosaccharide profiles of mAbs by hydrophilic interaction liquid chromatography (HILIC). (**a**) Nivolumab (human anti-PD-1 IgG4), (**b**) Mogamulizumab (humanized anti-CCR4 IgG1). The oligosaccharides were released with peptide-*N*-glycosidase F from the heavy chains of the mAbs in the SDS-PAGE gel bands and labeled with 2-aminobenzamide (2-AB) as previously described (Royle et al. 2006). The fluorescently labeled oligosaccharides were separated by ultraperformance liquid chromatography (UPLC) on a sub-2μm hydrophilic interaction-based stationary phase with a Waters Ethylene Bridged Hybrid (BEH) Glycan chromatography column (150 × 2.1 mm i.d., 1.7μm BEH particles) (Bones et al. 2010; Doherty et al. 2012). The oligosaccharide peaks were assigned in accordance with the previous study (Pucic et al. 2011). 2-AB-labeled oligosaccharides were digested using arrays of the following enzymes: *Arthrobacter ureafaciens* sialidase (ABS), bovine testis β-galactosidase (BTG), bovine kidney α-fucosidase (BKF), β-*N*-acetylglucosaminidase cloned from *Streptococcus pneumoniae*, expressed in *Escherichia coli* (GUH)

Various glycoengineering approaches have been developed to produce nonfucosylated mAbs because of the importance of ADCC for the treatment of cancers, inflammatory diseases, and infection. As of 2020, four nonfucosylated mAbs have been approved for treatment of cancers, asthma, or autoimmune diseases, and currently >30 glycoengineered mAbs are under clinical evaluation (Lu et al. 2020; Kaplon et al. 2020; Pereira et al. 2018). Recently, the endoglycosidase-based chemoenzymatic glycoengineering approach has been introduced for remodeling of IgG-Fc glycosylation with preassembled oligosaccharides. This method allows for the preparation of potentially any homogeneous IgG-Fc glycoform, including those bearing fully sialylated, bisected, multi-antennary complex-type, oligomannose-type, or hybrid-type oligosaccharides (Wang et al. 2019). Furthermore, in vitro glycoengineering approaches are also exploited for the preparation of ADCs with tightly controlled drug-antibody ratios (DARs) and homogeneity. Production of selective IgG glycoforms via glycoengineering is a promising means to tailor the clinical efficacy of mAbs. This review provides an overview of the current understanding of glycobiology of natural IgG and mAbs that may be exploited to develop next-generation antibody therapeutics with optimized activities via glycoengineering.

15.2 Influence of Fc Glycosylation on the Structure and Function of IgG-Fc

Glycosylation of IgG-Fc has been shown to be essential for optimal expression of biological activities mediated through FcγRs (FcγRI, FcγRIIa/b/c, FcγRIIIa/b) and the C1q component of complement, and deglycosylated IgG is deficient in binding or activating these effector ligands. The IgG-Fc crystal structure reveals that the oligosaccharides are integral to the protein structure and sequestered within the internal space enclosed by the two C_H2 domains (Deisenhofer 1981). The electron density map provides coherent diffraction for the mono-galactosylated oligosaccharide and allows the possibility of >70 non-covalent interactions with 14 amino acid residues of the C_H2 domain (Fig. 15.2) (Deisenhofer 1981; Padlan 1990).

Following successive truncation of non-reducing Fc sugar residues, the thermal stability of the C_H2 domains was progressively decreased while the C_H3 domains

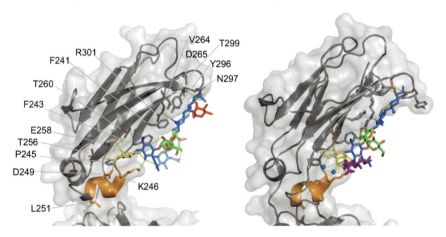

Fig. 15.2 Contacts between the Asn297-linked oligosaccharide and amino acid residues in the C_H2 domain of human IgG-Fc. Crystal structures of the C_H2 domains of monogalactosylated IgG-Fc (PDB ID code: 1Fc1) (**a**) and disialylated IgG-Fc (PDB ID code: 5GSQ) (**b**). GlcNAc, fucose, mannose, galactose, and sialic acid are shown in blue, red, green, yellow, and purple, respectively. Hydrogen bonds are shown with yellow dashed lines. Note that the terminal galactose is hydrogen-bonded with Asp249 (**a**) while the terminal sialic acid and galactose interact with Asp249 via water-mediated hydrogen bonds (water: blue sphere, **b**). The α-helix from Lys246 to Leu251 at the C_H2/C_H3 interface is shown in orange. The molecular models were produced with PyMOL (The PyMOL Molecular Graphics System, Version 2.1.0, Schrödinger, LLC)

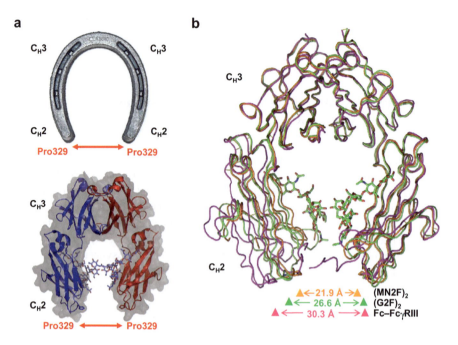

Fig. 15.3 The IgG-Fc horseshoe model (**a**) and superimposed image of IgG-Fc glycoforms (G2F)$_2$ and (MN2F)$_2$ and IgG-Fc complexed with a recombinant soluble form of FcγRIII (sFcγRIII) (**b**). The G2F oligosaccharides are shown in green sticks. The Pro329 residues located in the FG loop of the C$_H$2 domains of (G2F)$_2$, (MN2F)$_2$ glycoforms, and IgG-Fc complexed with sFcγRIII are indicated by green, orange, and pink arrowheads, respectively (PDB ID codes: 1H3V, 1H3T, and 1E4K) (Krapp et al. 2003; Sondermann et al. 2000)

were unchanged as revealed by differential scanning microcalorimetry. The fully galactosylated (G2F)$_2$ Fc glycoform exhibited a higher enthalpy change (ΔH) for the unfolding of the C$_H$2 domains than the (G0F)$_2$ glycoform, indicating that terminal galactose residues can confer stability on the C$_H$2 domains (Ghirlando et al. 1999). Removal of terminal GlcNAc from the (G0F)$_2$ glycoform resulted in a marked reduction of thermal stability, due to loss of the interactions with the side chain residues of Phe243, Lys246, and Thr260. Subsequent removal of the branch mannose residues, yielding the trisaccharide GlcNAc-GlcNAc-Man, did not result in further loss of stability but binding to a soluble form of FcγRIIb was markedly reduced (Mimura et al. 2000, 2001).

Tracing the α-carbon chain amino acid sequence of the IgG-Fc, as determined by X-ray crystallography, yields a "horseshoe" structure (Fig. 15.3a), in which the C$_H$3 domains form the "head" and the C$_H$2 domains are distanced from each other to accommodate the oligosaccharide. Removal of the oligosaccharide results in partial closure of this "open" structure with consequent loss of FcγR and C1 binding activities. By contrast, the horseshoe-shaped Fc opens further upon complex formation with an FcγR (Sondermann et al. 2000; Kiyoshi et al. 2015); it is presumed;

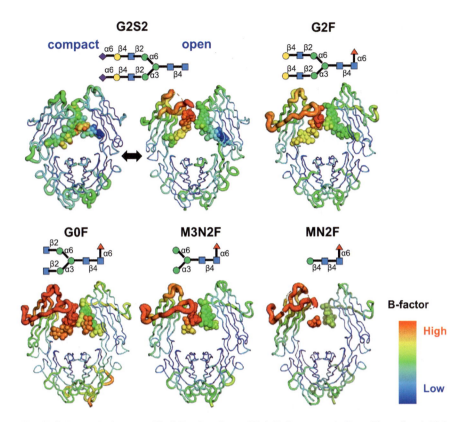

Fig. 15.4 Dynamic features of IgG-Fc glycoforms. High B-factors are indicated by red and thick tubes in a series of truncated glycoforms of IgG1-Fc (PDB ID codes: 5GSQ, 1H3V, 1H3X, 1H3U, and 1H3T). The shorter the oligosaccharides on the C_H2 domains the higher the B-factors, indicating increased mobility of the atoms in the C_H2 domains bearing short oligosaccharides. The molecular models were produced with PyMOL (The PyMOL Molecular Graphics System, Version 2.1.0, Schrödinger, LLC)

therefore, that the different IgG-Fc glycans modulate the open conformation(s) of the IgG-Fc to afford optimal interactions/binding to effector molecules. This notion is supported by the crystal structures of the Fc glycoforms bearing sequentially truncated glycans (($G2F)_2$, $(G0F)_2$, $(M3N2F)_2$ and $(MN2F)_2$, G: galactose; M: mannose; N: GlcNAc; F: fucose) in which the distance between the Pro329-Pro329 Cα residues are equated with the extent of openness; thus the distance between the Pro residues of the $(G2F)_2$ glycoform is measured as ~26.6 Å whereas for the $(MN2F)_2$ glycoform the distance is ~21.9 Å (Krapp et al. 2003) (Fig. 15.3b). Truncated IgG–Fc glycoforms showed progressive increases in the B-factors for the hinge proximal region of the C_H2 domains, as evidence of progressive structural disorder (Fig. 15.4) (Krapp et al. 2003). Progressive truncation of the terminal sugar residues results in decreased binding affinity to FcγRIIb and C1q (Mimura et al. 2000, 2001), and FcγRIII (Yamaguchi et al. 2006); these effector ligands engage amino acid residues within the lower hinge and hinge proximal region of the C_H2 domain (Shields et al. 2001; Idusogie et al. 2000; Sarmay et al. 1992; Lund et al. 1990, 1991). It is interesting to note that subtle modulations of the IgG-Fc structure can result in profound impacts on effector ligand engagement, as determined ex vivo. Importantly, the FcγRIII receptor has been shown to bear five oligosaccharide moieties that, individually, modulate IgG-Fc engagement (Ferrara et al. 2011; Patel et al. 2018; Shibata-Koyama et al. 2009). Therefore, the glycan profiles of the IgG-Fc and each FcγR, in concert, can exert profound regulation of outcomes in vivo.

15.2.1 Fucosylation of IgG-Fc

The increased binding affinity to FcγRIII and mediation of ADCC for nonfucosylated G0/G1/G2 glycoforms in comparison with G0F/G1F/G2F glycoforms of IgG1 has been exploited to increase the efficacy of approved antibody therapeutics. Additionally, human IgG1 binds more strongly to NK cells homozygous for the FcγRIIIa-Val158 allotype than to those expressing FcγRIIIa-Phe158 allotype (Koene et al. 1997; Wu et al. 1997): nonfucosylated IgG exhibits higher binding affinity to the recombinant exo-domains of both FcγRIIIa-Val158 and FcγRIIIa-Phe158, 50-fold and 30-fold, respectively (Shields et al. 2002; Ferrara et al. 2006b). The enhanced FcγRIIIa binding of nonfucosylated Fc is dependent on the glycosylation status of FcγRIIIa, which has five potential glycosylation sites (Asn38, Asn45, Asn74, Asn162, and Asn169). Interestingly, the increased affinity for the nonfucosylated glycoform of IgG-Fc was negated when FcγRIIIa was not glycosylated at Asn162 (Ferrara et al. 2006b). The oligosaccharide attached at Asn162 is required for high-affinity binding to nonfucosylated Fc glycoforms while glycosylation of Asn45 has an inhibitory effect on the binding to nonfucosylated Fc (Shibata-Koyama et al. 2009); however, glycosylation at this site is required for expression of the protein in culture (Ferrara et al. 2011). In addition, individual glycoforms present at Asn162 also influence the interaction between the two components. The oligosaccharides of FcγRIIIa on NK cells from human donors are of complex- and oligomannose-type (Edberg and Kimberly 1997) whereas those of recombinant FcγRIIIa are of complex-type (Hayes et al. 2017). The affinity of FcγRIIIa bearing oligomannose-type oligosaccharides is ~twofold higher than FcγRIIIa with complex-type oligosaccharides (Ferrara et al. 2011). Increased affinity binding of nonfucosylated IgG is similarly observed for FcγRIIIb which is expressed on neutrophils and mediates degranulation and phagocytosis of antibody-opsonized targets (Subedi and Barb 2016; Dekkers et al. 2017).

The crystallographic studies of the complex between nonfucosylated IgG-Fc and glycosylated FcγRIIIa revealed unique interactions between the nonfucosylated oligosaccharide at Asn297 of IgG-Fc and the oligosaccharide at Asn162 of sFcγRIIIa (Ferrara et al. 2011; Mizushima et al. 2011) (Fig. 15.5a). Non-fucosylation of IgG-Fc increases carbohydrate–carbohydrate and

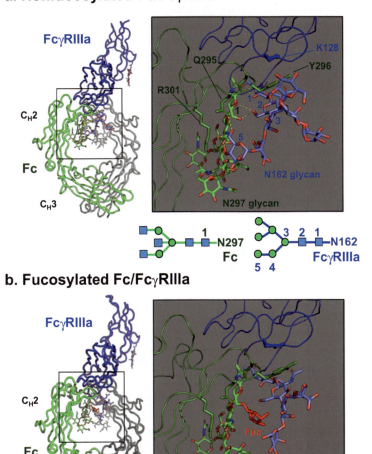

Fig. 15.5 Structures of nonfucosylated and fucosylated Fc fragments complexed with FcγRIIIa. Crystal structures of the complexes between nonfucosylated Fc and FcγRIIIa (PDB ID code: 3SGK) (**a**) and between fucosylated Fc and FcγRIIIa (PDB ID code: 3SGJ) (B) are shown (Ferrara et al. 2011). Hydrogen bonds between GlcNAc1 and GlcNAc2 of the Asn162 glycan of FcγRIIIa and GlcNAc1 of the nonfucosylated glycan of the Fc and between Man5 of FcγRIIIa and Gln295 of the Fc are observed as yellow dashed lines. Tyr296 of the Fc makes contacts with Man3 and Lys128 of FcγRIIIa. The fucose residue of the Fc glycan is shown in red (**b**). The molecular models were produced with PyMOL (The PyMOL Molecular Graphics System, Version 2.1.0, Schrödinger, LLC)

carbohydrate–protein interactions between FcγRIIIa and IgG-Fc, thereby stabilizing complex formation. The crystal structure of the fucosylated Fc–glycosylated FcγRIIIa complex showed that the core fucose residue is oriented toward the secondary GlcNAc residue of the oligosaccharide at Asn162 of sFcγRIIIa, indicating steric inhibition of the glycosylated FcγRIIIa binding by Fc fucosylation (Fig. 15.5b). Thus, high-affinity IgG-Fc/FcγRIIIa binding requires an interaction of sugar residues attached at Asn162 with surface structures of the nonfucosylated IgG-Fc glycoform and due to the asymmetry of the IgG-Fc/FcγRIIIa interaction non-fucosylation of one heavy chain should be sufficient for tight binding.

Although most studies were conducted with IgG1 subclass proteins, increased ADCC was also demonstrated for nonfucosylated IgG3 and IgG4 antibodies; some activity also being observed for IgG2 (Niwa et al. 2005). Increased ADCC activity was also reported for a nonfucosylated glycoform of C_H1/C_L-deleted fusion protein and could, presumably, be extended to IgG-Fc fusion proteins (Natsume et al. 2005).

15.2.2 Galactosylation of IgG-Fc

Early studies of endogenous (polyclonal) serum-derived IgG showed it to be comprised of 22% G0(F), 40% G1(F), and 38% G2(F) (Jefferis et al. 1990); however, a similar analysis of monoclonal IgG, isolated from serum collected from patients with multiple myeloma showed each sample to express a restricted glycoform profile and highly variable proportions of each glycoform (Jefferis et al. 1990). A possible relevance of IgG-Fc galactosylation to function was first posited following the demonstration that IgG isolated from the sera of patients diagnosed with rheumatoid arthritis exhibited a deficit in galactosylation relative to that isolated from the sera of individuals in normal health; however, conflicting claims were reported mainly due to the fact that, at that time: (1) the role of fucosylation was not appreciated and, (2) that ~20% of serum IgG bears additional glycosylation site(s) within the Fab region (Jefferis 2017a). Subsequently, the glycoform profile of serum IgG has been shown to vary depending on age, sex, pregnancy, and disease (Parekh et al. 1985, 1988; Cheng et al. 2020; Ercan et al. 2012; Gudelj et al. 2018). The oligosaccharide profile of recombinant IgG produced in cell culture varies depending on the production platform employed and may be altered by precise manipulation of the composition of the culture medium, temperature, culture period, etc. (Raju and Jordan 2012).

Employing glycoengineered G0 and G2 IgG glycoforms of known fucosylation levels, it has been established that galactosylation of IgG-Fc has positive effects on C1q and FcγRIIIa binding and activation (Wada et al. 2019; Aoyama et al. 2019; Kurogochi et al. 2015; Dekkers et al. 2017; Thomann et al. 2016; Houde et al. 2010). Although the increase of FcγRIIIa binding by Fc galactosylation is subtle (~twofold) as compared with prominent enhancement by non-fucosylation of IgG-Fc, the positive effect of galactosylation is independent of non-fucosylation (Houde et al.

2010; Thomann et al. 2016). In addition, Fc galactosylation is also shown to increase affinities to FcγRIIa/b (Subedi and Barb 2016; Thomann et al. 2015).

Although the molecular basis for the beneficial effects of galactosylation is not fully elucidated, the terminal galactose residue(s) contribute to the stability of IgG-Fc structure, as shown by an increase of the enthalpy for the unfolding of the C_H2 domains (Ghirlando et al. 1999; Wada et al. 2019) and lowered deuterium uptake in the hydrophobic surface of the C_H2 domain spanning Phe241 to Met252 (Aoyama et al. 2019). Furthermore, the crystal structure of the $(G2F)_2$ Fc glycoform shows the open conformation of the two C_H2 domains of the horseshoe-shaped Fc, which may favor FcγR binding (Fig. 15.3b) (Krapp et al. 2003; Sondermann et al. 2000).

Galactosylated IgG is increased in sera during pregnancy and preferentially transported to the fetus via the placenta (Kibe et al. 1996; Williams et al. 1995). While placental transport of maternal IgG is mediated by the neonatal Fc receptor (FcRn), numerous studies have suggested that noncanonical placental Fc receptors including FcγRIIb and FcγRIIIa expressed on placental endothelial cells and trophoblast cells, respectively, are involved in placental IgG transport (Martinez et al. 2019; Ishikawa et al. 2015; Wilcox et al. 2017; Jennewein et al. 2019).

15.2.3 Sialylation of IgG-Fc

The oligosaccharide of human serum IgG-Fc is hypo-sialylated at ~15% in α(2-6)-linkage, being mostly monosialylated on α(1-3)-arm and sparsely disialylated (Mimura et al. 2018; Arnold et al. 2006). The paucity of sialylation is presumed to reflect the intimate integration of the oligosaccharides within the IgG-Fc structure such that the steric/spatial requirements of the α(2-6)-sialyltransferase cannot be met (Jefferis 2017a). In fact, it is not easy to prepare homogeneous disialylated IgG by the in vitro glycoengineering approach using a combination of β(1-4)-galactosyltransferase and α(2-6)-sialyltransferase (Thomann et al. 2015).

Structural analyses of highly sialylated Fc revealed that the sialic acid residues are highly dynamic and free of strong interaction with the protein moiety of the Fc (Crispin et al. 2013; Barb et al. 2009, 2012; Ahmed et al. 2014; Chen et al. 2017). The terminal sialic acid on the α(1-6)-arm projects away from the protein surface in a solvent-exposed manner (Fig. 15.2b, shown in purple) while both the terminal sialic acid and the galactose residue on the α(1-3)-arm are not visible (PDB ID codes: 4Q6Y and 5GSQ).

The crystallographic studies revealed conformational heterogeneity in disialylated Fcs, which adopt open and compact conformations per asymmetric unit in the crystal (Fig. 15.4) (Ahmed et al. 2014; Chen et al. 2017). In the open conformer, the terminal sialic acid residue on the α(1-6)-arm interacts through water-mediated hydrogen bonds with the α-helix at the C_H2/C_H3 domain interface (Fig. 15.2b, shown in orange) while in the compact conformer the oligosaccharides mutually interact and the sialic acid residue on the α(1-6)-arm is free of interaction with the C_H2/C_H3 domain. For nonfucosylated IgG, α(2-6)-sialylation slightly decreases FcγRIIIa binding and ADCC although these activities are several-fold higher than those of native IgG (Wada et al. 2019; Dekkers et al. 2017). On the other hand, for α(2-3) sialylated IgG the terminal α(2-3) sialic acid on the α(1-6)-arm is shown to weaken the interaction between the Fc oligosaccharide and the C_H2 domain possibly due to steric hindrance (Zhang et al. 2019; Kuhne et al. 2019), which is consistent with the reduced ADCC for α(2-3)-sialylated IgG (Scallon et al. 2007).

There are contradictory reports concerning biological consequences of α(2-6) sialylation of IgG-Fc. This relates to attempts to elucidate the mechanism(s) by which high doses of intravenous immunoglobulin (IVIG) mediate an anti-inflammatory activity in various human autoimmune/inflammatory diseases. In a mouse arthritis model a prophylactic use of α(2-6) sialylated IVIG was effective in the prevention of joint inflammation (Kaneko et al. 2006), and subsequently, the C-type lectin receptor SIGN-R1 in mice, or DC-SIGN in humans, was reported to engage sialylated Fc, which resulted in upregulated expression of FcγRIIb on macrophages, attenuating autoantibody-initiated inflammation (Anthony et al. 2011). However, it has been asserted that caution should be exercised when extrapolating from mouse models to humans, since the tissue distribution of SIGN-R1 and DC-SIGN differs. Other studies showed that the efficacy of IVIG to ameliorate mouse models of autoimmune diseases was independent of Fc sialylation or FcγRIIb (Leontyev et al. 2012; Guhr et al. 2011; Campbell et al. 2014; Othy et al. 2014; Yu et al. 2013) and that DC-SIGN does not interact with human IgG-Fc, regardless of glycosylation status (Yu et al. 2013; Temming et al. 2019). As usual for apparently contradictory scientific reports, both outcomes may be valid but critically dependent on precise experimental protocols and animal models (Schwab and Nimmerjahn 2014).

15.3 Impact of Glycosylation of mAbs and Fc-Fusion Proteins on Pharmacokinetics

The catabolic half-life of human IgG1, IgG2, and IgG4 is ~23 days, the longest of any serum protein, while for IgG3 it varies between allotypes and is ~7 or ~23 days in IgG3-Arg435 or IgG3-His435-expressing individuals, respectively (Stapleton et al. 2011; Braster et al. 2017; Vidarsson et al. 2014). IgG antibodies are protected from degradation in lysosomes through the FcRn recycling mechanism (Roopenian and Akilesh 2007). FcRn interacts with IgG at the C_H2/C_H3 interface with high affinity at an acidic pH (<6.5) but not at a physiological pH (Burmeister et al. 1994). Although FcRn binding to IgG is presumed to be independent of Fc glycosylation, longer retention times have been noted by FcRn affinity chromatography for galactosylated and sialylated IgG glycoforms than more truncated glycoforms during elution with a linear pH gradient from pH 5.5 to 8.8 (Cymer et al. 2017; Wada et al. 2019); however, an association between differential retention patterns of IgG glycoforms in an FcRn column and pharmacokinetics remains unknown. Other receptors that are known to bind and clear proteins with specific glycans include the asialoglycoprotein receptor that binds to terminal galactose residues of N-glycans (Ashwell and Harford 1982; Stockert 1995) and the mannose receptor that recognizes terminal mannose or GlcNAc sugars (Lee et al. 2002).

Oligomannose-type glycoforms are found at low proportions when produced from Chinese hamster ovary (CHO) and murine cells (Mimura et al. 2009; Goetze et al. 2011; Zhang et al. 2016). It should be noted that while oligomannose-type IgG glycoforms exhibit higher ADCC activity than fucosylated complex-type IgG glycoforms, shorter half-lives have been demonstrated for the former than the latter (Kanda et al. 2007b; Liu et al. 2011; Alessandri et al. 2012; Liu 2015). When therapeutic IgG1 or IgG2 antibody was administered in human subjects, the relative abundance of IgG glycoforms with terminal galactose or GlcNAc remained constant during 34 days after injection while oligomannose-type IgG glycoforms were selectively cleared more rapidly at lower intravenous doses (Goetze et al. 2011). There might be an association between the number of oligomannose glycans on IgG-Fc and in vivo clearance. In the investigation of the impact of IgG-Fc oligomannose glycan pairing on antibody clearance, it was demonstrated that IgG1 and IgG2 mAbs with both symmetrical and asymmetrical pairings exhibited similarly fast clearance in humans, independently of the extent of mannosylation (Liu and Flynn 2016). Thus, the presence of oligomannose-type glycoforms may compromise the efficacy of antibody therapeutics through enhanced clearance and/or possible immunogenicity elicited by uptake of immune complexes via the mannose receptor on macrophages/dendritic cells and the activation of the mannan-binding lectin pathway (Jefferis 2017b; Arnold et al. 2006).

The Fab is also glycosylated in 15–20% of polyclonal human IgG and the Fab oligosaccharides can be of highly galactosylated and sialylated complex-type (Mimura et al. 2007; Holland et al. 2006) or of oligomannose-type, depending on the location of the glycosylation site in the V_H region (Gala and Morrison 2004; Wright et al. 1991; Radcliffe et al. 2007). However, it is important to emphasize that IgG-Fab glycosylation has not been shown to compromise clearance rates, at least in mouse models (Huang et al. 2006).

On the other hand, it has been reported that the clearance rate of Fc-fusion proteins can be influenced by the fusion partner and its attached oligosaccharides (Liu 2015). Fc-fusion proteins may also have N-glycosylation sites in the fusion partner and the attached oligosaccharides can be highly galactosylated and sialylated, in contrast to those of the Fc portion. Although the catabolic half-lives of some Fc-fusion proteins are shorter than intact IgG molecules, possibly reflecting reduced affinity of Fc-fusion proteins to FcRn (Suzuki et al. 2010), the quantities of total sialic acids in the fusion partner of etanercept are positively correlated with a catabolic half-life in rats (Liu 2015).

15.4 Glycosylation of Total IgG and Antigen-Specific IgG in Autoimmunity

An association of increased levels of agalactosylated IgG in serum with disease activity in rheumatoid arthritis has raised the question of whether the agalactosylated glycoforms are causative (proinflammatory) or a bystander phenomenon (Parekh et al. 1985; Ercan et al. 2010; Stambuk et al. 2020). In fact, galactosylation of IgG has been shown to increase affinities for FcγRIIIa and C1q and results in increased ADCC and CDC activities (Dekkers et al. 2017), respectively, while agalactosylated IgG-containing immune complexes may activate the lectin pathway of complement via mannan-binding lectin (Malhotra et al. 1995; Banda et al. 2008). Vidarsson and coworkers have recently proposed differential effects of certain IgG-Fc glycoforms on disease activity of autoimmunity between antigen-specific IgG (auto- or alloantibodies) and nonspecific, total IgG (irrelevant bulk antibodies) (Dekkers et al. 2018). Galactosylated (G2) glycoforms of total IgG antibodies are presumed to occupy FcγRs, raising activation thresholds and preventing antigen-specific IgG from engagement with FcγRs including FcγRIIIa/b during remission. Upon relapse, however, agalactosylated (G0) glycoforms of total IgG are dissociated from FcγRs, lowering the threshold and allowing antigen-specific antibodies to provoke inflammation (Fig. 15.6).

Oligosaccharide profiles of antigen-specific IgG (e.g., ACPA and anti-red blood cell autoantibodies) and total IgG have been analyzed and compared before and after relapse in patients with autoimmune diseases (Sonneveld et al. 2017, 2018; Rombouts et al. 2015; Kemna et al. 2017). Notably, in Granulomatosis with polyangiitis glycosylation of total IgG differed from that of proteinase 3-antineutrophil cytoplasmic antibody (PR3-ANCA) IgG1 (Wuhrer et al. 2015; Kemna et al. 2017). For example, it has been reported that during relapse the oligosaccharide profiles of total IgG exhibited significantly decreased galactosylation/bisection and increased fucosylation while those of PR3-ANCA showed decreased galactosylation/fucosylation. On the other hand, in non-relapsing patients, the oligosaccharide profiles of total IgG were not significantly altered while those of PR3-ANCA showed decreased galactosylation/fucosylation and increased bisection (Kemna et al. 2017). Notably, a comparison of the oligosaccharide profiles of total IgG between relapsing and non-relapsing patients reveals higher levels of galactosylation, bisection, and non-fucosylation for the latter. These observations are consistent with the possible inhibitory activity of galactosylated species of total IgG, which may account for the association of low galactosylation levels of total IgG with severity in autoimmunity. Even if an increase in affinity of IgG for FcγRs by galactosylation is subtle, synergistic effect of galactosylation and non-fucosylation on affinity to FcγRs may overcome high-avidity of immune complexes. Although the proportion of plasma IgG bearing at least one nonfucosylated oligosaccharide is estimated to be as low as ~10% (Kapur et al. 2014), oligosaccharide analysis of IgG1 isolated from FcγRIIIa on human NK cells revealed that >50% of the oligosaccharides were nonfucosylated (Patel et al. 2019). This suggests that nonfucosylated IgG

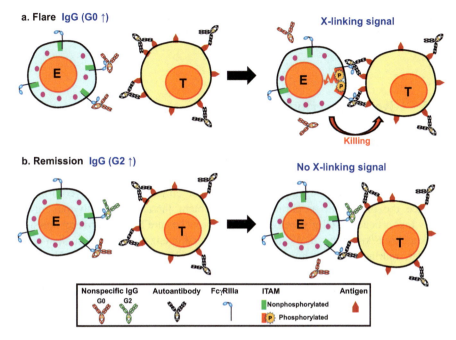

Fig. 15.6 Hypothetical role for glycoforms of plasma IgG in blockade of FcγRs on immune effector cells. (**a**) Nonspecific agalactosylated IgGs (G0) cannot compete with high-avidity autoantibody-antigen complexes on a target cell for FcγR binding, resulting in cross-linking FcγRs in effector cells during a flare. (**b**) During remission, nonspecific galactosylated IgGs (G2) occupy FcγRs and suppress the generation of cross-linking signals via FcγRs. E, effector cell. T, target cell

substantially occupies FcγRIIIa/b on circulating NK cells and neutrophils, potentially suppressing the recognition of autoantibody-opsonized target cells by NK cells and neutrophils. In addition, galactosylation can increase FcγRIIa binding, therefore, this provides a clue to enhancement of the anti-inflammatory activity of total IgG (i.e., IVIG) via glycoengineering.

IVIG is a therapeutic preparation of polyclonal IgG, derived from pooled plasma of thousands of healthy donors, used for the treatment of autoimmune/inflammatory diseases, including Kawasaki Disease and Guillain–Barré syndrome. The anti-inflammatory activity of IVIG is shown to reside in the Fc from a clinical study on the treatment of immune thrombocytopenic purpura with the Fc fragments of IVIG (Debre et al. 1993), and multiple mechanisms of action have been proposed for the efficacy of IVIG, including FcγR blockade and FcγRIIb upregulation. As anti-inflammatory activity of IVIG requires Fc glycosylation and large quantities (1–3 g of IgG/kg of body weight), it is postulated that the active constituent of IVIG can be certain IgG-Fc glycoforms such as sialylated glycoforms (Nimmerjahn and Ravetch 2008). The precise mechanism of action of IVIG in autoimmune diseases remains elusive and the issue as to whether sialylated glycoforms exert anti-inflammatory effect remains an open question. As stated above, it is postulated that in autoimmune diseases, including rheumatoid arthritis, the alterations in glycosylation of total IgG that modulate ADCC and ADCP play a key role in the disease activity.

15.5 Glycoengineering

Glycoproteins are comprised of heterogeneous glycoforms, and separating them into individual glycoforms has been challenging due to the attachment of diverse glycan structures as well as variations in glycosylation site occupancy. Glycoengineering approaches have been developed to generate or enrich desirable glycoforms to enhance the biological properties of glycoprotein therapeutics with maximized safety and efficacy. Cell glycoengineering approaches are based on overexpression and knockout/knockdown of certain glycoprocessing enzymes in mammalian (e.g., hamster, murine, rat, and human) (Yang et al. 2015; Goh et al. 2014; Yamane-Ohnuki et al. 2004) and non-mammalian host cells (e.g., insect, plant, and yeast) (Li et al. 2006; Le et al. 2016; Strasser et al. 2008; Mabashi-Asazuma et al. 2014), knockout of genes of enzymes involved in sugar-nucleotide biosynthesis and loss-of-function mutations in the Golgi sugar-nucleotide transporter genes (Kanda et al. 2007a; Kelly et al. 2018).

For the last decade, a variety of host cell engineering approaches have been directed toward the preparation of fucose-deficient IgG to enhance the affinity for FcγRIIIa and ADCC activity. Among common approaches are the modifications of the fucosylation pathway in CHO cell lines, including (1) knockout/knockdown of α1,6-fucosyltransferase (FUT8) gene using sequential homologous recombination (Potelligent® technology, Kyowa Hakko Kirin) (Yamane-Ohnuki et al. 2004), co-expression of β1,4-N-acetylglucosaminyltransferase-III (GnT-III) and β-mannosidase-II (Man-II) (GlycoMAb® technology, Roche) that localize the GnT-III to the early Golgi compartment, resulting in increased addition of bisecting GlcNAc and increased inhibition of the addition of fucose (Ferrara et al. 2006a); (2) disruption of the de novo GDP-fucose biosynthetic pathway in CHO cells (GlymaxX® technology, ProBioGen); (3) inhibitors of cellular protein fucosylation (fluorinated fucose analogs) (Okeley et al. 2013; Dekkers et al. 2016). Several glycoengineered, therapeutic IgG antibodies lacking core fucose or with low fucose content have entered the clinic, including mogamulizumab, benralizumab, obinutuzumab, etc.

15.5.1 Chemoenzymatic Glycoengineering

The in vitro chemoenzymatic glycan remodeling approach by combined use of endo-β-N-acetylglucosaminidase (ENGase), glycosynthase, and sugar oxazoline has a capability to prepare potentially any desired homogeneous glycoform of

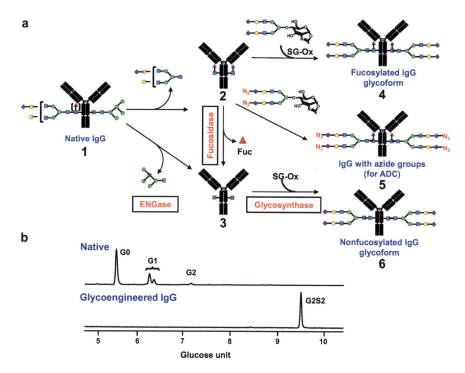

Fig. 15.7 Endoglycosidase-catalyzed glycan remodeling of IgG-Fc. (**a**) Schematic representation of the chemoenzymatic glycoengineering method. (**b**) UPLC-HILIC profiles of native (top) and G2S2 oligosaccharide (bottom) released from IgG1 mAb mogamulizumab glycoengineered with EndoM-N175Q (Tokyo Chemical Industry, Japan). *ENGase* endo-β-*N*-acetylglucosaminidase, *SG-Ox* sialoglycan oxazoline, *ADC* antibody-drug conjugate

IgG-Fc. Details on the approach are provided by a recent review by Wang et al. (2019). The approach involves cleavage of N-glycan(s) at the chitobiose core leaving the innermost GlcNAc or Fuc-GlcNAc with ENGase, together with fucosidase digestion where necessary, and transfer of a preassembled glycan to the innermost GlcNAc of the acceptor protein with ENGase-based glycosynthase (Fig. 15.7). The approach utilizes highly active glycan oxazolines, the mimics of the transition state, as donor substrates (Kobayashi et al. 1996), and transglycosylation with the synthetic glycan oxazoline proceeds in both a stereo- and regiospecific manner (Li et al. 2005). Various structures of glycan oxazoline are synthesized by endoglycosidase digestion of intact or exoglycosidase-digested sialylglycopeptide (SGP) from egg yolk and condensation with 2-chloro-1,3-dimethyl-imidazolinium chloride (DMC) in the presence of triethylamine (Sun et al. 2014; Noguchi et al. 2009). Homogeneous IgG glycoforms prepared by this method include fully sialylated diantennary and triantennary complex-type, hybrid-type, and oligomannose-type glycoforms (Lin et al. 2015; Li et al. 2016).

This method is based on the discovery of the transglycosylation activity of various ENGases, including EndoA from *Arthrobacter protophormiae* (Takegawa et al. 1995, 1997), EndoM from *Mucor hiemalis* (Yamamoto et al. 1994; Fujita et al. 2004), and EndoD from *Streptococcus pneumoniae* (Fan et al. 2012; Muramatsu et al. 2001) of the glycoside hydrolase (GH)-85 family; and Endo-S from *Streptococcus pyogenes* (Huang et al. 2012), EndoS2 from *Streptococcus pyogenes* NZ131 of serotype M49 (Li et al. 2016), and EndoF3 from *Elizabethkingia meningoseptica* (Giddens et al. 2016) of the GH18 family. To date, numerous mutants of ENGases have been generated as glycosynthases that abolish the hydrolytic activity on the transglycosylation products and improve the transglycosylation efficiency although different ENGases have distinct substrate specificity and limitations. EndoM acts on both the complex-type and oligomannose-type glycans whereas EndoA or EndoS acts on the oligomannose-type or the complex-type, respectively.

The first ENGase-catalyzed remodeling of IgG glycoforms was performed by deglycosylation of yeast-expressed human IgG1-Fc with EndoH and transglycosylation with EndoA of synthetic glycan-oxazolines including $Man_3GlcNAc$-oxazoline to the innermost GlcNAc residue of IgG-Fc (Wei et al. 2008). Furthermore, EndoA was found to efficiently catalyze transglycosylation to IgG-Fc of truncated glycan-oxazolines including bisecting GlcNAc-containing $Man_3GlcNAc$-oxazoline, which showed a higher affinity of the bisected $Man_3GlcNAc_2$-bearing Fc for FcγRIIIa, relative to the counterpart Fc glycoform without bisecting GlcNAc (Zou et al. 2011). In addition, glycosynthase mutants from EndoD (EndoD-N332Q) and EndoM (EndoM-N175Q) were reported to exhibit the transglycosylation activity to IgG and ribonuclease A (Fan et al. 2012; Huang et al. 2009); however, the former does not transfer full-length complex-type glycan to the Fc while the latter does not use fucosylated Fc as a substrate, which limits the applicability of the chemoenzymatic approach. The discovery of EndoS mutants D233A and D233Q represents a breakthrough of this method as these glycosynthase mutants catalyze the transfer of full-length complex-type glycan-oxazoline to both nonfucosylated and fucosylated GlcNAc-IgG with high efficiency (Huang et al. 2012). Furthermore, the D184M and D184Q mutants of EndoS2 have broader substrate specificity than EndoS-D233Q (Li et al. 2016), acting on the complex-type, hybrid-type, and oligomannose-type glycans. EndoF3-D165A and D165Q can transfer both di- and triantennary oligosaccharides to intact IgG-Fc (Giddens et al. 2016). According to recent protocols, transglycosylation of glycan oxazoline to IgG (~5 mg/ml) can be achieved by EndoS-D233Q or EndoS2-D184M within 4 h at 30 °C at a glycan oxazoline/IgG molar ratio of 40 (Tang et al. 2017; Li et al. 2018).

The structural integrity of transglycosylated IgG products is presumed to be maintained. Comparison of crystal structures of glycoengineered Fc glycoform and native Fc revealed no major difference in the conformation between the transferred and native oligosaccharides and in the orientation of hydrophobic side chains adjacent to each oligosaccharide (Fig. 15.2) (Ahmed et al. 2014; Chen et al. 2017). With regard to the functionality, rituximab and trastuzumab with the Fc glycans defucosylated via the chemoenzymatic glycoengineering exhibited enhanced FcγRIIIa binding and ADCC, as observed for nonfucosylated IgG variants produced in FUT8-knockout CHO host cells (Li et al. 2018; Lin et al. 2015; Chen et al. 2017;

Kurogochi et al. 2015; Huang et al. 2012; Liu et al. 2018). It should be noted that, currently, only this chemoenzymatic glycoengineering approach allows for the preparation of homogeneous disialylated IgG-Fc at high efficiency, which is superior to conventional approaches that combine βGal4T1 and ST6Gal1 treatment (Thomann et al. 2015). Recently, homogeneous IgG glycoforms prepared by this method have been utilized to investigate the stability and functionality of individual IgG glycoforms (Lin et al. 2015; Ahmed et al. 2014; Wada et al. 2019; Li et al. 2017). Another breakthrough is the discovery of α1,6-fucosidases that can remove core fucose from the innermost GlcNAc of IgG-Fc (Tang et al. 2017; Tsai et al. 2017). This chemoenzymatic glycoengineering approach has also been exploited to ADCs because of its robust, stable, and site-specific conjugation.

15.5.2 Antibody-Drug Conjugates Via Glycosylation

The Fc oligosaccharide is exploited as an attractive reaction site to prepare ADCs. Most of ADCs approved for clinical use are based on the conjugation of cytotoxic drugs to Lys (trastuzumab emtansine (Kadcyla®), inotuzumab ozogamicin (Besponsa®)) and Cys residues (brentuximab vedotin (Adcetris®)), resulting in a variable distribution of DAR. It has been demonstrated that random conjugation has a negative impact on therapeutic efficacy. To circumvent the structural heterogeneity, various methods have been developed that utilize the Fc oligosaccharide as ADC conjugation site (Toftevall et al. 2020) (Fig. 15.8). Zhou and coworkers have developed a glycosylation site-specific method consisting of enzymatic remodeling of the Fc oligosaccharide with a combination of β4GalT1 with ST6Gal1, periodate oxidation of the terminal sialic acids to generate aldehyde groups and conjugation of aminooxy functionalized cytotoxic agents via oxime ligation (Zhou et al. 2014). This method converts diantennary complex-type oligosaccharides to monosialylated forms (>94%) and incorporates ~1.6 cytotoxic agents per antibody molecule. However, periodate treatment can also oxidize Met residues (Chap. 1, Fig. 15.1), and the DAR is generally low. van Geel and coworkers at SynAffix have developed the GlycoConnect™ technology which consists of oligosaccharide release with EndoS2 and the transfer of azido group-containing GalNAc (*N*-azidoacetylgalactosamine) to the primary GlcNAc residue by β4GalT Y289L, which serves as a chemical handle for conjugation with a cytotoxic agent (Fig. 15.8a) (van Geel et al. 2015). The applicability of this technology is shown by using different isotypes and linker–drug combinations. Huang and coworkers have recently reported an alternative approach via EndoS-catalyzed Fc oligosaccharide remodeling in which native Fc oligosaccharides are replaced with N_3-modified sialylated complex-type oligosaccharides as a chemical handle, by deglycosylation with ENGase and reglycosylation with glycosynthase (e.g., EndoS-D233Q) (Figs. 15.7a and 15.8b) (Tang et al. 2017). The approaches using the Fc oligosaccharides have some advantages, including the controlled DAR as 2.0, the stability of

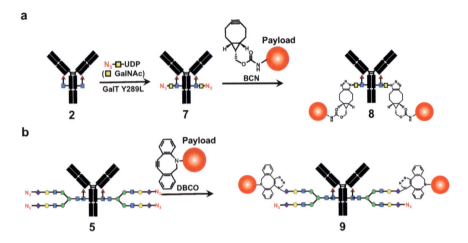

Fig. 15.8 Chemoenzymatic syntheses of glycosylation site-specific antibody-drug conjugates (ADCs). (**a**) Transfer of azido-functionalized GalNAc to the core GlcNAc of IgG (Structure **2** in Fig. 15.7) with GalT (Y289L) and site-specific conjugation of bicyclo[6.1.0]nonyne (BCN)-tagged drug payload to the azide-group by click chemistry (van Geel et al. 2015). (**b**) Site-specific conjugation of dibenzoazacyclooctyne (DBCO)-tagged drug payload to the azido-functionalized N-glycans of IgG (Structure **5** in Fig. 15.7) via a copper-free "click" reaction, yielding ADC. Synthesis of DBCO-tagged cytotoxic agent is described in Tang et al. (2017)

the conjugates in the circulation, and homogeneity of the predefined ADC structure (Tang et al. 2016, 2017; Wang et al. 2019).

IgM antibody is also considered as a promising platform for ADC due to the presence of 51 oligosaccharides in a pentameric IgM antibody including J-chain (Moh et al. 2019). By using ST6Gal1 and azide-functionalized CMP-sialic acid, conjugation of a drug candidate having dibenzoazacyclooctyne (DBCO) group to azide-sialic labeled oligosaccharides on IgM is achieved by copper-free click chemistry, with a DAR of 8–10.

15.5.3 Glycoengineering of Antibodies for Treatment of Cancers, Inflammation, and Infectious Diseases

Glycoengineering has been applied to optimize the effector functions of therapeutic IgG antibodies for the treatment of cancers, inflammatory, and infectious diseases. Numerous glycoengineered antibodies, including fucose-deficient or aglycosylated IgG antibodies, have entered the clinic or have been evaluated under clinical trials.

15.5.3.1 Cancers

As of 2020, ~20 mAbs in clinical development have been glycoengineered to deplete core fucose for cancer treatment (Kaplon et al. 2020; Pereira et al. 2018). The first approved nonfucosylated therapeutic antibody is humanized anti-CC chemokine receptor 4 (CCR4) IgG1 antibody mogamulizumab (Poteligeo®) for treatment of CCR4-positive adult T-cell leukemia or relapsed peripheral T-cell lymphoma (Ishida et al. 2012). Other nonfucosylated IgG antibodies include humanized anti-CD20 IgG1 obinutuzumab (Gazyva®, approved in 2013, USA) and primatized anti-EGFR IgG1 imgatuzumab (Gonzalez-Nicolini et al. 2016). Mogamulizumab is produced in $FUT8^{-/-}$ CHO cells (Potelligent® technology) and completely devoid of core fucose (Fig. 15.1b). Obinutuzumab with a low fucose content is produced in CHO-K1 cells engineered to overexpress GnT-III and Golgi α-mannosidase II (GlycoMab® technology) and exhibits superior antitumor activities to the reference antibody rituximab (Sehn et al. 2012, 2015). Fucose depletion of the licensed antitumor therapeutic IgG antibodies rituximab and trastuzumab has been shown to enhance ADCC activities (Iida et al. 2006; Junttila et al. 2010). In fact, nonfucosylated anti-CD20 antibody obinutuzumab combined with cytotoxic agent chlorambucil has shown better outcome than rituximab in patients with chronic lymphocytic leukemia despite infusion-related reactions and neutropenia (Goede et al. 2014).

The efficacy of anticancer therapeutic antibodies is influenced by FcγRIIIa-Val158Phe polymorphism, with patients homozygous for FcγRIIIa-Val158 showing higher clinical responses compared to FcγRIIIa-Phe158 carriers (Overdijk et al. 2014). Nonfucosylated glycoforms of anticancer therapeutic antibodies have some benefits, with regard to (1) the ability to elicit potent ADCC in individuals irrespective of the FcγRIIIa-Val158Phe allotype (Niwa et al. 2004), (2) the ability to mediate ADCC in the presence of large excess of plasma IgG (Nechansky et al. 2007), (3) lower doses compared with fucosylated mAbs (e.g., only 1 mg/kg for mogamulizumab) (Iida et al. 2006), and (4) the ability to mediate monocyte and macrophage phagocytic and cytotoxic activities (Herter et al. 2014). Obinutuzumab-induced ADCC is not negatively influenced by KIR/HLA interactions (Terszowski et al. 2014). Furthermore, the enhanced efficacy of obinutuzumab is explained by the ability to induce higher IFN-γ secretion from NK cells than rituximab upon interaction with the antibody-opsonized tumor cells irrespective of FcγRIIIa-Val158Phe polymorphism (Capuano et al. 2017). It is suggested that, besides short-term cytotoxic properties, obinutuzumab-experienced NK cells promote dendritic cell maturation and long-lasting T cell responses via increased IFNγ secretion.

15.5.3.2 Autoimmune and Inflammatory Diseases

Glycoengineered mAb therapeutics for autoimmune/inflammatory diseases currently approved or under clinical evaluation also have low fucose contents.

Nonfucosylated humanized anti-IL-5Rα IgG1 benralizumab (Fasenra®) produced in FUT8$^{-/-}$ CHO cells (Potelligent® technology) has been approved for the treatment of severe, uncontrolled asthma, with eosinophilic phenotype (Wang et al. 2017a). Complete depletion of eosinophils and their early progenitors in the bone marrow has been reported in eosinophilic asthmatic patients who received benralizumab (Sehmi et al. 2018), and its clinical efficacy was comparable to the anti-IL-5 antibody mepolizumab (Bourdin et al. 2018). Nonfucosylated humanized anti-CD19 inebilizumab (Uplizna®) has received its first global approval in 2020 for treatment of neuromyelitis optica spectrum disorder, an autoimmune, demyelinating disease of the central nervous system, which is indicated for adult patients who are seropositive for anti-aquaporin-4 autoantibodies (Frampton 2020). This antibody therapeutic is also under clinical evaluation for kidney transplant desensitization, myasthenia gravis, and IgG4-related disease through depletion of CD19-expressing B cells by ADCC.

Roledumab (LFB-R593) is a human IgG1 anti-rhesus (Rh)D mAb with a low fucose content produced from rat YB2/0 cells by EMABling® technology (Yver et al. 2012), and its safety and efficacy are being evaluated in the Phase 2/3 NCT02287896 study for the prevention of feto-maternal alloimmunization in RhD-negative women, as a substitute for human polyclonal anti-RhD antibodies. Interestingly, plasma anti-RhD IgG antibodies (RhIG) from hyperimmunized healthy anti-D donors are found to have lowered Fc fucosylation (47% for females) compared with normal IgG (93%) (Kapur et al. 2015). Fc fucosylation levels of RhIG from manufacturers are variably lowered (56%–91%), which may inversely correlate with the efficiency of clearance of D+ red blood cells by FcγRs on macrophages and NK cells from maternal circulation. Although the precise mechanism of action of RhIG is not known, high-affinity binding of nonfucosylated RhIG to FcγRIIIa and enhanced ADCC may be related to successful prevention of maternal immunization with RhD-positive red blood cells.

15.5.3.3 Infection

Neutralizing mAbs specific for viruses including influenza A and respiratory syncytial virus (RSV) are known to confer potent protection against infection. Humanized anti-RSV IgG1 mAb palivizumab (Synagis®) is recommended for prophylaxis of high-risk neonates during bronchiolitis seasons; however, the therapeutic efficacy of the antibody for the treatment of RSV infection has not been demonstrated (Alansari et al. 2019). Interestingly, a glycovariant of palivizumab produced in a genetically engineered plant host bearing mostly nonfucosylated, nongalactosylated glycans showed increased FcγRIIIa binding and decreased RSV titers in the lungs of cotton rats, compared with the licensed fucosylated palivizumab (Hiatt et al. 2014). Thus, the induction of ADCC is important for successful protection against viral infection, which is also noted for influenza A (He et al. 2016; Jegaskanda et al. 2019). Anti-hemagglutinin stalk domain antibodies are efficient at induction of ADCC, in contrast to anti-neuraminidase antibodies, but the latter has an additive effect on

ADCC with the former. For antibodies against HIV, dengue virus, and SARS-CoV-2, the oligosaccharide profiles of the specific antibodies from the infected subjects are characterized by elevated levels of nonfucosylated IgG glycoforms. For HIV, this was observed for those who had a longer disease-free survival which may correlate with antibody-mediated cellular viral inhibition (Ackerman et al. 2013); however, for dengue virus and SARS-CoV-2, it may cause antibody-dependent enhancement of infection (Wang et al. 2017b; Weber and Oxenius 2014; Bournazos et al. 2020; Chakraborty et al. 2021; Larsen et al. 2020). It remains unknown what factors determine the outcome in the presence of nonfucosylated antibodies against these infections. Nonetheless, glycoengineering to deplete core fucose of IgG for enhanced ADCC activity may provide an opportunity to improve the treatment of viral infection.

15.6 Conclusion

The improved clinical efficacy of nonfucosylated IgGs through enhanced FcγRIIIa binding and ADCC activity has boosted the development of the glycoengineered mAbs for the treatment of cancers and chronic inflammatory diseases. A protective role of ADCC against infectious diseases has also been noted including RSV, dengue virus, influenza virus, and HIV. A nonfucosylated variant of the broadly neutralizing human IgG1 anti-HIV mAb b12 exhibits greater in vitro ADCC activity against HIV-infected cells compared to fucosylated b12 (Moldt et al. 2012); however, no enhanced protection with nonfucosylated b12 was observed against a vaginal simian-human immunodeficiency virus challenge in rhesus macaques, compared to fucosylated b12. These findings suggest the requirement of FcγR-mediated activities other than FcγRIIIa-mediated ADCC for protection against viral infection.

The long-standing question of whether agalactosylated IgG is proinflammatory whereas sialylated and/or galactosylated IgG are anti-inflammatory in autoimmune diseases has been approached by observation of differential Fc oligosaccharide profiles between total IgG and antigen-specific IgG antibodies. Galactosylation of total IgG may raise the threshold for FcγR activation of immune effector cells through improved binding affinity for FcγRs, and the opposite may be true for agalactosylation of total IgG. Additionally, it seems likely that circulating nonfucosylated IgGs dominantly occupy FcγRIIIa/b, presumably raising the threshold for activation of NK cells and neutrophils. Increased levels of fucosylated IgG have been reported for total IgG in patients with rheumatoid arthritis (Gornik et al. 1999) and Granulomatosis with polyangiitis (Kemna et al. 2017), which may reflect a decrease of circulating nonfucosylated IgG as a result of binding to FcγRIIIa/b on immune cells that expand during a flare. If it is the case, nonfucosylated, galactosylated IVIG may serve as promising anti-inflammatory therapeutics for autoimmune diseases, inhibiting FcγRs-mediated immune cell activation through FcγRIIIa/b blockade (Mimura, Y. et al., manuscript in preparation).

The introduction of chemoenzymatic glycoengineering method employing glycosynthase mutants allows remodeling of IgG glycoforms at high efficiency and flexibility. This approach has several advantages over conventional in vitro glycoengineering approaches, with respect to a reaction time and the capabilities for defucosylation, sialylation, bisection, and branching. It should be noted that the longer the reaction time, the higher the risk of spontaneous degradations (e.g., deamidation and oxidation, see Chap. 1). Transfer of pre-defined oligosaccharides to IgG-Fc *en bloc* can mostly be achieved within a few hours, maintaining homogeneity of a glycoproteoform. Thus, this novel chemoenzymatic glycoengineering approach has been employed in numerous studies on the structure/function of IgG-Fc glycoforms (Wada et al. 2019; Aoyama et al. 2019; Li et al. 2017; Ahmed et al. 2014; Chen et al. 2017) and opens a new avenue to glycoform remodeling for therapeutic purposes. Overall, the oligosaccharides of IgG-Fc are the fertile ground for the development of next-generation antibody therapeutics with improved efficacy.

Acknowledgments Pauline M Rudd acknowledges Waters Corporation for research funding and donation of equipment.

Compliance with Ethical Standards

Conflict of Interest Yusuke Mimura, Radka Saldova, Yuka Mimura-Kimura, Pauline M Rudd, and Roy Jefferis declare that they have no conflict of interest.

Ethical Approval This chapter does not contain any studies with human participants or animals performed by any of the authors.

References

Ackerman ME, Crispin M, Yu X, Baruah K, Boesch AW, Harvey DJ, Dugast AS, Heizen EL, Ercan A, Choi I, Streeck H, Nigrovic PA, Bailey-Kellogg C, Scanlan C, Alter G (2013) Natural variation in Fc glycosylation of HIV-specific antibodies impacts antiviral activity. J Clin Invest 123(5):2183–2192. https://doi.org/10.1172/JCI65708

Ahmed AA, Giddens J, Pincetic A, Lomino JV, Ravetch JV, Wang LX, Bjorkman PJ (2014) Structural characterization of anti-inflammatory immunoglobulin G Fc proteins. J Mol Biol 426 (18):3166–3179. https://doi.org/10.1016/j.jmb.2014.07.006

Alansari K, Toaimah FH, Almatar DH, El Tatawy LA, Davidson BL, Qusad MIM (2019) Monoclonal antibody treatment of RSV bronchiolitis in young infants: a randomized trial. Pediatrics 143(3). https://doi.org/10.1542/peds.2018-2308

Alessandri L, Ouellette D, Acquah A, Rieser M, Leblond D, Saltarelli M, Radziejewski C, Fujimori T, Correia I (2012) Increased serum clearance of oligomannose species present on a human IgG1 molecule. MAbs 4(4):509–520. https://doi.org/10.4161/mabs.20450

Anthony RM, Kobayashi T, Wermeling F, Ravetch JV (2011) Intravenous gammaglobulin suppresses inflammation through a novel T(H)2 pathway. Nature 475(7354):110–113. https://doi.org/10.1038/nature10134

Aoyama M, Hashii N, Tsukimura W, Osumi K, Harazono A, Tada M, Kiyoshi M, Matsuda A, Ishii-Watabe A (2019) Effects of terminal galactose residues in mannose alpha1-6 arm of Fc-glycan on the effector functions of therapeutic monoclonal antibodies. MAbs 11(5):826–836. https://doi.org/10.1080/19420862.2019.1608143

Arnold JN, Wormald MR, Sim RB, Rudd PM, Dwek RA (2006) The impact of glycosylation on the biological function and structure of human immunoglobulins. Annu Rev Immunol 25:21–50

Ashwell G, Harford J (1982) Carbohydrate-specific receptors of the liver. Annu Rev Biochem 51:531–554. https://doi.org/10.1146/annurev.bi.51.070182.002531

Banda NK, Wood AK, Takahashi K, Levitt B, Rudd PM, Royle L, Abrahams JL, Stahl GL, Holers VM, Arend WP (2008) Initiation of the alternative pathway of murine complement by immune complexes is dependent on N-glycans in IgG antibodies. Arthritis Rheum 58(10):3081–3089. https://doi.org/10.1002/art.23865

Barb AW, Brady EK, Prestegard JH (2009) Branch-specific sialylation of IgG-Fc glycans by ST6Gal-I. Biochemistry 48(41):9705–9707. https://doi.org/10.1021/bi901430h

Barb AW, Meng L, Gao Z, Johnson RW, Moremen KW, Prestegard JH (2012) NMR characterization of immunoglobulin G Fc glycan motion on enzymatic sialylation. Biochemistry 51 (22):4618–4626. https://doi.org/10.1021/bi300319q

Beck A, Liu H (2019) Macro- and micro-heterogeneity of natural and recombinant IgG antibodies. Antibodies (Basel) 8(1). https://doi.org/10.3390/antib8010018

Bones J, Mittermayr S, O'Donoghue N, Guttman A, Rudd PM (2010) Ultra performance liquid chromatographic profiling of serum N-glycans for fast and efficient identification of cancer associated alterations in glycosylation. Anal Chem 82(24):10208–10215. https://doi.org/10.1021/ac102860w

Bourdin A, Husereau D, Molinari N, Golam S, Siddiqui MK, Lindner L, Xu X (2018) Matching-adjusted indirect comparison of benralizumab versus interleukin-5 inhibitors for the treatment of severe asthma: a systematic review. Eur Respir J 52(5). https://doi.org/10.1183/13993003.01393-2018

Bournazos S, Gupta A, Ravetch JV (2020) The role of IgG Fc receptors in antibody-dependent enhancement. Nat Rev Immunol 20(10):633–643. https://doi.org/10.1038/s41577-020-00410-0

Braster R, Grewal S, Visser R, Einarsdottir HK, van Egmond M, Vidarsson G, Bogels M (2017) Human IgG3 with extended half-life does not improve Fc-gamma receptor-mediated cancer antibody therapies in mice. PLoS One 12(5):e0177736. https://doi.org/10.1371/journal.pone.0177736

Burmeister WP, Huber AH, Bjorkman PJ (1994) Crystal structure of the complex of rat neonatal Fc receptor with Fc. Nature 372(6504):379–383. https://doi.org/10.1038/372379a0

Campbell IK, Miescher S, Branch DR, Mott PJ, Lazarus AH, Han D, Maraskovsky E, Zuercher AW, Neschadim A, Leontyev D, McKenzie BS, Kasermann F (2014) Therapeutic effect of IVIG on inflammatory arthritis in mice is dependent on the Fc portion and independent of sialylation or basophils. J Immunol 192(11):5031–5038. https://doi.org/10.4049/jimmunol.1301611

Capuano C, Pighi C, Molfetta R, Paolini R, Battella S, Palmieri G, Giannini G, Belardinilli F, Santoni A, Galandrini R (2017) Obinutuzumab-mediated high-affinity ligation of FcgammaRIIIA/CD16 primes NK cells for IFNgamma production. Onco Targets Ther 6(3): e1290037. https://doi.org/10.1080/2162402X.2017.1290037

Chakraborty S, Gonzalez J, Edwards K, Mallajosyula V, Buzzanco AS, Sherwood R, Buffone C, Kathale N, Providenza S, Xie MM, Andrews JR, Blish CA, Singh U, Dugan H, Wilson PC, Pham TD, Boyd SD, Nadeau KC, Pinsky BA, Zhang S, Memoli MJ, Taubenberger JK, Morales T, Schapiro JM, Tan GS, Jagannathan P, Wang TT (2021) Proinflammatory IgG Fc structures in patients with severe COVID-19. Nat Immunol 22(1):67–73. https://doi.org/10.1038/s41590-020-00828-7

Chen CL, Hsu JC, Lin CW, Wang CH, Tsai MH, Wu CY, Wong CH, Ma C (2017) Crystal structure of a homogeneous IgG-Fc glycoform with the N-glycan designed to maximize the antibody dependent cellular cytotoxicity. ACS Chem Biol 12(5):1335–1345. https://doi.org/10.1021/acschembio.7b00140

Cheng HD, Tirosh I, de Haan N, Stockmann H, Adamczyk B, McManus CA, O'Flaherty R, Greville G, Saldova R, Bonilla FA, Notarangelo LD, Driessen GJ, Holm IA, Rudd PM, Wuhrer M, Ackerman ME, Nigrovic PA (2020) IgG Fc glycosylation as an axis of humoral immunity in childhood. J Allergy Clin Immunol 145(2):710–713 e719. https://doi.org/10.1016/j.jaci.2019.10.012

Cobb BA (2020) The history of IgG glycosylation and where we are now. Glycobiology 30 (4):202–213. https://doi.org/10.1093/glycob/cwz065

Crispin M, Yu X, Bowden TA (2013) Crystal structure of sialylated IgG Fc: implications for the mechanism of intravenous immunoglobulin therapy. Proc Natl Acad Sci USA 110(38):E3544–E3546. https://doi.org/10.1073/pnas.1310657110

Cymer F, Schlothauer T, Knaupp A, Beck H (2017) Evaluation of an FcRn affinity chromatographic method for IgG1-type antibodies and evaluation of IgG variants. Bioanalysis 9 (17):1305–1317. https://doi.org/10.4155/bio-2017-0109

De Leoz MLA, Duewer DL, Fung A, Liu L, Yau HK, Potter O, Staples GO, Furuki K, Frenkel R, Hu Y, Sosic Z, Zhang P, Altmann F, Gru Nwald-Grube C, Shao C, Zaia J, Evers W, Pengelley S, Suckau D, Wiechmann A, Resemann A, Jabs W, Beck A, Froehlich JW, Huang C, Li Y, Liu Y, Sun S, Wang Y, Seo Y, An HJ, Reichardt NC, Ruiz JE, Archer-Hartmann S, Azadi P, Bell L, Lakos Z, An Y, Cipollo JF, Pucic-Bakovic M, Stambuk J, Lauc G, Li X, Wang PG, Bock A, Hennig R, Rapp E, Creskey M, Cyr TD, Nakano M, Sugiyama T, Leung PA, Link-Lenczowski-P, Jaworek J, Yang S, Zhang H, Kelly T, Klapoetke S, Cao R, Kim JY, Lee HK, Lee JY, Yoo JS, Kim SR, Suh SK, de Haan N, Falck D, Lageveen-Kammeijer GSM, Wuhrer M, Emery RJ, Kozak RP, Liew LP, Royle L, Urbanowicz PA, Packer NH, Song X, Everest-Dass A, Lattova E, Cajic S, Alagesan K, Kolarich D, Kasali T, Lindo V, Chen Y, Goswami K, Gau B, Amunugama R, Jones R, Stroop CJM, Kato K, Yagi H, Kondo S, Yuen CT, Harazono A, Shi X, Magnelli PE, Kasper BT, Mahal L, Harvey DJ, O'Flaherty R, Rudd PM, Saldova R, Hecht ES, Muddiman DC, Kang J, Bhoskar P, Menard D, Saati A, Merle C, Mast S, Tep S, Truong J, Nishikaze T, Sekiya S, Shafer A, Funaoka S, Toyoda M, de Vreugd P, Caron C, Pradhan P, Tan NC, Mechref Y, Patil S, Rohrer JS, Chakrabarti R, Dadke D, Lahori M, Zou C, Cairo C, Reiz B, Whittal RM, Lebrilla CB, Wu L, Guttman A, Szigeti M, Kremkow BG, Lee KH, Sihlbom C, Adamczyk B, Jin C, Karlsson NG, Ornros J, Larson G, Nilsson J, Meyer B, Wiegandt A, Komatsu E, Perreault H, Bodnar ED, Said N, Francois YN, Leize-Wagner E, Maier S, Zeck A, Heck AJR, Yang Y, Haselberg R, Yu YQ, Alley W, Leone JW, Yuan H, Stein SE (2020) NIST interlaboratory study on glycosylation analysis of monoclonal antibodies: comparison of results from diverse analytical methods. Mol Cell Proteomics 19(1):11–30. https://doi.org/10.1074/mcp.RA119.001677

de Taeye SW, Rispens T, Vidarsson G (2019) The ligands for human IgG and their effector functions. Antibodies (Basel) 8(2). https://doi.org/10.3390/antib8020030

Debre M, Bonnet MC, Fridman WH, Carosella E, Philippe N, Reinert P, Vilmer E, Kaplan C, Teillaud JL, Griscelli C (1993) Infusion of Fc gamma fragments for treatment of children with acute immune thrombocytopenic purpura. Lancet 342(8877):945–949

Deisenhofer J (1981) Crystallographic refinement and atomic models of a human Fc fragment and its complex with fragment B of protein A from *Staphylococcus aureus* at 2.9- and 2.8-A resolution. Biochemistry 20(9):2361–2370

Dekkers G, Plomp R, Koeleman CA, Visser R, von Horsten HH, Sandig V, Rispens T, Wuhrer M, Vidarsson G (2016) Multi-level glyco-engineering techniques to generate IgG with defined Fc-glycans. Sci Rep 6:36964. https://doi.org/10.1038/srep36964

Dekkers G, Treffers L, Plomp R, Bentlage AEH, de Boer M, Koeleman CAM, Lissenberg-Thunnissen SN, Visser R, Brouwer M, Mok JY, Matlung H, van den Berg TK, van Esch WJE, Kuijpers TW, Wouters D, Rispens T, Wuhrer M, Vidarsson G (2017) Decoding the human immunoglobulin G-glycan repertoire reveals a spectrum of Fc-receptor- and complement-mediated-effector activities. Front Immunol 8:877. https://doi.org/10.3389/fimmu.2017.00877

Dekkers G, Rispens T, Vidarsson G (2018) Novel concepts of altered immunoglobulin G galactosylation in autoimmune diseases. Front Immunol 9:553. https://doi.org/10.3389/fimmu.2018.00553

Doherty M, McManus CA, Duke R, Rudd PM (2012) High-throughput quantitative N-glycan analysis of glycoproteins. Methods Mol Biol 899:293–313. https://doi.org/10.1007/978-1-61779-921-1_19

Edberg JC, Kimberly RP (1997) Cell type-specific glycoforms of Fc gamma RIIIa (CD16): differential ligand binding. J Immunol 159(8):3849–3857

Ercan A, Cui J, Chatterton DE, Deane KD, Hazen MM, Brintnell W, O'Donnell CI, Derber LA, Weinblatt ME, Shadick NA, Bell DA, Cairns E, Solomon DH, Holers VM, Rudd PM, Lee DM (2010) Aberrant IgG galactosylation precedes disease onset, correlates with disease activity, and is prevalent in autoantibodies in rheumatoid arthritis. Arthritis Rheum 62(8):2239–2248. https://doi.org/10.1002/art.27533

Ercan A, Barnes MG, Hazen M, Tory H, Henderson L, Dedeoglu F, Fuhlbrigge RC, Grom A, Holm IA, Kellogg M, Kim S, Adamczyk B, Rudd PM, Son MB, Sundel RP, Foell D, Glass DN, Thompson SD, Nigrovic PA (2012) Multiple juvenile idiopathic arthritis subtypes demonstrate proinflammatory IgG glycosylation. Arthritis Rheum 64(9):3025–3033. https://doi.org/10.1002/art.34507

Fan SQ, Huang W, Wang LX (2012) Remarkable transglycosylation activity of glycosynthase mutants of endo-D, an endo-beta-N-acetylglucosaminidase from *Streptococcus pneumoniae*. J Biol Chem 287(14):11272–11281. https://doi.org/10.1074/jbc.M112.340497

Ferrara C, Brunker P, Suter T, Moser S, Puntener U, Umana P (2006a) Modulation of therapeutic antibody effector functions by glycosylation engineering: influence of Golgi enzyme localization domain and co-expression of heterologous beta1, 4-N-acetylglucosaminyltransferase III and Golgi alpha-mannosidase II. Biotechnol Bioeng 93(5):851–861

Ferrara C, Stuart F, Sondermann P, Brunker P, Umana P (2006b) The carbohydrate at FcgammaRIIIa Asn-162. An element required for high affinity binding to non-fucosylated IgG glycoforms. J Biol Chem 281(8):5032–5036

Ferrara C, Grau S, Jager C, Sondermann P, Brunker P, Waldhauer I, Hennig M, Ruf A, Rufer AC, Stihle M, Umana P, Benz J (2011) Unique carbohydrate-carbohydrate interactions are required for high affinity binding between FcgammaRIII and antibodies lacking core fucose. Proc Natl Acad Sci USA 108(31):12669–12674. https://doi.org/10.1073/pnas.1108455108

Frampton JE (2020) Inebilizumab: first approval. Drugs 80(12):1259–1264. https://doi.org/10.1007/s40265-020-01370-4

Fujita K, Kobayashi K, Iwamatsu A, Takeuchi M, Kumagai H, Yamamoto K (2004) Molecular cloning of Mucor hiemalis endo-beta-N-acetylglucosaminidase and some properties of the recombinant enzyme. Arch Biochem Biophys 432(1):41–49. https://doi.org/10.1016/j.abb.2004.09.013

Gala FA, Morrison SL (2004) V region carbohydrate and antibody expression. J Immunol 172(9):5489–5494

Ghirlando R, Lund J, Goodall M, Jefferis R (1999) Glycosylation of human IgG-Fc: influences on structure revealed by differential scanning micro-calorimetry. Immunol Lett 68(1):47–52

Giddens JP, Lomino JV, Amin MN, Wang LX (2016) Endo-F3 glycosynthase mutants enable chemoenzymatic synthesis of core-fucosylated triantennary complex type glycopeptides and glycoproteins. J Biol Chem 291(17):9356–9370. https://doi.org/10.1074/jbc.M116.721597

Goede V, Fischer K, Busch R, Engelke A, Eichhorst B, Wendtner CM, Chagorova T, de la Serna J, Dilhuydy MS, Illmer T, Opat S, Owen CJ, Samoylova O, Kreuzer KA, Stilgenbauer S, Dohner H, Langerak AW, Ritgen M, Kneba M, Asikianus E, Humphrey K, Wenger M, Hallek M (2014) Obinutuzumab plus chlorambucil in patients with CLL and coexisting conditions. N Engl J Med 370(12):1101–1110. https://doi.org/10.1056/NEJMoa1313984

Goetze AM, Liu YD, Zhang Z, Shah B, Lee E, Bondarenko PV, Flynn GC (2011) High-mannose glycans on the Fc region of therapeutic IgG antibodies increase serum clearance in humans. Glycobiology 21(7):949–959. https://doi.org/10.1093/glycob/cwr027

Goh JS, Liu Y, Chan KF, Wan C, Teo G, Zhang P, Zhang Y, Song Z (2014) Producing recombinant therapeutic glycoproteins with enhanced sialylation using CHO-gmt4 glycosylation mutant cells. Bioengineered 5(4):269–273. https://doi.org/10.4161/bioe.29490

Gonzalez-Nicolini V, Herter S, Lang S, Waldhauer I, Bacac M, Roemmele M, Bommer E, Freytag O, van Puijenbroek E, Umana P, Gerdes CA (2016) Premedication and chemotherapy agents do not impair imgatuzumab (GA201)-mediated antibody-dependent cellular cytotoxicity and combination therapies enhance efficacy. Clin Cancer Res 22(10):2453–2461. https://doi.org/10.1158/1078-0432.CCR-14-2579

Gornik I, Maravic G, Dumic J, Flogel M, Lauc G (1999) Fucosylation of IgG heavy chains is increased in rheumatoid arthritis. Clin Biochem 32(8):605–608. https://doi.org/10.1016/s0009-9120(99)00060-0

Gudelj I, Lauc G, Pezer M (2018) Immunoglobulin G glycosylation in aging and diseases. Cell Immunol 333:65–79. https://doi.org/10.1016/j.cellimm.2018.07.009

Guhr T, Bloem J, Derksen NI, Wuhrer M, Koenderman AH, Aalberse RC, Rispens T (2011) Enrichment of sialylated IgG by lectin fractionation does not enhance the efficacy of immunoglobulin G in a murine model of immune thrombocytopenia. PLoS One 6(6):e21246. https://doi.org/10.1371/journal.pone.0021246

Hayes JM, Frostell A, Karlsson R, Muller S, Martin SM, Pauers M, Reuss F, Cosgrave EF, Anneren C, Davey GP, Rudd PM (2017) Identification of Fc gamma receptor glycoforms that produce differential binding kinetics for rituximab. Mol Cell Proteomics 16(10):1770–1788. https://doi.org/10.1074/mcp.M117.066944

He W, Tan GS, Mullarkey CE, Lee AJ, Lam MM, Krammer F, Henry C, Wilson PC, Ashkar AA, Palese P, Miller MS (2016) Epitope specificity plays a critical role in regulating antibody-dependent cell-mediated cytotoxicity against influenza A virus. Proc Natl Acad Sci USA 113(42):11931–11936. https://doi.org/10.1073/pnas.1609316113

Herter S, Birk MC, Klein C, Gerdes C, Umana P, Bacac M (2014) Glycoengineering of therapeutic antibodies enhances monocyte/macrophage-mediated phagocytosis and cytotoxicity. J Immunol 192(5):2252–2260. https://doi.org/10.4049/jimmunol.1301249

Hiatt A, Bohorova N, Bohorov O, Goodman C, Kim D, Pauly MH, Velasco J, Whaley KJ, Piedra PA, Gilbert BE, Zeitlin L (2014) Glycan variants of a respiratory syncytial virus antibody with enhanced effector function and in vivo efficacy. Proc Natl Acad Sci USA 111(16):5992–5997. https://doi.org/10.1073/pnas.1402458111

Holland M, Yagi H, Takahashi N, Kato K, Savage CO, Goodall DM, Jefferis R (2006) Differential glycosylation of polyclonal IgG, IgG-Fc and IgG-Fab isolated from the sera of patients with ANCA-associated systemic vasculitis. Biochim Biophys Acta 1760(4):669–677

Houde D, Peng Y, Berkowitz SA, Engen JR (2010) Post-translational modifications differentially affect IgG1 conformation and receptor binding. Mol Cell Proteomics 9(8):1716–1728. https://doi.org/10.1074/mcp.M900540-MCP200

Huang L, Biolsi S, Bales KR, Kuchibhotla U (2006) Impact of variable domain glycosylation on antibody clearance: an LC/MS characterization. Anal Biochem 349(2):197–207

Huang W, Li C, Li B, Umekawa M, Yamamoto K, Zhang X, Wang LX (2009) Glycosynthases enable a highly efficient chemoenzymatic synthesis of N-glycoproteins carrying intact natural N-glycans. J Am Chem Soc 131(6):2214–2223. https://doi.org/10.1021/ja8074677

Huang W, Giddens J, Fan SQ, Toonstra C, Wang LX (2012) Chemoenzymatic glycoengineering of intact IgG antibodies for gain of functions. J Am Chem Soc 134(29):12308–12318. https://doi.org/10.1021/ja3051266

Idusogie EE, Presta LG, Gazzano-Santoro H, Totpal K, Wong PY, Ultsch M, Meng YG, Mulkerrin MG (2000) Mapping of the C1q binding site on rituxan, a chimeric antibody with a human IgG1 Fc. J Immunol 164(8):4178–4184

Iida S, Misaka H, Inoue M, Shibata M, Nakano R, Yamane-Ohnuki N, Wakitani M, Yano K, Shitara K, Satoh M (2006) Nonfucosylated therapeutic IgG1 antibody can evade the inhibitory effect of serum immunoglobulin G on antibody-dependent cellular cytotoxicity through its high binding to FcgammaRIIIa. Clin Cancer Res 12(9):2879–2887. https://doi.org/10.1158/1078-0432.CCR-05-2619

Ishida T, Joh T, Uike N, Yamamoto K, Utsunomiya A, Yoshida S, Saburi Y, Miyamoto T, Takemoto S, Suzushima H, Tsukasaki K, Nosaka K, Fujiwara H, Ishitsuka K, Inagaki H, Ogura M, Akinaga S, Tomonaga M, Tobinai K, Ueda R (2012) Defucosylated anti-CCR4 monoclonal antibody (KW-0761) for relapsed adult T-cell leukemia-lymphoma: a multicenter phase II study. J Clin Oncol 30(8):837–842. https://doi.org/10.1200/JCO.2011.37.3472

Ishikawa T, Takizawa T, Iwaki J, Mishima T, Ui-Tei K, Takeshita T, Matsubara S, Takizawa T (2015) Fc gamma receptor IIb participates in maternal IgG trafficking of human placental endothelial cells. Int J Mol Med 35(5):1273–1289. https://doi.org/10.3892/ijmm.2015.2141

Jefferis R (2016) Glyco-engineering of human IgG-Fc to modulate biologic activities. Curr Pharm Biotechnol 17(15):1333–1347. https://doi.org/10.2174/1389201017666161029225929

Jefferis R (2017a) Antibody posttranslational modifications. In: Liu C, Morrow KJJ (eds) Biosimilars of monoclonal antibodies: a practical guide to manufacturing, preclinical, and clinical development. Wiley, New York, pp 155–199. https://doi.org/10.1002/9781118940648

Jefferis R (2017b) Characterization of biosimilar biologics: the link between structure and functions. In: Endrenyi L, Declerck P, Chow S-C (eds) Drugs and the pharmaceutical sciences, vol 216. CRC, Boca Raton, pp 109–149

Jefferis R (2017c) Recombinant proteins and monoclonal antibodies. Adv Biochem Eng Biotechnol. https://doi.org/10.1007/10_2017_32

Jefferis R, Lund J, Mizutani H, Nakagawa H, Kawazoe Y, Arata Y, Takahashi N (1990) A comparative study of the N-linked oligosaccharide structures of human IgG subclass proteins. Biochem J 268(3):529–537

Jegaskanda S, Vanderven HA, Tan HX, Alcantara S, Wragg KM, Parsons MS, Chung AW, Juno JA, Kent SJ (2019) Influenza virus infection enhances antibody-mediated NK cell functions via type I interferon-dependent pathways. J Virol 93(5). https://doi.org/10.1128/JVI.02090-18

Jennewein MF, Goldfarb I, Dolatshahi S, Cosgrove C, Noelette FJ, Krykbaeva M, Das J, Sarkar A, Gorman MJ, Fischinger S, Boudreau CM, Brown J, Cooperrider JH, Aneja J, Suscovich TJ, Graham BS, Lauer GM, Goetghebuer T, Marchant A, Lauffenburger D, Kim AY, Riley LE, Alter G (2019) Fc glycan-mediated regulation of placental antibody transfer. Cell 178(1):202–215. e214. https://doi.org/10.1016/j.cell.2019.05.044

Junttila TT, Parsons K, Olsson C, Lu Y, Xin Y, Theriault J, Crocker L, Pabonan O, Baginski T, Meng G, Totpal K, Kelley RF, Sliwkowski MX (2010) Superior in vivo efficacy of afucosylated trastuzumab in the treatment of HER2-amplified breast cancer. Cancer Res 70(11):4481–4489. https://doi.org/10.1158/0008-5472.CAN-09-3704

Kanda Y, Imai-Nishiya H, Kuni-Kamochi R, Mori K, Inoue M, Kitajima-Miyama K, Okazaki A, Iida S, Shitara K, Satoh M (2007a) Establishment of a GDP-mannose 4,6-dehydratase (GMD) knockout host cell line: a new strategy for generating completely non-fucosylated recombinant therapeutics. J Biotechnol 130(3):300–310. https://doi.org/10.1016/j.jbiotec.2007.04.025

Kanda Y, Yamada T, Mori K, Okazaki A, Inoue M, Kitajima-Miyama K, Kuni-Kamochi R, Nakano R, Yano K, Kakita S, Shitara K, Satoh M (2007b) Comparison of biological activity among nonfucosylated therapeutic IgG1 antibodies with three different N-linked Fc oligosaccharides: the high-mannose, hybrid, and complex types. Glycobiology 17(1):104–118

Kaneko Y, Nimmerjahn F, Ravetch JV (2006) Anti-inflammatory activity of immunoglobulin G resulting from Fc sialylation. Science 313(5787):670–673. https://doi.org/10.1126/science.1129594

Kaplon H, Muralidharan M, Schneider Z, Reichert JM (2020) Antibodies to watch in 2020. MAbs 12(1):1703531. https://doi.org/10.1080/19420862.2019.1703531

Kapur R, Einarsdottir HK, Vidarsson G (2014) IgG-effector functions: "the good, the bad and the ugly". Immunol Lett 160(2):139–144. https://doi.org/10.1016/j.imlet.2014.01.015

Kapur R, Della Valle L, Verhagen OJ, Hipgrave Ederveen A, Ligthart P, de Haas M, Kumpel B, Wuhrer M, van der Schoot CE, Vidarsson G (2015) Prophylactic anti-D preparations display variable decreases in Fc-fucosylation of anti-D. Transfusion 55(3):553–562. https://doi.org/10.1111/trf.12880

Kelly RM, Kowle RL, Lian Z, Strifler BA, Witcher DR, Parekh BS, Wang T, Frye CC (2018) Modulation of IgG1 immunoeffector function by glycoengineering of the GDP-fucose biosynthesis pathway. Biotechnol Bioeng 115(3):705–718. https://doi.org/10.1002/bit.26496

Kemna MJ, Plomp R, van Paassen P, Koeleman CAM, Jansen BC, Damoiseaux J, Cohen Tervaert JW, Wuhrer M (2017) Galactosylation and sialylation levels of IgG predict relapse in patients with PR3-ANCA associated vasculitis. EBioMedicine 17:108–118. https://doi.org/10.1016/j.ebiom.2017.01.033

Kibe T, Fujimoto S, Ishida C, Togari H, Okada S, Nakagawa H, Tsukamoto Y, Takahashi N (1996) Glycosylation and placental transport of immunoglobulin G. J Clin Biochem Nutr 21:57–63. https://doi.org/10.3164/jcbn.21.57

Kiyoshi M, Caaveiro JM, Kawai T, Tashiro S, Ide T, Asaoka Y, Hatayama K, Tsumoto K (2015) Structural basis for binding of human IgG1 to its high-affinity human receptor FcgammaRI. Nat Commun 6:6866. https://doi.org/10.1038/ncomms7866

Kobayashi S, Kiyosada T, Shoda S-I (1996) Synthesis of artificial chitin: irreversible catalytic behavior of a glycosyl hydrolase through a transition state analogue substrate. J Am Chem Soc 118(51):13113–13114. https://doi.org/10.1021/ja963011u

Koene HR, Kleijer M, Algra J, Roos D, von dem Borne AE, de Haas M (1997) Fc gammaRIIIa-158V/F polymorphism influences the binding of IgG by natural killer cell Fc gammaRIIIa, independently of the Fc gammaRIIIa-48L/R/H phenotype. Blood 90(3):1109–1114

Krapp S, Mimura Y, Jefferis R, Huber R, Sondermann P (2003) Structural analysis of human IgG-Fc glycoforms reveals a correlation between glycosylation and structural integrity. J Mol Biol 325(5):979–989

Kuhne F, Bonnington L, Malik S, Thomann M, Avenal C, Cymer F, Wegele H, Reusch D, Mormann M, Bulau P (2019) The impact of immunoglobulin G1 Fc sialylation on backbone amide H/D exchange. Antibodies (Basel) 8(4). https://doi.org/10.3390/antib8040049

Kurogochi M, Mori M, Osumi K, Tojino M, Sugawara S, Takashima S, Hirose Y, Tsukimura W, Mizuno M, Amano J, Matsuda A, Tomita M, Takayanagi A, Shoda S, Shirai T (2015) Glycoengineered monoclonal antibodies with homogeneous glycan (M3, G0, G2, and A2) using a chemoenzymatic approach have different affinities for FcgammaRIIIa and variable antibody-dependent cellular cytotoxicity activities. PLoS One 10(7):e0132848. https://doi.org/10.1371/journal.pone.0132848

Larsen MD, de Graaf EL, Sonneveld ME, Plomp HR, Nouta J, Hoepel W, Chen HJ, Linty F, Visser R, Brinkhaus M, Sustic T, de Taeye SW, Bentlage AEH, Toivonen S, Koeleman CAM, Sainio S, Kootstra NA, Brouwer PJM, Geyer CE, Derksen NIL, Wolbink G, de Winther M, Sanders RW, van Gils MJ, de Bruin S, Vlaar APJ, Amsterdam UMCC-bsg, Rispens T, den Dunnen J, Zaaijer HL, Wuhrer M, Ellen van der Schoot C, Vidarsson G (2020) Afucosylated IgG characterizes enveloped viral responses and correlates with COVID-19 severity. Science. https://doi.org/10.1126/science.abc8378

Le NP, Bowden TA, Struwe WB, Crispin M (2016) Immune recruitment or suppression by glycan engineering of endogenous and therapeutic antibodies. Biochim Biophys Acta 1860 (8):1655–1668. https://doi.org/10.1016/j.bbagen.2016.04.016

Lee SJ, Evers S, Roeder D, Parlow AF, Risteli J, Risteli L, Lee YC, Feizi T, Langen H, Nussenzweig MC (2002) Mannose receptor-mediated regulation of serum glycoprotein homeostasis. Science 295(5561):1898–1901. https://doi.org/10.1126/science.1069540

Leontyev D, Katsman Y, Ma XZ, Miescher S, Kasermann F, Branch DR (2012) Sialylation-independent mechanism involved in the amelioration of murine immune thrombocytopenia using intravenous gammaglobulin. Transfusion 52(8):1799–1805. https://doi.org/10.1111/j.1537-2995.2011.03517.x

Li B, Zeng Y, Hauser S, Song H, Wang LX (2005) Highly efficient endoglycosidase-catalyzed synthesis of glycopeptides using oligosaccharide oxazolines as donor substrates. J Am Chem Soc 127(27):9692–9693. https://doi.org/10.1021/ja051715a

Li H, Sethuraman N, Stadheim TA, Zha D, Prinz B, Ballew N, Bobrowicz P, Choi BK, Cook WJ, Cukan M, Houston-Cummings NR, Davidson R, Gong B, Hamilton SR, Hoopes JP, Jiang Y, Kim N, Mansfield R, Nett JH, Rios S, Strawbridge R, Wildt S, Gerngross TU (2006) Optimization of humanized IgGs in glycoengineered Pichia pastoris. Nat Biotechnol 24(2):210–215

Li T, Tong X, Yang Q, Giddens JP, Wang LX (2016) Glycosynthase mutants of endoglycosidase S2 show potent transglycosylation activity and remarkably relaxed substrate specificity for antibody glycosylation remodeling. J Biol Chem 291(32):16508–16518. https://doi.org/10.1074/jbc.M116.738765

Li T, DiLillo DJ, Bournazos S, Giddens JP, Ravetch JV, Wang LX (2017) Modulating IgG effector function by Fc glycan engineering. Proc Natl Acad Sci USA 114(13):3485–3490. https://doi.org/10.1073/pnas.1702173114

Li C, Li T, Wang LX (2018) Chemoenzymatic defucosylation of therapeutic antibodies for enhanced effector functions using bacterial alpha-fucosidases. Methods Mol Biol 1827:367–380. https://doi.org/10.1007/978-1-4939-8648-4_19

Lin CW, Tsai MH, Li ST, Tsai TI, Chu KC, Liu YC, Lai MY, Wu CY, Tseng YC, Shivatare SS, Wang CH, Chao P, Wang SY, Shih HW, Zeng YF, You TH, Liao JY, Tu YC, Lin YS, Chuang HY, Chen CL, Tsai CS, Huang CC, Lin NH, Ma C, Wu CY, Wong CH (2015) A common glycan structure on immunoglobulin G for enhancement of effector functions. Proc Natl Acad Sci USA 112(34):10611–10616. https://doi.org/10.1073/pnas.1513456112

Liu L (2015) Antibody glycosylation and its impact on the pharmacokinetics and pharmacodynamics of monoclonal antibodies and Fc-fusion proteins. J Pharm Sci 104(6):1866–1884. https://doi.org/10.1002/jps.24444

Liu YD, Flynn GC (2016) Effect of high mannose glycan pairing on IgG antibody clearance. Biologicals 44(3):163–169. https://doi.org/10.1016/j.biologicals.2016.02.003

Liu L, Stadheim A, Hamuro L, Pittman T, Wang W, Zha D, Hochman J, Prueksaritanont T (2011) Pharmacokinetics of IgG1 monoclonal antibodies produced in humanized Pichia pastoris with specific glycoforms: a comparative study with CHO produced materials. Biologicals 39(4):205–210. https://doi.org/10.1016/j.biologicals.2011.06.002

Liu CP, Tsai TI, Cheng T, Shivatare VS, Wu CY, Wu CY, Wong CH (2018) Glycoengineering of antibody (Herceptin) through yeast expression and in vitro enzymatic glycosylation. Proc Natl Acad Sci USA 115(4):720–725. https://doi.org/10.1073/pnas.1718172115

Lu RM, Hwang YC, Liu IJ, Lee CC, Tsai HZ, Li HJ, Wu HC (2020) Development of therapeutic antibodies for the treatment of diseases. J Biomed Sci 27(1):1. https://doi.org/10.1186/s12929-019-0592-z

Lund J, Tanaka T, Takahashi N, Sarmay G, Arata Y, Jefferis R (1990) A protein structural change in aglycosylated IgG3 correlates with loss of huFc gamma R1 and huFc gamma R111 binding and/or activation. Mol Immunol 27(11):1145–1153

Lund J, Winter G, Jones PT, Pound JD, Tanaka T, Walker MR, Artymiuk PJ, Arata Y, Burton DR, Jefferis R et al (1991) Human Fc gamma RI and Fc gamma RII interact with distinct but overlapping sites on human IgG. J Immunol 147(8):2657–2662

Mabashi-Asazuma H, Kuo CW, Khoo KH, Jarvis DL (2014) A novel baculovirus vector for the production of nonfucosylated recombinant glycoproteins in insect cells. Glycobiology 24(3):325–340. https://doi.org/10.1093/glycob/cwt161

Malhotra R, Wormald MR, Rudd PM, Fischer PB, Dwek RA, Sim RB (1995) Glycosylation changes of IgG associated with rheumatoid arthritis can activate complement via the mannose-binding protein. Nat Med 1(3):237–243

Martinez DR, Fong Y, Li SH, Yang F, Jennewein MF, Weiner JA, Harrell EA, Mangold JF, Goswami R, Seage GR 3rd, Alter G, Ackerman ME, Peng X, Fouda GG, Permar SR (2019) Fc characteristics mediate selective placental transfer of IgG in HIV-infected women. Cell 178(1):190–201 e111. https://doi.org/10.1016/j.cell.2019.05.046

Mimura Y, Jefferis R (2021) Human IgG glycosylation in inflammation and inflammatory disease. In: Barchi JJ (ed) Comprehensive glycoscience, vol 5, 2nd edn. Elsevier, London, pp 215–232

Mimura Y, Church S, Ghirlando R, Ashton PR, Dong S, Goodall M, Lund J, Jefferis R (2000) The influence of glycosylation on the thermal stability and effector function expression of human IgG1-Fc: properties of a series of truncated glycoforms. Mol Immunol 37(12–13):697–706

Mimura Y, Sondermann P, Ghirlando R, Lund J, Young SP, Goodall M, Jefferis R (2001) Role of oligosaccharide residues of IgG1-Fc in Fc gamma RIIb binding. J Biol Chem 276 (49):45539–45547

Mimura Y, Ashton PR, Takahashi N, Harvey DJ, Jefferis R (2007) Contrasting glycosylation profiles between Fab and Fc of a human IgG protein studied by electrospray ionization mass spectrometry. J Immunol Methods 326(1–2):116–126

Mimura Y, Jefferis R, Mimura-Kimura Y, Abrahams J, Rudd PM (2009) Glycosylation of therapeutic IgGs. In: An Z (ed) Therapeutic monoclonal antibodies: from the bench to the clinic. Wiley, Hoboken, pp 67–89

Mimura Y, Katoh T, Saldova R, O'Flaherty R, Izumi T, Mimura-Kimura Y, Utsunomiya T, Mizukami Y, Yamamoto K, Matsumoto T, Rudd PM (2018) Glycosylation engineering of therapeutic IgG antibodies: challenges for the safety, functionality and efficacy. Protein Cell 9 (1):47–62. https://doi.org/10.1007/s13238-017-0433-3

Mizushima T, Yagi H, Takemoto E, Shibata-Koyama M, Isoda Y, Iida S, Masuda K, Satoh M, Kato K (2011) Structural basis for improved efficacy of therapeutic antibodies on defucosylation of their Fc glycans. Genes Cells 16(11):1071–1080. https://doi.org/10.1111/j.1365-2443.2011. 01552.x

Moh ESX, Sayyadi N, Packer NH (2019) Chemoenzymatic glycan labelling as a platform for site-specific IgM-antibody drug conjugates. Anal Biochem 584:113385. https://doi.org/10.1016/j. ab.2019.113385

Moldt B, Shibata-Koyama M, Rakasz EG, Schultz N, Kanda Y, Dunlop DC, Finstad SL, Jin C, Landucci G, Alpert MD, Dugast AS, Parren PW, Nimmerjahn F, Evans DT, Alter G, Forthal DN, Schmitz JE, Iida S, Poignard P, Watkins DI, Hessell AJ, Burton DR (2012) A nonfucosylated variant of the anti-HIV-1 monoclonal antibody b12 has enhanced FcgammaRIIIa-mediated antiviral activity in vitro but does not improve protection against mucosal SHIV challenge in macaques. J Virol 86(11):6189–6196. https://doi.org/10.1128/ JVI.00491-12

Muramatsu H, Tachikui H, Ushida H, Song X, Qiu Y, Yamamoto S, Muramatsu T (2001) Molecular cloning and expression of endo-beta-N-acetylglucosaminidase D, which acts on the core structure of complex type asparagine-linked oligosaccharides. J Biochem 129(6):923–928

Natsume A, Wakitani M, Yamane-Ohnuki N, Shoji-Hosaka E, Niwa R, Uchida K, Satoh M, Shitara K (2005) Fucose removal from complex-type oligosaccharide enhances the antibody-dependent cellular cytotoxicity of single-gene-encoded antibody comprising a single-chain antibody linked the antibody constant region. J Immunol Methods 306(1–2):93–103. https://doi.org/10.1016/j. jim.2005.07.025

Nechansky A, Schuster M, Jost W, Siegl P, Wiederkum S, Gorr G, Kircheis R (2007) Compensation of endogenous IgG mediated inhibition of antibody-dependent cellular cytotoxicity by glyco-engineering of therapeutic antibodies. Mol Immunol 44(7):1815–1817. https://doi.org/10. 1016/j.molimm.2006.08.013

Nimmerjahn F, Ravetch JV (2008) Anti-inflammatory actions of intravenous immunoglobulin. Annu Rev Immunol 26:513–533. https://doi.org/10.1146/annurev.immunol.26.021607.090232

Niwa R, Shoji-Hosaka E, Sakurada M, Shinkawa T, Uchida K, Nakamura K, Matsushima K, Ueda R, Hanai N, Shitara K (2004) Defucosylated chimeric anti-CC chemokine receptor 4 IgG1 with enhanced antibody-dependent cellular cytotoxicity shows potent therapeutic activity to T-cell leukemia and lymphoma. Cancer Res 64(6):2127–2133. https://doi.org/10.1158/0008-5472.can-03-2068

Niwa R, Natsume A, Uehara A, Wakitani M, Iida S, Uchida K, Satoh M, Shitara K (2005) IgG subclass-independent improvement of antibody-dependent cellular cytotoxicity by fucose removal from Asn297-linked oligosaccharides. J Immunol Methods 306(1–2):151–160

Noguchi M, Tanaka T, Gyakushi H, Kobayashi A, Shoda S (2009) Efficient synthesis of sugar oxazolines from unprotected N-acetyl-2-amino sugars by using chloroformamidinium reagent in water. J Org Chem 74(5):2210–2212. https://doi.org/10.1021/jo8024708

Okeley NM, Alley SC, Anderson ME, Boursalian TE, Burke PJ, Emmerton KM, Jeffrey SC, Klussman K, Law CL, Sussman D, Toki BE, Westendorf L, Zeng W, Zhang X, Benjamin DR, Senter PD (2013) Development of orally active inhibitors of protein and cellular fucosylation. Proc Natl Acad Sci USA 110(14):5404–5409. https://doi.org/10.1073/pnas.1222263110

Othy S, Topcu S, Saha C, Kothapalli P, Lacroix-Desmazes S, Kasermann F, Miescher S, Bayry J, Kaveri SV (2014) Sialylation may be dispensable for reciprocal modulation of helper T cells by intravenous immunoglobulin. Eur J Immunol 44(7):2059–2063. https://doi.org/10.1002/eji.201444440

Overdijk MB, Verploegen S, Bleeker WK, Parren PWHI (2014) Role of IgG Fc receptors in monoclonal antibody therapy of cancer. In: Ackerman M, Nimmerjahn F (eds) Antibody Fc: linking adaptive and innate immunity. Academic, Amsterdam, pp 239–255

Padlan EA (1990) X-ray diffraction studies of antibody constant regions. In: Metzger H (ed) Fc receptors and the action of antibodies. American Society for Microbiology, Washington, DC, pp 12–30

Parekh RB, Dwek RA, Sutton BJ, Fernandes DL, Leung A, Stanworth D, Rademacher TW, Mizuochi T, Taniguchi T, Matsuta K et al (1985) Association of rheumatoid arthritis and primary osteoarthritis with changes in the glycosylation pattern of total serum IgG. Nature 316(6027):452–457

Parekh R, Roitt I, Isenberg D, Dwek R, Rademacher T (1988) Age-related galactosylation of the N-linked oligosaccharides of human serum IgG. J Exp Med 167(5):1731–1736

Patel KR, Roberts JT, Subedi GP, Barb AW (2018) Restricted processing of CD16a/Fc gamma receptor IIIa N-glycans from primary human NK cells impacts structure and function. J Biol Chem 293(10):3477–3489. https://doi.org/10.1074/jbc.RA117.001207

Patel KR, Nott JD, Barb AW (2019) Primary human natural killer cells retain proinflammatory IgG1 at the cell surface and express CD16a glycoforms with donor-dependent variability. Mol Cell Proteomics 18(11):2178–2190. https://doi.org/10.1074/mcp.RA119.001607

Pereira NA, Chan KF, Lin PC, Song Z (2018) The "less-is-more" in therapeutic antibodies: afucosylated anti-cancer antibodies with enhanced antibody-dependent cellular cytotoxicity. MAbs 10(5):693–711. https://doi.org/10.1080/19420862.2018.1466767

Pucic M, Knezevic A, Vidic J, Adamczyk B, Novokmet M, Polasek O, Gornik O, Supraha-Goreta S, Wormald MR, Redzic I, Campbell H, Wright A, Hastie ND, Wilson JF, Rudan I, Wuhrer M, Rudd PM, Josic D, Lauc G (2011) High throughput isolation and glycosylation analysis of IgG-variability and heritability of the IgG glycome in three isolated human populations. Mol Cell Proteomics 10(10):M111 010090. https://doi.org/10.1074/mcp.M111.010090

Radcliffe CM, Arnold JN, Suter DM, Wormald MR, Harvey DJ, Royle L, Mimura Y, Kimura Y, Sim RB, Inoges S, Rodriguez-Calvillo M, Zabalegui N, de Cerio AL, Potter KN, Mockridge CI, Dwek RA, Bendandi M, Rudd PM, Stevenson FK (2007) Human follicular lymphoma cells contain oligomannose glycans in the antigen-binding site of the B-cell receptor. J Biol Chem 282(10):7405–7415

Raju TS, Jordan RE (2012) Galactosylation variations in marketed therapeutic antibodies. MAbs 4 (3):385–391. https://doi.org/10.4161/mabs.19868

Rombouts Y, Ewing E, van de Stadt LA, Selman MH, Trouw LA, Deelder AM, Huizinga TW, Wuhrer M, van Schaardenburg D, Toes RE, Scherer HU (2015) Anti-citrullinated protein antibodies acquire a pro-inflammatory Fc glycosylation phenotype prior to the onset of rheumatoid arthritis. Ann Rheum Dis 74(1):234–241. https://doi.org/10.1136/annrheumdis-2013-203565

Roopenian DC, Akilesh S (2007) FcRn: the neonatal Fc receptor comes of age. Nat Rev Immunol 7 (9):715–725

Royle L, Radcliffe CM, Dwek RA, Rudd PM (2006) Detailed structural analysis of N-glycans released from glycoproteins in SDS-PAGE gel bands using HPLC combined with exoglycosidase array digestions. Methods Mol Biol 347:125–143

Rudd PM, Dwek RA (1997) Glycosylation: heterogeneity and the 3D structure of proteins. Crit Rev Biochem Mol Biol 32(1):1–100

Sarmay G, Lund J, Rozsnyay Z, Gergely J, Jefferis R (1992) Mapping and comparison of the interaction sites on the Fc region of IgG responsible for triggering antibody dependent cellular cytotoxicity (ADCC) through different types of human Fc gamma receptor. Mol Immunol 29 (5):633–639

Scallon BJ, Tam SH, McCarthy SG, Cai AN, Raju TS (2007) Higher levels of sialylated Fc glycans in immunoglobulin G molecules can adversely impact functionality. Mol Immunol 44 (7):1524–1534. https://doi.org/10.1016/j.molimm.2006.09.005

Schwab I, Nimmerjahn F (2014) Role of sialylation in the anti-inflammatory activity of intravenous immunoglobulin – F(ab′)(2) versus Fc sialylation. Clin Exp Immunol 178(Suppl 1):97–99. https://doi.org/10.1111/cei.12527

Sehmi R, Lim HF, Mukherjee M, Huang C, Radford K, Newbold P, Boulet LP, Dorscheid D, Martin JG, Nair P (2018) Benralizumab attenuates airway eosinophilia in prednisone-dependent asthma. J Allergy Clin Immunol 141(4):1529–1532. e1528. https://doi.org/10.1016/j.jaci.2018.01.008

Sehn LH, Assouline SE, Stewart DA, Mangel J, Gascoyne RD, Fine G, Frances-Lasserre S, Carlile DJ, Crump M (2012) A phase 1 study of obinutuzumab induction followed by 2 years of maintenance in patients with relapsed CD20-positive B-cell malignancies. Blood 119 (22):5118–5125. https://doi.org/10.1182/blood-2012-02-408773

Sehn LH, Goy A, Offner FC, Martinelli G, Caballero MD, Gadeberg O, Baetz T, Zelenetz AD, Gaidano G, Fayad LE, Buckstein R, Friedberg JW, Crump M, Jaksic B, Zinzani PL, Padmanabhan Iyer S, Sahin D, Chai A, Fingerle-Rowson G, Press OW (2015) Randomized phase II trial comparing obinutuzumab (GA101) with rituximab in patients with relapsed CD20+ indolent B-cell non-Hodgkin Lymphoma: final analysis of the GAUSS study. J Clin Oncol 33(30):3467–3474. https://doi.org/10.1200/JCO.2014.59.2139

Shibata-Koyama M, Iida S, Okazaki A, Mori K, Kitajima-Miyama K, Saitou S, Kakita S, Kanda Y, Shitara K, Kato K, Satoh M (2009) The N-linked oligosaccharide at Fc gamma RIIIa Asn-45: an inhibitory element for high Fc gamma RIIIa binding affinity to IgG glycoforms lacking core fucosylation. Glycobiology 19(2):126–134. https://doi.org/10.1093/glycob/cwn110

Shields RL, Namenuk AK, Hong K, Meng YG, Rae J, Briggs J, Xie D, Lai J, Stadlen A, Li B, Fox JA, Presta LG (2001) High resolution mapping of the binding site on human IgG1 for Fc gamma RI, Fc gamma RII, Fc gamma RIII, and FcRn and design of IgG1 variants with improved binding to the Fc gamma R. J Biol Chem 276(9):6591–6604

Shields RL, Lai J, Keck R, O'Connell LY, Hong K, Meng YG, Weikert SH, Presta LG (2002) Lack of fucose on human IgG1 N-linked oligosaccharide improves binding to human Fcgamma RIII and antibody-dependent cellular toxicity. J Biol Chem 277(30):26733–26740

Sondermann P, Huber R, Oosthuizen V, Jacob U (2000) The 3.2-A crystal structure of the human IgG1 Fc fragment-Fc gammaRIII complex. Nature 406(6793):267–273

Sonneveld ME, de Haas M, Koeleman C, de Haan N, Zeerleder SS, Ligthart PC, Wuhrer M, van der Schoot CE, Vidarsson G (2017) Patients with IgG1-anti-red blood cell autoantibodies show aberrant Fc-glycosylation. Sci Rep 7(1):8187. https://doi.org/10.1038/s41598-017-08654-y

Sonneveld ME, Koeleman CAM, Plomp HR, Wuhrer M, van der Schoot CE, Vidarsson G (2018) Fc-glycosylation in human IgG1 and IgG3 is similar for both total and anti-red-blood cell anti-K antibodies. Front Immunol 9:129. https://doi.org/10.3389/fimmu.2018.00129

Stambuk T, Klasic M, Zoldos V, Lauc G (2020) N-glycans as functional effectors of genetic and epigenetic disease risk. Mol Aspects Med 100891. https://doi.org/10.1016/j.mam.2020.100891

Stapleton NM, Andersen JT, Stemerding AM, Bjarnarson SP, Verheul RC, Gerritsen J, Zhao Y, Kleijer M, Sandlie I, de Haas M, Jonsdottir I, van der Schoot CE, Vidarsson G (2011)

Competition for FcRn-mediated transport gives rise to short half-life of human IgG3 and offers therapeutic potential. Nat Commun 2:599. https://doi.org/10.1038/ncomms1608

Stockert RJ (1995) The asialoglycoprotein receptor: relationships between structure, function, and expression. Physiol Rev 75(3):591–609. https://doi.org/10.1152/physrev.1995.75.3.591

Strasser R, Stadlmann J, Schahs M, Stiegler G, Quendler H, Mach L, Glossl J, Weterings K, Pabst M, Steinkellner H (2008) Generation of glyco-engineered Nicotiana benthamiana for the production of monoclonal antibodies with a homogeneous human-like N-glycan structure. Plant Biotechnol J 6(4):392–402. https://doi.org/10.1111/j.1467-7652.2008.00330.x

Subedi GP, Barb AW (2016) The immunoglobulin G1 N-glycan composition affects binding to each low affinity Fc gamma receptor. MAbs 8(8):1512–1524. https://doi.org/10.1080/19420862.2016.1218586

Sun B, Bao W, Tian X, Li M, Liu H, Dong J, Huang W (2014) A simplified procedure for gram-scale production of sialylglycopeptide (SGP) from egg yolks and subsequent semi-synthesis of Man3GlcNAc oxazoline. Carbohydr Res 396:62–69. https://doi.org/10.1016/j.carres.2014.07.013

Suzuki T, Ishii-Watabe A, Tada M, Kobayashi T, Kanayasu-Toyoda T, Kawanishi T, Yamaguchi T (2010) Importance of neonatal FcR in regulating the serum half-life of therapeutic proteins containing the Fc domain of human IgG1: a comparative study of the affinity of monoclonal antibodies and Fc-fusion proteins to human neonatal FcR. J Immunol 184(4):1968–1976. https://doi.org/10.4049/jimmunol.0903296

Takegawa K, Tabuchi M, Yamaguchi S, Kondo A, Kato I, Iwahara S (1995) Synthesis of neoglycoproteins using oligosaccharide-transfer activity with endo-beta-N-acetylglucosaminidase. J Biol Chem 270(7):3094–3099

Takegawa K, Yamabe K, Fujita K, Tabuchi M, Mita M, Izu H, Watanabe A, Asada Y, Sano M, Kondo A, Kato I, Iwahara S (1997) Cloning, sequencing, and expression of Arthrobacter protophormiae endo-beta-N-acetylglucosaminidase in *Escherichia coli*. Arch Biochem Biophys 338(1):22–28. https://doi.org/10.1006/abbi.1996.9803

Tang F, Yang Y, Tang Y, Tang S, Yang L, Sun B, Jiang B, Dong J, Liu H, Huang M, Geng MY, Huang W (2016) One-pot N-glycosylation remodeling of IgG with non-natural sialylglycopeptides enables glycosite-specific and dual-payload antibody-drug conjugates. Org Biomol Chem 14(40):9501–9518. https://doi.org/10.1039/c6ob01751g

Tang F, Wang LX, Huang W (2017) Chemoenzymatic synthesis of glycoengineered IgG antibodies and glycosite-specific antibody-drug conjugates. Nat Protoc 12(8):1702–1721. https://doi.org/10.1038/nprot.2017.058

Temming AR, Dekkers G, van de Bovenkamp FS, Plomp HR, Bentlage AEH, Szittner Z, Derksen NIL, Wuhrer M, Rispens T, Vidarsson G (2019) Human DC-SIGN and CD23 do not interact with human IgG. Sci Rep 9(1):9995. https://doi.org/10.1038/s41598-019-46484-2

Terszowski G, Klein C, Stern M (2014) KIR/HLA interactions negatively affect rituximab- but not GA101 (obinutuzumab)-induced antibody-dependent cellular cytotoxicity. J Immunol 192 (12):5618–5624. https://doi.org/10.4049/jimmunol.1400288

Thomann M, Schlothauer T, Dashivets T, Malik S, Avenal C, Bulau P, Ruger P, Reusch D (2015) In vitro glycoengineering of IgG1 and its effect on Fc receptor binding and ADCC activity. PLoS One 10(8):e0134949. https://doi.org/10.1371/journal.pone.0134949

Thomann M, Reckermann K, Reusch D, Prasser J, Tejada ML (2016) Fc-galactosylation modulates antibody-dependent cellular cytotoxicity of therapeutic antibodies. Mol Immunol 73:69–75. https://doi.org/10.1016/j.molimm.2016.03.002

Toftevall H, Nyhlen H, Olsson F, Sjogren J (2020) Antibody conjugations via glycosyl remodeling. Methods Mol Biol 2078:131–145. https://doi.org/10.1007/978-1-4939-9929-3_9

Tsai TI, Li ST, Liu CP, Chen KY, Shivatare SS, Lin CW, Liao SF, Lin CW, Hsu TL, Wu YT, Tsai MH, Lai MY, Lin NH, Wu CY, Wong CH (2017) An effective bacterial fucosidase for glycoprotein remodeling. ACS Chem Biol 12(1):63–72. https://doi.org/10.1021/acschembio.6b00821

van Geel R, Wijdeven MA, Heesbeen R, Verkade JM, Wasiel AA, van Berkel SS, van Delft FL (2015) Chemoenzymatic conjugation of toxic payloads to the globally conserved N-glycan of native mAbs provides homogeneous and highly efficacious antibody-drug conjugates. Bioconjug Chem 26(11):2233–2242. https://doi.org/10.1021/acs.bioconjchem.5b00224

Varki A, Cummings RD, Esko JD, Stanley P, Hart GW, Aebi M, Darvill AG, Kinoshita T, Packer NH, Prestegard JH, Schnaar RL, Seeberger PH (2017) Essentials of glycobiology, 3rd edn. Cold Spring Harbor Laboratory Press, New York

Vidarsson G, Dekkers G, Rispens T (2014) IgG subclasses and allotypes: from structure to effector functions. Front Immunol 5:520. https://doi.org/10.3389/fimmu.2014.00520

Wada R, Matsui M, Kawasaki N (2019) Influence of N-glycosylation on effector functions and thermal stability of glycoengineered IgG1 monoclonal antibody with homogeneous glycoforms. MAbs 11(2):350–372. https://doi.org/10.1080/19420862.2018.1551044

Wang B, Yan L, Yao Z, Roskos LK (2017a) Population pharmacokinetics and pharmacodynamics of benralizumab in healthy volunteers and patients with asthma. CPT Pharmacometrics Syst Pharmacol. https://doi.org/10.1002/psp4.12160

Wang TT, Sewatanon J, Memoli MJ, Wrammert J, Bournazos S, Bhaumik SK, Pinsky BA, Chokephaibulkit K, Onlamoon N, Pattanapanyasat K, Taubenberger JK, Ahmed R, Ravetch JV (2017b) IgG antibodies to dengue enhanced for FcgammaRIIIA binding determine disease severity. Science 355(6323):395–398. https://doi.org/10.1126/science.aai8128

Wang LX, Tong X, Li C, Giddens JP, Li T (2019) Glycoengineering of antibodies for modulating functions. Annu Rev Biochem 88:433–459. https://doi.org/10.1146/annurev-biochem-062917-012911

Weber SS, Oxenius A (2014) Antibody-dependent cellular phagocytosis and its impact on pathogen control. In: Ackerman M, Nimmerjahn F (eds) Antibody Fc: linking adaptive and innate immunity. Academic, Amsterdam, pp 29–47

Wei Y, Li C, Huang W, Li B, Strome S, Wang LX (2008) Glycoengineering of human IgG1-Fc through combined yeast expression and in vitro chemoenzymatic glycosylation. Biochemistry 47(39):10294–10304. https://doi.org/10.1021/bi800874y

Wilcox CR, Holder B, Jones CE (2017) Factors affecting the FcRn-mediated transplacental transfer of antibodies and implications for vaccination in pregnancy. Front Immunol 8:1294. https://doi.org/10.3389/fimmu.2017.01294

Williams PJ, Arkwright PD, Rudd P, Scragg IG, Edge CJ, Wormald MR, Rademacher TW (1995) Short communication: selective placental transport of maternal IgG to the fetus. Placenta 16(8):749–756. https://doi.org/10.1016/0143-4004(95)90018-7

Wright A, Tao MH, Kabat EA, Morrison SL (1991) Antibody variable region glycosylation: position effects on antigen binding and carbohydrate structure. EMBO J 10(10):2717–2723

Wu J, Edberg JC, Redecha PB, Bansal V, Guyre PM, Coleman K, Salmon JE, Kimberly RP (1997) A novel polymorphism of FcgammaRIIIa (CD16) alters receptor function and predisposes to autoimmune disease. J Clin Invest 100(5):1059–1070. https://doi.org/10.1172/JCI119616

Wuhrer M, Stavenhagen K, Koeleman CA, Selman MH, Harper L, Jacobs BC, Savage CO, Jefferis R, Deelder AM, Morgan M (2015) Skewed Fc glycosylation profiles of anti-proteinase 3 immunoglobulin G1 autoantibodies from granulomatosis with polyangiitis patients show low levels of bisection, galactosylation, and sialylation. J Proteome Res 14(4):1657–1665. https://doi.org/10.1021/pr500780a

Yamaguchi Y, Nishimura M, Nagano M, Yagi H, Sasakawa H, Uchida K, Shitara K, Kato K (2006) Glycoform-dependent conformational alteration of the Fc region of human immunoglobulin G1 as revealed by NMR spectroscopy. Biochim Biophys Acta 1760(4):693–700. https://doi.org/10.1016/j.bbagen.2005.10.002

Yamamoto K, Kadowaki S, Fujisaki M, Kumagai H, Tochikura T (1994) Novel specificities of Mucor hiemalis endo-beta-N-acetylglucosaminidase acting complex asparagine-linked oligosaccharides. Biosci Biotechnol Biochem 58(1):72–77

Yamane-Ohnuki N, Kinoshita S, Inoue-Urakubo M, Kusunoki M, Iida S, Nakano R, Wakitani M, Niwa R, Sakurada M, Uchida K, Shitara K, Satoh M (2004) Establishment of FUT8 knockout

Chinese hamster ovary cells: an ideal host cell line for producing completely defucosylated antibodies with enhanced antibody-dependent cellular cytotoxicity. Biotechnol Bioeng 87 (5):614–622

Yang Z, Wang S, Halim A, Schulz MA, Frodin M, Rahman SH, Vester-Christensen MB, Behrens C, Kristensen C, Vakhrushev SY, Bennett EP, Wandall HH, Clausen H (2015) Engineered CHO cells for production of diverse, homogeneous glycoproteins. Nat Biotechnol 33(8):842–844. https://doi.org/10.1038/nbt.3280

Yu X, Vasiljevic S, Mitchell DA, Crispin M, Scanlan CN (2013) Dissecting the molecular mechanism of IVIg therapy: the interaction between serum IgG and DC-SIGN is independent of antibody glycoform or Fc domain. J Mol Biol 425(8):1253–1258. https://doi.org/10.1016/j.jmb.2013.02.006

Yver A, Homery MC, Fuseau E, Guemas E, Dhainaut F, Quagliaroli D, Beliard R, Prost JF (2012) Pharmacokinetics and safety of roledumab, a novel human recombinant monoclonal anti-RhD antibody with an optimized Fc for improved engagement of FCgammaRIII, in healthy volunteers. Vox Sang 103(3):213–222. https://doi.org/10.1111/j.1423-0410.2012.01603.x

Zhang P, Woen S, Wang T, Liau B, Zhao S, Chen C, Yang Y, Song Z, Wormald MR, Yu C, Rudd PM (2016) Challenges of glycosylation analysis and control: an integrated approach to producing optimal and consistent therapeutic drugs. Drug Discov Today 21(5):740–765. https://doi.org/10.1016/j.drudis.2016.01.006

Zhang Z, Shah B, Richardson J (2019) Impact of Fc N-glycan sialylation on IgG structure. MAbs 11 (8):1381–1390. https://doi.org/10.1080/19420862.2019.1655377

Zhou Q, Stefano JE, Manning C, Kyazike J, Chen B, Gianolio DA, Park A, Busch M, Bird J, Zheng X, Simonds-Mannes H, Kim J, Gregory RC, Miller RJ, Brondyk WH, Dhal PK, Pan CQ (2014) Site-specific antibody-drug conjugation through glycoengineering. Bioconjug Chem 25 (3):510–520. https://doi.org/10.1021/bc400505q

Zou G, Ochiai H, Huang W, Yang Q, Li C, Wang LX (2011) Chemoenzymatic synthesis and Fcgamma receptor binding of homogeneous glycoforms of antibody Fc domain. Presence of a bisecting sugar moiety enhances the affinity of Fc to FcgammaIIIa receptor. J Am Chem Soc 133(46):18975–18991. https://doi.org/10.1021/ja208390n

Chapter 16
Glycosylation of Plant-Produced Immunoglobulins

Kathrin Göritzer and Richard Strasser

Contents

16.1	Introduction to N-Glycan Processing in Plants	521
16.2	Introduction to O-Glycan Biosynthesis in Plants	524
16.3	Engineering of N-Glycan Processing Pathways in Plants	525
16.4	Engineering of O-Glycosylation Biosynthesis Pathways in Plants	526
16.5	Glycosylation of Plant-Produced IgGs	527
16.6	Glycosylation of Plant-Produced IgAs	530
16.7	Glycosylation of Plant-Produced IgEs	533
16.8	Glycosylation of Plant-Produced IgMs	534
16.9	Conclusion and Outlook	535
References		535

Abstract Many economically important protein-based therapeutics like monoclonal antibodies are glycosylated. Due to the recognized importance of this type of posttranslational modification, glycoengineering of expression systems to obtain highly active and homogenous therapeutics is an emerging field. Although most of the monoclonal antibodies on the market are still produced in mammalian expression platforms, plants are emerging as an alternative cost-effective and scalable production platform that allows precise engineering of glycosylation to produce targeted human glycoforms at large homogeneity. Apart from producing more effective antibodies, pure glycoforms are required in efforts to link biological functions to specific glycan structures. Much is already known about the role of IgG1 glycosylation and this antibody class is the dominant recombinant format that has been expressed in plants. By contrast, little attention has been paid to the glycoengineering of recombinant IgG subtypes and the other four classes of human immunoglobulins (IgA, IgD, IgE, and IgM). Except for IgD, all these

---

K. Göritzer
St. George's University of London, London, UK

R. Strasser (✉)
University of Natural Resources and Life Sciences Vienna, Vienna, Austria
e-mail: richard.strasser@boku.ac.at

© The Author(s), under exclusive license to Springer Nature Switzerland AG 2021
M. Pezer (ed.), *Antibody Glycosylation*, Experientia Supplementum 112,
https://doi.org/10.1007/978-3-030-76912-3_16

antibody classes have been expressed in plants and the glycosylation has been analyzed in a site-specific manner. Here, we summarize the current data on glycosylation of plant-produced monoclonal antibodies and discuss the findings in the light of known functions for these glycans.

Keywords Antibody · Glycan · Glycoengineering · Glycoprotein · *Nicotiana benthamiana* · Plant biotechnology · Recombinant protein

Abbreviations

ADCC	Antibody-dependent cellular cytotoxicity
AG	Arabinogalactan
AGP	Arabinogalactan proteins
ALG	Asn-linked glycosylation
CHO	Chinese hamster ovary
CNX	Calnexin
CRISPR/Cas9	Clustered Regularly Interspaced Short Palindromic Repeats/CRISPR-associated-9
CRT	Calreticulin
EPO	Erythropoietin
ER	Endoplasmic reticulum
ERAD	ER-associated degradation
Fc	Fragment crystallizable
FUT	Fucosyltransferase
GALT	Galactosyltransferase
GMII	Golgi mannosidase II
GnGn	GlcNAc$_2$Man$_3$GlcNAc$_2$ N-glycan
GnGnXF	GlcNAc$_2$XylFucMan$_3$GlcNAc$_2$ N-glycan
GnT	*N*-acetylglucosaminyltransferase
HEK	Human embryonic kidney
HEXO	*N*-acetylhexosaminidase
Hyp	Hydroxyproline
Ig	Immunoglobulin
JC	Joining chain
MMXF	Man$_3$XylFucGlcNAc$_2$
MNS	Mannosidase
MUC1	Mucin 1
OST	Oligosaccharyltransferase
P4H	Prolyl-4-hydroxylase
pIgR	Polymeric immunoglobulin receptor
SC	Secretory component
SIgA	Secretory IgA
ST	Sialyltransferase

| STT3 | Staurosporine and temperature sensitive 3 |
| XylT | Xylosyltransferase |

16.1 Introduction to N-Glycan Processing in Plants

N-Glycosylation of secretory proteins is initiated by the en bloc transfer of a preassembled oligosaccharide (Glc$_3$Man$_9$GlcNAc$_2$) precursor in the lumen of the endoplasmic reticulum (ER). The assembly of the lipid-linked oligosaccharide precursor involves multiple Asn-linked glycosylation (ALG) enzymes that are all conserved in plants (Strasser 2016). The Glc$_3$Man$_9$GlcNAc$_2$ moiety is transferred by the oligosaccharyltransferase (OST) complex to asparagine residues in the sequence Asn-X-Ser/Thr (X can be any amino acid except proline) of newly synthesized polypeptides. In budding yeast, mammals, and plants, OST is a multimeric membrane-bound protein complex consisting of one catalytically active subunit (STT3) and several different non-catalytic subunits that mediate interactions with the translocation channel and ribosome or might be required for glycosylation of specific sites (Shrimal and Gilmore 2019). Mammals harbor two different OST complexes. While the STT3A complex interacts with the translocon and mediates co-translational glycosylation, the STT3B complex catalyzes posttranslational glycosylation of proteins and glycosylates sites that have been skipped by the STT3A complex. Plants have also two catalytic subunits, termed STT3A and STT3B (Koiwa et al. 2003) that likely form two distinct heteromeric OST complexes (Niu et al. 2020). However, the function of individual OST subunits appears different in plants (Farid et al. 2013; Castilho et al. 2018) and our current understanding of the role of the STT3A and STT3B complexes in N-glycosylation of plant proteins is still limited (Jeong et al. 2018).

Once the oligosaccharide has been transferred by the OST complex, the N-glycan is subjected to stepwise processing. Removal of the terminal α1,2-linked glucose by α-glucosidase I and subsequent removal of the first α1,3-linked glucose by α-glucosidase II result in a glycan structure that can be recognized by the lectin chaperones calnexin (CNX) and calreticulin (CRT) that promote protein folding (Strasser 2018). In this ER-quality control process, misfolded glycoproteins are subjected to several rounds of interaction with CNX/CRT and monitoring of their folding status. Proteins that have acquired their native conformation are released from the CNX/CRT cycle and allowed to exit the ER to downstream compartments. Terminally misfolded proteins are recognized by a poorly understood process and directed towards ER-associated degradation (ERAD) to prevent the accumulation or secretion of potentially harmful proteins. The basic biological functions of the glycan-dependent ER quality control process and clearance mechanism of aberrant glycoproteins are conserved in plants.

Correctly folded and assembled secretory glycoproteins leave the ER and transit through the Golgi apparatus where they encounter multiple glycosidases and glycosyltransferases that process oligomannosidic N-glycans to complex N-glycans (Strasser 2016). While in budding yeast and mammals, the first trimming reaction is catalyzed by an ER-resident α-mannosidase, MNS3 the corresponding plant enzyme is primarily located in the *cis*-Golgi (Schoberer et al. 2019). MNS3 catalyzes the trimming of a single α1,2-linked mannose from the middle branch of the $Man_9GlcNAc_2$ N-glycan to form $Man_8GlcNAc_2$. Subsequently, three additional mannose residues are cleaved off by Golgi α-mannosidase I. MNS1 and MNS2 are two functionally redundant Golgi α-mannosidases in *Arabidopsis thaliana* (Liebminger et al. 2009). The resulting $Man_5GlcNAc_2$ structure is used by β1,2-*N*-acetylglucosaminyltransferase I (GnTI) to initiate hybrid and complex N-glycan formation (von Schaewen et al. 1993; Strasser et al. 1999). The transfer of the GlcNAc residue to the α1,3-linked mannose by GnTI is required for further N-glycan modifications in the Golgi. In the next processing steps, Golgi α-mannosidase II (GMII) removes the α1,6- and α-1,3-linked mannose residues and β1,2-*N*-acetylglucosaminyltransferase II (GnTII) attaches a single GlcNAc to the α1,6-linked mannose to generate the complex N-glycan $GlcNAc_2Man_3GlcNAc_2$ (GnGn) (Fig. 16.1). Until this step, the processing reactions are conserved between mammals and plants. Subsequently, a β1,2-linked xylose and a core α1,3-linked fucose are attached to GnGn to generate $GlcNAc_2XylFucMan_3GlcNAc_2$ (GnGnXF) the predominant complex N-glycan found on glycoproteins in plants (Wilson et al. 2001; Léonard et al. 2004; Strasser et al. 2004). Of note, the substrate specificities and the overlapping Golgi localization of the corresponding enzymes allow an alternative order of processing from $GlcNAcMan_5GlcNAc_2$ to GnGnXF with the β1,2-xylosyltransferase (XylT) activity preceding trimming by GMII (Strasser 2016). Further modifications of complex N-glycans are catalyzed by β1,3-galactosyltransferase 1 (GALT1) and α1,4-fucosyltransferase (FUT13). The two *trans*-Golgi resident enzymes generate the Lewis-a carbohydrate epitope [Fucα1,4 (Galβ1,3)GlcNAc-R] that is ubiquitously found in plants, but occurs only on a very limited number of plant glycoproteins (Fitchette-Lainé et al. 1997; Wilson et al. 2001; Strasser et al. 2007). Truncated $Man_3XylFucGlcNAc_2$ (MMXF) N-glycans are generated from GnGnXF by removal of terminal GlcNAc residues. This reaction is catalyzed either by the vacuolar β-*N*-acetylhexosaminidase 1 (HEXO1) or by an apoplast located HEXO3 that is the major contributor to the formation of truncated N-glycans on secreted glycoproteins (Liebminger et al. 2011; Shin et al. 2017).

Apart from the formation of Lewis-a structures, no further complex N-glycan modifications have been described. Plants lack *N*-acetylglucosaminyltransferases for the formation of tri- or tetra-antennary N-glycans and the attachment of a bisecting GlcNAc. Common mammalian complex N-glycan modifications that are found on immunoglobulins such as core α1,6-fucose and terminal β1,4-galactose have not been described (Strasser et al. 2009). Moreover, plants lack the biosynthesis pathway for CMP-sialic acid, a Golgi CMP-sialic acid transporter as well as α2,3- and α2,6-sialyltransferases (ST) that catalyze the transfer of CMP-Neu5Ac to complex N-glycans in the Golgi (Zeleny et al. 2006; Castilho et al. 2010). Due to the absence of these mammalian-type complex N-glycan modifications, the N-glycan heterogeneity on plant-produced glycoproteins is clearly reduced which is an enormous

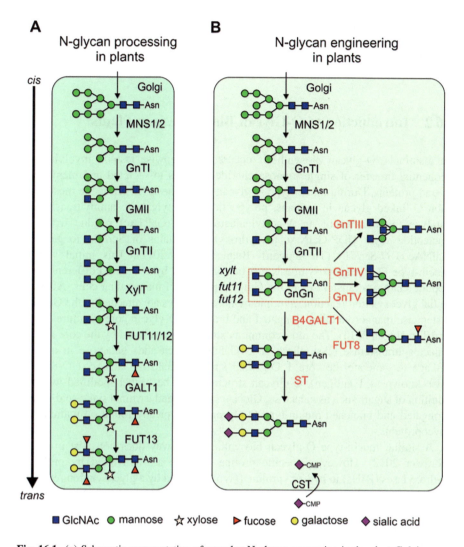

Fig. 16.1 (a) Schematic representation of complex N-glycan processing in the plant Golgi apparatus. Golgi-α-mannosidase I (MNS1/2), N-acetylglucosaminyltransferase I (GnTI), Golgi-α-mannosidase II (GMII), N-acetylglucosaminyltransferase II (GnTII), β1,2-xylosyltransferase (XylT), core α1,3-fucosyltransferase (FUT11/12), β1,3-galactosyltransferase (GALT1), and α1,4-fucosyltransferase (FUT13) are indicated. (b) N-glycan engineering approaches to produce defined homogenous complex N-glycans on plant-produced recombinant antibodies: the generation of *xylt*, *fut11 fut12* knockouts results in the formation of the GnGn structure which can serve as acceptor substrate for N-acetylglucosaminyltransferase III (GnTIII), IV (GnTIV), and V (GnTV), core α1,6-fucosyltransferase (FUT8), β1,4-galactosyltransferase (B4GALT1), and α2,6-sialyltransferases (ST). Sialylation in plants requires the co-expression of the Golgi CMP-sialic acid transporter (CST) and proteins for CMP-sialic acid biosynthesis (not shown)

advantage for approaches aiming at the generation of defined homogenous N-glycans for different applications or glycan structure function studies (Schoberer and Strasser 2018) (Fig. 16.1).

16.2 Introduction to O-Glycan Biosynthesis in Plants

In mammals, O-glycan biosynthesis occurs in a stepwise fashion involving the sequential transfer of single monosaccharide residues to secreted and membrane-bound proteins. During mucin-type O-glycan biosynthesis, which is the most common O-linked glycan in humans, polypeptide *N*-acetylgalactosaminyltransferases catalyze the transfer of an *N*-acetylgalactosamine (GalNAc) residue from the nucleotide sugar UDP-GalNAc to hydroxyl side chains of Ser/Thr to generate GalNAcα1-*O*-Ser/Thr (Tn antigen) (Bennett et al. 2012). This initial step in mucin-type O-glycan biosynthesis can be carried out by one of 20 different polypeptide GalNAc-transferases in the Golgi apparatus of mammalian cells. After the initial glycosylation reaction, multiple monosaccharides are attached to the Golgi in a stepwise manner yielding elongated and branched O-glycan core structures (Tran and Ten Hagen 2013). The most common extension is catalyzed by the core 1 β1,3-galactosyltransferase (T-synthase or C1GalT1), which adds galactose in a β1,3-linkage to generate the core 1 O-glycan structure Galβ1–3GalNAcα1-*O*-Ser/Thr (also known as T antigen). O-Glycan structures can be further modified with the addition of sugars such as galactose, GlcNAc, fucose, and terminal sialic acid. These elongated and branched mucin-type O-glycans are typically found on mammalian glycoproteins.

A similar mucin-type O-glycan biosynthesis pathway does not exist in plants (Strasser 2012). However, specific proline residues are converted by prolyl-4-hydroxylases (P4Hs) to hydroxyproline (Hyp) followed by a glycosylation reaction. Two major types of O-glycans are attached to glycoproteins with Hyp residues. Unbranched chains composed of up to five arabinose (Ara) residues are added to clusters of Hyp residues in proteins such as extensins, whereas complex arabinogalactans (AGs) are attached to clustered noncontiguous Hyp residues on arabinogalactan proteins (AGPs) (Ellis et al. 2010). The O-glycosylation of AGPs is initiated by a set of Hyp-galactosyltransferases that add a single galactose to a Hyp residue in the Golgi of plants. The AGP glycan structures are not well characterized but include incorporation of multiple galactose residues and additional modifications with arabinose, xylose, fucose, or glucuronic acid. On extensins, a single galactose can be attached to a Ser residue next to a Hyp repeat that is not further modified or elongated with other sugar residues.

16.3 Engineering of N-Glycan Processing Pathways in Plants

Initial attempts to engineer the N-glycan processing pathway in plants aimed to prevent the formation of complex or truncated N-glycans with β1,2-xylose and core α1,3-fucose. Both N-glycan modifications have been associated with an increased risk for immunogenicity and adverse allergic reactions in humans (Bardor et al. 2003; Jin et al. 2008; Paulus et al. 2011). The potential immunogenicity of the plant-specific N-glycans has been discussed extensively in the context of molecular farming, and there is still an ongoing debate whether β1,2-xylose and core α1,3-fucose modifications have an adverse effect when present on recombinant therapeutic proteins (Grabowski et al. 2014; Ward et al. 2014; Piron et al. 2015; Shaaltiel and Tekoah 2016; Rup et al. 2017). Furthermore, unwanted regulatory concerns make their elimination desirable.

Pioneering work in *A. thaliana* demonstrated that plants tolerate the complete removal of β1,2-xylose and core α1,3-fucose residues without any adverse effects on plant growth or development (Strasser et al. 2004). Based on this study, gene silencing approaches were successfully applied to almost completely remove these N-glycan modifications in the aquatic plant *Lemna minor* and in *Nicotiana benthamiana* (Cox et al. 2006; Strasser et al. 2008). *N. benthamiana* is currently used by academic groups and companies worldwide as a transient expression system for monoclonal antibodies, Fc-fusion proteins, virus-like particles, and antigens used for therapy, prophylaxis, and diagnostics (Stoger et al. 2014; Lomonossoff and D'Aoust 2016; Margolin et al. 2020; Sainsbury 2020). Multiplex CRISPR/Cas9 genome editing was recently used to generate *N. benthamiana* deficient in plant-specific core α1,3-fucosyltransferase and β1,2-xylosyltransferase activities (Jansing et al. 2019). Consistent with previous findings for *A. thaliana* (Strasser et al. 2004), no obvious phenotype was described for this multiple knockout lines highlighting that *N. benthamiana* plants tolerate the removal of plant-specific complex N-glycans very well. Anti-HIV IgG antibodies produced in the ΔXT/FT knockdown line (Strasser et al. 2008) or in the recently described knockout line (Jansing et al. 2019) displayed primarily GnGn N-glycans (Fig. 16.1). This human-type N-glycan structure is the preferred base for engineering glycan extensions and introduction of mammalian-type complex N-glycan modifications (Montero-Morales and Steinkellner 2018). GnGn N-glycans can serve as acceptor substrates for the attachment of β1,4-linked galactose (Strasser et al. 2009), branching (Castilho et al. 2011b; Nagels et al. 2011), bisecting GlcNAc (Castilho et al. 2015), core α1,6-fucosylation (Castilho et al. 2011a) or the formation of immunomodulatory helminth N-glycans carrying Galβ1–4(Fucα1–3)GlcNAc (Lewis X), or GalNAcβ1–4GlcNAc (LDN) structures (Wilbers et al. 2017). The formation of complex N-glycans with β1,4-galactose in plants paves the way for subsequent sialylation which has been achieved by transient or stable expression of the mammalian sialylation machinery (Castilho et al. 2010; Kallolimath et al. 2016). Using these approaches, complex N-glycan branches were capped with single α2,3- or α2,6-linked sialic acid residues or further extended with α2,8-linked polysialic acid (Kallolimath et al. 2016).

Other glycoengineering approaches are intended to eliminate interfering complex N-glycan modifications. The Lewis-a epitope formation interferes with β1,4-galactosylation and potentially presents an immunogenic epitope when highly abundant on recombinant therapeutic proteins. Lewis-a structures have, for example, been detected on recombinant human erythropoietin (EPO) produced in *N. benthamiana* and *Physcomitrella patens* (Weise et al. 2007; Castilho et al. 2011b). Knockout of the GALT1 orthologue in *P. patens* prevented the formation of the Lewis-a epitope on recombinant EPO (Parsons et al. 2012). Transient knockdown of HEXO3 in *N. benthamiana* enriched the amount of GnGn-containing N-glycans on recombinant glycoproteins (Shin et al. 2017) and depletion of a specific β-galactosidase from the apoplast of *N. benthamiana* prevented the removal of β1,4-galactose from recombinant glycoproteins (Kriechbaum et al. 2020). Complete knockout of the endogenous plant genes coding for these glycosyl hydrolases will further improve the *N. benthamiana* expression system resulting in the formation of recombinant glycoproteins with highly homogeneous glycans.

Besides differences in N-glycan processing, some recombinant proteins expressed in plants are underglycosylated (Van Droogenbroeck et al. 2007; Hamorsky et al. 2015; Castilho et al. 2018; Göritzer et al. 2019; Montero-Morales et al. 2019; Stelter et al. 2020). The reduced N-glycosylation efficiency is caused by yet unknown differences in the function of the plant OST complex. For some recombinant proteins including antibodies, the underglycosylation of N-glycosylation sites can be overcome by co-expression of a single subunit OST from *Leishmania major* (LmSTT3D) (Castilho et al. 2018; Montero-Morales et al. 2019; Göritzer et al. 2020).

16.4 Engineering of O-Glycosylation Biosynthesis Pathways in Plants

Despite the huge differences between mammalian and plant-type O-glycans, comparatively little attempts have been directed toward the production of human-type O-glycans in plants. The analysis of plant-produced recombinant proteins carrying mucin-type O-glycosylation sites revealed the presence of Hyp as well as several pentose residues corresponding to unbranched arabinose chains found on plant extensins (Karnoup et al. 2005; Pinkhasov et al. 2011; Castilho et al. 2012; Yang et al. 2012; Dicker et al. 2016; Göritzer et al. 2017). Hyp residues are not found on human proteins such as IgA1 or EPO. The presence of the arabinose chain may cause adverse effects and bears the risk of an unwanted immune response against plant-produced therapeutic proteins. Therefore, one aim of plant O-glycan engineering approaches is the elimination of specific P4H activities to prevent Hyp formation and subsequent plant-specific glycosylation. In *P. patens*, knockout of a single P4H completely abolished the production of Hyp on recombinant EPO (Parsons et al. 2013). Plants like *A. thaliana* or *N. benthamiana* have numerous potential P4H candidates that could be involved in the hydroxylation of proline on recombinant proteins (Velasquez et al. 2011). Consequently, the removal of the plant-specific modification will likely require the knockout of several P4H genes coding for enzymes with similar substrate specificities.

In addition to the removal of the unwanted native O-glycosylation repertoire of plants, other engineering strategies aimed to introduce mucin-type O-glycosylation. Human polypeptide GalNAc-transferase 2 has, for example, been transiently expressed in *N. benthamiana* to initiate O-GalNAc formation on different recombinant glycoproteins including peptides derived from human mucin 1 (MUC1), EPO-Fc, or IgA1 (Pinkhasov et al. 2011; Castilho et al. 2012; Yang et al. 2012; Dicker et al. 2016). On the single O-glycosylation site of human EPO-Fc, the core 1 structure could be generated by expression of the *Drosophila melanogaster* core 1 β1,3-galactosyltransferase. Co-expression of the mammalian CMP-sialic acid biosynthesis pathway, the CMP-sialic acid transporter, and the corresponding sialyltransferases led to the production of IgA1 or EPO-Fc with sialylated mucin-type O-glycans (Castilho et al. 2012; Dicker et al. 2016). For the generation of defined O-glycan structures on recombinant proteins, the absence of an endogenous mucin-type O-glycan biosynthesis pathway is of great advantage as it allows the stepwise modification of O-glycans with only the desired monosaccharides.

16.5 Glycosylation of Plant-Produced IgGs

In wild-type plant-produced IgGs, the Fc-resident GnGn N-glycan (Fig. 16.1) is commonly modified with β1,2-xylose and core α1,3-fucose residues to produce GnGnXF which is not present in mammals. In glycoengineered *N. benthamiana*, IgG with humanized complex-type GnGn N-glycans as major glycoforms were produced (Strasser et al. 2008; Jansing et al. 2019). Besides the conserved N-glycosylation sites on the Fc portion (Fig. 16.2), additional carbohydrate chains can be linked to the hypervariable regions of IgG. For instance, up to 25% of IgG molecules isolated from the serum of healthy individuals as well as several therapeutic monoclonal antibodies like cetuximab have been reported to carry N-glycans on their variable domains which exhibit site-specific differences compared to the Fc-resident N-glycan on the same molecule. While the Fc N-glycan of such antibodies is less modified, the N-glycan in the variable region is more exposed and displays extensive processing (Teh et al. 2014). Similarly, such IgG1 antibodies produced in plants were shown to carry up to 30% α1,3-fucose residues in the Fab resident N-glycan revealing a leaky knockdown in the ΔXT/FT plants (Castilho et al. 2015).

On some plant-produced IgGs low amounts of truncated structures have been detected (Strasser et al. 2008; Stelter et al. 2020). Depending on the IgG idiotype expressed, also small amounts of not fully processed oligomannosidic structures can

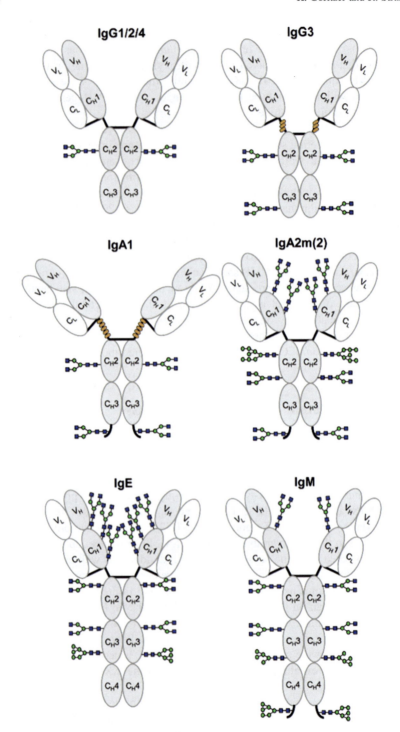

Fig. 16.2 Schematic illustration of the structure and glycosylation sites of IgGs, IgAs, IgE, and IgM. The light chain is colored in light gray and the heavy chain in dark gray. N-glycans found in

occur resulting from different secretion efficiency to the apoplast due to ER retention of potentially incompletely folded IgGs (Westerhof et al. 2014). An issue that has only been tackled recently are the differences in N-glycosylation occupancy of plant- and mammalian-produced glycoproteins. While the single N-glycosylation site present in the human IgG Fc region is almost 100% glycosylated when expressed in mammalian cells, 10–30% of plant-produced IgG is underglycosylated at this site due to yet unknown features of the plant OST complex (Castilho et al. 2018; Stelter et al. 2020). However, transient expression of the single-subunit OST from *L. major* successfully increases the N-glycan occupancy on the IgG Fc site. Using these different strategies, plants can produce IgGs with very little microheterogeneity carrying a homogenous glycosylation profile with mostly GnGn N-glycans.

Glycoengineering of IgG has focused mainly on the elimination of core fucose from the N-glycan in the Fc region of the heavy chain as major contributions to antibody activities have been assigned to that N-glycan residue (Umaña et al. 1999; Shinkawa et al. 2003; Yamane-Ohnuki et al. 2004; Junttila et al. 2010). However, co-expression of the responsible mammalian core α1,6-fucosyltransferase in glycoengineered ΔXT/FT plants facilitated the generation of IgGs with and without fucose while retaining an otherwise identical N-glycosylation profile (Forthal et al. 2010; Castilho et al. 2011a). This led to a series of studies of plant-produced IgG showing that the absence of fucose increases the affinity for FcγRIII receptor binding and improved antibody-dependent cellular cytotoxicity (ADCC) on natural killer cells (Jez et al. 2012; Loos and Steinkellner 2012; Qiu et al. 2014; Marusic et al. 2018; Stelter et al. 2020). A similar glycosylation-dependent mechanism has an impact on antibody-dependent cellular phagocytosis (ADCP) by macrophages and influences the receptor-mediated effector function of virus-neutralizing antibodies (Forthal et al. 2010; Lai et al. 2014; Hayes et al. 2017). Furthermore, it has been suggested that α1,6-linked fucose could contribute to antibody-dependent enhancement (ADE) of infection and therefore plant-produced IgGs with GnGn could be safer and more efficacious antibody-based therapeutics against dengue virus and other ADE-prone viral diseases (Dent et al. 2016; Hurtado et al. 2020). The success of afucosylated IgG antibodies produced in plants is highlighted with the case of ZMapp, an antibody cocktail for treatment of Ebola virus infections, which was used during the Ebola outbreak in 2014/2015 (Qiu et al. 2014). Core fucose-free monoclonal antibody 13F6 which is one of the ZMAPP components displayed clearly enhanced potency against Ebola virus compared to 13F6 variants with core fucose (Castilho et al. 2011a).

⬅

Fig. 16.2 (continued) the constant domains of the different antibody classes are indicated with symbols that are drawn according to the nomenclature from the Consortium for Functional Glycomics (http://www.functionalglycomics.org/). For each site, the predominant N-glycan structure (complex GnGn or oligomannosidic) found on ΔXT/FT *N. benthamiana* produced recombinant antibody is indicated. Potential O-glycosylation sites are marked in the hinge region of IgG3 and IgA1 (orange ellipse)

Capping of both branches (>80%) of the IgG Fc glycan (Strasser et al. 2009; Stelter et al. 2020) with β1,4-galactosylated structures could be achieved by targeting human β1,4-galactosyltransferase (B4GALT1) to a late Golgi compartment in *N. benthamiana* ΔXT/FT (Strasser et al. 2009; Castilho et al. 2011a; Jez et al. 2012). This is an improvement compared to CHO cell-produced IgG that frequently carries galactose residues only on one branch. Employing these glycoengineering approaches in plants it could be shown that β1,4-galactosylation can, although not improving interaction with Fcγ receptors, enhance neutralization activity of two anti-HIV antibodies (Strasser et al. 2009; Stelter et al. 2020).

The final and most complex step of human complex N-glycan processing is terminal sialylation. These negatively charged residues at the nonreducing end of N-glycans reduce protein turnover by preventing the exposure of galactose, GlcNAc, or mannose to lectin receptors like the asialoglycoprotein receptor (Ashwell and Morell 1974). In addition, there is also long-standing evidence that IgG molecules can have anti-inflammatory activity in autoimmune diseases and recent studies indicate that this activity is associated with the presence of sialic acid (Kaneko et al. 2006; Raju and Lang 2014; Wang and Ravetch 2019). The synthesis of sialylated N-glycans in plants involves the coordinated co-expression of several mammalian proteins acting in different subcellular compartments at different stages of the N-glycosylation pathway (Castilho et al. 2010). While the *in planta* sialylation works quite well for proteins like EPO, IgG sialylation is only possible in the presence of a core fucose residue (Castilho et al. 2015; Kallolimath et al. 2020).

The plant-based production of IgG1 to IgG4 subtypes has been reported recently and all of them display a quite similar N-glycan profile when expressed in glycoengineered *N. benthamiana* (Kallolimath et al. 2020). IgG3 has a second N-glycosylation site in the CH3 domain (Fig. 16.2) and an extended hinge region that is very likely modified with Hyp and plant-specific O-glycans. When produced in *N. benthamiana*, IgG3 displayed degradation products of the heavy chain. Whether the cleavage takes place in the extended hinge region remains to be shown. Modification of the hinge region with human mucin-type O-glycans might be a valuable strategy to reduce the proteolytic vulnerability of the hinge region of recombinant IgG3 produced in plants. Altogether, glycoengineering in plant-based systems provides a reliable platform to generate human IgG antibodies with a controlled glycosylation pattern.

16.6 Glycosylation of Plant-Produced IgAs

IgAs are increasingly gaining attention as possible biopharmaceuticals for treatment of infectious diseases and cancer, especially in mucosal settings due to their unique structural and functional properties. The two IgA isotypes (IgA1 and IgA2) carry two to five N-glycosylation sites on the α-(heavy) chain. In addition, the IgA1 hinge region is elongated and modified with up to six O-linked glycans (Yoo and Morrison 2005). In serum, IgA occurs mostly as its monomeric structural unit, however, it can be further assembled into dimers through incorporation of the joining chain (JC), a small polypeptide with a single biantennary complex N-glycan, which along with the IgA tailpiece N-glycan contributes to correct dimer formation (Atkin et al. 1996; Yoo et al. 1999; Göritzer et al. 2020). Newly synthesized dimeric IgA can associate with the pIgR receptor that is expressed as integral membrane protein on the basolateral side of epithelial cells lining mucosal surfaces, after which it is transported across the epithelium and released into the lumen. At the luminal side, pIgR is cleaved and a part referred to as secretory component (SC) remains attached thereby forming SIgA (Johansen et al. 2001; Mostov et al. 1984). The secretory component is a hydrophilic and highly glycosylated polypeptide carrying seven N-glycosylation sites, which protect SIgA from degradation and can interact with various host cell receptors and pathogens (Brandtzaeg 2013).

The N- and O-glycans attached to IgA in circulation are very heterogenous and their function is often not well understood. Therefore, generating recombinant monomeric and multimeric IgA variants bearing well-defined glycans is challenging but desired to study their contribution to IgA function. Furthermore, aberrant glycosylation such as galactose-deficient IgA1 O-glycans that are involved in the pathogenesis of IgA nephropathy should be avoided in therapeutic settings to reduce the risk of adverse side effects like the formation of anti-glycan antibodies (Novak et al. 2011).

Successful functional expression of a fully assembled recombinant secretory IgA (CaroRX™) to prevent dental caries has first been shown in transgenic *N. tabacum* (Ma et al. 1995). More recently, the production of IgA variants in different plant species displaying no apparent difference in assembly, integrity, and functionality compared to mammalian-produced IgA has been reported (Karnoup et al. 2005; Paul et al. 2014; Göritzer et al. 2017; Dicker et al. 2016). The glycosylation efficiency is essentially the same in plant and mammalian expression systems with an almost complete occupancy of N-glycans on all sites of IgA except the one in the C-terminal tailpiece (Göritzer et al. 2017). The N-glycan diversity found on plant-produced recombinant IgAs is, however, reduced compared to mammalian-derived IgA, with biantennary complex-type structures like $GlcNAc_1Man_3GlcNAc_2$ and GnGn as major glycoforms. In contrast, tri- and tetra-antennary structures, bisecting GlcNAc, and capping with sialic acid, which can be detected on mammalian-derived IgA (Royle et al. 2003; Mattu et al. 1998), are missing in plants. Furthermore, it has been reported that IgA transiently produced in *N. benthamiana* displays variable amounts of oligomannosidic structures indicating inefficient secretion (Paul et al. 2014; Westerhof et al. 2014; Göritzer et al. 2017). Site-specific N-glycan analysis revealed major differences between the individual N-glycosylation sites of each IgA subtype. These distinct features are conserved among the different IgA subtypes and expression systems, although their glycosylation repertoire is very different. The most pronounced difference is the complete lack of $\alpha 1,6$- and $\alpha 1,3$-linked core fucose in the CH2-resident N-glycosylation site in all IgA isotypes expressed in mammalian cells and wild-type plants, respectively (Göritzer et al. 2017; Dicker et al. 2016). N-glycans found on dimeric IgA produced in plants are similar but display a shift from paucimannosidic structures to more processed structures (GnGn) compared to their monomeric counterparts (Göritzer et al. 2020). This trimming likely occurs in a post-Golgi compartment by β-hexosaminidases and further differences between monomeric and dimeric IgAs can be explained by changes in the accessibility of the N-glycans due to dimer formation and incorporation of the JC. Like the JC of mammalian-derived IgA, the N-glycosylation site is fully occupied but displays high amounts of oligomannosidic N-glycans that are not commonly detected on the JC of dimeric IgA produced in mammalian cells (Paul et al. 2014; Göritzer et al. 2020). In humans, the SC of mucosal IgA is heavily glycosylated with branched complex N-glycans carrying high levels of sialic acid and the seven putative sites are occupied in varying degrees (Huang et al. 2015). A comprehensive site-specific and quantitative N-glycan analysis of the SC incorporated in plant-produced IgA is still lacking. Partial analysis revealed differing data on the N-glycan profile with either mostly complex-type structures or the presence of oligomannosidic structures indicating different subcellular trafficking routes of distinct IgA variants (Paul et al. 2014; Westerhof et al. 2014; Dicker et al. 2016).

The most significant difference between plant and mammalian expression hosts is the modification of the proline-rich hinge region. O-glycans found on IgA1 produced in mammalian cells are a combination of mucin-type core structures with a maximal occupation of six out of nine potential O-glycosylation sites (Göritzer et al. 2017; Royle et al. 2003). On the hinge region of plant-produced recombinant IgA1 the conversion of proline residues located next to O-glycosylation sites to Hyp and the presence of additional pentoses, presumably representing attached arabinose chains has been detected in different plant-based systems (Karnoup et al. 2005; Göritzer et al. 2017). These protein modifications increase the heterogeneity of plant-produced proteins, impede a detailed site-specific analysis of engineered O-glycan analysis, and may have adverse properties that affect the functionality or immunogenicity of therapeutic IgA.

One of the most important steps toward humanizing IgA1 antibodies produced in plants is the successful modification of the hinge-region with disialylated mucin-type core 1 O-glycans that largely resemble the human serum glycoform (Dicker et al. 2016). The recently completed sequencing of the glycoengineered *N. benthamiana* ΔXT/FT line (Schiavinato et al. 2019) allows now thorough mining of P4H candidates responsible for the conversion to Hyp to set up genome editing approaches for their elimination. Other shortcomings of plant-produced IgA such as underglycosylation of the IgA tailpiece as well as the presence of paucimannosidic structures could be counteracted applying similar glycoengineering approaches as described for plant-produced IgG. It is possible to overcome the reduced glycosylation efficiency by co-expression of the single subunit OST from *L. major* (LmSTT3D) (Göritzer et al. 2020; Castilho et al., 2018). A higher occupancy of the tailpiece N-glycan also increased the efficiency of dimeric IgA assembly in plants. Furthermore, through co-expression of human GnTII, the ratio of fully processed structures with two terminal GlcNAc residues (GnGn) can be substantially increased in *N. benthamiana* ΔXT/FT. In a recent study, the N-glycan core structure of monomeric IgA produced in the *N. benthamiana* ΔXT/FT could be further extended carrying terminally galactosylated and sialylated N-glycans with high homogeneity at each N-glycosylation sites of all IgA isotypes through co-expression of the respective mammalian glycosylation enzymes (Göritzer et al. 2019). Additionally, the generation of monomeric IgA variants carrying mainly truncated paucimannosidic glycans could be achieved through overexpression of two β-hexosaminidases targeted to the *trans*-Golgi and apoplast. The produced glycovariants were then used to investigate the influence of distinct glycoforms on conformational and thermal stability as well as binding to FcαRI, the main IgA receptor. Consistent with data from human serum IgA, no effect on FcαRI binding was observed for the plant-produced IgA glycoforms (Mattu et al. 1998; Göritzer et al. 2019). On the other hand, a recent study reported that removal of terminal sialylation from serum IgA1 increases its pro-inflammatory capacities and distinct site-specific glycan modifications could play a role for effector functions (Steffen et al. 2020). Further studies with glycoengineered plant-produced monomeric and polymeric IgAs will contribute to shed light on the function of distinct IgA glycan modifications.

16.7 Glycosylation of Plant-Produced IgEs

IgE is the least abundant serum antibody and a central player in the allergic response. IgE antibodies directed toward allergens lead to symptoms of allergy through binding to the high-affinity IgE receptor FcεRI. The IgE structure differs from IgG in that IgEs contain four constant domains compared to three constant domains in IgG classes (Arnold et al. 2007) (Fig. 16.2). IgE is the most heavily N-glycosylated antibody with seven N-glycosylation sites distributed across the constant domain of the human ε-(heavy) chain. Five sites are predominately occupied by complex N-glycans containing core fucose and different levels of sialic acids (Arnold et al. 2004; Plomp et al. 2014; Shade et al. 2015; Montero-Morales et al. 2017). Asn383 is not glycosylated on recombinant, myeloma or serum IgE and Ans394 carries exclusively oligomannosidic N-Glycans. The N-glycan at this site corresponds to Asn297 from the IgG1 heavy chain and occupies the cavity between two Fc domains (Wurzburg et al. 2000). While removal of the N-glycan at Asn394 impairs effector functions (Shade et al. 2015; Jabs et al. 2018), specific modifications of the other N-glycans did not result in altered FcεRI binding on mast cells (Montero-Morales et al. 2019).

A recombinant human IgE antibody targeting HER2 has been transiently expressed in *N. benthamiana* and compared to the same antibody produced in HEK293 cells (Montero-Morales et al. 2017). Like the human cell-derived variant, plant-produced IgE carried complex N-glycans at the same N-glycosylation sites, Asn383 was not occupied and Asn394 was modified with oligomannosidic N-glycans. When produced in the glycoengineered ΔXT/FT line, the majority of the N-glycans on these sites correspond to GnGn. N-glycosylation sites Asn140, Asn168, Asn265, and Asn394 were essentially fully glycosylated. By contrast, Asn218 and Asn371 displayed underglycosylation with 18–48% occupancy compared to 75–90% in human cell-derived IgE (Castilho et al. 2018). Co-expression of the single subunit OST LmSTT3D increased the occupancy at both sites and resulted in more than 60% glycosylation of Asn383 with a complex-type N-glycan. Moreover, transient expression of recombinant IgE in *N. benthamiana* capable of protein sialylation resulted in N-glycans with terminal sialic acid ranging from 45 to 78% (Montero-Morales et al. 2019). The sialylation content of IgEs differs in individuals with specific allergies and allergic reactions may be attenuated by reduced levels of sialylated IgEs (Shade et al. 2020). Recombinant IgE variants with different amounts of sialic acid are therefore valuable for the characterization of distinct IgE functions.

16.8 Glycosylation of Plant-Produced IgMs

IgMs are the first antibodies produced during a humoral immune response and the third most abundant antibody subclass in humans. IgMs are heavily glycosylated oligomers containing five N-glycosylation sites on each IgM μ-(heavy) chain (Asn171, Asn332, Asn395, Asn402, and Asn563) (Arnold et al. 2007) (Fig. 16.2). In human serum, IgMs circulate mainly as pentamers consisting of 10μ-chains, 10 light chains, and a single JC that are linked by disulfide bridges. Together with the single N-glycosylation site in the JC, a pentameric IgM has 51 potential N-glycosylation sites. In addition, IgM can occur as a hexamer with 60 potential N-glycosylation sites. Like IgE, the μ-chain has four domains in the constant region. Asn171 is in the CH1 domain, Asn332 in the CH2 domain, Asn395 as well as Asn402 are located in the CH3 domain. Asn563 is located in the C-terminal tailpiece region which is required for JC incorporation and multimerization (Wiersma et al. 1998). On human serum, IgM and recombinant pentameric IgM, Asn171, Asn332, and Asn395 carry predominately biantennary complex N-glycans with different degrees of sialylation (Loos et al. 2014; Pabst et al. 2015; Moh et al. 2016; Chandler et al. 2019; Hennicke et al. 2020). By contrast, Asn402 displays mainly $Man_5GlcNAc_2$ structures and Asn563 $Man_6GlcNAc_2$ to $Man_8GlcNAc_2$ oligomannosidic N-glycans. While sites Asn171, Asn332, Asn395, and Asn402 are typically fully occupied with N-glycans, there is some variation in the glycosylation efficiency of Asn563. On human serum-derived IgM or recombinantly produced IgM, full glycosylation (Loos et al. 2014; Pabst et al. 2015) as well as reduced N-glycosylation efficiency with only 17–60% occupancy were reported for Asn563 (Arnold et al. 2005; Moh et al. 2016; Chandler et al. 2019).

Previous studies have shown that IgM N-glycans are functionally important. Abolishing N-glycosylation impacts IgM secretion (Sitia et al. 1984) and immunomodulatory effects such as the internalization of IgM by T cells (Colucci et al. 2015) or complement activation (Wright et al. 1990; Gadjeva et al. 2008). On the other hand, distinct glycan modifications appear dispensable for the binding to the human Fcμ receptor (Lloyd et al. 2017).

Transient co-expression of the μ, light, and joining chains in *N. benthamiana* resulted in the expression of a functional IgM with a high proportion of hexamers

(Loos et al. 2014). The type of N-glycans found on plant-produced IgM resembled that of recombinant IgM derived from human cells. Glycosylation sites Asn171, Asn332, Asn395 carried more than 50% of complex GnGn N-glycans when expressed in the glycoengineered ΔXT/FT line. On sites Asn402 and Asn563 96% of oligomannosidic N-glycans were detected. Upon co-expression of the pathway for *in planta* protein sialylation, complex N-glycans with mono- and disialylated structures were present on sites Asn171, Asn332, and Asn395 (Loos et al. 2014). Although the N-glycosylation efficiency at site Asn563 of plant-produced IgM was not reported, it is likely that the site in the tailpiece is incompletely glycosylated. As described for plant-produced dimeric IgA, the reduced N-glycan occupancy may affect the JC incorporation and leads to the higher proportion of hexameric IgM in plants (Loos et al. 2014).

16.9 Conclusion and Outlook

In the last couple of years, a comprehensive glycosylation analysis of all antibody subclasses (except IgD) produced in plants has been performed. Overall, the analysis revealed that the type of N-glycans (complex vs. oligomannosidic) are conserved when expressed in plants. Differences are found due to the simplified N-glycan processing pathway, the sometimes reduced N-glycosylation efficiency and the completely missing mucin-type O-glycosylation pathway. *N. benthamiana* plants are amenable to glycoengineering that resulted in the production of different recombinant antibodies with quite homogenous human-like glycans. These tailored structures are essential to investigate the biological function of distinct glycan modifications and make plants an attractive platform for the generation of recombinant antibodies with diverse activities and applications (Wang and Ravetch 2019).

Compliance with Ethical Standards

Funding This study was funded by the Austrian Science Fund (FWF Project P31920-B32).

Ethical Approval This chapter is a review of previously published accounts, as such, no animal or human studies were performed.

Conflict of Interest Kathrin Göritzer declares that she has no conflict of interest. Richard Strasser declares that he has no conflict of interest.

References

Arnold JN, Radcliffe CM, Wormald MR, Royle L, Harvey DJ, Crispin M, Dwek RA, Sim RB, Rudd PM (2004) The glycosylation of human serum IgD and IgE and the accessibility of identified oligomannose structures for interaction with mannan-binding lectin. J Immunol 173 (11):6831–6840. https://doi.org/10.4049/jimmunol.173.11.6831

Arnold JN, Wormald MR, Suter DM, Radcliffe CM, Harvey DJ, Dwek RA, Rudd PM, Sim RB (2005) Human serum IgM glycosylation: identification of glycoforms that can bind to mannan-binding lectin. J Biol Chem 280(32):29080–29087. https://doi.org/10.1074/jbc.M504528200

Arnold JN, Wormald MR, Sim RB, Rudd PM, Dwek RA (2007) The impact of glycosylation on the biological function and structure of human immunoglobulins. Annu Rev Immunol 25:21–50. https://doi.org/10.1146/annurev.immunol.25.022106.141702

Ashwell G, Morell AG (1974) The role of surface carbohydrates in the hepatic recognition and transport of circulating glycoproteins. Adv Enzymol Relat Areas Mol Biol 41(0):99–128. https://doi.org/10.1002/9780470122860.ch3

Atkin JD, Pleass RJ, Owens RJ, Woof JM (1996) Mutagenesis of the human IgA1 heavy chain tailpiece that prevents dimer assembly. J Immunol 157(1):156–159

Bardor M, Faveeuw C, Fitchette A, Gilbert D, Galas L, Trottein F, Faye L, Lerouge P (2003) Immunoreactivity in mammals of two typical plant glyco-epitopes, core alpha(1,3)-fucose and core xylose. Glycobiology 13(6):427–434

Bennett EP, Mandel U, Clausen H, Gerken TA, Fritz TA, Tabak LA (2012) Control of mucin-type O-glycosylation: a classification of the polypeptide GalNAc-transferase gene family. Glycobiology 22(6):736–756. https://doi.org/10.1093/glycob/cwr182

Brandtzaeg P (2013) Secretory IgA: designed for anti-microbial defense. Front Immunol 4:222

Castilho A, Strasser R, Stadlmann J, Grass J, Jez J, Gattinger P, Kunert R, Quendler H, Pabst M, Leonard R, Altmann F, Steinkellner H (2010) In planta protein sialylation through overexpression of the respective mammalian pathway. J Biol Chem 285(21):15923–15930. https://doi.org/10.1074/jbc.M109.088401

Castilho A, Bohorova N, Grass J, Bohorov O, Zeitlin L, Whaley K, Altmann F, Steinkellner H (2011a) Rapid high yield production of different glycoforms of Ebola virus monoclonal antibody. PLoS One 6(10):e26040. https://doi.org/10.1371/journal.pone.0026040

Castilho A, Gattinger P, Grass J, Jez J, Pabst M, Altmann F, Gorfer M, Strasser R, Steinkellner H (2011b) N-glycosylation engineering of plants for the biosynthesis of glycoproteins with bisected and branched complex N-glycans. Glycobiology 21(6):813–823. https://doi.org/10.1093/glycob/cwr009

Castilho A, Neumann L, Daskalova S, Mason HS, Steinkellner H, Altmann F, Strasser R (2012) Engineering of sialylated mucin-type O-glycosylation in plants. J Biol Chem 287(43):36518–36526. https://doi.org/10.1074/jbc.M112.402685

Castilho A, Gruber C, Thader A, Oostenbrink C, Pechlaner M, Steinkellner H, Altmann F (2015) Processing of complex N-glycans in IgG Fc-region is affected by core fucosylation. MAbs 7(5):863–870. https://doi.org/10.1080/19420862.2015.1053683

Castilho A, Beihammer G, Pfeiffer C, Göritzer K, Montero-Morales L, Vavra U, Maresch D, Grünwald-Gruber C, Altmann F, Steinkellner H, Strasser R (2018) An oligosaccharyltransferase from Leishmania major increases the N-glycan occupancy on recombinant glycoproteins produced in *Nicotiana benthamiana*. Plant Biotechnol J 16(10):1700–1709. https://doi.org/10.1111/pbi.12906

Chandler KB, Mehta N, Leon DR, Suscovich TJ, Alter G, Costello CE (2019) Multi-isotype glycoproteomic characterization of serum antibody heavy chains reveals isotype- and subclass-specific. Mol Cell Proteomics 18(4):686–703. https://doi.org/10.1074/mcp.RA118.001185

Colucci M, Stöckmann H, Butera A, Masotti A, Baldassarre A, Giorda E, Petrini S, Rudd PM, Sitia R, Emma F, Vivarelli M (2015) Sialylation of N-linked glycans influences the immunomodulatory effects of IgM on T cells. J Immunol 194(1):151–157. https://doi.org/10.4049/jimmunol.1402025

Cox K, Sterling J, Regan J, Gasdaska J, Frantz K, Peele C, Black A, Passmore D, Moldovan-Loomis C, Srinivasan M, Cuison S, Cardarelli P, Dickey L (2006) Glycan optimization of a human monoclonal antibody in the aquatic plant *Lemna minor*. Nat Biotechnol 24(12):1591–1597

Dent M, Hurtado J, Paul AM, Sun H, Lai H, Yang M, Esqueda A, Bai F, Steinkellner H, Chen Q (2016) Plant-produced anti-dengue virus monoclonal antibodies exhibit reduced antibody-dependent enhancement of infection activity. J Gen Virol 97(12):3280–3290. https://doi.org/10.1099/jgv.0.000635

Dicker M, Tschofen M, Maresch D, König J, Juarez P, Orzaez D, Altmann F, Steinkellner H, Strasser R (2016) Transient glyco-engineering to produce recombinant IgA1 with defined N- and O-Glycans in plants. Front Plant Sci 7:18. https://doi.org/10.3389/fpls.2016.00018

Ellis M, Egelund J, Schultz CJ, Bacic A (2010) Arabinogalactan-proteins: key regulators at the cell surface? Plant Physiol 153(2):403–419. https://doi.org/10.1104/pp.110.156000

Farid A, Malinovsky FG, Veit C, Schoberer J, Zipfel C, Strasser R (2013) Specialized roles of the conserved subunit OST3/6 of the oligosaccharyltransferase complex in innate immunity and tolerance to abiotic stresses. Plant Physiol 162(1):24–38. https://doi.org/10.1104/pp.113.215509

Fitchette-Lainé A, Gomord V, Cabanes M, Michalski J, Saint Macary M, Foucher B, Cavelier B, Hawes C, Lerouge P, Faye L (1997) N-glycans harboring the Lewis a epitope are expressed at the surface of plant cells. Plant J 12(6):1411–1417

Forthal DN, Gach JS, Landucci G, Jez J, Strasser R, Kunert R, Steinkellner H (2010) Fc-glycosylation influences Fcγ receptor binding and cell-mediated anti-HIV activity of monoclonal antibody 2G12. J Immunol 185(11):6876–6882. https://doi.org/10.4049/jimmunol.1002600

Gadjeva MG, Rouseva MM, Zlatarova AS, Reid KB, Kishore U, Kojouharova MS (2008) Interaction of human C1q with IgG and IgM: revisited. Biochemistry 47(49):13093–13102. https://doi.org/10.1021/bi801131h

Göritzer K, Maresch D, Altmann F, Obinger C, Strasser R (2017) Exploring site-specific N-glycosylation of HEK293 and plant-produced human IgA isotypes. J Proteome Res 16(7):2560–2570. https://doi.org/10.1021/acs.jproteome.7b00121

Göritzer K, Turupcu A, Maresch D, Novak J, Altmann F, Oostenbrink C, Obinger C, Strasser R (2019) Distinct Fcα receptor. J Biol Chem 294(38):13995–14008. https://doi.org/10.1074/jbc.RA119.009954

Göritzer K, Goet I, Duric S, Maresch D, Altmann F, Obinger C, Strasser R (2020) Efficient. Front Chem 8:346. https://doi.org/10.3389/fchem.2020.00346

Grabowski GA, Golembo M, Shaaltiel Y (2014) Taliglucerase alfa: an enzyme replacement therapy using plant cell expression technology. Mol Genet Metab 112(1):1–8. https://doi.org/10.1016/j.ymgme.2014.02.011

Hamorsky KT, Kouokam JC, Jurkiewicz JM, Nelson B, Moore LJ, Husk AS, Kajiura H, Fujiyama K, Matoba N (2015) N-glycosylation of cholera toxin B subunit in *Nicotiana benthamiana*: impacts on host stress response, production yield and vaccine potential. Sci Rep 5:8003. https://doi.org/10.1038/srep08003

Hayes JM, Frostell A, Karlsson R, Müller S, Martín SM, Pauers M, Reuss F, Cosgrave EF, Anneren C, Davey GP, Rudd PM (2017) Identification of Fc gamma receptor glycoforms that produce differential binding kinetics for rituximab. Mol Cell Proteomics 16(10):1770–1788. https://doi.org/10.1074/mcp.M117.066944

Hennicke J, Schwaigerlehner L, Grünwald-Gruber C, Bally I, Ling WL, Thielens N, Reiser JB, Kunert R (2020) Transient pentameric IgM fulfill biological function-effect of expression host and transfection on IgM properties. PLoS One 15(3):e0229992. https://doi.org/10.1371/journal.pone.0229992

Huang J, Guerrero A, Parker E, Strum JS, Smilowitz JT, German JB et al (2015) Site-specific glycosylation of secretory immunoglobulin A from human colostrum. J Proteome Res 14(3):1335–1349

Hurtado J, Acharya D, Lai H, Sun H, Kallolimath S, Steinkellner H, Bai F, Chen Q (2020) In vitro and in vivo efficacy of anti-chikungunya virus monoclonal antibodies produced in wild-type and glycoengineered *Nicotiana benthamiana* plants. Plant Biotechnol J 18(1):266–273. https://doi.org/10.1111/pbi.13194

Jabs F, Plum M, Laursen NS, Jensen RK, Mølgaard B, Miehe M, Mandolesi M, Rauber MM, Pfützner W, Jakob T, Möbs C, Andersen GR, Spillner E (2018) Trapping IgE in a closed conformation by mimicking CD23 binding prevents and disrupts FcεRI interaction. Nat Commun 9(1):7. https://doi.org/10.1038/s41467-017-02312-7

Jansing J, Sack M, Augustine SM, Fischer R, Bortesi L (2019) CRISPR/Cas9-mediated knockout of six glycosyltransferase genes in *Nicotiana benthamiana* for the production of recombinant proteins lacking β-1,2-xylose and core α-1,3-fucose. Plant Biotechnol J 17(2):350–361. https://doi.org/10.1111/pbi.12981

Jeong IS, Lee S, Bonkhofer F, Tolley J, Fukudome A, Nagashima Y, May K, Rips S, Lee SY, Gallois P, Russell WK, Jung HS, von Schaewen A, Koiwa H (2018) Purification and characterization of *Arabidopsis thaliana* oligosaccharyltransferase complexes from the native host: a protein super-expression system for structural studies. Plant J 94(1):131–145. https://doi.org/10.1111/tpj.13847

Jez J, Antes B, Castilho A, Kainer M, Wiederkum S, Grass J, Rüker F, Woisetschläger M, Steinkellner H (2012) Significant impact of single N-glycan residues on the biological activity of Fc-based antibody-like fragments. J Biol Chem 287(29):24313–24319. https://doi.org/10.1074/jbc.M112.360701

Jin C, Altmann F, Strasser R, Mach L, Schähs M, Kunert R, Rademacher T, Glössl J, Steinkellner H (2008) A plant-derived human monoclonal antibody induces an anti-carbohydrate immune response in rabbits. Glycobiology 18(3):235–241

Johansen FE, Braathen R, Brandtzaeg P (2001) The J chain is essential for polymeric Ig receptor-mediated epithelial transport of IgA. J Immunol 167(9):5185–5192

Junttila TT, Parsons K, Olsson C, Lu Y, Xin Y, Theriault J, Crocker L, Pabonan O, Baginski T, Meng G, Totpal K, Kelley RF, Sliwkowski MX (2010) Superior in vivo efficacy of afucosylated trastuzumab in the treatment of HER2-amplified breast cancer. Cancer Res 70(11):4481–4489. https://doi.org/10.1158/0008-5472.CAN-09-3704

Kallolimath S, Castilho A, Strasser R, Grünwald-Gruber C, Altmann F, Strubl S, Galuska CE, Zlatina K, Galuska SP, Werner S, Thiesler H, Werneburg S, Hildebrandt H, Gerardy-Schahn R, Steinkellner H (2016) Engineering of complex protein sialylation in plants. Proc Natl Acad Sci USA 113(34):9498–9503. https://doi.org/10.1073/pnas.1604371113

Kallolimath S, Hackl T, Gahn R, Grünwald-Gruber C, Zich W, Kogelmann B, Lux A, Nimmerjahn F, Steinkellner H (2020) Expression profiling and glycan engineering of IgG subclass 1-4 in. Front Bioeng Biotechnol 8:825. https://doi.org/10.3389/fbioe.2020.00825

Kaneko Y, Nimmerjahn F, Ravetch JV (2006) Anti-inflammatory activity of immunoglobulin G resulting from Fc sialylation. Science 313(5787):670–673. https://doi.org/10.1126/science.1129594

Karnoup AS, Turkelson V, Anderson WH (2005) O-linked glycosylation in maize-expressed human IgA1. Glycobiology 15(10):965–981. https://doi.org/10.1093/glycob/cwi077

Koiwa H, Li F, McCully M, Mendoza I, Koizumi N, Manabe Y, Nakagawa Y, Zhu J, Rus A, Pardo J, Bressan R, Hasegawa P (2003) The STT3a subunit isoform of the Arabidopsis oligosaccharyltransferase controls adaptive responses to salt/osmotic stress. Plant Cell 15(10):2273–2284

Kriechbaum R, Ziaee E, Grünwald-Gruber C, Buscaill P, van der Hoorn RAL, Castilho A (2020) BGAL1 depletion boosts the level of β-galactosylation of N- and O-glycans in *N. benthamiana*. Plant Biotechnol J 18(7):1537–1549. https://doi.org/10.1111/pbi.13316

Lai B, Hasenhindl C, Obinger C, Oostenbrink C (2014) Molecular dynamics simulation of the crystallizable fragment of IgG1-insights for the design of Fcabs. Int J Mol Sci 15(1):438–455. https://doi.org/10.3390/ijms15010438

Léonard R, Kolarich D, Paschinger K, Altmann F, Wilson I (2004) A genetic and structural analysis of the N-glycosylation capabilities of rice and other monocotyledons. Plant Mol Biol 55(5):631–644

Liebminger E, Hüttner S, Vavra U, Fischl R, Schoberer J, Grass J, Blaukopf C, Seifert G, Altmann F, Mach L, Strasser R (2009) Class I alpha-mannosidases are required for N-glycan processing and root development in *Arabidopsis thaliana*. Plant Cell 21(12):3850–3867. https://doi.org/10.1105/tpc.109.072363

Liebminger E, Veit C, Pabst M, Batoux M, Zipfel C, Altmann F, Mach L, Strasser R (2011) {beta}-N-acetylhexosaminidases HEXO1 and HEXO3 are responsible for the formation of paucimannosidic N-glycans in *Arabidopsis thaliana*. J Biol Chem 286(12):10793–10802. https://doi.org/10.1074/jbc.M110.178020

Lloyd KA, Wang J, Urban BC, Czajkowsky DM, Pleass RJ (2017) Glycan-independent binding and internalization of human IgM to FCMR, its cognate cellular receptor. Sci Rep 7:42989. https://doi.org/10.1038/srep42989

Lomonossoff GP, D'Aoust MA (2016) Plant-produced biopharmaceuticals: a case of technical developments driving clinical deployment. Science 353(6305):1237–1240. https://doi.org/10.1126/science.aaf6638

Loos A, Steinkellner H (2012) IgG-Fc glycoengineering in non-mammalian expression hosts. Arch Biochem Biophys. https://doi.org/10.1016/j.abb.2012.05.011

Loos A, Gruber C, Altmann F, Mehofer U, Hensel F, Grandits M, Oostenbrink C, Stadlmayr G, Furtmüller PG, Steinkellner H (2014) Expression and glycoengineering of functionally active heteromultimeric IgM in plants. Proc Natl Acad Sci USA 111(17):6263–6268. https://doi.org/10.1073/pnas.1320544111

Ma JK, Hiatt A, Hein M, Vine ND, Wang F, Stabila P et al (1995) Generation and assembly of secretory antibodies in plants. Science 268(5211):716–719

Margolin EA, Strasser R, Chapman R, Williamson AL, Rybicki EP, Meyers AE (2020) Engineering the plant secretory pathway for the production of next-generation pharmaceuticals. Trends Biotechnol 38(9):1034–1044. https://doi.org/10.1016/j.tibtech.2020.03.004

Marusic C, Pioli C, Stelter S, Novelli F, Lonoce C, Morrocchi E, Benvenuto E, Salzano AM, Scaloni A, Donini M (2018) N-glycan engineering of a plant-produced anti-CD20-hIL-2 immunocytokine significantly enhances its effector functions. Biotechnol Bioeng 115(3):565–576. https://doi.org/10.1002/bit.26503

Mattu TS, Pleass RJ, Willis AC, Kilian M, Wormald MR, Lellouch AC, Rudd PM, Woof JM, Dwek RA (1998) The glycosylation and structure of human serum IgA1, Fab, and Fc regions and the role of N-glycosylation on Fcα receptor interactions. J Biol Chem 273(4):2260–2272. https://doi.org/10.1074/jbc.273.4.2260

Moh ES, Lin CH, Thaysen-Andersen M, Packer NH (2016) Site-specific N-glycosylation of recombinant pentameric and hexameric human IgM. J Am Soc Mass Spectrom 27(7):1143–1155. https://doi.org/10.1007/s13361-016-1378-0

Montero-Morales L, Steinkellner H (2018) Advanced plant-based glycan engineering. Front Bioeng Biotechnol 6:81. https://doi.org/10.3389/fbioe.2018.00081

Montero-Morales L, Maresch D, Castilho A, Turupcu A, Ilieva KM, Crescioli S, Karagiannis SN, Lupinek C, Oostenbrink C, Altmann F, Steinkellner H (2017) Recombinant plant-derived human IgE glycoproteomics. J Proteome 161:81–87. https://doi.org/10.1016/j.jprot.2017.04.002

Montero-Morales L, Maresch D, Crescioli S, Castilho A, Ilieva KM, Mele S, Karagiannis SN, Altmann F, Steinkellner H (2019) Glycan engineering and functional activities of IgE antibodies. Front Bioeng Biotechnol 7:242. https://doi.org/10.3389/fbioe.2019.00242

Mostov KE, Friedlander M, Blobel G (1984) The receptor for transepithelial transport of IgA and IgM contains multiple immunoglobulin-like domains. Nature 308(5954):37–43

Nagels B, Van Damme EJ, Pabst M, Callewaert N, Weterings K (2011) Production of complex multiantennary N-Glycans in *Nicotiana benthamiana* plants. Plant Physiol 155(3):1103–12. https://doi.org/10.1104/pp.110.168773

Niu G, Shao Z, Liu C, Chen T, Jiao Q, Hong Z (2020) Comparative and evolutionary analyses of the divergence of plant oligosaccharyltransferase STT3 isoforms. FEBS Open Bio 10(3):468–483. https://doi.org/10.1002/2211-5463.12804

Novak J, Moldoveanu Z, Julian BA, Raska M, Wyatt RJ, Suzuki Y et al (2011) Aberrant glycosylation of IgA1 and anti-glycan antibodies in IgA nephropathy: role of mucosal immune system. Adv Otorhinolaryngol 72:60–63

Pabst M, Küster SK, Wahl F, Krismer J, Dittrich PS, Zenobi R (2015) A microarray-matrix-assisted laser desorption/ionization-mass spectrometry approach for site-specific protein N-glycosylation analysis, as demonstrated for human serum immunoglobulin M (IgM). Mol Cell Proteomics 14(6):1645–1656. https://doi.org/10.1074/mcp.O114.046748

Parsons J, Altmann F, Arrenberg CK, Koprivova A, Beike AK, Stemmer C, Gorr G, Reski R, Decker EL (2012) Moss-based production of asialo-erythropoietin devoid of Lewis A and other plant-typical carbohydrate determinants. Plant Biotechnol J 10(7):851–861. https://doi.org/10.1111/j.1467-7652.2012.00704.x

Parsons J, Altmann F, Graf M, Stadlmann J, Reski R, Decker EL (2013) A gene responsible for prolyl-hydroxylation of moss-produced recombinant human erythropoietin. Sci Rep 3:3019. https://doi.org/10.1038/srep03019

Paul M, Reljic R, Klein K, Drake PM, van Dolleweerd C, Pabst M et al (2014) Characterization of a plant-produced recombinant human secretory IgA with broad neutralizing activity against HIV. mAbs 6(6):1585–1597

Paulus KE, Mahler V, Pabst M, Kogel KH, Altmann F, Sonnewald U (2011) Silencing β1,2-xylosyltransferase in transgenic tomato fruits reveals xylose as constitutive component of Ige-binding epitopes. Front Plant Sci 2:42. https://doi.org/10.3389/fpls.2011.00042

Pinkhasov J, Alvarez ML, Rigano MM, Piensook K, Larios D, Pabst M, Grass J, Mukherjee P, Gendler SJ, Walmsley AM, Mason HS (2011) Recombinant plant-expressed tumour-associated MUC1 peptide is immunogenic and capable of breaking tolerance in MUC1.Tg mice. Plant Biotechnol J 9(9):991–1001. https://doi.org/10.1111/j.1467-7652.2011.00614.x

Piron R, Santens F, De Paepe A, Depicker A, Callewaert N (2015) Using GlycoDelete to produce proteins lacking plant-specific N-glycan modification in seeds. Nat Biotechnol 33 (11):1135–1137. https://doi.org/10.1038/nbt.3359

Plomp R, Hensbergen PJ, Rombouts Y, Zauner G, Dragan I, Koeleman CA, Deelder AM, Wuhrer M (2014) Site-specific N-glycosylation analysis of human immunoglobulin e. J Proteome Res 13(2):536–546. https://doi.org/10.1021/pr400714w

Qiu X, Wong G, Audet J, Bello A, Fernando L, Alimonti JB, Fausther-Bovendo H, Wei H, Aviles J, Hiatt E, Johnson A, Morton J, Swope K, Bohorov O, Bohorova N, Goodman C, Kim D, Pauly MH, Velasco J, Pettitt J, Olinger GG, Whaley K, Xu B, Strong JE, Zeitlin L, Kobinger GP (2014) Reversion of advanced Ebola virus disease in nonhuman primates with ZMapp. Nature 514(7520):47–53. https://doi.org/10.1038/nature13777

Raju TS, Lang SE (2014) Diversity in structure and functions of antibody sialylation in the Fc. Curr Opin Biotechnol 30:147–152. https://doi.org/10.1016/j.copbio.2014.06.014

Royle L, Roos A, Harvey DJ, Wormald MR, van Gijlswijk-Janssen D, el Redwan RM et al (2003) Secretory IgA N- and O-glycans provide a link between the innate and adaptive immune systems. J Biol Chem 278(22):20140–20153

Rup B, Alon S, Amit-Cohen BC, Brill Almon E, Chertkoff R, Tekoah Y, Rudd PM (2017) Immunogenicity of glycans on biotherapeutic drugs produced in plant expression systems-the taliglucerase alfa story. PLoS One 12(10):e0186211. https://doi.org/10.1371/journal.pone.0186211

Sainsbury F (2020) Innovation in plant-based transient protein expression for infectious disease prevention and preparedness. Curr Opin Biotechnol 61:110–115. https://doi.org/10.1016/j.copbio.2019.11.002

Schiavinato M, Strasser R, Mach L, Dohm JC, Himmelbauer H (2019) Genome and transcriptome characterization of the glycoengineered *Nicotiana benthamiana* line ΔXT/FT. BMC Genomics 20(1):594

Schoberer J, Strasser R (2018) Plant glyco-biotechnology. Semin Cell Dev Biol 80:133–141. https://doi.org/10.1016/j.semcdb.2017.07.005

Schoberer J, König J, Veit C, Vavra U, Liebminger E, Botchway SW, Altmann F, Kriechbaumer V, Hawes C, Strasser R (2019) A signal motif retains Arabidopsis ER-α-mannosidase I in the cis-Golgi and prevents enhanced glycoprotein ERAD. Nat Commun 10(1):3701. https://doi.org/10.1038/s41467-019-11686-9

Shaaltiel Y, Tekoah Y (2016) Plant specific N-glycans do not have proven adverse effects in humans. Nat Biotechnol 34(7):706–708. https://doi.org/10.1038/nbt.3556

Shade KT, Platzer B, Washburn N, Mani V, Bartsch YC, Conroy M, Pagan JD, Bosques C, Mempel TR, Fiebiger E, Anthony RM (2015) A single glycan on IgE is indispensable for initiation of anaphylaxis. J Exp Med 212(4):457–467. https://doi.org/10.1084/jem.20142182

Shade KC, Conroy ME, Washburn N, Kitaoka M, Huynh DJ, Laprise E, Patil SU, Shreffler WG, Anthony RM (2020) Sialylation of immunoglobulin E is a determinant of allergic pathogenicity. Nature 582(7811):265–270. https://doi.org/10.1038/s41586-020-2311-z

Shin YJ, Castilho A, Dicker M, Sádio F, Vavra U, Grünwald-Gruber C, Kwon TH, Altmann F, Steinkellner H, Strasser R (2017) Reduced paucimannosidic N-glycan formation by suppression of a specific β-hexosaminidase from *Nicotiana benthamiana*. Plant Biotechnol J 15(2):197–206. https://doi.org/10.1111/pbi.12602

Shinkawa T, Nakamura K, Yamane N, Shoji-Hosaka E, Kanda Y, Sakurada M, Uchida K, Anazawa H, Satoh M, Yamasaki M, Hanai N, Shitara K (2003) The absence of fucose but not the presence of galactose or bisecting N-acetylglucosamine of human IgG1 complex-type oligosaccharides shows the critical role of enhancing antibody-dependent cellular cytotoxicity. J Biol Chem 278(5):3466–3473. https://doi.org/10.1074/jbc.M210665200

Shrimal S, Gilmore R (2019) Oligosaccharyltransferase structures provide novel insight into the mechanism of asparagine-linked glycosylation in prokaryotic and eukaryotic cells. Glycobiology 29(4):288–297. https://doi.org/10.1093/glycob/cwy093

Sitia R, Rubartelli A, Hämmerling U (1984) The role of glycosylation in secretion and membrane expression of immunoglobulins M and A. Mol Immunol 21(8):709–719. https://doi.org/10.1016/0161-5890(84)90023-3

Steffen U, Koeleman CA, Sokolova MV, Bang H, Kleyer A, Rech J, Unterweger H, Schicht M, Garreis F, Hahn J, Andes FT, Hartmann S, Hahn M, Mahajan A, Paulsen F, Hoffmann M, Lochnit G, Muñoz LE, Wuhrer M, Falck D, Herrmann M, Schett G (2020) IgA subclasses have different effector functions associated with distinct glycosylation profiles. Nat Commun 11(1):120. https://doi.org/10.1038/s41467-019-13992-8

Stelter S, Paul MJ, Teh AY, Grandits M, Altmann F, Vanier J, Bardor M, Castilho A, Allen RL, Ma JK (2020) Engineering the interactions between a plant-produced HIV antibody and human Fc receptors. Plant Biotechnol J 18(2):402–414. https://doi.org/10.1111/pbi.13207

Stoger E, Fischer R, Moloney M, Ma JK (2014) Plant molecular pharming for the treatment of chronic and infectious diseases. Annu Rev Plant Biol 65:743–768. https://doi.org/10.1146/annurev-arplant-050213-035850

Strasser R (2012) Challenges in O-glycan engineering of plants. Front Plant Sci 3:218. https://doi.org/10.3389/fpls.2012.00218

Strasser R (2016) Plant protein glycosylation. Glycobiology 26(9):926–939. https://doi.org/10.1093/glycob/cww023

Strasser R (2018) Protein quality control in the endoplasmic reticulum of plants. Annu Rev Plant Biol 69:147–172. https://doi.org/10.1146/annurev-arplant-042817-040331

Strasser R, Mucha J, Schwihla H, Altmann F, Glössl J, Steinkellner H (1999) Molecular cloning and characterization of cDNA coding for beta1,2N-acetylglucosaminyltransferase I (GlcNAc-TI) from *Nicotiana tabacum*. Glycobiology 9(8):779–785

Strasser R, Altmann F, Mach L, Glössl J, Steinkellner H (2004) Generation of *Arabidopsis thaliana* plants with complex N-glycans lacking beta1,2-linked xylose and core alpha1,3-linked fucose. FEBS Lett 561(1–3):132–136

Strasser R, Bondili J, Vavra U, Schoberer J, Svoboda B, Glössl J, Léonard R, Stadlmann J, Altmann F, Steinkellner H, Mach L (2007) A unique beta1,3-galactosyltransferase is indispensable for the biosynthesis of N-glycans containing Lewis a structures in *Arabidopsis thaliana*. Plant Cell 19(7):2278–2292

Strasser R, Stadlmann J, Schähs M, Stiegler G, Quendler H, Mach L, Glössl J, Weterings K, Pabst M, Steinkellner H (2008) Generation of glyco-engineered *Nicotiana benthamiana* for the production of monoclonal antibodies with a homogeneous human-like N-glycan structure. Plant Biotechnol J 6(4):392–402

Strasser R, Castilho A, Stadlmann J, Kunert R, Quendler H, Gattinger P, Jez J, Rademacher T, Altmann F, Mach L, Steinkellner H (2009) Improved virus neutralization by plant-produced anti-HIV antibodies with a homogeneous {beta}1,4-galactosylated N-glycan profile. J Biol Chem 284(31):20479–20485

Teh AY, Maresch D, Klein K, Ma JK (2014) Characterization of VRC01, a potent and broadly neutralizing anti-HIV mAb, produced in transiently and stably transformed tobacco. Plant Biotechnol J 12(3):300–311. https://doi.org/10.1111/pbi.12137

Tran DT, Ten Hagen KG (2013) Mucin-type O-glycosylation during development. J Biol Chem 288(10):6921–6929. https://doi.org/10.1074/jbc.R112.418558

Umaña P, Jean-Mairet J, Moudry R, Amstutz H, Bailey JE (1999) Engineered glycoforms of an antineuroblastoma IgG1 with optimized antibody-dependent cellular cytotoxic activity. Nat Biotechnol 17(2):176–180. https://doi.org/10.1038/6179

Van Droogenbroeck B, Cao J, Stadlmann J, Altmann F, Colanesi S, Hillmer S, Robinson DG, Van Lerberge E, Terryn N, Van Montagu M, Liang M, Depicker A, De Jaeger G (2007) Aberrant localization and underglycosylation of highly accumulating single-chain Fv-Fc antibodies in transgenic Arabidopsis seeds. Proc Natl Acad Sci USA 104(4):1430–1435. https://doi.org/10.1073/pnas.0609997104

Velasquez SM, Ricardi MM, Dorosz JG, Fernandez PV, Nadra AD, Pol-Fachin L, Egelund J, Gille S, Harholt J, Ciancia M, Verli H, Pauly M, Bacic A, Olsen CE, Ulvskov P, Petersen BL, Somerville C, Iusem ND, Estevez JM (2011) O-glycosylated cell wall proteins are essential in root hair growth. Science 332(6036):1401–1403. https://doi.org/10.1126/science.1206657

von Schaewen A, Sturm A, O'Neill J, Chrispeels M (1993) Isolation of a mutant Arabidopsis plant that lacks N-acetyl glucosaminyl transferase I and is unable to synthesize Golgi-modified complex N-linked glycans. Plant Physiol 102(4):1109–1118

Wang TT, Ravetch JV (2019) Functional diversification of IgGs through Fc glycosylation. J Clin Invest 129(9):3492–3498. https://doi.org/10.1172/JCI130029

Ward BJ, Landry N, Trépanier S, Mercier G, Dargis M, Couture M, D'Aoust MA, Vézina LP (2014) Human antibody response to N-glycans present on plant-made influenza virus-like particle (VLP) vaccines. Vaccine 32(46):6098–6106. https://doi.org/10.1016/j.vaccine.2014.08.079

Weise A, Altmann F, Rodriguez-Franco M, Sjoberg ER, Bäumer W, Launhardt H, Kietzmann M, Gorr G (2007) High-level expression of secreted complex glycosylated recombinant human erythropoietin in the Physcomitrella Delta-fuc-t Delta-xyl-t mutant. Plant Biotechnol J 5(3):389–401. https://doi.org/10.1111/j.1467-7652.2007.00248.x

Westerhof LB, Wilbers RH, van Raaij DR, Nguyen DL, Goverse A, Henquet MG, Hokke CH, Bosch D, Bakker J, Schots A (2014) Monomeric IgA can be produced in planta as efficient as IgG, yet receives different N-glycans. Plant Biotechnol J 12(9):1333–1342. https://doi.org/10.1111/pbi.12251

Wiersma EJ, Collins C, Fazel S, Shulman MJ (1998) Structural and functional analysis of J chain-deficient IgM. J Immunol 160(12):5979–5989

Wilbers RH, Westerhof LB, van Noort K, Obieglo K, Driessen NN, Everts B, Gringhuis SI, Schramm G, Goverse A, Smant G, Bakker J, Smits HH, Yazdanbakhsh M, Schots A, Hokke CH (2017) Production and glyco-engineering of immunomodulatory helminth glycoproteins in plants. Sci Rep 7:45910. https://doi.org/10.1038/srep45910

Wilson I, Zeleny R, Kolarich D, Staudacher E, Stroop C, Kamerling J, Altmann F (2001) Analysis of Asn-linked glycans from vegetable foodstuffs: widespread occurrence of Lewis a, core alpha1,3-linked fucose and xylose substitutions. Glycobiology 11(4):261–274

Wright JF, Shulman MJ, Isenman DE, Painter RH (1990) C1 binding by mouse IgM. The effect of abnormal glycosylation at position 402 resulting from a serine to asparagine exchange at residue 406 of the mu-chain. J Biol Chem 265(18):10506–10513

Wurzburg BA, Garman SC, Jardetzky TS (2000) Structure of the human IgE-Fc C epsilon 3-C epsilon 4 reveals conformational flexibility in the antibody effector domains. Immunity 13 (3):375–385. https://doi.org/10.1016/s1074-7613(00)00037-6

Yamane-Ohnuki N, Kinoshita S, Inoue-Urakubo M, Kusunoki M, Iida S, Nakano R, Wakitani M, Niwa R, Sakurada M, Uchida K, Shitara K, Satoh M (2004) Establishment of FUT8 knockout Chinese hamster ovary cells: an ideal host cell line for producing completely defucosylated antibodies with enhanced antibody-dependent cellular cytotoxicity. Biotechnol Bioeng 87 (5):614–622. https://doi.org/10.1002/bit.20151

Yang Z, Drew DP, Jørgensen B, Mandel U, Bach SS, Ulvskov P, Levery SB, Bennett EP, Clausen H, Petersen BL (2012) Engineering mammalian mucin-type O-glycosylation in plants. J Biol Chem 287(15):11911–11923. https://doi.org/10.1074/jbc.M111.312918

Yoo EM, Morrison SL (2005) IgA: an immune glycoprotein. Clin Immunol 116(1):3–10

Yoo EM, Coloma MJ, Trinh KR, Nguyen TQ, Vuong LU, Morrison SL et al (1999) Structural requirements for polymeric immunoglobulin assembly and association with J chain. J Biol Chem 274(47):33771–33777

Zeleny R, Kolarich D, Strasser R, Altmann F (2006) Sialic acid concentrations in plants are in the range of inadvertent contamination. Planta 224(1):222–227

Chapter 17
The Rapidly Expanding Nexus of Immunoglobulin G N-Glycomics, Suboptimal Health Status, and Precision Medicine

Alyce Russell and Wei Wang

Contents

17.1	IgG N-Glycans in the Context of Biomedical Research	547
17.2	IgG N-Glycans as an Intermediate Phenotype of Health Status	549
17.3	Defining the Grey Area Between Health and Chronic Disease	551
17.4	Unravelling IgG Fc N-Glycan Inter- and Intra-population Variability	552
	17.4.1 Genes	553
	17.4.2 Gene Expression	554
	17.4.3 Clinical Indicators	555
17.5	The IgG N-Glycome in the Precision Medicine Framework	557
References		559

Abstract Immunoglobulin G is a prevalent glycoprotein, whose downstream immune responses are partially mediated by the N-glycans within the fragment crystallisable domain. Collectively termed the N-glycome, it is considered a complex intermediate phenotype: an amalgamation of genetic predisposition, environmental exposure, and health behaviours over the life-course. Thus, the immunoglobulin G N-glycome may provide an indication of health status on the spectrum from health to disease and infirmary. Although variability exists within and between populations, composition of the immunoglobulin G N-glycome remains stable over short periods of time. This underscores the potential of harnessing the immunoglobulin G N-glycome as an ideal tool for preclinical disease risk prediction, stratification, and prognosis through the development of precise dynamic biomarkers.

Keywords Immunoglobulin G N-glycome · Biomarker · Precision medicine · Suboptimal health

---

A. Russell · W. Wang (✉)
Centre for Precision Health, Edith Cowan University, Joondalup, Australia

School of Medical and Health Sciences, Edith Cowan University, Joondalup, Australia
e-mail: a.russell@ecu.edu.au; wei.wang@ecu.edu.au

© The Author(s), under exclusive license to Springer Nature Switzerland AG 2021
M. Pezer (ed.), *Antibody Glycosylation*, Experientia Supplementum 112,
https://doi.org/10.1007/978-3-030-76912-3_17

Abbreviations

A/G Ratio	Android/Gynoid Ratio
AD	Alzheimer's disease
ADCC	Antibody-dependent cell cytotoxicity
BMI	Body mass index
BP	Blood pressure
CCA	Canonical correlation analysis
C_H	Constant heavy
C_L	Constant light
CRP	C-reactive protein
Fab	Fragment antigen binding
FBG	Fasting blood glucose
Fc	Fragment crystallisable
FcR	Fragment crystallisable receptor
gQTL	Quantitative trait loci of N-glycosylation
GWAS	Genome-wide association study
HDL	High-density lipoprotein
IBD	Inflammatory bowel disease
ICD	International Classification of Diseases
IgG	Immunoglobulin G
LDL	Low-density lipoprotein
MBL	Mannose-binding lectin
MetS	Metabolic syndrome
PD	Parkinson's disease
PTM	Post-translational modification
RA	Rheumatoid arthritis
SHS	Suboptimal health status
SLE	Systemic lupus erythematosus
T2DM	Type-2 diabetes mellitus
TC	Total cholesterol
TG	Triglycerides
UPLC	Ultra-performance liquid chromatography
V_H	Variable heavy
V_L	Variable light
WHO	World Health Organisation
WHR	Waist-to-hip ratio
WHtR	Waist-to-height ratio

17.1 IgG N-Glycans in the Context of Biomedical Research

The premise of the Human Genome Project, a collaborative international effort launched in the 1990s, was to explore the genetic underpinnings of a plethora of human phenotypes and identify causal genes of congenital and chronic diseases (Collins and McKusick 2001). Although most phenotypes were inevitably too complex to be explained by genetics alone, the Human Genome Project set precedence for ethical considerations with open-source databases and research. Importantly, the Consortium elucidated few genes than first thought, with approximately 24,000 identified instead of the hypothesised 100,000 (Salzberg 2018). These outcomes lend support to the emerging field of epigenetics, which are heritable modifications to gene expression rather than the genome itself, and the importance of post-translational modifications (PTM), among others, in the construction of the complex phenotypes in the human repertoire.

Protein N-glycosylation, a process involving the addition of complex branching carbohydrate moieties (known as N-glycans) to proteins, is an important PTM that affects many biological processes, including those underlying human disease (Lauc et al. 2013; Russell et al. 2018). The full significance of the timing, presence, and function of the diverse range of glycoproteins is the focus of abundant cutting-edge research, not least because of the realisation that the complexity of the human glycoproteome is several orders of magnitude greater than its proteome (Lauc et al. 2013). Protein N-glycosylation is important in many vital biological processes such as cell adhesion, protein folding, molecular trafficking and clearance, receptor activation, and signal transduction (Russell et al. 2018; Li et al. 2019). More than half of the plasma proteins are N-glycosylated, including the antibody immunoglobulin G (IgG), whose N-glycans are crucial for conformation of the fragment crystallisable (Fc) region, which mediates downstream immune responses (Russell et al. 2018).

IgG is an important effector glycoprotein linking the innate and adaptive branches of the immune system. The protein portion of IgG consists of four polypeptide chains: two identical light chains, which may be kappa or lambda light chains, and two identical heavy chains (Fig. 17.1a) (Vidarsson et al. 2014). These are held together by intra-peptide disulphide bonds that cause the formation of loops and link the anti-parallel β-sheets in the tertiary structure of IgG (Vidarsson et al. 2014). The IgG glycoproteins contain two functionally distinct regions: the fragment antigen-binding (Fab) and the Fc. The Fab contains the variable (V_L and V_H) and constant (C_L and C_H1) light and heavy chain domains. These form the antigen-binding cleft, with the highly variable polypeptide sequences complementing specific target antigens (Vidarsson et al. 2014), whereas the Fc consists of constant heavy chain domains (C_H2 and C_H3) and mediates key effector functions (Pincetic et al. 2014).

The IgG Fc orchestrates various immune responses, which are either independent from antigen-binding (e.g. anti-inflammatory activity of intravenous immunoglobulins) or triggered by antigen recognition and dependent on the affinity for a number

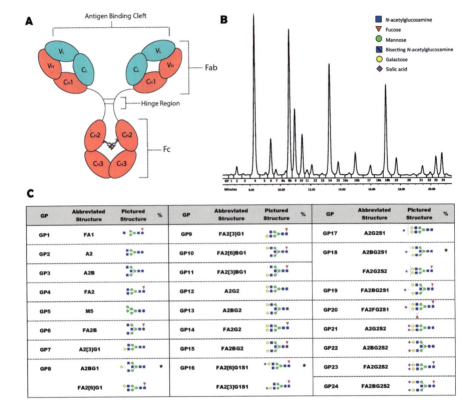

Fig. 17.1 The immunoglobulin G (IgG) N-glycome. (**a**) IgG contains two functionally distinct regions, linked via a protein scaffold hinge: the fragment antigen-binding (Fab) and the fragment crystallisable (Fc). Within the IgG Fab exists the antigen-binding cleft, responsible for antigen recognition (Russell et al. 2018). (**b**) The IgG Fc has a branching N-glycan within each constant heavy 2 (C_H2) domain. Collectively known as the IgG N-glycome, these may be analysed using ultra-performance liquid chromatography (UPLC), which separates the N-glycans by retention time and allows the estimation of abundance (Trbojevic-Akmacic et al. 2017). (**c**) These UPLC-separated N-glycan peaks contain unique N-glycan structures. The glycoforms prevalent within each peak are depicted in terms of abbreviated and pictured structure, with * indicating the major structure (Pučić et al. 2011). F—core (if the first letter) or antennary fucose, A2—biantennary, B—bisecting N-acetylglucosamine, Gx—galactose, Sx—sialic acid

of activating and inhibitory Fc receptors (FcRs) and complement factors (Russell et al. 2018; Quast et al. 2017). These key immune responses include pathogen clearance, antibody-dependent cell cytotoxicity (ADCC), and complement-initiated inflammation, all with both beneficial and detrimental effects depending on the premise of the IgG glycoprotein's activity (Russell et al. 2018). For example, during primary bacterial infection, IgG can initiate opsonisation through complement activation and phagocytosis of the bacterial cells by macrophages, monocytes, and neutrophils, as well as neutralise endotoxins and exotoxins (Subedi and Barb 2015; Krause et al. 2002; Ioan-Facsinay et al. 2002). These well-established beneficial effects have been harnessed for monoclonal antibody therapy in immunodeficient individuals (Schwab and Nimmerjahn 2013). On the contrary, there are examples where these effects are detrimental. In rheumatoid arthritis (RA) patients, IgG is thought to tandemly bind synovial cells and mannose-binding lectin (MBL), resulting in the initiation of the lectin complement cascade and secondary damage of surrounding tissues within the synovial joints (Fujita 2002; Quast and Lünemann 2014; Malhotra et al. 1995).

The IgG Fc N-glycans alter the glycoprotein's affinity for several FcRs and complement factors, which may be generalised as eliciting anti-inflammatory or pro-inflammatory responses (Russell et al. 2018; Ahmed et al. 2014). These immune responses are partially mediated by the conserved N-glycan within the C_H2 domain (Fig. 17.1a). Though most of the IgG Fc N-glycans are complex-type moieties (Russell et al. 2018), each IgG has two C_H2 chains (Fig. 17.1a). Thus, it assimilates two Fc N-glycans that often differ and interact with each other, leading to greater variability in downstream effector response (Nimmerjahn et al. 2007; Dekkers et al. 2017).

17.2 IgG N-Glycans as an Intermediate Phenotype of Health Status

IgG N-glycosylation is complex, mediated by glycosyltransferases and glycosylhydrolases that add and remove monosaccharides, respectively, from the maturing N-glycan structures within the endoplasmic reticulum and Golgi apparatus (Russell et al. 2018). The availability of these specialised enzymes is determined by the expression of hundreds of glycogenes within the producing plasma cell (Russell et al. 2017, 2018; Adua et al. 2017a; Vučković et al. 2015). Variations to the IgG N-glycans that appear structurally minute, however, can significantly alter its affinity to several FcRs and complement factors (summarised in Fig. 17.2). Numerous quantitative trait loci of N-glycosylation (gQTLs) have shown a clear directional effect with either increases or decreases in the relative abundance of certain IgG glycoforms (Lauc et al. 2013; Wang et al. 2011; Klarić et al. 2020). Though polymorphisms have been identified within these gQTLs, they cannot fully explain IgG N-glycome heterogeneity, even in very large population-based association studies. In fact, the post-translational timing of protein N-glycosylation indicates that the competing genetic underpinnings of the cell and the "cellular environment", not genetics alone, drive IgG N-glycome variability (Russell et al. 2018; Adua et al. 2017a; Wahl et al. 2018).

Several endogenous and exogenous factors are known to alter the cellular environment. These include cytokines and other immune mediators released by an array of leucocytes, not limited to those within the antibody-producing plasma cell (Russell et al. 2018; Wang et al. 2011; Horvat et al. 2011; Johnson et al. 2013; Rabinovich Gabriel and Croci 2012), and an array of complex phenotypes, including

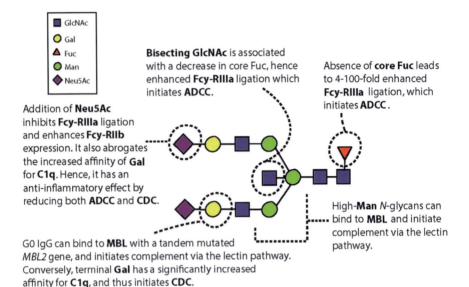

Fig. 17.2 Altered IgG N-glycosylation and its downstream effects. This figure was originally published in Russell et al. (2018) and is licensed under a Creative Commons Attribution 4.0 International License. No changes were made. GlcNAc—N-acetylglucosamine. Gal—galactose, Fuc—core fucose, Man—mannose, Neu5Ac—N-acetylneuraminic acid (sialic acid), ADCC—antibody-dependent cell cytotoxicity, CDC—complement-dependent cytotoxicity

biological and chronological age (Krištić et al. 2014; Yu et al. 2016), sex (Ercan et al. 2017), hormones levels (Ercan et al. 2017; Engdahl et al. 2017; Chen et al. 2012), and disease presence (Russell et al. 2017, 2018; Adua et al. 2017a; Vučković et al. 2015). Given its potential to identify and monitor health status, the IgG N-glycome has been implemented in several population-based studies exploring the biological mechanisms of altered disease states. A shift towards certain IgG N-glycan profiles has been reported for a number of diseases and conditions, including RA (Troelsen et al. 2012; Sebastian et al. 2016), metabolic syndrome (MetS), and type 2 diabetes mellitus (T2DM) (Lauc et al. 2013; Li et al. 2019; Lemmers et al. 2017; Adua et al. 2018; Liu et al. 2018a, 2019), inflammatory bowel disease (IBD) (Lauc et al. 2013), systemic lupus erythematosus (SLE) (Vučković et al. 2015), hypertension (Wang et al. 2016; Liu et al. 2018b), various cancers (Lauc et al. 2013; Meany and Chan 2011), and neurological disorders such as Alzheimer's disease (AD) and progressive mild cognitive impairment (Lundström et al. 2013), multiple sclerosis (Wuhrer et al. 2015), and Parkinson's disease (PD) (Russell et al. 2017). These studies report disease-specific shifts, underscoring the plausible biological importance of the IgG N-glycome in disease pathogenesis.

We consider the IgG N-glycome a complex "intermediate phenotype"; an amalgamation of genetic predisposition, environmental exposure, and health behaviours over the life-course, which provides an indication of health status on the spectrum from health to disease (Russell et al. 2018, 2019a). Although variability exists, IgG

N-glycome composition is considered a predesigned outcome of the producing plasma cell (Russell et al. 2018; Vidarsson et al. 2014) and is stable over short periods, with current estimates over 6–12 months (Pučić et al. 2011; Adua et al. 2018). This makes it an ideal target for measuring health status in the context of biomarker research and the shift towards precision medicine, which aims to diagnose early, delay further development of disease, monitor treatment efficacy, and improve quality of life through prevention and tailored interventions (Wang et al. 2014).

17.3 Defining the Grey Area Between Health and Chronic Disease

Health is the optimum level of functional or metabolic efficiency, in the absence of illness, stress, injury, or pain (Wang and Yan 2012). The World Health Organisation (WHO) defines health as "a state of complete physical, mental, and social well-being and not merely the absence of disease or infirmary" (Sartorius 2006). Though it has persisted as the most widely accepted definition of health since its dissemination in 1946, controversy has arisen due to the definition lacking operational value and the limitation created by the word "complete" (Sartorius 2006; Callahan 1973; Jadad and O'Grady 2008). A broader approach may be achieved by instead focusing on health status, which is the indication of a person's combined state of physical, mental, and social well-being (Wang and Yan 2012). Indeed, other classification systems, such as the International Classification of Diseases (ICD), place greater emphasis on identifying and measuring components of health status, which may be in different states of well-being (Wang and Yan 2012).

There has been considerable effort into defining the grey area between complete health and disease. For example, Suboptimal Health Status (SHS) has been defined as a physical state whereby no specific clinically diagnosable conditions are present (Yan et al. 2012), and focuses on identifying a decline in vitality, physiological function, and capacity for adaptation (Yan et al. 2009). SHS recognises a person's perception of health complaints, general weakness, chronic fatigue, and low energy, which may elucidate suboptimal or less-than-ideal components of health status requiring intervention (Wang et al. 2014). Early intervention is consistently regarded as the most effective method of preventing chronic disease and can be achieved through prescribing modified health behaviours or treating mild complaints from the perspectives of precision medicine.

Our initial efforts were directed at developing the Suboptimal Health Status Questionnaire-25 (SHSQ-25), which we have since validated in multiple populations, including Africans, Asians, and Caucasians (Wang et al. 2014; Wang and Yan 2012; Yan et al. 2012; Adua et al. 2017b, 2020; Anto et al. 2019; Kupaev et al. 2016). The SHSQ-25 incorporates Traditional Chinese Medicine principles, and assesses health status from the perspective of five domains: (1) the cardiovascular system, (2) the immune system, (3) the digestive system, (4) fatigue, and

(5) mental health. Subsequent research has highlighted the relationship between total and domain-specific SHS scores and many objective health indicators, including cardiometabolic risk factors such as triglycerides (TG), C-reactive protein (CRP), insulin, and alkaline phosphatase, as well as endothelial dysfunction, blood pressure (BP), obesity, plasma cortisol and glucocorticoid receptor α/β (Yan et al. 2015).

More recently, there have been investigations into the association of the total and domain-specific SHS scores and the IgG Fc N-glycans, and how these relate to complex phenotypes (Sebastian et al. 2016; Yan et al. 2012; Anto et al. 2019). Preliminary results suggest an interplay between the complexity of the IgG N-glycome, metabolic risk factors, and SHS (Liu et al. 2018c, 2019; Adua et al. 2017b). These profiles may hold the key to understanding the underlying biological mechanisms of SHS and are under further investigation (Yan et al. 2009; Lu et al. 2011). It is envisaged these findings will pave the way for preclinical disease risk prediction, stratification, and prognosis through the development of precise dynamic biomarkers, as well as identify potentially modifiable health behaviours or pharmaceutical targets for early intervention.

17.4 Unravelling IgG Fc N-Glycan Inter- and Intra-population Variability

The IgG N-glycome presents itself as a promising biomarker. Although variability exists within and between populations (Pučić et al. 2011; Štambuk et al. 2020), including between Chinese minority groups (Liu et al. 2018b), its composition is considered a predesigned and relatively stable outcome of the producing plasma cell (Russell et al. 2018; Vidarsson et al. 2014). Therefore, a plethora of research is underway to determine the IgG N-glycome's efficacy in elucidating overall health status, identifying the risk of developing disorders, tracking progression of phenotypic disorders, and indicating who may respond to certain therapies or whether a currently prescribed therapy is effective.

Our ongoing research demonstrates the discriminatory utility of the IgG N-glycome in various population studies of chronic disease and its related phenotypes, including RA (Troelsen et al. 2012; Sebastian et al. 2016), MetS and T2DM (Lauc et al. 2013; Li et al. 2019; Lemmers et al. 2017; Adua et al. 2018; Liu et al. 2018a, 2019), SLE (Vučković et al. 2015), hypertension (Wang et al. 2016; Liu et al. 2018b), central adiposity (Russell et al. 2019b), ischaemic stroke (Liu et al. 2018d), and PD (Russell et al. 2017). Although promising results are evident, considerable IgG N-glycome heterogeneity exists. Thus, they also underscore the need to identify what is driving altered IgG N-glycosylation, even among individuals with the same phenotypes. To bridge the gap in knowledge, we have explored genetic and other factors that may alter the cellular environment and, therefore, IgG Fc effector responses. Notably, these factors may be incorporated into IgG N-glycome models to improve their precision.

17.4.1 Genes

Several loci, termed gQTLs, are associated with aberrant N-glycosylation and have clear directional effects in the relative abundance of certain N-glycosylation features. The gQTL *HNF1α* was the first found to be associated with aberrant plasma glycosylation, a gene whose product is hepatocyte nuclear factor 1α (HNF1a). HNF1a is a master regulator of the expression of *FUT6* and *FUT8*. These gQTLs encode fucosyltransferases that influence multiple stages in fucosylation (Wang et al. 2011). Particularly for IgG, *FUT8* associates with core fucosylation in European Caucasian populations (Lauc et al. 2013; Wang et al. 2011). IgG glycoforms lacking core fucose have a 4- to 100-fold enhanced affinity for Fcγ-RIIIa, which initiates ADCC (Russell et al. 2018; Vidarsson et al. 2014).

Many other gQTLs have been identified, including the *ST6GAL1*, *B4GALT1*, and *MGAT3* genes that encode glycosyltransferases instrumental in adding sialic acid (Neu5Ac), galactose (Gal), and bisecting N-acetylglucosamine (GlcNAc) monosaccharides, respectively, to the branching N-glycans (Lauc et al. 2013; Wang et al. 2011; Bondt et al. 2014). Whereas genes previously associated with other diseases, such as autoimmune diseases and haematological cancers, have later been identified to have pleiotropy with IgG N-glycosylation and include *IL6ST-ANKRD55*, *IKZF1*, *ABCF2-SMARCD3*, *SUV420H1*, *SMARCB1-DERL3*, and *SYNGR1-TAB1-MGAT3-CACNA11* (Lauc et al. 2013; Menni et al. 2013). Moreover, five novel gQTLs were recently identified and validated within European Caucasians: *IGH*, *ELL2*, *HLA-B-C*, *AZI1*, and *FUT6-FUT3* (Shen et al. 2017a).

We recently confirmed two gQTLs using an Australian Caucasian cohort; *ST6GAL1* and *MGAT3* (Russell 2020). These gQTLs were previously identified in Caucasians from four European populations, two islands in Croatia (Vis and Korcula), the Orkney Islands in the United Kingdom, and Sweden (Lauc et al. 2013). *ST6GAL1* encodes sialyltransferase-6 that catalyses the transfer of Neu5Ac to the branching N-glycan moieties (Kuhn et al. 2013). This glycosyltransferase is localised in the membrane of the Golgi apparatus and is specific to B cells in the later stages, following differentiation into antibody-producing plasma cells (Kuhn et al. 2013). Mutations within *ST6GAL1* are consistently acknowledged in genome-wide association studies (GWAS) of IgG N-glycosylation (Lauc et al. 2013; Wahl et al. 2018; Shen et al. 2017a). Indeed, this gQTL associates with the relative abundance of various sialylated IgG glycoforms. Polymorphisms within *ST6GAL1* have pleiotropy to T2DM (Lauc et al. 2013). Interestingly, altered sialylation of IgG has been implicated in T2DM among European Caucasians (Lemmers et al. 2017), Han Chinese (Wu et al. 2020), and Uyghur Chinese (Liu et al. 2019). Thus, this finding may have biological significance.

The gQTL containing *SYNGR1-TAB1-MGAT3* is another commonly acknowledged gQTL (Lauc et al. 2013; Wahl et al. 2018; Shen et al. 2017a). The strongest associate genetic polymorphism we identified was within the *MGAT3* genomic region; a gene encoding mannosyl (β-1,4)-glycoprotein β-1,4-*N*-acetylglucosaminyltransferase which adds bisecting GlcNAc to the branching IgG glycoforms. This polymorphism associated with increased relative abundance of bisect-type IgG N-glycans among fucosylated disialylated N-glycans (Russell 2020), which has been implicated in E-cadherin, EGF-, Wnt- and integrin- cancer-associated signalling pathways (Kohler et al. 2016). Altered MGAT3 has pleiotropy with Alzheimer's disease (Fiala et al. 2011) and inflammatory bowel disease (Klasic et al. 2018). *SYNGR1* and *TAB1* are shown to be in linkage disequilibrium with *MGAT3* and encode synaptogyrin-1 and TGF-beta-activated kinase-1, respectively. Genetic polymorphisms within *SYNGR1* have pleiotropy with schizophrenia and bipolar disorder (Verma et al. 2005; Iatropoulos et al. 2009), whereas TAB1 may have pleiotropy with colorectal cancer (Gong et al. 2018).

Though associated with various N-glycosylation features, the vast repertoire of gQTLs cannot fully explain IgG N-glycome heterogeneity, even in very large association studies. The post-translational timing of N-glycosylation supports the competing effects of genetic predisposition of an antibody-producing plasma cell and its cellular environment. Certainly, this gene–environment interaction drives the variability among IgG N-glycome profiles within and between populations (Russell et al. 2018; Adua et al. 2017a; Wahl et al. 2018).

17.4.2 Gene Expression

The IgG N-glycome is both genetically and epigenetically regulated (Russell et al. 2018; Adua et al. 2017a; Wahl et al. 2018), and cytokines and other immune factors released by an array of leucocytes, not only those factors within the antibody-producing plasma cell, associate with IgG N-glycome heterogeneity (Wang et al. 2011; Horvat et al. 2011; Johnson et al. 2013; Rabinovich Gabriel and Croci 2012). Therefore, we recently analysed whole blood RNA-Seq to explore the genetic influence on the heterogeneity of the IgG N-glycome, in terms of the transcribed genes and not just genetic polymorphisms (Russell 2020). The decision to explore whole blood mRNA rather than targeting B-cell lymphocyte and plasma cell mRNA transpired since both endogenous and exogenous factors are known to alter IgG N-glycome composition and in turn, impact the inflammatory properties of IgG.

An integrative model explored how IgG N-glycome heterogeneity could be explained by the differential expression of whole blood cells via contemporaneously analysing both datasets (14,544 expressed genes and 24 IgG N-glycan peaks; see Fig. 17.1b, c). Our optimised model explicated that 58.1% of IgG N-glycome variability was explained by variable whole blood gene expression, while only 11.4% of the mRNA variability was explained by the IgG N-glycome. The biological validity of these findings was confirmed through interrogating gene ontology (GO) enrichment terms, which represent biosynthesis and degradation-related intracellular activity as well as several IgG Fc downstream effector responses (Russell 2020).

The IgG Fc binds distinct FcR and complement factors, and these are cell, spatial, and immune response specific (Russell et al. 2018; Pincetic et al. 2014).

Additionally, FcR specificity may be influenced by several factors, including the IgG Fc N-glycans (Russell et al. 2018; Quast et al. 2017). Since whole blood gene expression was considered, our results provide further evidence that cytokines and other factors extrinsic to the differentiating B-cell and antibody-producing plasma cell may mediate IgG N-glycosylation. On the contrary, the IgG N-glycans associate with very specific downstream immune responses due to IgG Fc specificity (Wang et al. 2011; Horvat et al. 2011; Johnson et al. 2013; Rabinovich Gabriel and Croci 2012). Hence, there exists a complex interplay in the epigenetic regulation of IgG N-glycosylation.

17.4.3 Clinical Indicators

Clinical indicators are arguably the most studied determinants of IgG N-glycome variability, with many studies arising from our research team (Yu et al. 2016; Sebastian et al. 2016; Wang et al. 2016; Liu et al. 2018d) and others (Krištić et al. 2014; Ercan et al. 2017; Baković et al. 2013; Keser et al. 2017). Although genetics plays a key role in the baseline risk of the chronic and altered relative abundances of IgG N-glycan features, the N-glycosylation of IgG Fc is regulated by several endogenous and exogenous factors, including routinely measured blood and clinical indicators. In this sense, the IgG N-glycome is malleable as it is reliant on the expression levels of the glycosyltransferases and glycosylhydrolases, as well as the abundance of sugar nucleotide donors, during biosynthesis.

An increase in pro-inflammatory IgG glycoforms associates with unfavourable levels of several cardiometabolic indicators, such as TG, total cholesterol (TC), low-density lipoprotein (LDL) and high-density lipoprotein (HDL) (Plomp et al. 2017), body mass index (BMI) and central adiposity (Russell et al. 2019b; Perkovic et al. 2014), fasting blood glucose (FBG) (Lemmers et al. 2017; Ge et al. 2018), BP (Wang et al. 2016), and CRP (Plomp et al. 2017). Very recently, IgG N-glycome heterogeneity was linked with the 10-year atherosclerotic cardiovascular disease risk score, which is a sex- and race-specific risk assessment derived using a combination of these clinical indicators (Menni et al. 2018).

We further underscored the association between pro-inflammatory IgG N-glycans and poorer health status using canonical correlation analysis (CCA); a multivariate form of correlation analysis that simultaneously associates all the IgG N-glycans with the multiple clinical indicators (Russell et al. 2019a). CRP had the strongest independent association with increased relative abundance of pro-inflammatory IgG glycoforms (Fig. 17.3). Importantly, CRP is a non-specific marker of systemic inflammation whose activity ranges from acute-phase reactions through low-grade chronic inflammation (Segman and Stein 2015), whereas the IgG N-glycome has unique signatures for various phenotypes. Both CRP and the IgG N-glycome are biomarkers of the cardiometabolic cluster of disorders, including hypertension, cardiovascular disease, ischaemic stroke, MetS, and T2DM (Wang et al. 2016; Liu et al. 2018b, d; Menni et al. 2018; Segman and Stein 2015). However, it is prudent to

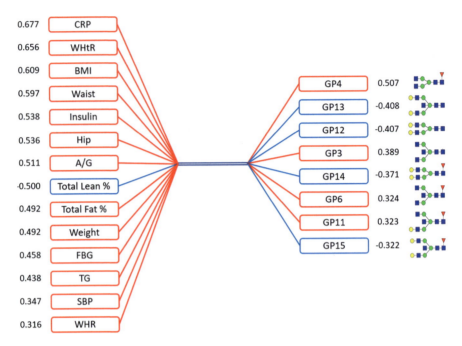

Fig. 17.3 Canonical structures of IgG N-glycan peaks (GPs) and clinical factors in the first canonical set. An absolute value of canonical loadings greater than 0.3 was considered significant. All variables are sorted by the absolute value of the canonical loadings. Positive relationships are in red, while negative relationships are in blue. No changes were made. *A/G* android-to-gynoid ratio (measured using DEXA), *BMI* body mass index, *CRP* C-reactive protein, *DEXA* dual-energy X-ray absorptiometry, *FBG* fasting blood glucose, *Hip* hip circumference, *SBP* systolic blood pressure, *TG* triglycerides, *Total Fat %* percentage of fat mass (total body), *Total Lean %* percentage of lean mass (total body), *Waist* waist circumference, *WHR* waist-to-hip ratio, *WHtR* waist-to-height ratio. Reprinted with permission from Russell et al. (2019a)

underscore the novelty of the IgG N-glycome, which has unique signatures for many phenotypes when compared with CRP, known for its non-specific indication of the presence of inflammation. Thus, although both pro-inflammatory IgG N-glycans and CRP are present in low-grade chronic inflammation, the IgG N-glycome elucidates more specific information about the underlying biology of preclinical and clinical diseases (Russell et al. 2017; Lemmers et al. 2017; Wang et al. 2016; Menni et al. 2018; Bondt et al. 2013; Dekkers et al. 2018).

CRP positively associates with adipokines released from central body fat, particularly visceral fat; an endocrine organ in its own right (Fontana et al. 2007). Increased central adiposity compared with hip fat (as measured with dual energy X-ray absorptiometry-derived android-to-gynoid ratio; A/G ratio) or height (waist-to-height ratio; WHtR) are strongly associated with pro-inflammatory IgG glycoforms, more so than BMI (Russell et al. 2019b). Moreover, visceral body fat produces interleukin-6 (IL-6), to a greater degree than subcutaneous fat (Panagiotakos et al. 2005), and associates with increased levels of an array of systemic inflammatory markers (Hsieh et al. 2014; Gaens et al. 2015), including CRP (Fontana et al. 2007). Chronic, subclinical inflammation has been linked to obesity (Festa et al. 2001; Alissa et al. 2016), and central adiposity is more detrimental than total body fat percentage; obese individuals (according to BMI guidelines) with low A/G ratios are less likely to develop MetS (Koster et al. 2010).

Central adiposity is a major risk factor of disorders such as MetS (Koster et al. 2010), cardiovascular disease (Shen et al. 2017b), and AD (Whitmer et al. 2008), while the IgG N-glycome has utility in identifying disease presence and severity (Lemmers et al. 2017; Wang et al. 2016; Liu et al. 2018d; Koster et al. 2010; Shen et al. 2017b; Novokmet et al. 2014). CCA further validated previously identified univariate associations, evidencing four measures of central adiposity to be moderately to strongly associated with the IgG N-glycome, namely WHtR, BMI, waist circumference, and waist-to-hip ratio (WHR) (Fig. 17.3). Indeed, IgG N-glycosylation heterogeneity may be an immunological response to centrally located body fat (Russell et al. 2019b). However, it remains to be determined whether increases in central adiposity cause the increase in pro-inflammatory IgG glycoforms, or whether it is reverse causation or bidirectional (Russell et al. 2019b).

17.5 The IgG N-Glycome in the Precision Medicine Framework

The recently initiated Human Glycome Project aims to follow in the footsteps of the Human Genome Project and elucidate biological regulation within the human body, leveraged by the collaborative effort of world leaders in glycobiology from several countries (Bennett 2019; Wang 2019). Moving beyond laboratory- and population-based studies, there has been a surge in effort focused on translating glycobiology research into the clinical practice (Fig. 17.4). Currently, clinical tests are emerging in Europe and China that quantify biological age using the dynamic IgG N-glycome, following results from several studies (Krištić et al. 2014; Yu et al. 2016; Vanhooren et al. 2010; Ruhaak et al. 2010). The premise of "GlycanAge" is to predict biological age, thought to combine genetic predisposition with the effects of several health behaviours and environmental exposures over the life-course. Therefore, an individual may use this to identify potential discordance from their chronological age. It is an important initiative in the context of precision medicine. However, it remains to be determined how individuals will truly benefit from the knowledge of subclinical morbidity risk or discordant biological age, particularly if much higher than chronological age.

Variation among several clinical indicators, particularly those related to cardiometabolic health, associate with the heterogeneity of IgG N-glycosylation, and may represent instrumental variables of health behaviours and environmental exposure (Lauc et al. 2013; Li et al. 2019; Liu et al. 2018a, d; Wang et al. 2016; Russell et al. 2019a, b; Menni et al. 2018). By virtue of this gene–environment

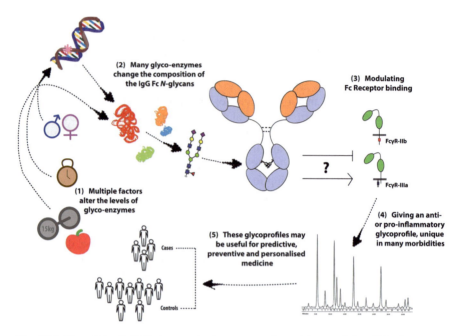

Fig. 17.4 Use of the IgG N-glycome in the context of precision medicine. This figure was originally published in Russell et al. (2018) and is licensed under a Creative Commons Attribution 4.0 International License. No changes were made

interplay, it may be possible to remodel IgG N-glycan composition by modifying certain negating health behaviours associated with poorer health outcomes (Russell 2020). Lowering or avoiding certain health behaviours is heavily promoted as a vector to improve overall health status and increase disease-free life. Aside from obesity and hypertension, other prominent modifiable health behaviours include physical inactivity, daily smoking, excessive alcohol consumption, inadequate consumption of fruits and vegetables, and consumption of whole milk (AIHW 2012). In fact, these seven health behaviours contribute to three-quarters of the total burden of chronic disease in Australia (AIHW 2012), as well as decline in cardiometabolic health, especially in middle-aged and elderly individuals (Díaz-Redondo et al. 2015). If shown to be causative, an adjustment in health behaviours may be prescribed to alter IgG N-glycome composition and improve health outcomes.

We have done preliminary analyses on three of these health behaviours: excessive alcohol consumption, current smoking status, and physical inactivity. There was no evidence of an effect between IgG N-glycosylation and the level of self-reported physical activity (Russell 2020). However, binge drinking and excessive alcohol consumption were associated with an increased relative abundance of pro-inflammatory IgG glycoforms (Russell 2020). Also, several IgG N-glycans were associated with current smoking status compared with those who have never smoked (Russell 2020). Smoking was previously implicated in a European-based study (Wahl et al. 2018). Wahl et al. (2018) identified the potential mechanism whereby the association between smoking and IgG N-glycosylation may be partially mediated by altered DNA methylation at several gQTLs (Wahl et al. 2018). Importantly, we found no difference in IgG N-glycan profiles when comparing those that have never smoked to ex-smokers (Wahl et al. 2018). This may support the positive effects on health status after quitting smoking, reflected in other health research (Knuchel-Takano et al. 2018). Though changes to these modifiable health behaviours make an ideal intervention strategy, it should be noted an interplay between these health behaviours, mental status (particularly untreated mental disorders), and addiction complicates the prospect of intervening (Verdurmen et al. 2005; Teesson et al. 2010).

In conclusion, we consider the IgG N-glycome a complex intermediate phenotype, which may be harnessed as a biomarker for measuring underlying biological processes on the spectrum from health to disease and infirmary. IgG N-glycome composition varies within and between populations, is stable over short periods, and associates with various subclinical and clinical phenotypes. We further suggest it may be possible to remodel IgG N-glycan composition through modifying certain negating health behaviours. Whether an adjustment in modifiable health behaviours may improve overall health status and have a downstream impact on the IgG N-glycome is yet to be validated. However, the composition of the IgG N-glycome may be pertinent to the pathophysiology of existent subclinical or clinical phenotypes, making it an ideal target for measuring health status in the context of precision medicine.

Acknowledgment This book chapter was partially supported by Australia–China International Collaborative Grant (NHMRC APP1112767-NSFC 81561128020). Parts of the data and presented figures in this chapter have previously been published in our earlier articles, which are referenced.

Disclosure of Interests The two authors declare they have no conflict of interest.

Ethical Approval The studies discussed in this book chapter have ethical approval from each institution. All procedures performed in the studies involving human participants were in accordance with the ethical standards of the institutional research committees and with the 1964 Helsinki declaration and its later amendments.

References

Adua E, Russell A, Roberts P, Wang Y, Song M, Wang W (2017a) Innovation analysis on postgenomic biomarkers: glycomics for chronic diseases. OMICS 21(4):183–196

Adua E, Roberts P, Wang WJEJ (2017b) Incorporation of suboptimal health status as a potential risk assessment for type II diabetes mellitus: a case-control study in a Ghanaian population. EPMA J 8(4):345–355

Adua E, Anto EO, Roberts P, Kantanka OS, Aboagye E, Wang W (2018) The potential of N-glycosylation profiles as biomarkers for monitoring the progression of type II diabetes mellitus towards diabetic kidney disease. J Diabetes Metab Disord 17(2):233–246

Adua E, Afrifa-Yamoah E, Frimpong K, Adama E, Karthigesu SP, Anto EO et al (2020) Construct validity of the suboptimal health status Questionnaire-25 in a Ghanaian population

Ahmed AA, Giddens J, Pincetic A, Lomino JV, Ravetch JV, Wang L-X et al (2014) Structural characterization of anti-inflammatory immunoglobulin G Fc proteins. J Mol Biol 426 (18):3166–3179

AIHW (2012) Risk factors contributing to chronic disease. Canberra, AIHW

Alissa EM, Maisa'a M, Alama NA, Ferns GA (2016) Role of omentin-1 and C-reactive protein in obese subjects with subclinical inflammation. J Clin Transl Endocrinol 3:7–11

Anto EO, Roberts P, Coall D, Turpin CA, Adua E, Wang Y et al (2019) Integration of suboptimal health status evaluation as a criterion for prediction of preeclampsia is strongly recommended for healthcare management in pregnancy: a prospective cohort study in a Ghanaian population. EPMA J 10(3):211–226

Baković MP, Selman MHJ, Hoffmann M, Rudan I, Campbell H, Deelder AM et al (2013) High-throughput IgG fc N-glycosylation profiling by mass spectrometry of glycopeptides. J Proteome Res 12(2):821–831

Bennett H (2019) Life is sweet. New Scientist 241(3223):34–37

Bondt A, Selman MHJ, Deelder AM, Hazes JMW, Willemsen SP, Wuhrer M et al (2013) Association between galactosylation of immunoglobulin G and improvement of rheumatoid arthritis during pregnancy is independent of sialylation. J Proteome Res 12(10):4522–4531

Bondt A, Rombouts Y, Selman MHJ, Hensbergen PJ, Reiding KR, Hazes JMW et al (2014) Immunoglobulin G (IgG) fab glycosylation analysis using a new mass spectrometric high-throughput profiling method reveals pregnancy-associated changes. Mol Cell Proteomics 13 (11):3029–3039

Callahan D (1973) The WHO definition of 'Health'. Hast Cent Stud 1(3):77–87

Chen G, Wang Y, Qiu L, Qin X, Liu H, Wang X et al (2012) Human IgG Fc-glycosylation profiling reveals associations with age, sex, female sex hormones and thyroid cancer. J Proteome 75 (10):2824–2834

Collins FS, McKusick VA (2001) Implications of the human genome project for medical science. JAMA 285(5):540–544

Dekkers G, Treffers L, Plomp R, Bentlage AEH, de Boer M, Koeleman CAM et al (2017) Decoding the human immunoglobulin G-glycan repertoire reveals a spectrum of fc-receptor- and complement-mediated-effector activities. Front Immunol 8(877)

Dekkers G, Rispens T, Vidarsson G (2018) Novel concepts of altered immunoglobulin g galactosylation in autoimmune diseases. Front Immunol 9(553)

Díaz-Redondo A, Giráldez-García C, Carrillo L, Serrano R, García-Soidán FJ, Artola S et al (2015) Modifiable risk factors associated with prediabetes in men and women: a cross-sectional analysis of the cohort study in primary health care on the evolution of patients with prediabetes (PREDAPS-study). BMC Fam Pract 16(1):5

Engdahl C, Raufer J, Harre U, Bondt A, Pfeifle R, Krönke G et al (2017) SAT0019 Estrogen influences the sialylation profile and inflammatory properties of antibodies – a potential explanation for the sex differences and increased risk for ra in postmenopausal women. Ann Rheum Dis 76(Suppl 2):775

Ercan A, Kohrt WM, Cui J, Deane KD, Pezer M, Yu EW et al (2017) Estrogens regulate glycosylation of IgG in women and men. JCI Insight 2(4)

Festa A, D'Agostino Jr R, Williams K, Karter A, Mayer-Davis E, Tracy R et al (2001) The relation of body fat mass and distribution to markers of chronic inflammation. Int J Obes Relat Metab Disord 25(10)

Fiala M, Mahanian M, Rosenthal M, Mizwicki MT, Tse E, Cho T et al (2011) MGAT3 mRNA: a biomarker for prognosis and therapy of Alzheimer's disease by vitamin D and curcuminoids. J Alzheimers Dis 25(1):135–144

Fontana L, Eagon JC, Trujillo ME, Scherer PE, Klein S (2007) Visceral fat adipokine secretion is associated with systemic inflammation in obese humans. Diabetes 56(4):1010–1013

Fujita T (2002) Evolution of the lectin–complement pathway and its role in innate immunity. Nat Rev Immunol 2(5):346–353

Gaens KH, Ferreira I, Van De Waarenburg MP, van Greevenbroek MM, Van Der Kallen CJ, Dekker JM et al (2015) Protein-bound plasma Nε-(carboxymethyl) lysine is inversely associated with central obesity and inflammation and significantly explain a part of the central obesity-related increase in inflammation: the Hoorn and CODAM studies. Arterioscler Thromb Vasc Biol 35(12):2707–2713

Ge S, Wang Y, Song M, Li X, Yu X, Wang H et al (2018) Type 2 diabetes mellitus: integrative analysis of multiomics data for biomarker discovery. OMICS 22(7):514–523

Gong H, Fang L, Li Y, Du J, Zhou B, Wang X et al (2018) miR873 inhibits colorectal cancer cell proliferation by targeting TRAF5 and TAB1. Oncol Rep 39(3):1090–1098

Horvat T, Zoldoš V, Lauc G (2011) Evolutional and clinical implications of the epigenetic regulation of protein glycosylation. Clin Epigenetics 2(2):425–432

Hsieh C-J, Wang P-W, Chen T-Y (2014) The relationship between regional abdominal fat distribution and both insulin resistance and subclinical chronic inflammation in non-diabetic adults. Diabetol Metab Syndr 6(1):49

Iatropoulos P, Gardella R, Valsecchi P, Magri C, Ratti C, Podavini D et al (2009) Association study and mutational screening of SYNGR1 as a candidate susceptibility gene for schizophrenia. Psychiatr Genet 19(5):237–243

Ioan-Facsinay A, de Kimpe SJ, Hellwig SMM, van Lent PL, Hofhuis FMA, van Ojik HH et al (2002) FcγRI (CD64) contributes substantially to severity of arthritis, hypersensitivity responses, and protection from bacterial infection. Immunity 16(3):391–402

Jadad AR, O'Grady L (2008) How should health be defined? BMJ 337:a2900

Johnson JL, Jones MB, Ryan SO, Cobb BAJT (2013) The regulatory power of glycans and their binding partners in immunity. Trends Immunol 34(6):290–298

Keser T, Vučković F, Barrios C, Zierer J, Wahl A, Akinkuolie AO et al (2017) Effects of statins on the immunoglobulin G glycome. Biochim Biophys Acta Gen Subj 1861(5):1152–1158

Klarić L, Tsepilov YA, Stanton CM, Mangino M, Sikka TT, Esko T et al (2020) Glycosylation of immunoglobulin G is regulated by a large network of genes pleiotropic with inflammatory diseases. Sci Adv 6(8):eaax0301

Klasic M, Markulin D, Vojta A, Samarzija I, Birus I, Dobrinic P et al (2018) Promoter methylation of the MGAT3 and BACH2 genes correlates with the composition of the immunoglobulin G glycome in inflammatory bowel disease. Clin Epigenetics 10:75

Knuchel-Takano A, Hunt D, Jaccard A, Bhimjiyani A, Brown M, Retat L et al (2018) Modelling the implications of reducing smoking prevalence: the benefits of increasing the UK tobacco duty escalator to public health and economic outcomes. Tob Control 27(e2):e124–e1e9

Kohler RS, Anugraham M, López MN, Xiao C, Schoetzau A, Hettich T et al (2016) Epigenetic activation of MGAT3 and corresponding shift in N-glycans bisecting GlcNAc shortens the survival of cancer patients. Oncotarget 7(32):51674–51686

Koster A, Stenholm S, Alley DE, Kim LJ, Simonsick EM, Kanaya AM et al (2010) Body fat distribution and inflammation among obese older adults with and without metabolic syndrome. Obesity 18(12):2354–2361

Krause I, Wu R, Sherer Y, Patanik M, Peter J, Shoenfeld Y (2002) In vitro antiviral and antibacterial activity of commercial intravenous immunoglobulin preparations–a potential role for adjuvant intravenous immunoglobulin therapy in infectious diseases. Transfusion Med 12(2):133–139

Krištić J, Vučković F, Menni C, Klarić L, Keser T, Becceheli I et al (2014) Glycans are a novel biomarker of chronological and biological ages. J Gerontol A Biol Sci Med Sci 69(7):779–789

Kuhn B, Benz J, Greif M, Engel AM, Sobek H, Rudolph MG (2013) The structure of human alpha-2,6-sialyltransferase reveals the binding mode of complex glycans. Acta Crystallogr D Biol Crystallogr 69(Pt 9):1826–1838

Kupaev V, Borisov O, Marutina E, Yan Y-X, Wang WJEJ (2016) Integration of suboptimal health status and endothelial dysfunction as a new aspect for risk evaluation of cardiovascular disease. EPMA J 7(1):19

Lauc G, Huffman JE, Pučić M, Zgaga L, Adamczyk B, Mužinić A et al (2013) Loci associated with N-glycosylation of human immunoglobulin G show pleiotropy with autoimmune diseases and haematological cancers. PLoS One 9(1):e1003225

Lemmers RF, Vilaj M, Urda D, Agakov F, Šimurina M, Klaric L et al (2017) IgG glycan patterns are associated with type 2 diabetes in independent European populations. Biochim Biophys Acta Gen Subj 1861(9):2240–2249

Li X, Wang H, Russell A, Cao W, Wang X, Ge S et al (2019) Type 2 diabetes mellitus is associated with the immunoglobulin G N-glycome through putative proinflammatory mechanisms in an Australian population. OMICS (Ahead of print)

Liu D, Chu X, Wang H, Dong J, Ge S-Q, Zhao Z-Y et al (2018a) The changes of immunoglobulin G N-glycosylation in blood lipids and dyslipidaemia. J Transl Med 16(1):235

Liu JN, Dolikun M, Štambuk J, Trbojević-Akmačić I, Zhang J, Wang H et al (2018b) The association between subclass-specific IgG Fc N-glycosylation profiles and hypertension in the Uygur, Kazak, Kirgiz, and Tajik populations. J Hum Hypertens

Liu D, Chu X, Wang H, Dong J, Ge S-Q, Zhao Z-Y et al (2018c) The changes of immunoglobulin GN-glycosylation in blood lipids and dyslipidaemia. J Transl Med 16(1):235

Liu D, Zhao Z, Wang A, Ge S, Wang H, Zhang X et al (2018d) Ischemic stroke is associated with the pro-inflammatory potential of N-glycosylated immunoglobulin G. J Neuroinflammation 15(1):123

Liu J, Dolikun M, Štambuk J, Trbojević-Akmačić I, Zhang J, Zhang J et al (2019) Glycomics for type 2 diabetes biomarker discovery: promise of immunoglobulin G subclass-specific fragment crystallizable N-glycosylation in the Uyghur population. OMICS: J Integr Biol 23(12):640–648

Lu J-P, Knezevic A, Wang Y-X, Rudan I, Campbell H, Zou Z-K et al (2011) Screening novel biomarkers for metabolic syndrome by profiling human plasma N-glycans in Chinese Han and Croatian populations. J Proteome Res 10(11):4959–4969

Lundström SL, Yang H, Lyutvinskiy Y, Rutishauser D, Herukka S-K, Soininen H et al (2013) Blood plasma IgG Fc glycans are significantly altered in Alzheimer's disease and progressive mild cognitive impairment. J Alzheimers Dis 38(3):567–579

Malhotra R, Wormald MR, Rudd PM, Fischer PB, Dwek RA, Sim RB (1995) Glycosylation changes of IgG associated with rheumatoid arthritis can activate complement via the mannose-binding protein. Nat Med 1(3):237–243

Meany DL, Chan DW (2011) Aberrant glycosylation associated with enzymes as cancer biomarkers. Clin Proteomics 8(7):10.1186

Menni C, Keser T, Mangino M, Bell JT, Erte I, Akmačić I et al (2013) Glycosylation of immunoglobulin g: role of genetic and epigenetic influences. PLoS One 8(12):e82558

Menni C, Gudelj I, MacDonald-Dunlop E, Mangino M, Zierer J, Bešić E et al (2018) Glycosylation profile of immunoglobulin G is cross-sectionally associated with cardiovascular disease risk score and subclinical atherosclerosis in two independent cohorts. Circ Res 122(11):1555–1564

Nimmerjahn F, Anthony RM, Ravetch JV (2007) Agalactosylated IgG antibodies depend on cellular Fc receptors for in vivo activity. PNAS 104(20):8433–8437

Novokmet M, Lukić E, Vučković F, Đurić Ž, Keser T, Rajšl K et al (2014) Changes in IgG and total plasma protein glycomes in acute systemic inflammation. Scientific Rep 4:4347

Panagiotakos DB, Pitsavos C, Yannakoulia M, Chrysohoou C, Stefanadis C (2005) The implication of obesity and central fat on markers of chronic inflammation: the ATTICA study. Atherosclerosis 183(2):308–315

Perkovic MN, Bakovic MP, Kristic J, Novokmet M, Huffman JE, Vitart V et al (2014) The association between galactosylation of immunoglobulin G and body mass index. Prog Neuro-Psychoph 48:20–25

Pincetic A, Bournazos S, DiLillo DJ, Maamary J, Wang TT, Dahan R et al (2014) Type I and type II Fc receptors regulate innate and adaptive immunity. Nat Immunol 15(8):707–716

Plomp R, Ruhaak LR, Uh H-W, Reiding KR, Selman M, Houwing-Duistermaat JJ et al (2017) Subclass-specific IgG glycosylation is associated with markers of inflammation and metabolic health. Sci Rep 7(1):12325

Pučić M, Knežević A, Vidič J, Adamczyk B, Novokmet M, Polašek O et al (2011) High throughput isolation and glycosylation analysis of IgG–variability and heritability of the IgG glycome in three isolated human populations. Mol Cell Proteomics 10(10):M111.010090

Quast I, Lünemann JD (2014) Fc glycan-modulated immunoglobulin G effector functions. J Clin Immunol 34(1):51–55

Quast I, Peschke B, Lünemann JD (2017) Regulation of antibody effector functions through IgG Fc N-glycosylation. Cell Mol Life Sci 74(5):837–847

Rabinovich Gabriel A, Croci DO (2012) Regulatory circuits mediated by lectin-glycan interactions in autoimmunity and cancer. Immunity 36(3):322–335

Ruhaak LR, Uh H-W, Beekman M, Koeleman C, Hokke CH, Westendorp R et al (2010) Decreased levels of bisecting GlcNAc glycoforms of IgG are associated with human longevity. PLoS One 5(9):e12566

Russell A (2020) Quantifying the heterogeneity of the immunoglobulin G N-glycome in an ageing Australian population: the Busselton healthy ageing study: Edith Cowan University

Russell A, Šimurina M, Garcia M, Novokmet M, Wang Y, Rudan I et al (2017) The N-glycosylation of immunoglobulin G as a novel biomarker of Parkinson's disease. Glycobiology 27(5):501–510

Russell A, Adua E, Ugrina I, Laws S, Wang W (2018) Unravelling immunoglobulin G Fc N-glycosylation: a dynamic marker potentiating predictive, preventive and personalised medicine. Int J Mol Sci 19(2):390

Russell AC, Kepka A, Trbojević-Akmačić I, Ugrina I, Song M, Hui J et al (2019a) Why not use the immunoglobulin G N-glycans as predictor variables in disease biomarker-phenotype association studies? A multivariate analysis. OMICS (in press)

Russell AC, Kepka A, Trbojević-Akmačić I, Ugrina I, Song M, Hui J et al (2019b) Increased central adiposity is associated with pro-inflammatory immunoglobulin G N-glycans. Immunobiology 224(1):110–115

Salzberg SL (2018) Open questions: how many genes do we have? BMC Biol 16(1):94

Sartorius N (2006) The meanings of health and its promotion. Croat Med J 47(4):662–664

Schwab I, Nimmerjahn F (2013) Intravenous immunoglobulin therapy: how does IgG modulate the immune system? Nat Rev Immunol 13(3):176–189

Sebastian A, Alzain MA, Asweto CO, Song H, Cui L, Yu X et al (2016) Glycan biomarkers for rheumatoid arthritis and its remission status in Han Chinese patients. OMICS 20(6):343–351

Segman RH, Stein MB (2015) C-reactive protein: a stress diathesis marker at the crossroads of maladaptive behavioral and cardiometabolic sequelae. Am J Psychiatry 172(4):307–309

Shen X, Klarić L, Sharapov S, Mangino M, Ning Z, Wu D et al (2017a) Multivariate discovery and replication of five novel loci associated with immunoglobulin G N-glycosylation. Nat Commun 8(1):447

Shen S, Lu Y, Qi H, Li F, Shen Z, Wu L et al (2017b) Waist-to-height ratio is an effective indicator for comprehensive cardiovascular health. Sci Rep 7

Štambuk J, Nakić N, Vučković F, Pučić-Baković M, Razdorov G, Trbojević-Akmačić I et al (2020) Global variability of the human IgG glycome. Aging 12(15):15222

Subedi GP, Barb AW (2015) The structural role of antibody N-glycosylation in receptor interactions. Structure 23(9):1573–1583

Teesson M, Hall W, Slade T, Mills K, Grove R, Mewton L et al (2010) Prevalence and correlates of DSM-IV alcohol abuse and dependence in Australia: findings from the 2007 National Survey of mental health and wellbeing. Addiction 105(12):2085–2094

Trbojevic-Akmacic I, Ugrina I, Lauc G (2017) Comparative analysis and validation of different steps in glycomics studies. Methods Enzymol 586:37–55

Troelsen LN, Jacobsen S, Abrahams JL, Royle L, Rudd PM, Narvestad E et al (2012) IgG glycosylation changes and MBL2 polymorphisms: associations with markers of systemic inflammation and joint destruction in rheumatoid arthritis. J Rheumatol 39(3):463–469

Vanhooren V, Dewaele S, Libert C, Engelborghs S, De Deyn PP, Toussaint O et al (2010) Serum N-glycan profile shift during human ageing. Exp Gerontol 45(10):738–743

Verdurmen J, Monshouwer K, Dorsselaer SV, Bogt TT, Vollebergh W (2005) Alcohol use and mental health in adolescents: interactions with age and gender-findings from the Dutch 2001 health behaviour in school-aged children survey. J Stud Alcohol 66(5):605–609

Verma R, Kubendran S, Das SK, Jain S, Brahmachari SK (2005) SYNGR1 is associated with schizophrenia and bipolar disorder in southern India. J Hum Genet 50(12):635–640

Vidarsson G, Dekkers G, Rispens T (2014) IgG subclasses and allotypes: from structure to effector functions. Front Immunol 5

Vučković F, Krištić J, Gudelj I, Teruel M, Keser T, Pezer M et al (2015) Association of systemic lupus erythematosus with decreased immunosuppressive potential of the IgG glycome. Arthritis Rheumatol 67(11):2978–2989

Wahl A, van den Akker E, Klaric L, Štambuk J, Benedetti E, Plomp R et al (2018) Genome-wide association study on immunoglobulin G glycosylation patterns. Front Immunol 9(277)

Wang W (2019) Glycomics research in China: the current state of the art. OMICS: J Integr Biol 23 (12):601–602

Wang W, Yan Y (2012) Suboptimal health: a new health dimension for translational medicine. Clin Transl Med 1(1):28

Wang J, Balog CI, Stavenhagen K, Koeleman CA, Scherer HU, Selman MH et al (2011) Fc-glycosylation of IgG1 is modulated by B-cell stimuli. Mol Cell Prot 10(5):M110. 004655

Wang W, Russell A, Yan Y (2014) Traditional Chinese medicine and new concepts of predictive, preventive and personalized medicine in diagnosis and treatment of suboptimal health. EPMA J 5(1):4

Wang Y, Klaric L, Yu X, Thaqi K, Dong J, Novokmet M et al (2016) The association between glycosylation of immunoglobulin G and hypertension: a multiple ethnic cross-sectional study. Medicine 95(17):e3379

Whitmer R, Gustafson D, Barrett-Connor E, Haan M, Gunderson E, Yaffe K (2008) Central obesity and increased risk of dementia more than three decades later. Neurology 71(14):1057–1064

Wu Z, Li H, Liu D, Tao L, Zhang J, Liang B et al (2020) IgG glycosylation profile and the glycan score are associated with type 2 diabetes in independent Chinese populations: a case-control study

Wuhrer M, Selman MH, McDonnell LA, Kümpfel T, Derfuss T, Khademi M et al (2015) Pro-inflammatory pattern of IgG1 Fc glycosylation in multiple sclerosis cerebrospinal fluid. J Neuroinflamm 12(1):235

Yan Y-X, Liu Y-Q, Li M, Hu P-F, Guo A-M, Yang X-H et al (2009) Development and evaluation of a questionnaire for measuring suboptimal health status in urban Chinese. J Epidemiol 19 (6):333–341

Yan YX, Dong J, Liu YQ, Yang XH, Li M, Shia G et al (2012) Association of suboptimal health status and cardiovascular risk factors in Urban Chinese Workers. J Urban Health 89(2):329–338

Yan Y-X, Dong J, Liu Y-Q, Zhang J, Song M-S, He Y et al (2015) Association of suboptimal health status with psychosocial stress, plasma cortisol and mRNA expression of glucocorticoid receptor α/β in lymphocyte. Stress 18(1):29–34

Yu X, Wang Y, Kristic J, Dong J, Chu X, Ge S et al (2016) Profiling IgG N-glycans as potential biomarker of chronological and biological ages: a community-based study in a Han Chinese population. Medicine 95(28)

Chapter 18
Glycosylation of Antigen-Specific Antibodies: Perspectives on Immunoglobulin G Glycosylation in Vaccination and Immunotherapy

Pranay Bharadwaj and Margaret E. Ackerman

Contents

18.1	Introduction	567
18.2	Antibody Effector Functions	567
18.3	Immunomodulatory Antibody Activities	570
18.4	Induction and Regulation of Antigen-Specific Antibodies	571
18.5	Importance of Ab Glycosylation	571
18.6	Typical Serum IgG Fc Glycan Composition	572
18.7	Variations in Ab Glycoprofiles	573
18.8	ASA Glycosylation in Infectious Disease	573
18.9	ASA Glycosylation in Allo/Autoimmunity	575
18.10	In vivo Fc Glycan Programming	578
18.11	Summary and Future Perspectives	579
References		579

Abstract Exciting developments have been made in understanding antibody-mediated immunity, deepening understanding of antibody effector functions increasingly recognized as critical mechanisms of action beyond antigen recognition, and significantly broadening the evidence base for the importance of these effector mechanisms across diverse infectious and autoimmune diseases. Because these activities critically depend on the specific glycoforms present on a conserved site of the IgG Fc domain, relationships between the Fc glycosylation profiles of antigen-specific antibody pools and outcomes in infectious and autoimmune disease have

P. Bharadwaj
Department of Microbiology and Immunology, Geisel School of Medicine, Dartmouth College, Hanover, NH, USA

M. E. Ackerman (✉)
Department of Microbiology and Immunology, Geisel School of Medicine, Dartmouth College, Hanover, NH, USA

Thayer School of Engineering, Dartmouth College, Hanover, NH, USA
e-mail: Margaret.E.Ackerman@Dartmouth.edu

begun to be defined, pointing to the key role of this posttranslational modification as a biomarker and mechanistic modifier of antibody-mediated immunity. Here we summarize studies evaluating the profiles and activities of antigen-specific antibodies elicited by infection and vaccination as well as within the context of allo- and autoimmunity, and consider current approaches to rational modification of Fc glycans in vivo.

Keywords Immunoglobulin · Antibody · Fc domain · Glycosylation · Vaccine · Allergy · Autoimmunity · IgG · Effector function

Abbreviations

ACPA	Anti-citrullinated protein antibodies
ADCC	Antibody-dependent cellular cytotoxicity
ADCP	Antibody-dependent cellular phagocytosis
ADE	Antibody-dependent enhancement
AMI	Antibody-mediated immunity
ASA	Antigen-specific antibody
CDC	Complement-dependent cytotoxicity
COVID-19	Coronavirus disease 2019
CSR	Class switch recombination
DC	Dendritic cell
DENV	Dengue virus
EndoS	Endoglycosidase S
Fab	Fragment antigen-binding
Fc	Fragment crystallizable
FcR	Fc receptor
FNAIT	Fetal or neonatal allo-immune thrombocytopenia
Fv	Fragment variable
GlcNAc	N-acetylglucosamine
HPA-1a	Human platelet antigen 1a
IdeS	IgG digesting enzyme S
IgG	Immunoglobulin G
IVIg	Intravenous immunoglobulin
K	Kell
mAb	Monoclonal antibody
MAC	Membrane attack complex
MBL	Mannose-binding lectin
MHC	Major histocompatibility complex
NK	Natural killer
HIV	Human immunodeficiency virus
RA	Rheumatoid arthritis
RhD	Rhesus D

RSV   Respiratory syncytial virus
SARS-CoV-2   Severe acute respiratory syndrome coronavirus 2

18.1 Introduction

Evidence of the critical importance of effector functions to antibody-mediated immunity (AMI) has been accumulating across studies of diverse infectious and autoimmune diseases. The ability to study and clinically leverage AMI was made a much simpler task after Kohler and Milstein's (Köhler and Milstein 1975) discovery of an approach to generate consistent, reliable, and reproducible monoclonal antibody (mAb) preparations, which represented challenges to the use of polyclonal sera prevalent at that time. Advanced molecular methods in antibody cloning (Wang et al. 2019; Hunter et al. 2019; Kim et al. 2014; Winzeler and Wang 2013; Chon and Zarbis-Papastoitsis 2011) and engineering (Bruggeman et al. 2018; Dekkers et al. 2018; Crooks et al. 2018; Dekkers et al. 2016) have complemented the discovery of diverse Fc receptors (FcR) (Bournazos and Ravetch 2017; Castro-Dopico and Clatworthy 2016; Wu et al. 2014; Hirvinen et al. 2013; Nimmerjahn and Ravetch 2008a; Lazar et al. 2006; Hogarth 2002) and development of elegant knockout mouse models (Walsh et al. 2016; Verkoczy 2017; Stackowicz et al. 2020) to enable further basic science exploration and to support therapeutic optimization of AMI. In parallel, higher resolution means of profiling serum antibodies have accompanied these advances and greatly expanded the ability to interrogate polyclonal responses in serum and tissue. The high throughput of many of these profiling approaches has now turned the heterogeneity observed among polyclonal samples into a strength, providing means to interrogate the features and activities of antibodies that are associated with AMI.

18.2 Antibody Effector Functions

Antibodies play an important role in both effecting and regulating an immune response. They have the capacity to either amplify or dampen an inflammatory immune response based on their specificity, affinity, titer, isotype, and glycosylation profile. While the antigen-binding fragment (Fab) domain confers antigen specificity, the crystallizable fragment (Fc) domain is responsible for linking antigen recognition to downstream effector functions (Schroeder and Cavacini 2010). Antibodies can neutralize pathogens by directly binding through the Fab domain and occluding the binding of the pathogen or its toxins to cognate receptors. Such Fab-mediated antibody action is complemented by Fc domain engagement of complement proteins and FcR (Fig. 18.1). There are broadly two categories of FcRs—activating and inhibitory—that are ubiquitously expressed on human hematopoietic

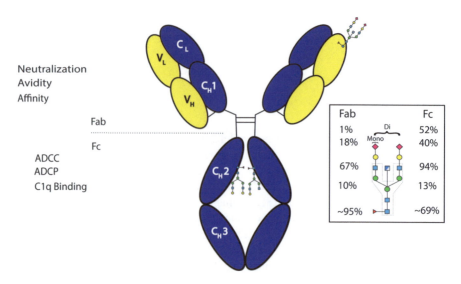

Fig. 18.1 Antibody glycosylation sites in IgG. Glycosylation sites can be present in the Fab region (15–25% of IgGs), but are always present in the Fc region. There are distinct qualitative differences among the glycans commonly found at these sites, however, the importance of the glycans in mediating function via respective antibody domains is a shared characteristic. These glycans share the core heptasaccharide (dotted line in the inset) to which extensions of specific sugars are attached. Fc glycans tend to be heavily fucosylated whereas the Fab glycans have generally been observed to exhibit relatively high levels of sialylation. Adapted from Bondt et al. (2014)

cells and play a key role in orchestrating potent antibody-mediated effector immune responses in the context of both protective immunity to pathogens and pathogenic immune responses to self (Fig. 18.2).

The most widely studied Fc-mediated antibody effector functions are antibody-dependent cell-mediated cytotoxicity (ADCC) (Worley et al. 2018), antibody-dependent cellular phagocytosis (ADCP) (Gerber and Mosser 2001), and complement-dependent cytotoxicity (CDC) (Goldberg and Ackerman 2020). These activities are induced by engagement of the FcγRs on innate effector cells, or by soluble complement cascade initiators, such as C1q or Mannose Binding Lectin (MBL), by the Fc domain of antibodies that are bound to a target antigen. ADCC is characterized by FcγR engagement that causes the release of cytotoxic granules that contain perforin and granzyme, resulting in the killing of target cells (Smyth et al. 2005). FcγRIIIA-expressing Natural Killer (NK) cells are widely considered to be an important contributor to ADCC and are often assayed in vitro. However, in vivo, neutrophils, monocytes, and macrophages are also capable of driving ADCC and have been found to make important contributions to antibody mechanism of action (Smyth et al. 2005; van Erp et al. 2019).

ADCP or opsonophagocytosis is the uptake of immune complexes or antibody-coated antigens by phagocytic cells including monocytes, macrophages, dendritic cells (DCs), and others that express FcγRI, FcγRII, and/or FcαRI, each of which can

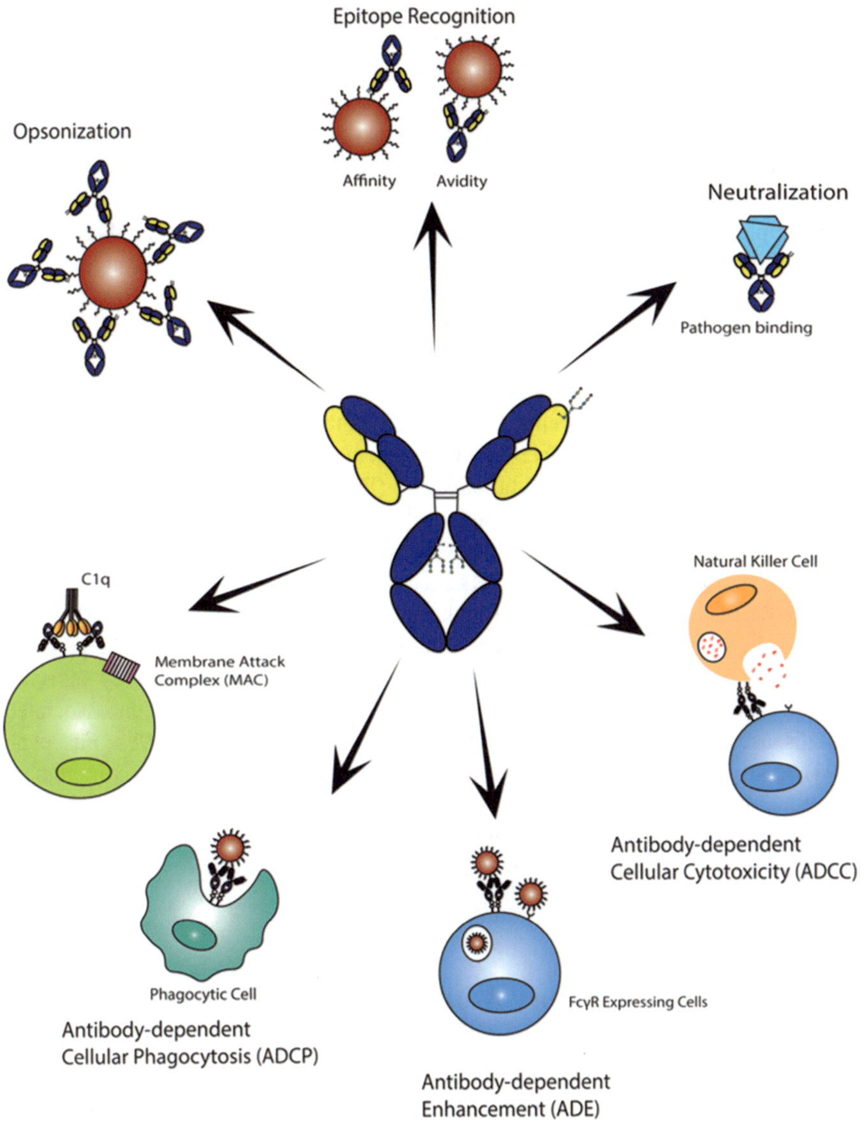

Fig. 18.2 Antibody mechanisms of action. While the Fab domain functions are directly driven by antigen recognition, Fc-mediated functions result from recruitment of various components of the innate immune system

mediate immune complex uptake (Li and Kimberly 2014). ADCP mediates clearance of immune complexes by trafficking them to the lysosomes for degradation and antigen processing for presentation on Major Histocompatibility Complex (MHC) molecules on the cell surface (Mantegazza et al. 2013). Previous work on influenza virus has shown that ADCP contributes to protection from infection in mice (Huber et al. 2001; He et al. 2017) and potentially plays a role in recovery from severe infections in humans (Vanderven et al. 2017; Ana-Sosa-Batiz et al. 2016). Associations between ADCP and improved outcomes in other disease settings such as HIV (Barouch et al. 2013, 2015), West Nile Virus (Vogt et al. 2011) in humans, and Respiratory Syncytial Virus (RSV) (Bukreyev et al. 2012) have also been recently established.

Besides ADCC and ADCP, antibodies can also induce complement activation. The complement cascade contributes to pathogen elimination either directly, by means of complement-dependent cytotoxicity (CDC), or indirectly, through phagocytic clearance of complement-coated targets and the induction of an inflammatory response (Goldberg and Ackerman 2020; van Erp et al. 2019; Grafals and Thurman 2019; Casadevall and Pirofski 2012). The complement cascade consists of a large number of distinct plasma proteins that react with one another to opsonize pathogens, inducing a series of inflammatory responses that help to fight infection (Noris and Remuzzi 2013). A number of complement proteins are proteases that are themselves activated by proteolytic cleavage (Dunkelberger and Song 2010). The terminal complement components assemble into the membrane attack complex (MAC), resulting in lysis of the pathogen-infected cell. Complement has been shown to have both protective and pathogenic effects in various disease conditions. In HIV (Barouch et al. 2013, 2015; Pittala et al. 2019), influenza (Co et al. 2014; Wu et al. 2015), and vaccinia (Benhnia et al. 2009) infection, antibody-mediated CDC has been shown to correlate or mechanistically contribute to antibody antiviral activity. Alternatively, complement-mediated activation has also been associated with disease severity (Nascimento et al. 2009; Churdboonchart et al. 1983; Füst et al. 1994). Lastly, the binding of complement-coated immune complexes to complement receptor 2 on B cells is reported to lower the B cell activation threshold, thereby promoting long-lived adaptive immunity and higher antibody levels (van Erp et al. 2019; Hebell et al. 1991; Gonzalez et al. 2010).

18.3 Immunomodulatory Antibody Activities

In contrast, anti-inflammatory effects of antibodies can help in alleviating severe immune damage. Based on this concept, administering intravenous immunoglobulin (IVIg) to treat inflammatory conditions such as autoimmune disease has found an important clinical application (Bayry 2016). While the underpinnings of IVIg mechanisms have yet to be clearly elucidated (Schwab and Nimmerjahn 2013), different mechanisms of action such as neonatal Fc receptor blockade resulting in accelerated clearance of autoantibodies (Li and Kimberly 2014), direct interaction with the inhibitory FcγRIIb (Nagelkerke and Kuijpers 2015), or occlusion of activating receptors and tempering the inflammatory effector responses (Nimmerjahn and Ravetch 2008b) have been proposed. However, since IVIg treatment is used to treat various diseases, it is likely that the mode of action differs per clinical setting.

18.4 Induction and Regulation of Antigen-Specific Antibodies

During an immune response, B cells are stimulated to mature and to undergo class switch recombination (CSR) resulting in genetic modification of the IgH locus and selection of the antibody isotype and subclass to be secreted (Stavnezer and Schrader 2014). Just as B cells undergo rounds of somatic hypermutation over the course of affinity maturation as they migrate in and out of regions in the germinal center, CSR can occur in rounds with repeated switching to downstream types (Mesin et al. 2016). This heterogeneity in the amino acid sequence of both variable fragment (Fv) and Fc regions is coupled to further functional diversification via incorporation of one of the >30 possible glycoforms (Jennewein and Alter 2017) in the conserved N-linked glycosylation motif. While multiple isotypes are glycosylated in the Fc, we will focus on glycosylation in the context of the four IgG subclasses (Vidarsson et al. 2014).

In the past 20 years, the role of antibody glycosylation as an important parameter modulating the potency of effector functions has been firmly established through advances in monoclonal antibody research and development, as well as in studies of natural immune responses in the context of infectious and autoimmune disease. Here we focus on recent research considering the glycosylation of antigen-specific antibodies in these settings.

18.5 Importance of Ab Glycosylation

The Fc domain contains a consensus N-linked glycosylation site that is typically occupied by a heptasaccharide core structure consisting of four N-acetylglucosamine (GlcNAc) and three mannose moieties that form a biantennary complex (Liu 2015). Additional glycosylation features such as fucose, galactose, sialic acid, and GlcNAc can be added later to the core structure to produce over 30 distinct glycovariants. As both heavy chains are glycosylated, a single IgG molecule can have a diverse array of glycosylation heterogeneity (Jefferis 2009). Nuclear magnetic resonance (NMR) studies have shown that variability in the glycans at this conserved position has a profound effect on the hinge region conformation (Yamaguchi et al. 2006). Similarly, interactions with Fcγ and other IgG and glycan receptors are entirely dependent on or modified by glycan composition and conformation, thus the type of glycan occupying this site modifies antibody effector function (Saunders and Conceptual 2019). Unlike genetically templated factors that impact IgG activity, such as Fv sequence and Fc subclass, antibody glycosylation is remarkably varied, resulting in a high level of microheterogeneity that facilitates the fine tuning of antibody function (Alter et al. 2018a). These dynamic changes in antibody glycosylation can have a subtle or profound effect in their interactions or downstream functions.

18.6 Typical Serum IgG Fc Glycan Composition

Given the importance of IgG Fc glycans, the composition of serum antibodies has been evaluated in a number of populations, providing insight into changes associated with age, sex, hormone levels, and disease status. Nonetheless, "typical" compositions have been articulated among healthy individuals (Fig. 18.1), and deviations from this profile suggest active processes regulating this posttranslational modification at multiple levels.

Serum IgG Fc is typically overwhelmingly fucosylated (>90%) (Gudelj et al. 2018). However, skewed glycosylation variants, produced by chemoenzymatic modifications or expressed in engineered cells, have been produced that lack this fucose moiety, and as a result exhibit significantly improved effector function. For example, an afucosylated form of an anti-CD20 IgG1 showed a 50-fold improvement in binding to FcγRIIIa and enhanced ADCC activity (Shields et al. 2002). Later, structural studies found that the fucose on the Fc glycan clashes with a GlcNAc$_2$ group of an FcγRIIIa glycan, thereby providing a structural rationale to the improved ADCC activity of afucosylated antibody (Ferrara et al. 2011).

About 10% of all circulating IgGs in healthy human adults exhibit bisected Fc glycans (Gudelj et al. 2018), which have been shown previously to relate to ADCC activity (Hodoniczky et al. 2005). However, this amplification in ADCC, caused by the increased engagement of the FcγRIII, is believed predominantly to be due to the indirect role of bisection in decreasing fucosylation, rather than a direct consequence of its presence in the antibody structure (Shinkawa et al. 2002).

Similarly, agalactosylated, monogalactosylated, and digalactosylated glycan structures account for approximately 35%, 35%, and 15% of circulating IgG Fc-glycans, respectively (Gudelj et al. 2018). A prominent bias towards agalactosylated antibodies has been observed in people with active autoimmune and inflammatory diseases (Parekh et al. 1989; Tomana et al. 1992; Rademacher et al. 1994; Decker et al. 2016), however, a clear consensus on cause or consequence is yet to be achieved (Alter et al. 2018a). Furthermore, there are conflicting reports on the role of galactosylated antibodies in mediating proinflammatory activities, with some reports observing the presence of galactosylation on the IgGs to enhance the ADCC and complement binding (C1q) in vitro (Nimmerjahn et al. 2007; Peschke et al. 2017; Thomann et al. 2015; Tsuchiya et al. 1989), while others have noted a dampening of an inflammatory response by highly galactosylated immune complexes (Karsten et al. 2012). A lack of correlation between the presence or absence of galactosylation on IgGs and corresponding in vivo activity has also been reported (Nimmerjahn et al. 2007), suggesting that the consequences of variable galactosylation may be best investigated per disease model and per antibody.

Lastly, approximately 10% of circulating IgG Fc is sialylated (Gudelj et al. 2018). Sialylated IgG Fc is associated with an anti-inflammatory profile of antibodies in mouse models, in which neuraminidase-treated, asialylated pooled human IgG (IVIg) has been observed to abrogate the normally anti-inflammatory activity of IVIg (Kaneko et al. 2006). However, this mechanism of action remains controversial in humans. Discrepant observations have been made as to the ability of IgG to interact with the candidate receptor proposed on the basis of mouse studies (Anthony et al. 2008; Temming et al. 2019), and sialylated IgG has shown slightly elevated binding to activating FcγR and C1q, and associated effector functions (Dekkers et al. 2017; Subedi and Barb 2016), which would suggest a greater inflammatory capacity.

18.7 Variations in Ab Glycoprofiles

Deviations from these "typical" profiles have been associated with diverse physiological and immunological states. For example, changes in total serum IgG Fc glycosylation are observed in early life (Cheng et al. 2019), in adolescence (Gudelj et al. 2018; de Haan et al. 2016), and in association with hormonal status (Ercan et al. 2017), as well as more gradual changes during immune senescence (Krištić et al. 2013), across a broad range of glycoforms and constituent sugar moieties. In the context of ongoing inflammation, such as observed in chronic infection (Moore et al. 2005) or autoimmunity (Parekh et al. 1985), global IgG Fc glycosylation is often modified, showing reduced galactose and sialic acid content (Lastra et al. 2009).

Beyond approaches to evaluate these global changes, the role of Fc glycans in antibody function has also motivated the development of robust methods to define the glycosylation profiles of antigen-specific antibodies (ASA) purified from serum. Early questions about ASA fractions related to whether they are typically composed of IgG Fc glycovariants with similar prevalence to those observed for total serum IgG, and if not, whether glycoprofiles vary by pathogen, antigen, and epitope specificity.

18.8 ASA Glycosylation in Infectious Disease

In the context of responses to the HIV envelope protein among chronically infected individuals, HIV envelope glycoprotein-specific antibodies were found to exhibit reduced galactosylation, fucosylation, and sialylation (Ackerman et al. 2013), even when compared to global serum IgG Fc glycan profiles that were shifted in these same directions as compared to uninfected and acutely infected individuals (Moore et al. 2005). Among ASA, galactosylation levels correlated with Ab-dependent inhibition of viral infection and replication and were consistent with glycosyltransferase and glycosidase expression in peripheral B cells (Ackerman et al. 2013). Perhaps surprisingly, these global and HIV-specific plasma IgG Fc glycan changes were not resolved by either antiretroviral drug therapy or in the context of spontaneous virus control. Subsequent studies have shown the contribution of HIV-specific IgG glycans to predicting HIV-specific antibody effector functions (Alter et al. 2018b) and vaccine efficacy (Vaccari et al. 2016; Ackerman et al. 2018).

These and other early studies have firmly established that ASA can differ from total serum IgG in their glycosylation states. As methods for analysis of ASA have advanced, analysis of ASA targeting different proteins has become increasingly feasible but not yet common. To the extent studies have addressed multiple target antigens, there has been some evidence for consistent glycoforms across distinct specificities and other cases in which different antigen-specificities, or even different epitope-specificities within the same protein have shown distinct profiles. For example, in tuberculosis, distinct ASA IgG Fc glycan profiles for two different antigen types were reported to show similar glycan profiles to each other, but with striking decreases in fucose and increases in galactose, sialic acid, and bisecting GlcNac as compared to total serum IgG Fc (Lu et al. 2020). In contrast, Wang et al. reported that the abundance of sialylation and fucosylation among influenza hemagglutinin-specific (HA) IgG differed depending on specificity of the Fab domain. Antibodies to the HA globular head were significantly more sialylated and fucosylated than those directed against the HA stem domain (Wang and Ravetch 2019), though it may be important to keep in mind that the globular head functions as a sialic acid-binding protein.

One of the most interesting examples of the effect of ASA Fc glycosylation comes from the setting of flavivirus infection. This family of viruses has been associated with a phenomenon called Antibody-Dependent Enhancement (ADE), in which virus-specific antibodies increase infection of FcγR-bearing target cells. Among these, dengue is a mosquito-borne pathogen caused by four distinct but closely related dengue virus (DENV) types. Recovery from infection is believed to typically provide immunity against infection from the same type. However, cross-type immunity is partial and temporary. Subsequent (secondary) infection by another serotype is associated with an increased risk of developing severe dengue via ADE (Katzelnick et al. 2017; Guzman et al. 2013). While prior work has shown that waning antibody titer is associated with severe disease upon secondary exposure (Katzelnick et al. 2017), recent work has highlighted the potential importance and clinical impact of the glycosylation of dengue-specific antibodies. As perhaps the most elegant setting in which to evaluate ADE, severe disease of neonates is associated with the level of passively transferred maternal dengue-specific antibody that is afucosylated, resulting in dengue hemorrhagic fever or dengue shock syndrome (Wang et al. 2017; Thulin et al. 2020; Khandia et al. 2018). This potent ADE response is thought to manifest via non-neutralizing, dengue-specific antibodies that exhibit increased affinity to the activating FcγRIIIA receptor.

In the context of coronavirus disease 2019 (COVID-19), Fc glycans of IgG antibodies to Severe Acute Respiratory Syndrome Coronavirus 2 (SARS-CoV-2) envelope spike and nucleocapsid proteins differ from those of total serum IgG (Larsen et al. 2020). In this study, these profiles were observed to differ between spike and capsid, and multiple studies have observed that spike-specific IgG Fc afucosylation is correlated to disease severity (Larsen et al. 2020; Chakraborty et al. 2020), with some evidence that they may contribute to pathology via inducing inflammatory responses from macrophages (Hoepel et al. 2020). Global serum IgG glycans have also been reported to diverge according to COVID-19 severity, with decreased bisecting GlcNAc observed in multiple cohorts (Petrovic et al. 2020). The role of IgG Fc glycosylation of ASA remains to be defined in many more infectious disease settings. Like in COVID-19, HIV, and other settings in which ASAs have been profiled, intriguing observations regarding differences in global IgG Fc glycosylation abound—such as in meningococcal sepsis (Haan 2018), visceral leishmaniasis (Haan 2018; Gardinassi et al. 2014), and tuberculosis (Lu et al. 2016, 2020)—and have been found to relate to disease status or outcomes.

18.9 ASA Glycosylation in Allo/Autoimmunity

Rheumatoid arthritis (RA) is a common systemic inflammatory autoimmune disease in which joint synovium is affected by a dysregulated immune system. RA is typically associated with serological evidence of systemic autoimmunity as indicated by the presence of autoantibodies in serum and synovial fluid (Coutant 2019; Song and Kang 2010). Instead of being characterized by specific reactivity to a particular autoantigen, RA is associated with antibodies reactive against a wide spectrum of autoantigens, which can make the etiology of disease progression in RA patients very different. Among various autoantigens targeted in RA, anti-citrullinated protein antibodies (ACPA) have been identified as a useful marker in diagnosis (Coutant 2019) and predicting whether undifferentiated arthritis will progress to RA (Forslind et al. 2004). ACPA are associated with an increased risk of developing bone erosions (Rönnelid et al. 2005; Rycke et al. 2004), suggesting their potential to contribute to joint pathology. Like total serum IgG, long known to show decreased galactosylation, ACPA are observed to exhibit further reduction in sialylation and galactosylation (Scherer et al. 2010; Ohmi et al. 2016), though there is some evidence that the IgG subclasses may differ from each other in this regard (Lundström et al. 2014). Reinforcing the controversy regarding the potentially conflicting roles of sialylated IgG in different species, but supporting the role of glycoengineered ASA as therapeutic interventions, sialylated ACPA have been shown to reduce arthritis pathology in a mouse model (Ohmi et al. 2016).

Whether ACPA are a cause or consequence of RA status remains controversial, but they have been reported to activate effector cells via FcγR (Clavel et al. 2008), whose allotypic and copy number variation have sometimes but not always been observed to associate with RA status and severity (Thabet et al. 2009; Kastbom and Ahmadi 2005; Nieto et al. 2000; Radstake et al. 2003). Further, several longitudinal studies have observed that galactosylation and sialylation levels of ACPAs decreased shortly before symptom onset in patients who had ACPA but no evidence of RA at baseline (Pfeifle et al. 2017; Harre et al. 2015; Rombouts et al. 2015), suggesting the potential value of measuring the level of ACPA galactosylation/sialylation as a biomarker to predict the risk of progression from pre-clinical disease to chronic inflammatory disease. Beyond differences between ACPA and total IgG Fc glycosylation, differences in ACPA Fc glycan profiles have also been noted between individuals with and without rheumatoid factor, and between serum and synovial fluid (Scherer et al. 2010). While some have interpreted these differences to potentially relate to active alteration of Fc glycans in affected joints, the lack of differences in total serum and synovial fluid IgG1 agalactosylation suggests that alternative mechanisms may be at play. To this end, ACPA-secreting plasma cells have been reported to exist in synovial fluid (Rodríguez-Bayona et al. 2007), suggesting the possibility that differences in systemic versus synovial ACPA Fc glycosylation may be driven by differences associated with plasma cells in the synovium and elsewhere.

Beyond these alterations in Fc glycosylation, ACPA have more recently been reported to exhibit striking glycosylation of their Fab domains. Unlike total IgG, a majority of ACPA variable domains are glycosylated (Lloyd et al. 2018; Hafkenscheid et al. 2019; Hafkenscheid et al. 2017). Unlike their Fc domains, these APCA Fab glycans are overwhelmingly sialylated (Hafkenscheid et al. 2017). Variable domain glycosylation has also been reported to modify antigen binding among ACPA (Rombouts et al. 2016), suggesting the potential for antibody glycosylation in both variable and crystallizable domains to contribute to RA pathogenesis.

Functional consequences of variations in the profile of ASA have also been reported in fetal or neonatal allo-immune thrombocytopenia (FNAIT). In this disease condition, fetal allo-antigens induce production of maternal antibodies that are then transported across the placenta and drive lysis of fetal cells. While allotypic variation of a variety of maternal fetal antigens is possible, the best studied is that of rhesus D (RhD) antigen incompatibility. Curiously, this incompatibility, which resulted in hemolytic disease in 1% of babies born through the 1940s, 40% of which would die as a result (Bowman 2003), is treated by administration of IVIG from RhD-sensitized donors. While like IVIG used in other indications, the precise mechanisms of this intervention remain unclear; prevention of sensitization, immunomodulatory effects, and accelerated clearance of endogenous maternal IgG have all been proposed as candidate mediators. To this end, the RhD-specific antibodies in at least one commercial product show increased galactosylation and sialylation relative to the entire mixture of antibodies in that product (Winkler et al. 2013), suggesting their potential immunosuppressive character. Evaluations of the mechanism of action have been hampered by the difficulty in recapitulating protective effects of polyclonal RhD IgG with monoclonal antibodies. The difference in the effect of polyclonal versus monoclonal antibody infusions may relate to differential glycosylation of RhD-specific fraction or entire pool, differences in affinity and avidity, altered red blood cell clearance capacity, or other factors, but have led to observations of alternatively enhanced or inhibited maternal sensitization, leaving many unanswered questions (Kumpel 2007; Kumpel et al. 1995). To this end, it has been recently reported that RhD-specific monoclonal antibodies varied in their ability to clear RhD+ target cells and prevent alloimmunization, dependent on their fucosylation status and associated ADCC activity (Kumpel et al. 2020). Similarly, in the context of seropositive mothers, IgG Fc fucosylation of RhD-specific antibodies have been found to correlate with ADCC activity and low fetal neonatal hemoglobin levels (Kapur et al. 2014a).

Despite questions as to mechanism, RhD+ serum IgG has all but eliminated pregnancy loss and neonatal death from RhD incompatibility in much of the world. In contrast, other less frequently observed incompatibilities have no effective preventative interventions. For a number of these antigens, maternal antibody titer is a poor indicator of pathology, and in some of these settings, variation in ASA-Fc glycosylation has been investigated for its predictive value. Here, more mixed results as to the importance of glycosylation profiles of fetal antigen-specific antibodies have been observed. As compared to RhD-specific antibodies, those recognizing red blood cell antigens K, c, and E were less distinct from total plasma IgG Fc glycans than those recognizing RhD, but nonetheless, afucosylation of Kell (K)-specific antibodies and high galactosylation and sialylation of anti-c antibodies were correlated with severe anemia of the fetus (Sonneveld et al. 2016a). In a small follow up study of maternal K-specific antibodies, IgG1 and IgG3 fractions were shown to exhibit similar glycoform prevalences, and while the previously observed relationship between afucosylation and disease severity did not meet an arbitrary significance threshold of $p = 0.05$, galactose content was shown to correlate with disease severity (Sonneveld et al. 2018).

Beyond red blood cell alloantigens, Fc glycoforms of human platelet antigen 1a (HPA-1a)-specific antibodies have been analyzed. Like other maternal alloantibody responses, HPA-1a-specific antibodies show markedly decreased levels of fucosylation as compared to total serum IgG1 (Kapur et al. 2014b). These significantly less fucosylated anti-HPA-1a antibodies showed enhanced phagocytosis of platelets on account of higher binding affinity to FcγRIIIa and FcγRIIIb, but not to FcγRIIa, compared with antibodies with a high amount of Fc fucose. Most critically, the extent of HPA-1a-specific antibody Fc fucosylation was shown to correlate with clinical disease severity. In a follow-up study, stability of ASA Fc glycans was defined and correlations between bleeding severity and fucose, galactose, and antibody titer were observed (Sonneveld et al. 2016b). Similarly, Jo1 anti-histidyl tRNA synthetase autoantibodies, which are observed in idiopathic inflammatory myopathy and anti-synthetase syndrome, have demonstrated similar reductions in galactose, sialic acid, and fucose, with glycoprofiles relating to disease status (Fernandes-Cerqueira et al. 2018).

Collectively, auto- and alloimmune responses have supported the importance of Fc glycans of ASA to diverse antigens. These observations have motivated investigation of deglycosylated IgGs to prevent FNAIT (Bakchoul et al. 2013), and sialylated ACPA to treat RA (Ohmi et al. 2016). While similar evaluation of alloantibodies in the setting of organ transplant has proven challenging, the role of effector functions is well established, with assessment of complement deposition associated with transplant- or donor-specific antibodies (DSA) forming part of the basis for evaluation of suitability of transplant (Zeevi et al. 2013; Mohan et al. 2012; Stegall et al. 2011; Lefaucheur et al. 2010), and enzymatic Fc restriction of serum IgG showing potential in reducing transplant loss associated with DSA positive organ recipients (Jordan et al. 2017).

18.10 In vivo Fc Glycan Programming

The importance of IgG Fc glycans to Ab biology in vivo has motivated a number of interventions that take advantage of this dependence. Beyond glycoengineering of therapeutic antibodies to optimize their activity, sophisticated new approaches are being explored to control antibody activity. These include leveraging B cell-independent sialylation (Jones et al. 2016) by administration of exogenous galactosyl and sialyltransferase in order to accomplish in vivo sialylation and thereby ameliorate autoimmune disease (Pagan et al. 2018). Similarly, changes in sialyltransferase expression induced by estrogen therapy suggest alternatives to exogenous enzyme therapy (Engdahl et al. 2018).

As opposed to extending IgG Fc glycans, glycan restriction is also being employed toward the same goal of reducing autoimmunity. Glycosidase therapy, most notably EndoS from *S. pyogenes*, the same organism that expresses the IgG protease IdeS used to disarm HLA alloantibodies in kidney transplant, has been investigated in diverse autoimmune conditions in animal models. These settings include IgG-driven thrombocytopenia purpura (Collin et al. 2008), collagen autoimmunity (Hirose et al. 2012), anti-neutrophil cytoplasmic autoantibody-mediated glomerulonephritis (van Timmeren et al. 2010), and autoimmune hemolysis (Allhorn et al. 2010). Challenges to clinical translation remain, including the consequences of globally eliminating effector function non-specifically, as well as the induction of anti-enzyme antibodies, but recent translation of the Fc protease IdeS suggests that these barriers may be surmountable (Collin and Bjorck 2017).

Other possibilities, such as the ability to vaccinate to drive specific inflammatory or anti-inflammatory antibody responses, also exist. A future in which allergen therapy leverages B cell transcriptional programs to not only undergo CSR toward less inflammatory IgG4 molecules but also toward anti-inflammatory glycans comes to mind, as has been shown to lessen allergic reactions in a mouse model using a recombinant glycoengineered antibody (Epp et al. 2017). To this end, Vestrheim et al. considered four distinct bacterial and viral vaccines and observed that the IgG subclass that dominated the response exhibited a temporal increase in galactosylation and sialylation for most vaccinees (Vestrheim et al. 2014). Other studies have observed this effect only within the ASA fraction (Selman et al. 2012).

With a more nuanced perspective, Larsen et al. compared and contrasted ASA targeting enveloped and non-enveloped viral pathogens and found decreased fucose content that is consistent with responses to infection by enveloped viruses, though to varying extents (Larsen et al. 2020). Natural infection, at least in the case of Hepatitis B Virus, was found to better induce afucosylated IgG1 as compared to immunization with a protein subunit vaccine. In contrast, attenuated Mumps virus vaccination induced a similar level of IgG1 afucosylation as natural infection. A study considering HIV-specific IgG Fc glycans observed that vaccination was able to overcome the normally observed variations in total serum IgG associated with geography (Mahan et al. 2016). ASA showed similar glycosylation patterns for a given vaccine, but distinct vaccine regimens resulted in distinct ASA glycosylation profiles.

These and complementary observations related to difference in induction of the IgG subclasses mediated by distinct antigen, pathogen, or vaccine stimuli suggest the existence of "rules" regulating the CSR and glycosylation processes in B cells. While refined insight into these pathways continues to develop, using an in vitro B-cell culture system resembling the in vivo T-cell-dependent antibody production, Wang et al. showed that B-cells secreted variably glycosylated IgG1 when stimulated with TLR ligands, metabolites, and cytokines (Wang et al. 2011). Indeed, because the antibody Fc domain itself can regulate responses by antigen-presenting cells and B-cells, manipulation of Fc glycans in the context of immune complex vaccines has been used to intentionally influence subsequent Ab induction/maturation (Lofano et al. 2018).

18.11 Summary and Future Perspectives

Distinctly different global IgG and ASA Fc profiles have been observed in both infectious disease and auto- and alloimmune settings. Studying the Fc glycosylation profile of ASA presents an excellent opportunity to understand the mechanistic underpinnings and the in vivo regulation of the diverse adaptive immune processes that define protective and pathological humoral responses. To this end, many unanswered questions remain.

Acknowledgments The authors thank Chanc VanWinkle Orzell for copyediting the manuscript.

Compliance with Ethical Standards

Funding This study was funded in part by the National Institutes of Health NIAID and NIGMS (grant number R01AI131975).

Conflict of Interest PB declares he has no conflict of interest. MEA declares that she has no conflict of interest.

Ethical Approval This chapter does not contain any studies with human participants or animals performed by any of the authors.

References

Ackerman ME et al (2013) Natural variation in Fc glycosylation of HIV-specific antibodies impacts antiviral activity. J Clin Invest 123:2183–2192. https://doi.org/10.1172/JCI65708

Ackerman ME et al (2018) Route of immunization defines multiple mechanisms of vaccine-mediated protection against SIV. Nat Med 24:1590–1598. https://doi.org/10.1038/s41591-018-0161-0

Allhorn M et al (2010) The IgG-specific endoglycosidase EndoS inhibits both cellular and complement-mediated autoimmune hemolysis. Blood 115:5080–5088. https://doi.org/10.1182/blood-2009-08-239020

Alter G, Ottenhoff THM, Joosten SA (2018a) Antibody glycosylation in inflammation, disease and vaccination. Semin Immunol 39:102–110. https://doi.org/10.1016/j.smim.2018.05.003

Alter G et al (2018b) High-resolution definition of humoral immune response correlates of effective immunity against HIV. Mol Syst Biol 14:e7881. https://doi.org/10.15252/msb.20177881

Ana-Sosa-Batiz F et al (2016) Influenza-specific antibody-dependent phagocytosis. PLoS One 11: e0154461. https://doi.org/10.1371/journal.pone.0154461

Anthony RM, Wermeling F, Karlsson MCI, Ravetch JV (2008) Identification of a receptor required for the anti-inflammatory activity of IVIG. Proc Natl Acad Sci 105:19571–19578. https://doi.org/10.1073/pnas.0810163105

Bakchoul T et al (2013) Inhibition of HPA-1a alloantibody-mediated platelet destruction by a deglycosylated anti–HPA-1a monoclonal antibody in mice: toward targeted treatment of fetal-alloimmune thrombocytopenia. Blood 122:321–327. https://doi.org/10.1182/blood-2012-11-468561

Barouch DH et al (2013) Protective efficacy of a global HIV-1 mosaic vaccine against heterologous SHIV challenges in rhesus monkeys. Cell 155:531–539. https://doi.org/10.1016/j.cell.2013.09.061

Barouch DH et al (2015) Protective efficacy of adenovirus/protein vaccines against SIV challenges in rhesus monkeys. Science 349:320–324. https://doi.org/10.1126/science.aab3886

Bayry J (2016) Lupus pathogenesis: role of IgE autoantibodies. Cell Res 26:271–272. https://doi.org/10.1038/cr.2016.12

Benhnia MR-E-I et al (2009) Heavily isotype-dependent protective activities of human antibodies against vaccinia virus extracellular virion antigen B5ᵥ. J Virol 83:12355–12367. https://doi.org/10.1128/jvi.01593-09

Bondt A et al (2014) Immunoglobulin G (IgG) Fab glycosylation analysis using a new mass spectrometric high-throughput profiling method reveals pregnancy-associated changes. Mol Cell Proteomics 13:3029–3039. https://doi.org/10.1074/mcp.M114.039537

Bournazos S, Ravetch JV (2017) Fcγ receptor function and the design of vaccination strategies. Immunity 47:224–233. https://doi.org/10.1016/j.immuni.2017.07.009

Bowman J (2003) Thirty-five years of Rh prophylaxis. Transfusion 43:1661–1666. https://doi.org/10.1111/j.0041-1132.2003.00632.x

Bruggeman CW et al (2018) IgG Glyco-engineering to improve IVIg potency. Front Immunol 9:2442. https://doi.org/10.3389/fimmu.2018.02442

Bukreyev A, Yang L, Collins PL (2012) The secreted G protein of human respiratory syncytial virus antagonizes antibody-mediated restriction of replication involving macrophages and complement. J Virol 86:10880–10884. https://doi.org/10.1128/jvi.01162-12

Casadevall A, Pirofski, L.-a. (2012) Immunoglobulins in defense, pathogenesis, and therapy of fungal diseases. Cell Host Microbe 11:447–456. https://doi.org/10.1016/j.chom.2012.04.004

Castro-Dopico T, Clatworthy MR (2016) Fcγ receptors in solid organ transplantation. Curr Transplant Rep 3:284–293. https://doi.org/10.1007/s40472-016-0116-7

Chakraborty S et al (2020) Proinflammatory IgG Fc structures in patients with severe COVID-19. Nat Immunol. https://doi.org/10.1038/s41590-020-00828-7

Cheng HD et al (2019) IgG Fc glycosylation as an axis of humoral immunity in childhood. J Allergy Clin Immunol 145:710–713.e719. https://doi.org/10.1016/j.jaci.2019.10.012

Chon JH, Zarbis-Papastoitsis G (2011) Advances in the production and downstream processing of antibodies. New Biotechnol 28:458–463. https://doi.org/10.1016/j.nbt.2011.03.015

Churdboonchart V, Futrakul P, Bhamarapravati N (1983) Crossed immunoelectrophoresis for the detection of split products of the third complement in dengue hemorrhagic fever: I. Observations in patients' plasma*. Am J Tropical Med Hyg 32:569–576. https://doi.org/10.4269/ajtmh.1983.32.569

Clavel C et al (2008) Induction of macrophage secretion of tumor necrosis factor α through Fcγ receptor IIa engagement by rheumatoid arthritis–specific autoantibodies to citrullinated proteins complexed with fibrinogen. Arthritis Rheum 58:678–688. https://doi.org/10.1002/art.23284

Co MDT et al (2014) Relationship of preexisting influenza hemagglutination inhibition, complement-dependent lytic, and antibody-dependent cellular cytotoxicity antibodies to the development of clinical illness in a prospective study of A(H1N1)pdm09 influenza in children. Viral Immunol 27:375–382. https://doi.org/10.1089/vim.2014.0061

Collin M, Bjorck L (2017) Toward clinical use of the IgG specific enzymes IdeS and EndoS against antibody-mediated diseases. Methods Mol Biol 1535:339–351. https://doi.org/10.1007/978-1-4939-6673-8_23

Collin M, Shannon O, Bjorck L (2008) IgG glycan hydrolysis by a bacterial enzyme as a therapy against autoimmune conditions. Proc Natl Acad Sci USA 105:4265–4270. https://doi.org/10.1073/pnas.0711271105

Coutant F (2019) Pathogenic effects of anti-citrullinated peptide antibodies in rheumatoid arthritis – role for glycosylation. Joint Bone Spine 86:562–567. https://doi.org/10.1016/j.jbspin.2019.01.005

Crooks ET et al (2018) Glycoengineering HIV-1 Env creates 'supercharged' and 'hybrid' glycans to increase neutralizing antibody potency, breadth and saturation. PLoS Pathog 14:e1007024. https://doi.org/10.1371/journal.ppat.1007024

de Haan N, Reiding KR, Driessen G, van der Burg M, Wuhrer M (2016) Changes in healthy human IgG Fc-glycosylation after birth and during early childhood. J Proteome Res 15:1853–1861. https://doi.org/10.1021/acs.jproteome.6b00038

Decker Y et al (2016) Abnormal galactosylation of immunoglobulin G in cerebrospinal fluid of multiple sclerosis patients. Mult Scler J 22:1794–1803. https://doi.org/10.1177/1352458516631036

Dekkers G et al (2016) Multi-level glyco-engineering techniques to generate IgG with defined Fc-glycans. Sci Rep 6:36964. https://doi.org/10.1038/srep36964

Dekkers G et al (2017) Decoding the human immunoglobulin G-glycan repertoire reveals a Spectrum of Fc-receptor- and complement-mediated-effector activities. Front Immunol 8:877. https://doi.org/10.3389/fimmu.2017.00877

Dekkers G, Rispens T, Vidarsson G (2018) Novel concepts of altered immunoglobulin G galactosylation in autoimmune diseases. Front Immunol 9:553. https://doi.org/10.3389/fimmu.2018.00553

Dunkelberger JR, Song W-C (2010) Complement and its role in innate and adaptive immune responses. Cell Res 20:34–50. https://doi.org/10.1038/cr.2009.139

Engdahl C et al (2018) Estrogen induces St6gal1 expression and increases IgG sialylation in mice and patients with rheumatoid arthritis: a potential explanation for the increased risk of rheumatoid arthritis in postmenopausal women. Arthritis Res Ther 20:84. https://doi.org/10.1186/s13075-018-1586-z

Epp A et al (2017) Sialylation of IgG antibodies inhibits IgG-mediated allergic reactions. J Allergy Clin Immunol. https://doi.org/10.1016/j.jaci.2017.06.021

Ercan A et al (2017) Estrogens regulate glycosylation of IgG in women and men. Jci Insight 2:e89703. https://doi.org/10.1172/jci.insight.89703

Fernandes-Cerqueira C et al (2018) Patients with anti-Jo1 antibodies display a characteristic IgG Fc-glycan profile which is further enhanced in anti-Jo1 autoantibodies. Sci Rep 8:17958. https://doi.org/10.1038/s41598-018-36395-z

Ferrara C et al (2011) Unique carbohydrate–carbohydrate interactions are required for high affinity binding between FcγRIII and antibodies lacking core fucose. Proc Natl Acad Sci 108:12669–12674. https://doi.org/10.1073/pnas.1108455108

Forslind K et al (2004) Prediction of radiological outcome in early rheumatoid arthritis in clinical practice: role of antibodies to citrullinated peptides (anti-CCP). Ann Rheum Dis 63:1090. https://doi.org/10.1136/ard.2003.014233

Füst G et al (1994) Neutralizing and enhancing antibodies measured in complement-restored serum samples from HIV-1-infected individuals correlate with immunosuppression and disease. AIDS 8:603–610. https://doi.org/10.1097/00002030-199405000-00005

Gardinassi LG et al (2014) Clinical severity of visceral leishmaniasis is associated with changes in immunoglobulin G Fc N-glycosylation. MBio 5:e01844–e01814. https://doi.org/10.1128/mbio.01844-14

Gerber JS, Mosser DM (2001) Stimulatory and inhibitory signals originating from the macrophage Fcγ receptors. Microbes Infect 3:131–139. https://doi.org/10.1016/s1286-4579(00)01360-5

Goldberg BS, Ackerman ME (2020) Antibody-mediated complement activation in pathology and protection. Immunol Cell Biol 98:305–317. https://doi.org/10.1111/imcb.12324

Gonzalez SF et al (2010) Complement-dependent transport of antigen into B cell follicles. J Immunol 185:2659–2664. https://doi.org/10.4049/jimmunol.1000522

Grafals M, Thurman JM (2019) The role of complement in organ transplantation. Front Immunol 10:2380. https://doi.org/10.3389/fimmu.2019.02380

Gudelj I, Lauc G, Pezer M (2018) Immunoglobulin G glycosylation in aging and diseases. Cell Immunol 333:65–79. https://doi.org/10.1016/j.cellimm.2018.07.009

Guzman MG, Alvarez M, Halstead SB (2013) Secondary infection as a risk factor for dengue hemorrhagic fever/dengue shock syndrome: an historical perspective and role of antibody-dependent enhancement of infection. Arch Virol 158:1445–1459. https://doi.org/10.1007/s00705-013-1645-3

Haan N (2018) Differences in IgG Fc glycosylation are associated with outcome of pediatric *Meningococcal sepsis*. MBio 9:e00546–e00518. https://doi.org/10.1128/mbio.00546-18

Hafkenscheid L et al (2017) Structural analysis of variable domain glycosylation of anti-citrullinated protein antibodies in rheumatoid arthritis reveals the presence of highly sialylated glycans. Mol Cell Proteomics 16:278–287. https://doi.org/10.1074/mcp.m116.062919

Hafkenscheid L et al (2019) N-linked glycans in the variable domain of IgG anti–citrullinated protein antibodies predict the development of rheumatoid arthritis. Arthritis Rheumatol 71:1626–1633. https://doi.org/10.1002/art.40920

Harre U et al (2015) Glycosylation of immunoglobulin G determines osteoclast differentiation and bone loss. Nat Commun 6:6651. https://doi.org/10.1038/ncomms7651

He W et al (2017) Alveolar macrophages are critical for broadly-reactive antibody-mediated protection against influenza A virus in mice. Nat Commun 8:846. https://doi.org/10.1038/s41467-017-00928-3

Hebell T, Ahearn JM, Fearon DT (1991) Suppression of the immune response by a soluble complement receptor of B lymphocytes. Science 254:102–105. https://doi.org/10.1126/science.1718035

Hirose M et al (2012) Enzymatic autoantibody glycan hydrolysis alleviates autoimmunity against type VII collagen. J Autoimmun 39:304–314. https://doi.org/10.1016/j.jaut.2012.04.002

Hirvinen M et al (2013) Fc-gamma receptor polymorphisms as predictive and prognostic factors in patients receiving oncolytic adenovirus treatment. J Transl Med 11:1–12. https://doi.org/10.1186/1479-5876-11-193

Hodoniczky J, Zheng YZ, James DC (2005) Control of recombinant monoclonal antibody effector functions by Fc N-glycan remodeling in vitro. Biotechnol Prog 21:1644–1652. https://doi.org/10.1021/bp050228w

Hoepel W et al (2020) Anti-SARS-CoV-2 IgG from severely ill COVID-19 patients promotes macrophage hyper-inflammatory responses. Biorxiv 2020.2007.2013.190140. https://doi.org/10.1101/2020.07.13.190140

Hogarth PM (2002) Fc receptors are major mediators of antibody based inflammation in autoimmunity. Curr Opin Immunol 14:798–802. https://doi.org/10.1016/s0952-7915(02)00409-0

Huber VC, Lynch JM, Bucher DJ, Le J, Metzger DW (2001) Fc receptor-mediated phagocytosis makes a significant contribution to clearance of influenza virus infections. J Immunol 166:7381–7388. https://doi.org/10.4049/jimmunol.166.12.7381

Hunter M, Yuan P, Vavilala D, Fox M (2019) Optimization of protein expression in mammalian cells. Curr Protoc Protein Sci 95:e77. https://doi.org/10.1002/cpps.77

Jefferis R (2009) Glycosylation as a strategy to improve antibody-based therapeutics. Nat Rev Drug Discov 8:226–234. https://doi.org/10.1038/nrd2804

Jennewein MF, Alter G (2017) The immunoregulatory roles of antibody glycosylation. Trends Immunol 38:358–372. https://doi.org/10.1016/j.it.2017.02.004

Jones MB et al (2016) B-cell-independent sialylation of IgG. Proc Natl Acad Sci USA 113:7207–7212. https://doi.org/10.1073/pnas.1523968113

Jordan SC et al (2017) IgG endopeptidase in highly sensitized patients undergoing transplantation. New Engl J Med 377:442–453. https://doi.org/10.1056/nejmoa1612567

Kaneko Y, Nimmerjahn F, Ravetch JV (2006) Anti-inflammatory activity of immunoglobulin G resulting from Fc sialylation. Science 313:670–673. https://doi.org/10.1126/science.1129594

Kapur R et al (2014a) Low anti-RhD IgG-Fc-fucosylation in pregnancy: a new variable predicting severity in haemolytic disease of the fetus and newborn. Br J Haematol 166:936–945. https://doi.org/10.1111/bjh.12965

Kapur R et al (2014b) A prominent lack of IgG1-Fc fucosylation of platelet alloantibodies in pregnancy. Blood 123:471–480. https://doi.org/10.1182/blood-2013-09-527978

Karsten CM et al (2012) Anti-inflammatory activity of IgG1 mediated by Fc galactosylation and association of FcγRIIB and dectin-1. Nat Med 18:1401–1406. https://doi.org/10.1038/nm.2862

Kastbom A, Ahmadi A (2005) Söderkvist, P. & Skogh, T. The 158V polymorphism of fc gamma receptor type IIIA in early rheumatoid arthritis: increased susceptibility and severity in male patients (the Swedish TIRA*TIRA is a Swedish acronym for 'early invention in rheumatoid arthritis' and is a multicentre cooperation between rheumatology units in southeastern Sweden. project). Rheumatology 44:1294–1298. https://doi.org/10.1093/rheumatology/kei010

Katzelnick LC et al (2017) Antibody-dependent enhancement of severe dengue disease in humans. Science eaan6836. https://doi.org/10.1126/science.aan6836

Khandia R et al (2018) Modulation of dengue/Zika virus pathogenicity by antibody-dependent enhancement and strategies to protect against enhancement in Zika virus infection. Front Immunol 9:597. https://doi.org/10.3389/fimmu.2018.00597

Kim H-Y, Stojadinovic A, Izadjoo MJ (2014) Affinity maturation of monoclonal antibodies by multi-site-directed mutagenesis. Methods Mol Biol Clifton N J 1131:407–420. https://doi.org/10.1007/978-1-62703-992-5_24

Köhler G, Milstein C (1975) Continuous cultures of fused cells secreting antibody of predefined specificity. Nature 256:495–497. https://doi.org/10.1038/256495a0

Krištić J et al (2013) Glycans are a novel biomarker of chronological and biological ages. J Gerontol Ser 69:779–789. https://doi.org/10.1093/gerona/glt190

Kumpel BM (2007) Efficacy of RhD monoclonal antibodies in clinical trials as replacement therapy for prophylactic anti-D immunoglobulin: more questions than answers. Vox Sang 93:99–111. https://doi.org/10.1111/j.1423-0410.2007.00945.x

Kumpel BM et al (1995) Human Rh D monoclonal antibodies (BRAD-3 and BRAD-5) cause accelerated clearance of Rh D+ red blood cells and suppression of Rh D immunization in Rh D-volunteers. Blood 86:1701–1709

Kumpel BM et al (2020) Anti-D monoclonal antibodies from 23 human and rodent cell lines display diverse IgG Fc-glycosylation profiles that determine their clinical efficacy. Sci Rep 10:1464. https://doi.org/10.1038/s41598-019-57393-9

Larsen MD et al (2020) Afucosylated immunoglobulin G responses are a hallmark of enveloped virus infections and show an exacerbated phenotype in COVID-19. Biorxiv 2020.2005.2018.099507. https://doi.org/10.1101/2020.05.18.099507

Lastra GC, Thompson SJ, Lemonidis AS, Elson CJ (2009) Changes in the galactose content of IgG during humoral immune responses. Autoimmunity 28:25–30. https://doi.org/10.3109/08916939808993842

Lazar GA et al (2006) Engineered antibody Fc variants with enhanced effector function. Proc Natl Acad Sci USA 103:4005–4010. https://doi.org/10.1073/pnas.0508123103

Lefaucheur C et al (2010) Preexisting donor-specific HLA antibodies predict outcome in kidney transplantation. J Am Soc Nephrol 21:1398–1406. https://doi.org/10.1681/asn.2009101065

Li X, Kimberly RP (2014) Targeting the Fc receptor in autoimmune disease. Expert Opin Ther Tar 18:335–350. https://doi.org/10.1517/14728222.2014.877891

Liu L (2015) Antibody glycosylation and its impact on the pharmacokinetics and pharmacodynamics of monoclonal antibodies and Fc-fusion proteins. J Pharm Sci 104:1866–1884. https://doi.org/10.1002/jps.24444

Lloyd KA et al (2018) Variable domain N-linked glycosylation and negative surface charge are key features of monoclonal ACPA: implications for B-cell selection. Eur J Immunol 48:1030–1045. https://doi.org/10.1002/eji.201747446

Lofano G et al (2018) Antigen-specific antibody Fc glycosylation enhances humoral immunity via the recruitment of complement. Sci Immunol 3:eaat7796. https://doi.org/10.1126/sciimmunol.aat7796

Lu LL et al (2016) A functional role for antibodies in tuberculosis. Cell 167:433–443.e414. https://doi.org/10.1016/j.cell.2016.08.072

Lu LL et al (2020) Antibody Fc glycosylation discriminates between latent and active tuberculosis. J Infect Dis. https://doi.org/10.1093/infdis/jiz643

Lundström SL et al (2014) IgG antibodies to cyclic Citrullinated peptides exhibit profiles specific in terms of IgG subclasses, Fc-Glycans and a Fab-peptide sequence. PLoS One 9:e113924. https://doi.org/10.1371/journal.pone.0113924

Mahan AE et al (2016) Antigen-specific antibody glycosylation is regulated via vaccination. PLoS Pathog 12:e1005456. https://doi.org/10.1371/journal.ppat.1005456

Mantegazza AR, Magalhaes JG, Amigorena S, Marks MS (2013) Presentation of phagocytosed antigens by MHC class I and II. Traffic 14:135–152. https://doi.org/10.1111/tra.12026

Mesin L, Ersching J, Victora GD (2016) Germinal center B cell dynamics. Immunity 45:471–482. https://doi.org/10.1016/j.immuni.2016.09.001

Mohan S et al (2012) Donor-specific antibodies adversely affect kidney allograft outcomes. J Am Soc Nephrol 23:2061–2071. https://doi.org/10.1681/asn.2012070664

Moore JS et al (2005) Increased levels of galactose-deficient IgG in sera of HIV-1-infected individuals. AIDS 19:381–389. https://doi.org/10.1097/01.aids.0000161767.21405.68

Nagelkerke SQ, Kuijpers TW (2015) Immunomodulation by IVIg and the role of Fc-gamma receptors: classic mechanisms of action after all? Front Immunol 5:674. https://doi.org/10.3389/fimmu.2014.00674

Nascimento EJM et al (2009) Alternative complement pathway deregulation is correlated with dengue severity. PLoS One 4:e6782. https://doi.org/10.1371/journal.pone.0006782

Nieto A et al (2000) Involvement of Fcγ receptor IIIA genotypes in susceptibility to rheumatoid arthritis. Arthritis Rheum 43:735–739. https://doi.org/10.1002/1529-0131(200004)43:4<735::aid-anr3>3.0.co;2-q

Nimmerjahn F, Ravetch JV (2008a) Fcgamma receptors as regulators of immune responses. Nat Rev Immunol 8:34–47. https://doi.org/10.1038/nri2206

Nimmerjahn F, Ravetch JV (2008b) Anti-inflammatory actions of intravenous immunoglobulin. Annu Rev Immunol 26:513–533. https://doi.org/10.1146/annurev.immunol.26.021607.090232

Nimmerjahn F, Anthony RM, Ravetch JV (2007) Agalactosylated IgG antibodies depend on cellular Fc receptors for in vivo activity. Proc Natl Acad Sci 104:8433–8437. https://doi.org/10.1073/pnas.0702936104

Noris M, Remuzzi G (2013) Overview of complement activation and regulation. Semin Nephrol 33:479–492. https://doi.org/10.1016/j.semnephrol.2013.08.001

Ohmi Y et al (2016) Sialylation converts arthritogenic IgG into inhibitors of collagen-induced arthritis. Nat Commun 7:11205. https://doi.org/10.1038/ncomms11205

Pagan JD, Kitaoka M, Anthony RM (2018) Engineered sialylation of pathogenic antibodies in vivo attenuates autoimmune disease. Cell 172:564–577 e513. https://doi.org/10.1016/j.cell.2017.11.041

Parekh RB et al (1985) Association of rheumatoid arthritis and primary osteoarthritis with changes in the glycosylation pattern of total serum IgG. Nature 316:452–457. https://doi.org/10.1038/316452a0

Parekh R et al (1989) A comparative analysis of disease-associated changes in the galactosylation of serum IgG. J Autoimmun 2:101–114. https://doi.org/10.1016/0896-8411(89)90148-0

Peschke B, Keller CW, Weber P, Quast I, Lünemann JD (2017) Fc-Galactosylation of human immunoglobulin gamma isotypes improves C1q binding and enhances complement-dependent cytotoxicity. Front Immunol 8:646. https://doi.org/10.3389/fimmu.2017.00646

Petrovic T et al (2020) Composition of the immunoglobulin G glycome associates with the severity of COVID-19. Glycobiology. https://doi.org/10.1093/glycob/cwaa102

Pfeifle R et al (2017) Regulation of autoantibody activity by the IL-23–TH17 axis determines the onset of autoimmune disease. Nat Immunol 18:104–113. https://doi.org/10.1038/ni.3579

Pittala S et al (2019) Antibody Fab-Fc properties outperform titer in predictive models of SIV vaccine-induced protection. Mol Syst Biol 15:e8747. https://doi.org/10.15252/msb.20188747

Rademacher TW, Williams P, Dwek RA (1994) Agalactosyl glycoforms of IgG autoantibodies are pathogenic. Proc Natl Acad Sci 91:6123–6127. https://doi.org/10.1073/pnas.91.13.6123

Radstake TRDJ et al (2003) Role of Fcgamma receptors IIA, IIIA, and IIIB in susceptibility to rheumatoid arthritis. J Rheumatol 30:926–933

Rodríguez-Bayona B, Pérez-Venegas JJ, Rodríguez C, Brieva JA (2007) CD95-mediated control of anti-citrullinated protein/peptides antibodies (ACPA)-producing plasma cells occurring in rheumatoid arthritis inflamed joints. Rheumatology 46:612–616. https://doi.org/10.1093/rheumatology/kel395

Rombouts Y et al (2015) Anti-citrullinated protein antibodies acquire a pro-inflammatory Fc glycosylation phenotype prior to the onset of rheumatoid arthritis. Ann Rheum Dis 74:234. https://doi.org/10.1136/annrheumdis-2013-203565

Rombouts Y et al (2016) Extensive glycosylation of ACPA-IgG variable domains modulates binding to citrullinated antigens in rheumatoid arthritis. Ann Rheum Dis 75:578. https://doi.org/10.1136/annrheumdis-2014-206598

Rönnelid J et al (2005) Longitudinal analysis of citrullinated protein/peptide antibodies (anti-CP) during 5 year follow up in early rheumatoid arthritis: anti-CP status predicts worse disease activity and greater radiological progression. Ann Rheum Dis 64:1744. https://doi.org/10.1136/ard.2004.033571

Rycke LD et al (2004) Rheumatoid factor and anticitrullinated protein antibodies in rheumatoid arthritis: diagnostic value, associations with radiological progression rate, and extra-articular manifestations. Ann Rheum Dis 63:1587. https://doi.org/10.1136/ard.2003.017574

Saunders, Conceptual O (2019) Approaches to modulating antibody effector functions and circulation half-life. Front Immunol 10:1296. https://doi.org/10.3389/fimmu.2019.01296

Scherer HU et al (2010) Glycan profiling of anti–citrullinated protein antibodies isolated from human serum and synovial fluid. Arthritis Rheum 62:1620–1629. https://doi.org/10.1002/art.27414

Schroeder HW, Cavacini L (2010) Structure and function of immunoglobulins. J Allergy Clin Immunol 125:S41–S52. https://doi.org/10.1016/j.jaci.2009.09.046

Schwab I, Nimmerjahn F (2013) Intravenous immunoglobulin therapy: how does IgG modulate the immune system? Nat Rev Immunol 13:176–189. https://doi.org/10.1038/nri3401

Selman MHJ et al (2012) Changes in antigen-specific IgG1 Fc N-glycosylation upon influenza and tetanus vaccination. Mol Cell Proteomics 11:M111.014563. https://doi.org/10.1074/mcp.M111.014563

Shields RL et al (2002) Lack of Fucose on human IgG1 N-linked oligosaccharide improves binding to human FcγRIII and antibody-dependent cellular toxicity. J Biol Chem 277:26733–26740. https://doi.org/10.1074/jbc.m202069200

Shinkawa T et al (2002) The absence of fucose but not the presence of galactose or bisecting N-acetylglucosamine of human IgG1 complex-type oligosaccharides shows the critical role of enhancing antibody-dependent cellular cytotoxicity. J Biol Chem 278:3466–3473. https://doi.org/10.1074/jbc.m210665200

Smyth MJ et al (2005) Activation of NK cell cytotoxicity. Mol Immunol 42:501–510. https://doi.org/10.1016/j.molimm.2004.07.034

Song YW, Kang EH (2010) Autoantibodies in rheumatoid arthritis: rheumatoid factors and anticitrullinated protein antibodies. Qjm Int J Med 103:139–146. https://doi.org/10.1093/qjmed/hcp165

Sonneveld ME et al (2016a) Antigen specificity determines anti-red blood cell IgG-Fc alloantibody glycosylation and thereby severity of haemolytic disease of the fetus and newborn. Br J Haematol. https://doi.org/10.1111/bjh.14438

Sonneveld ME et al (2016b) Glycosylation pattern of anti-platelet IgG is stable during pregnancy and predicts clinical outcome in alloimmune thrombocytopenia. Br J Haematol. https://doi.org/10.1111/bjh.14053

Sonneveld ME et al (2018) Fc-glycosylation in human IgG1 and IgG3 is similar for both total and anti-red-blood cell anti-K antibodies. Front Immunol 9:129. https://doi.org/10.3389/fimmu.2018.00129

Stackowicz J, Jönsson F, Reber LL (2020) Mouse models and tools for the in vivo study of neutrophils. Front Immunol 10:3130. https://doi.org/10.3389/fimmu.2019.03130

Stavnezer J, Schrader CE (2014) IgH chain class switch recombination: mechanism and regulation. J Immunol 193:5370–5378. https://doi.org/10.4049/jimmunol.1401849

Stegall MD et al (2011) Terminal complement inhibition decreases antibody-mediated rejection in sensitized renal transplant recipients: terminal complement inhibition decreases antibody-mediated rejection. Am J Transplant 11:2405–2413. https://doi.org/10.1111/j.1600-6143.2011.03757.x

Subedi GP, Barb AW (2016) The immunoglobulin G1 N-glycan composition affects binding to each low affinity Fc gamma receptor. MAbs 8:1512–1524. https://doi.org/10.1080/19420862.2016.1218586

Temming AR et al (2019) Human DC-SIGN and CD23 do not interact with human IgG. Sci Rep 9:9995. https://doi.org/10.1038/s41598-019-46484-2

Thabet MM et al (2009) Contribution of Fcγ receptor IIIA gene 158V/F polymorphism and copy number variation to the risk of ACPA-positive rheumatoid arthritis. Ann Rheum Dis 68:1775. https://doi.org/10.1136/ard.2008.099309

Thomann M et al (2015) In vitro Glycoengineering of IgG1 and its effect on Fc receptor binding and ADCC activity. PLoS One 10:e0134949. https://doi.org/10.1371/journal.pone.0134949

Thulin NK et al (2020) Maternal anti-dengue IgG fucosylation predicts susceptibility to dengue disease in infants. Cell Rep 31:107642. https://doi.org/10.1016/j.celrep.2020.107642

Tomana M, Schrohenloher RE, Reveille JD, Arnett FC, Koopman WJ (1992) Abnormal galactosylation of serum IgG in patients with systemic lupus erythematosus and members of families with high frequency of autoimmune diseases. Rheumatol Int 12:191–194. https://doi.org/10.1007/bf00302151

Tsuchiya N et al (1989) Effects of galactose depletion from oligosaccharide chains on immunological activities of human IgG. J Rheumatol 16:285–290

Vaccari M et al (2016) Adjuvant-dependent innate and adaptive immune signatures of risk of SIVmac251 acquisition. Nat Med 22:762–770. https://doi.org/10.1038/nm.4105

van Erp EA, Luytjes W, Ferwerda G, van Kasteren PB (2019) Fc-mediated antibody effector functions during respiratory syncytial virus infection and disease. Front Immunol 10:548. https://doi.org/10.3389/fimmu.2019.00548

van Timmeren MM et al (2010) IgG glycan hydrolysis attenuates ANCA-mediated glomerulonephritis. J Am Soc Nephrol 21:1103–1114. https://doi.org/10.1681/ASN.2009090984

Vanderven HA et al (2017) Fc functional antibodies in humans with severe H7N9 and seasonal influenza. Jci Insight 2:e92750. https://doi.org/10.1172/jci.insight.92750

Verkoczy L (2017) Chapter five humanized immunoglobulin mice models for HIV vaccine testing and studying the broadly neutralizing antibody problem. Adv Immunol 134:235–352. https://doi.org/10.1016/bs.ai.2017.01.004

Vestrheim AC et al (2014) A pilot study showing differences in glycosylation patterns of IgG subclasses induced by pneumococcal, meningococcal, and two types of influenza vaccines. Immun Inflamm Dis 2:76–91. https://doi.org/10.1002/iid3.22

Vidarsson G, Dekkers G, Rispens T (2014) IgG subclasses and Allotypes: from structure to effector functions. Front Immunol 5:520. https://doi.org/10.3389/fimmu.2014.00520

Vogt MR et al (2011) Poorly neutralizing cross-reactive antibodies against the fusion loop of West Nile virus envelope protein protect in vivo via Fcγ receptor and complement-dependent effector mechanisms. J Virol 85:11567–11580. https://doi.org/10.1128/jvi.05859-11

Walsh NC et al (2016) Humanized mouse models of clinical disease. Annu Rev Pathol Mech Dis 12:187–215. https://doi.org/10.1146/annurev-pathol-052016-100332

Wang TT, Ravetch JV (2019) Functional diversification of IgGs through Fc glycosylation. J Clin Invest 129:3492–3498. https://doi.org/10.1172/jci130029

Wang J et al (2011) Fc-glycosylation of IgG1 is modulated by B-cell stimuli. Mol Cell Proteomics 10:M110.004655. https://doi.org/10.1074/mcp.m110.004655

Wang TT et al (2017) IgG antibodies to dengue enhanced for FcγRIIIA binding determine disease severity. Science 355:395–398. https://doi.org/10.1126/science.aai8128

Wang Q et al (2019) Design and production of bispecific antibodies. Antibodies 8:43. https://doi.org/10.3390/antib8030043

Winkler A, Berger M, Ehlers M (2013) Anti-rhesus D prophylaxis in pregnant women is based on sialylated IgG antibodies. F1000research 2(169). https://doi.org/10.12688/f1000research.2-169.v1

Winzeler A, Wang JT (2013) Culturing hybridoma cell lines for monoclonal antibody production. Cold Spring Harbor Protocols 2013:pdb.prot074914. https://doi.org/10.1101/pdb.prot074914

Worley MJ et al (2018) Neutrophils mediate HIV-specific antibody-dependent phagocytosis and ADCC. J Immunol Methods 457:41–52. https://doi.org/10.1016/j.jim.2018.03.007

Wu J et al (2014) Functional Fcgamma receptor polymorphisms are associated with human allergy. PLoS One 9:e89196. https://doi.org/10.1371/journal.pone.0089196

Wu Y et al (2015) A potent broad-spectrum protective human monoclonal antibody crosslinking two haemagglutinin monomers of influenza A virus. Nat Commun 6:7708. https://doi.org/10.1038/ncomms8708

Yamaguchi Y et al (2006) Glycoform-dependent conformational alteration of the Fc region of human immunoglobulin G1 as revealed by NMR spectroscopy. Biochim Biophys Acta Bba – Gen Subj 1760:693–700. https://doi.org/10.1016/j.bbagen.2005.10.002

Zeevi A et al (2013) Persistent strong anti-HLA antibody at high titer is complement binding and associated with increased risk of antibody-mediated rejection in heart transplant recipients. J Hear Lung Transplant 32:98–105. https://doi.org/10.1016/j.healun.2012.09.021

Printed by Printforce, the Netherlands